Advances in

Heterocyclic Chemistry

Volume 21

Editorial Advisory Board

R. A. Abramovitch
A. Albert
A. T. Balaban
S. Gronowitz
T. Kametani
C. W. Rees
Yu. N. Sheinker
H. A. Staab
M. Tišler

Advances in
HETEROCYCLIC CHEMISTRY

Edited by

A. R. KATRITZKY

A. J. BOULTON

School of Chemical Sciences
University of East Anglia
Norwich, England

Volume 21

Academic Press · New York San Francisco London · 1977
A Subsidiary of Harcourt Brace Jovanovich, Publishers

COPYRIGHT © 1977, BY ACADEMIC PRESS, INC.
ALL RIGHTS RESERVED.
NO PART OF THIS PUBLICATION MAY BE REPRODUCED OR
TRANSMITTED IN ANY FORM OR BY ANY MEANS, ELECTRONIC
OR MECHANICAL, INCLUDING PHOTOCOPY, RECORDING, OR ANY
INFORMATION STORAGE AND RETRIEVAL SYSTEM, WITHOUT
PERMISSION IN WRITING FROM THE PUBLISHER.

ACADEMIC PRESS, INC.
111 Fifth Avenue, New York, New York 10003

United Kingdom Edition published by
ACADEMIC PRESS, INC. (LONDON) LTD.
24/28 Oval Road, London NW1

LIBRARY OF CONGRESS CATALOG CARD NUMBER: 62–13037

ISBN 0–12–020621–8

PRINTED IN THE UNITED STATES OF AMERICA

Contents

CONTRIBUTORS vii

PREFACE ix

Pyrrolodiazines with a Bridgehead Nitrogen

D. E. KUHLA AND J. G. LOMBARDINO

I. Introduction 2
II. Pyrrolo[1,2-a]pyrimidines 3
III. Pyrrolo[1,2-c]pyrimidines 25
IV. Pyrrolo[1,2-a]pyrazines 31
V. Pyrrolo[1,2-b]pyridazines 50

The Thienopyridines

JOHN M. BARKER

I. Introduction 65
II. Synthesis 67
III. Electron Distribution 89
IV. Reactions 92
V. Physical Properties 107
VI. Biological Activity 114
VII. Dyestuffs 117

Tellurophene and Related Compounds

FRANCESCO FRINGUELLI, GIANLORENZO MARINO, AND ALDO TATICCHI

I. Introduction and Scope of the Review 120
II. Molecular Structure and Physical Properties of Tellurophene . . 120
III. Synthesis and Chemical Reactivity of Tellurophene and Its Derivatives . 142
IV. Other Tellurium Heterocycles 152
V. Appendix. Tables of Tellurophene Derivatives and Related Compounds . 166
Note Added in Proof 172

New Developments in the Chemistry of Oxazolones

ROBERT FILLER AND Y. SHYAMSUNDER RAO

I. Introduction 176
II. 2-Oxazolin-5-ones 176

III. 3-Oxazolin-5-ones	191
IV. 2-Oxazolin-4-ones	198
V. 4-Oxazolin-2-ones	202
VI. 3-Oxazolin-2-ones	205

The Chemistry of Isoxazolidines

YOSHITO TAKEUCHI AND FUMIO FURUSAKI

I. Introduction	208
II. Synthetic Methods	209
III. Physical Properties	234
IV. Spectroscopic Properties	237
V. Chemical Properties	241
VI. Uses of Isoxazolidines	248

(2 + 2)-Cycloaddition and (2 + 2)-Cycloreversion Reactions of Heterocyclic Compounds

D. N. REINHOUDT

I. Introduction	254
II. Thermal (2 + 2)-Cycloadditions	257
III. Photochemical (2 + 2)-Cycloadditions	285
IV. (2 + 2)-Cycloreversion Reactions	310
Appendix	321

Recent Advances in Tetrazole Chemistry

R. N. BUTLER

I. Introduction	324
II. Physicochemical Studies	325
III. Studies of Synthesis and Mechanism	354
IV. Azidoazomethine Tetrazole Isomerism—Fused Tetrazoloheterocycles	402
V. Conclusion	424
Notes Added in Proof	425

The Chemistry of 1,2-Dioxetanes

WALDEMAR ADAM

I. Introduction	438
II. Preparation	439
III. Characterization	448
IV. Dioxetanes as Reactive Intermediates in Oxygenations	451
V. Chemical Properties	463

CUMULATIVE INDEX OF TITLES	483

Contributors

Numbers in parentheses indicate the pages on which the authors' contributions begin.

WALDEMAR ADAM, *Department of Chemistry, University of Puerto Rico, Rio Piedras, Puerto Rico* (437)

JOHN M. BARKER, *Trent Polytechnic, Nottingham, England* (65)

R. N. BUTLER, *Chemistry Department, University College, Galway, Ireland* (323)

ROBERT FILLER, *Department of Chemistry, Illinois Institute of Technology, Chicago, Illinois* (175)

FRANCESCO FRINGUELLI, *Istituto di Chimica Organica, Università di Perugia, Perugia, Italy* (119)

FUMIO FURUSAKI, *Central Research Laboratory, Mitsubishi Chemical Industries Ltd., Kamoshida-cho, Midori-ku, Yokohama, Japan* (207)

D. E. KUHLA, *Pfizer Central Research, Groton, Connecticut* (1)

J. G. LOMBARDINO, *Pfizer Central Research, Groton, Connecticut* (1)

GIANLORENZO MARINO, *Istituto di Chimica Organica, Università di Perugia, Perugia, Italy* (119)

Y. SHYAMSUNDER RAO, *Department of Chemistry, Kennedy-King College, Chicago, Illinois* (175)

D. N. REINHOUDT, *Koninklijke-Shell-Laboratorium, Amsterdam, The Netherlands* (253)

YOSHITO TAKEUCHI, *Department of Chemistry, College of General Education, The University of Tokyo, Komaba, Meguro-ku, Tokyo, Japan* (207)

ALDO TATICCHI, *Istituto di Chimica Organica, Università di Perugia, Perugia, Italy* (119)

Preface

Four of the reviews in the present volume survey complete topics: the tellurophenes (Fringuelli, Marino, and Taticchi), the 1,2-dioxetanes (Adam), the thienopyridines (Barker), and the isoxazolidines (Takeuchi and Furusaki). The first two fields are relatively new with little of significance reported before the present decade, and while the other two have a longer history, neither has previously been completely reviewed.

Two other chapters, one covering the tetrazoles (including fused tetrazoles) (Butler), and one entitled "Pyrrolodiazines with a Bridgehead Nitrogen" (i.e., the pyrrolo[1,2]-pyridazines, -pyrimidines, and -pyrazines) (Kuhla and Lombardino), update fields which have been reviewed elsewhere. Another, by Filler and Rao, surveys development in the oxazolone field since it was reviewed in this Series eleven years ago.

The remaining chapter, by Reinhoudt, deals with (2 + 2)-cycloaddition *and* (2 + 2)-cycloreversions. This chapter, concerned with the formation and rupture of membered heterocycles, is structured as to reaction type rather than heterocyclic ring.

A. J. BOULTON
A. R. KATRITZKY

Pyrrolodiazines with a Bridgehead Nitrogen

D. E. KUHLA AND J. G. LOMBARDINO

Pfizer Central Research, Groton, Connecticut

I. Introduction	2
II. Pyrrolo[1,2-*a*]pyrimidines (**1**)	3
A. Syntheses	3
1. Syntheses Starting with a Preformed Pyrimidine Ring	4
2. Syntheses Starting with the Pyrrole Portion of the Molecule Preformed	6
3. Syntheses Starting with Acyclic Precursors	14
B. Reactions	17
C. Spectral Properties	19
D. 2,3,4,6,7,8-Hexahydropyrrolo[1,2-*a*]pyrimidine (DBN)	21
1. Synthesis	21
2. Reactions	21
3. Utility	22
III. Pyrrolo[1,2-*c*]pyrimidines (**2**)	25
A. Syntheses	25
B. Reactions	28
C. Spectral Properties	30
D. Natural Products	30
IV. Pyrrolo[1,2-*a*]pyrazines (**3**)	31
A. Syntheses	31
B. Reactions	33
C. Spectral Properties	34
D. Octahydro Derivatives	34
1. Octahydropyrrolo[1,2-*a*]pyrazines	34
2. Mono-oxo Octahydropyrrolo[1,2-*a*]pyrazines	38
3. 1,4-Dioxo-octahydropyrrolo[1,2-*a*]pyrazines	40
4. Other Oxo Derivatives	49
V. Pyrrolo[1,2-*b*]pyridazines (**4**)	50
A. Syntheses	50
1. Syntheses Starting with a Preformed Pyridazine Ring . . .	50
2. Syntheses Starting with a Preformed Pyrrole Ring . . .	56
3. Synthesis via Molecular Rearrangement	57
B. Reactions	58
1. Electrophilic Substitution	58
2. Functional Group Transformations	60
C. Spectral Properties	61

I. Introduction

The general class of pyrrolodiazines with a bridgehead nitrogen atom encompasses four distinct heterocyclic systems (**1–4**). These compounds were last reviewed in a systematic fashion by Mosby,[1] who, while covering the literature through 1957, found 35 references to these ring systems. A brief review[2] restricted to pyrrolopyrimidines [including pyrrolo[1,2-*a*]pyrimidines (**1**) and pyrrolo[1,2-*c*]pyrimidines (**2**)] was published after this present work was essentially complete. Consequently, an attempt has been made to complement and supplement both of these earlier reviews.

The present review, which surveys the primary literature until mid-1975 and *Chemical Abstracts* Subject Indexes to Volume 81, has been restricted to the four heterocyclic systems (**1–4**) pictured below together with their approved *Chemical Abstracts* names and numbering systems. Other names which have been applied to these compounds in different chemical journals are given in parentheses; however, these names are considered ambiguous and should not be used in place of the proper approved name. Throughout this review, the numbering and naming systems applied in the literature have all been changed to conform to the latest *Chemical Abstracts* format. Fused benzo derivatives of these ring systems are not discussed since their chemistry usually differs greatly from that observed with the simpler bicyclic systems of compounds **1–4**.

Examples of the completely unsaturated 10 π-electron heteroaromatic ring system are known for all the above heterocycles. The chemical reactivity of these heterocyclic systems can be viewed as a microcosm of heterocyclic chemistry. Although little explored, the literature on these compounds is already replete with examples of imaginative synthetic entry, nucleophilic and electrophilic substitution reactions, mechanistically intriguing molecular rearrangements and compilations of spectral data, which now allow structure assignment to new reaction products with relative certainty.

In addition, many examples of partially or completely saturated pyrrolo[1,2-*a*]pyrimidines (**1**) and pyrrolo[1,2-*a*]pyrazines (**3**) are known, and a wide variety of interesting biological activities have been attributed to these systems.

Finally, there are numerous references in the literature to 2,3,4,6,7,8-hexahydropyrrolo[1,2-*a*]pyrimidine (DBN), which is an important industrial catalyst because of its nonnucleophilic, yet strongly basic, properties.

---

[1] W. L. Mosby, *in* "Heterocyclic Systems with Bridgehead Nitrogen Atoms," Part I. Wiley (Interscience), New York, 1961.

[2] V. Amarnath and R. Madhav, *Synthesis*, 837 (1974).

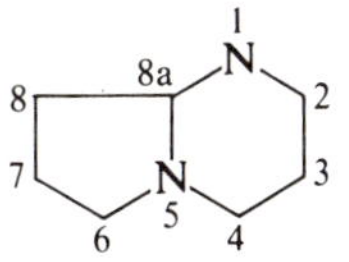

Pyrrolo[1,2-*a*]pyrimidine
(not 1-azaindolizine
or 1,5-diazabicyclo-
[4.3.0]nonane)

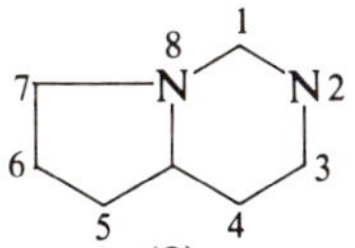

(1)

Pyrrolo[1,2-*c*]pyrimidine
(not 2-azaindolizine,
6-azapyrrocoline, or
1,3-diazabicyclo[4.3.0]nonane)

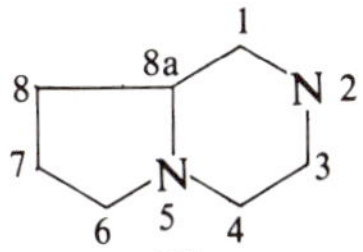

(2)

Pyrrolo[1,2-*a*]pyrazine
(not 7-azapyrrocoline,
polyazaindolizine,
1,4-diazabicyclo[4.3.0]nonane, or
1,2-trimethylenepiperazine)

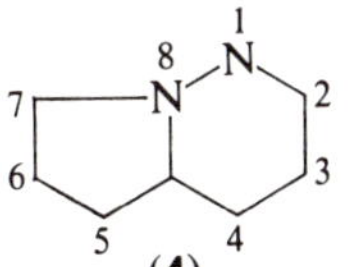

(3)

Pyrrolo[1,2-*b*]pyridazine
(not 5-azapyrrocoline,
1,2-diazabicyclo[4.3.0]nonane, or
azaindolizine)

(4)

This review attempts to survey the pertinent world literature on these four heterocyclic systems in order systematically to compare and constrast their chemistry and to compile spectral data of value in making structural assignments for derivatives of these compounds.

II. Pyrrolo[1,2-*a*]pyrimidines (1)

A. SYNTHESES

To systematize and clarify the many types of syntheses reported for **1**, methods for preparing this ring system are divided into three categories:

1. Syntheses starting with the pyrimidine portion of the bicyclic system preformed.
2. Syntheses starting with the pyrrole portion of the molecule preformed.
3. Syntheses employing only acyclic precursors.

In addition, each subsection starts with the syntheses that afford the most unsaturated ring system and proceeds through syntheses of octahydro derivatives.

1. Syntheses Starting with a Preformed Pyrimidine Ring

Kobayashi et al.[3] treated 4-methoxypyrimidine (**5**) with tetracyanoethylene oxide (**6**) generating ylide (**7**), which reacted readily with dimethylacetylenedicarboxylate (DMAD) in a 2 + 3-dipolar cycloaddition reaction to give dimethyl 6-cyano-2-methoxypyrrolo[1,2-*a*]-pyrimidine-7,8-dicarboxylate (**8**) [Eq. (1)]. NMR spectroscopy was used to rule out structures **9** and **10** as the product from the above reaction. The coupled ($J_{3,4}$ = 7.5 Hz) protons on carbon atoms 3 and 4 are readily apparent in the NMR spectrum of **8**, ruling out the pyrrolo[1,2-*c*]-pyrimidine (**9**). Conversion of the 6-cyano group in **8** to a carbomethoxy group leads to a shift to lower magnetic field for the H-4 proton, thus ruling out structure **10**, which has a methoxy group in this position. This low-field shift of a proton (H-4 in this case) by a peri-ester group has been documented by a number of workers in this area.[3-6]

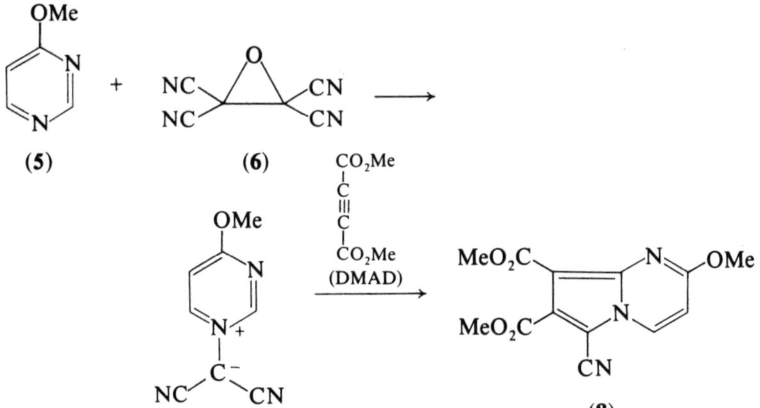

[3] Y. Kobayashi, T. Katsuma, and K. Morinaga, *Chem. Pharm. Bull.* **19**, 2106 (1971).
[4] R. M. Acheson and D. A. Robinson, *J. Chem. Soc. C*, 1633 (1968).
[5] R. M. Acheson and M. W. Foxton, *J. Chem. Soc. C*, 2218 (1966).
[6] R. L. Letsinger and R. Lasco, *J. Org. Chem.* **21**, 764 (1956).

Lown and Matsumoto,[7] in a comprehensive study on the reactions of diphenylcyclopropenone (11) with nitrogen heterocycles, prepared 6-hydroxy-2,7,8-triphenylpyrrolo[1,2-*a*]pyrimidine (13) by allowing 4-phenylpyrimidine (12) to react with 11 [Eq. (2)]. This synthesis can also be viewed as a 2,3-dipolar cycloaddition reaction in which 11 serves as a very reactive 1,3-dipole and the 4-phenylpyrimidine (12) is a dipolarophile. Note that in both of these syntheses the pyrrole portion of the resulting bicyclic system is completely substituted.

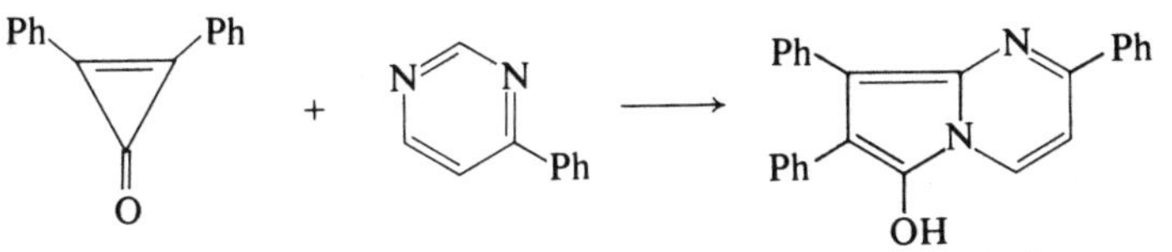

During their work on the total synthesis of diazasteroids, Burckhalter and Abramson[8] investigated, as a model, the synthesis of 2-(2-ethoxycarbonylethyl)-3,4,5,6-tetrahydropyrimidine hydrochloride (16) via the condensation of 1,3-propanediamines (14) with ethyl 3-ethoxycarbonylpropionimidate hydrochloride (15). When 16 was heated above its melting point, and the melt was allowed to sublime, 2,3,4,6,7,8-hexahydro-6-oxopyrrolo[1,2-*a*]pyrimidine hydrochloride (17) was isolated in excellent yield [Eq. (3)].

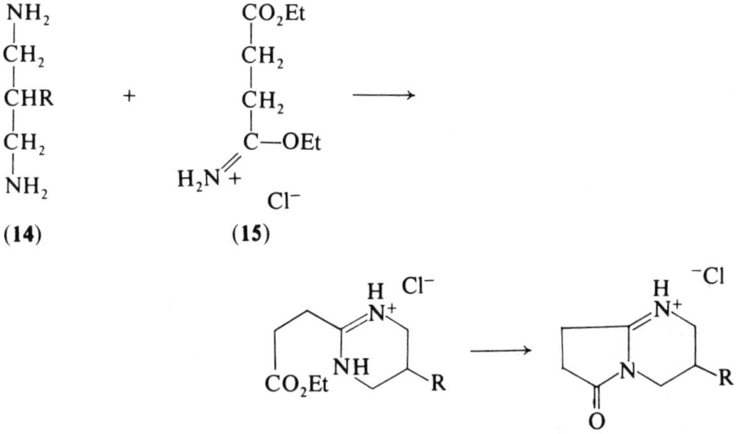

---

[7] J. W. Lown and K. Matsumoto, *Can. J. Chem.* **49**, 1165 (1971).
[8] J. H. Burckhalter and H. N. Abramson, *Chem. Commun.*, 805 (1966).

2. Syntheses Starting with the Pyrrole Portion of the Molecule Preformed

Syntheses of completely unsaturated pyrrolo[1,2-a]pyrimidines that begin with the pyrrole portion of the bicyclic system preformed all lead to products in which the pyrrole portion of this molecule is heavily substituted. This is due to the instability of the required 2-aminopyrrole starting material unless it is heavily substituted, usually with an electron-withdrawing group in the 3 position.

Shvedov et al.[9] have found what is probably the most general synthetic route to the completely unsaturated pyrrolo[1,2-a]pyrimidine system reported to date. These workers condense substituted 2-amino-3-cyanopyrroles (**18**) with a variety of β-diketones (**19**) affording the substituted 8-cyanopyrrolo[1,2-a]pyrimidines (**20**) in good overall yield.

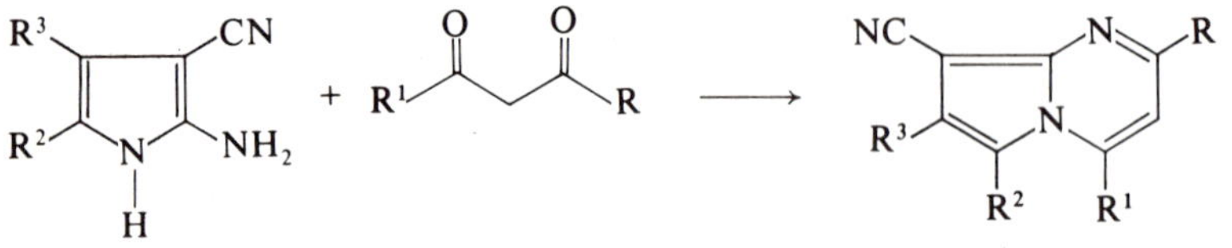

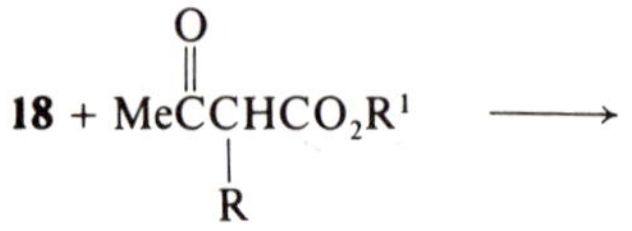

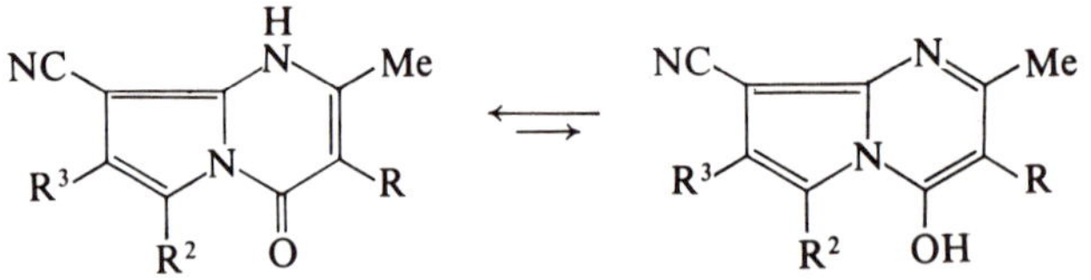

SCHEME 1

When β-ketoesters (**21**) are substituted for the β-diketones, 4-hydroxy-8-cyanopyrrolo[1,2-a]pyrimidines (**22a**) are obtained. Infrared (IR) spectroscopy indicates that these materials exist predominantly in the keto form (**22b**) (Scheme 1). A related, but much less general, synthesis was reported a few years later by Sowell and Blanton.[10] These workers alkylate 2-amino-3-cyano-4,5-dimethylpyrrole (**23**) with phenacyl

---
[9] V. I. Shvedov, J. A. Kharizomenova, L. B. Altukhova, and A. N. Grinev, *Khim. Geterotsikl. Soedin.* **3**, 428 (1970) [*CA* **73**, 25403 (1970)].
[10] J. W. Sowell and C. D. Blanton, *J. Heterocycl. Chem.* **10**, 287 (1973).

bromides, giving the monoalkylated products (**24**). Upon treating **24** with base, the substituted pyrrolo[1,2-*a*]pyrimidines (**25**) are obtained in excellent yields. The authors postulate the mechanistic course pictured in Scheme 2 to account for the products obtained, although the 2-amino-3-cyanopyrrole by-product was never isolated.

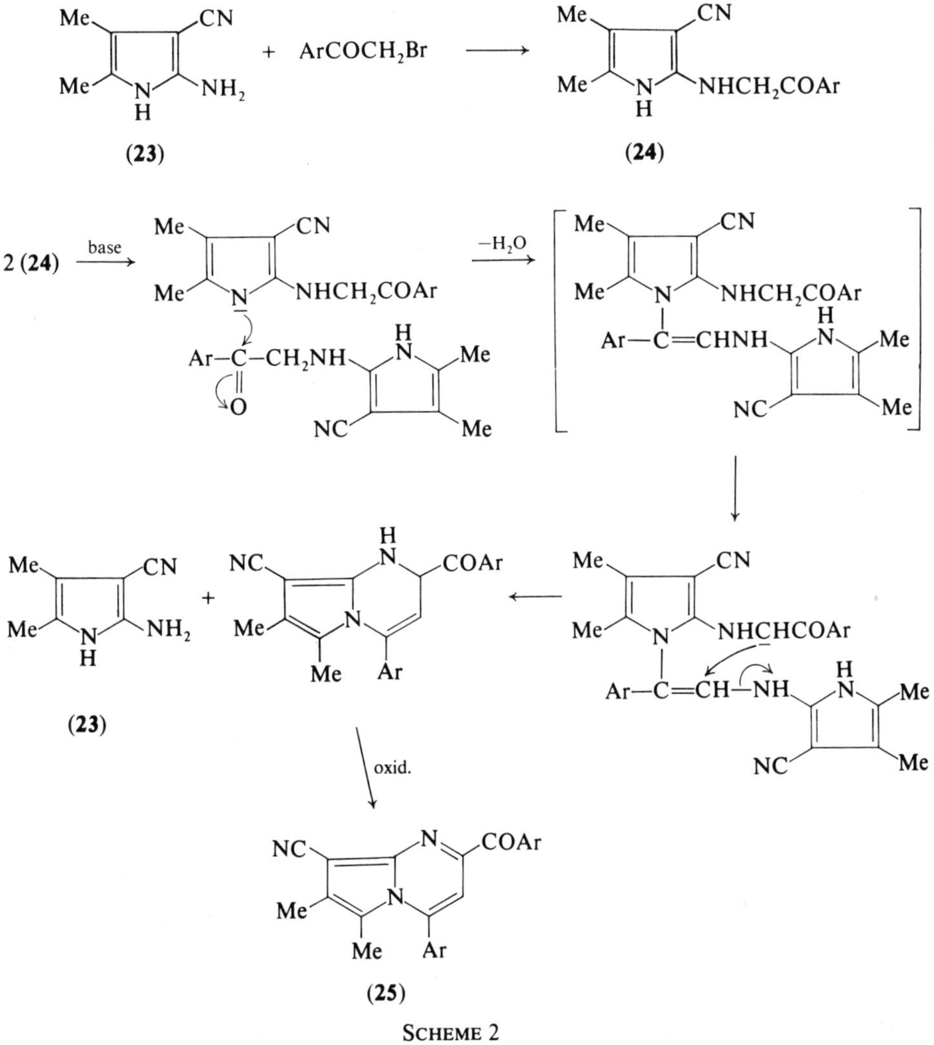

SCHEME 2

The only other example of the completely unsaturated pyrrolo[1,2-*a*]-pyrimidine nucleus reported is the hexachloro analog (**27**) prepared by

Beck et al.[11,12] by high-temperature chlorination of β-cyanoethylpyrrolidine (26) followed by heating *in vacuo* [Eq. (4)]. Two isomeric

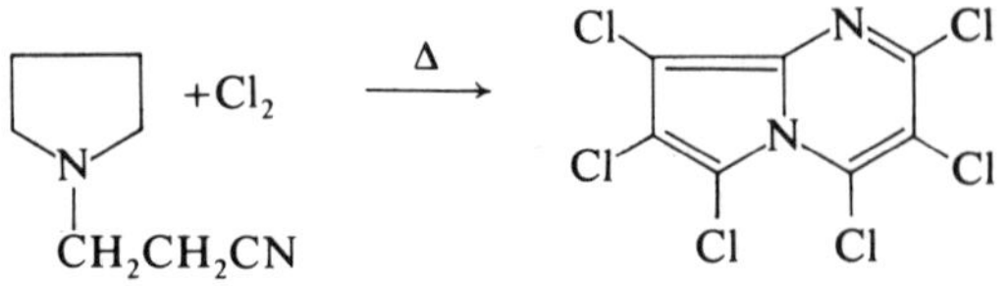

$$\text{(26)} \xrightarrow{+Cl_2, \Delta} \text{(27)} \quad (4)$$

pentachloro analogs were also isolated in very low yield, but structure assignment remains indefinite. The parent unsubstituted pyrrolo[1,2-*a*]-pyrimidine (28) has not yet been reported, nor have simple mono- or di-substituted homologs of 28.

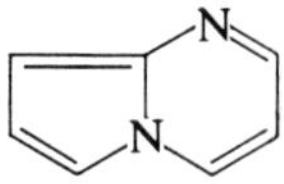

(28)

LeBerre and Renault have utilized 2-amino-1-pyrroline (30) in condensation reactions with α,β-unsaturated carbonyl derivatives and β-dicarbonyl compounds to prepare a wide variety of partially saturated pyrrolo[1,2-*a*]pyrimidines in which the pyrrole portion of the molecule is always completely saturated. For example, condensation of the aminopyrroline (30) with substituted malonic esters (29) affords[13,14] 2-hydroxy-4,6,7,8-tetrahydropyrrolo[1,2-*a*]pyrimidin-4-one (31a). Spectral data indicate the betaine form (31b) to be preferred [Eq. (5)].

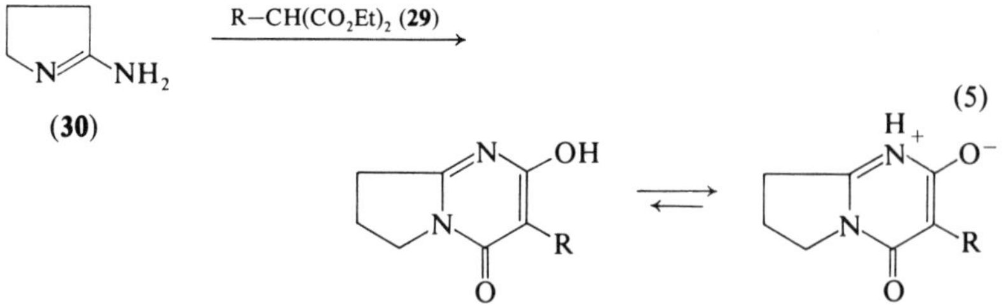

$$\text{(30)} \xrightarrow{R-CH(CO_2Et)_2, (29)} \text{(31a)} \rightleftharpoons \text{(31b)} \quad (5)$$

[11] G. Beck, H. Holtschmidt, and H. Heitzer, *Justus Liebigs Ann. Chem.* **731**, 45 (1970).
[12] G. Beck, H. Holtschmidt, and H. Heitzer, *Angew Chem., Int. Ed. Engl.* **13**, 210 (1974).
[13] A. Etienne, A. LeBerre, and C. Renault, *C.R. Hebd. Seances Acad. Sci., Ser. C* **262**, 365 (1966).
[14] A. LeBerre and C. Renault, *Bull. Soc. Chim. Fr.* **9**, 3133 (1969).

Reaction of **30** with α,β-unsaturated aldehydes or ketones[15,16] gives 2-hydroxy-2,3,4,6,7,8-hexahydryopyrrolo[1,2-a]pyrimidines (**32**), which can be dehydrated to afford the tetrahydro derivative **33** [Eq. (6)]. The

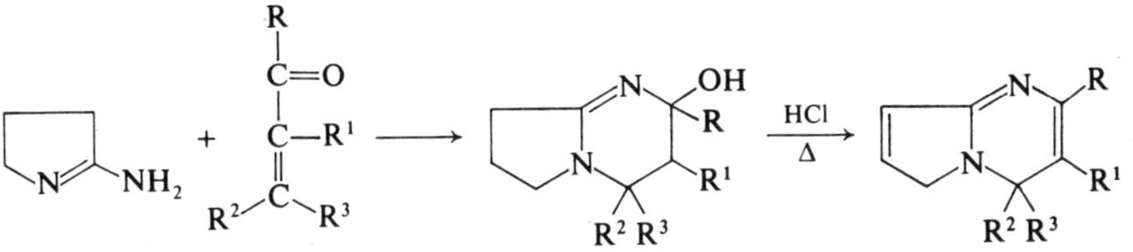

2-amino-1-pyrroline (**30**) also reacts with β-keto esters yielding 2,6,7,8-tetrahydropyrrolo[1,2-a]pyrimidin-2-one (**34**) or -4-one (**35**) depending on the conditions employed [Eq. (7)]. However, when **30** is condensed with α,β-unsaturated esters, only 2,3,4,6,7,8-hexahydropyrrolo[1,2-a]-pyrimidin-2-one (**36**) is isolated[17] [Eq. (8)].

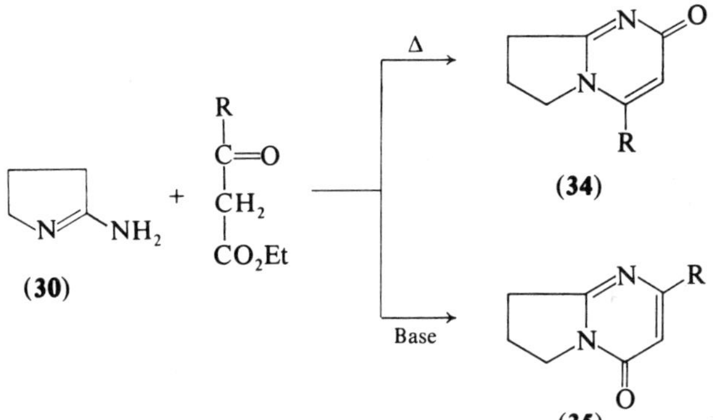

The other possible isomer, 2,3,4,6,7,8-hexahydropyrrolo[1,2-a]-pyrimidin-4-one (**39**), not observed in the above reaction, can be prepared readily by treating the imidate **37** with a variety of β-amino acid esters (**38**) [Eq. (9)].[17]

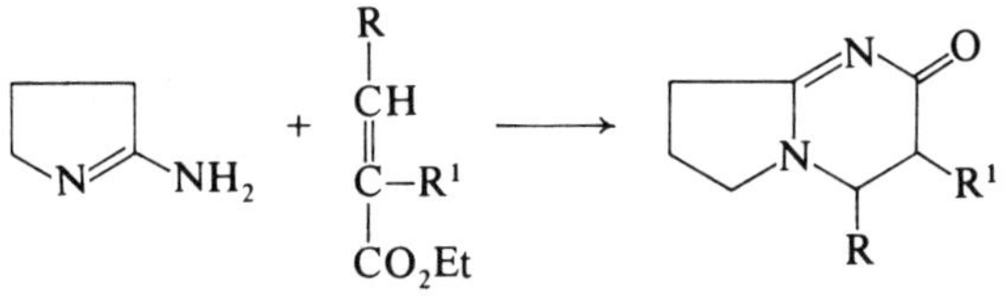

[15] A. LeBerre and C. Renault, *C.R. Hebd. Seances Acad. Sci., Ser. C* **265**, 249 (1967).
[16] A. LeBerre and C. Renault, *Bull. Soc. Chim. Fr.* **9**, 3146 (1969).
[17] A. LeBerre and C. Renault, *Bull. Soc. Chim. Fr.* **9**, 3139 (1969).

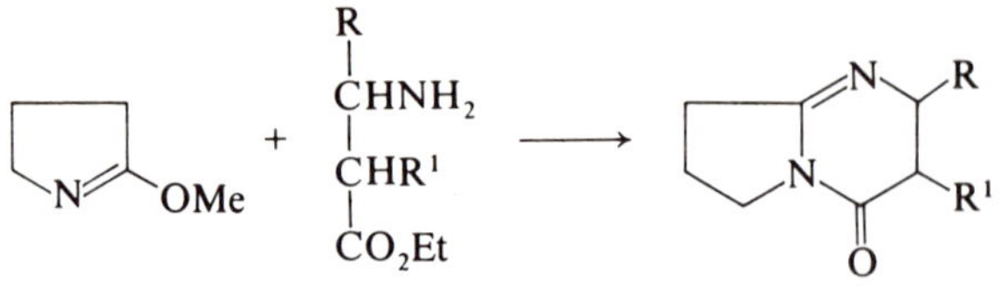

[Structures for Eq. (9): compounds (37), (38), (39)]

$$\text{(37)} + \text{(38)} \longrightarrow \text{(39)} \quad (9)$$

Other workers have extended this synthetic approach of LeBerre and Renault. Golubushina and Chuiguk,[18] for example, allowed 2-amino-1-pyrroline hydrochloride and sodium perchlorate to react with a series of β-diketones. They obtained the 7,8-dihydro-6H-pyrrolo[1,2-a]-pyrimidinium perchlorates (40) in fair yields [Eq. (10)]. A Japanese

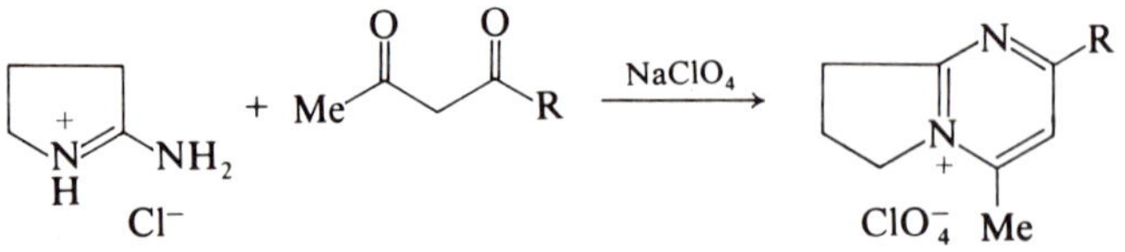

$$\text{(30)} + \text{Me-CO-CH}_2\text{-CO-R} \xrightarrow{\text{NaClO}_4} \text{(40)} \quad (10)$$

patent[19] describes the reaction of 30 with diethyl ethoxymethylenemalonate (41) giving 3-carboethoxy-4,6,7,8-tetrahydropyrrolo[1,2-a]-pyrimidin-4-one (42) [Eq. (11)].

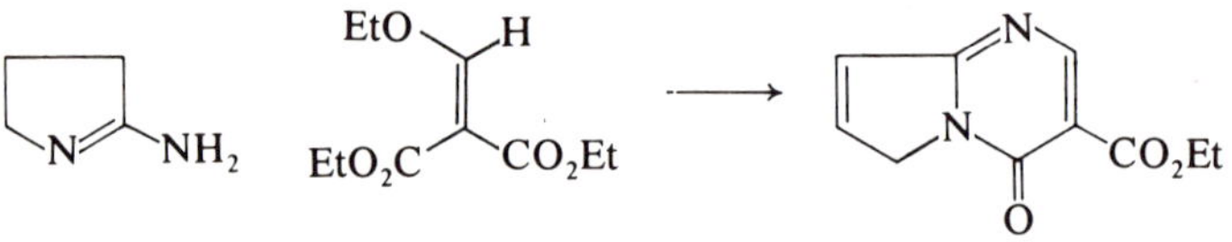

$$\text{(30)} + \text{(41)} \longrightarrow \text{(42)} \quad (11)$$

Finally, reaction[20] of compound 30 with α-acetyl-γ-butyrolactone (43) affords two isomeric pyrrolo[1,2-a]pyrimidine derivatives (44 and 45) [Eq. (12)].

---

[18] G. M. Golubushina and V. A. Chuiguk, *Khim. Geterotsikl. Soedin.* **3**, 419 (1972) [*CA* **77**, 88425 (1972)].

[19] I. Agata, S. Noguchi, and K. Tanaka, Japanese Patent 73 34,897 (1973) [*CA* **79**, P42537 (1973)].

[20] H-J. Willenbrock, H. Wamhoff, and F. Korte, *Justus Liebigs Ann. Chem.* 103, (1973).

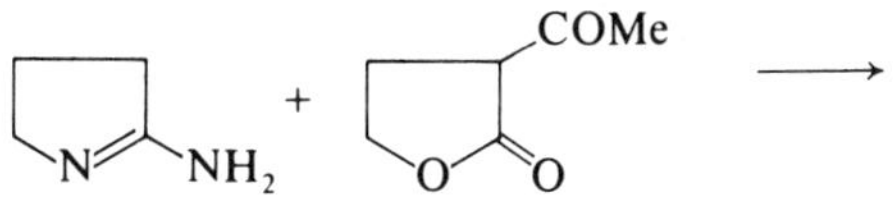

Sec. II.A] PYRROLODIAZINES WITH A BRIDGEHEAD NITROGEN

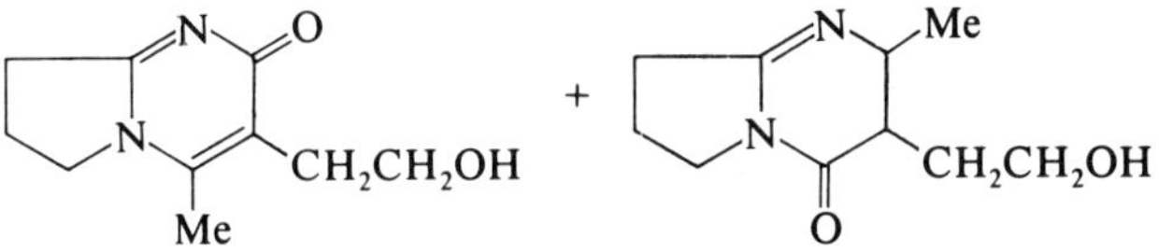

Bormann[21,22] has developed an alternative synthetic method for preparing pyrrolo[1,2-a]pyrimidinones similar in structure to those synthesized by LeBerre and Renault. Heating imidate **37** with a variety of substituted 2-azetidinones (β-lactams) (**46**) gives 2,3,4,6,7,8-hexahydropyrrolo[1,2-a]pyrimidin-4-ones (**48**). The intermediates (**47**) are relatively labile and are isolated and characterized only when R^1 and R^2 are both alkyl groups [Eq. (13)].

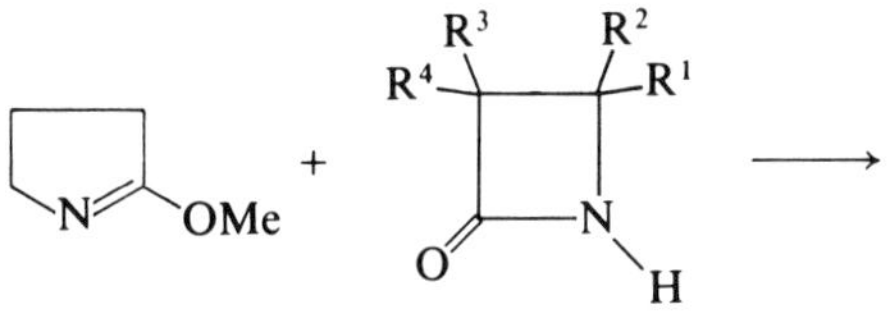

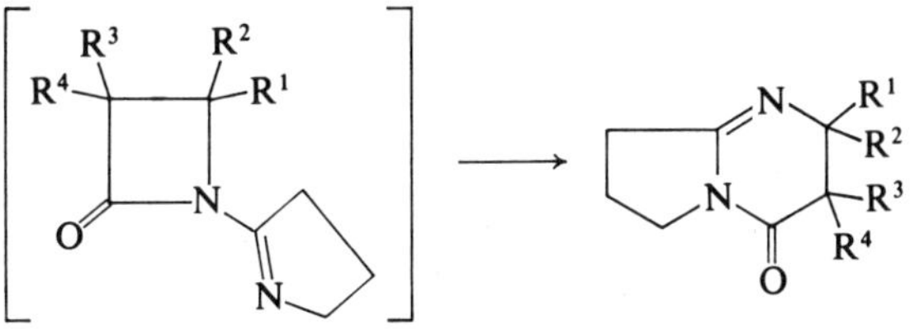

Hirai et al.[23] and Bormann[24] have extended the above synthesis by starting with a 2-azetidinone containing a potential leaving group (e.g., R^1 = OMe, O_2CR, or SC_6H_4Cl) in the 4-position (**49**). In this case a further degree of unsaturation is introduced into the product, and a

---

[21] D. Bormann, *Chem. Ber.* **103**, 1797 (1970).
[22] D. Bormann, German Patent 1,803,785 (1970) [*CA* **73**, P35400 (1970)].
[23] K. Hirai, H. Matsuda, and Y. Kishida, *Chem. Pharm. Bull.* **21**, 1305 (1973).
[24] D. Bormann, *Chem. Ber.* **107**, 270 (1974).

4,5,7,8-tetrahydropyrrolo[1,2-a]pyrimidin-4-one (**52**) is isolated [Eq. (14)]. The reaction presumably proceeds via intermediates **50** and **51**, which are similar to the corresponding structures **47** and **48** in Eq. (13).

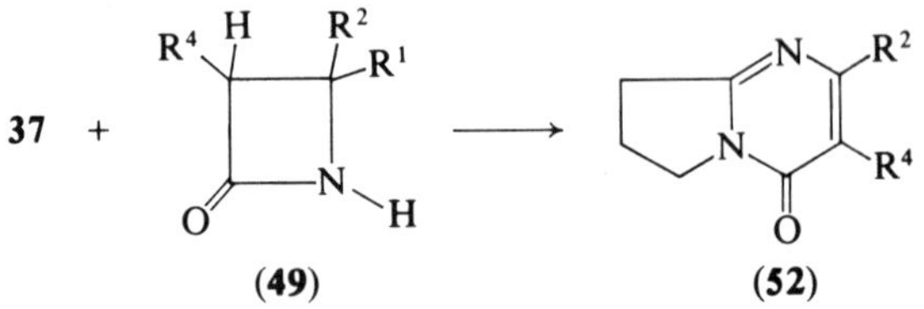

$$37 + (49) \longrightarrow (52) \qquad (14)$$

The final major group of pyrrolo[1,2-a]pyrimidine syntheses, which begin with an intact five-membered ring, employ N-substituted 2-pyrrolidinones as starting materials.

Sugasawa et al.[25] in a series of model studies directed toward the total synthesis of camptothecin, allowed 2-pyrrolidinone (**53**) to react with diketene (**54**) followed by ammonium acetate in acetic acid to afford 2-methyl-4,6,7,8-tetrahydropyrrolo[1,2-a]pyrimidin-4-one (**56**). Presumably, keto imide **55** is an intermediate in this reaction [Eq. (15)].

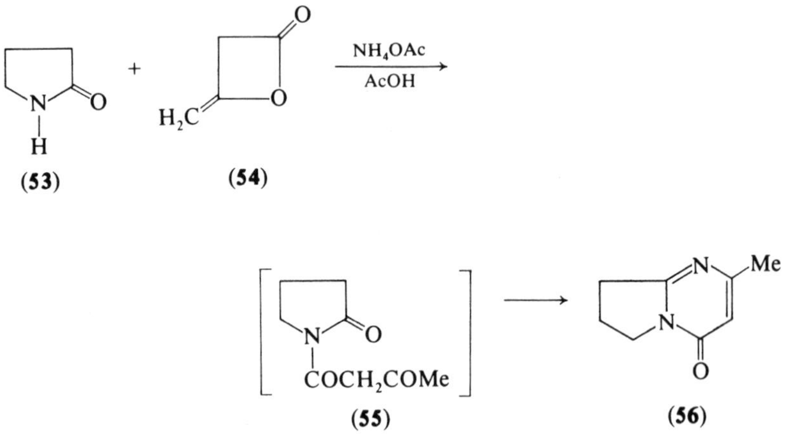

$$(53) + (54) \xrightarrow{NH_4OAc}_{AcOH}$$

$$[(55)] \longrightarrow (56) \qquad (15)$$

Möhrle and Engelsing[26] report a similar transformation. Aminolactam **57** when heated in xylene with an acid catalyst, eliminates water, and gives 2-phenyl-2,3,4,6,7,8-hexahydropyrrolo[1,2-a]pyrimidine (**58**) [Eq. (16)].

[25] T. Sugasawa, T. Toyoda, and K. Sasakura, *Chem. Pharm. Bull.* **22**, 771 (1974).
[26] H. Möhrle and R. Engelsing, *Arch. Pharm.* (*Weinheim, Ger.*) **306**, 325 (1973).

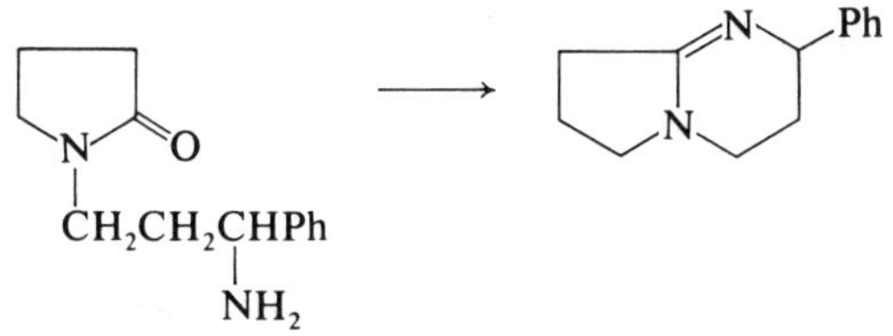

(57) → (58) (16)

A series of 6-alkyl-2,3,4,6,7,8-hexahydropyrimidines (63) have been prepared[27] by the route shown in Eq. (17). This synthesis is directly analogous to the industrial method[28,29] used to prepare 2,3,4,6,7,8-hexahydropyrrolo[1,2-a]pyrimidine (DBN), a very useful strong base and industrial catalyst (see below). These compounds were tested[27] for antibacterial and fungistatic activity. A number of derivatives of (63) were potent fungistatic agents, but only one (R = heptyl) was at all active as an antibacterial agent.

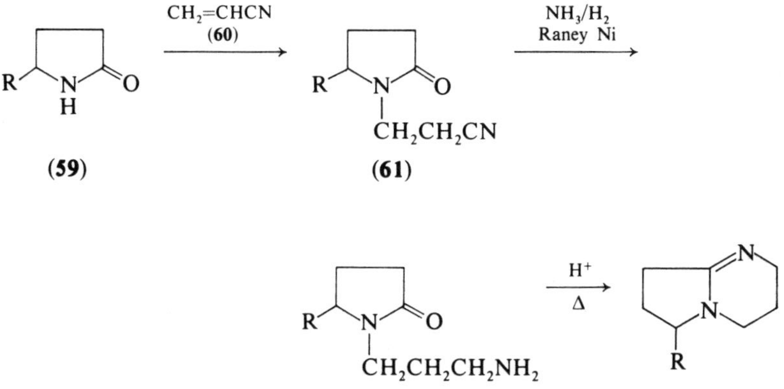

(59) → (61) → (62) → (63) (17)

Shvedov et al.[30] allowed substituted 2-amino-3-carboethoxythiophenes (64) to react with 2-pyrrolidones (65) to give the corresponding thienopyrimidine derivative (66). Heating 66 in the presence of Raney nickel causes loss of sulfur, and pyrrolo[1,2-a]pyrimidine derivatives (67) are obtained [Eq. (18)].

[27] V. A. Sedavkina, N. A. Morozova, A. A. Rechinskaya, and L. K. Kulikova, Khim. Farm. Zh. **8**, 21 (1974) [CA **80**, 82870 (1974)].
[28] W. Reppe, Justus Liebigs Ann. Chem. **596**, 210 (1955).
[29] H. Oediger, H. Kabbe, F. Möller, and K. Eiter, Chem. Ber. **99**, 2012 (1966).
[30] V. I. Shvedov, I. A. Kharizomenova, and A. N. Grinev, Khim. Geterotsikl. Soedin. **6**, 765 (1975).

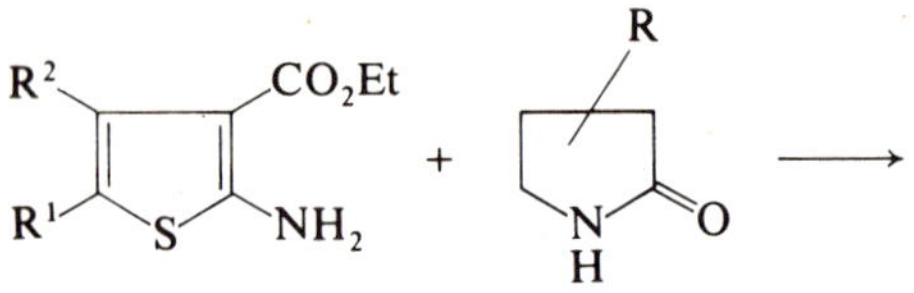

[Structures: (64), (65) → (66) → Raney Ni → (67)] (18)

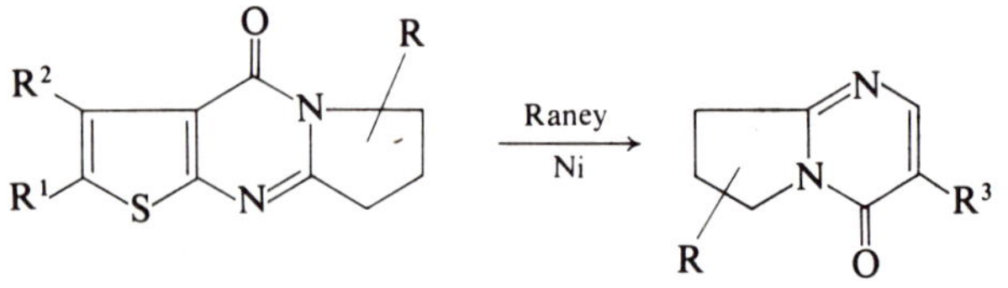

In an extremely interesting microbial transformation,[31] β-cyanoethylpyrrolidine derivative **68** is converted into 3,3-diphenyloctahydropyrrolo[1,2-a]pyrimidin-2-one (**69**) [Eq. (19)].

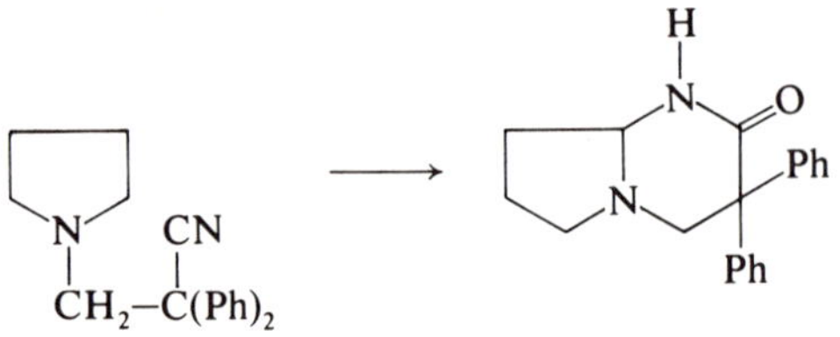

[Structures: (68) → (69)] (19)

3. Syntheses Starting with Acyclic Precursors

This section details syntheses of the pyrrolo[1,2-a]pyrimidine ring system in which only acyclic starting materials are used. In most cases, however, a substituted pyrrole or pyrimidine monocyclic structure is probably an intermediate in this process. These react further under the conditions employed to give the observed bicyclic product. Wollweber et al.[32] have provided experimental evidence for just such an intermediate (see below).

[31] R. J. Theriault, T. H. Longfield, and H. E. Zaugg, *Biochemistry* **11**, 385 (1972).
[32] H. Wollweber, J. Kurz, and W. Nägele, *Arch. Pharm. (Weinheim, Ger.)* **304**, 774 (1971).

Bortnick and Fegley[33-36] report that the condensation of γ-ketonitriles (**70**) with 1,3-propanediamine (**71**) produces 2,3,4,8-tetrahydropyrrolo-[1,2-a]pyrimidines (**72**) [Eq. (20)]. A few years after these initial reports,

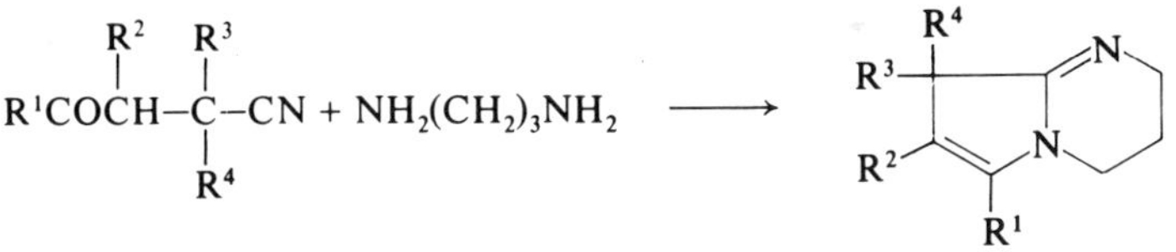

$$R^1COCH-\underset{R^4}{\underset{|}{C}}-CN + NH_2(CH_2)_3NH_2 \longrightarrow \text{(72)} \qquad (20)$$

(**70**)   (**71**)   (**72**)

DeBenneville and Niederhauwer[37] proposed the exocyclic unsaturated structure (**73**) as the product obtained in the above condensation where $R^1 = CH_3$. Since definitive spectral data are lacking in the original references,[33-36] it is impossible to know which structure is correct for this reaction product.

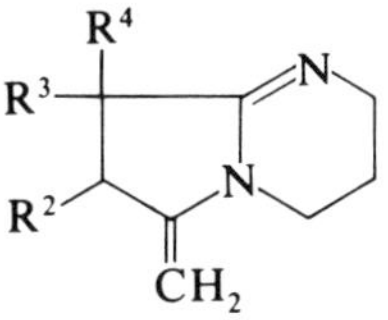

(**73**)

Almost simultaneously, workers at Geigy A.-G.[38] and Oeberli and Houlihan[39] and Houlihan[40-42] reported the condensation of γ-benzoylpropionic acids (**74**) with 1,3-diaminopropane (**71**) to give substituted 8a-phenyl-1,2,3,4,6,7,8a-octahydropyrrolo[1,2-a]pyrimidin-6-one (**75**) [Eq. (21)].

---

[33] N. M. Bortnick and M. F. Fegley, U.S. Patent 2,984,665 (1961) [*CA* **55**, P2479 (1961)].

[34] N. M. Bortnick and M. F. Fegley, U.S. Patent 2,993,049 (1961) [*CA* **56**, 476 (1962)].

[35] N. M. Bortnick and M. F. Fegley, U.S. Patent 2,993,046 (1961) [*CA* **56**, P4779 (1962)].

[36] N. M. Bortnick and M. F. Fegley, U.S. Patent 2,984,666 (1961) [*CA* **56**, P3483 (1962)].

[37] P. L. DeBenneville and W. D. Niederhauwer, U.S. Patent 3,267,082 (1966) [*CA* **65**, P18721 (1966)].

[38] J. R. Geigy A.-G., Belgian Patent 659,530 (1965) [*CA* **64**, P6665 (1966)].

[39] P. Aeberli and W. J. Houlihan, *J. Org. Chem.* **34**, 165 (1969).

[40] W. J. Houlihan, U.S. Patent 3,526,660 (1970) [*CA* **73**, P120339 (1970)].

[41] W. J. Houlihan, U.S. Patent 3,334,099 (1967) [*CA* **69**, P96769 (1968)].

[42] W. J. Houlihan, U.S. Patent 3,526,626 (1970) [*CA* **73**, P98976 (1970)].

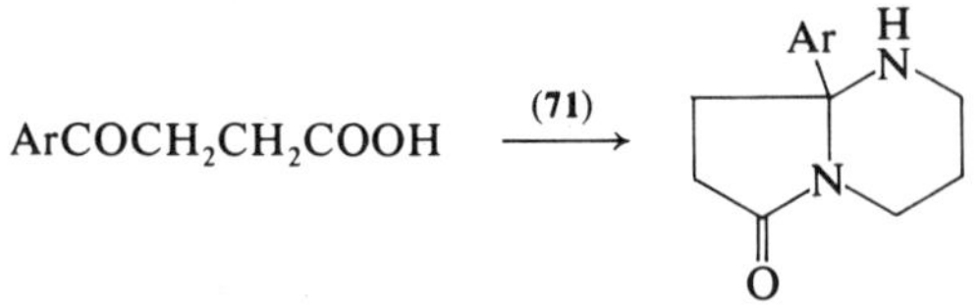

ArCOCH$_2$CH$_2$COOH $\xrightarrow{(71)}$ [structure **75**] (21)

(74)   (75)

Wollweber[43] has extended this reaction to aliphatic γ-keto acids. The products he obtained are identical to **75** except that the 8a-aryl substituent has been replaced with an aliphatic group. Wollweber et al.[32] have also studied the course of the above reaction using NMR spectroscopy. Monitoring the condensation of **76** with **77**, they observe the "aminal ester" (**78**) intermediate after 15 minutes. This intermediate slowly disappears as product **79** is formed [Eq. (22)].

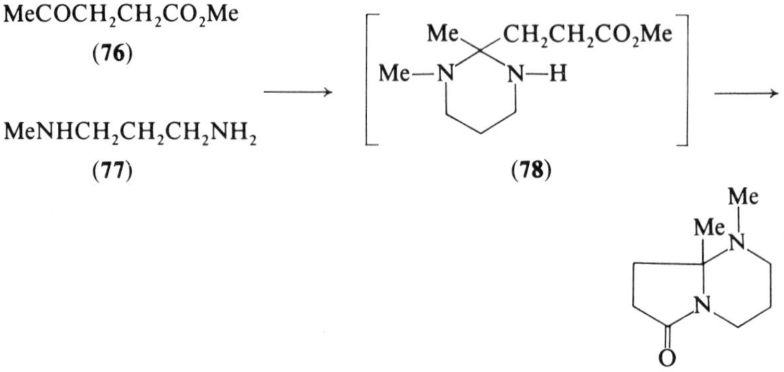

MeCOCH$_2$CH$_2$CO$_2$Me
(76)

MeNHCH$_2$CH$_2$CH$_2$NH$_2$
(77)

$\longrightarrow$ [**78**] $\longrightarrow$ [**79**] (22)

Shemyakin and co-workers[44–47] have shown that the cyclopeptide **80**, when heated in inert solvents, undergoes transannular amide–amide interaction with conversion via the azacyclol **81** into 2,3,4,6,7,8-hexahydropyrrolo[1,2-a]pyrimidin-4-one (**82**) [Eq. (23)]. This appears to be a general reaction of medium ring cyclopeptides.

[43] H. Wollweber, *Angew. Chem., Int. Ed. Engl.* **8**, 69 (1969).
[44] V. K. Antonov, Z. E. Agadzhanyan, T. R. Telesnina, M. M. Shemyakin, G. G. Dvoryantseva, and Yu. N. Shlinker, *Tetrahedron Lett.*, 727 (1964).
[45] M. M. Shemyakin, V. K. Antonov, A. M. Shkrob, V. I. Shchelokov, and Z. E. Agadzhanyan, *Tetrahedron* **21**, 3537 (1965).
[46] V. K. Antonov, Z. E. Agadzhanyan, T. R. Telesnina, and M. M. Shemyakin, *Zh. Obshch. Khim.* **35**, 2231 (1965) [*CA* **64**, 11310 (1966)].
[47] Yu. V. Denisov, V. A. Puchkov, N. S. Vulfson, Z. E. Agadzhanyan, V. K. Antonov, and M. M. Shemyakin, *J. Gen. Chem. USSR* **38**, 743 (1968) [*CA* **69**, 77718 (1968)].

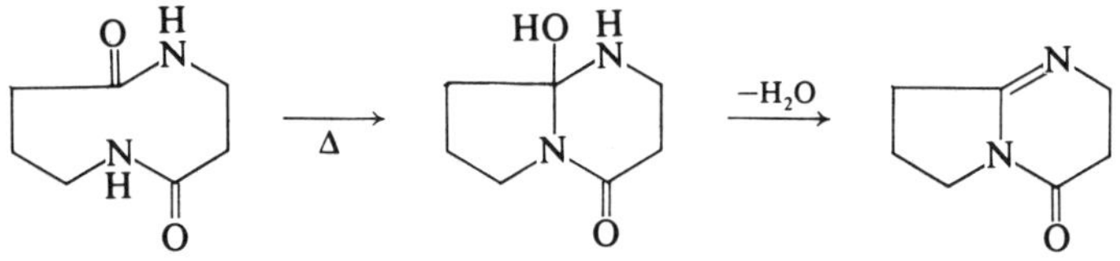

Sec. II.B] PYRROLODIAZINES WITH A BRIDGEHEAD NITROGEN 17

(80) → (81) $-H_2O$ → (82) (23)

B. Reactions

Only a very few reactions, detailing straightforward functional group transformations, such as etherification and nitrile hydrolysis, have been reported for derivatives of the completely unsaturated pyrrolo[1,2-a]-pyrimidine nucleus (Scheme 3). In addition, hydrolysis of a substituted 8-cyanopyrrolo[1,2-a]pyrimidine (25) to 4-aryl-2-aroyl-8-carboxamido-6,7-dimethylpyrrolo[1,2-a]pyrimidine (85) has been reported.[10]

Ref.

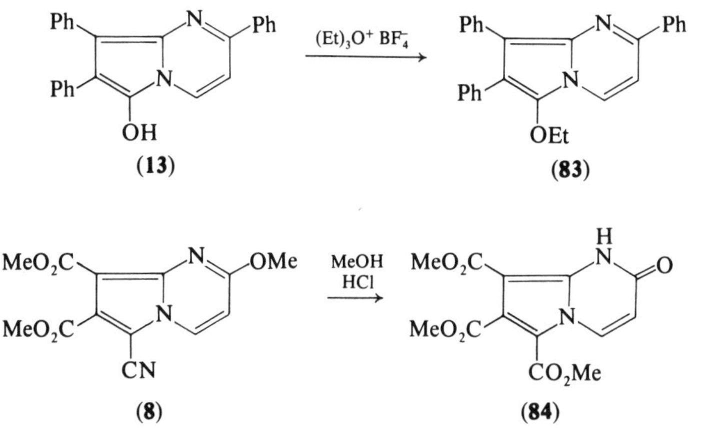

(13) $(Et)_3O^+ BF_4^-$ → (83)   7

(8) MeOH HCl → (84)   3

SCHEME 3

Noteworthy is the stability of the heteroaromatic nucleus to strong acid reaction conditions. The fact that no reports have appeared on either electrophilic or nucleophilic substitution reactions on this ring system is probably due to the total lack of appropriate starting materials; all completely unsaturated examples of this ring system synthesized to date are at least tetrasubstituted.

Etienne et al.[13] and LeBerre and Renault[14-17] have subjected a number of partially unsaturated pyrrolo[1,2-a]pyrimidinones to catalytic hydrogenation. The results are shown in Scheme 4.

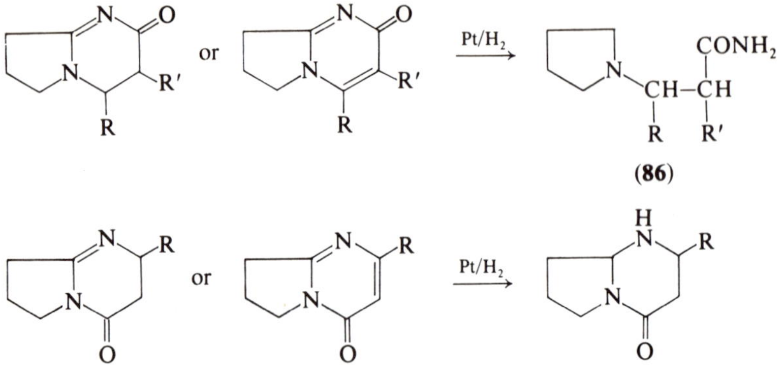

(86) (87)

SCHEME 4

This same group[14] also converted the 2-hydroxyl group in 2-hydroxy-4,6,7,8-tetrahydropyrrolo[1,2-*a*]pyrimidin-4-one (**31b**) into the 2-chloro (**88**) and 2-methoxy (**89**) analogs. In addition, reaction of **31b** with methyl iodide gives betaine **90** (Scheme 5). Patents have recently appeared[48,49] describing the conversion of **31b** into thiophosphate

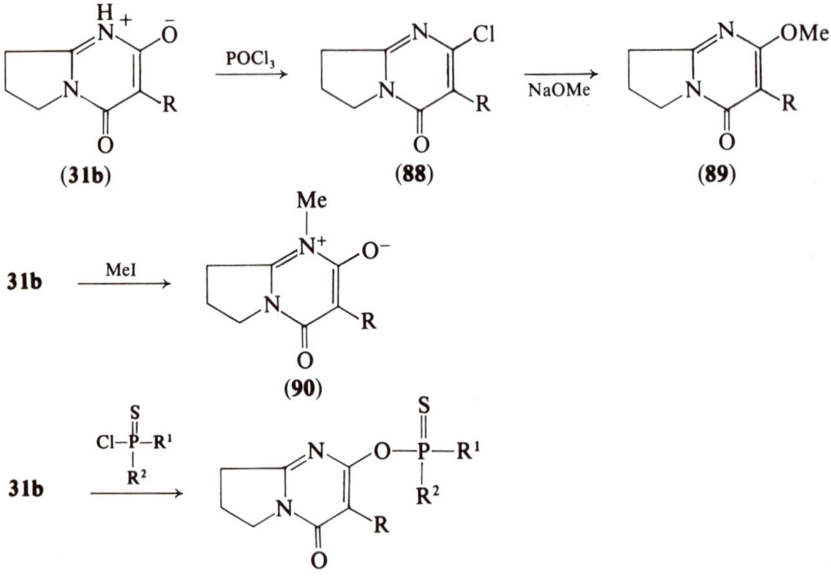

(31b) (88) (89) (90) (91)

SCHEME 5

[48] J. Perronnet and A. Poittevin, German Patent 2,245,386 (1973) [*CA* **79**, P78842 (1973)].

[49] J. Perronnet, A. Poittevin, and L. Taliani, French Patent 2,197,513 (1974) [*CA* **82**, P112101 (1975)].

derivatives (**91**) (Scheme 5). These compounds are reported to have insecticidal and/or acaricidal activity.

Bortnick and Fegley[33-36] have described the reaction of 2,3,4,8-tetrahydropyrrolo[1,2-a]pyrimidines (**72**) with a variety of protic reagents. In all cases addition occurred across the 6,7-double bond affording hexahydro derivatives (**92**) [Eq. (24)].

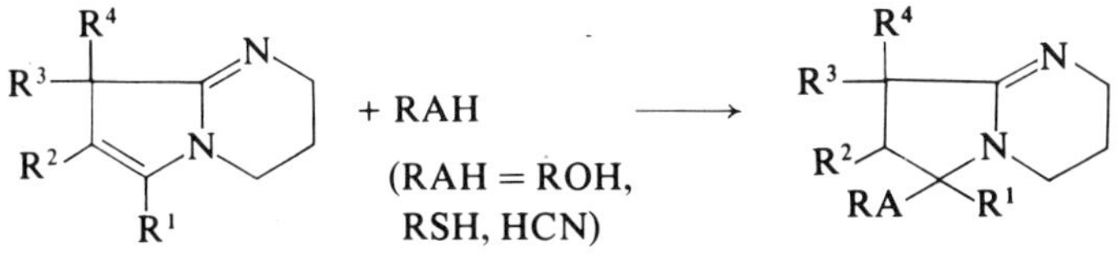

The 6-cyano derivative (**93**), formed in the above reaction can be hydrogenated to the 6-aminomethyl analog **94** in good overall yield [Eq. (25)]. Finally, the reduction of lactam **75** (Ar=Ph) to 8a-phenyl-1,2,3,4,6,7,8,8a-octahydropyrrolo[1,2-a]pyrimidine (**95**) has been accomplished[42] using lithium aluminum hydride [Eq. (26)]. The starting material (**75**) in the above reaction is reported to have sedative activity whereas the product (**95**) exhibits antiinflammatory properties.

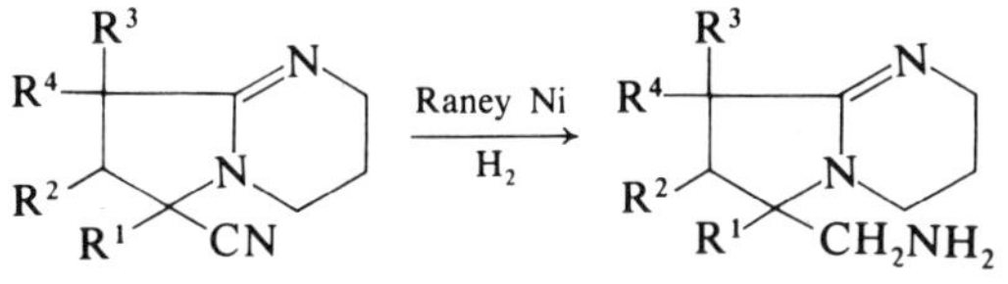

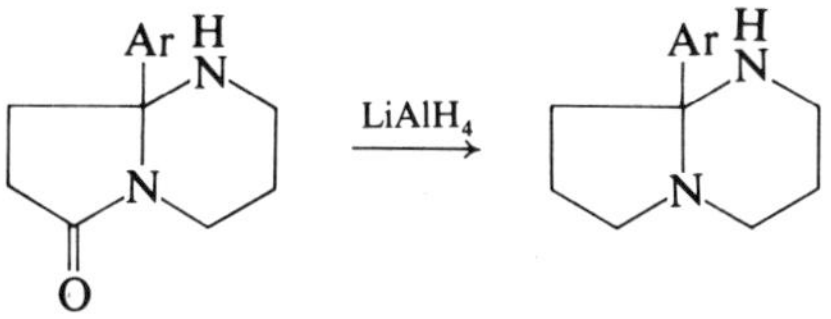

C. Spectral Properties

The NMR and UV spectra of a few selected pyrrolo[1,2-a]pyrimidines with varying degrees of unsaturation are presented in Table I. Additional NMR and UV data are available.[3,7,10,16,21,39] Substituted

TABLE I
SPECTRAL PROPERTIES OF SOME PYRROLO[1,2-a]PYRIMIDINES

Structure	Ultraviolet spectra (solvent) λ_{max}, nm (ϵ)	Nuclear magnetic resonance spectra (solvent) δ	References
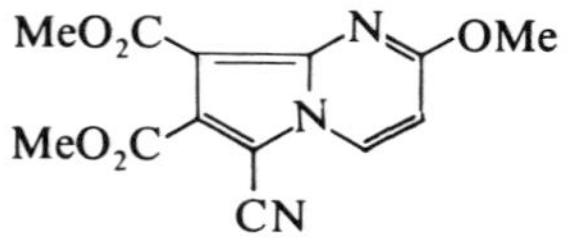 MeO₂C, MeO₂C, OMe, CN structure	(CH₃CN) 241 (34,700) 247 (35,800) 269 (12,800) 298 (7,740) 330 (3,640)	(CDCl₃) 8.28 (d, H4, $J_{3,4}$ = 7.5 Hz) 6.62 (d, H3) 4.08 (2-OCH_3) 3.98 and 3.92 (s, 7,8-CO_2CH_3)	3
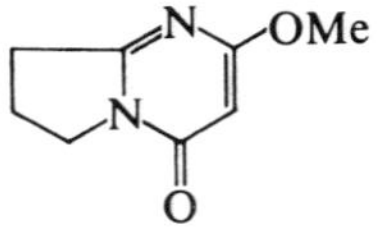 OMe pyrrolopyrimidinone	(EtOH) 247 (26,900) 264 (38,940)	(CDCl₃) 5.53 (s, H3) 4.11 (t, H6) 3.85 (s, 2-OCH₃) 3.10 (t, H8) 2.26 (m, H7)	14
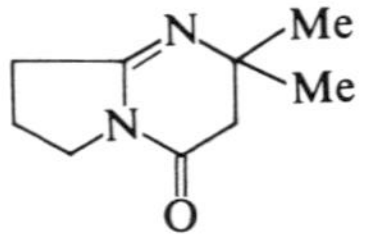 2,2-Me₂ pyrrolopyrimidinone		(CDCl₃) 3.78 (t, H6) 2.65 (m, H8) 2.32 (s, H3) 2.09 (m, H7) 1.21 (s, 2,2-(CH₃)₂)	21

pyrrolo[1,2-a]pyrimidines give a major parent ion in the mass spectrum, and some of the more common fragmentation pathways have also been discussed.[24,39,47]

Theoretical molecular orbital interpretations of the electronic spectra of pyrrolo[1,2-a]pyrimidines have also been reported.[50,51]

D. 2,3,4,6,7,8-Hexahydropyrrolo[1,2-a]pyrimidine (DBN)

1. Synthesis

Reppe[28] reported one of the first authenticated syntheses of the pyrrolo[1,2-a]pyrimidine nucleus with his preparation of DBN (**98**) in 1955. *N*-Cyanoethylation of 2-pyrrolidinone with acrylonitrile gives β-cyanoethylpyrrolidin-2-one (**96**). High-pressure hydrogenation of **96** in the presence of Raney cobalt and ammonia affords *N*-(3-aminopropyl)-pyrrolidone (**97**), which cyclizes under the reaction conditions to yield DBN (**98**) [Eq. (27)]. Others[29,52] have substituted Raney nickel for the Raney cobalt catalyst originally used by Reppe. Patents have also appeared describing the conversion of the amine **97** to **98** using acids[53] or metal oxides[54] at elevated temperatures.

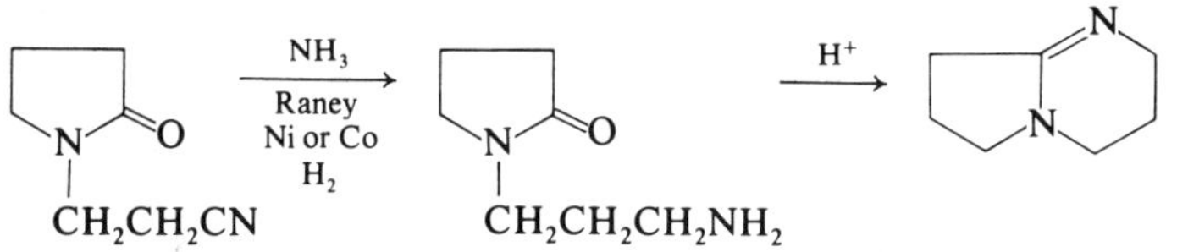

2. Reactions

Iminodicarboxylic acid derivatives such as **99** or **100** react with DBN to afford tricyclic uracil derivatives **101** which are found[55] to be useful as plant protective agents [Eq. (28)]. DBN has also been reported[56] to react

---

[50] V. Galasso, G. DeAlti, and A. Bigotto, *Theoret. Chim. Acta* **9**, 222 (1968).
[51] A. Gumba and G. Arvini, *Gazz. Chim. Ital.* **98**, 167 (1968).
[52] T. Hashimoto, K. Nakatani, S. Suzuki, H. Diago, K. Sugiura, and K. Fujino, Japanese Patent 71 26,516 (1971) [*CA* **75**, P140882 (1971)].
[53] Farbenfabriken Bayer A.-G., French Patent 1,491,791 (1967) [*CA* **69**, P67412 (1968)].
[54] S. Hashimoto, K. Nakatani, S. Suzuki, H. Daigo, and K. Sugiura, U.S. Patent 3,761,436 (1973) [*CA* **79**, P137185 (1973)].
[55] K. Ley, G. Aichinger, A. Botta, H. Hagemann, and E. Niemers, German Patent 2,126,148 (1972) [*CA* **78**, P111360 (1973)].
[56] R. Richter, *Chem. Ber.* **101**, 3002 (1968).

with arylisocyanates to afford tricyclic compounds of type **102** [Eq. (29)].

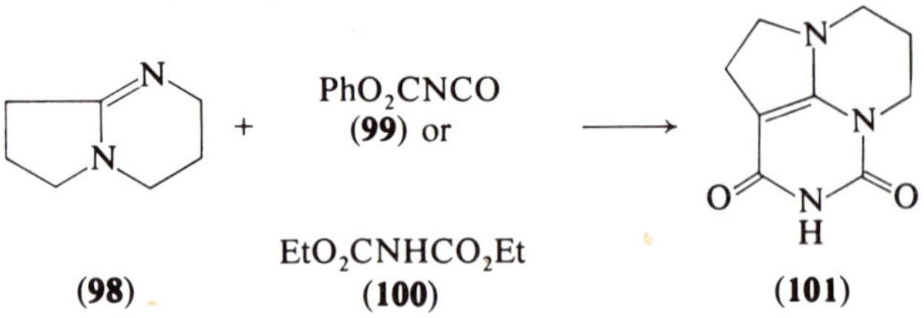

$$(98) + \text{PhO}_2\text{CNCO} \; (99) \text{ or } \text{EtO}_2\text{CNHCO}_2\text{Et} \; (100) \longrightarrow (101) \quad (28)$$

$$(98) + 2 \; \text{ArNCO} \longrightarrow (102) \quad (29)$$

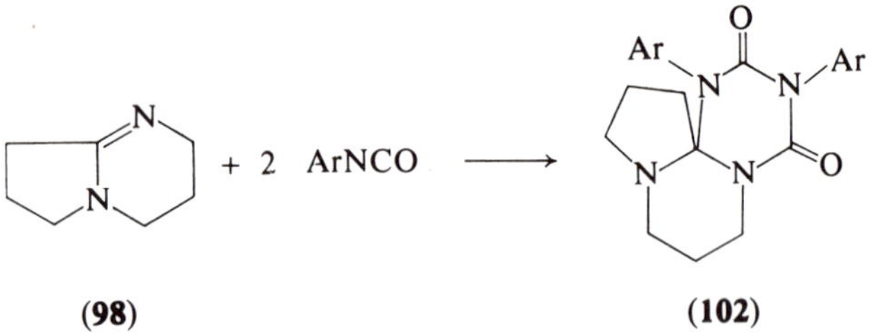

Jarvis and Tong[57] have recently reported on a new class of vinyl betaines prepared by allowing DBN to react with 2,3-diphenylthiirene-1,1-dioxide (**103**) [Eq. (30)]. The yellow betaine **104** prepared initially is oxidized to the colorless sulfobetaine **105** with *m*-chloroperbenzoic acid [Eq. (30)].

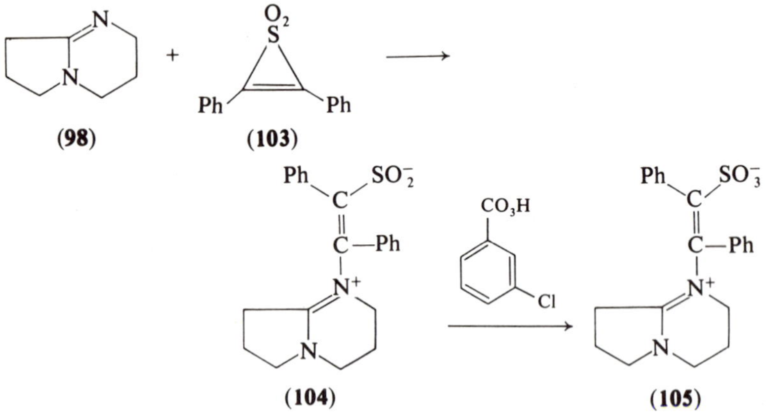

$$(98) + (103) \longrightarrow (104) \xrightarrow{\text{CO}_3\text{H}} (105) \quad (30)$$

3. Utility

a. DBN as a Strong Base. DBN was first effectively used as a base for a dehydrohalogenation sequence in a synthesis of vitamin A.[58] As a

[57] B. Jarvis and W. Tong, *Synthesis*, 102 (1975).
[58] H. Oediger, F. Möller, and K. Eiter, *Synthesis*, 591 (1972).

result of the excellent results obtained in this initial application, DBN was rapidly exploited by a number of chemists and has become one of the preferred reagents for dehydrohalogenations of relatively sensitive molecules. The use of DBN (and other bicyclic amidines) as strong, nonnucleophilic bases in organic synthesis was reviewed in 1972;[58] therefore, only a few examples of its utility as a base, all taken from the recent literature, will be presented (Scheme 6).

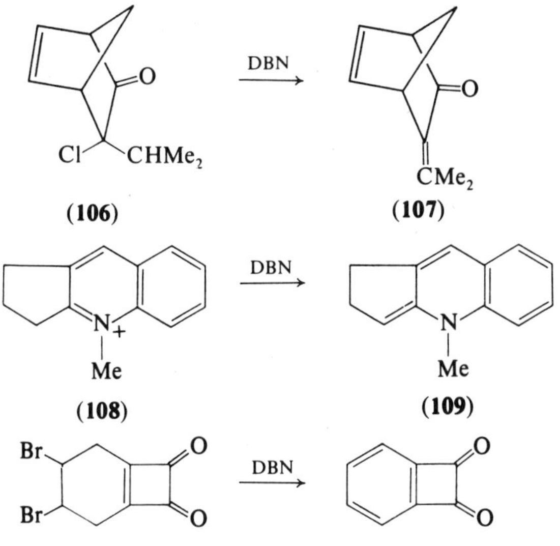

(106) (107) (108) (109) (110) (111)

SCHEME 6

b. DBN in Medicinal Products. DBN itself has been patented[62] as an inhibitor of indoleamine-*N*-methyltransferase.

Combining DBN with cretyl bromide (**112**) in refluxing benzene gives the expected quaternary salt (**113**) [Eq. (31)], which has antibacterial activity.[63]

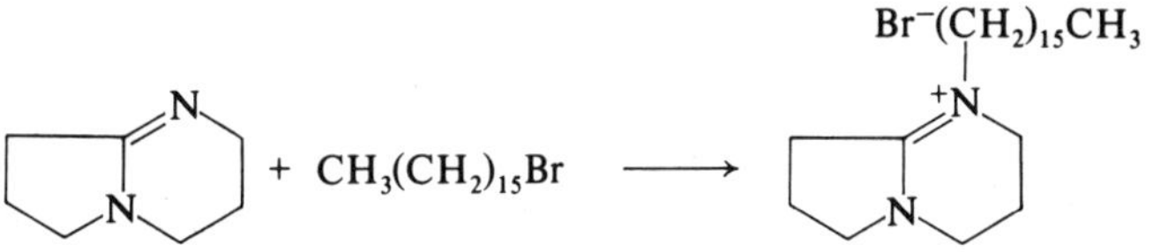

(**98**) (**112**) (**113**) (31)

[59] R. C. DeSelms and F. Delay, *J. Am. Chem. Soc.* **95**, 274 (1973).
[60] J. J. Eisch, F. J. Gadek, and G. Gupta, *J. Org. Chem.* **38**, 431 (1973).
[61] T. Kowar and E. LeGoff, *Synthesis*, 212 (1973).
[62] L. R. Mandel, U.S. Patent, 3,749,781 (1973) [*CA* **79**, P96967 (1973)].
[63] D. J. Ellis and D. Rammler, U.S. Patent 3,652,564 (1972) [*CA* **76**, P153771 (1972)].

In a similar fashion 3-β-bromoandrost-5-en-17-one (**114**) affords the salt **115** when heated in the presence of DBN [Eq. (32)]. This quaternary steroid is also found[64] to have antibacterial activity.

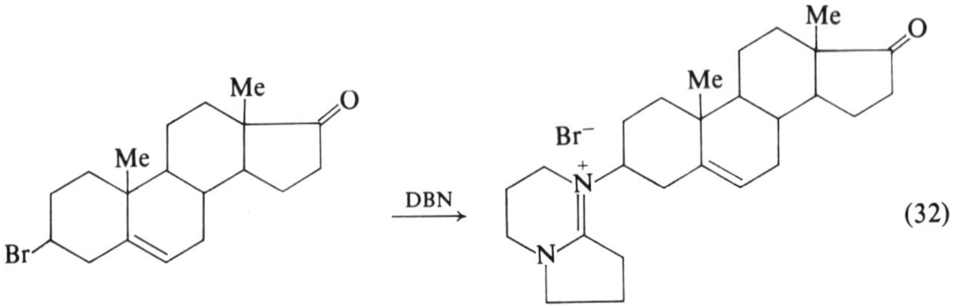

McFarland et al.,[65] in studying structure–activity relationships around the potent anthelmintic agent pyrantel (**116**), prepared 2,3,4,6,7,8-hexahydro-8-(2-thenylidene)pyrrolo[1,2-a]pyrimidine (**118**) by condensing DBN with 2-thiophenecarboxaldehyde (**117**) in the presence of methyl formate [Eq. (33)]. However, **118** was inactive as an anthelmintic agent.

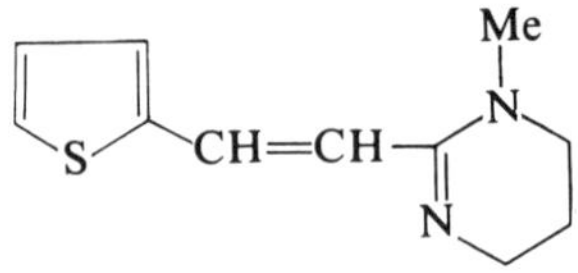

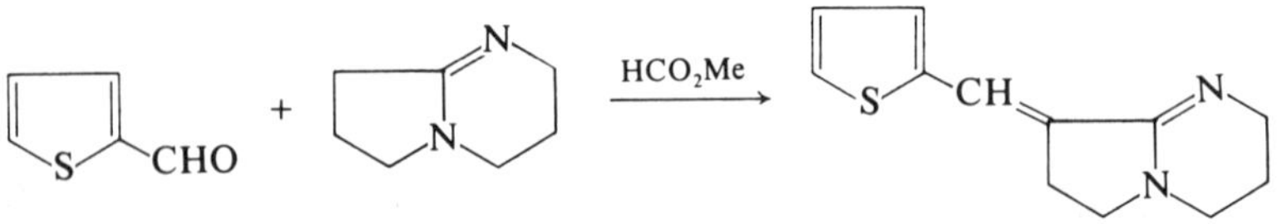

c. *DBN as an Industrial Catalyst.* DBN has been shown to catalyze the reaction of polyhydroxy compounds with polyisocyanates, giving

---

[64] D. J. Ellis and D. Rammler, U.S. Patent 3,553,211 (1971) [*CA* **74**, 88224 (1971)].
[65] J. W. McFarland, L. H. Conover, H. L. Howes, J. E. Lynch, D. R. Chisholm, W. C. Austin, R. L. Cornwell, J. C. Danilewicz, W. Courtney, and D. H. Morgan, *J. Med. Chem.* **12**, 1066 (1969).

polyurethanes and polyurethane foams.[66–70] DBN has also been patented as a catalyst for hardening epoxy resins,[71–73] as a corrosion inhibitor,[74,75] as a vulcanization accelerator,[76] as a desensitizer for leuco dye copying papers,[77] and as a catalyst in ethyleneimine manufacture.[78] In addition, copper salts of DBN are reported[79] to catalyze the oxidative coupling of 2,6-disubstituted phenols to polyaryl ethers.

III. Pyrrolo[1,2-c]pyrimidines (2)

A. Syntheses

The first derivative of the pyrrolo[1,2-c]pyrimidine system was isolated by Herz[80] in unspecified yield from treatment of the acrylic acid (**119**) or its azlactone with sodium hydroxide as shown in Eq. (34). The product, 1-phenylpyrrolo[1,2-c]pyrimidine-3-carboxylic acid (**120**), was then converted into a methyl ester.

[66] C. A. Aufdermarsh and A. W. Fogiel, British Patent 1,287,150 (1972) [*CA* **73**, P4585 (1970)].
[67] T. Hashimoto, K. Nakatani, S. Suzuki, H. Daigo, and I. Fujino, Japanese Patent 70 35,071 (1970) [*CA* **74**, P64964 (1971)].
[68] S. Hashimoto, K. Nakatani, S. Suzuki, H. Daigo, and I. Fujino, Japanese Patent 70 40,554, (1970) [*CA* **74**, P126482 (1971)].
[69] S. Hashimoto, K. Nakatani, S. Suzuki, H. Daigo, and I. Fujino, Japanese Patent 70 40,553 (1970) [*CA* **74**, P126484 (1971)].
[70] S. Hashimoto, K. Nakatani, and H. Daigo, U.S. Patent 3,769,244 (1973) [*CA* **80**, P84247 (1974)].
[71] T. Hashimoto, K. Nakatani, and S. Suzuki, Japanese Patent 73 17,880 (1973) [*CA* **81**, P4435 (1974)].
[72] T. Hashimoto, K. Nakatani, and S. Suzuki, Japanese Patent 72 01,115 (1972) [*CA* **77**, P115411 (1972)].
[73] S. Hashimoto, K. Nakatani, and S. Suzuki, U.S. Patent 3,662,540 (1971) [*CA* **77**, P6463 (1972)].
[74] S. Hashimoto and K. Hirose, U.S. Patent 3,625,859 (1971) [*CA* **73**, P6558 (1970)].
[75] T. Hashimoto and K. Hirose, Japanese Patent 71 02,169 (1971) [*CA* **75**, P100462 (1971)].
[76] J. M. Bowman, British Patent 1,300,782 (1972) [*CA* **76**, P47156 (1972)].
[77] A. Miyamoto, H. Matsukawa, A. Watanabe, German Patent 2,328,312 (1973) [*CA* **80**, P151274 (1974)].
[78] F. Matsuda, T. Takahashi, and N. Oogiya, Japanese Patent, 74 24,945 (1974) [*CA* **80**, P145994 (1974)].
[79] H. Wieden and U. Bahr, British Patent 1,134,613 (1968) [*CA* **70**, P29589 (1969)].
[80] W. Herz, *J. Am. Chem. Soc.* **71**, 3982 (1949).

Preparation of 1-oxo-1,5,6,7-tetrahydropyrrolo[1,2-c]pyrimidine (121) and its 3-methyl derivative (122) has been accomplished by Schuett and Rapoport[81] from the corresponding 2-aminoethylpyrroles as in Scheme 7. Spectral and other data on compound 121 indicate it to be identical to a product isolated from degradation (phosphorus/hydriodic acid) of the natural product saxitoxin (see Section III,D).

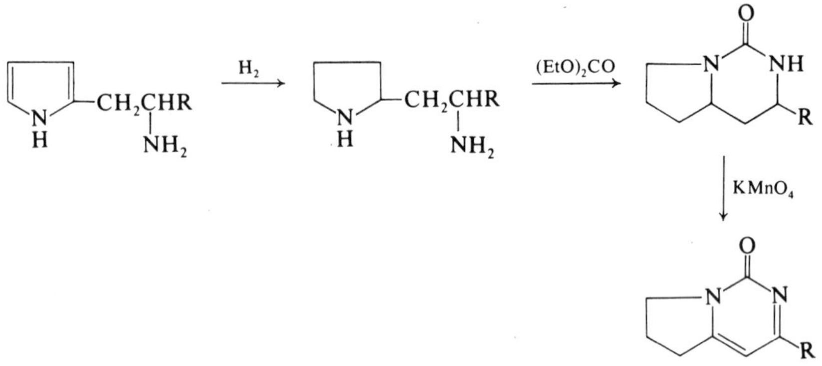

SCHEME 7

The Chichibabin cyclization procedure for quaternary salts has been successfully applied [Eq. (35)] by Boekelheide and Kertelj[82] to certain phenacylpyrimidinium bromides to produce fair yields of 3-methyl-6-phenylpyrrolo[1,2-c]pyrimidine (123).

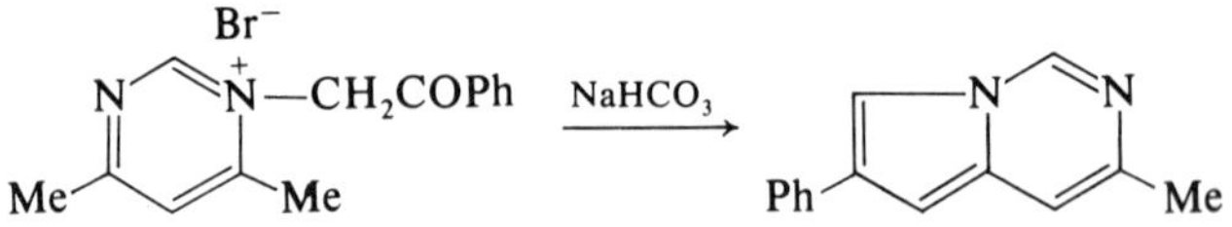

Wibberley and his co-workers[83,84] have examined several synthetic routes to derivatives of 2. For example, an intramolecular aldol-type cyclization [Eq. (36)] of the pyrimidine (124) in acetic anhydride produces 3,6,7-triphenyl-5-acetyl-pyrrolo[1,2-c]pyrimidine (125) in unspecified yield.

[81] W. Schuett and H. Rapoport, *J. Am. Chem. Soc.* **84**, 2266 (1962).
[82] V. Boekelheide and S. S. Kertelj, *J. Org. Chem.* **28**, 3212 (1963).
[83] T. Melton, J. Taylor, and D. G. Wibberley, *Chem. Commun,* 151 (1965).
[84] J. Taylor and D. G. Wibberley, *J. Chem. Soc. C,* 2693 (1968).

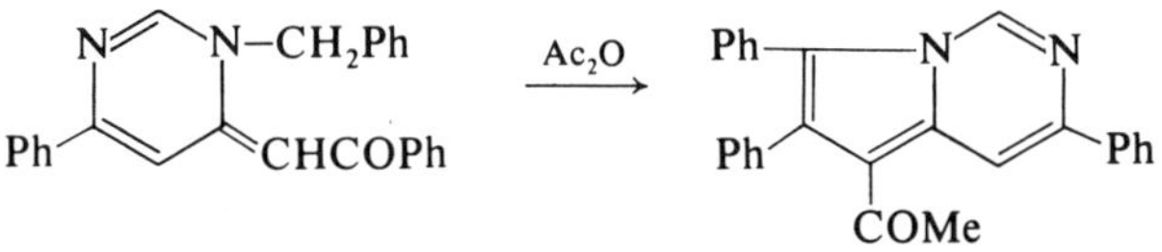

Sec. III.A] PYRROLODIAZINES WITH A BRIDGEHEAD NITROGEN

$$\text{(124)} \xrightarrow{Ac_2O} \text{(125)} \quad (36)$$

After probing four possible synthetic routes to derivatives of **2**, best results were obtained[84] from the Chichibabin procedure shown in Eq. (37) (e.g., the preparation of the 3,6-diphenyl derivative **126** in 98% yield). A model synthetic procedure and tabulation of yields for five analogs of **126** is available.[2]

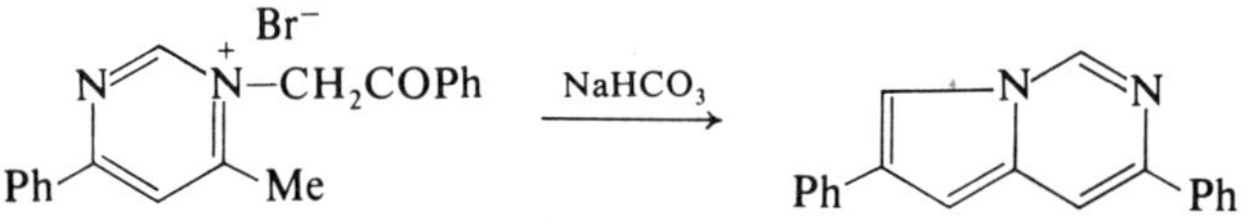

$$\xrightarrow{NaHCO_3} \text{(126)} \quad (37)$$

The completely unsubstituted pyrrolo[1,2-c]pyrimidine **127** has been synthesized from 4-methylpyrimidine and butyl glyoxalate in seven synthetic steps[85] (Scheme 8). This last step proceeds in only 9% yield, probably a reflection of the poor nucleophilicity of the pyrimidine nitrogen atom. Compound **127** melts at 36°–39°, is highly volatile, and darkens slowly in air.

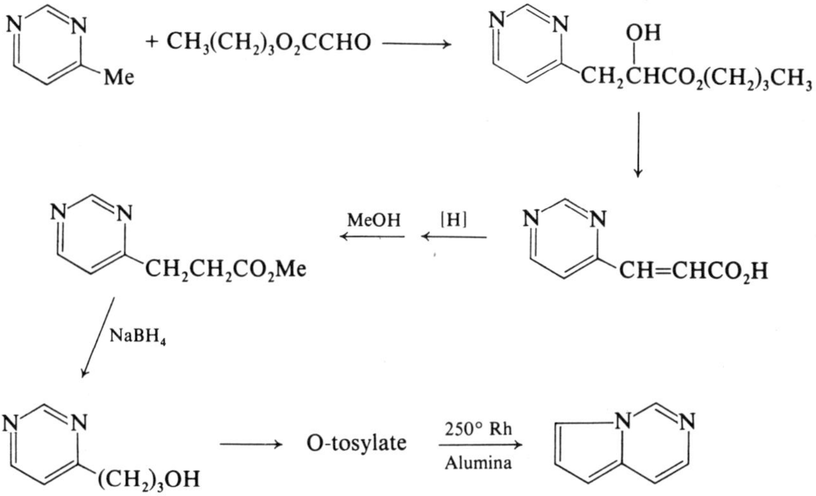

SCHEME 8

[85] J. L. Wong, M. S. Brown, and H. Rapoport, *J. Org. Chem.* **30**, 2398 (1965).

A German patent[86] claims the preparation of 7-oxo-1,2,3,4,4a,5,6,7-octahydro-4a-methyl-pyrrolo[1,2-c]pyrimidine (**128**) from angelicalactone and 1,3-propylenediamine in inert solvents at elevated temperatures [Eq. (38)]. However, the assigned structure is clearly incorrect and the product of this reaction is surely a pyrrolo[1,2-a]pyridine analogous to **75** (p. 16).

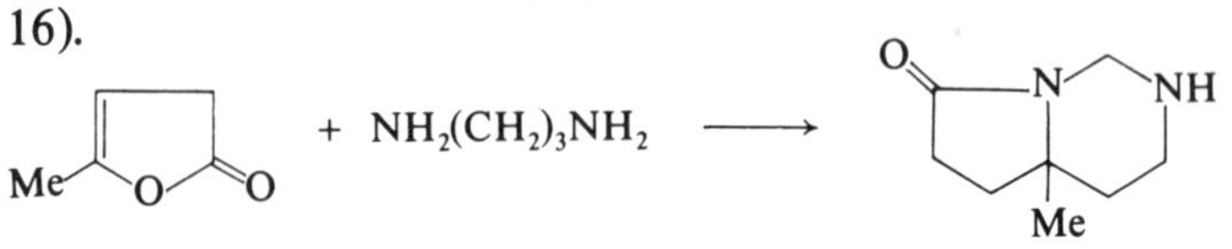

$$\text{Me-furanone} + NH_2(CH_2)_3NH_2 \longrightarrow \text{(128)} \quad (38)$$

(**128**)

B. Reactions

The relatively few reactions reported for pyrrolo[1,2-c]pyrimidines are of three main types—electrophilic attack, addition reactions, and ring opening reactions.

Electrophilic reactions on compound **2** produce 7-substituted derivatives. Thus, nitrosation, acylation, diazonium coupling, Reimer–Tiemann conditions or phosgenation of a substituted **2**, produce[84] the corresponding 7-substituted derivatives shown in Eq. (39)

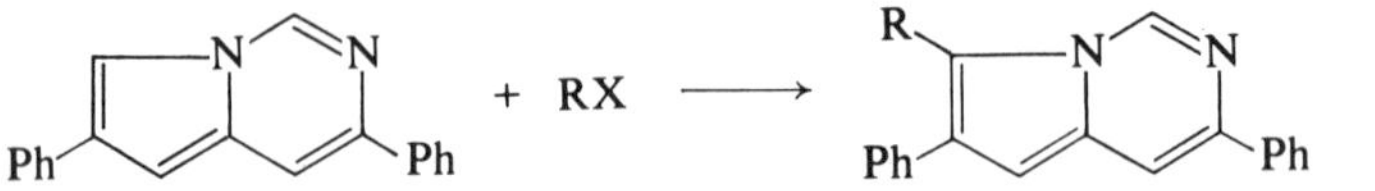

$$\text{(126)} + RX \longrightarrow \text{product} \quad (39)$$

(**126**)   R = NO, COMe; N = NPh, CHO, COOH

in yields of 48–82%. The loss of the 7-H NMR resonance at δ 7.11 after an electrophilic reaction on **126** is evidence for the position of the newly introduced R group. Bromination of **126** using one equivalent of N-bromosuccinimide produces minor amounts of 5-bromo- and 7-bromo products with major quantities of the 5,7-dibromo product (**126a**).[87] However, use of two equivalents of brominating agent produces only the 5,7-dibromo compound (**126a**). Bromine in chloroform converts compound **126** to **126a** perbromide salt, which on treatment with base gives the dibromo derivative (**126a**); treatment of the perbromide salt with an alcohol produces 4-alkoxy-5,7-dibromo-3,6-diphenylpyrrolo[1,2-c]-pyrimidines (**126b**).

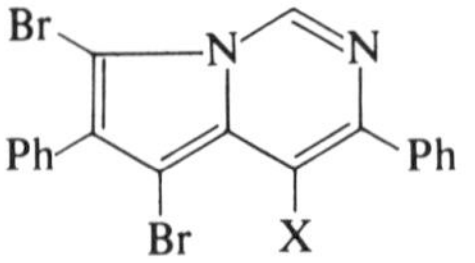

(**126a**) X = H
(**126b**) X = OMe or OEt

---

[86] German Patent 1,802,468 (1968) [*CA* **73**, P255432 (1970)].
[87] W. J. Irwin, D. G. Wibberley, and G. Cooper, *J. Chem. Soc. C*, 3870 (1971).

A single addition reaction has been reported for a pyrrolo[1,2-c]-pyrimidine. Dimethyl acetylenedicarboxylate has been added to the 3-methyl-6-phenyl derivative **123** to produce the 5-azacycl[3.2.2]azine **129** in good yield[82] [Eq. (40)].

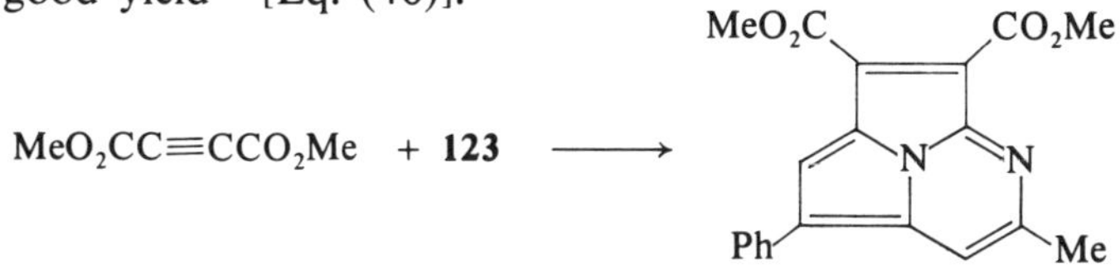

$$\text{MeO}_2\text{CC} \equiv \text{CCO}_2\text{Me} + \mathbf{123} \longrightarrow \mathbf{(129)} \quad (40)$$

In the course of a slow catalytic hydrogenation of 7-nitroso-3,6-diphenylpyrrolo[1,2-c]pyrimidine (**130**), a product resulting from pyrrole ring cleavage was isolated.[88,89] The suggestion was made by Irwin and Wibberley[88,89] that an intermediate nitrene derived from the nitroso group in **130** isomerizes to yield compound **131** [Eq. (41)]. To support this hypothesis, the proposed nitrene was generated by another route by converting the nitroso compound **130** into the corresponding 7-amino derivative.[88,89] Oxidation of this amine, a reaction also capable of proceeding via a nitrene intermediate, was then shown to yield the same nitrile (**131**).

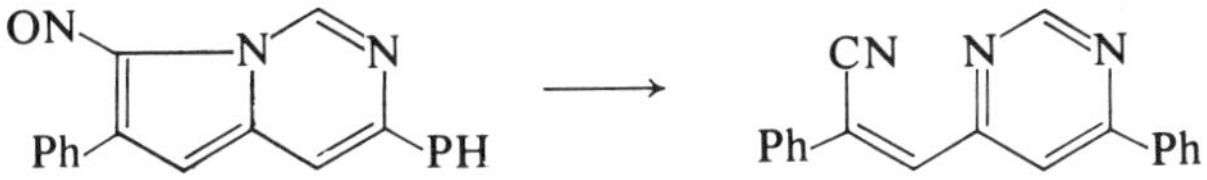

$$\mathbf{(130)} \longrightarrow \mathbf{(131)} \quad (41)$$

4-Cyanopyrrolo[1,2-c]pyrimidine (**132**) is reported to form from pyrrole-2-acetonitrile and N,N-dimethyl-(3-dimethylamino-2-aza-alkylidene)ammonium perchlorate [Eq. (42)] in quinoline solution.[90]

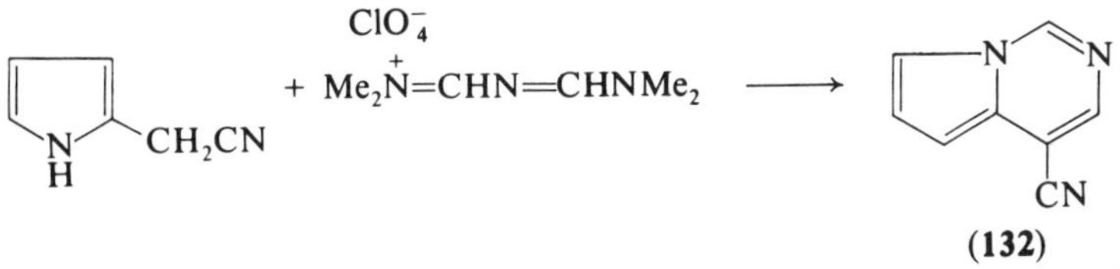

$$+ \text{Me}_2\overset{+}{\text{N}}=\text{CHN}=\text{CHNMe}_2 \quad \text{ClO}_4^- \longrightarrow \mathbf{(132)} \quad (42)$$

However, no physical data beyond a melting point were quoted. If confirmed, this process would appear to have the potential for preparing a number of substituted pyrrolo[1,2-c]pyrimidines from appropriately 2-substituted pyrroles.

---

[88] W. J. Irwin and D. G. Wibberley, *Chem. Commun.*, 878 (1968).
[89] W. J. Irwin and D. G. Wibberley, *J. Chem. Soc. C*, 3237 (1971).
[90] C. Jutz, R. M. Wagner, and H.-G. Löbering, *Angew. Chem., Int. Ed. Engl.* **13**, 737 (1974).

Polarographic reduction of pyrrolo[1,2-c]pyrimidines proceeds through two distinct 1-electron steps to form the dihydro derivative.[91]

C. Spectral Properties

A compilation of the spectral properties of two pyrrolo[1,2-c]-pyrimidines is presented in Table II. Theoretical molecular orbital interpretations of the electronic spectra of pyrrolo[1,2-c]pyrimidines have also been reported.[51]

TABLE II

Spectral Properties of Some Pyrrolo[1,2-c]pyrimidines

Structure	Ultraviolet spectra (solvent) λ_{max}, nm (ϵ)	Nuclear magnetic resonance (solvent) δ	Infrared spectra (vehicle) μm	References
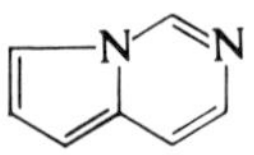 (pyrrolo[1,2-c]pyrimidine)	(EtOH) 229 (27,400) 272 (5,450) 283 (5,900) 345 (1,080)	(DCCl$_3$) 8.84 (s, H1) 6.94 (t, H6) 6.50 (d, H5) 7.37 (m, H3, 4, 7)	(CHCl$_3$) 3.52, 6.92, 7.24 7.41, 8.71, 9.1 9.31, 11.85	85
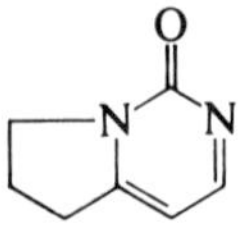 (1-oxo-tetrahydropyrrolo[1,2-c]pyrimidine)	(CH$_3$OH) 302 (6,100) (MeOH/acid) 307 (7,200)	(CDCl$_3$) 8.43 (H3)[a] 6.40 (H4)	(CHCl$_3$) 6.02–6.21 (broad) HCl salt: s.89, 6.21	81

[a] Proton assignments were not made by the authors of the original article. These assignments have been made by the authors of this review article and should be considered as tentative.

D. Natural Products

Two naturally occurring compounds have been shown to contain a pyrrolo[1,2-c]pyrimidine ring system. Saxitoxin, a paralytic poison isolated from certain shellfish, has been investigated by several groups.[81,92,93] It has a complex structure that has been degraded through exhaustive reduction to 3-methyl-1-oxo-1,5,6,7-tetrahydropyrrolo[1,2-c]-pyrimidine (122). Synthesis of the isomeric 4-methyl derivative by Irino,[93] and demonstration of its nonequivalence to the product isolated

[91] A. V. Lizogub, Z. N. Timofeeva, M. L. Aleksandrova, A. V. El'tsov, and E. G. Petrova, *Zh. Obshch. Khim.* **43**, 2291 (1973) [*CA* **80**, 47906 (1974)].
[92] J. L. Wong, R. Oesterlin, and H. Rapoport, *J. Am. Chem. Soc.* **93**, 7344 (1974).
[93] R. R. Irino, *Sci. Aerospace Rep.* **3**, 2499 (1965) [*CA* **66**, 94980 (1967)].

from the natural source, helps establish the identity of the degradation product **122**.

Seeds of the *Plantago arenaria* plant yield an alkaloid,[94] called "arenaine" (**133**) whose structure was assigned by Rabaron *et al*.[95] mainly on the basis of ^{13}C nuclear magnetic spectroscopy.

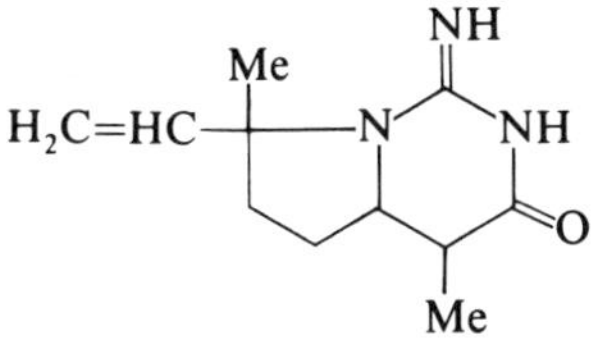

(**133**)

IV. Pyrrolo[1,2-*a*]pyrazines (3)

Over 120 references to this heterocyclic ring system has appeared since Mosby[1] reviewed the literature and found 28 references prior to 1957. The octahydro derivatives, in addition to being called 1,4-diazabicyclo[4.3.0]nonanes or 1,2-trimethylenepiperazines, are also known as anhydrides or lactams of proline with a second amino acid or as diketopiperazines. The disproportionately large number of octahydro derivatives of this ring system will be treated separately in Section IV,D.

A. SYNTHESES

Only a few new syntheses of the fully unsaturated pyrrolo[1,2-*a*]-pyrazine system have appeared since the literature was last reviewed.[1] Acheson and Foxton[5] obtained low yields of the triesters **134** and **135** [Eq. (43)] from the corresponding pyrazines and dimethyl acetylenedi-

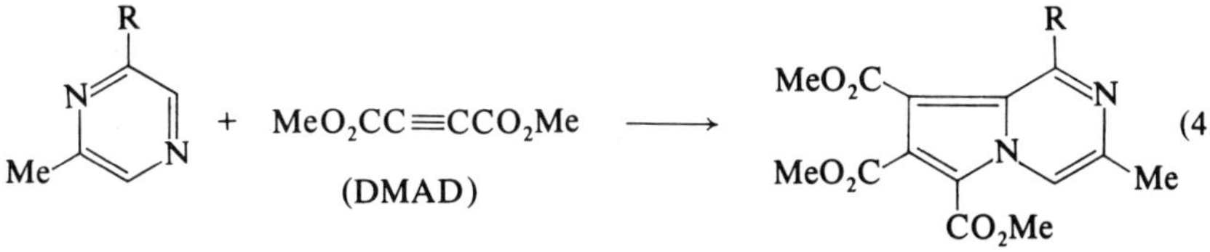

R = H, Me

(**134**) R = H
(**135**) R = Me

(43)

[94] J. Peyroux, M. H. Mehri, M. Plat, P. Rossignol, and G. Valette, *Ann. Pharm. Fr.* **30**, 51 (1972).
[95] A. Rabaron, M. Koch, M. Plat, J. Peyroux, E. Wenkert, and D. Cochran, *J. Am. Chem. Soc.* **93**, 6270 (1971).

carboxylate (DMAD). A more versatile synthesis of substituted pyrrolo-[1,2-a]pyrazines has been described.[96,97] 2-Pyrrole aldehyde sodium salt is alkylated with an α-halocarbonyl compound followed by cyclization of the resultant product in the presence of ammonium acetate [Eq. (44)].

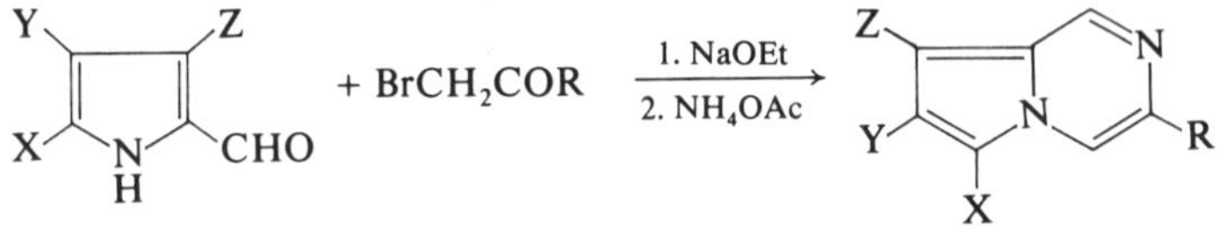

$$\underset{\underset{H}{X}}{\overset{Y}{\bigvee}}\overset{Z}{\underset{CHO}{\bigvee}} + BrCH_2COR \xrightarrow[\text{2. NH}_4\text{OAc}]{\text{1. NaOEt}} \text{(136)} \quad (44)$$

Four examples of compound **136** were prepared[97] by this synthetic procedure. The possibility of utilizing other 2-acylpyrroles in order to obtain 1-substituted derivatives of **136** is also mentioned.[96]

In a broad study of 1,3-dipolar additions to various nitrogen heterocycles, Sasaki et al.[98] obtained very low yields (4–8%) of the diesters **137** and **138** when 3-methylpyrazinium N-phenacylide was treated with DMAD [Eq. (45)].

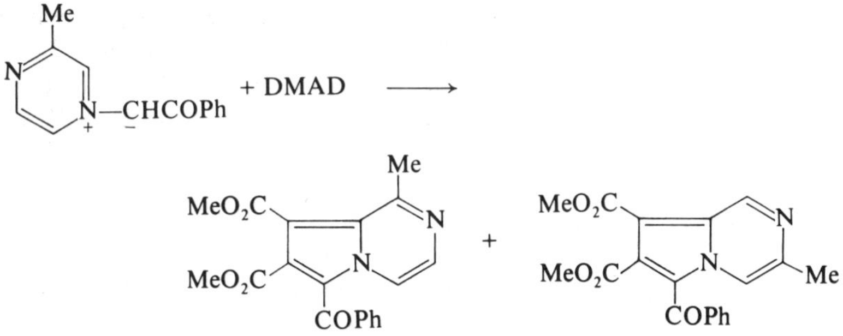

$$\text{pyrazinium ylide} + DMAD \longrightarrow \text{(137)} + \text{(138)} \quad (45)$$

Another pyrazinium ylide (**139**), however, produces[3] satisfactory yields of the 6-cyano compound (**140**) [Eq. (46)] when combined with DMAD.

Lown and Matsumoto[7,99] have examined the reaction of pyrazine and methylpyrazines with either diphenylcyclopropenone[7] or diphenylcyclopropenethione.[99] In the case of the ketone and a pyrazine derivative, a 6-

---

[96] V. I. Shvedov, L. B. Altukhova, A. V. Bocharnikova, and A. N. Grinev, USSR Patent 237,153 (1969) [CA **71**, 13142 (1969)].

[97] V. I. Shvedov, L. B. Altukhova, and A. N. Grinev, Khim. Geterotsikl. Soedin., 1048 (1970) [CA **74**, 125628 (1971)].

[98] T. Sasaki, K. Kanematsu, Y. Yukimoto, and S. Ochiai, J. Org. Chem. **36**, 813 (1971).

[99] J. W. Lown and K. Matsumoto, Can. J. Chem. **49**, 3119 (1971).

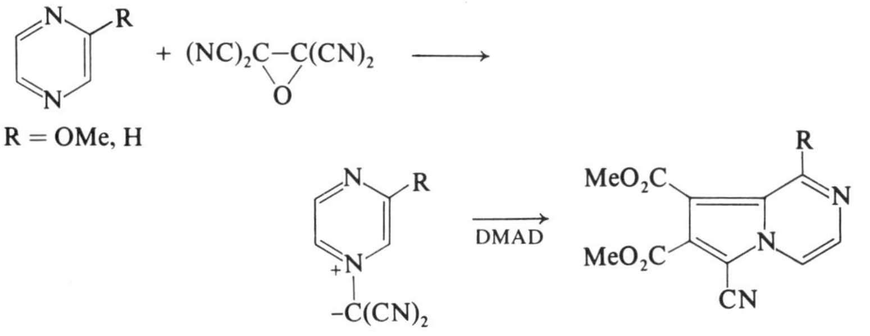

Sec. IV.B] PYRROLODIAZINES WITH A BRIDGEHEAD NITROGEN 33

$$\text{(pyrazine-R)} + (NC)_2C-C(CN)_2 \longrightarrow$$
$$\underset{O}{\diagdown\diagup}$$

R = OMe, H

$$\text{(139)} \xrightarrow{\text{DMAD}} \text{(140)} \quad (46)$$

hydroxy-7,8-diphenylpyrrolo[1,2-*a*]pyrazine (**141**) is obtained [Eq. (47)] in high yield. Full spectra data are given[7] for compound **141** and for a related product obtained from unsubstituted pyrazine itself.

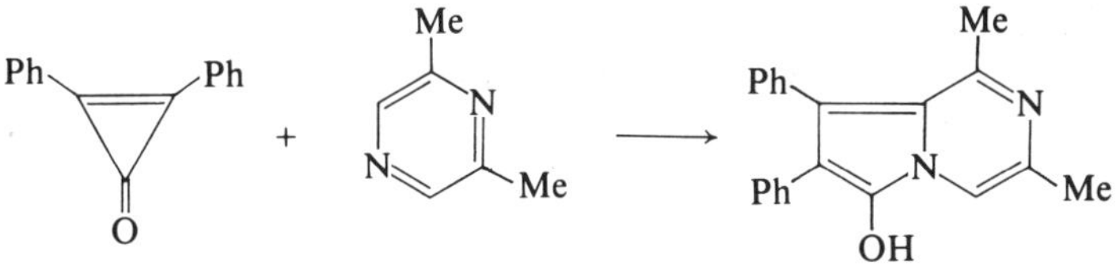

$$\text{(Ph-cyclopropenone + Me-pyrazine)} \longrightarrow \text{(141)} \quad (47)$$

In the case of the thione **142** and pyrazine derivatives, a 6-mercapto-7,8-diphenylpyrrolo[1,2-*a*]pyrazine (**143**) is obtained [Eq. (48)] in yields of 25–75%, depending on the substituents on the starting pyrazine.[99] A further complication of this thione reaction is the variable amounts of the disulfide derived from **143**, which forms[99] readily in the presence of oxygen.

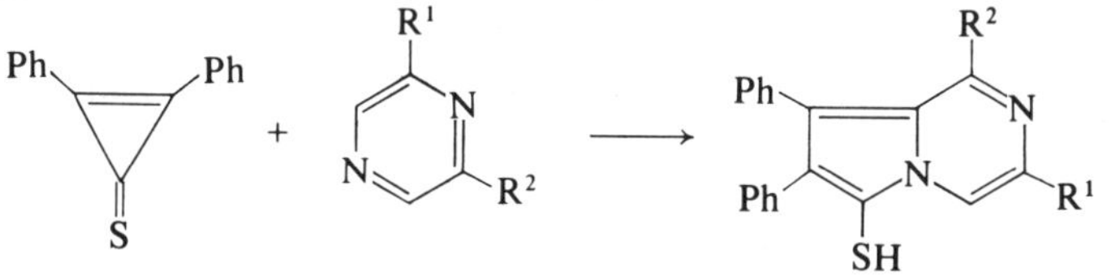

$$\text{(142)} + \text{(pyrazine } R^1, R^2\text{)} \longrightarrow \text{(143)} \quad (48)$$

(**142**) $R^1, R^2 = H, Me$ (**143**)

B. REACTIONS

The only reported studies of reactions of pyrrolo[1,2-*a*]pyrazines are those of Paudler and his co-workers, who examined the effects of both an electrophilic[100] and a nucleophilic[101] reagent on this ring system.

[100] W. W. Paudler and D. E. Dunham, *J. Heterocycl. Chem.* **2**, 410 (1965).
[101] W. W. Paudler, C. I. P. Chao, and L. S. Helmick, *J. Heterocycl. Chem.* **9**, 1157 (1972).

Bromination of the unsubstituted pyrrolo[1,2-a]pyrazine (**144**) in ethanol–water [Eq. (49)] produces 6,8-dibromopyrrolo[1,2-a]pyrazine (**145**) in good yield.[100]

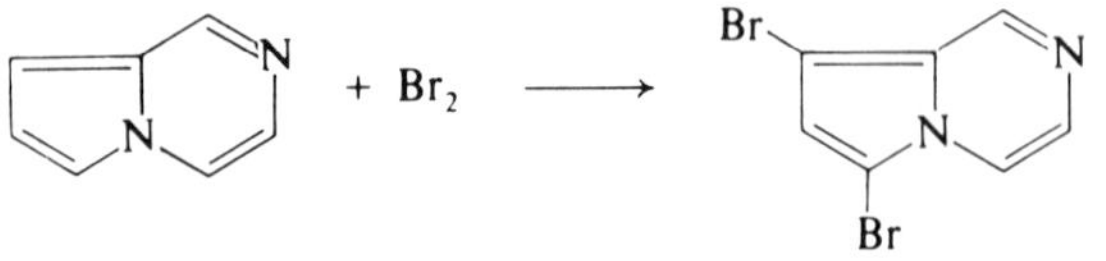

$$\text{(144)} + Br_2 \longrightarrow \text{(145)} \quad (49)$$

(**144**)   (**145**)

The nucleophilic reagent phenyllithium, on the other hand, adds to the 1-position of compound **144**, and the resulting dihydro derivative is air oxidized[101] to 1-phenylpyrrolo[1,2-a]pyrazine (**146**) [Eq. (50)]. Other examples of nucleophilic additions to pyrrolo[1,2-a]pyrazines should be examined in order to delineate the scope of this approach for preparing 1-substituted derivatives of this ring system.

$$\mathbf{144} + PhLi \longrightarrow \text{(146)} \quad (50)$$

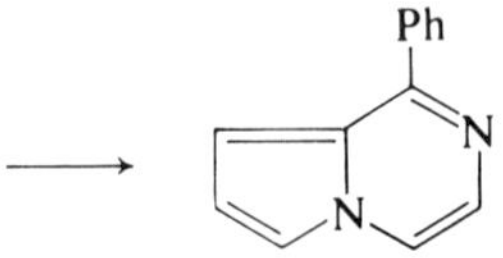

(**146**)

C. Spectral Properties

A compilation of a few representative pyrrolo[1,2-a]pyrazines and their spectral properties are given in Table III. In addition, the electronic spectrum of this ring system has been calculated[50] and interpreted[51] using molecular orbital methods.

D. Octahydro Derivatives

1. *Octahydropyrrolo[1,2-a]pyrazines*

Several synthetic routes have been utilized for the preparation of the completely saturated derivatives of pyrrolo[1,2-a]pyrazines. Stoll and Petrzilka[102] found that reduction of the dioxo compound **147** (a "leucylproline lactam," see Section D,3) with lithium aluminum hydride (LAH) [Eq. (51)] gives 3-isobutyloctahydropyrrolo[1,2-a]pyrazine (**148**).

---

[102] A. Stoll and T. Petrzilka, U.S. Patent 2,673,850 (1954) [*CA* **49**, 4032 (1955)].

TABLE III
SPECTRAL PROPERTIES OF SOME PYRROLO[1,2-a]PYRAZINES

Structure	Ultraviolet spectra (solvent) λ_{max}, nm(ϵ)	Nuclear magnetic resonance (solvent) δ	Infrared spectra (vehicle) μm	References
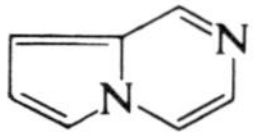 (pyrrolo[1,2-a]pyrazine)		(CDCl$_3$) 8.91 (H1, $J_{1,4}$ = 1.55 Hz) 7.89 (H4, $J_{3,4}$ = 5.5 Hz) 7.58 (H3, $J_{4,3}$ = 5.5 Hz) 7.46 (H6, $J_{6,8}$ = 1.45 Hz) 6.97 (H7, $J_{6,7}$ = 2.55 Hz) 6.85 (H8, $J_{3,8}$ = 1.00 Hz; $J_{8,7}$ = 4.50 Hz)		100
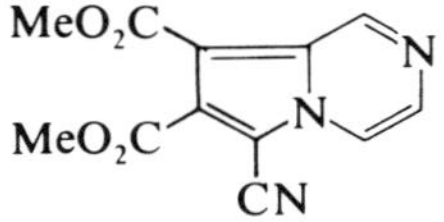 (MeO$_2$C, MeO$_2$C, CN substituted)	(CH$_3$CN) 216 (18,800) 244 (38,600) 320 (8050)	(CDCl$_3$) 9.62 ($J_{1,4}$ = 2.0, H1) 8.23 (J = 4.8 Hz, H3) 8.16 (J = 4.8 Hz, H4)	(KBr) 4.47 (C=N) 5.70 (C=O)	3
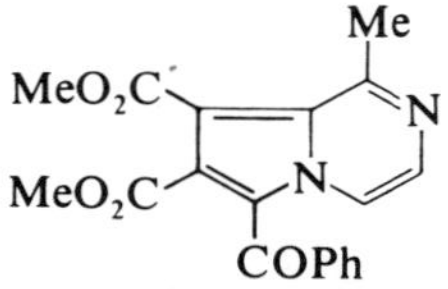 (Me, MeO$_2$C, MeO$_2$C, COPh substituted)		(CDCl$_3$) 8.86 (d, H4, J = 5.5 Hz)[a] 7.86 (d, H3, J = 5.5 Hz)[a] 7.54 (m, C$_6$H$_5$) 2.84 (s, 1-CH$_3$)		98

[a] Proton assignments were not made by the authors of the original article. These assignments have been made by the authors of this review article and should be considered tentative.

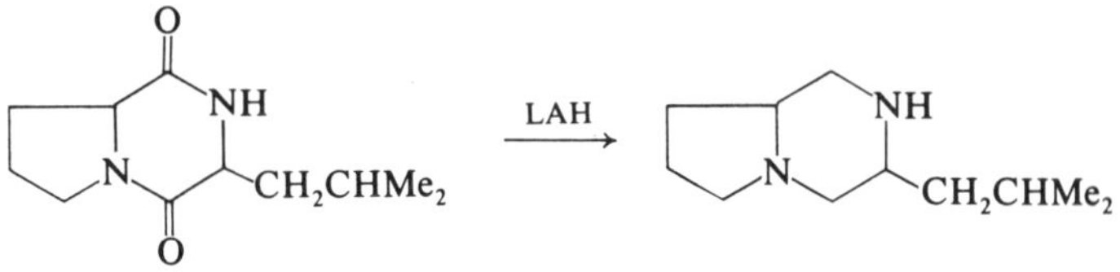

(51)

Similarly, Paquette and Scott[103] prepared the 3- and 1-oxo derivatives **149** and **150** (Scheme 9) and reduced both to the same octahydropyrrolo[1,2-a]pyrazine (**151**). The preparation of compound **150** by the route shown in Scheme 9 had been previously accomplished by Freed and Day[104] and later by Casagrande et al.[105]

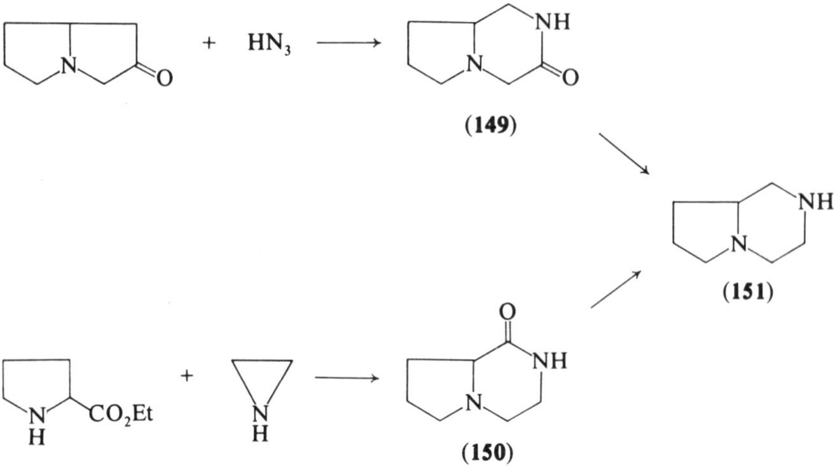

SCHEME 9

Another synthetic route to 2-substituted octahydropyrrolo[1,2-a]-pyrazines (**153**) involves the cyclization of the dichloropyrrole (**152**) with various amines[104,106,107] [Eq. (52)] in yields of 26–80%. In the case of compound **153** where R = CH_2Ph, catalytic debenzylation[104] yields the parent compound **151**.

[103] L. A. Paquette and M. K. Scott, *J. Org. Chem.* **33**, 2379 (1968).
[104] M. E. Freed and A. R. Day, *J. Org. Chem.* **25**, 2108 (1960).
[105] C. Casagrande, A. Galli, R. Ferrini, and G. Miragoli, *Arzneim.-Forsch.* **21**, 808 (1971).
[106] M. E. Freed, U.S. Patent 3,164,598 (1965) [*CA* **62**, 9154 (1965)].
[107] M. E. Freed, U.S. Patent 3,531,485 (1970) [*CA* **73**, 120672 (1970)].

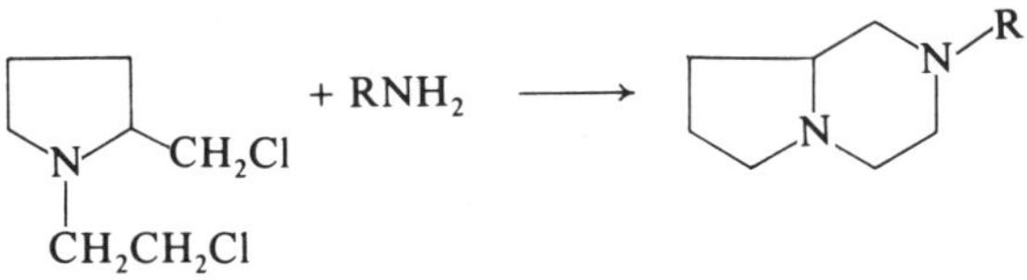

Sec. IV.D] PYRROLODIAZINES WITH A BRIDGEHEAD NITROGEN

[structure 152] + RNH$_2$ ⟶ [structure 153]   (52)

(152)   (153)

Ponomarev and Skvortsov[108,109] report the two-step preparation of octahydropyrrolo[1,2-a]pyrazine (151) in modest yield from tetrahydrofurfural and ethylenediamine [Eq. (53)].

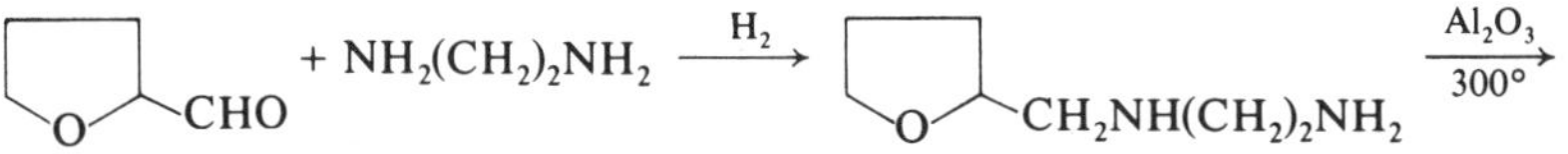

[structure] + NH$_2$(CH$_2$)$_2$NH$_2$ $\xrightarrow{H_2}$ [structure with CH$_2$NH(CH$_2$)$_2$NH$_2$] $\xrightarrow[300°]{Al_2O_3}$ 151   (53)

Combination of the secondary amine 151 with various phenothiazine derivatives produces compounds with several pharmacological activities. For example, two synthetic routes[105,110] are reported to yield the phenothiazines 154a,b, which are claimed to exhibit neuroleptic activity. The structurally similar aminoacylphenothiazines (155) are claimed[111,112] to be coronary dilating agents.

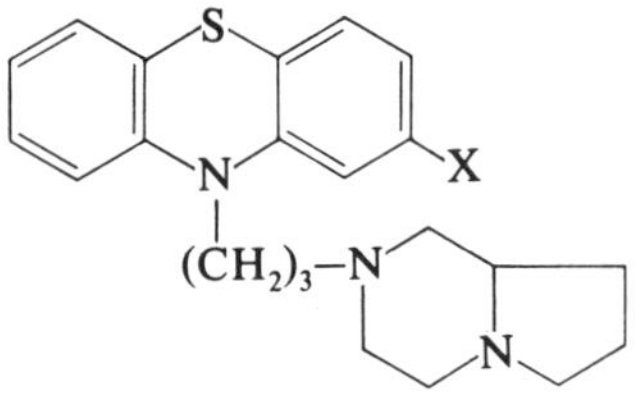

(154) a: X = H
      b: X = CF$_3$

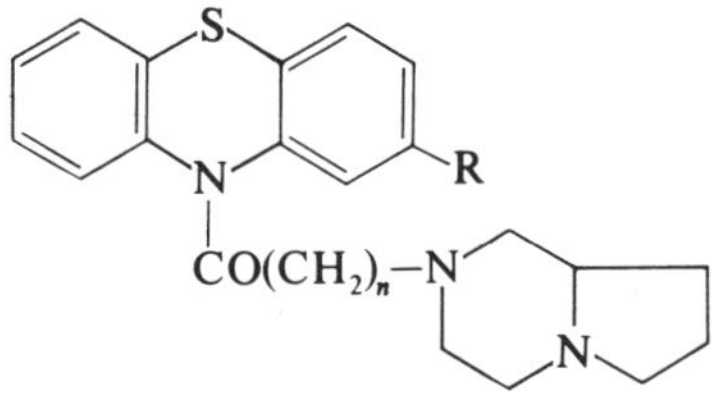

(155) R = H, Cl, CF$_3$

Alkylation of 151 produces biologically active compounds of another type. Several 2-benzhydryl derivatives of 151 are claimed[106] as anti-

---

[108] A. A. Ponomarev and I. M. Skvortsov, *Dokl. Akad. Nauk. SSSR* **148**, 860 (1963) [*CA* **59**, 3918 (1963)].

[109] A. A. Ponomarev and I. M. Skvortsov, *Metody Poluch. Khim. Reakt. Prep.*, 5 (1967) [*CA* **70**, 87778 (1969)].

[110] V. V. Zukusov, A. P. Skoldinov, K. S. Raevskii, A. M. Likhosherstov, and L. S. Nazarova, USSR Patent 367,102 [*CA* **79**, 5354 (1973)].

[111] A. M. Likhosherstov, L. S. Nazarova, A. P. Skoldinov, G. A. Markova, and N. V. Kaverina, German Patent 2,210,382 (1973) [*CA* **79**, 126514 (1973)].

[112] A. M. Likhosherstov, L. S. Nazarova, A. P. Skoldinov, G. A. Markova, and N. V. Kaverina, British Patent 1,338,363 (1973) [*CA* **80**, 83008 (1974)].

histaminic and anticholinergic agents, and 2-alkylation of **151** with other alkyl groups produces **156** and **157** with antiarrhythmic activity.[113,114]

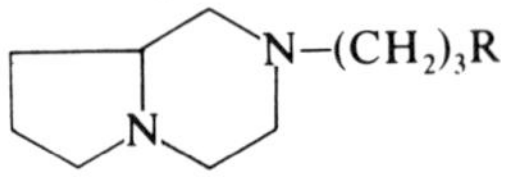

(**156**) R = CO—⟨C6H4⟩—F

(**157**) R = CPh$_2$CONR$_2^1$

2. Mono-oxo Octahydropyrrolo[1,2-a]pyrazines

Likhosherstov et al.[115–117] have prepared 1-oxo-octahydropyrrolo-[1,2-a]pyrazine (**150**) from dihalocarboxylic acids or esters (**158**) and ethylenediamine [Eq. (54)] in good yields. Three examples of this synthetic route are known[115] and various substituted examples of **150** are clearly accessible depending on the dihaloacid or diamine employed in this versatile method.

Cl(CH$_2$)$_3$CHBrCO$_2$Et + NH$_2$CH$_2$CH$_2$NH$_2$ ⟶ (**150**)   (54)

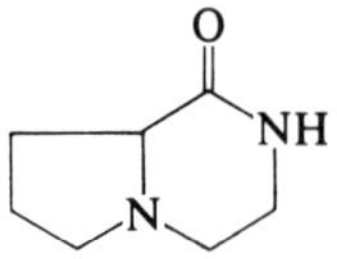

(**158**)

Treatment of 1,4-dioxo-octahydro[1,2-a]pyrazines (**159**) with Meerwein's reagent produces[118] 1-ethoxy-4-oxo-3,4,6,7,8,8a-hexahydropyrrolo[1,2-a]pyrazines (**160**) [Eq. (55)]. These compounds were

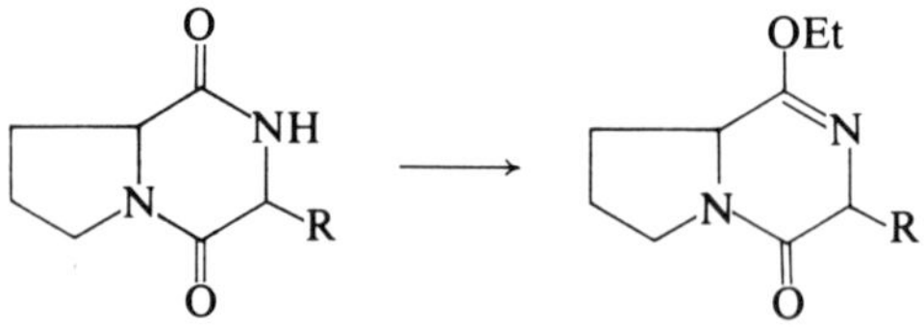

(**159**) ⟶ (**160**)   (55)

---

[113] V. V. Zakusov, A. P. Skoldinov, A. M. Likhosherstov, K. S. Raevskii, and L. S. Nazarova, USSR Patent 352,877 [*CA* **78**, 58053 (1973)].

[114] C. Lovell, German Patent 2,226,063 (1972) [*CA* **78**, 58458 (1973)].

[115] A. M. Likhosherstov, L. S. Nazarova, and A. P. Skoldinov, *Zh. Org. Khim.* **6**, 1729 (1970) [*CA* **73**, 109724 (1970)].

[116] A. M. Likhosherstov, L. S. Nazarova, and A. P. Skoldinov, French Patent 1,510,781 (1968) [*CA* **70**, 68427 (1969)].

[117] Sci. Res. Inst. Pharmacol. Chemother., Moscow, British Patent 1,144,749 (1969) [*CA* **70**, 106558 (1969)].

Sec. IV.D] PYRROLODIAZINES WITH A BRIDGEHEAD NITROGEN 39

then combined[118] with anthranilic acids to produce tetracyclic quinazolone derivatives.

Kato[119] has reduced the methiodide salt (**161**) of compound **150** with Li/NH$_3$ to produce the nine-membered amino lactam **162**. Further reduction of **162** gives the diazonine **163** [Eq. (56)].

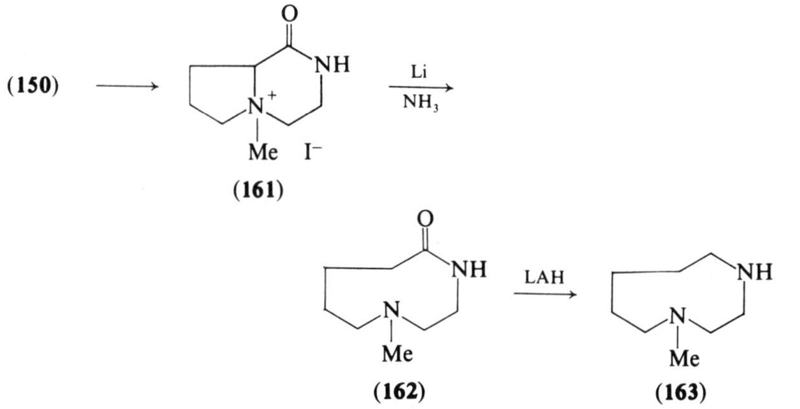

Morimoto and Watanabe[120-122] have described a process for nitrating the mono-oxo compound **165**, prepared via Beckmann rearrangement of the oxime (**164**), to the 6- and 7-nitro (**166**, **167**) [Eq. (57)] derivatives. These products are claimed to be protozoacidal, trichomonacidal, and antidiabetic.[122]

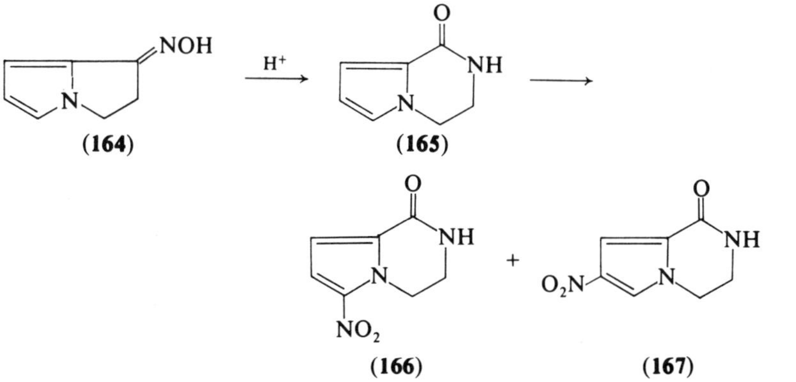

[118] S. Rajappa and B. G. Advani, *Tetrahedron* **29**, 1299 (1973).
[119] H. Kato, *J. Pharm. Soc. Jpn.* **95**, 830 (1975).
[120] A. Morimoto and T. Watanabe, Japanese Patent 74 27,879 (1974) [*CA* **82**, 125416 (1975)].
[121] A. Morimoto and T. Watanabe, Japanese Patent 74 27,877 (1974) [*CA* **82**, 156398 (1975)].
[122] A. Morimoto and T. Watanabe, Japanese Patent 74 27,878 (1974) [*CA* **82**, 140180 (1975)].

3. 1,4-Dioxo-octahydropyrrolo[1,2-a]pyrazines

By far the greatest number of literature references have been made to this particular class of octahydropyrrolo[1,2-*a*]pyrazines (**168**). The reason for this high rate of citation is due to the fact that cyclodehydration of proline with another amino acid gives rise to compound **168** [Eq. (58)]. Derivatives of **168** are often referred to as "proline lactams" or

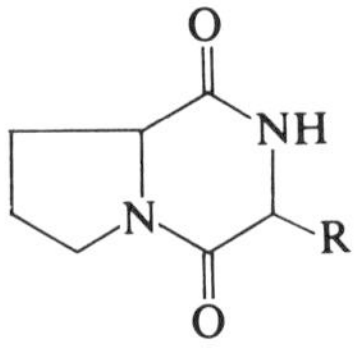

(**168**)

"proline anhydrides," as derivatives of diketopiperazine or as 1,4-diazabicyclo[4.3.0]nonane-2,5-diones. Many reports are concerned with the direct synthesis of examples of compound **168**. In addition, derivatives of **168** have been isolated from peptides, fermentations, and other natural sources, such as ergot alkaloids. These general classifications will be treated separately in Sections III,D,3a and b, respectively.

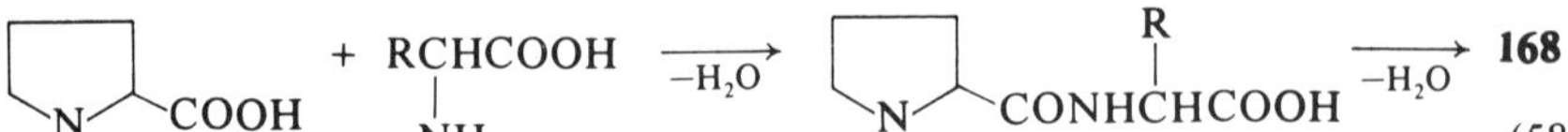

(58)

a. Syntheses. The bimolecular, cyclodehydration of proline [Eq. (59)] to a 1,4-dioxopyrrolo[1,2-*a*]pyrazine derivative (**169**) is carried out in glycol solution at elevated temperature.[123] Several dipeptide methyl esters are conveniently cyclized to derivatives of **168** by heating their formate salts and removing formic acid azeotropically.[124] A related technique utilizes *p*-nitrophenyl esters of proline peptides.[125]

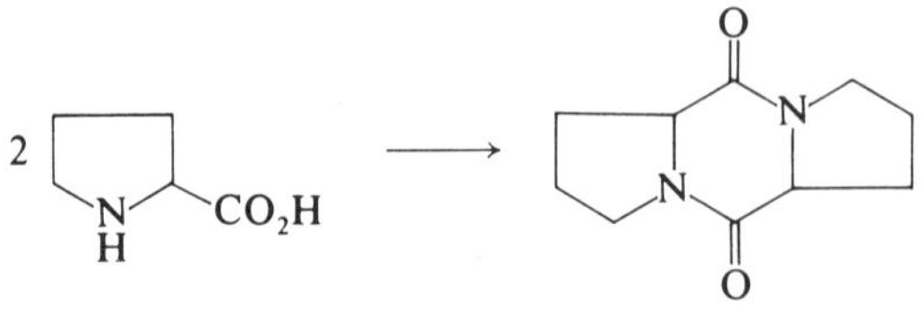

(59)

(**169**)

---

[123] N. A. Poddubnaya and G. I. Lavrenova, *Ves. Moskov. Univ. Ser. Mat. Mekh. Astron., Fiz., Khim.* **13**, No. 3, 165 (1958) [*CA* **53**, 11396 (1959)].
[124] D. E. Nitecki, B. Halpern, and J. W. Westley, *J. Org. Chem.* **33**, 864 (1968).
[125] G. Lucente and P. Frattesi, *Tetrahedron Lett.*, 4283 (1972).

An alternative route[126] to these types of compounds utilizes a β-chloroalanyl proline amide (**170**), which is cyclized by a strongly basic, sterically hindered amine as in Eq. (60)

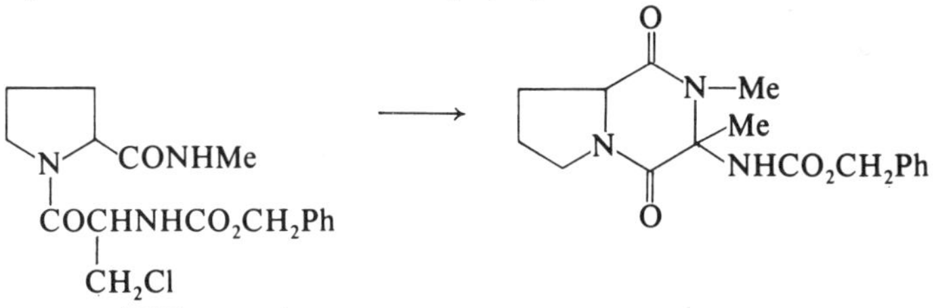

$$\text{(170)} \longrightarrow \text{product} \quad (60)$$

Several workers have examined the preparation of "proline anhydrides" from a variety of peptides [see Eq. (58)]. Thus, Vicar et al.[127] prepared a number of derivatives of compound **168**. "Prolylvaline anhydride" (**168**, R = CHMe$_2$) has been reported by several laboratories.[128-130] "Prolylhistidine anhydride" (**171**) has been isolated[131] from cleavage of the tripeptide His-Pro-Phe-OCH$_3$. "Leucylproline anhydride" (**168**, R = CH$_2$CHMe$_2$) is prepared[132] from N-carbobenzyloxy-L-leucine azide and proline benzyl ester in methanolic hydrochloric acid. "Prolyl-N-methylphenylalanine" (**172**) has been isolated[133]

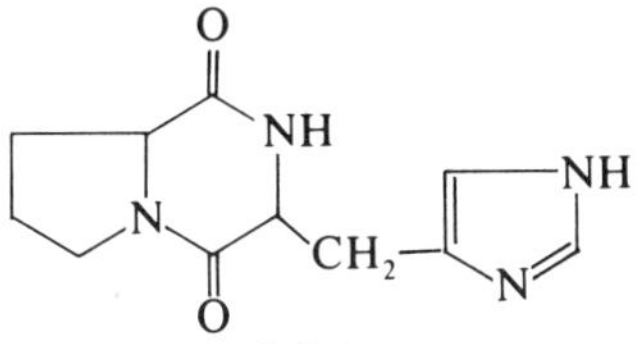

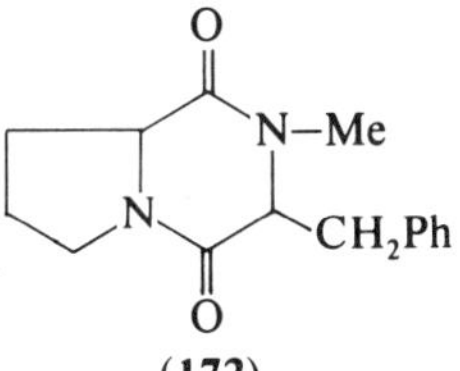

[126] U. Schmidt, A. Perco, and E. Ohler, *Chem. Ber.* **107**, 2816 (1974).
[127] J. Vicar, J. Smolikova, and K. Blaha, *Collect. Czech. Chem. Commun.* **37**, 4060 (1972) [*CA* **78**, 84810 (1973)].
[128] R. A. Boissonnas, S. Guttmann, R. L. Huguenin, P. A. Jaquenoud, and E. Sandrin, *Helv. Chim. Acta* **41**, 1867 (1958).
[129] B. F. Gisin and R. B. Merrifield, *J. Am. Chem. Soc.* **94**, 3102 (1972).
[130] A. B. Mauger, *J. Chem. Soc. D*, 39 (1971).
[131] R. H. Mazur and J. M. Schlatter, *J. Org. Chem.* **28**, 1025 (1963).
[132] R. E. Neuman and E. L. Smith, *J. Biol. Chem.* **193**, 97 (1951).
[133] M. C. Khosla, R. R. Semby, and F. M. Bumpus, *J. Am. Chem. Soc.* **94**, 4721 (1972).

from an attempt to synthesize an angiotensin derivative using the solid-state peptide synthesis method. Optically active **174** is made[134] by catalytic reduction of the unsaturated derivative **173** [Eq. (61)]. In several examples,[134] this reduction proceeds in good yield with high induction of optical activity in the products.

$$\text{(173)} \longrightarrow \text{(174)} \tag{61}$$

(173)   (174)

"Prolylglycine anhydride" (**168**, R = H) has been isolated by many workers[135–140] from a variety of different peptide syntheses. Solid-phase peptide syntheses are also prone to produce dioxopyrrolo[1,2-a]-pyrazines.[141] Radiolabeled examples of **168** have been made[142] as potential precursors of ergotoxin alkaloids. A series of N-methylated derivatives of **168** was prepared,[143] and their NMR properties were studied. Tripeptides containing a prolyl group have been cyclized[144] to **168**, R = CH$_2$Ph, in basic solution. Cyclization of N-acylated prolylamides affords [Eq. (62)] a convenient synthesis[145] of the 3-hydroxy derivative (**175**) of a 1,4-dioxopyrrolo[1,2-a]pyrazine. The hydroxy group in **175** has been replaced by alkylthio groups.[145]

---

[134] H. Poisel and U. Schmidt, *Chem. Ber.* **106**, 3408 (1973).
[135] M. Goodman and K. C. Stueben, *J. Am. Chem. Soc.* **84**, 1279 (1962).
[136] E. Wuensch, Peptides, *Proc. Eur. Symp., 5th*, 89 (1962) [*CA* **62**, 5334 (1965)].
[137] H. N. Rydon and P. W. G. Smith, *J. Chem. Soc.*, 3642 (1956).
[138] K. T. Poroshin, T. D. Kozarenko, and V. A. Shibnev, *Izv. Akad. Nauk SSSR Otd. Khim. Nauk.*, 1129 (1958) [*CA* **53**, 3077 (1959)].
[139] N. S. Andreeva, V. I. Iveronova, T. D. Kozarenko, K. T. Poroshin, V. A. Shibnev, and N. E. Shutskever, *Izv. Akad. Nauk. SSSR, Otd. Khim. Nauk.*, 376 (1958) [*CA* **52**, 12762 1958)].
[140] W. Grassmann, H. Hormann, A. Nordwig, and E. Wunsch, *Z. Physiol. Chem.* **316**, 287 (1959) [*CA* **54**, 8958 (1960)].
[141] M. Rothe and J. Mazanek, *Justus Liebigs Ann. Chem.*, 439 (1974).
[142] S. Johne and D. Groger, *Pharmazie* **29**, 181 (1974).
[143] A. B. Mauger, R. B. Desai, I. Rittner, and W. J. Rzeszotarski, *J. Chem. Soc., Perkin Trans. 1*, 2146 (1972).
[144] G. Lucente, G. Fiorentini, and D. Rossi, *Gazz. Chim. Ital.* **101**, 109 (1971) [*CA* **75**, 64266 (1971)].
[145] J. Haeusler, U. Schmidt, A. Perco, and E. Oehler, *Chem. Ber.* **107**, 2804 (1974).

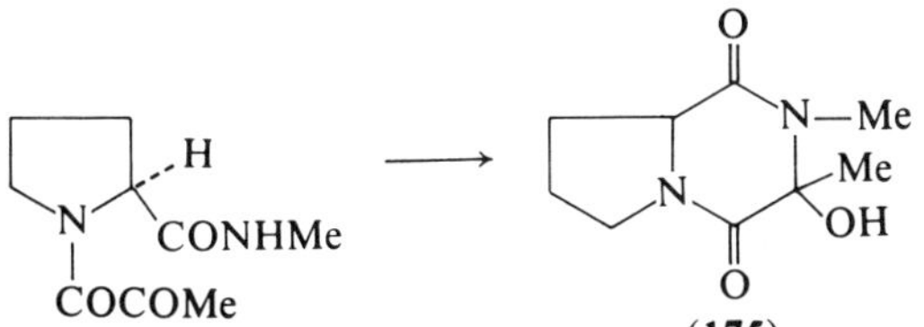

(62)

(175)

The relative selectivity for nucleophilic reaction at one of the three amide bonds in **176** has been examined by Bruice and co-workers.[146,147] Primary and secondary amines generally yield the deacylated compound **177** while oxygen bases (e.g., formate, acetate, hydroxide ion) provide the ring-cleaved product **178** [Eq. (63)].

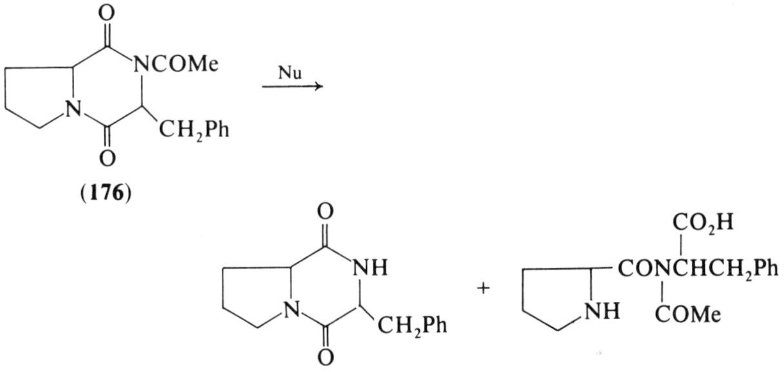

(176)

(177)   (178)   (63)

Spectral studies on 1,4-dioxopyrrolo[1,2-a]pyrazines abound. Infrared absorption properties of a variety of derivatives of **168** have been compiled[148,149] and related to the conformations of the piperazinedione ring. Gas-chromatographic techniques for separating diastereomeric mixtures of derivatives of **168** have been described.[150] Mass spectral fragmentation patterns for a series of dioxopiperazines, including derivatives of **168**, has been reported.[151] Numerous nuclear

---

[146] T. C. Bruice and D. M. McMahon, *Biochemistry* **11**, 1273 (1972).
[147] J. E. Dixon and T. C. Bruice, *J. Am. Chem. Soc.* **94**, 2052 (1972).
[148] J. Vicar, J. Smolikova, and K. Blaha, *Collect. Czech. Chem. Commun.* **38**, 1957 (1973) [*CA* **79**, 126784 (1973)].
[149] Yu. N. Chirgadze, *Opt. Spektrosk., Akad. Nauk. SSSR, Otd. Fiz. Mat. Nauk, Sb. Statei* **2**, 242 (1963) [*CA* **60**, 117 (1964)].
[150] A. B. Mauger, *J. Chromatog.* **37**, 315 (1968).
[151] N. S. Vul'fson, V. A. Puchkov, Yu. V. Denisov, B. V. Rozynov, V. N. Bochkarev, M. M. Shemyakin, Yu. A. Ovchinnikov, and V. K. Antonov, *Khim. Geterotsikl. Soedin.*, 614 (1966) [*CA* **66**, 28272 (1967)].

magnetic resonance studies[152-157] on 1,4-dioxopyrrolo[1,2-a]pyrazines have addressed the question of steric configuration of substituents on this ring system. Carbon-13 NMR studies[158] on ^{13}C-enriched cyclic dipeptides produce simplified spectra and allow conformational determinations to be made. An X-ray examination of the crystal structure of **168**, R = CH$_2$CHMe$_2$, is available.[159]

Shemyakin and co-workers[45,160-162] have examined the tautomeric forms of 2-(α-hydroxyacyl) derivatives of **168** [i.e., compound **179** in Eq. (64)]. The cyclol **180** is detected by spectra on solutions of **179**;

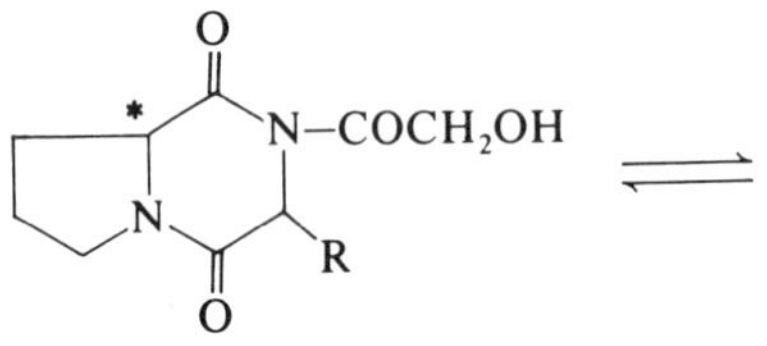

(**179**)

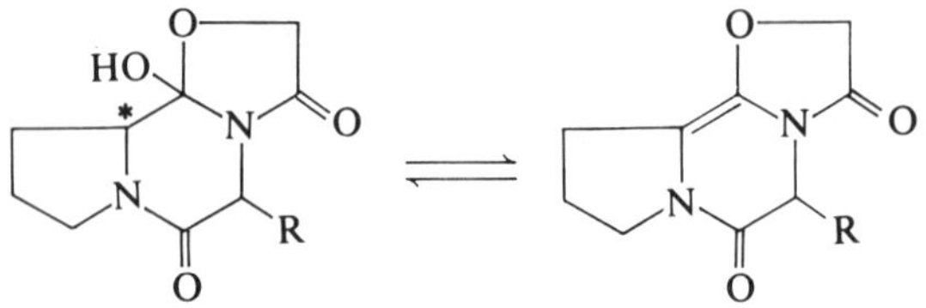

(64)

(**180**)   (**181**)

reversible dehydration of compound **180** to **181** is suggested to explain the loss of optical purity at the bridgehead carbon atom in **179**. Lucente

[152] J. W. Westley, V. A. Close, D. N. Nitecki, and B. Halpern, *Anal. Chem.* **40**, 1888 (1968).
[153] I. Z. Siemion, *Justus Liebigs Ann. Chem.* **748**, 88 (1971).
[154] J. Vicar, M. Budesinsky, and K. Blaha, *Collect. Czech. Chem. Commun.* **38**, 1940 (1973) [*CA* **79**, 126783 (1973)].
[155] I. Z. Siemion and A. Sucharda-Sobczyk, *Rocz. Chem.* **46**, 1257 (1972) [*CA* **77**, 140487 (1972)].
[156] K. Blaha, M. Budesinsky, I. Fric, J. Smolikova, and J. Vicar *Tetrahedron Lett.*, 1437 (1972).
[157] I. Z. Siemion, *Org. Magn. Reson.* **3**, 545 (1971).
[158] S. Tran Dinh and S. Fermandjian, *J. Phys. (Paris) Colloq.*, 45 (1973) [*CA* **81**, 131777 (1974)].
[159] I. L. Karle, *J. Am. Chem. Soc.* **94**, 81 (1972).
[160] M. M. Shemyakin, V. V. Antonov, A. M. Shkrob, Yu. N. Sheinker, and L. B. Senyavina, *Tetrahedron Lett.*, 701 (1962).
[161] V. K. Antonov, A. M. Shkrob, and M. M. Shemyakin, *Peptides, Proc. Eur. Symp., 5th.* 221 (1962) [*CA* **62**, 4242 (1965)].
[162] V. K. Antonov, A. M. Shkrob, and M. M. Shemyakin, *Zh. Obshch. Khim.* **37**, 2225 (1967) [*CA* **69**, 10416 (1968)].

and co-workers[163,164] presented evidence for a related base-catalyzed interconversion of **182** to **183** where a nitrogen atom, rather than an oxygen atom, acts as a nucleophile toward the 1-oxo function in compound **182** [Eq. (65)).

(65)

(**182**)   (**183**)

Liberek et al.[165] found that base-catalyzed mutarotation of various "cyclo dipeptides" (i.e., **168**) proceeds as expected except for the "cyclo-Pro-Pro" compound **169**. The authors attributed this lack of reactivity for **169** to a conformation unfavourable to mutarotation. Another study[166] of various cyclic dipeptides and their cis–trans isomerization in ethanolic sodium ethoxide indicated that the ratio of cis–trans isomers depended on the R group in compounds of type **168**. Again, compound **169** showed no tendency to isomerize and was in the cis form at equilibrium. The stability of the cis form of **169** toward base was confirmed[167] while the strained trans-**169** was converted to the racemic cis isomer.

b. Natural Products. A number of microbiological transformations have yielded 1,4-dioxopyrrolo[1,2-*a*]pyrazines as side products. Thus, "leucylproline anhydride" (**168**, R = CH_2CHMe_2) is found[168] in a fermentation designed to transform steroids. This same product has been detected from other fermentations.[169–172] Other fermentation conditions

[163] G. Lucente, A. Romeo, and G. Zanotti, *Experientia* **31**, 17 (1975).
[164] F. Conti, G. Lucente, A. Romeo, and G. Zanotti, *Int. J. Peptide Res.* **5**, 353 (1973) [*CA* **80**, 60181 (1974)].
[165] B. Liberek, A. Chodup, J. Ciarkowski, and M. Kozlowska, *Rocz. Chem.* **48**, 361 (1974) [*CA* **81**, 50019 (1974)].
[166] C. Eguchi and A. Kakuta, *J. Am. Chem. Soc.* **96**, 3985 (1974).
[167] J. Vicar and K. Blaha, *Collect. Czech. Chem. Commun.* **38**, 3307 (1973) [*CA* **80**, 71090 (1974)].
[168] S. H. Eppstein, P. D. Meister, D. H. Peterson, H. C. Murray, H. M. Leigh, D. A. Lyttle, L. M. Reineke, and A. Weintraub, *J. Am. Chem. Soc.* **75**, 408 (1953).
[169] G. Leudemann, W. Charney, A. Woyciesjes, E. Pettersen, W. D. Peckham, M. J. Gentles, H. Marshall, and H. L. Herzog, *J. Org. Chem.* **26**, 4128 (1961).
[170] B. T. Lingappa, M. Prasad, Y. Lingappa, D. F. Hunt, and K. Biemann, *Science* **163**, 192 (1969).
[171] T. Ohashi, H. Takahashi, and M. Abe, *Nippon Nogei Kagaku Kaishi* **46**, 535 (1972) [*CA* **78**, 39411 (1973)].
[172] I. Belic, A. Cimerman, and H. Socic, *Mikrobiologija* **9**, 251 (1972) [*CA* **81**, 164988 (1974)].

produce compound **168**, R = H,[173,174] R = CHMe$_2$,[175] or R = CH$_2$C$_6$H$_4$OH-*p*.[176]

Miscellaneous fermentation products containing more complex molecules incorporating a 1,4-dioxopyrrolo[1,2-*a*]pyrazine have also been isolated.[177–181] A culture of *Streptomyces* has been observed by three different laboratories[182–185] to produce **168**, R = CH$_2$CHMe$_2$. The synthesis[186] and degradation[187] of actinomycins have involved "valylproline anhydride" (**168**, R = CHMe$_2$). Gramicidin C also yields **168**, R = CH$_2$Ph, on acid hydrolysis in an autoclave.[188] An active principle has been isolated[189,190] from *Streptomyces* cultures that acts as a rice plant seed germination promoter. Hydrolysis of this active factor produces "prolylvaline anhydride" (**168**, R = CHMe$_2$). Radiolabeled **168**, R = tryptophyl, has been employed[192] to examine the biosynthetic pathway to brevianamide A.

A fungus affecting the growth of silkworms produces a number of derivatives of **168**.[191,193,194] Similarly, various examples of compound

---

[173] G. R. Pettit, R. B. Von Dreele, G. Bolliger, P. M. Traxler, and P. Brown, *Experientia* **29**, 521 (1973).
[174] E. P. White, *N.Z. J. Sci.* **15**, 178 (1972).
[175] Y-S. Chen, *Bull. Agr. Chem. Soc. Jpn.* **24**, 372 (1960) [*CA* **55**, 680 (1961)].
[176] T. Tatsuno, M. Sato, Y. Kubota, Y. Kubota, and H. Tsunoda, *Chem. Pharm. Bull.* **19**, 1498 (1971) [*CA* **75**, 116967 (1971)].
[177] M. Abe, T. Fukuhara, S. Ohmomo, M. Hori, and T. Tabuchi, *Nippon Nogei Kagaku Kaishi* **44**, 573 (1970) [*CA* **74**, 136416 (1971)].
[178] P. S. Steyn, *Tetrahedron Lett.*, 3331 (1971).
[179] P. S. Steyn, *Tetrahedron* **29**, 107 (1973).
[180] M. Abe, T. Ohashi, S. Ohmomo, and T. Tabuchi, *Nippon Nogei Kagaku Kaishi* **45**, 6 (1971) [*CA* **75**, 60162 (1971)].
[181] A. J. Birch and R. A. Russell, *Tetrahedron* **28**, 2999 (1972).
[182] Y. Koaze, *Bull. Agr. Chem. Soc. Jpn.* **24**, 449 (1960) [*CA* **55**, 2792 (1961)].
[183] Y. Koaze, *Bull. Agr. Chem. Soc. Jpn.* **24**, 530 (1960) [*CA* **55**, 3714 (1961)].
[184] J. L. Johnson, W. G. Jackson, and T. E. Eble, *J. Am. Chem. Soc.* **73**, 2947 (1951).
[185] L. Ettlinger, E. Gaumann, R. Hutter, W. Keller-Schierlein, F. Kradolfer, L. Neipp, V. Prelog, P. Reusser, and H. Zahner, *Monatsh. Chem.* **88**, 989 (1957).
[186] J. Meinhofer, *J. Am. Chem. Soc.* **92**, 3771 (1970).
[187] H. Brockmann, G. Bohnsack, and C. H. Suling, *Angew. Chem.* **68**, 66 (1956).
[188] N. A. Poddubnaya and M. I. Kiselev, *Zhr. Obshch. Khim.* **26**, 1508 (1956) [*CA* **50**, 14538 (1956)].
[189] Y. Koaze, *Bull. Agr. Chem. Soc. Jpn.* **21**, 197 (1957) [*CA* **51**, 15718 (1957)].
[190] Y. Koaze, *Bull. Agr. Chem. Soc. Jpn.* **22**, 91 (1958) [*CA* **52**, 20460 (1958)].
[191] Y. Kodaira, *J. Fac. Textile Sci. Technol. Shinshu Univ., Ser. E*, 1 (1961) [*CA* **59**, 5522 (1963)].
[192] J. Baldas, A. J. Birch, and R. A. Russell, *J. Chem. Soc., Perkin Trans. 1* 50 (1974).
[193] Y. Kodaira, *Agr. Biol. Chem. (Tokyo)* **25**, 261 (1961) [*CA* **55**, 12678 (1961)].
[194] A. Butenandt, P. Karlson, and W. Zillig, *Hoppe-Seylers Z. Physiol. Chem.* **288**, 279 (1951).

168 have been isolated from peptone.[195-198] Telomycin, a complex polypeptide, has been degraded by Sheehan et al.[199,200] to the 1,4-dioxopyrrolo[1,2-a]pyrazine **184**.

(**184**)

"Leucylproline anhydride" (**168**, R = CH_2CHMe_2) has been identified as a bitter-tasting component of aged sake.[201]

Finally, and not surprisingly, complex protein hydrolyzates[202] and blood hydrolyzates[203] are found to contain 1,4-dioxopyrrolo[1,2-a]-pyrazines. In phenylketonuric patients, levels of compound **171** in urine are found to be elevated.[204]

In a series of publications, workers from the Sandoz company have described the synthesis or isolation of ergot alkaloids that contain 1,4-dioxopyrrolo[1,2-a]pyrazines such **185** or **186**.

Acid hydrolysis[205] and hydrazine cleavage[206] of dihydroergotamine produces compound **177**. An initial assignment[207] for the structure of pyroergotamine was later corrected[208] to compound **187** ("N-pyruvoyl-phenylalanylproline lactam") and an authentic synthesis of **187** has

[195] T. Kosuge and H. Kamiya, *Chem. Pharm. Bull* **10**, 154 (1962) [*CA* **58**, 2582 (1963)].
[196] T. Kosuge and H. Kamiya, *Nature (London)* **188**, 1112 (1960).
[197] S. Tamura, A. Suzuki, A. Yasuo, and N. Otake, *Agr. Biol. Chem. (Tokyo)* **28**, 650 (1964) [*CA* **62**, 5471 (1965)].
[198] L. A. Mitscher, M. P. Kunstmann, J. H. Martin, W. W. Andres, R. H. Evans, K. J. Sax, and E. L. Patterson, *Experientia* **23**, 796 (1967).
[199] J. C. Sheehan, P. E. Drummond, J. N. Gardner, K. Maeda, D. Mania, S. Nakamura, A. K. Sen, and J. A. Stock, *J. Am. Chem. Soc.* **85**, 2867 (1963).
[200] J. C. Sheehan, D. Mania, S. Nakamura, J. A. Stock, and K. Maeda, *J. Am. Chem. Soc.* **90**, 462 (1968).
[201] K. Takahashi, M. Tadenuma, K. Kitamoto, and S. Sato, *Agr. Biol. Chem.* **38**, 927 (1974) [*CA* **81**, 76421 (1974)].
[202] I. Chmielewska, B. Bulhak, and K. Toczko, *Acta Biol. Pol.* **14**, 409 (1967) [*CA* **68**, 46264 (1968)].
[203] I. Chmielewska, B. Bulhak, and K. Toczko, *Bull. Acad. Pol. Sci., Ser. Sci. Biol.* **15**, 719 (1967) [*CA* **68**, 103216 (1968)].
[204] T. L. Perry, K. C. Richardson, S. Hansen, and A. J. D. Friesen, *J. Biol. Chem.* **240**, 4540 (1965).
[205] A. Stoll and A. Hofmann, *Helv. Chim. Acta* **33**, 1705 (1950).
[206] A. Stoll, T. Petrzilka and B. Becker, *Helv. Chim. Acta* **33**, 57 (1950).
[207] A. Stoll, A. Hoffmann and T. Petrzilka, *Helv. Chim. Acta* **34**, 1544 (1951).
[208] M. Green and E. A. C. Lucken, *Helv. Chim. Acta* **44**, 1417 (1961).

been reported.[209] Compound **186** has been shown to be stable in 0.1 N NaOH for 24 hours with no evidence of opening of the cyclol ring.[210]

(185)

(186)

(187)

Such a cyclol structure is a part of the ergotamine structure.[211,212] Compound **185** has been isolated[213] from a fermentation residue.

The important pharmacological activities, such as antiserotonin effects, of many of the ergot derivatives has led to many patents being issued.[214-224] Detailed synthetic procedures are contained in these patents for the multistep synthesis of dioxopyrrolo[1,2-a]diazines such as **186** from derivatives of α-methylproline and phenylalanine. In addition to the pharmacologically active (hypotensive, antiserotonin, etc.)

[209] C. A. Grob and W. Meier, *Helv. Chim. Acta* **39**, 776 (1956).
[210] H. Ott, A. J. Frey, and A. Hofmann, *Tetrahedron* **19**, 1675 (1963).
[211] A. Hofmann, H. Ott, R. Griot, P. A. Stadler, and A. J. Frey, *Helv. Chim. Acta* **46**, 2506 (1963).
[212] P. A. Stadler, A. J. Frey, H. Ott, and A. Hoffmann, *Helv. Chim. Acta* **47**, 1911 (1964).
[213] P. Stuetz, R. Brunner, and P. A. Stadler, *Experientia* **29**, 936 (1973).
[214] P. Stadler, H. Hauth, G. Wersin, S. Guttmann, A. Hofmann, P. Stutz, and H. Willems, British Patent 1,238,348 (1971) [*CA* **75**, 88815 (1971); **73**, 13103 (1970)].
[215] S. Guttmann and R. Huguenin, German Patent 2,029,447 (1971) [*CA* **74**, 88191 (1971)].
[216] A. Hofmann and P. Stadler, French Patent 6822 (1969) [*CA* **74**, 88192 (1971)].
[217] P. Stadler, A. Hofmann and F. Troxler, French Patent 1,583,797 (1969) [*CA* **73**, 77459 (1970)]; Swiss Patent 550,801 (1974) [*CA* **81**, 152488 (1975)].
[218] Sandoz Ltd., Belgian Patent 617,449 (1962) [*CA* **58**, 13969 (1963)].
[219] Sandoz Ltd., Belgian Patent 617,451 (1962) [*CA* **58**, 14029 (1963)].
[220] S. Guttmann and R. Huguenin, German Patent 1,931,081 (1970) [*CA* **72**, 133049 (1970)].
[221] Sandoz Ltd., Neth. Appl. 6,511,933 (1966) [*CA* **65**, 7230 (1966)].
[222] Sandoz Ltd., Neth. Appl. 6,513,688 (1966) [*CA* **65**, 8980 (1966)].
[223] Sandoz Ltd., Neth. Appl. 6,609,202 (1967) [*CA* **67**, 82297 (1967)].
[224] P. Stadler, H. Hauth, G. Wersin, S. Guttman, A. Hofmann, P. Stutz, and H. Willems, British Patent 1,230,614 (1971) [*CA* **73**, 131033 (1970)].

lysergic acid derivatives discussed above, other dioxopyrrolo[1,2-*a*]-pyrazines are found to have sedative properties.[225]

4. *Other Oxo Derivatives*

A 1,6-dioxopyrrolo[1,2-*a*]pyrazine (**189**) has been made[226] by hydrolysis of the nitrile **188** [Eq. (66)] derived from an imidazo[1,2-*a*]-pyrrole. An intermediate amino acid was also isolated[36] from this reaction and then cyclized to compound **189**. Harnden[226–228] has described the synthesis of 1,4,6-troxopyrrolo[1,2-*a*]pyrazines (**191**) from 3,6-dioxopiperazine-2-propionic acid (**190**) in trifluoroacetic acid anhydride solvent [Eq. (67)]. The requisite starting compound **190** is prepared by cyclodehydration of glycyl-L-glutamic acid; the two-step conversion to **191** proceeds in 79% overall yield.

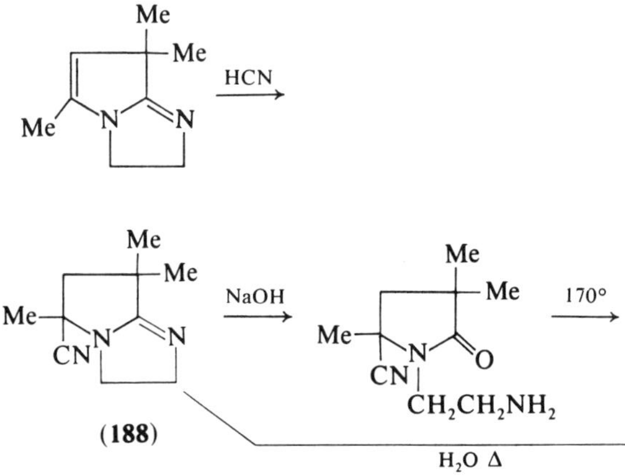

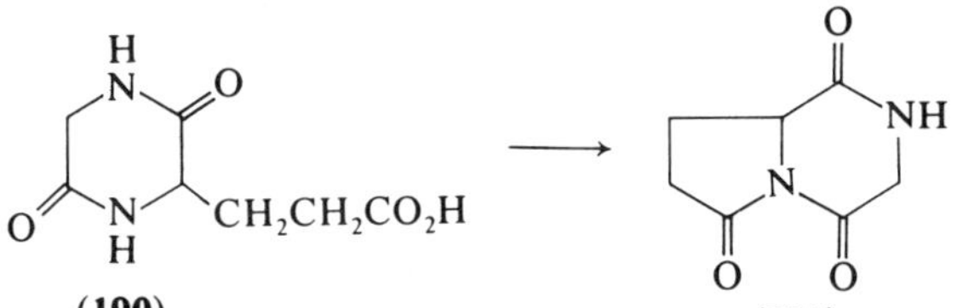

[225] L. Fontanella and L. Mariani, German Patent 2,354,046 (1974).
[226] M. R. Harnden, *J. Chem. Soc. C,* 2341 (1967).
[227] M. R. Harnden, *J. Heterocycl. Chem.* **5**, 307 (1968).
[228] M. R. Harnden, U.S. Patent 3,563,992 (1971) [*CA* **75**, 20443 (1971)].

V. Pyrrolo[1,2-b]pyridazines (4)

A. Syntheses

1. *Syntheses Starting with a Preformed Pyridazine Ring*

Letsinger and Lasco[6] reported the first authenticated synthesis of the pyrrolo[1,2-b]pyridazine ring system. These authors demonstrate that, in methanol solution, pyridazines (**192**) would react readily with dimethyl acetylenedicarboxylate to give 5,6,7-tricarbomethoxypyrrolo-[1,2-b]pyridazines (**193**) [Eq. (68)]. In contrast, it was found that when

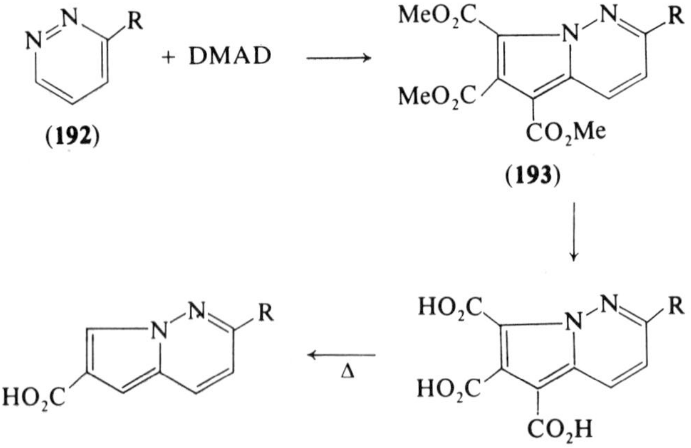

ether was used as solvent the reaction afforded only tars. Triester **193** was hydrolyzed to give tricarboxylic acid **194**, which decarboxylated when heated to afford the 2-substituted (or unsubstituted) pyrrolo[1,2-b]pyridazine-6-carboxylic acid (**195**). This synthesis is analogous to an earlier indolizine synthesis, reported by Diels and Meyer,[229] which utilizes pyridines and DMAD as starting materials. In addition, the complete parallelism of the chemical transformations to those observed in the earlier indolizine work[229] was used by Letsinger and Lasco[6] for their initial structure assignments.

In order to conclusively demonstrate the structure of the new ring system, Letsinger and Lasco[6] synthesized **195**, R = Me in an unequivocal fashion from 3,6-dimethylpyridazine and ethyl bromopyruvate [Eq. (69)].

A few years after this initial report, Acheson and Foxton[5] reinvestigated some of Letsinger's work and showed that the dimethyl acetylenedicarboxylate reaction with pyridazine was very solvent dependent. In a

---

[229] O. Diels and R. Meyer, *Justus Liebigs Ann. Chem.* **513**, 129 (1934).

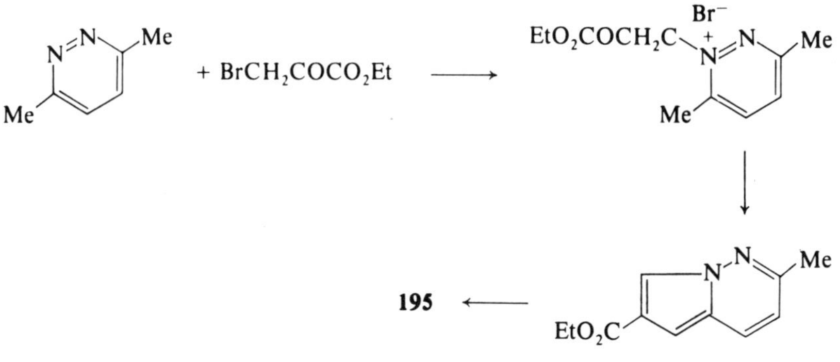

protic solvent, such as methanol, one obtained the expected pyrrolo-[1,2-b]pyridazines. However, when acetonitrile was the medium the major product was the 1:2 molar adduct having the pyridopyridazine structure **197** [Eq. (70)]. Acheson[230] has proposed a mechanism for the

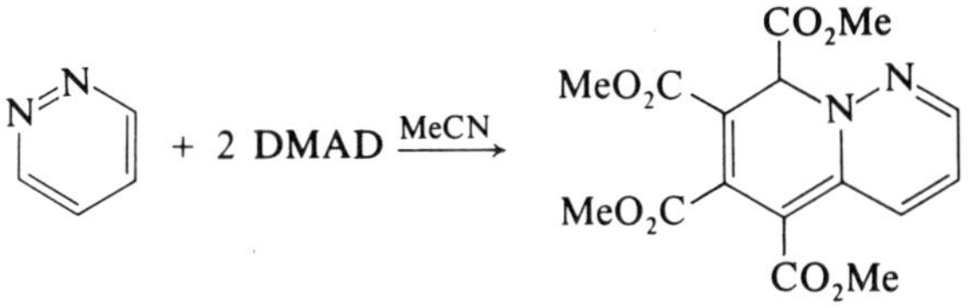

formation of compounds similar to **193** which is extended to the pyridazine case in Scheme 10. The synthesis of 6-carboethoxypyrrolo-[1,2-b]pyridazine (**196**) from 3,6-dimethylpyridazine and ethyl bromopyruvate initially used by Letsinger and Lasco[6] as a structure proof, has been extended into a general synthetic method by Fraser,[231] who prepared a variety of alkyl- and aryl-substituted pyrrolo[1,2-b]pyridazines (**198**) by combining α-haloketones with 3,6-dimethylpyridazine [Eq. (71)].

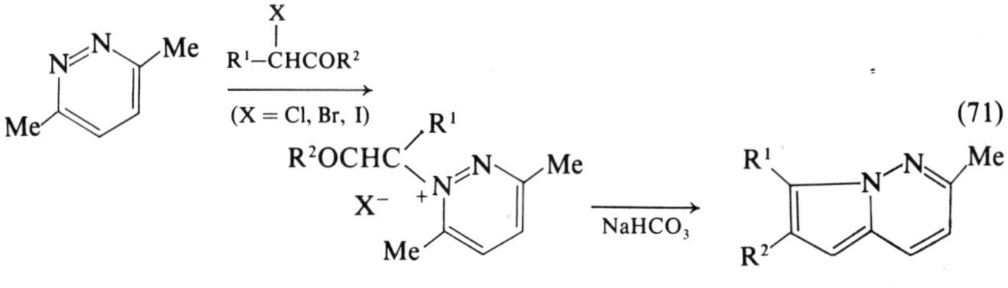

[230] R. M. Acheson, *Adv. Heterocycl. Chem.* **1**, 125 (1963).
[231] M. Fraser, *J. Org. Chem.* **36**, 3087 (1971).

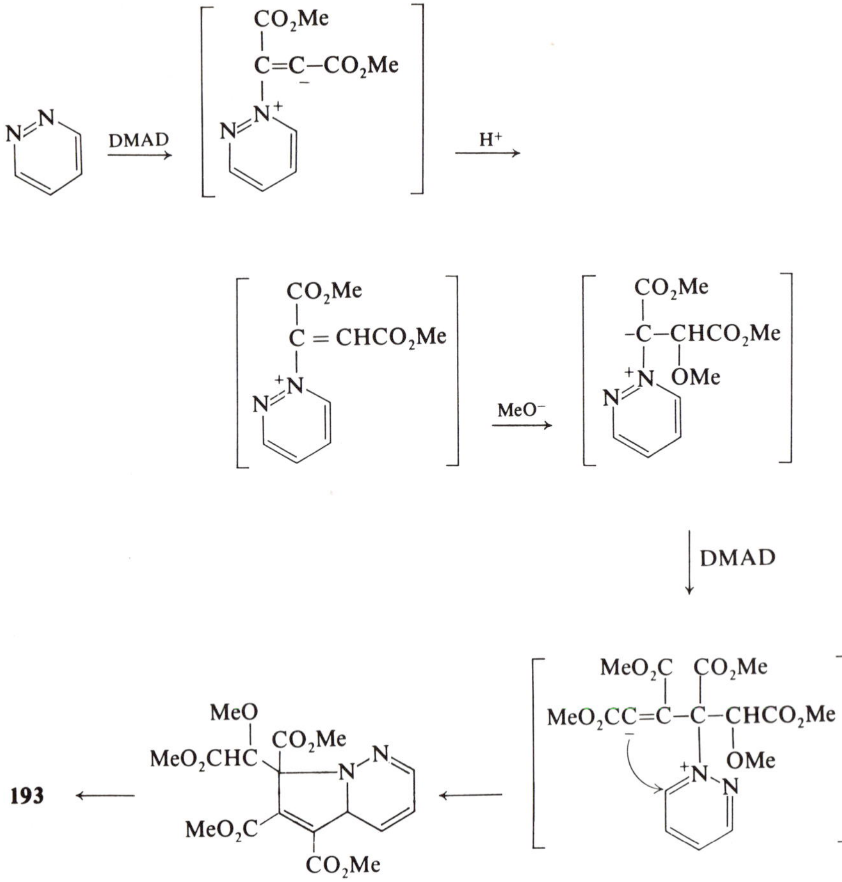

SCHEME 10

Another major group of syntheses of the pyrrolo[1,2-*b*]pyridazine ring system employing pyridazines as starting materials proceeds via the intermediacy of a stabilized 1,3-dipole, followed by a dipolar cycloaddition reaction of the 1,3-dipoles to a variety of acetylenes or olefins. Thus, two groups[3,98] have studied the reaction of tetracyanoethylene oxide with pyridazines, affording ylides of type **199** in yields ranging from 30 to 85%. These ylides react readily with acetylenes to produce substituted 7-cyanopyrrolo[1,2-*b*]pyridazines (**200**) [Eq. (72)]. Sasaki *et al.*[98] have also observed that, when 3,6-dialkoxypyridazines were employed in the above reaction, compounds of type **201** were formed, but only in very poor yield (ca. 10%) [Eq. (73)].

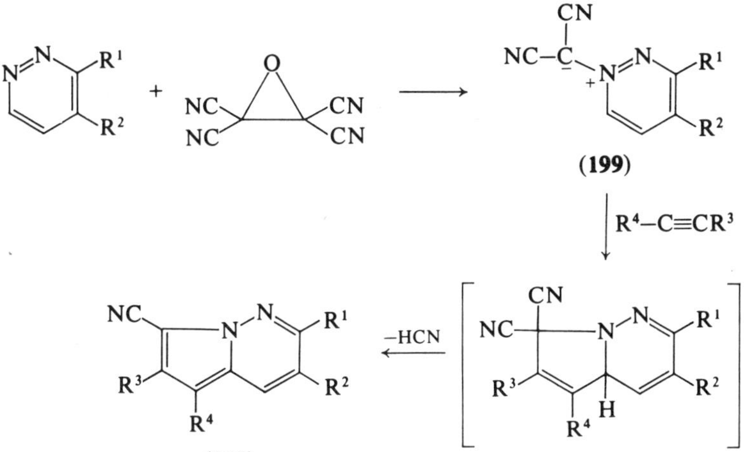

Masaki et al.[232] effectively extended the synthetic utility of this 1,3-dipolar cycloaddition approach to pyrrolo[1,2-b]pyridazines by preparing a wide variety of pyridazinium ylides (202), followed by *in situ* reaction with acetylenes to give 7,8-dihydropyrrolo[1,2-b]pyridazines (203). These dihydro derivatives could not be isolated but instead were oxidized directly with chloranil to substituted pyrrolo[1,2-b]pyridazines (204) [Eq. (74)]. Yields in this reaction range from 9 to 85% and were directly related to the stabilizing influence of the R group on the intermediate ylide (202); for example, when R = H and R^1 = CO_2Me, the yield of 204 was 9.5%; however, under identical reaction conditions, when R = PhCO and R^1 = CO_2Me, the yield of 204 was 82%.

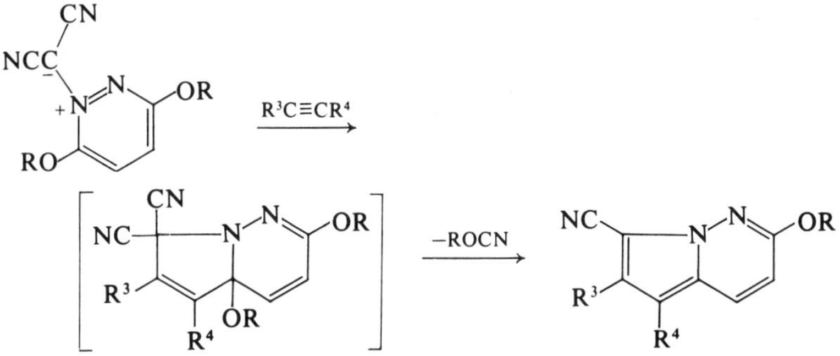

[232] Y. Masaki, H. Otsuka, Y. Nakayama, and M. Hioki, *Chem. Pharm. Bull.* **21**, 2780 (1973).

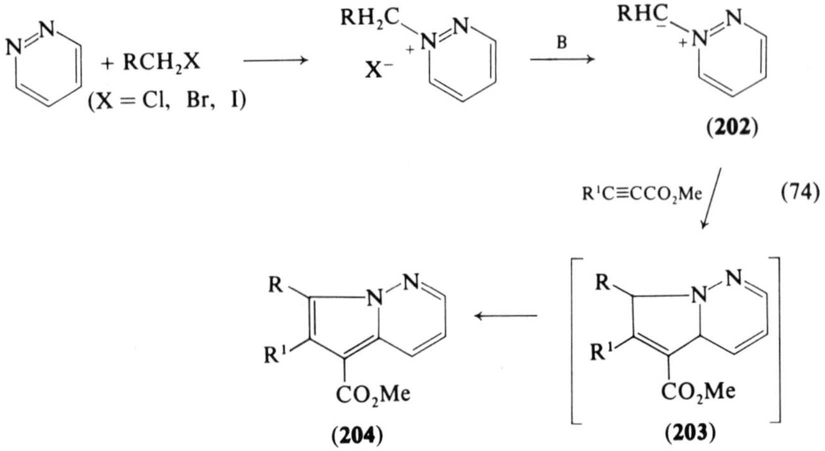

In a related effort, Petrovanii et al.[233,234] showed that when stabilized ylide **205** was generated in the absence of a dipolarophile only a dimer (**206**) could be isolated. Compound **205** was also shown to react with maleic anhydride to afford a 4a,5,6,7-tetrahydropyrrolo[1,2-b]-pyridazine (**207**) [Eq. (75)].

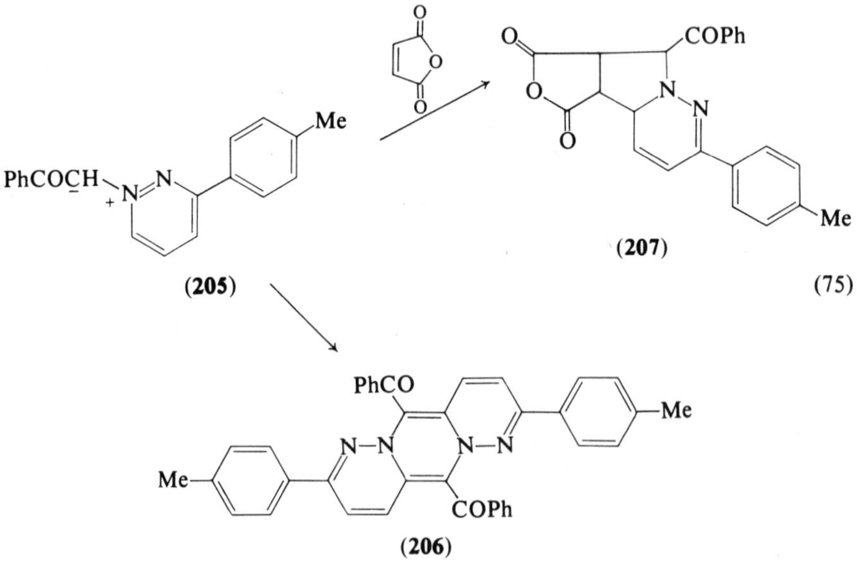

---

[233] M. Petrovanii, E. Stefanescu, and J. Druta, *Rev. Roum. Chim.* **16**, 1107 (1971) [*CA* **75**, 98524 (1971)].

[234] M. Petrovanii, E. Stefanescu, and J. Druta, *An. Stiint. Univ. "Al. I Cuza" Iasi, Sect. Ic* **19**, 175–81 (1973) [*CA* **80**, 108471 (1974)].

Farnum et al.[235] also utilized a 1,3-dipolar cycloaddition in an attempt to prepare derivatives of the pyrazolo[1,2-a]pyridazine **209**. The 2,6-dimethyl ylide **208** was prepared, and combination of it with DMAD did not afford the desired **209** but instead produced the pyrrolo[1,2-b]-pyridazine (**210**) [Eq. (76)]. Although a methyl group is apparently lost during this comparatively mild process which affords **210**, no mechanistic speculation was presented by the authors as to how the above transformation might have occurred.

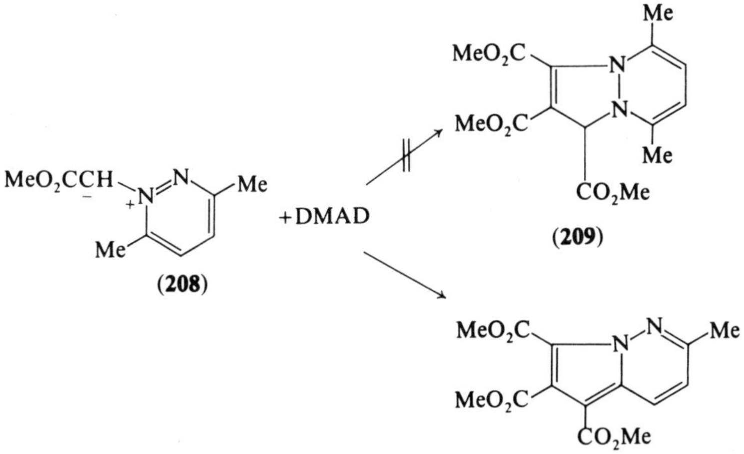

Finally, Lown and Matsumoto[7] combined diphenylcyclopropenone (**211**) with pyridazine to give 5,6-diphenyl-7-hydroxypyrrolo[1,2-b]-pyridazine (**212**) [Eq. (77)]. This reaction can also be viewed as a 1,3-dipolar cycloaddition with the diphenylcyclopropenone serving as a very reactive 1,3-dipole and pyridazine acting as the dipolarophile. When this reaction was repeated with diphenylcyclopropenethione (**213**),[99] 6,7-

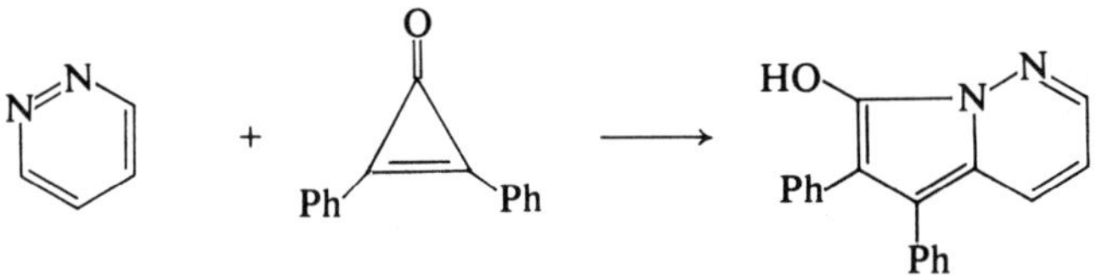

---

[235] D. G. Farnum, R. J. Alaimo, and J. M. Dunston, *J. Org. Chem.* **32**, 1130 (1967).

diphenyl-8-thioxopyrazolo[1,2-a]pyridazine (**214**) was the only product isolated [Eq. (78)].

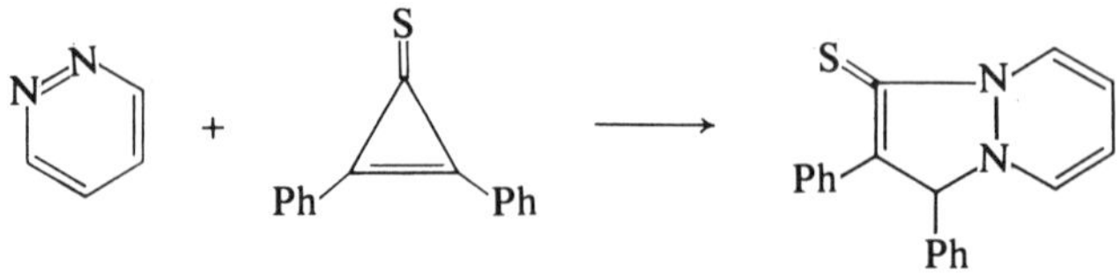

$$\text{(213)} \quad + \quad \underset{Ph \quad Ph}{\overset{S}{\triangle}} \quad \longrightarrow \quad \text{(214)} \qquad (78)$$

Most of the above synthetic routes, utilizing pyridazine or substituted pyridazines as starting materials, suffer from the fact that the resulting pyrrolo[1,2-b]pyridazines are heavily substituted in the pyrrole portion of the molecule, usually with electron withdrawing groups. Syntheses starting with a preformed pyrrole ring obviate this problem (see below) and appear to be of greater synthetic utility.

2. Syntheses Starting with a Preformed Pyrrole Ring

The most versatile method for preparing pyrrolo[1,2-b]pyridazines appears to be the reaction of β-dicarbonyl compounds, or their synthons, with N-aminopyrroles. Flitsch and Kramer[236,237] prepared the completely unsubstituted compound, pyrrolo[1,2-b]pyridazine (**215**),

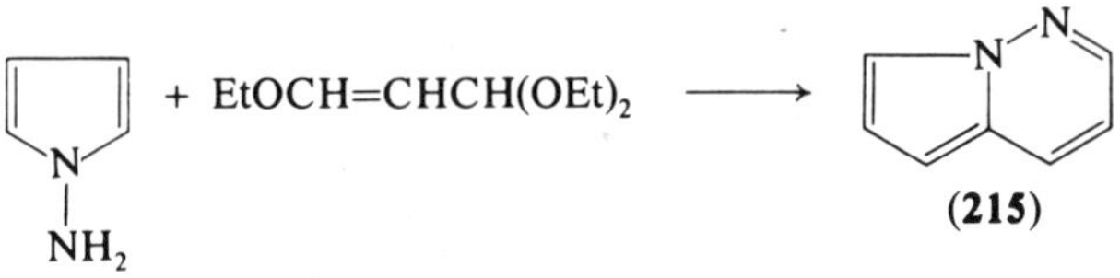

$$\underset{NH_2}{\text{pyrrole-N}} + \text{EtOCH=CHCH(OEt)}_2 \longrightarrow \text{(215)} \qquad (79)$$

from β-ethoxyacrolein diethylacetal and N-aminopyrrole using this method [Eq. (79)]. Phenylmalondialdehyde reacts[238] in a completely analogous fashion, producing 3-phenylpyrrolo[1,2-b]pyridazine (**216**) [Eq. (80)].

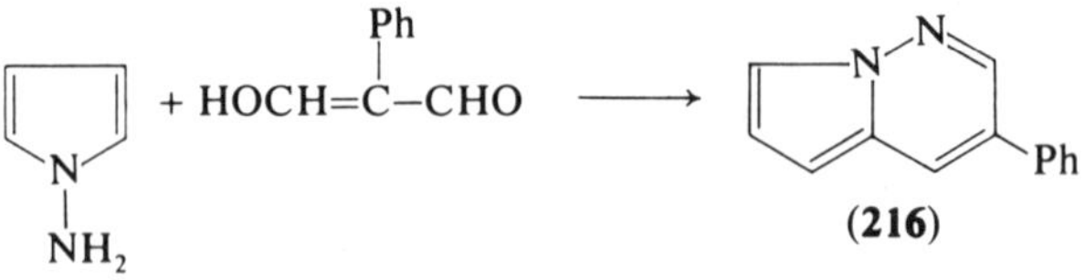

$$\underset{NH_2}{\text{pyrrole-N}} + \text{HOCH=}\underset{Ph}{\overset{|}{C}}\text{-CHO} \longrightarrow \text{(216)} \qquad (80)$$

---

[236] W. Flitsch and U. Kramer, *Tetrahedron Lett.*, 1479 (1968).
[237] W. Flitsch and U. Kramer, *Justus Liebigs Ann. Chem.* **735**, 35 (1970).
[238] G. Cappola, G. E. Hardtmann, and B. S. Huegi, *J. Heterocycl. Chem.* **11**, 51 (1974).

Symmetrical β-diketones afford 2,4-disubstituted pyrrolo[1,2-b]-pyridazines in good yield.[236,237] For example, 2,4-pentanedione yields 2,4-dimethylpyrrolo[1,2-b]pyridazine (217) [Eq. (81)].

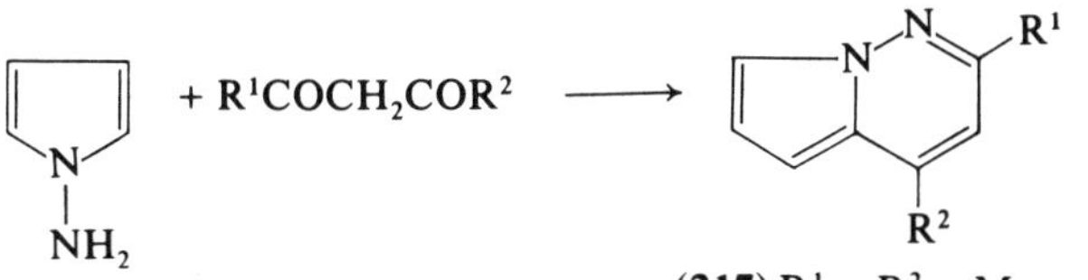

$$\text{pyrrole-NH}_2 + R^1COCH_2COR^2 \longrightarrow \text{product} \quad (81)$$

(217) $R^1 = R^2 = Me$
(218) $R^1 = Me, R^2 = Ph$
(219) $R = Ph, R^2 = H$
(220) $R^1 = H, R^2 = Ph$

Unsymmetrical β-dicarbonyl compounds give variable results.[236,237] 2-Methyl-4-phenylpyrrolo[1,2-b]pyridazine (218) is the only product isolated from the reaction of benzoylacetone with N-aminopyrrole. However, benzoylacetaldehyde produces a mixture of 2- and 4-phenyl-pyrrolo[1,2-b]pyridazines 219 and 220 [Eq. (81)]. 1,2-Dihydro derivatives (221) are prepared via a closely related transformation[236,237] [Eq. (82)].

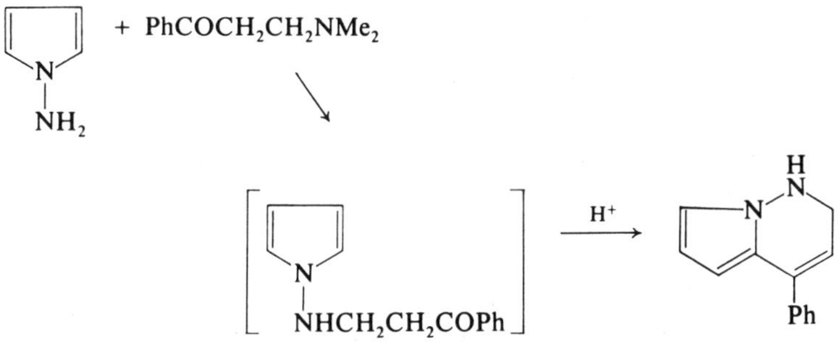

$$\text{pyrrole-NH}_2 + PhCOCH_2CH_2NMe_2 \longrightarrow [\text{pyrrole-NHCH}_2CH_2COPh] \xrightarrow{H^+} \text{(221)} \quad (82)$$

(221)

3. Synthesis via Molecular Rearrangement

Moore et al.[239] report an interesting rearrangement of the diaza-bicyclo[4.2.0]ketone 222 to the pyrrolo[1,2-b]pyridazine derivative 223 in almost quantitative yield. The reaction appears to be catalyzed by acid, base, or heat. The course of this rearrangement, as visualized by Moore, is as in Scheme 11.

[239] J. A. Moore, R. C. Gearhart, O. S. Rothenberger, P. C. Thorstenson, and R. H. Wood, J. Org. Chem. 37, 3774 (1972).

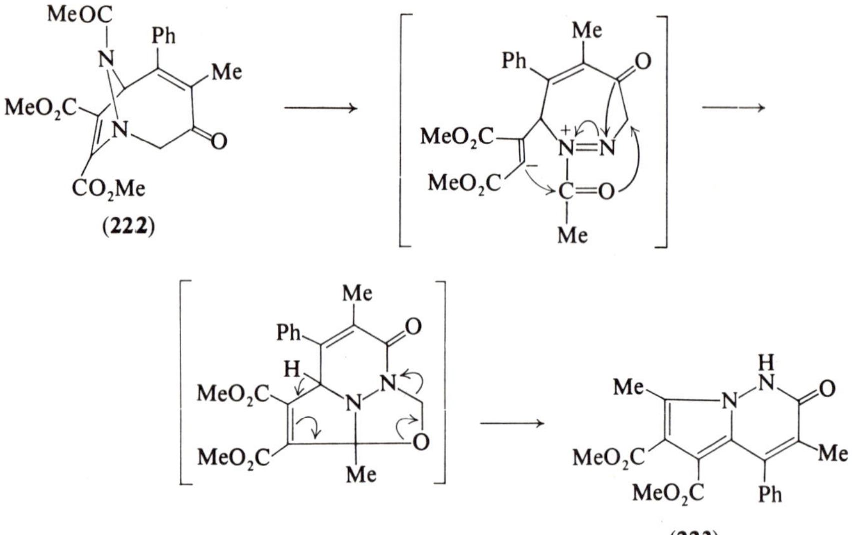

SCHEME 11

B. REACTIONS

Only two basic types of reactions have been reported for the pyrrolo-[1,2-b]pyridazine ring system: a variety of electrophilic substitution reactions and some rather straightforward functional group manipulations of substituted ring systems.

1. *Electrophilic Substitution*

Mild electrophilic reagents[240,241] afford only monosubstituted derivatives; for example, nitrosation, azo coupling, acylation with trifluoroacetic anhydride or a Vilsmeier reaction on **217** give 7-substituted derivatives (**224**) (E= —NO, —N=NAr, CF$_3$CO, or —CHO [Eq. (83)].

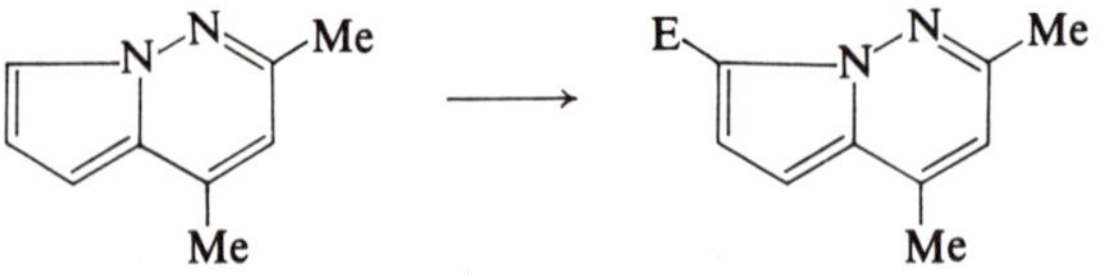

[240] M. Zupan, N. Stanovnik, and M. Tisler, *J. Heterocycl. Chem.* **8**, 1 (1971).
[241] M. Fraser, *J. Org. Chem.* **37**, 3027 (1972).

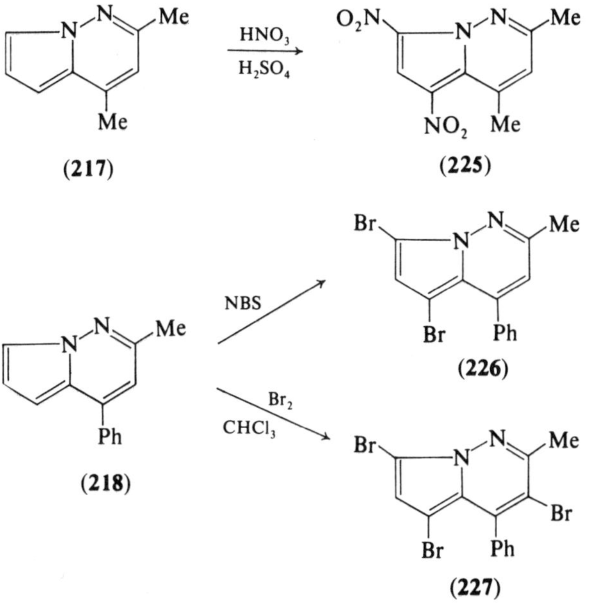

SCHEME 12

Stronger electrophiles[240] lead to di-, tri-, or tetra-substituted derivatives. Thus, nitration or bromination yield polysubstituted derivatives (Scheme 12). In the case of disubstituted products (**225, 226**) a 5,7-substitution pattern is observed. The trisubstituted compound **227** has also apparently been prepared by Flitsch and Kramer,[236,237] who originally assigned the structure as the 5,6,7-tribromo derivative. The weight of available evidence favors the 3,5,7-tribromo structure **227** for this material, as proposed by Zupan et al.[240]

Two groups[236,237,240] have demonstrated that compound **228**, resulting from electrophilic addition of DMAD to the 7-position of the 2,4-dimethylpyrrolo[1,2-b]pyridazine (**217**), can be cyclized under acidic conditions to give the tricyclic diazacyclopentindene derivative (**229**)

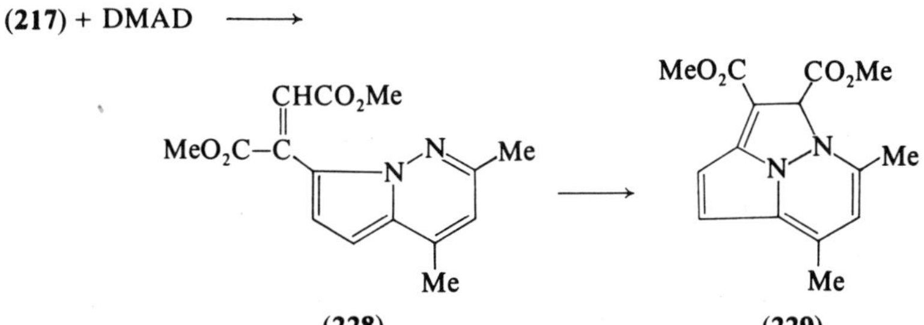

(**217**) + DMAD ⟶

(84)

[Eq. (84)]. All electrophilic substitution reactions reported to date for the pyrrolo[1,2-b]pyridazine nucleus are in complete accord with the calculated charge densities which show the 7 and 5 positions to have the greatest electron densities of the seven carbon atoms comprising this bicyclic ring system.[231]

In contrast, a very unusual feature of this ring system is its propensity to protonate on carbon rather than on the π-equivalent nitrogen at the 1-position.[231,236,237,241] Only a very few heteroaromatic systems containing a π-equivalent nitrogen atom protonate in this manner.[241] These results are in complete disagreement with the expected site of protonation based on the calculated charge densities of the ring atoms which show the N-1 nitrogen atom to have the highest electron density. Fraser[231,241] and others[236,237] have shown that protonation usually occurs at C-7 to give cation **230**; if however C-7 is substituted as in compound **231** then some protonation can occur at C-5, affording cation **232** (Scheme 13). In contrast, pyrrolo[1,2-c]pyrimidine (**123**) and

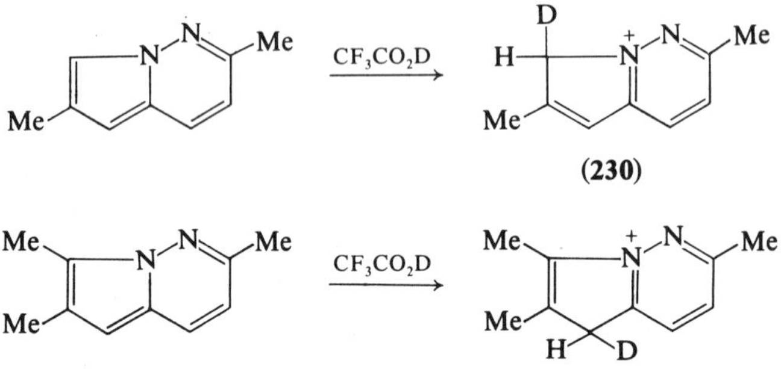

SCHEME 13

pyrrolo[1,2-a]pyrazine (**144**) protonate in the expected manner on the π-equivalent (nonbridgehead) nitrogen.[231]

2. Functional Group Transformations

Some standard transformations that can be carried out on substituted pyrrolo[1,2-b]pyridazines are illustrated in Scheme 14.

As indicated previously in Eq. (68), triester **193** can be hydrolyzed to the tricarboxylic acid **194** which can then be decarboxylated to **195**. It has also been reported[6] that pyrrolo[1,2-b]pyridazines of type **195** could not be quaternized under conditions that usually convert pyridines or pyridazines into their quaternary derivatives.

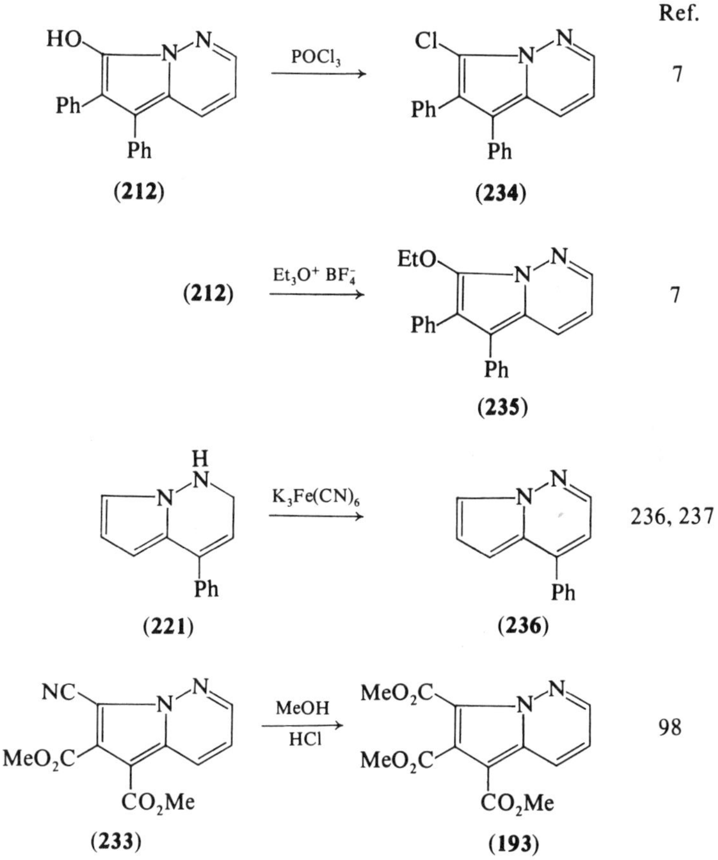

SCHEME 14

C. SPECTRAL PROPERTIES

A compilation of the spectral properties of various pyrrolo[1,2-b]-pyridazines is presented in Table IV. As a first approximation, the nuclear magnetic resonance (NMR) spectra of the pyrrolo[1,2-b]-pyridazines can be considered to be the spectrum to be expected from combination of the spectrum of a substituted pyrrole with that of a pyridazine. Enough NMR data have appeared in the literature (including a detailed analysis of all coupling constants) that the substitution patterns of many unknown derivatives could be assigned after a careful analysis of their proton magnetic resonance spectrum. Table IV together with other published[7,98,99,231,232,237,238,240,241] spectral data on this ring system should be useful for this purpose.

TABLE IV

SPECTRAL PROPERTIES OF SOME PYRROLO[1,2-b]PYRIDAZINES

Structure	Ultraviolet spectra (solvent) λ_{max}, nm (ε)	NMR spectra (solvent) δ	Infrared spectra (vehicle) μm	References
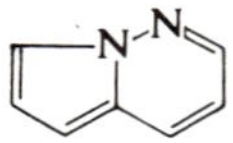 (pyrrolo[1,2-b]pyridazine)	(EtOH) 233 (25,120) 240 (25,120) 282 (1585) 292 (1000) 370 (1995)	$(CS_2)^a$ 7.79 (m, H2, $J_{2,3}$ = 4.5 Hz) 7.52 (dd, H7, $J_{5,7}$ = 1.0 Hz) 7.51 (dd, H4, $J_{2,4}$ = 2.0 Hz) 6.67 (dd, H6, $J_{6,7}$ = 2.5 Hz) 6.32 (m, H5, $J_{5,6}$ = 4.0 Hz) 6.31 (dd, H3, $J_{3,4}$ = 9.0 Hz)		237
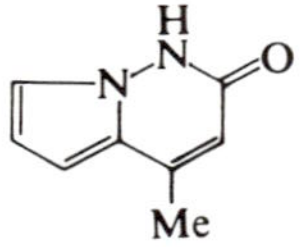 (2-oxo-4-methyl)	(EtOH) 241 (12,590) 298 (1585) 340 (3160)	$(DMSO-d_6)^a$ 10.91 (s, NH) 7.44 (t, H7, $J_{5,7}$ = 1.5 Hz) 6.54 (dd, H6, $J_{6,7}$ = 2.0 Hz) 6.37 (dd, H5, $J_{5,6}$ = 4.0 Hz) 6.12 (s, H3) 2.31 (s, 4-CH$_3$)	(KBr) 3.39 6.09 6.29	237
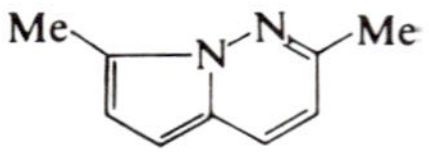 (2,6-dimethyl)	(EtOH) 235 (sh) (13,180) 242 (17,780) 250 (17,380) 310 (br) (10,235) 370 (br) (16,995)	$(CDCl_3)$ 7.47 (d, H4, $J_{3,4}$ = 9.0 Hz) 7.46 (s, H7) 6.31 (d, H3) 6.24 (s, H5) 2.42 (s, 2-CH$_3$) 2.34 (s, 6-CH$_3$)	(thin film) 6.02 6.47 7.75 9.01 12.50	231
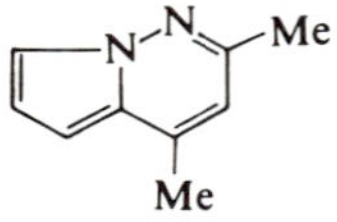 (2,4-dimethyl)		(CD_3COCD_3) 7.81 (dd, H7, $J_{5,7}$ = 1.8 Hz) 6.83 (dd, H6, $J_{6,7}$ = 2.7 Hz) 6.53 (dd, H5, $J_{5,6}$ = 4.5 Hz) 6.35 (s, H3) 2.35 (s, 2-CH$_3$ and 4-CH$_3$)		240

[a] Proton assignments for various chemical shifts were not made by the authors of the original article. These assignments have been made by the authors of this review article and should be considered as tentative.

Numerous workers[7,98,99,232] have reported that a variety of substituted pyrrolo[1,2-b]pyridazines give a visible parent ion in the mass spectrum. No reports have yet appeared on a more detailed analysis of the mass spectra of this ring system.

Theoretical molecular orbital interpretations of the electronic spectra of pyrrolo[1,2-b]pyridazines have also been reported.[50,51]

Acknowledgments

The authors wish to thank Dr. Beryl Dominy of the Pfizer Technical Information Department for help in carrying out computer-assisted searches of the literature on the title compounds and Mrs. Eileen Hartie and her co-workers of the Pfizer Word Processing Group for typing this manuscript.

The Thienopyridines

JOHN M. BARKER

Trent Polytechnic, Nottingham, England

I. Introduction	65
II. Synthesis	67
A. The Quinoline Isosteres	67
1. Syntheses Involving Formation of the Pyridine Ring	67
2. Syntheses Involving Formation of the Thiophene Ring	74
3. Thieno[3,4-*b*]pyridines	77
B. The Isoquinoline Isosteres	78
1. Syntheses Involving Formation of the Pyridine Ring	78
2. Syntheses Involving Formation of the Thiophene Ring	85
3. Thieno[3,4-*c*]pyridines	86
III. Electron Distribution	89
IV. Reactions	92
A. Electrophilic Substitution	92
1. The Quinoline Isosteres	92
2. The Isoquinoline Isosteres	94
B. Nucleophilic Substitution	96
C. Halogen–Metal Interconversion	98
D. Reactions of Active Methyl Groups	99
E. Oxidation	99
F. Reduction	102
G. Hydroxy Compounds	102
H. Amino Compounds	105
I. Miscellaneous Reactions	106
V. Physical Properties	107
A. Base Strengths	107
B. Spectroscopic Properties	108
1. Ultraviolet Spectra	108
2. Infrared Spectra	108
3. Nuclear Magnetic Resonance Spectra	110
4. Mass Spectra	112
C. Other Physical Properties	114
VI. Biological Activity	114
VII. Dyestuffs	117

I. Introduction

This report covers the literature up to the end of June 1975, and refers to one or two papers published in July and August of that year.

The six possible thienopyridine systems, the thieno[x,y-z]pyridines, fall into two groups, those that are analogs of quinoline, the [b]-fused systems **1–3**, and those that are analogs of isoquinoline, the [c]-fused systems, **4–6**.

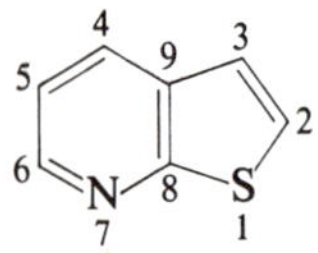

(1) [2,3-b]   (2) [3,2-b]   (3) [3,4-b]

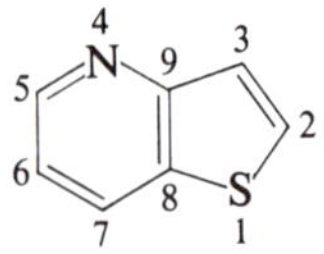

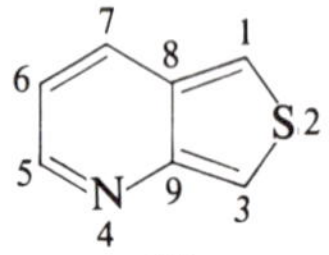

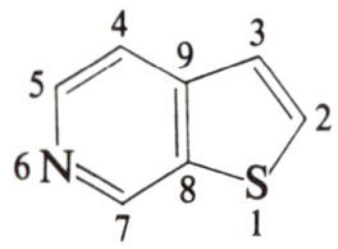

(4) [2,3-c]   (5) [3,2-c]   (6) [3,4-c]

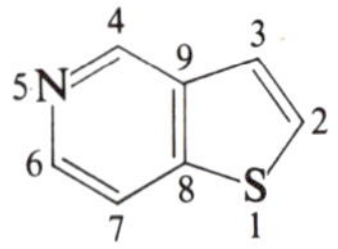

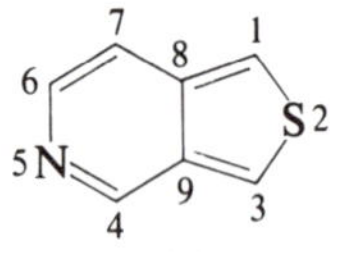

The earliest report (1913) of the preparation of a thienopyridine was by Steinkopf,[1,2] who applied the Skraup synthesis to 2-aminothiophene to obtain a low yield of **1**. During the 35 years that followed, there were very few publications in this field, and those that did appear were mostly concerned with approaches to thiophene analogs of indigo dyes.[3-6] In the period 1950–1965 a certain amount of work on thienopyridines was reported, including the first syntheses of **2**,[7] **4**,[8] and **5**.[8] Judging from the number of entries in *Chemical Abstracts,* interest began to increase in the period 1966–1969 (20 entries) and has been maintained at a comparatively high level during the 1970s (1970, 14 entries; 1971, 33; 1972, 18; 1973, 16; 1974, 18). There are two main explanations for this. First, there is the obvious theoretical interest in the behavior of systems that contain, fused together, a π-excessive and a π-deficient ring. Second, the search for pharmacologically active substances has led to the synthesis of analogs of various quinolines and isoquinolines in which the benzene ring is replaced by a thiophene nucleus.

[1] W. Steinkopf and G. Lutzkendorf, *Justus Liebigs Ann. Chem.* **403**, 45 (1914).
[2] W. Steinkopf, and G. Lutzkendorf, *Chem. Ztg.* **36**, 379 (1913).
[3] M. Colonna, *Gazz. Chim. Ital.* **70**, 154 (1940).
[4] E. Koenigs and H. Geissler, *Ber.* **57B**, 2076 (1924).
[5] E. Plazek and E. Sucharda, *Rocz. Chim.* **7**, 187 (1927).
[6] E. Koenigs and H. Kantrowitz, *Ber.* **60**, 2097 (1927).
[7] L. H. Klemm and D. Reid, *J. Org. Chem.* **25**, 1816 (1960).
[8] W. Herz and Lin Tsai, *J. Am. Chem. Soc.* **75**, 5122 (1953).

Unlike their benzene counterparts, thienopyridines do not occur widely in nature. The basic components of a shale oil of high sulfur content were found to contain certain methylthieno[2,3-*b*]- and -[3,2-*b*]-pyridines,[9] but no other natural occurrence of the compounds has been reported.

Very little work has so far been reported on thieno[3,4-*b*]- and -[3,4-*c*]-pyridines **3** and **6**; the parent substances were first prepared in 1970,[10] but they were found to be much less stable than the other four isomers.

Thienopyridine chemistry has previously been reviewed, in rather less detail than in the present work, by S. W. Schneller.[11]

II. Synthesis

Synthetic approaches to thienopyridines are conveniently considered under two headings, according to which heterocyclic ring is constructed. Generally syntheses involving formation of the pyridine ring have been adaptations of classical quinoline and isoquinoline syntheses.

In the account which follows the [2,3-] and [3,2-] fused systems will be considered separately from the much less common [3,4-] fused systems.

A. The Quinoline Isosteres

THIENO[2,3-*b*]- AND -[3,2-*b*]PYRIDINES

1. *Syntheses Involving Formation of the Pyridine Ring*

   a. The Skraup Reaction. The preparation of thieno[2,3-*b*]pyridine from 2-aminothiophene has already been mentioned;[1,2] in fact, Steinkopf did not employ the free amine (which is rather unstable) but used the tin double salt $(C_4H_3S-\overset{+}{N}H_3)_2SnCl_6^{2-}$ obtained directly by reduction of 2-nitrothiophene. This is still common practice in work involving aminothiophenes; in this review such salts will be represented as, for example, **7**:

(**7**)

[9] M. Pailer and W. Jiresch, *Monatsh.* **100**, 121 (1969).
[10] L. H. Klemm, W. O. Johnson, and D. V. White, *J. Heterocycl. Chem.* **7**, 463 (1970).
[11] S. W. Schneller, *Int. J. Sulphur Chem.* **7**, 309 (1972).

It is well known that acrolein is an intermediate in the Skraup synthesis of quinoline, so the use of other $\alpha\beta$-unsaturated carbonyl compounds can be considered at this point. Russian workers[12] reported that methyl vinyl ketone reacted with 2-aminothiophene double salt to give 4-methylthieno[2,3-b]pyridine [Eq. (1)] but later Klemm[13] showed that a

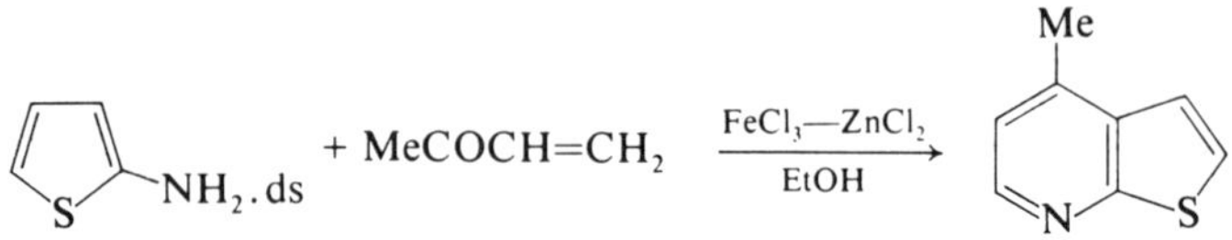

$$\text{thiophene-NH}_2 \cdot \text{ds} + \text{MeCOCH=CH}_2 \xrightarrow[\text{EtOH}]{\text{FeCl}_3\text{—ZnCl}_2} \text{4-methylthieno[2,3-b]pyridine} \quad (1)$$

smaller quantity of the 6-methyl isomer is also produced in this reaction. In similar fashion 3-aminothiophene double salt gives a mixture of products, that arising from Michael addition (**8**) predominating over that from the Schiff's base (**9**).

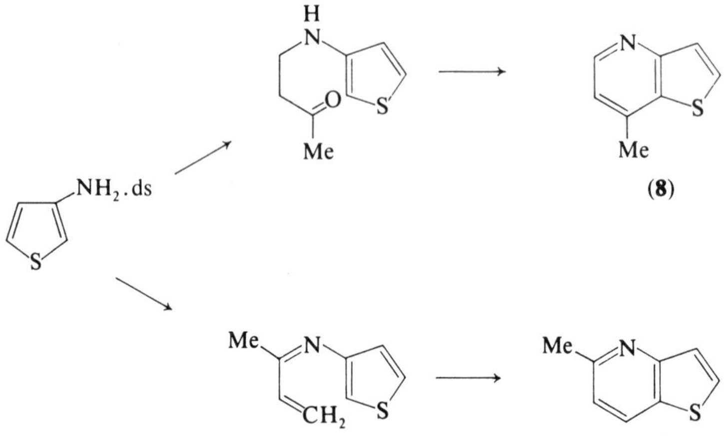

(41%) (**8**)

(9%) (**9**)

b. Reactions of Aminothiophenes with 1,3-Dicarbonyl Compounds. The first application of this approach to the thienopyridine system was reported by Emerson, Holly, and Klemm,[14] who cyclized the Schiff's base (**10**) to obtain excellent yields (80%) of 4,6-dimethylthieno[2,3-b]-pyridine [Eq. (2)].

[12] V. G. Zhiryakov and P. I. Abramenko, *Zh. Vsesoyuz. Obshch. D. T. Mendeleeva* **5**, 707 (1960) [*CA* **55**, 11416 (1961)].

[13] L. H. Klemm, C. E. Klopfenstein, R. Zell, and D. R. McCoy, *J. Org. Chem.* **34**, 347 (1969).

[14] W. S. Emerson, F. W. Holly, and L. H. Klemm, *J. Am. Chem. Soc.* **63**, 2569 (1941).

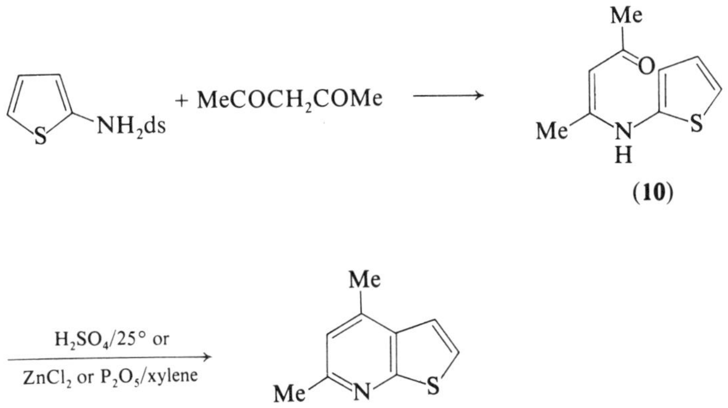

[Structure: 2-aminothiophene + MeCOCH₂COMe → compound (10)]

$$\xrightarrow{\text{H}_2\text{SO}_4/25° \text{ or } \text{ZnCl}_2 \text{ or } \text{P}_2\text{O}_5/\text{xylene}}$$

[4,6-dimethylthieno[2,3-b]pyridine] (2)

However, attempts to apply the sequence to a number of other 1,3-dicarbonyl compounds led only to tar formation. Abramenko[15] also reported the above reaction and, in addition, prepared the corresponding dimethylthieno[3,2-b]pyridine from the same dione and 3-aminothiophene double salt, using zinc chloride in ethanol.[16] Klemm[13] used zinc chloride in dioxane to effect cyclization of 10 and also prepared 4,5,6-trimethylthieno[3,2-b]pyridine via the 3-methylpentane-2,4-dione/2-aminothiophene Schiff's base. The acetals and ketals of 1,3-dicarbonyl compounds are also effective in this synthesis. Thus Klemm[13] has prepared the parent systems by condensation–cyclization of 2- and 3-aminothiophene double salts with malondialdehyde tetraethyl acetal (MTA) [Eq. (3)].

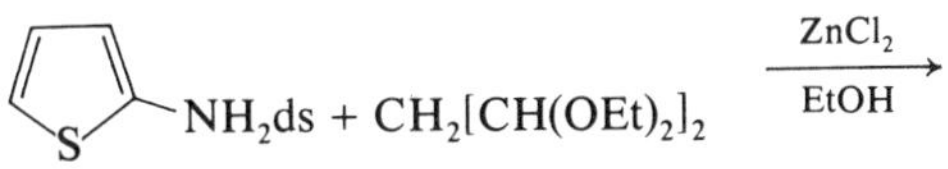

[Thiophene-NH₂ds + CH₂[CH(OEt)₂]₂] $\xrightarrow[\text{EtOH}]{\text{ZnCl}_2}$ **1** (44%)   (3)

The inaccessibility of 3-aminothiophene is a considerable practical problem, and recently Outurquin, Ah Kow, and Paulmier[17] have presented an interesting alternative synthesis of thieno[3,2-b]pyridine from 3-acetylthiophene via a Schmidt reaction [Eq. (4)].

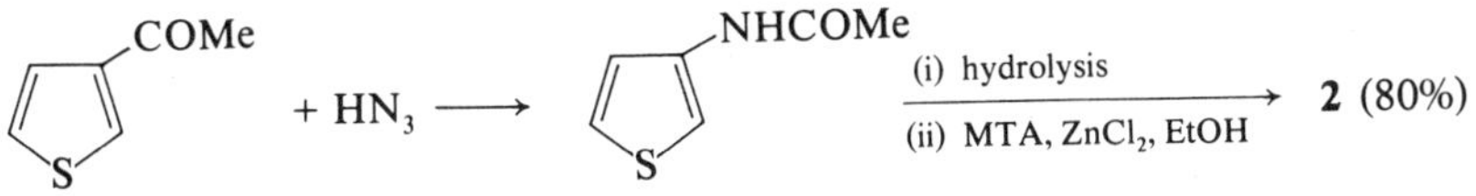

[3-acetylthiophene + HN₃ → 3-(NHCOMe)thiophene] $\xrightarrow[\text{(ii) MTA, ZnCl}_2, \text{EtOH}]{\text{(i) hydrolysis}}$ **2** (80%)   (4)

The presence of a deactivating group on the thiophene ring does not prevent cyclization; Klemm,[13] for example, showed that the mixture of

---

[15] P. I. Abramenko, *Khim. Geterotsikl. Soedin.* **7**, 468 (1971) [*CA* **76**, 25128 (1972)].
[16] P. I. Abramenko, *Khim. Geterotsikl. Soedin.* **3**, 368 (1967) [*CA* **67**, 116827 (1967)].
[17] F. Outurquin, G. Ah Kow, and C. Paulmier, *C.R. Hebd. Seances Acad. Sci., Paris* **277**, 29 (1973).

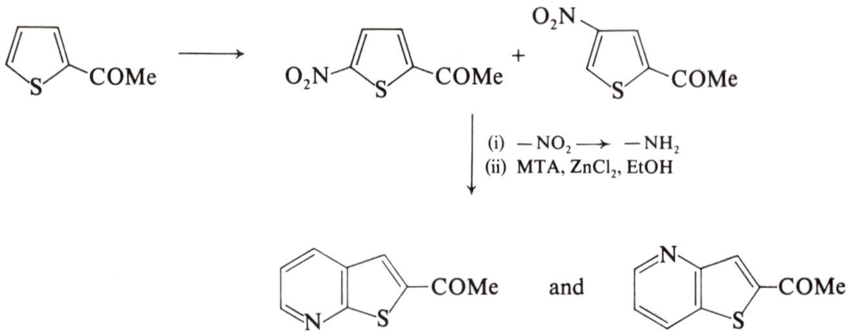

Scheme 1

(i) $-NO_2 \rightarrow -NH_2$
(ii) MTA, $ZnCl_2$, EtOH

SCHEME 1

amines formed on nitration-reduction of 2-acetylthiophene led to overall yields of 10% each of 2-acetylthieno[2,3-b]- and -[3,2-b]pyridines, which were separated by acid extraction and chromatography (Scheme 1). When no losses are to be expected through separation, cyclization onto a deactivated ring proceeds quite well, e.g. [Eq. (5)].[13]

$$\text{MeCO}-\text{[thiophene]}-NH_2 \cdot ds + MeCOCH_2COMe \longrightarrow \text{[product]} \quad (5)$$

(42%)

Zhiryakov and Abramenko[18] claimed that reaction of the diethylacetal diethylketal of 3-ketobutanal with 2-aminothiophene double salt led to 6-methylthieno[2,3-b]pyridine. The diethylacetal (**11**) of this dicarbonyl compound, however, was shown by Klemm[13] to yield mainly 5-acetylthieno[2,3-b]pyridine (**12**), only minor amounts of methylthienopyridines being formed (Scheme 2). Klemm presents evidence that the product reported by the Russian workers is actually the 4-methyl isomer. The mechanism suggested for the formation of **12** is given in Scheme 3.

Ethyl acetoacetate condenses with 3-aminothiophene, in the presence of acetic acid, to provide 5-methylthieno[3,2-b]pyridin-7(4H)-one (**13**), which can be converted into 5-methylthieno[3,2-b]pyridine by standard reactions (Scheme 4). The same compound was obtained from the double salt and paraldehyde.[15,19]

[18] V. G. Zhiryakov and P. I. Abramenko, *Khim. Geterotsikl. Soedin.* **1**, 334 (1965) [*CA* **63**, 13231 (1965)].

[19] P. I. Abramenko, *Zh. Vsesoyuz. Obshch. D. T. Mendeleeva* **17**, 478 (1972) [*CA* **77**, 152020 (1972)].

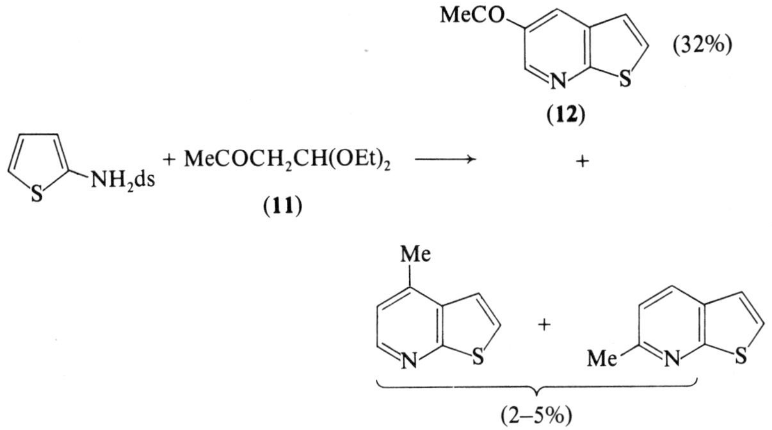

Sec. II.A] THE THIENOPYRIDINES 71

SCHEME 2

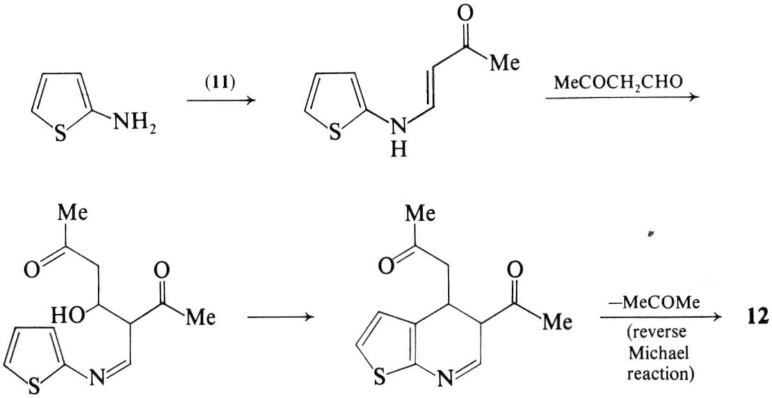

SCHEME 3

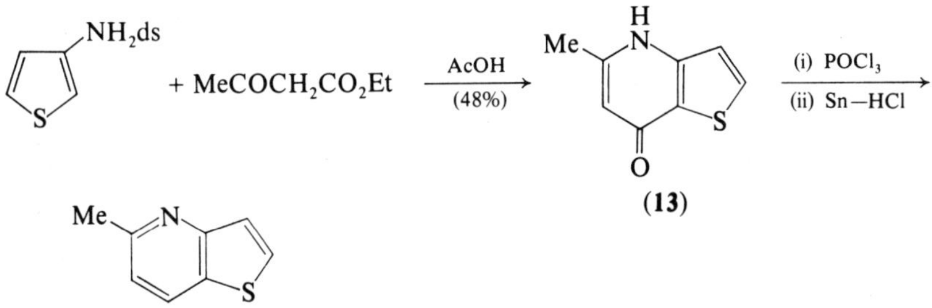

SCHEME 4

c. *Cyclizations of o-Aminocarbonylthiophenes.* This approach is summarized in Eqs. (6) and (7). Clearly the principal difficulty lies in the preparation of appropriately substituted thiophenes. In a comparatively early paper, Raich and Hamilton[20] described the multistep synthesis depicted in Scheme 5.

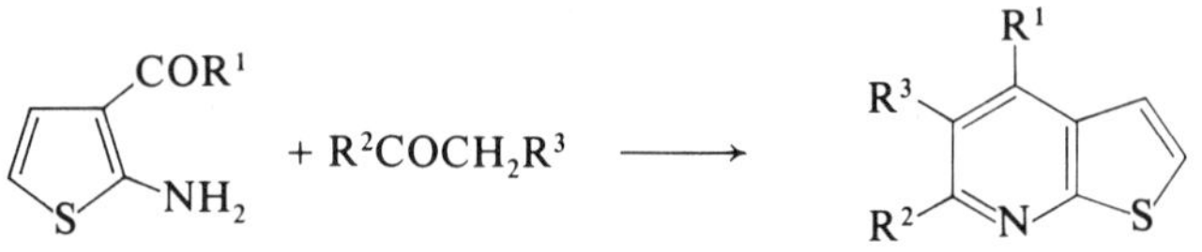

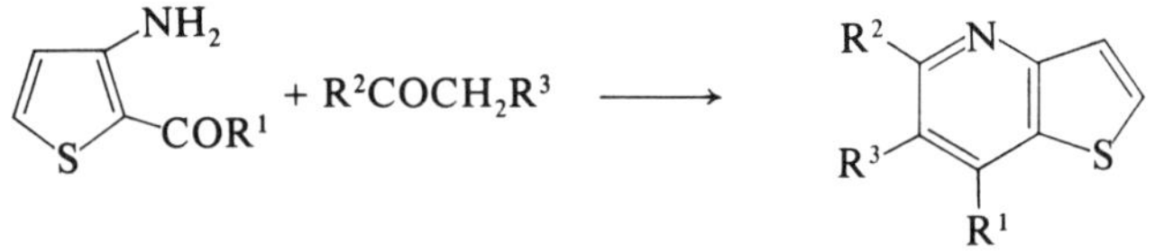

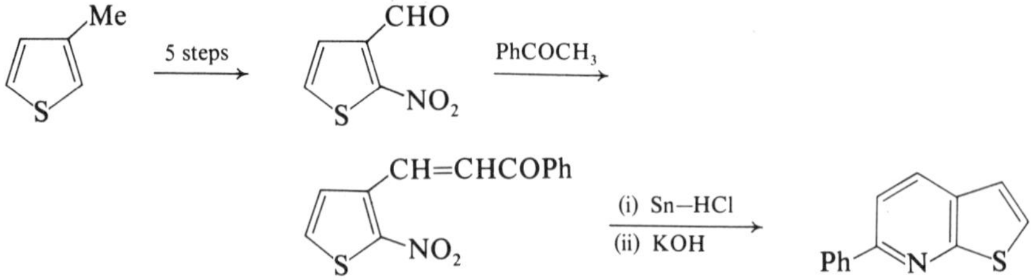

SCHEME 5

Shvedov *et al.*[21,22] carried out Vilsmeier reactions on various 4,5-disubstituted 2-acylthienylamines and cyclized the resulting 3-formyl compounds with a variety of substances containing active methylene groups; the process is summarized in Eq. (8) (R, R^1 were alkyl or cycloalkyl; R^3 was $-CN$, $-COR^4$, $-CO_2H$, $-CO_2Et$, or $-CONH_2$).

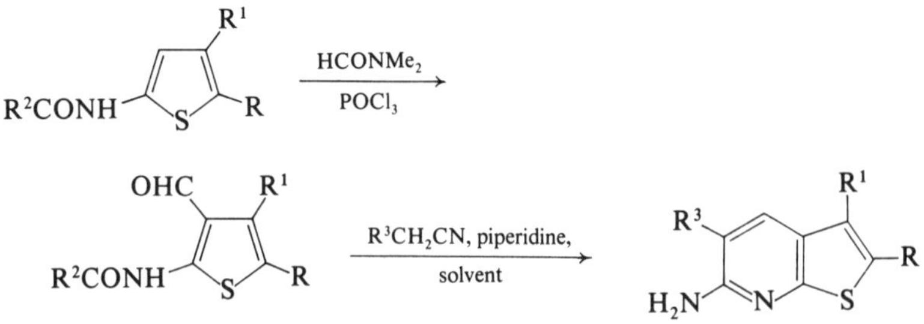

[20] W. J. Raich and C. S. Hamilton, *J. Am. Chem. Soc.* **79**, 3800 (1957).
[21] V. I. Shvedov, I. A. Kharizomenova, and A. N. Grinev, USSR Patent 364,613 [*CA* **78**, 159580 (1973)].
[22] V. I. Shvedov, I. A. Kharizomenova, and A. N. Grinev, *Khim. Geterotsikl. Soedin.* **10**, 58 (1974) [*CA* **80**, 95865 (1974)].

Perchloric acid has been used to effect condensation of 3-amino-2-benzoylthiophenes with ketones [Eq. (9)].[23]

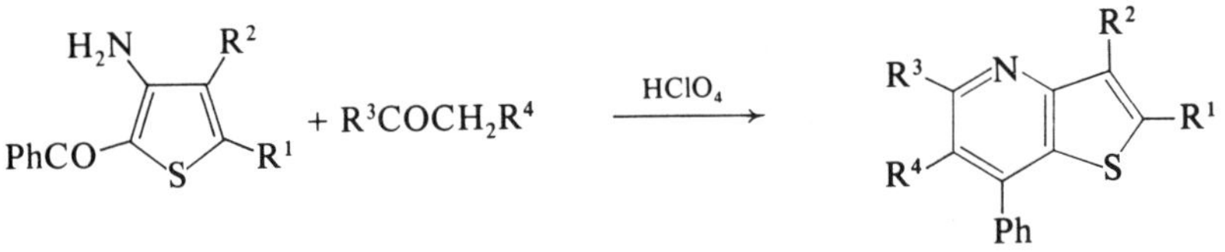

$$\text{[structure: 3-amino-2-benzoylthiophene with } R^1, R^2 \text{]} + R^3COCH_2R^4 \xrightarrow{HClO_4} \text{[thienopyridine product with } R^1, R^2, R^3, R^4, Ph\text{]} \quad (9)$$

(49–87%)

A variation, in which the acyl function on the thiophene ring is replaced by a thionoester group, leads to a thienopyridinethione

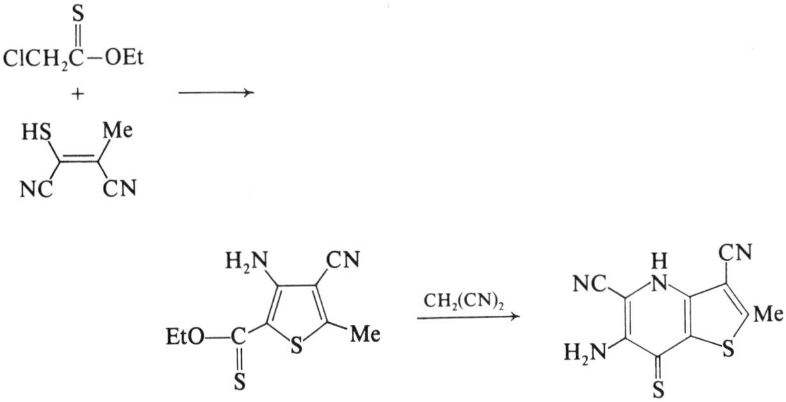

$$ClCH_2\overset{S}{\underset{\|}{C}}-OEt + \underset{NC}{\overset{HS}{\diagdown}}\underset{CN}{\overset{Me}{\diagup}} \longrightarrow$$

$$\text{[aminothiophene intermediate]} \xrightarrow{CH_2(CN)_2} \text{[thienopyridinethione } \mathbf{14}\text{]} \quad (10)$$

(14)

(14)[24] [Eq. (10)], and the sulfur-substituted compound (15) arose from reaction of 3-benzoyl-2-amino-4,5,6,7-tetrahydrobenzo[b]thiophene and 1,1-dimethylthio-2-nitroethylene[25] [Eq. (11)].

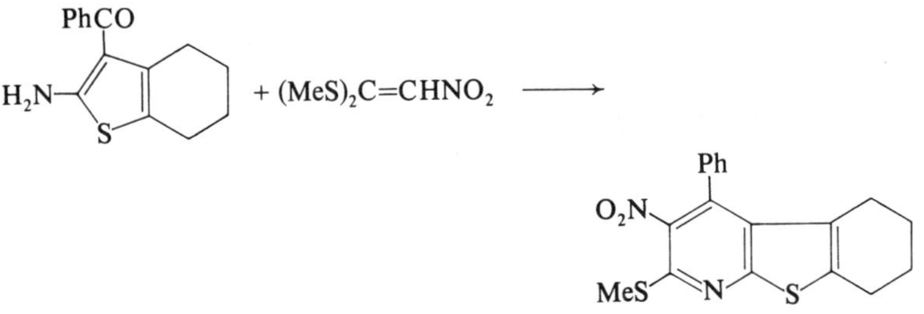

$$\text{[3-benzoyl-2-amino-tetrahydrobenzothiophene]} + (MeS)_2C=CHNO_2 \longrightarrow \text{[product } \mathbf{15}\text{]} \quad (11)$$

(15)

[23] H. Schaefer, K. Gewald, and H. Hartmann, *J. Prakt. Chem.* **316**, 169 (1974).
[24] K. Hartke and G. Goelz, *Justus Liebigs Ann. Chem.* 1644 (1973).
[25] H. Schaefer, B. Bartho, and K. Gewald, *Z. Chem.* **13**, 294 (1973).

d. *Miscellaneous Cyclizations.* An interesting synthesis of a rather elaborate thieno[2,3-*b*]pyridine has been published by Blechert, Gericke, and Winterfeldt.[26] The ring closure was followed by a hetero-Cope rearrangement [Eq. (12)].

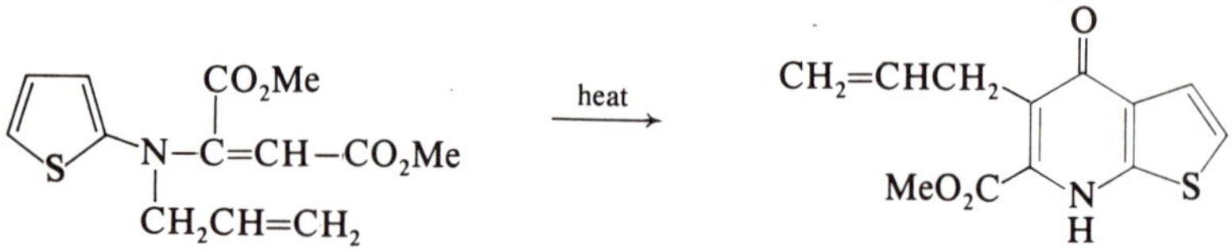

2. *Syntheses Involving Formation of the Thiophene Ring*

a. *Cyclization of Carboxypyridylthioacetic Acids.* An early report by Koenigs and Geissler[4] that 3-hydroxythieno[2, 3-*b*]pyridine was formed by treatment of 2-pyridylthioacetic acid with acetic anhydride was questioned by Chichibabin and Vorozhtov,[27] who showed that Koenigs and Geissler's product differed from authentic material prepared as in Scheme 6.

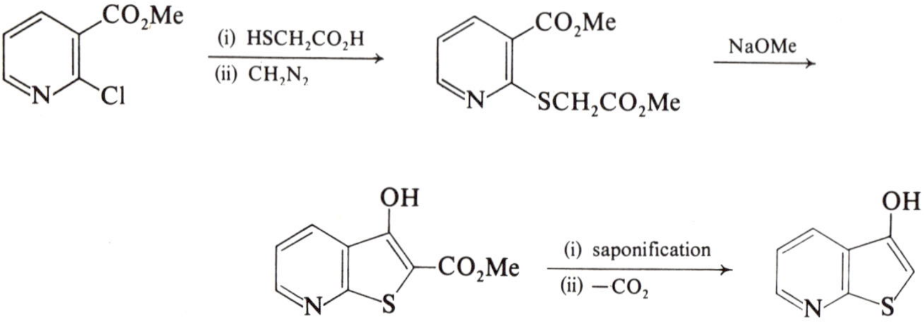

SCHEME 6

A new structure (**16**) was proposed for Koenigs and Geissler's substance, but this, too, was incorrect. Its true mesionic nature (**17**) was established some time later by Duffin and Kendall.[28]

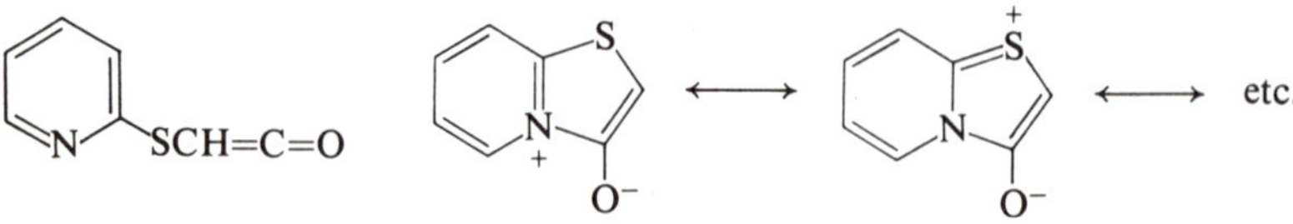

(**16**)   (**17**)

[26] S. Blechert, R. Gericke, and E. Winterfeldt, *Ber.* **106**, 368 (1973).
[27] A. E. Chichibabin and N. N. Vorozhtov, *Ber.* **66B**, 364 (1933).
[28] G. F. Duffin and J. D. Kendall, *J. Chem. Soc.*, 734 (1951).

Plazek and Sucharda[29] and Colonna[3] prepared thioindigo dyes from 2-pyridylacetic acids; it may be assumed that 3-hydroxythieno[2,3-b]-pyridines were intermediates, but they were not isolated. An improvement in yield of hydroxythienopyridines in the cyclization of carboxy-pyridylthioacetic acids with acetic anhydride is achieved if the intermediate acetoxy compound is isolated[30] (Scheme 7).

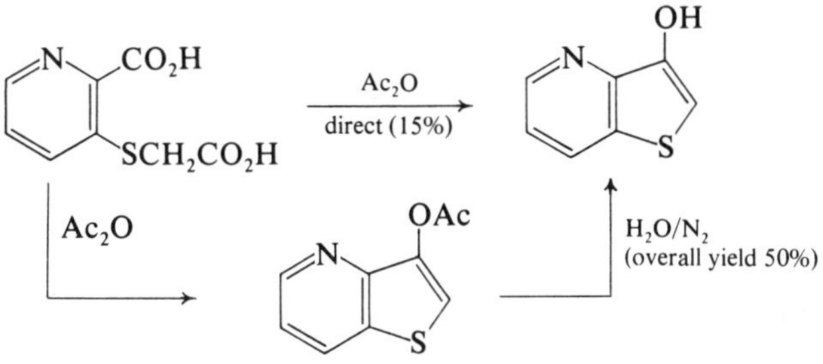

SCHEME 7

This method has also been employed by Zhiryakov and Abramenko[18] for the preparation of 3-hydroxy-6-methylthieno[2,3-b]-pyridine from 3-carboxy-6-methyl-2-pyridylthioacetic acid.

Related syntheses use cyanopyridylacetic acids. For example 3-amino-2-ethoxycarbonylthieno[2,3-b]pyridine is formed by the sequence of Eq. (13).[31]

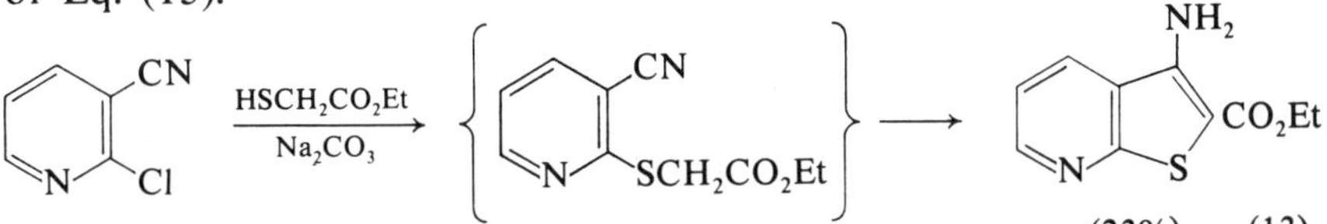

(23%)   (13)

The formation of more complex derivatives, by similar reactions of more highly substituted pyridines, has been reported.[32,33] Replacement of the ethoxycarbonyl function by other electron-withdrawing substituents ($-COMe$, $-CN$, $-NO_2$) is possible[34] [Eq. (14)].

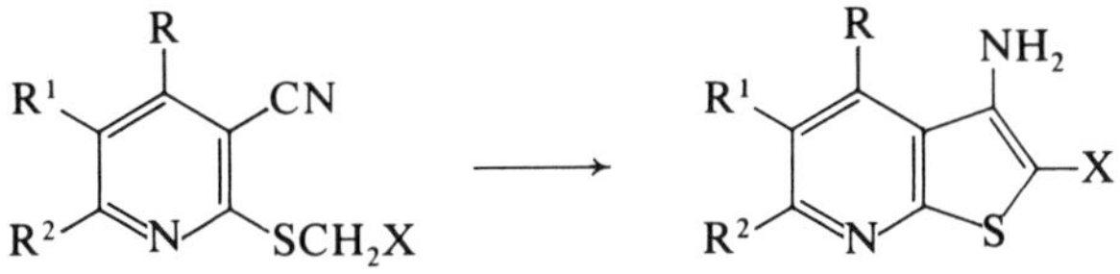

(14)

(X = COMe, CN, NO_2)

[29] E. Plazek and E. Sucharda, *Ber.* **59B**, 2282 (1926).
[30] J. T. Sheehan and G. J. Leitner, *J. Am. Chem. Soc.* **74**, 5501 (1952).
[31] S. W. Schneller and F. W. Clough, *J. Heterocycl. Chem.* **11**, 975 (1974).
[32] A. L. Cossey, R. L. N. Harris, J. L. Huppatz, and J. N. Philips, *Angew. Chem., Int. Ed. Engl.* **11**, 1099 (1972).
[33] K. Gewald and M. Hentschel, East German Patent 105,805 [*CA* **82**, 16813 (1975)].
[34] R. Niess and H. Eilingsfeld, German Patent 2,241,717 [*CA* **80**, 146133 (1974)].

b. High-Temperature Catalytic Methods. Klemm and his co-workers have developed high-temperature catalytic processes for the preparation of thienopyridines from the readily available vinylpyridines. The first report of such an approach was that of Hansch and Carpenter,[35] who were unsuccessful in their attempt to prepare thieno[2,3-*c*]pyridine by reaction of 4-vinylpyridine and hydrogen sulfide over an alumina catalyst at 600°. Klemm and Reid[7] found that an iron(II) sulfide–alumina catalyst caused the desired reaction to occur, albeit in very low yield. Thus 2-vinylpyridine gave thieno[3, 2-*b*]pyridine (1.6%)[7] and 3-vinylpyridine gave a mixture of thienopyridines [Eq. (15)].[36] Similar

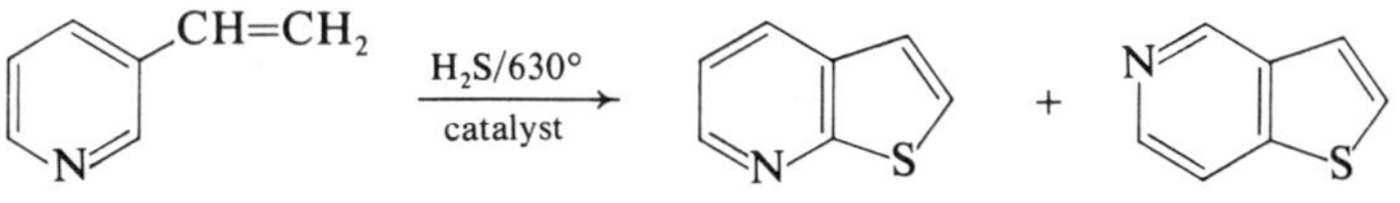

yields were obtained from 3-ethylpyridine. Application of the reaction to 2-methyl-5-ethyl- or -5-vinylpyridine led to the corresponding 6-methylthienopyridines.

It was found that much better yields are obtained if the vinylpyridine is first converted into the corresponding benzyl pyridylethyl sulfide (by reaction with benzylmercaptan), followed by pyrolysis of the latter over glass helices at high temperature with either nitrogen or hydrogen sulfide as carrier gas[37] [Eq. (16)].

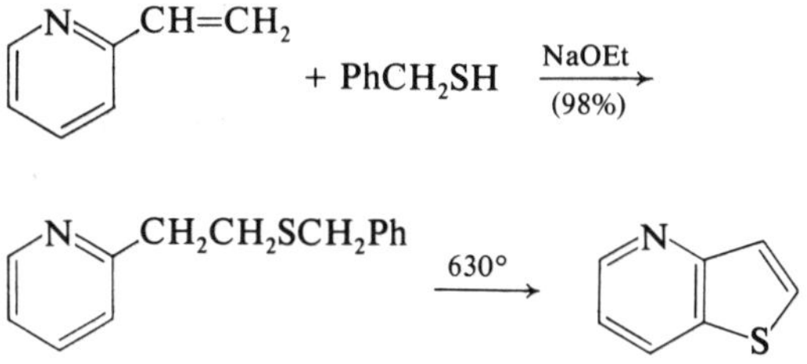

It is believed that this reaction involves thioethylpyridyl radicals, arising by homolysis of the sulfur–benzyl bond.

---

[35] C. Hansch and W. Carpenter, *J. Org. Chem.* **22**, 936 (1957).
[36] L. H. Klemm and D. R. McCoy, *J. Heterocycl. Chem.* **6**, 73 (1969).
[37] L. H. Klemm, J. Shabtai, D. R. McCoy, and W. K. T. Kiang, *J. Heterocycl. Chem.* **5**, 883 (1968).

c. *Miscellaneous Cyclizations.* Two interesting reactions leading to 2,3-dihydrothienopyridine derivatives have been recorded. The first[38] gave 6-*t*-butyl-2,3-dihydro-3,3-dimethylthieno[3,2-*b*]pyridine-1,1-diox-

ide (**18**) [Eq. (17)] and the second[39] led to 2,3-dihydro-2,2-diphenyl-thieno[2,3-*b*]pyridine (**19**) [Eq. (18)].

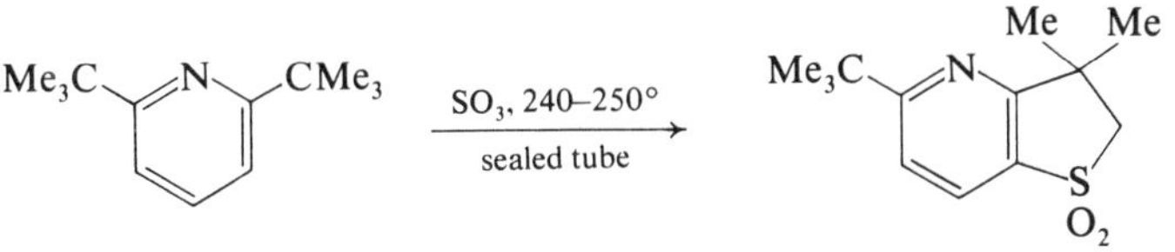

3. *Thieno[3,4-b]pyridines*

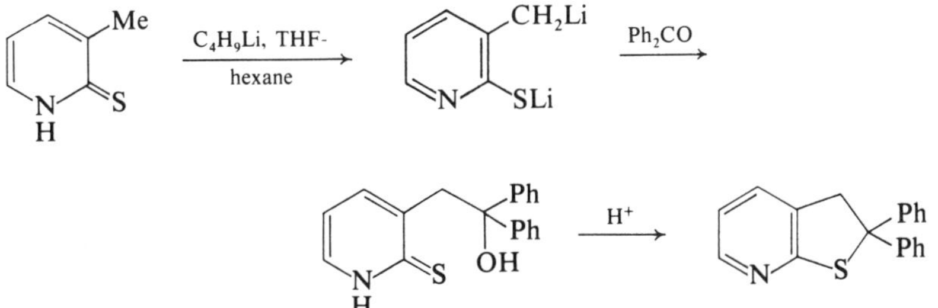

Very little work has been done in this area. The parent system has been prepared by Klemm and his colleagues[10,40,41] from 2,3-dimethyl-pyridine by the sequence of Scheme 8.

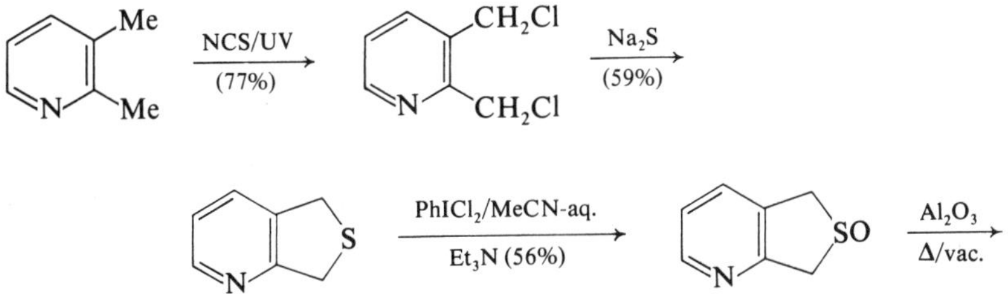

SCHEME 8

---

[38] H. C. van der Plas and T. H. Crawford, *J. Org. Chem.* **26**, 2611 (1961).
[39] R. C. Smith, S. Boatman, and C. R. Hauser, *J. Org. Chem.* **33**, 2083 (1968).
[40] L. H. Klemm, W. O. Johnson, and D. V. White, *J. Heterocycl. Chem.* **9**, 843 (1972).
[41] L. H. Klemm, W. O. Johnson, and D. V. White. U.S. Patent 3,709,894 [*CA* **78**, 72091 (1973)].

Spinner and Yeoh[42] have synthesized a number of 1,3-dihydro derivatives of the system. The starting point was dimethyl quinolinate N-oxide, which was converted by conventional reactions into the dichloro compound (20) and thence into 5-methoxy- and 5-hydroxy-1,3-dihydrothieno[3,4-b]pyridine (which exists predominantly in the 4H-5-one form: see Section IV,G.).

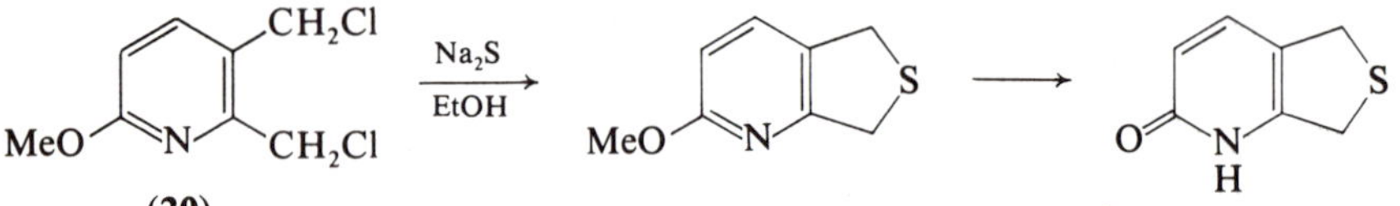

B. THE ISOQUINOLINE ISOSTERES

1. Syntheses Involving Formation of the Pyridine Ring

a. The Bischler–Napieralski Synthesis. In 1938 Barger and Easson[43] made the first reported attempt to prepare a thieno[3,2-c]pyridine by ring closure of N-formyl-2-(2-thienyl)ethylamine (21, R = H), but found that both phosphorus pentachloride and pentoxide failed to produce the desired compound. Later Herz[44] showed that the corresponding acetyl- and benzoylamines (21) were cyclized, in reasonable yields, by treatment

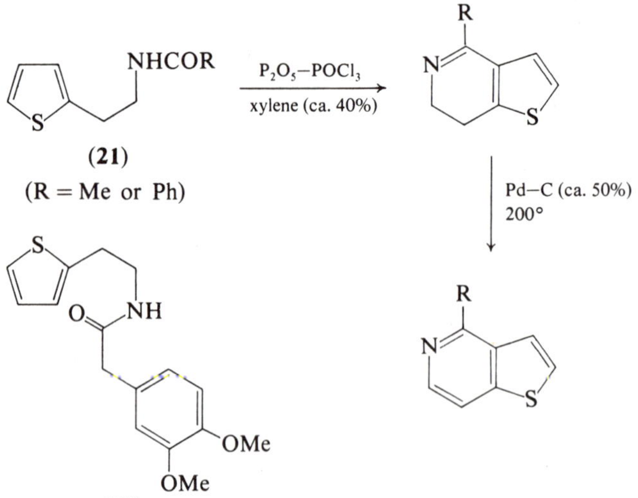

---

[42] E. Spinner and G. B. Yeoh, *J. Chem. Soc. B*, 289 (1971).
[43] G. Barger and A. P. T. Easson, *J. Chem. Soc.*, 2100 (1938).
[44] W. Herz, *J. Am. Chem. Soc.* **73**, 351 (1951).

with a mixture of phosphorus pentoxide and phosphoryl chloride in xylene. The resulting 4-substituted-6,7-dihydrothieno[3,2-c]pyridines were dehydrogenated to the fully aromatic systems [Eq. (19)]. Similar treatment of N-acyl-2-(3-thienyl)ethylamines led to dihydrothieno-[2,3-c]pyridines. Herz and Tsai[45] attempted to extend the reaction to the homoveratroylamide (22) in the hope of producing a papaverine-like substance, but the amide could not be cyclized under a variety of conditions. Herz also applied the reaction to the 5-methoxy-2-(2-thienyl) ethylamides (23). However, the conditions necessary to bring about the ring closure also caused demethylation of the rather labile methoxy group, the isolated product having the structure 24.

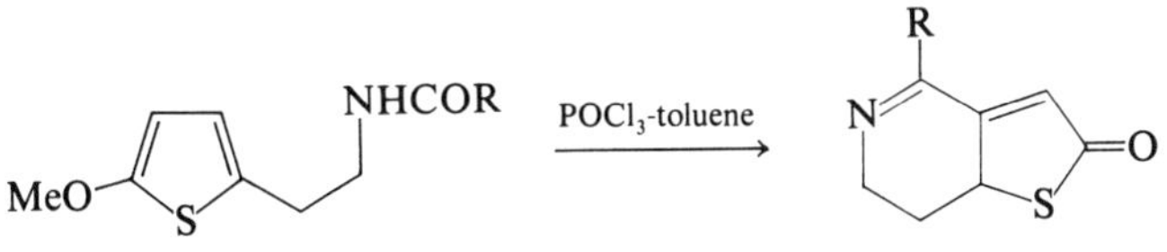

(23) → (24)
(R = Me or Ph)

The failure of 22 to cyclize seems to be due in some way to the presence of the benzyl methylene group, since Kametani and his co-workers[46] found that comparable aroylamines (21, R = 4-hydroxy-3-methoxyphenyl or 4-benzyloxy-3-methoxyphenyl) gave 1-aryl-6,7-dihydrothienol[3,2-c]pyridines when treated with phosphoryl chloride.

The same reagent has been used by Descamps and Binon[47] to cyclize amides of the type 21, in which R was phenyl, 4-methoxyphenyl, and 3,4,5-trimethoxyphenyl; yields were high (72–86%), and the dihydro compounds were successfully dehydrogenated with palladium–charcoal in xylene (yields 65–70%). Dressler and Joullié[48] have employed the Bischler–Napieralski synthesis to prepare more highly substituted thieno[3,2-c]pyridines (25) than those discussed above. The general reaction sequence used is shown in Scheme 9.

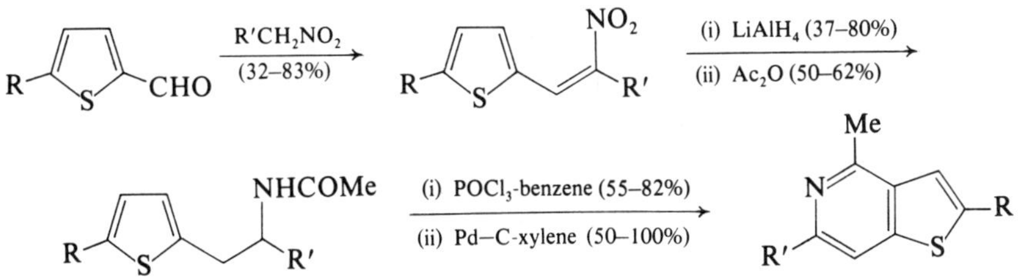

SCHEME 9

[45] W. Herz and Lin Tsai, *J. Am. Chem. Soc.* **77**, 3529 (1955).
[46] T. Kametani, I. Yuichi, and S. Aonuma, *J. Pharm. Soc. Jpn.* **74**, 1301 (1954).
[47] M. Descamps and F. Binon, *Bull. Soc. Chem. Belges* **71**, 579 (1962).
[48] M. L. Dressler and M. M. Joullié, *J. Heterocycl. Chem.* **7**, 1257 (1970).

Ott, in a series of patents,[49–52] has described the synthesis of both the [2,3-c]- and -[3,2-c]- fused isomers of **26** and the conversion of these substrates into the corresponding thienopyridobenzodiazepines (**27**).

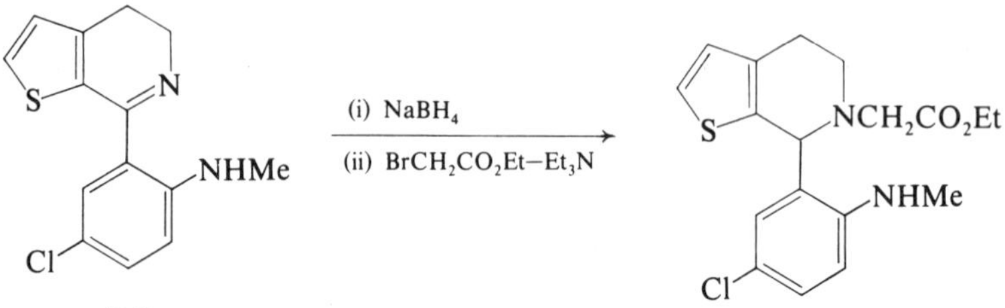

(**26**) $\xrightarrow{\text{(i) NaBH}_4 \text{ (ii) BrCH}_2\text{CO}_2\text{Et–Et}_3\text{N}}$ → product → (**27**)

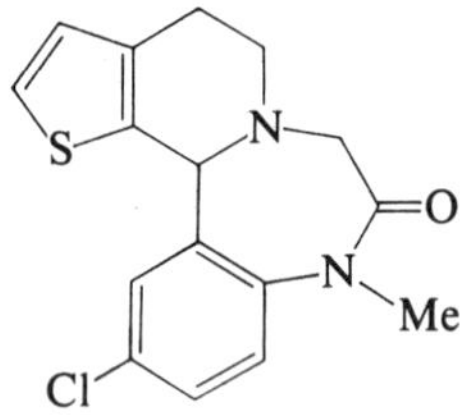

b. *The Pomeranz–Fritsch Synthesis.* This synthesis, involving acid-induced cyclization of the Schiff's base derived from aminoacetal and an aryl aldehyde or ketone, has been very little used in the thienopyridine field. When applied to the Schiff's bases (**28**) of thiophene-2-aldehyde, hydrolysis of the imino function (and subsequent polymerization of the hydrolysis products) competes successfully with the desired reaction.[8]

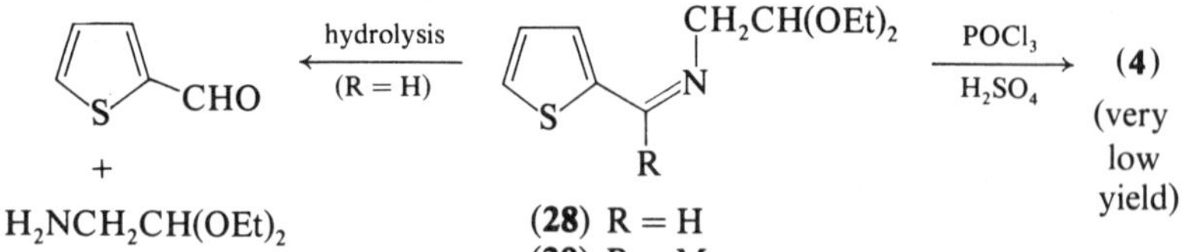

(**28**) R = H
(**29**) R = Me

Cyclization by phosphoryl chloride–polyphosphoric acid gives slightly better yields, but even this modification results in < 1% of pure

---

[49] H. Ott, French Patent 5,774 [*CA* **70**, 115191 (1969)].
[50] H. Ott, U.S. Patent 3,497,529 [*CA* **72**, 10076 (1970)].
[51] H. Ott, U.S. Patent 3,334,090 [*CA* **68**, 13010 (1968)].
[52] H. Ott, U.S. Patent 3,334,086 [*CA* **69**, 19228 (1968)].

Sec. II.B] THE THIENOPYRIDINES 81

products.[48] Apparently the Schiff's bases from thienyl ketones are less susceptible to hydrolysis, since the ketimine (**29**) provided 7-methylthieno[2,3-c]pyridine in 60% yield.[48]

c. The Pictet–Spengler Synthesis. Gronowitz[53] has successfully applied the Pictet–Spengler synthesis to the thienopyridine series. The overall route is exemplified by the preparation of thieno[3,2-c]pyridine (Scheme 10).

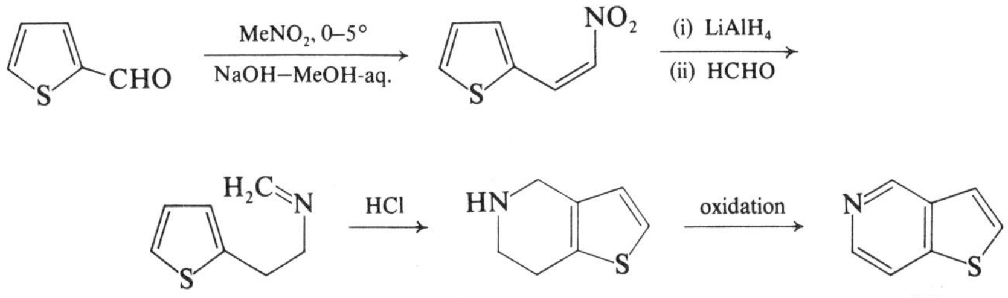

SCHEME 10

The final, oxidation stage presented some difficulties. Several conventional reagents failed; eventually it was found that $K_3Fe(CN)_6$–dioxane–water gave moderate yields of aromatized product. Application of the above reaction sequence to 4-bromo- and 5-chlorothiophen-2-aldehyde and to 4-bromothiophen-3-aldehyde yielded the appropriate halogenothienopyridine, but in the case of 5-bromothiophen-2-aldehyde reductive debromination occurred during the reduction of the nitrothienylethylene.

d. The Pictet–Gams Synthesis. The only report of the application of this method to the preparation of thienopyridines is by Herz and Tsai,[45] who obtained 4-methylthieno[3,2-c]pyridine by the route of Scheme 11. Cyclization of the homoveratroylamide of amine (**30**) under the same conditions proceeded in only extremely low yield.

e. Syntheses via the Cyclization of Isocyanates. Eloy and Deryckere have applied their synthesis of isocarbostyrils[54] (Scheme 12) to the preparation of thieno[2,3-c]- and -[3,2-c]pyridines.[55,56] The derived chlorocompounds were then reduced (by zinc–acetic acid) to the parent

---
[53] S. Gronowitz and E. Sandberg, *Ark. Kemi* **32**, 217 (1970).
[54] F. Eloy and A. M. Deryckere, *Helv. Chim. Acta* **52**, 1755 (1969).
[55] F. Eloy and A. M. Deryckere, *Bull. Soc. Chim. Belges* **79**, 301 (1970).
[56] A. M. Deryckere and F. Eloy, German Patent 2,059,386 [*CA* **75**, 76802 (1971)].

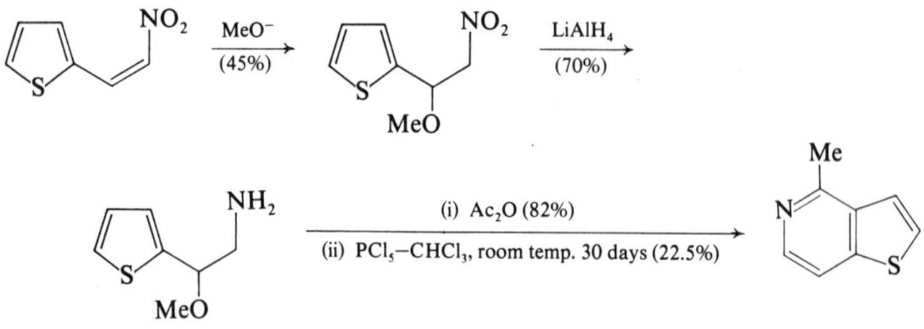

SCHEME 11

heterocyclic systems. Overall yields were good and the route is of particular interest since the intermediate 4-chlorothieno[2,3-c]- and 7-chlorothieno[3,2-c]pyridines can readily be converted into other derivatives by nucleophilic substitution of the halogen (see Section IV.B).

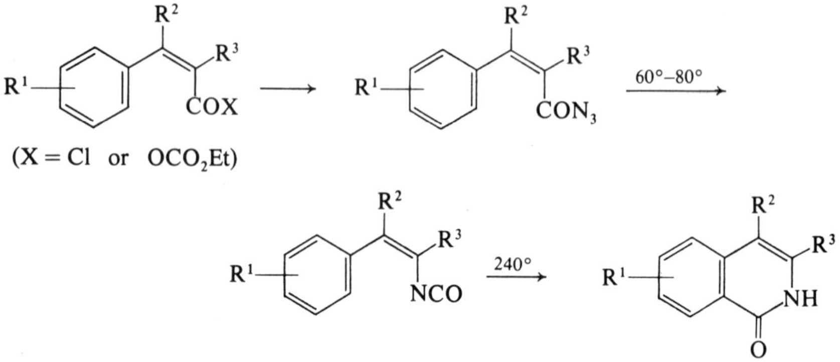

SCHEME 12

Very recently the formation of 4-thiomethyl-6,7-dihydrothieno[3,2-c]-pyridine (**32**) by cyclization of the isothiocyanate (**31**) has been reported.[57]

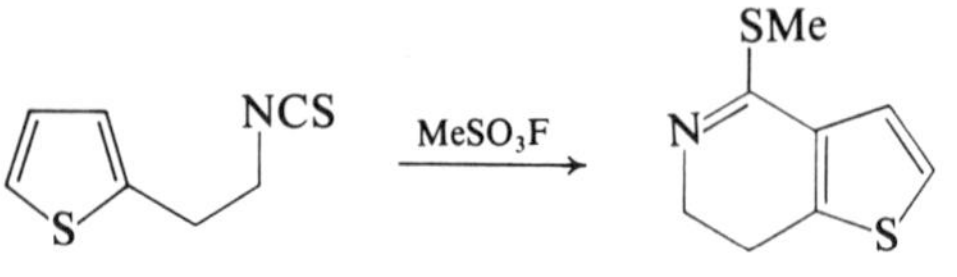

[57] W. M. Gittos, German Patent 2,318,399 [*CA* **80**, 14857 (1974)].

f. Synthesis from Pyrylium and Nitrilium Salts. Russian workers[58,59] have found that acetonylthiophenes yield pyrylium salts when allowed to react with anhydrides in the presence of 70% perchloric acid, and that these salts are converted into thienopyridines by treatment with ethanolic ammonia. For example 4-acetonyl-2-methylthiophene and acetic anhydride gave 2,5,7-trimethylthieno[2,3-*c*]pyrylium perchlorate (33), from which 2,5,7-trimethylthieno[2,3-*c*]pyridine (34) was obtained.

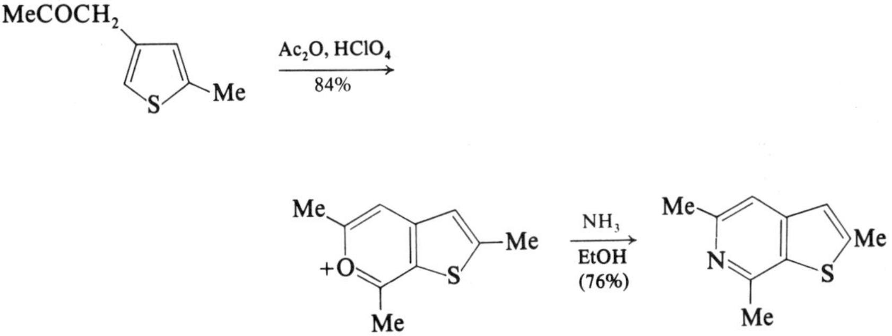

A number of variously alkyl-substituted thieno[2,3-*c*]- and -[3,2-*c*]-pyridines have been made in this way.

Reaction of 2-[2-chloroethyl]thiophene with the complex formed between a nitrile and tin(IV) chloride has been shown to give a dihydrothienopyridine[60] [Eq. (20)]. Yields with a wide variety of nitriles were low (8–17%) so that, in view of the necessity for a further step (dehydrogenation), the synthesis is less attractive than some of the alternatives.

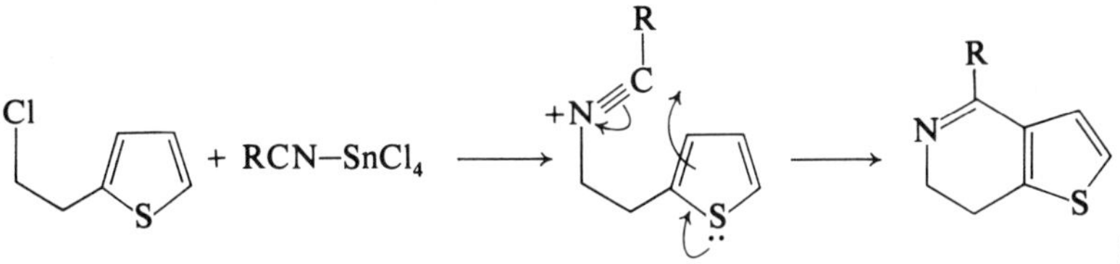

---

[58] L. V. Dulenko, G. N. Dorofeenko, S. N. Baranov, I. G. Katts, and V. I. Dulenko, *Khim. Geterotsikl. Soedin.* **7**, 320 (1971) [*CA* **76**, 14378 (1972)].

[59] S. V. Krivun, V. I. Dulenko, L. V. Dulenko, and G. N. Dorofeenko, *Dokl. Akad. Nauk SSSR* **166**, 359 (1966) [*CA* **64**, 11153 (1966)].

[60] M. Lora-Tamayo, R. Madronero, and M. Perez, *Ber.* **95**, 2188 (1962).

g. *Miscellaneous Syntheses*. The Beckmann rearrangement of oximes derived from thiophene analogs of indanones gives rise to dihydrothienopyridones.[61] For example 5-methylthieno[b]indan-4-one oxime (**35**) gave **36** (60%) with polyphosphoric acid.

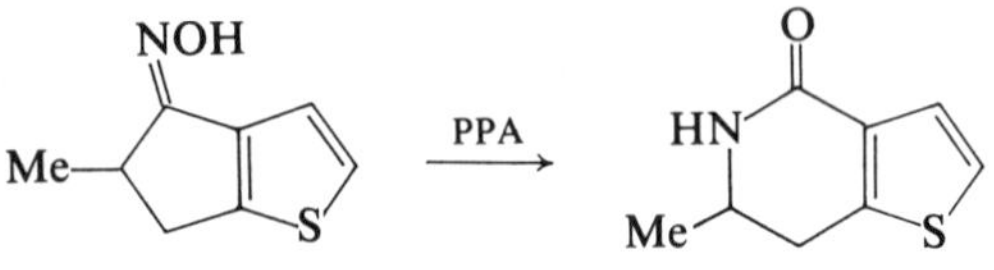

Tsuge[62] isolated 7-methyl-5-(2-thienyl)thieno[2,3-c]pyridine (**37**) from the pyrolysis products of 2-acetylthiophene ketazine; it is interesting that the thiophene side chain was selectively desulfurized by Raney nickel.

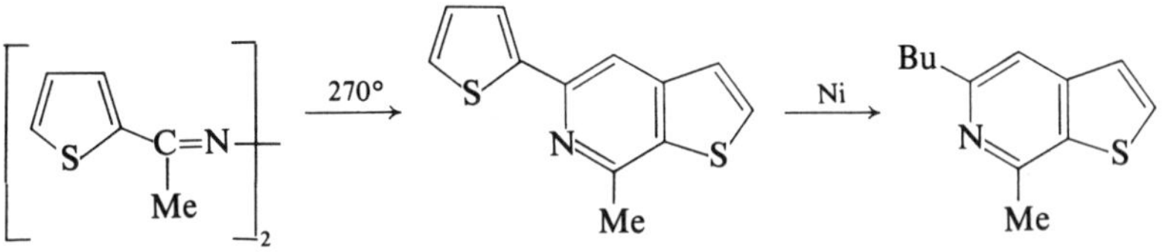

Recently Sandberg[63] has published an account of a new approach to thieno[2,3-c]- and -[3,2-c]pyridines. A noteworthy feature is the selective

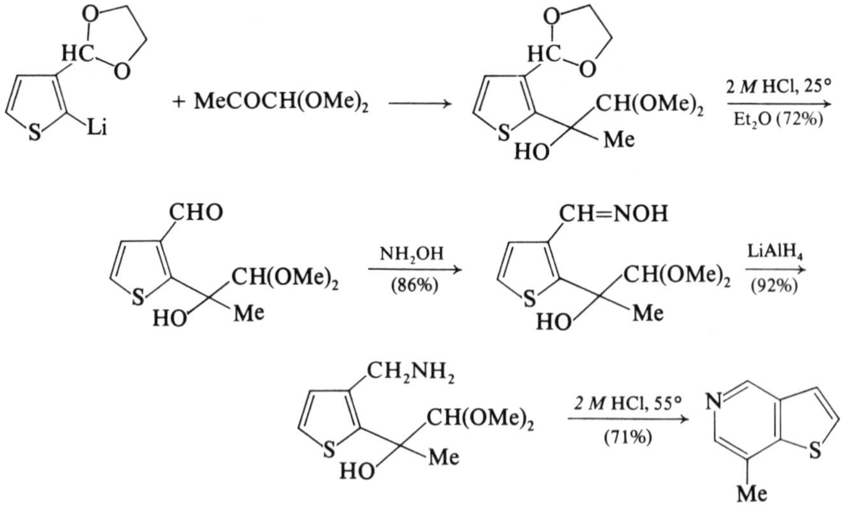

SCHEME 13

[61] K. Aparajithan, A. C. Thompson, and J. Sam, *J. Heterocycl. Chem.* **3**, 466 (1966).
[62] O. Tsuge, H. Watanabe, and H. Hokama, *Bull. Soc. Chem. Jpn.* **44**, 505 (1971).
[63] E. Sandberg, *Chem. Scripta* **2**, 241 (1972).

Sec. II.B] THE THIENOPYRIDINES 85

hydrolysis of an araldehyde ethylene acetal in the presence of an alkyl dimethyl acetal. The preparation of 7-methylthieno[3,2-c]pyridine serves to illustrate the route (Scheme 13). In a very recent publication, Ames and Ribeiro[64] described what could be a general method for the preparation of [c]-fused thienopyridones (and hence, therefore, thienopyridines) from o-bromothiophene carboxylic acids. Since most of the compounds made by these workers belong to the [3,4-c] series, the method will be described in Section II,B,3.

2. Syntheses Involving Formation of the Thiophene Ring

a. Cyclization of Carboxypyridylthioacetic Acids. The only report of the synthesis of a thiophene isoquinoline isostere by this method is due to Koenigs and Kantrowitz,[6] who obtained 3-hydroxy-4,6-dimethyl-thieno[3,2-c]pyridine (**39**) from the pyridine **38**.

$$\text{(38)} \xrightarrow[\text{(67\%)}]{\text{Ac}_2\text{O}} \text{(39)}$$

b. High-Temperature Catalytic Methods. Hansch and his colleagues prepared thieno[2,3-c]pyridine[65] and 6-methylthieno[3,2-c]pyridine[35] by cyclization-dehydrogenation, at 425° over a copper–chromium oxide catalyst, of 4-ethylpyridine-3-thiol and 5-ethyl-2-methylpyridine-4-thiol, respectively. Although the yields in the cyclization step were acceptable (50 and 20–25%), the intermediate pyridine thiols required multistage syntheses, so that overall yields were very low (< 5%); the failure of 4-vinylpyridine and hydrogen sulfide to form thieno[2,3-c]pyridine under the conditions employed by Hansch and Carpenter[35] has already been mentioned (Section II,A,2,b). Klemm *et al.*[66] found that this reaction could be brought about, under different conditions, but only in very low yield. The formation of thieno[3,2-c]pyridine (0.9%) together with the [2,3-b]-isomer (5.9%) from 3-vinylpyridine–hydrogen sulfide[36] has already been described (Section II,A,2,b). The indirect route from vinylpyridines, in which they are converted into benzyl pyridyl ethyl sulfides prior to thermolysis, was particularly effective for the synthesis of thieno[2,3-c]pyridine (**4**).[66]

[64] D. E. Ames and O. Ribeiro, *J. Chem. Soc., Perkin Trans. I*, 1390 (1975).
[65] C. Hansch and W. Carpenter, *J. Org. Chem.* **23**, 1924 (1958).
[66] L. H. Klemm, D. R. McCoy, J. Shabtai, and W. T. Kiang, *J. Heterocycl. Chem.* **6**, 813 (1969).

c. *Miscellaneous Cyclizations.* The cyclization of an acetonylthiopyridine, by hot polyphosphoric acid [Eq. (21)] has been reported.[67]

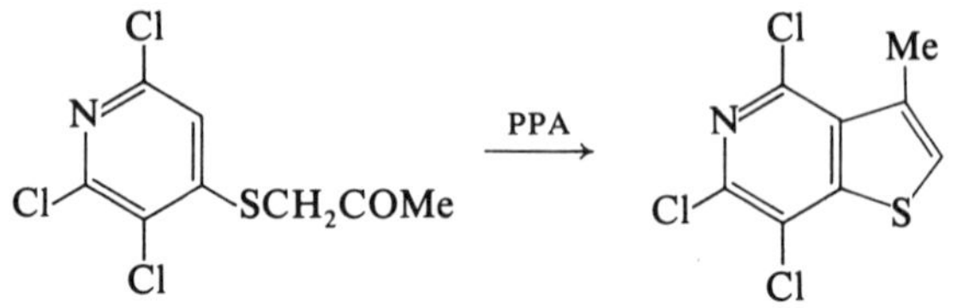

$$\text{(21)}$$

3. *Thieno[3,4-c]pyridines*

The first report of the preparation of a derivative of the system is due to Benary,[68] who prepared **40** by the reaction sequence shown. It would be interesting to discover whether the product has any spectroscopic characteristics of the fully aromatic enolic form (**41**).

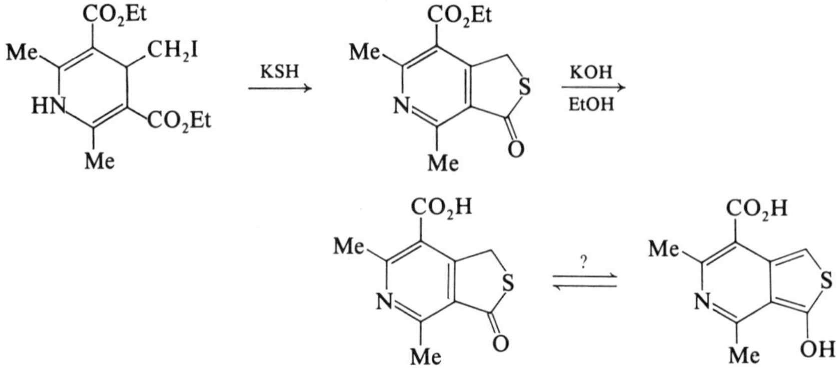

(**40**)   (**41**)

Klemm, Johnson, and White[10,40,41] synthesized the parent system from 3,4-dimethylpyridine by a route analogous to that already described for the [3,4-*b*]-isomer (Section II,A,3).

The method of Ames and Ribeiro,[64] already referred to briefly, would appear to offer an attractive route to a wide variety of hitherto inaccessible thieno[3,4-*c*]pyridines. The sodium salt of 4-bromothiophene-3-carboxylic acid reacts with carbanions, in the presence of copper or copper(II) acetate, to give condensation products (**42**) by displacement of bromide ion. With certain carbanions (e.g., those derived from PhCOCH$_2$CN, MeCOCH$_2$CO$_2$Et, PhCOCH$_2$COMe) deacylation (loss of PhCO, MeCO and MeCO, respectively) occurred simultaneously;

---

[67] E. Ager, B. Iddon, and H. Suschitzky, *J. Chem. Soc. C*, 193 (1970).
[68] E. Benary, *Ber.* **51B**, 572 (1918).

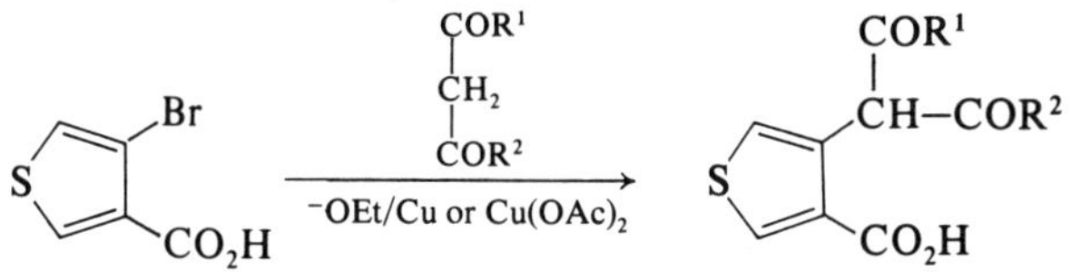

Sec. II.B] THE THIENOPYRIDINES

the diacyl products (42) can also be deacylated in a separate step by treatment with aqueous ammonia. The oxo-acids formed in these reactions cyclize with ammonium acetate (or with primary amine acetates) to form thieno[3,4-c]pyridin-4(5H)-ones, which have obvious potential for further transformation. The syntheses of 4-chloro-6-phenylthieno[3,4-c]pyridine (43) and 5-methyl-6-phenylthieno[3,4-c]-pyridin-4(5H)-one (44) depicted in Scheme 14 serve to illustrate the overall route.

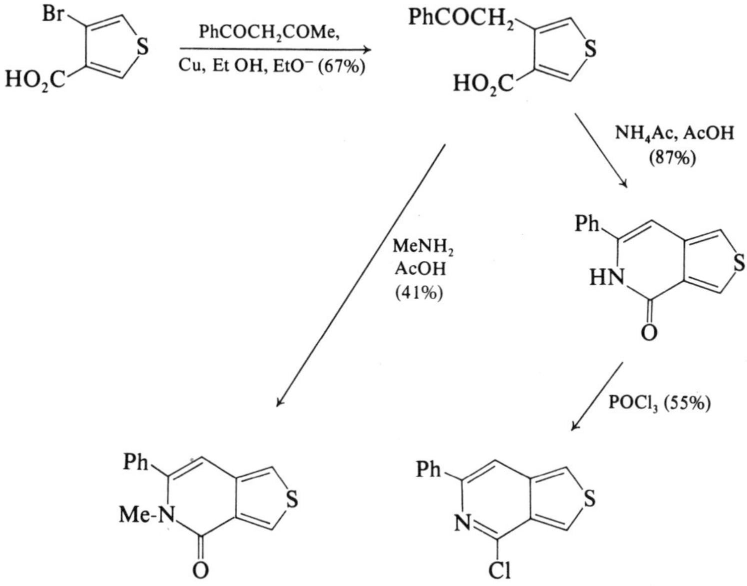

SCHEME 14

Most of the work reported on the thieno[3,4-c]pyridine system has been concerned with the preparation of substances related to pyridoxine and pyridoxal. Kreisky[69] claimed to have obtained 4-hydroxy-5-methyl-

[69] S. Kreisky, Monatsh. 89, 685 (1958).

1,3-dihydrothieno[3,4-c]pyridine (**46**) as a substance of m.p. 197–199° from the 3,4-bisbromomethylpyridine (**45a**) and KSH or thiourea,

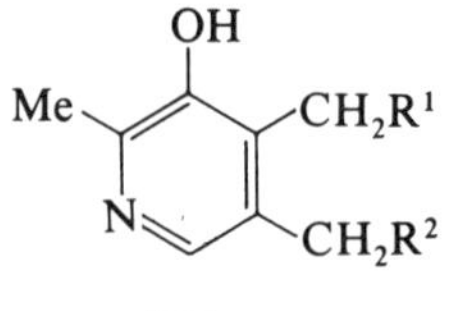

(**45**)

a: $R^1 = R^2 = Br$
b: $R^1 = SH, R^2 = OMe$
c: $R^1 = OMe, R^2 = SH$

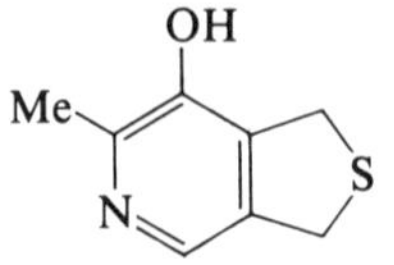

(**46**)

followed by ammonia, but later the identity of the product was disputed by Schmidt and Giesselmann,[70] who prepared **46** by acid-catalyzed cyclization of (**45b**) and (**45c**), and found it to differ from Kreisky's material, having m.p. 232°.

The synthesis of the fully aromatic analog (**48**) of **46**, from bis-(4-formyl-5-hydroxy-6-methyl-3-pyridylmethyl)disulfide (**47a**)[71] and from the corresponding diethylacetal (**47b**)[72] has been reported in patents.

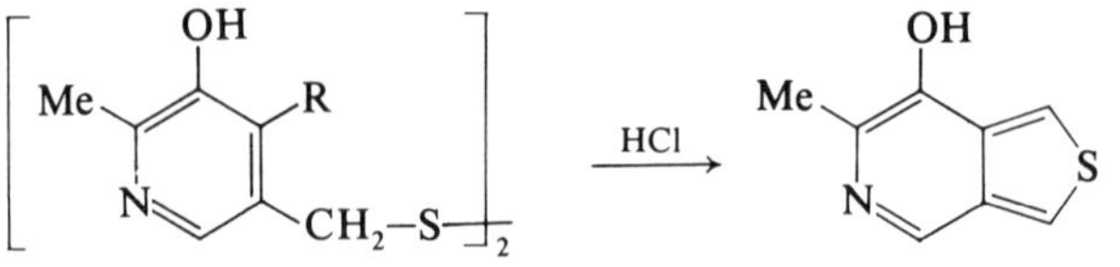

(**47**)

a: R = CHO; b: R = CH(OEt)$_2$

(**48**)

The 3-hydroxy-1,3-dihydrothieno[3,2-c]pyridine (**49**), prepared by Korytnyk and Ahrens[73] by the route of Scheme 15, can be regarded as a

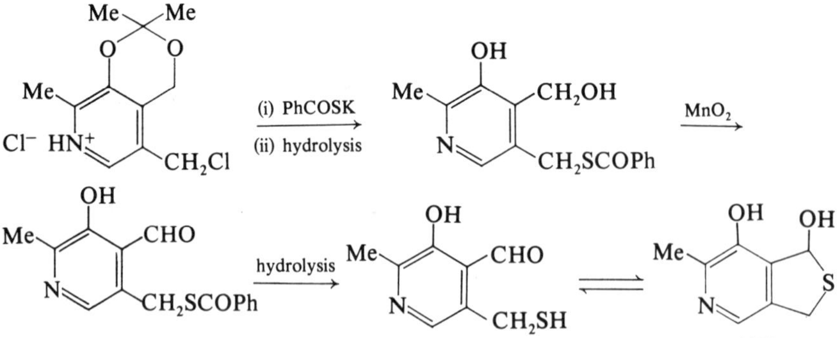

SCHEME 15

[70] U. Schmidt and G. Giesselmann, *Justus Liebigs Ann. Chem.* **657**, 162 (1962).
[71] E. Merck A-G., German Patent 1,193,049 [*CA* **63**, 2961 (1965)].
[72] E. Merck A-G., French Patent 3,091 [*CA* **62**, 16209 (1965)].
[73] W. Korytnyk and H. Ahrens, *J. Med. Chem.* **14**, 947 (1971).

thiohemiacetal. The hemiacetal ring was found to be stable only within a narrow pH range.

An interesting transformation of various 4-methylnicotinic acids, in which the activated 4-methyl group enters into the reaction of the carboxyl group with thionyl chloride, leads to a 1,3-dioxo-1,3-dihydrothieno[3,4-c]pyridine[74] [Eq. (22)].

III. Electron Distribution

One of the major points of interest in thienopyridines is the presence in them of a π-excessive and a π-deficient system. It is not surprising, therefore, that efforts have been made to calculate π-electron densities and to correlate them with experimental observations.

Klemm and his colleagues have used simple molecular orbital (MO) theory to obtain electron densities (q_r for position r) and have calculated superdelocalizability values for electrophilic (S_r^e) and nucleophilic (S_r^n) substitution for thieno[2,3-b]-,[13] -[3,2-b]-[75] -[3,4-b]-,[40] and -[3,4-c]-[40] pyridines. Similar calculations for the [2,3-c]-isomer have been conducted by Dressler and Jouillé.[48] The results are collected in Table I.

TABLE I

ELECTRON DENSITIES AND SUPERDELOCALIZABILITY VALUES OF THIENOPYRIDINES

	[2,3-b]			[3,2-b]			[3,4-b]			[2,3-c]			[3,4-c]		
Atom	q_r	S_r^e	S_r^n	q_r	S_r^e	S_r^n	q_r	S_r^e	S_r^n	S_r^e	S_r^n	q_r	S_r^e	S_r^n
1	1.32	3.10	0.49	1.31	3.16	0.55	1.08	1.06	0.59	3.36	0.56	1.22	2.25	0.57
2	1.09	1.20	0.78	1.09	1.22	0.80	1.13	1.52	0.86	1.19	0.75	1.13	1.47	0.82
3	1.21	2.21	0.54	1.22	2.17	0.49	1.24	2.20	0.52	2.31	0.52	1.21	2.31	0.63
4	0.94	0.96	1.19	1.32	1.60	0.95	1.33	1.60	0.94	1.28	0.83	0.90	1.04	1.28
5	1.06	1.13	0.71	0.92	0.91	1.15	0.93	0.93	1.17	1.29	0.84	1.33	1.44	0.79
6	0.93	0.92	1.15	1.06	1.14	0.72	1.06	1.13	0.71	1.54	0.84	1.01	1.29	0.87
7	1.32	1.59	0.94	0.94	0.95	1.18	0.94	0.98	1.22	1.14	1.11	1.04	1.26	0.84
8	1.05	1.18	0.76	1.08	1.05	0.63	1.08	1.06	0.59	1.04	0.59	1.07	1.06	0.64
9	1.08	1.02	0.60	1.05	1.14	0.73	1.05	1.11	0.70	1.08	0.63	1.09	0.98	0.56

[74] E. Wenkert, F. Haglid, and S. L. Mueller, *J. Org. Chem.* **34**, 247 (1969).
[75] L. H. Klemm, I. T. Barnish, R. A. Klemm, C. E. Klopfenstein, and D. R. McCoy, *J. Heterocycl. Chem.* **7**, 373 (1970).

It will be noted that in all five systems electrophilic substitution is expected to occur in the thiophene ring, β- to the sulfur atom when that position is free. Nucleophilic attack should occur in the pyridine ring, with a slight preference to the carbon γ- over that α- to the nitrogen atom where such a choice exists.

In a paper devoted to a study of nuclear magnetic resonance (NMR) spectra, Gronowitz and Sandberg[76] have examined the applicability of the relationship proton chemical shift = constant × π-electron charge of carbon atom to which proton attached (which holds for many aromatic systems) to thieno-[2,3-c]- and -[3,2-c]pyridines and to the derived pyridinium ions (50 and 51). It was found that a linear relationship did apply when π-electron charges were calculated by the ω-method of Janssen and Sandström,[77] but not when the simple Hückel method was employed. The results of the π-electron charge calculations (in which the value of ω was taken as 1.0), carried out by Gronowitz and Sandberg, are presented in Table II.

The most recent publication in this area is due to Helland and Skancke,[78] who have investigated the electronic structures of all six

TABLE II

π-ELECTRON CHARGES OF THIENO[2,3-c]- AND -[3,2-c]PYRIDINES AND -PYRIDINIUM IONS

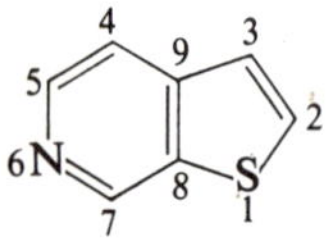

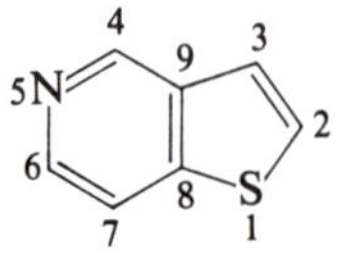

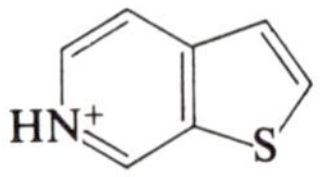

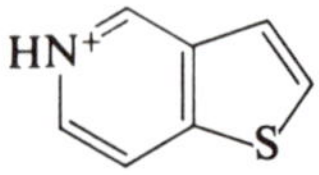

	π-Electron charge			
Atom	(4)	(5)	(50)	(51)
1	+0.117	+0.118	+0.126	+0.130
2	+0.007	−0.023	+0.036	+0.005
3	−0.039	−0.040	−0.032	−0.037
4	−0.001	+0.040	+0.011	+0.127
5	+0.026	−0.159	+0.094	−0.412
6	−0.152	+0.029	−0.422	+0.099
7	+0.033	−0.005	+0.114	+0.010
8	−0.003	+0.022	+0.015	+0.075
9	+0.007	−0.011	+0.057	+0.003

[76] S. Gronowitz and E. Sandberg, *Ark. Kemi* **32**, 269 (1970).
[77] M. J. Janssen and J. Sandström, *Tetrahedron* **20**, 2339 (1964).
[78] A. Helland and P. N. Skancke, *Acta Chem. Scand.* **26**, 2601 (1972).

thienopyridines by two methods, the π-electron approximation (Pariser–Parr–Pople, PPP) and the complete neglect of differential overlap (CNDO) method. The bond lengths of the systems, estimated by modification of known (or calculated) bond lengths in related molecules by taking mobile bond orders into consideration, are shown in Fig. 1.

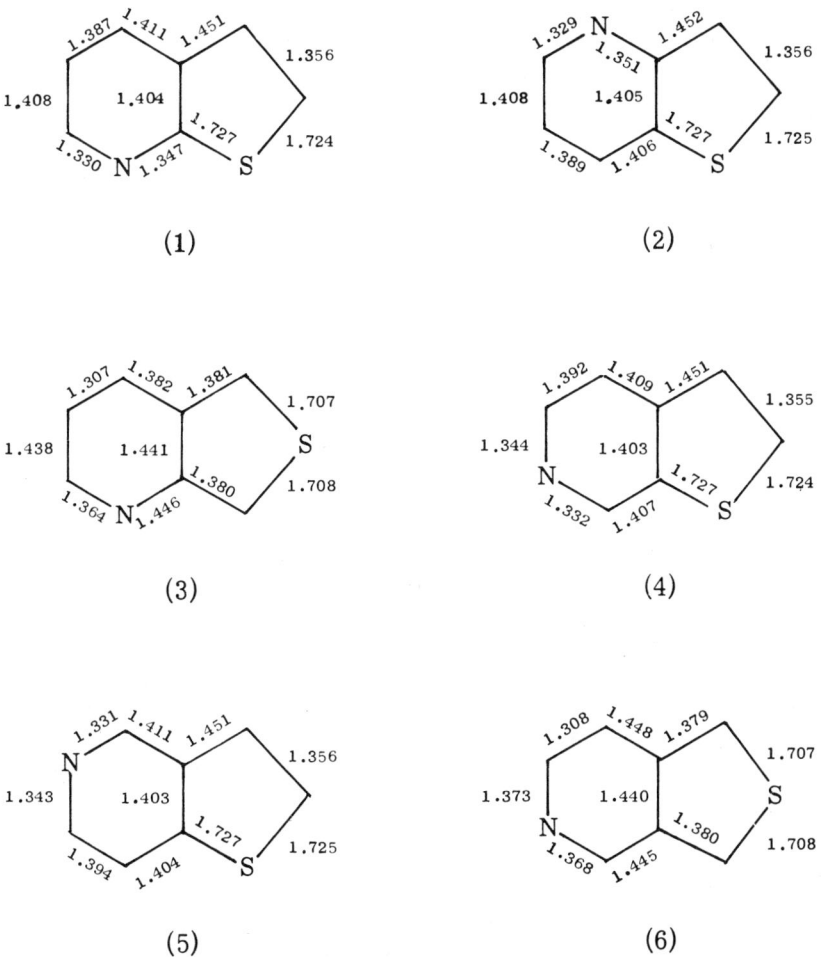

FIG. 1. Calculated bond lengths (Å) of thienopyridines.[78]

The electron charges at each ring position, for all six thienopyridines, obtained by the PPP and CNDO methods, are given in Helland and Skancke.[78]

Points of note are that both methods of calculation give a net negative charge on the nitrogen atom, a net positive charge on the carbons adjacent to it, and a net positive charge on sulfur. For the [2,3-]- and [3,2-]-fused systems, the PPP results indicate that, of the carbon atoms, C-2 (α- to the sulfur atom) has the highest net negative charge, which does not accord with the simpler Hückel calculations discussed above. In the [3,4-]-fused systems both sets of results indicate that those secondary carbon atoms α- to the sulfur atom have the greatest negative charges.

Using their CNDO results Helland and Skancke calculated indices of reactivity (frontier electron density, FED, for electrophilic substitution; frontier orbital density, FOD, for nucleophilic substitution; frontier radical density, FRD, for radical substitution) for the thienopyridines. It was indicated that the FED index has its highest value for C-3 in the [2,3-]- and [3,2-]-fused systems and for C-2 in the [3,4-]-fused isomers. As far as the former group is concerned, the predictions are in agreement with experimental observations (see Section IV,A.). Little experimental evidence is available for the [3,4-]-fused systems, but it seems highly probable that they would have a considerable tendency to undergo addition reactions at the 1,3-positions, since the product would contain a "normal" rather than a quinoid pyridine ring [Eq. (23)].

$$\text{[structure]} \xrightarrow{AB} \text{[structure]} \quad (23)$$

The FOD indices are greatest for C-2 of the [2,3-]- and]3,2-]-fused heterocycles. It seems highly improbable that nucleophilic substitution *would* occur in the thiophene ring rather than α or γ to the nitrogen atom, but experimental evidence is extremely sparse; in fact no systematic study of nucleophilic substitution of the parent fused heterocycles has been reported to date.

No evidence is available to confirm the predictions that free-radical substitution of [2,3-]- and [3,2]-fused systems will occur at C-3, of the [3,4-b]-isomer at C-1, and of thieno[3,4-c]pyridine at C-3.

IV. Reactions

A. ELECTROPHILIC SUBSTITUTION

1. *The Quinoline Isosteres*

The first report of an attempt to carry out an electrophilic substitution in this group was by Sheehan and Leitner,[30] who found that treatment of 3-hydroxythieno[3,2-b]pyridine with iodine monochloride gave a yellow monoiodo derivative in which the iodine atom was very labile. The structure of the product was not established, but it seems probable that it was 3-hydroxy-2-iodothieno[3,2-b]pyridine: the lability of the halogen could be due to its position α to the carbonyl function in the keto tautomer [Eq. (24)].

$$\underset{S}{\underset{N}{\bigodot}}\text{-OH} \xrightarrow{ICl} \underset{S}{\underset{N}{\bigodot}}\text{-OH, I} \rightleftarrows \underset{S}{\underset{N}{\bigodot}}\text{=O, I} \quad (24)$$

The only other work on electrophilic substitution of the quinoline isosteres has been carried out by Klemm and his group. NMR studies of the reaction of thieno[2,3-b]pyridine with deuteriosulfuric acid[13] at 98.5° showed that the proton at C-3 was replaced much more rapidly than that at C-2. Bromination of the same compound with an excess of bromine in carbon tetrachloride–water[13] led to a low yield (17%) of the 2,3-dibromo derivative: chlorination under similar conditions[79] gave both mono- and dichlorothieno[2,3-b]pyridine. Halogenation with elemental halogen in sulfuric acid–silver sulfate[79] gave moderate yields of the 3-halogenated heterocycle (**52**).

$$\mathbf{1} \xrightarrow[\text{AgSO}_4]{X_2\text{-H}_2\text{SO}_4} \underset{N \quad S}{\bigodot}\text{-X}$$

(**52**)
a: X = Cl (26%)
b: X = Br (47%)
c: X = I (22%)

In an attempt to effect reaction on an unprotonated species thieno-[2,3-b]pyridine was treated with bromine in chloroform in the presence of a buffer,[79] producing an improved yield (57%) of **52b**. Nitration of **52a–c** (mixed acid) gave rather poor yields (22–47%) of the 3-halogeno-2-nitrothieno[2,3-b]pyridine. Both of the quinoline isosteres[75] and also 5-ethylthieno[2,3-b]pyridine[80] were nitrated (mixed acid) at C-3; yields in all three cases were approximately 50%.

It is interesting to note that the pyridine ring in thieno[2,3-b]pyridine-N-oxide (**53**) is more reactive toward electrophiles than is the thiophene ring[81]; moreover, the orientation in nitration of this molecule depends on the reagent.

[79] L. H. Klemm, R. E. Merrill, F. H. W. Lee, and C. E. Klopfenstein, *J. Heterocycl. Chem.* **11**, 205 (1974).
[80] L. H. Klemm and H. Lund, *J. Heterocycl. Chem.* **10**, 871 (1973).
[81] L. H. Klemm, I. T. Barnish, and R. Zell, *J. Heterocycl. Chem.* **7**, 81 (1970).

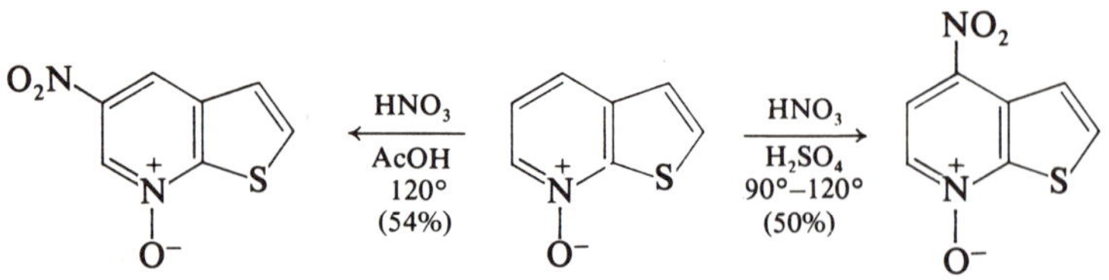

(54) ←[HNO₃ / AcOH / 120° / (54%)]— (53) —[HNO₃ / H₂SO₄ / 90°–120° / (50%)]→ (55)

Klemm has concluded that the 4-nitro derivative (55) results from electrophilic attack by NO_2^+ on the unprotonated substrate. A mechanism involving 1,3-dipolar addition to the nitrone moiety and subsequent electrophilic attack is proposed to account for the formation of the 5-nitro isomer (54).

2. *The Isoquinoline Isosteres*

That electrophilic attack on thienopyridines can occur on neutral or positively charged species receives support from the observations of Dressler and Joullié.[48] These workers found that, under conditions where complete protonation of nitrogen was not to be expected, 7-methylthieno[2,3-c]pyridine (56; pK_a 5.81) underwent the Friedel–Crafts reaction whereas 4-methylthieno[3,2-c]pyridine (57; pK_a 6.17) did not, which the authors ascribed to the effect of the stronger basicity of 57.

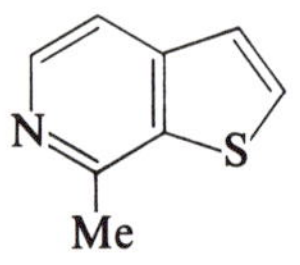

(56) 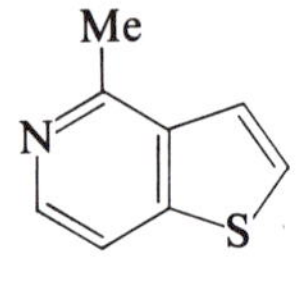   (57)

Both substances yielded the 3-sulfonic acid (15% fuming sulfuric acid) and the 3-nitro compound (fuming nitric–sulfuric acid); neither 56 nor 57 could be formylated by the Vilsmeier procedure.

A detailed study of electrophilic substitution of thieno[2,3-c]- and -[3,2-c]pyridines has been carried out by Gronowitz and Sandberg.[82] Deuterium exchange (by D_2SO_4), readily followed by NMR spectroscopy, was found to be temperature dependent. At 55° the signal due to the C-3 proton disappeared, that due to the C-2 proton remaining unchanged, but at 100° the latter was also substantially reduced. Nitration with fuming nitric–sulfuric acid gave excellent yields of the 3-nitro derivatives, but no nitration occurred at all if the sulfuric acid was

[82] S. Gronowitz and E. Sandberg, *Ark. Kemi* **32**, 249 (1970).

replaced by acetic acid or acetic anhydride. The general resistance to electrophilic attack was well illustrated by the conditions necessary to effect bromination. Treatment with bromine in carbon tetrachloride or in acetic acid caused no reaction; of several tried, the most effective agents were bromine in 48% hydrobromic acid or in thionyl chloride (giving yields of 60–80% of the 3-bromothienopyridine). Dibromoisocyanuric acid in sulfuric acid, a reagent usually effective for deactivated systems (e.g., nitrobenzene, *m*-dinitrobenzene) also gave reasonable yields of 3-bromo derivatives, which could be converted into the 2,3-dibromo compounds by treatment with the same reagent in fuming sulfuric acid.

Eloy and Deryckere[83] showed that a large excess of aluminum chloride catalyst was effective for the bromination of 4-chlorothieno-[3,2-*c*]pyridine, although no yields were quoted [Eq. (25)].

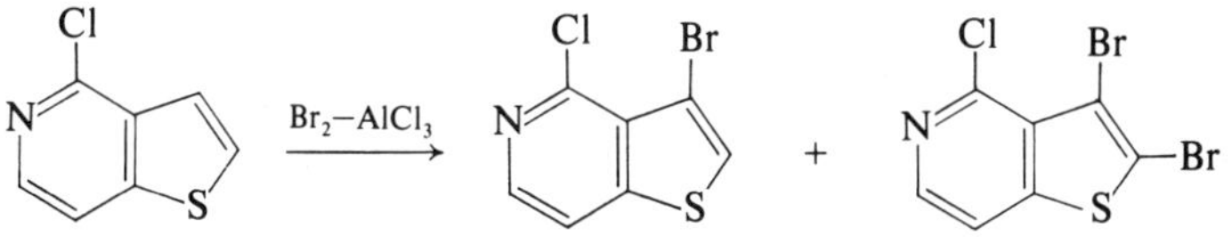

The same authors found that thieno[2,3-*c*]- and -[3,2-*c*]pyridones (e.g., **58**) yielded substitution products (**59**) when treated with hydrogen halides and hydrogen peroxide, or with nitric acid. The orientation of substitution and increased reactivity compared with the parent thienopyridine were ascribed to the +M effect of the hydroxy group in that tautomeric form; however, it is clear that the same results would be expected from the pyridone tautomer.

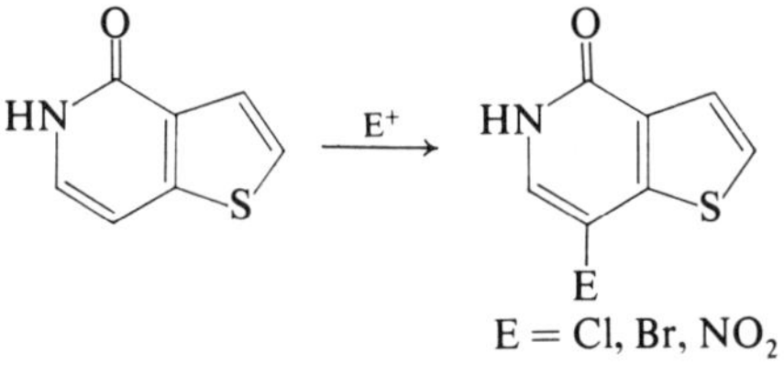

The results of the work on electrophilic substitution of thienopyridines confirm the general predictions of electron distribution calculations, i.e., that substitution should occur preferentially β to the sulfur atom. Although relative rate data are lacking, qualitatively it would seem that the systems are even more resistant to attack than the isosteric quinoline and isoquinoline.

[83] F. Eloy and A. Deryckere, *Bull. Soc. Chim. Belges* **79**, 407 (1970).

B. NUCLEOPHILIC SUBSTITUTION

Quinoline and isoquinoline undergo a set of reactions in which, in effect, a hydride ion is replaced by a strongly nucleophilic species. With organolithium compounds the initial adduct, on hydrolysis and mild oxidation, yields the corresponding alkyl or aryl derivative, and hydroxy or amino groups can be introduced by reaction with alkali metal (or alkaline earth) hydroxides or amides.

Only reactions of the first type have been reported for thienopyridines, and, in such reactions, metallation of the thiophene ring (α to the sulfur atom) is an obvious competitor. Thus Klemm et al.[13] found that treatment of thieno[2,3-b]pyridine with n-butyl lithium at 25°–35°, followed by mild oxidation, gave the 6-butyl derivative (47%) and an appreciable quantity of starting material. At a lower temperature methyl lithium yielded a product consisting of 6-methylthieno[2,3-b]pyridine (25%) and thieno[2,3-b]pyridine (75%); in both instances the recovered starting material could well have arisen from hydrolysis (during workup) of 2-lithiothieno[2,3-b]pyridine. Evidence for this view was obtained when the reaction mixture was treated with deuterium oxide, then water; in this case the recovered thieno[2,3-b]pyridine consisted of approximately equal amounts of the 2-D and 2-H isomers (Scheme 16).

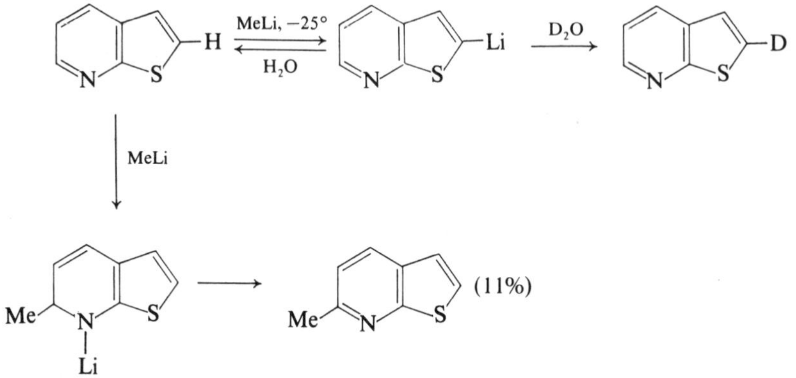

SCHEME 16

At still lower temperatures, addition to the azomethine bond is apparently unimportant in the quinoline isosteres, since Klemm and Merrill[84] obtained a good yield (66%) of thieno[2,3-b]pyridine-2-aldehyde by formylation of the lithio derivative, prepared at −70°. Although very little evidence is available, it appears that the situation with the isoquinoline analogs is different, for Gronowitz and Sandberg[82] found that

---

[84] L. H. Klemm and R. E. Merrill, *J. Heterocycl. Chem.* **11**, 355 (1974).

both thieno[2,3-c]- and -[3,2-c]pyridines were not metallated in the thiophene ring by n-butyl lithium at $-70°$, all reaction occurring by addition to the C=N bond.

Nucleophilic displacement of halogen at C-2 or C-4 of quinoline, or at C-1 of isoquinoline, takes place readily; halogens at other positions in these molecules are much more resistant to nucleophilic attack. Similar behavior is observed in halogenated thienopyridines.

In the [c]-fused series, derivatives in which a chlorine atom is α to the nitrogen are readily available from the pyridones produced by the synthesis of Eloy and Deryckere[55] [Eq. (26)].

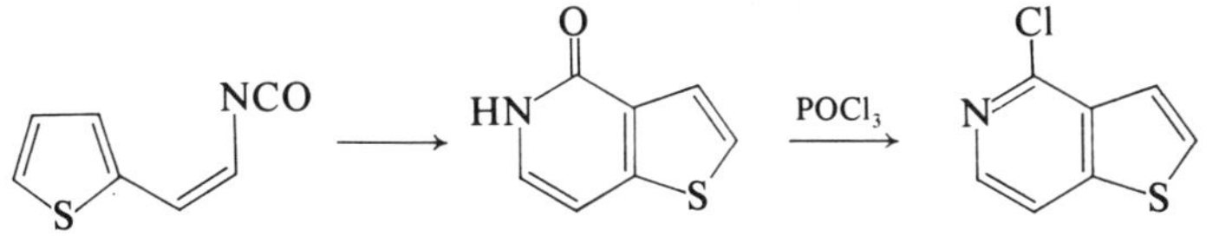

Such chlorine atoms are readily displaced by a variety of nucleophiles,[83] including alkoxides and primary and secondary amines. 7-Chlorothieno[2,3-c]pyridine (**60**) and 4-chlorothieno[3,2-c]pyridines have been used as intermediates in the preparation of substances of potential pharmacological activity.[85] For example, displacement of chloride ion from **60** by the anion derived from 3,4-dimethoxyphenylacetonitrile led to the benzylisoquinoline analog (**61**). Analogs (e.g., **62**)

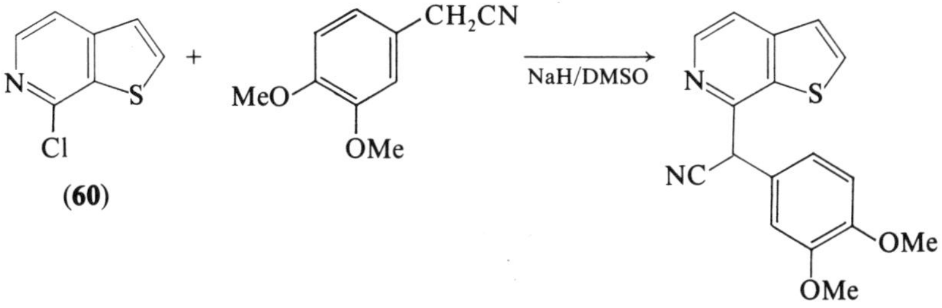

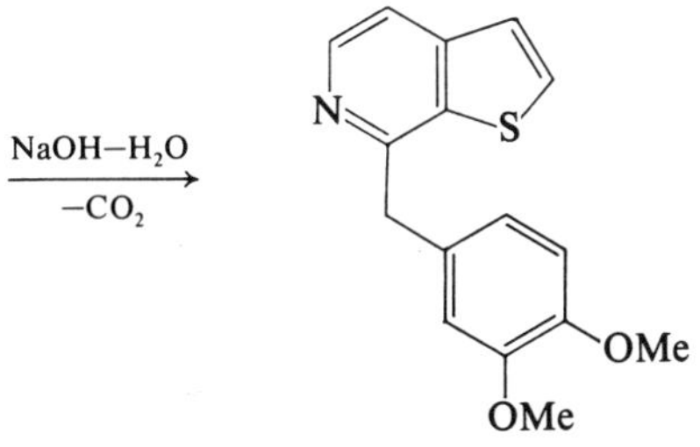

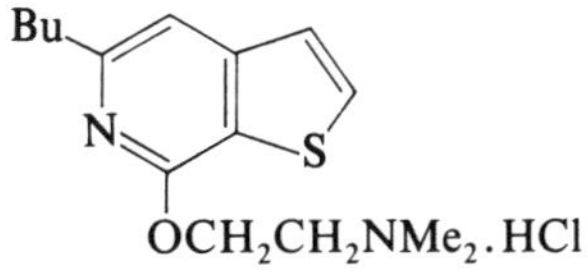

[85] F. Eloy and A. Deryckere, *Bull. Chim. Therap.* **4**, 466 (1969).

of the local anesthetic Quinisocaine were prepared from the appropriately substituted chlorothienopyridine by reaction with N,N-dimethylethoxide ion.

Reductive dehalogenation (by tin– or zinc–acid) of halogen atoms in these reactive positions occurs readily in both the [b]-[19] and [c]-fused[55,56] systems.

The comparative unreactivity toward nucleophilic substitution of halogens not α to the nitrogen atom was clearly demonstrated by the selective substitution of the 4-substituent in 4-chloro-7-halogenothieno-[3,2-c]pyridines (**63**; X = Cl or Br).[83]

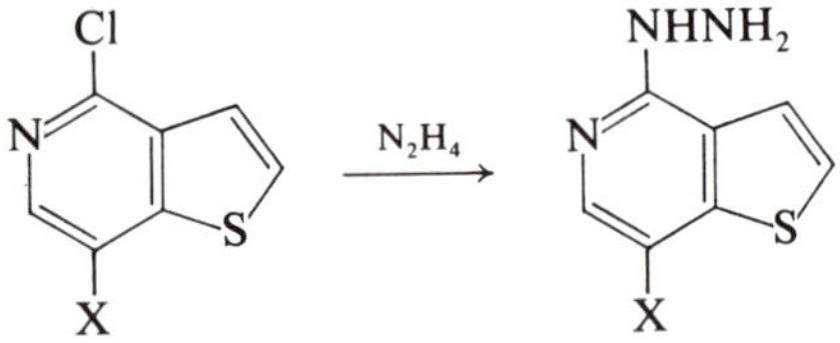

(**63**)

Under more forcing conditions, substitution of halogen at less reactive sites can be achieved; both 3-bromo-[79] and 5-bromothieno[2,3-b]-pyridine[86] gave the corresponding nitrile with copper(I) cyanide in refluxing dimethylformamide.

The halogen atom in 4-chloroquinoline-N-oxides is highly activated toward nucleophilic substitution. From the only examples reported for the thienopyridine analog[81] (**64**), it is clear that reaction does not occur particularly readily; it seems likely that electron release from the thiophene ring reduces the activating influence of the N-oxide group.

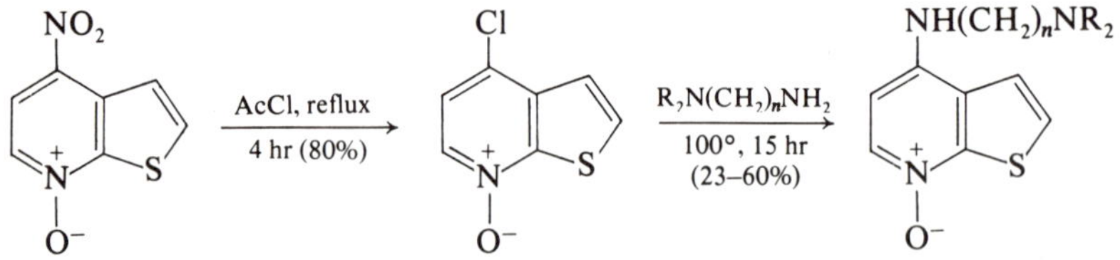

(**64**)

C. HALOGEN–METAL INTERCONVERSION

The complications associated with direct lithiation of thienopyridines have already been discussed, but lithio derivatives in both series are readily available by halogen–metal interchange at low temperature.

[86] L. H. Klemm and R. Zell, *J. Heterocycl. Chem.* **5**, 773 (1968).

Thieno[3,2-c]pyridine-3-carboxylic acid has been prepared from treatment of the 3-bromo compound with ethyl lithium at −70° followed by carbonation,[82] and Klemm and Merrill[84] have prepared 3-lithiothieno-[2,3-b]pyridine similarly and have used it to obtain a number of compounds (Scheme 17).

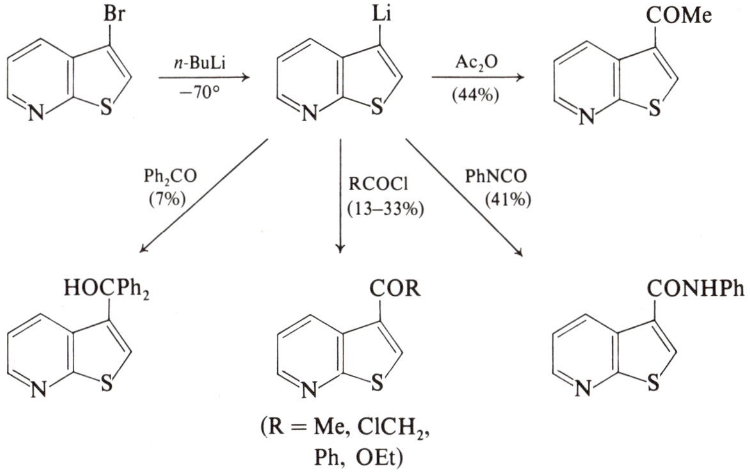

SCHEME 17

D. REACTIONS OF ACTIVE METHYL GROUPS

The reactivity of methyl groups in the 2- or 4-position of quinoline, and in the 1-position of isoquinoline, is also observed in the analogous thienopyridines. 7-Methylthieno[2,3-c]pyridine, for example, undergoes the typical reactions shown in Scheme 18.[48]

The dimer (**65**) was formed *in the presence of* benzaldehyde (or acetophenone), an indication that the reaction took place at a surface. In 4,6-dimethylthieno[2,3-b]pyridine, the isostere of 2,4-dimethylquinoline, both methyl groups are in active positions, and the compound forms a dibenzylidene derivative without difficulty.[14]

E. OXIDATION

In thienopyridines the possibility exists of oxidation at nitrogen (to the *N*-oxide) or at sulfur (to the sulfoxide or sulfone). Per-acids selectively oxidize the nitrogen atom in thieno[2,3-b]-[81] and -[3,2-b]pyridine[87] and

---

[87] L. H. Klemm, S. B. Mathur, R. Zell, and R. E. Merrill, *J. Heterocycl. Chem.* **8**, 931 (1971).

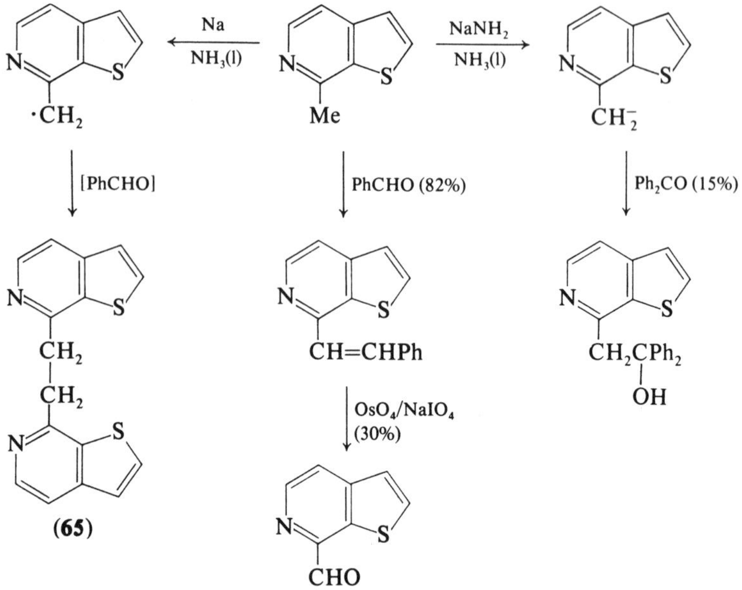

SCHEME 18

in various methyl-substituted thieno[2,3-c]- and -[3,2-c]pyridines,[48] but the preparation of sulfoxides and sulfones is less easily achieved. Iodobenzene dichloride converts nonaromatic cyclic sulfides (including 1,3-dihydrothieno[3,4-b]- and -[3,4-c]pyridines) into sulfoxides[10] and is also effective with certain aromatic tricyclic structures, e.g., benzothieno[3,2-b]pyridine (66).[87] Use of this reagent (or chlorine water,

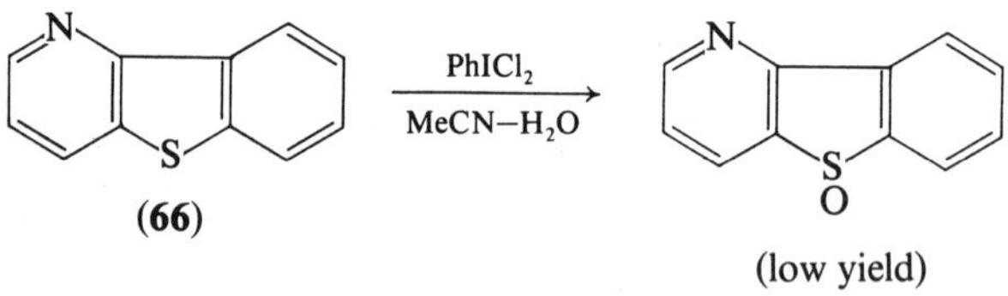

which often gives the same result) with thieno[2,3-b]pyridine led not to the desired sulfoxide, but instead to a low yield of the oxidation–addition product 2,3-dichloro-2,3-dihydrothieno[2,3-b]pyridine-S-oxide (67).[81]

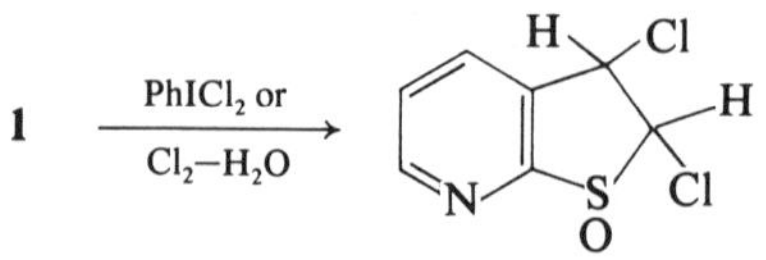

Klemm and Merrill[88] later devised a method for the preparation of the sulfones of thieno[2,3-*b*]-, -[3,2-*b*]-, and -[2,3-*c*]pyridines by reaction of the parent heterocycle with sodium hypochlorite and dilute hydrochloric acid. Thieno[2,3-*b*]pyridine sulfone (**68**) behaved as a dienophile toward anthracene, naphthacene, and furan and yielded 8-(3-pyridyl)quinoline on heating; Klemm proposes the mechanism of Scheme 19 for this interesting reaction.

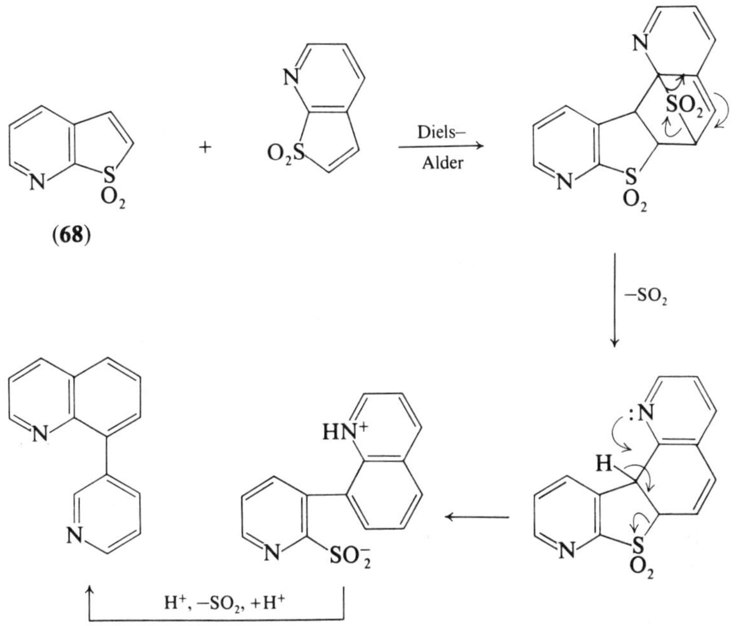

SCHEME 19

It would be expected that vigorous oxidation of a thienopyridine would result in destruction of the thiophene rather than the pyridine ring, and this was found to be the case with 5-butyl-7-methylthieno[2,3-*c*]-pyridine[62] [Eq. (27)].

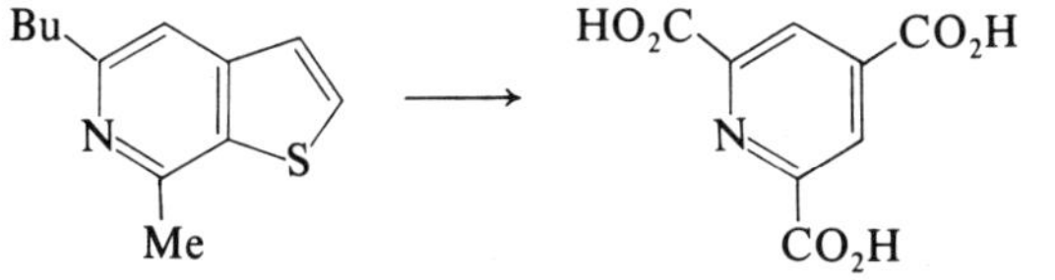

(27)

[88] L. H. Klemm and R. E. Merrill, *J. Heterocycl. Chem.* **9**, 293 (1972).

F. Reduction

In contrast to the quinoline and isoquinoline series, thienopyridines are apparently resistant to reduction by tin–hydrochloric acid, since the parent systems can be obtained by reductive dehalogenation of chloroderivatives. Quaternary salts are reduced to the N-alkyl-4,5,6,7-tetrahydrothienopyridine by sodium borohydride,[45,48,89–91] and the azomethine bond in dihydro derivatives (e.g. **69**) is reduced by lithium aluminum hydride.[47]

(69)

Reductive N-formylation, followed by hydrolysis has been employed as a route to 4,5,6,7-tetrahydrothieno[2,3-c]- and -[3,2-c]pyridines[89] [Eq. (28)]. N-oxide functions are reduced by iron and acetic acid.[81]

(28)

G. Hydroxy Compounds

Few hydroxythienopyridines have been described. A major point of interest is the extent to which the compounds exist as the hydroxy or keto tautomer. Derivatives in which the group is attached to the pyridine ring would be expected to resemble their quinoline or isoquinoline analogs, but, in view of the fact that hydroxythiophenes exist to some extent in keto forms, the genuinely phenolic properties of hydroxy groups on the benzene rings of the isosteric systems might not be reproduced in thienopyridines. 1-Hydroxyisoquinoline and 2- and 4-hydroxyquinoline exist almost exclusively in the keto forms, whereas 3-hydroxyquinoline and 4-hydroxyisoquinoline are extensively enolized; in 3-hydroxyisoquinoline the two forms are of comparable stability and which one predominates is dependent on the solvent. A similar pattern is

---

[89] F. Eloy and A. Deryckere, *Bull. Soc. Chim. Belges* **79**, 415 (1970).
[90] German Patent 2,404,308 [*CA* **81**, 136131 (1974)].
[91] J. P. Mattrand and F. Eloy, *Eur. J. Med. Chem.* **9**, 483 (1974).

observed in thienopyridines. The UV[55] and NMR[83] spectra of isocarbostyril (**70**) and its thieno[2,3-c]- (**71**) and -[3,2-c]pyridine (**58**) analogs are very similar; the 7-oxygenated-thieno[3,2-b]pyridine (**13**) is also formulated[19] as existing in the keto form.

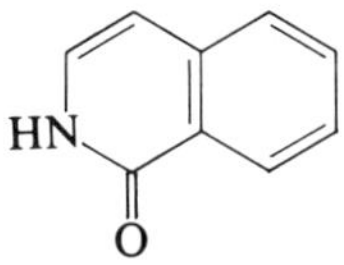

(**70**) 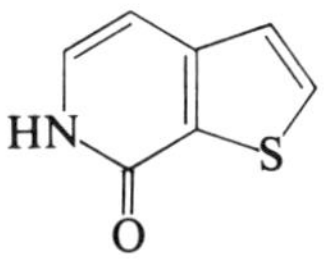   (**71**) 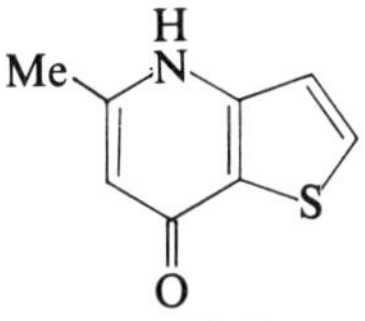   (**13**)

The infrared (IR) data quoted for 5-hydroxythieno[2,3-b]pyridine[86] (the isostere of 3-hydroxyisoquinoline) implies no contribution from the keto tautomer, and in numerous reports on studies of pyridoxine-type analogs,[69, 71–73, 92–94] which contain a 7-hydroxy-thieno[3,4-c]pyridine system, the oxygen function β to the nitrogen is invariably represented as a normal hydroxy group. In a paper devoted to ketoenol equilibria in 1,3-dihydro-5-hydroxythieno[3,2-b]pyridine derivatives[42] (**72–75**), it was reported that the keto forms were exclusively present in the solid state, but in dioxane the ratios **72:73** and **74:75** were 0.43:1 and 1.09:1, respectively. These facts were interpreted in terms of steric and electronic effects.

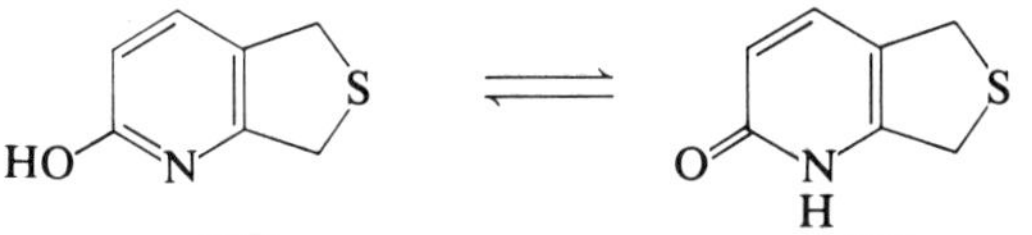

(**72**)   (**73**)

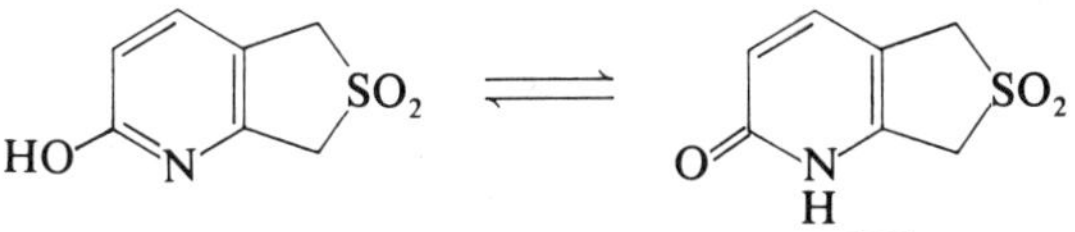

(**74**)   (**75**)

No 2-hydroxy derivative of a fully aromatic thienopyridine is known; the UV and IR spectra of the only 2-oxygenated compound (**24**) recorded to date[45] revealed no evidence of the hydroxy tautomer.

Nearly all the available evidence indicates that 3-hydroxythienopyridines exist in the enolic form. Sheehan and Leitner[30] found no

[92] R. Koch, H. Lagendorff, and H. Moenig, *Atomkernenergie* **11**, 209 (1966) [*CA* **65**, 9302 (1966)].
[93] H. Moenig and R. Koch, *Biophysik* **3**, 11 (1966).
[94] H. G. Kraft, L. Fiebig, and R. Hotovy, *Arzneim.-Forsch.* **11**, 922 (1961).

chemical or spectroscopic evidence for the keto form in 3-hydroxythieno[3,2-b]pyridine. The spectra of this compound closely resembled those of 8-hydroxyquinoline, and the substance formed a benzoate and a methyl ether. Koenigs and Kantrowitz[6] obtained an azo dye (presumably **77**) from 3-hydroxy-4,6-dimethylthieno[3,2-c]pyridine (**76**) and diazotized p-toluidine, but also found that **76** yielded a semicarbazone.

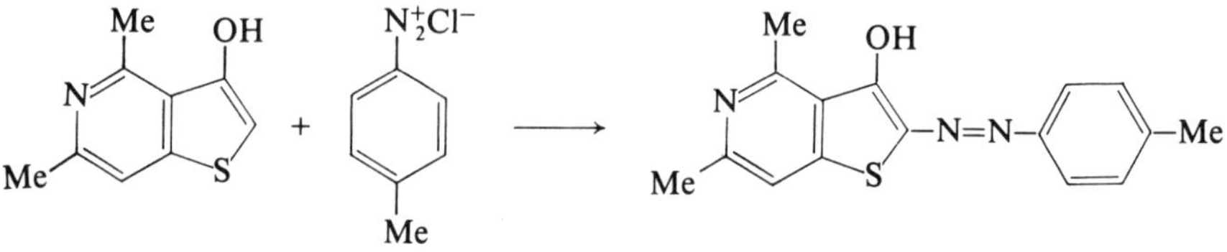

The latter substance was unusual in that it was colored and had a very high m.p. (> 300°), and it seems possible that it was not a simple carbonyl group derivative.

Early interest in hydroxythienopyridines centered on their potential use for the preparation of dyes related to indigo ("thioindigos").[3, 5, 6, 29] Oxidative coupling occurs on brief treatment with concentrated sulfuric acid at 200°–230°, or on exposure of an alkaline solution of the phenol to the air [Eq. (29)].

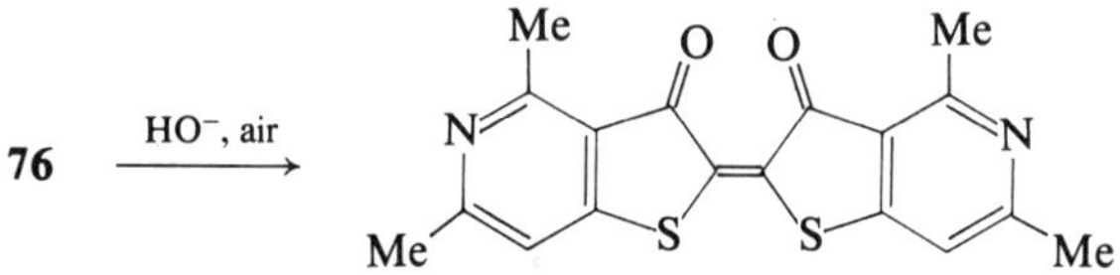

Nucleophilic displacement of the hydroxy group of 3-hydroxythieno[3,2-b]pyridine, by a variety of amines under rather forcing conditions, has been reported [Eq. (30)].[95, 96] The reactions proceeded in the presence of a catalytic quantity of potassium iodide, and it seems probable that the 3-iodo compound was an intermediate.

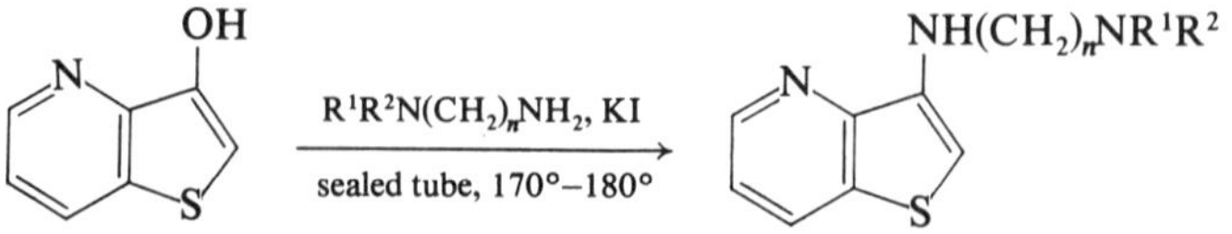

[95] J. T. Sheehan, J. Am. Chem. Soc. **74**, 5504 (1952).
[96] J. T. Sheehan, U.S. Patent 2,811,527 [CA **52**, 5482 (1958)].

H. Amino Compounds

The few known amino derivatives of the thienopyridines behave unexceptionally. As is the case with simple aminothiophenes, compounds in which a primary amino group is attached directly to the thiophene ring are unstable as the free bases but stable as *N*-acetyl derivatives.

The obvious route to 3-aminothienopyridines, i.e., reduction of the 3-nitro compound, is not always free of complications. Klemm *et al.*[75] found that treatment of 3-nitrothieno[2,3-*b*]pyridine with tin–hydrochloric acid gave the desired 3-amino derivative, but that iron–hydrochloric acid gave only the diamine (**78**) (Scheme 20). Metal–acid

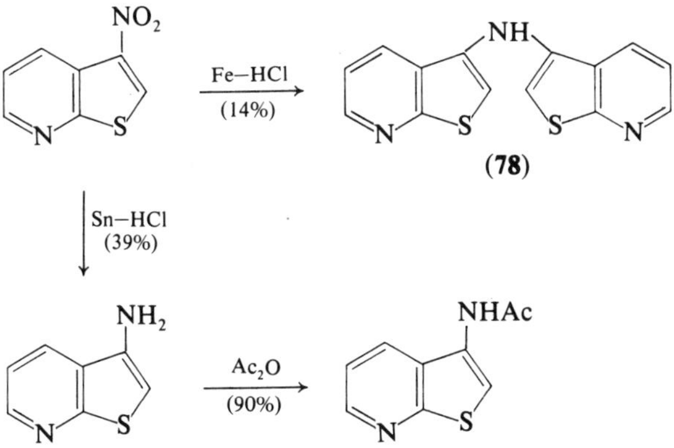

SCHEME 20

reduction of 3-nitrothieno[3,2-*b*]pyridine followed a different pattern; in this instance, treatment with tin–hydrochloric acid gave 3-amino-2-chlorothieno[3,2-*c*]pyridine (**79**). The acetylamine (**80**) was produced directly by reaction with iron–acetic acid.

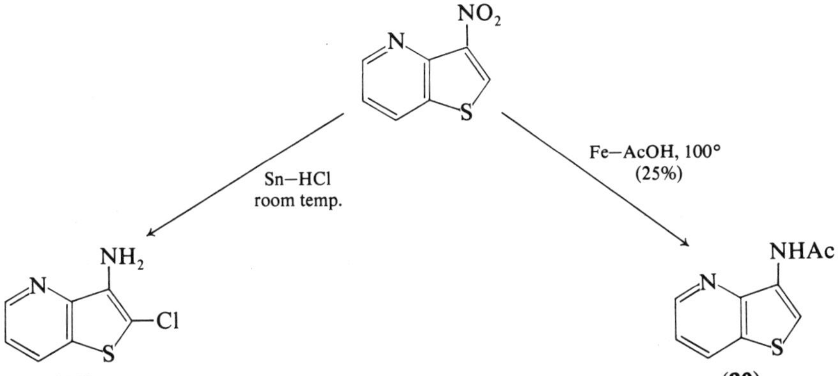

A more efficient method for the direct production of an acetylamino from a nitro compound, involving reaction of the latter with iron–acetic acid–acetic anhydride has been described;[80] the yield from a model compound, 5-ethyl-3-nitrothieno[2,3-b]pyridine, was in excess of 80%.

Nitro groups attached to the pyridine nucleus of thienopyridines are reduced without complications.

Both 4- and 5-nitrothieno[2,3-b]pyridine-N-oxide gave the amines on treatment with iron–acetic acid (67%) and tin–hydrochloric acid (41%), respectively, the N-oxide group also being removed during the reduction.[81] Klemm and Zell[86] had previously prepared 5-aminothieno[2,3-b]pyridine (81) from the 5-acetyl compound (via the oxime, Beckmann rearrangement, and hydrolysis) and had shown that it readily gave Schiff's bases with a variety of aldehydes. Under the same, or more forcing, conditions 4-aminothieno[2,3-b]pyridine (81) failed to react with these aldehydes.[81] The considerable difference in reactivity was ascribed to the decreased nucleophilicity of the 4-amino vs. the 5-amino group owing to the position of the former relative to the ring nitrogen.

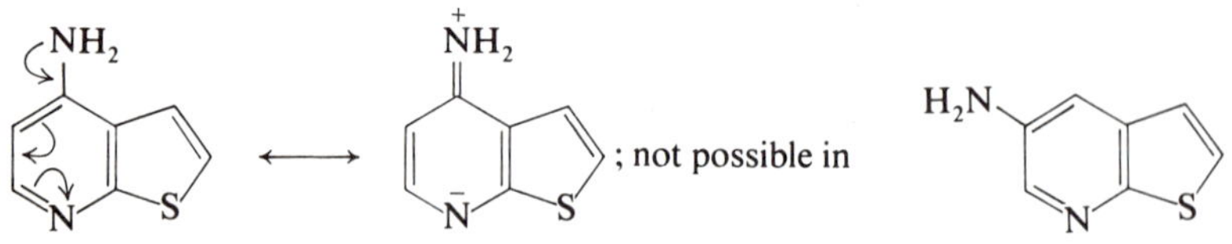

Schiff's bases were eventually prepared from 81; both these and those derived from 82 were readily reduced to secondary amines by sodium borohydride [Eq. (31)].

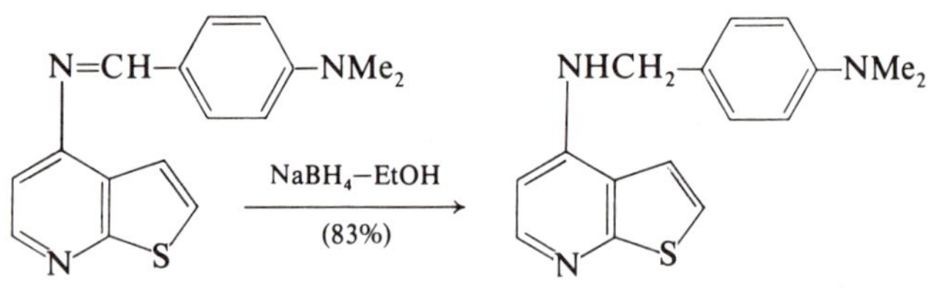

5-Aminothieno[2,3-b]pyridine (81) closely resembled the analogous 3-aminoquinoline, behaving as a typical aromatic primary amine.[86] It could be diazotized, and the diazonium salt yielded the 5-bromo- (48%), -chloro- (40%), -hydroxy- (65%), and -cyanothieno[2,3-b]pyridine (13%) under the usual conditions.

I. Miscellaneous Reactions

The foregoing account reveals that the substituents dicussed behave much like those of comparable quinolines and isoquinolines. Other groups also behave much as one might expect. For example, 5-acetylthieno[2,3-b]pyridine undergoes the Kindler and Wolff–Kishner reactions, forms an oxime, and gives thieno[2,3-b]pyridine-5-carboxylic acid on oxidation with sodium hypochlorite.[86]

Evidence for the existence of an aryne in the series was provided by the formation of a mixture of 4-amino- (**82**; 42%) and 5-aminothieno-[2,3-b]pyridine (**81**; 13%) on reaction of 5-bromothieno[2,3-b]pyridine with potassamide in liquid ammonia at $-70°$.[86] At $-35°$ only **82** was formed.

Deryckere and Eloy[97] have prepared some tricyclic compounds by cyclizations involving 4-hydrazinothieno[3,2-c]pyridine (**83**); interaction of this substance with ethylorthoformate gave the 1,2,4-triazole (**84**), and nitrous acid gave the tetrazole (**85**).

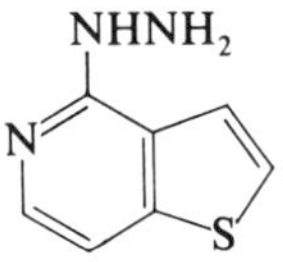

(**83**)

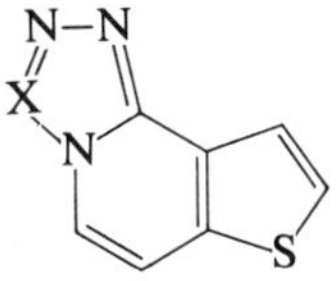

(**84**) X = CH
(**85**) X = N

V. Physical Properties

A. Base Strengths

The pK_a values for the four readily accessible thienopyridines have been determined; the data for these compounds, for quinoline, and for isoquinoline are collected in Table III.

TABLE III

pK_a Values (Water)

Compound	pK_a/temp.	Reference
Thieno[2,3-b]pyridine	2.75/20°	84
Thieno[3,2-b]pyridine	4.35/25°	98
Thieno[2,3-c]pyridine	5.58/25°	98
	5.57/20°	53
Thieno[3,2-c]pyridine	5.67/20°	53
Quinoline	4.94/20°	99
Isoquinoline	5.40/20°	100

[97] A. Deryckere and F. Eloy, German Patent 1,965,710 [*CA* **73**, 120639 (1970)].
[98] L. H. Klemm and R. D. Jacquot, *J. Electroanalyt. Chem. Interfacial Electrochem.* **45**, 181 (1973).
[99] A. Albert, R. Goldacre, and J. N. Phillips, *J. Chem. Soc.*, 2240 (1948).
[100] H. Osborn, K. Schofield, and L. Short, *J. Chem. Soc.*, 4191 (1956).

Klemm and Jacquot[98] have concluded that three factors have a bearing on the trends revealed by these figures: (a) the disposition of the nitrogen atom relative to the second ring (isoquinoline is a stronger base than quinoline and the same is true of their thienopyridine analogs); (b) the inductive effect of the sulfur atom (which attenuates rapidly as the sulfur atom becomes more remote from nitrogen); and (c) the resonance effect of the sulfur atom (orthoquinoid structures should be less important than the para-quinoid forms).

Factors (a) and (b) together explain the fact that the [c]-fused systems are stronger bases than those that are [b]-fused. A further point to note is that isoquinoline is less basic than the isosteric thienopyridines, but the reverse is true for quinoline. The introduction of methyl groups into the pyridine ring has the expected effect of increasing base strength; Dressler and Joullié[48] list pK_a values for a number of methyl derivatives in the [3,2-c]- and [2,3-c]-fused series.

B. Spectroscopic Properties

1. *Ultraviolet Spectra*

Details of the UV spectra of thienopyridines, quinoline, and isoquinoline are given in Table IV.

The spectra of the three [c]-fused systems show considerable similarity to that of isoquinoline; the thienopyridines have four main absorption bands, at ca. 225, 260–270, 280, and 295–300 nm (log ϵ_{max} values ca. 4.5, 3.5, 3.5, and 3.3, respectively), which parallel those at 217, ca. 265, 276, and 305 (log ϵ_{max} 4.8, 3.5, 3.4, and 3.4) given by the benzopyridine. There are also some features common to the spectra of quinoline and the thieno[b]pyridines, notably an intense absorption near 230 nm (log ϵ_{max} ca. 4.4), absorptions in the region 290–300 nm (log ϵ_{max} 3.4–3.9) and a generally less intense peak at ca. 315 nm.

The UV spectra of a number of substituted thienopyridines with fusion [2,3-b],[13,79,80] [3,2-b],[13] [2,3-c],[55] and [3,2-c],[47,55] and also the S,S-dioxides of the first three fusion types,[88] have been reported.

2. *Infrared Spectra*

The IR spectra of thienopyridines are apparently unremarkable, since no detailed discussion of them has appeared. In the [c]-fused systems the two adjacent hydrogen atoms and the isolated hydrogen atom of the pyridine ring give rise to out-of-plane deformation absorptions near 820 and 890 cm^{-1}, respectively,[48] and in thieno[3,4-b]pyridine it is reported that the three vicinal hydrogen atoms give bands at 775 and 800 cm^{-1}.[40]

TABLE IV
ULTRAVIOLET SPECTRA

		$\lambda_{max.}$ (nm) log $\epsilon_{max.}$					
Compound	Solvent	215–235	250–270	271–290	291–310	>310	References
Quinoline	EtOH	235 (4.5)			300 (3.4)	314 (3.5)	101
Thieno[3,2-b]pyridine	EtOH	230 (4.4)		278 (3.5)	290 (3.65)	315 (2.4)	7
Thieno[3,4-b]pyridine	EtOH	224 (4.3)		278 (3.7)	293 (3.8)	343 (3.5)	10
					296 (3.8)		
					306 (3.9)		
Isoquinoline	MeOH	217 (4.8)	258 (3.5)	276 (3.4)	305 (3.4)	319 (3.5)	55
			268 (3.6)				
Thieno[2,3-c]pyridine	i-C$_8$H$_{18}$	223 (4.2)			293 (3.7)		8
		227 (4.2)			299 (3.9)		
					304 (3.8)		
	MeOH	228 (4.4)	257 (3.0)		295 (3.8)		55
					305 (3.9)		
Thieno[3,2-c]pyridine	i-C$_8$H$_{18}$	222 (4.5)	256 (3.6)	281 (3.4)	291 (3.3)	337 (2.4)	8
	MeOH	223 (4.6)	266 (3.6)	278 (3.5)	293 (3.3)		55
Thieno[3,4-c]pyridine	EtOH	224 (4.3)	270 (3.35)	280 (3.5)	291 (3.3)	342 (3.5)	10

[101] J. M. Hearn, R. A. Morton, and J. C. E. Simpson, *J. Chem. Soc.*, 3318 (1951).

N-Oxides have a characteristic absorption in the region 1250–1280 cm^{-1},[79,81,87] sulfoxides in the region 1040–1050 cm^{-1};[87] sulfones give bands at 1300–1330 and 1165–1170 cm^{-1}.[86, 87]

3. Nuclear Magnetic Resonance Spectra

The bulk of work on thienopyridines has been carried out since the advent of NMR spectroscopy, and NMR data are available for many compounds in the [2,3-b][13,75,79–81,84–86,88], [3,2-b][13,75,88], and [2,3-c] and [3,2-c] series.[48,55,63,76,88] Details of the spectra of the parent heterocycles, all run at 60 MHz, are given in Table V.

The following observations may be made.

1. The proton(s) α to the nitrogen atom resonate at the lowest fields. In the isoquinoline isosteres, which contain two such protons, it is that proton which lies between the nitrogen atom and the ring junction which resonates at the lowest field of all.

2. Generally, the protons on the thiophene ring appear at the next lowest fields.

3. In the [2,3-]- and [3,2-]-fused systems, the proton α to the sulfur atom (H-2) resonates at lower field than that β (H-3) to it, and J_{23} is 5.5–6 Hz in the parent heterocycles and in many of their derivatives.

4. In the [3,2-b]-, [2,3-c]-, [3,2-c], and [3,4-b]-fused systems H-3 (on the thiophene ring) is coupled with H-7 (on the pyridine ring).

5. The coupling constants of protons β and γ to the nitrogen atom are greater in the quinoline (8.1–8.8 Hz) than in the isoquinoline (5.5–6.5 Hz) isosteres. The corresponding values for quinoline and isoquinoline are 8.2 and 6.0 Hz, respectively.

6. In the isoquinoline isosteres H-4 is coupled to H-7 ($J_{47} = 1.0$–1.2 Hz).

7. In the quinoline isosteres and their derivatives, the coupling constant between the pyridine ring protons α and β to the nitrogen atom is usually ca. 4.5 Hz (value for quinoline is 4.1 Hz).

In many cases it should be possible to locate a substituent from the NMR spectrum alone. The presence of a substituent on the thiophene ring is apparent not only from the disappearance of the appropriate resonance, but also from the effect on the signals from the remaining protons. If the substituent is at C-3, the H-2 signal becomes a singlet and (in the cases covered in paragraph 4 above) the H3–H7 coupling disappears; in addition, the position of the H-4 resonance may be changed significantly. A substituent at C-2, on the other hand, does not affect the H-4 resonance to any great extent, gives a simpler H-3 pattern, and does not affect the H3–H7 coupling.

TABLE V
NUCLEAR MAGNETIC RESONANCE SPECTRA

Thienopyridine	Solvent	Chemical shift (τ)							Coupling constant (Hz)										References
		H-1	H-2	H-3	H-4	H-5	H-6	H-7	13	23	26	37	45	46	47	56	57	67	
|2,3-b|	CCl$_4$	(S)	2.60	2.92	2.15	2.90	1.54	(N)	—	5.9	—	—	8.1	1.6	—	4.5	—	—	13
|3,2-b|	CCl$_4$	(S)	2.38	2.49	(N)	1.36	2.88	1.92	—	6.1	0.4	0.8	—	—	—	4.8	1.6	8.8	13
|3,4-b|	CDCl$_3$	2.34	(S)	1.98 2.22	(N)	1.25	3.10	2.22 1.98	3.2	—	—	0.7	—	—	—	3.7	1.8	8.5	40
|2,3-c|	CDCl$_3$	(S)	2.20	2.53	2.17	1.37	(N)	0.71	—	5.6	—	1.0	5.5	—	—	—	—	—	76
	CDCl$_3$	(S)	2.35	2.69	2.32	1.54	(N)	0.86	—	5.2	—	—	5.5	—	—	—	—	—	55
	CCl$_4$	(S)	2.40	2.74	2.44	1.58	(N)	0.94	—	—	—	—	—	—	—	—	—	—	48
	CCl$_4$	(S)	2.37	2.78	2.43	1.56	(N)	0.88	—	5.8	—	—	5.7	—	—	—	—	—	66
|3,2-c|	CDCl$_3$	(S)	(AB 2.48)	0.79	(N)	1.47	2.15	—	—	—	0.9	—	—	1.2	—	—	5.6	76
	CDCl$_3$	(S)	2.53	2.57	0.89	(N)	1.57	2.72	—	—	—	0.8	—	—	1.0	—	—	5.6	55
|3,4-c|	CDCl$_3$	1.98 2.36	(S)	2.36 1.98	0.88	(N)	1.96	2.62	—	—	—	—	—	—	—	—	—	—	40

TABLE VI

COUPLINGS IN THIENOPYRIDINES SUBSTITUTED IN THE PYRIDINE RING

Thienopyridine	Location of substituent	Coupling[a]						
		37	45	46	47	56	57	67
[2,3-b]	4	—	×	×	—	√	—	—
	5	—	×	√	—	×	—	—
	6	—	√	×	—	×	—	—
[3,2-b]	5	√	—	—	—	×	×	√
	6	√	—	—	—	×	√	×
	7	×	—	—	—	√	×	×
[2,3-c]	4	√	×	—	×	—	√	—
	5	√	×	—	√	—	×	—
	7	×	√	—	×	—	×	—
[3,2-c]	4	√	—	—	×	—	—	√
	6	√	—	—	√	—	—	×
	7	×	—	—	×	—	—	×

[a] —, one of the numbered positions is a heteroatom; ×, coupling not expected; √, coupling expected.

Table VI reveals the expected couplings for the [2,3-]- and [3,2-]-fused systems containing one substituent in the pyridine ring and shows that unambiguous location of such a substituent should be possible.

4. Mass Spectra

Mass spectral fragmentation data have been recorded for a few thienopyridine derivatives; the majority of the compounds studied have been substituted in the thiophene ring only.[79,80,82,84,88] Examination of these spectra reveals certain characteristic fragmentation patterns: (a) Substituents on the thiophene ring are lost to give the ion of the parent heterocycle (e.g., **86**; from monosubstituted compounds) or of the derived aryne (e.g., **87**; from disubstituted compounds). (b) Losses of HCN and CS from the parent ion, or from derived fragments, are common processes. (c) The molecular ion is usually very abundant. (d) The fragments $[C_3SH]^+$ (m/e 69) and $[C_5H_3]^+$ (m/e 63) appear in many of the spectra.

The essential features of the fragmentation patterns of thieno[2,3-b]-pyridine-2-aldehyde[84] and 2,3-dibromothieno[2,3-c]pyridine,[82] shown in Fig. 2, are typical.

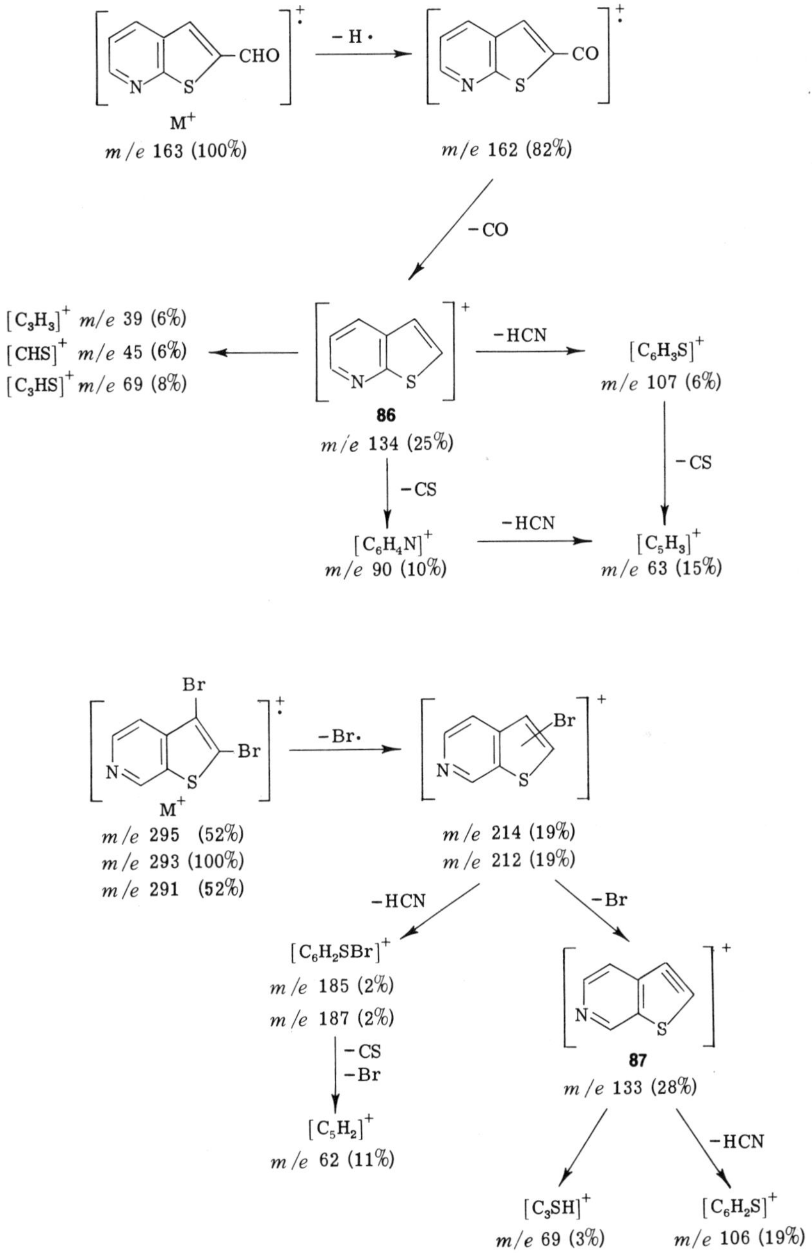

Sec. V.B] THE THIENOPYRIDINES 113

FIG. 2. Some mass spectral fragmentation patterns.

C. Other Physical Properties

Gronowitz and Sandberg[53] have determined the dipole moments of thieno[2,3-c]- (2.61 D) and -[3,2-c]pyridines (2.16 D); the values are fairly close to that of isoquinoline (2.54 D). Klemm and his colleagues have attempted to correlate the behavior of thienopyridines (among numerous other nitrogen heterocycles) on gas–liquid[102] and thin-layer chromatography[103] with the stereochemistry of the molecules and with their electronic interactions with the chromatographic system.

VI. Biological Activity

Most interest in biological activity has been centered on 4,5,6,7-tetrahydrothieno[2,3-c]- or -[3,2-c]pyridines. A considerable number of derivatives of these systems have been prepared by the sodium borohydride reduction of quaternary ammonium salts,[90,91] or by reaction of the 4,5,6,7-tetrahydro base with a suitable halide,[91,104] tosyl derivative,[105] epoxide,[106] or activated alkene.[107] An alternative, and very convenient, synthesis of 2-amino-3,6-substituted 4,5,6,7-tetrahydrothieno[2,3-c]-pyridines (89), used mostly by Nakanishi and his co-workers, involves reaction of an N-substituted 4-piperidone, a compound of the type 88 and sulfur, in the presence of morpholine. The group X can be CN,

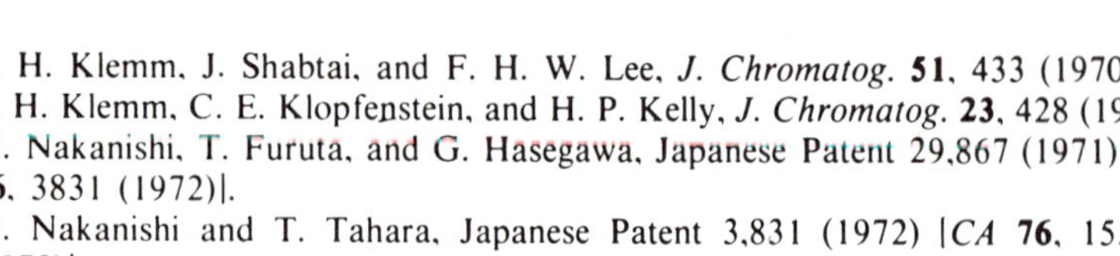

[102] L. H. Klemm, J. Shabtai, and F. H. W. Lee, *J. Chromatog.* **51**, 433 (1970).
[103] L. H. Klemm, C. E. Klopfenstein, and H. P. Kelly, *J. Chromatog.* **23**, 428 (1966).
[104] M. Nakanishi, T. Furuta, and G. Hasegawa, Japanese Patent 29,867 (1971) [*CA* **76**, 3831 (1972)].
[105] M. Nakanishi and T. Tahara, Japanese Patent 3,831 (1972) [*CA* **76**, 153727 (1972)].
[106] M. Nakanishi, T. Tahara, and T. Kobayakawa, Japanese Patent 4,071 (1972) [*CA* **76**, 140763 (1972)].
[107] M. Nakanishi, T. Kobayakawa, and T. Tahara, Japanese Patent 43,791 (1971) [*CA* **76**, 72497 (1972)].

CO_2R^1, $CONH_2$, COAr, $CSNH_2$, $CSNHR^2$, or $CSNR_2^3$, and R can be alkyl or alkaryl, carrying almost any further substituent(s). Numerous patents[31,108–117] record the preparation of very large numbers of compounds of this type, in which the groups R and X are present in various combinations. Further transformations, e.g., subsequent modification of the side chain[118,119] or of the amino group,[120] lead to yet more derivatives. Many of the compounds have been evaluated pharmacologically and have been found to show activity, for example against diabetes mellitus,[105,106,113] as analgesics and antiinflammants,[90,111,114,121] as sedatives,[114] and as anticoagulants.[90,121] One substance, 2-amino-6-benzyl-3-ethoxycarbonyl-4,5,6,7-tetrahydrothieno[2,3-c]pyridine (**89**; R = $PhCH_2$, X = CO_2Et), has been particularly closely studied, no doubt as a prelude to its introduction for clinical use. The compound is antipyretic, analgesic, and antiedematous.[122–125] Its effect on the organs

---

[108] M. Nakanishi, T. Tahara, H. Imamura, and H. Maruyama, Japanese Patent 39,350 (1971) [*CA* **76**, 59593 (1972)].

[109] M. Nakanishi and T. Tahara, Japanese Patent 4,072 (1972) [*CA* **76**, 153728 (1972)].

[110] M. Nakanishi, T. Tahara, H. Imamura, and H. Maruyama, Japanese Patent 21,032 (1971) [*CA* **75**, 129788 (1971)].

[111] M. Nakanishi, T. Tahara, H. Imamura, and H. Maruyama, Japanese Patent 38,338 (1970) [*CA* **74**, 87935 (1971)].

[112] M. Nakanishi, T. Tahara, O. Nakatsu, H. Imamura, and Y. Maruyama, German Patent 1,812,404 [*CA* **71**, 124402 (1969)].

[113] M. Nakanishi, T. Tobayakawa, and T. Tahara, German Patent 2,004,816 [*CA* **73**, 87908 (1970)].

[114] M. Nakanishi, T. Tahara, H. Imamura, and H. Maruyama, Japanese Patent 13,670 (1972) [*CA* **77**, 48429 (1972)].

[115] M. Nakanishi and T. Tahara, Japanese Patent 15,957 (1973) [*CA* **79**, 66340 (1973)].

[116] K. Eichenberger, P. Schmidt, and E. Schweiz, German Patent 1,937,459 [*CA* **72**, 100743 (1970)].

[117] I. Wellings, U.S. Patent 3,633,379 [*CA* **77**, 5440 (1972)].

[118] M. Nakanishi, T. Furuta, and G. Hasegawa, Japanese Patent 29,868 (1971) [*CA* **75**, 140816 (1971)].

[119] M. Nakanishi, T. Furuta, and G. Hasegawa, Japanese Patent 2,555 (1973) [*CA* **78**, 147933 (1973)].

[120] R. Aries, French Patent 2,168,180 [*CA* **80**, 47966 (1974)].

[121] M. Podesta, D. Aubert, and J. C. Ferrand, *Eur. J. Med. Chem.* **9**, 487 (1974).

[122] M. Nakanishi, H. Imamura, Y. Maruyama, and H. Hirosuke, *Yakugaku Zasshi* **90**, 272 (1970) [*CA* **73**, 54504 (1970)].

[123] M. Nakanishi, H. Imamura, and K. Goto, *Yakugaku Zasshi* **90**, 548 (1970) [*CA*, **73**, 43786 (1970)].

[124] M. Nakanishi, H. Imamura, and Y. Maruyama, *Arzneim.-Forsch.* **20**, 998 (1970).

[125] M. Nakanishi, H. Imamura, and Y. Maruyama, *Yakugaku Zasshi* **90**, 227 (1970) [*CA* **73**, 54505 (1970)].

of experimental animals has been examined in great detail,[126-136] and its absorption, distribution, metabolism and excretion in rats and mice has been followed by use of material labeled with ^{14}C at the benzyl carbon,[137-140] or with ^{35}S.[141,142] In both experimental animals and in man,[143] a major metabolic pathway involves debenzylation, the benzyl fragment appearing as hippuric acid (48–59%) or as benzoic acid (4–6%).

[126] T. Nanba, Y. Hamada, K. Izaki, and H. Imamura, *Yakugaku Zasshi* **90**, 1447 (1970) [*CA* **74**, 75010 (1971)].

[127] T. Namba, M. Takeuchi, Y. Hamada, J. Moriguchi, and H. Imamura, *Yakugaku Zasshi* **90**, 1439 (1970) [*CA* **74**, 75009 (1971)].

[128] M. Nakanishi, H. Imamura, and K. Goto, *Yakugaku Zasshi* **90**, 557 (1970) [*CA* **73**, 54262 (1970)].

[129] M. Nakanishi, H. Imamura, and K. Goto, *Biochem. Pharmacol.* **20**, 2116 (1971).

[130] M. Nakanishi, H. Imamura, and K. Goto, *Yakugaku Zasshi* **91**, 921 (1971) [*CA* **76**, 121562 (1972)].

[131] M. Nakanishi, H. Imamura, and Y. Maruyama, *Yakugaku Zasshi* **90**, 284 (1970) [*CA* **73**, 54506 (1970)].

[132] M. Nakanishi, H. Imamura, and Y. Maruyama, *Yakugaku Zasshi* **90**, 291 (1970) [*CA* **73**, 54507 (1970)].

[133] M. Nakanishi, H. Imamura, K. Ikegami, and K. Goto, *Arzneim.-Forsch.* **20**, 1004 (1970).

[134] H. Imamura, E. Matsui, Y. Kato, and T. Furuta, *Yakugaku Zasshi* **91**, 702 (1971) [*CA* **75**, 139199 (1971)].

[135] H. Imamura and E. Matsui, *Yakugaku Zasshi* **90**, 329 (1970) [*CA* **73**, 54510 (1970)].

[136] M. Murakami, K. Odake, M. Takase, and K. Yoshino, *Thromb. Diath. Haemorrh.* **27**, 252 (1972) [*CA* **77**, 73257 (1972)].

[137] H. Imamura, E. Matsui, and Y. Kato, *Yakugaku Zasshi* **90**, 302 (1970) [*CA* **73**, 54508 (1970)].

[138] H. Imamura, E. Matsui, and Y. Kato, *Yakugaku Zasshi* **90**, 317 (1970) [*CA* **73**, 54509 (1970)].

[139] H. Imamura, E. Matsui, and Y. Kato, *Yakugaku Zasshi* **90**, 296 (1970) [*CA* **73**, 64872 (1970)].

[140] Y. Kato, *Yakugaku Zasshi* **92**, 1140 (1970) [*CA* **78**, 37994 (1973)].

[141] Y. Kato, *Yakugaku Zasshi* **92**, 1152 (1972) [*CA* **78**, 37995 (1973)].

[142] H. Imamura, E. Matsui, Y. Kato, and T. Furuta, *Yakugaku Zasshi* **90**, 1126 (1970) [*CA* **73**, 129450 (1970)].

[143] E. Matsui, Y. Kato, T. Furuta, and H. Imamura, *Yakugaku Zasshi* **90**, 1156 (1970) [*CA* **73**, 129451 (1970)].

Nakanishi and Arimura[144] have prepared compounds in the thieno-
[3,4-c]pyridine series (91) analogous to those discussed above, from the
thioamides 90.

The group X was electron-withdrawing (e.g., CO_2R^4), R^3 was an acyl
group, R^1 ranged from alkyl to benzyloxycarbonyl, and R^2 could be an
alkyl or an aryl group or a heterocyclic residue.

There can be no doubt that many other thienopyridine derivatives
have been subjected to biological testing, but only very few reports of
the results of such work appear in the literature. Sheehan[95,96] prepared a
number of 3-substituted thieno[3,2-b]pyridines, including several of the
type $Ar-NH(CH_2)_nNR^1R^2$, but none of the compounds showed the
hoped-for antimalarial or antibacterial properties. Quaternary salts (92)
derived from 7-methylthieno[2,3-c]pyridine and 2-substituted 4-amino-
5-bromomethylpyridines are useful as coccidiostats.[145]

(92)

Several substances related to isoquinoline derivatives in clinical use
have been prepared,[46,49-52,85] but unfortunately no biological testing
results were published.

Some thieno[3,4-c]pyridine derivatives related to pyridoxine have
been studied; 1,3-dihydro-1,7-dihydroxy-6-methylthieno[3,4-c]pyridine
(49) was found to be inactive against mammary adenocarcinoma cells
grown in suspension.[73] A related compound (46), lacking the 1-hydroxy
group, was one of a number of substances employed in an electron spin
resonance study of the localization behavior and quantum yield of
radicals (produced by X-ray/UV irradiation); the object of the work was
to increase understanding of radiation damage to biological systems.[92]

VII. Dyestuffs

The early interest in thienopyridine analogs of indigo dyes has already
been mentioned; the dyes are apparently rather unstable. Zhiryakov and
Abramenko, in a series of papers and patents, have described the
preparation of numerous polymethine dyes containing thieno[2,3-b]-

---

[144] M. Nakanishi and K. Arimura, Japanese Patent 57,994 (1973) [CA 79, 11551 (1973)].

[145] E. F. Rogers, J. Hannah, and R. A. Dybas, German Patent 2,306,001 [CA 79, 126518 (1973)].

and -[3,2-b]pyridine systems.[146-150] The carbocyanine (93) and merocyanine (94) types are typical of the complex structures described by the Russian workers.

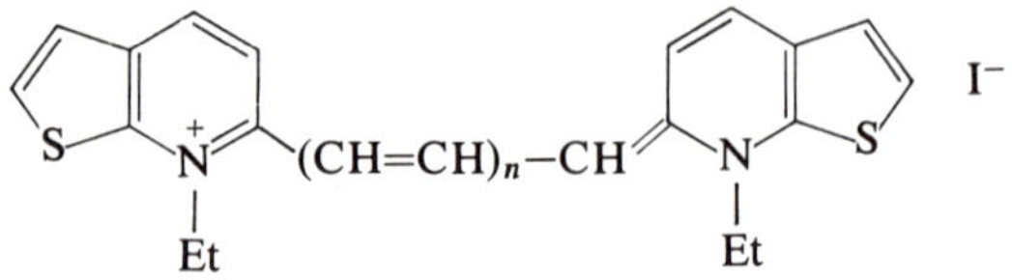

(93)

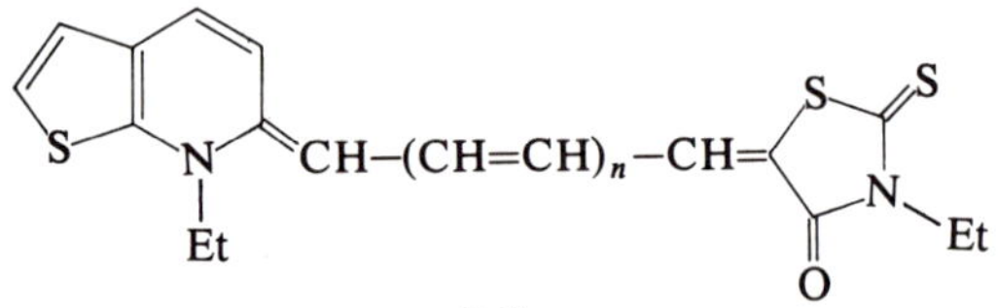

(94)

---

[146] V. G. Zhiryakov, P. I. Abramenko, and G. F. Kurepina, USSR Patent 159,726 (1963) [CA 61, 8449 (1964)].
[147] V. G. Zhiryakov and P. I. Abramenko, Zh. Obshch. Khim. 35, 150 (1965) [CA 62, 14861 (1965)].
[148] V. G. Zhiryakov, P. I. Abramenko, and N. I. Sennikova, USSR Patent 175,820 (1965) [CA 64, 8361 (1966)].
[149] V. G. Zhiryakov and P. I. Abramenko, Khim. Geterotsikl. Soedin. 3, 621 (1967); [CA 68, 96791 (1968)].
[150] V. G. Zhiryakov and P. I. Abramenko, Tr. Vses. Nauch.-Issled. Proekt. Zh. Khim., 1970, Abstr. No. 1, 18 [CA 74, 127522 (1971)].

# Tellurophene and Related Compounds

FRANCESCO FRINGUELLI, GIANLORENZO MARINO, AND ALDO TATICCHI

Istituto di Chimica Organica, Università di Perugia, Perugia, Italy

I. Introduction and Scope of the Review 120
II. Molecular Structure and Physical Properties of Tellurophene . . . 120
   A. Geometrical Parameters 120
   B. Physical Properties 123
     1. General 123
     2. Dipole Moment 124
     3. UV Spectrum 125
     4. Infrared and Raman Spectra 126
     5. Microwave Spectrum 129
     6. NMR Spectra 130
     7. Photoelectron Spectra 135
   C. Ground-State Aromaticity 137
   D. Conformational Aspects of Tellurophene Derivatives 140
III. Synthesis and Chemical Reactivity of Tellurophene and Its Derivatives . 142
   A. Synthesis of the Tellurophene Ring 142
   B. Chemical Properties of Tellurophene. Preparation of Tellurophene Derivatives 144
   C. Quantitative Studies of Reactivity 145
     1. Electrophilic Substitutions 145
     2. Side-Chain Reactions 148
     3. Ionization of Carboxylic Acids 150
   D. Molecular Complexes 151
IV. Other Tellurium Heterocycles 152
   A. Tetrahydrotellurophene 152
     1. Synthesis and Chemical Properties 152
     2. Physical Properties 153
   B. Benzo[b]tellurophene and Derivatives 155
     1. Synthesis 155
     2. Physical Properties 157
     3. Chemical Properties 159
     4. Quantitative Studies of Reactivity 160
   C. Dibenzotellurophene and Derivatives 161
     1. Synthesis and Chemical Reactivity 161
     2. Physical Properties 163
   D. Other Systems Containing the Tellurophene Nucleus . . . 164
V. Appendix: Tables of Tellurophene Derivatives and Related Compounds . 166
   Note Added in Proof 172

I. Introduction and Scope of the Review

(1)

Tellurophene (1) is a five-membered heteroaromatic compound, congener of furan, thiophene, and selenophene.

Since the first synthesis of a tellurophene derivative 15 years ago, numerous papers have been published on many aspects of tellurophene chemistry, which justify a review of this interesting aromatic system.

The term "related compounds" carries a 2-fold implication. "Related compounds" are the derivatives of tellurophene, i.e., substituted tellurophenes, hydrogenated tellurophenes, benzo- and dibenzotellurophene. Therefore this review reports the preparations and the physical, spectroscopic, and chemical properties of these compounds. But "related compounds" are also the congener rings: furan, thiophene, and selenophene, and the review also underlines similarities and differences in the behavior of these related molecules and gives particular emphasis to the *quantitative comparison* of their properties. Comparative studies are fundamental because they allow the influence of some basic parameters (electronegativity and mass of the heteroatom, ring geometry, etc.) on the physical and chemical behavior of this interesting series of molecules to be evaluated.

The literature has been covered until June 1975. Some papers published in Western journals after that date and some unpublished work is also reported.

II. Molecular Structure and Physical Properties of Tellurophene

A. Geometrical Parameters

The geometrical parameters of tellurophene were first determined by microwave spectrum analysis,[1] but assumptions were necessary to derive the complete molecular geometry, as a consequence of the small number of molecules of different isotopic constitution investigated. By analogy with the known geometries of the congener rings, it was assumed that the C(3)H(3) and C(2)H(2) bond lengths are 1.081 and 1.078 Å, respectively, and that these CH bonds bisect the corresponding external angles. All these assumptions except one—equal external angles at C(2)—are reasonable.

[1] R. D. Brown and J. G. Crofts, *Chem. Phys.* **1**, 217 (1973).

TABLE I

GEOMETRICAL PARAMETERS OF FURAN, THIOPHENE, SELENOPHENE, AND TELLUROPHENE, DERIVED FROM MICROWAVE STUDIES

Parameters	Furan[a]	Thiophene[b]	Selenophene[c]	Tellurophene[d]
Bond length (Å)				
X(1)C(2)	1.362	1.714	1.855	2.055
C(2)C(3)	1.361	1.370	1.369	1.375
C(3)C(4)	1.431	1.423	1.433	1.423[e]
C(2)H(2)	1.075	1.078	1.070	1.078[e]
C(3)H(3)	1.077	1.081	1.079	1.081[e]
Angle (degrees)				
C(2)X(1)C(5)	106.55	92.17	87.76	82.53
X(1)C(2)C(3)	110.68	111.47	111.56	110.81
C(2)C(3)C(4)	106.06	112.45	114.55	117.93
X(1)C(2)H(2)	115.92	119.85	121.73	124.59[f]
C(2)C(3)H(3)	126.00	123.28	122.59	121.04[f]

[a] See B. Bak, D. Christensen, W. B. Dixon, L. Hansen-Nygaard, J. Rastrup-Andersen, and M. Schottlander, *J. Mol. Spectrosc.* **9**, 124 (1962).
[b] See B. Bak, D. Christensen, L. Hansen-Nygaard, and J. Rastrup-Andersen, *J. Mol. Spectrosc.* **7**, 58 (1961); more recently the structure of thiophene has been determined by gas-phase electron diffraction: W. R. Harshbarger and S. H. Bauer, *Acta Crystallogr., Sect. B* **26**, 1010 (1970).
[c] See N. M. Pozdeev, O. B. Akulinin, A. A. Shapkin, and N. N. Magdesieva, *Dokl. Akad. Nauk. SSSR* **1851**, 384 (1969); the structure determined from R. D. Brown, F. R. Burden, and P. D. Godfrey, *J. Mol. Spectrosc.* **25**, 415 (1968) is more approximate because the microwave spectrum was analyzed for only two isotopic constitutions of the molecule.
[d] Data from Brown and Crofts.[1]
[e] Bond lengths are based on corresponding data from thiophene.
[f] External angles are assumed to be equal.

The geometrical parameters of the four congener rings, as derived from microwave studies, are summarized in Table I. An analysis of the data of Table I shows that the C(2)XC(5) angle decreases, the C(2)C(3)C(4) angle increases, and the X(1)C(2)C(3) angle remains approximately constant with increasing size of the heteroatom. The X(1)C(2) bond length increases along the series, whereas the C(2)C(3) and C(3)C(4) bond lengths have similar values. The results of these changes is that the shape of the molecule is progressively elongated, going from furan to tellurophene.

The analysis of the ¹H-NMR spectrum of tellurophene in nematic phase solution[2] has allowed determination of the ratios of the interproton distances and the whole ring geometry. Assuming the geometrical parameters derived from microwave spectra [except the external angles at C(2)] to be correct, the best compromise between the data from the liquid crystal and the microwave investigation is a TeC(2)H(2) angle of 126.5°.

The structural parameters of tellurophene have also been calculated using the regression equations of linear correlations between the proton coupling constants $J_{2,3}$ and the bond angles of furan, thiophene, selenophene, and pyrrole, and between the X(1)C(2) bond length of these heterocycles and the single bond covalent radius of the heteroatoms.[3] The values calculated using this procedure (Table II) agree well with those derived from microwave spectra, nematic-phase NMR spectra, and X-ray data.

TABLE II

GEOMETRICAL PARAMETERS OF TELLUROPHENE
DERIVED FROM REGRESSION ANALYSIS[a]

Bond length (Å)		Angle (degrees)	
TeC(2)	2.046	C(2)TeC(5)	82.82
C(2)C(3)	1.371[b]	TeC(2)C(3)	111.83
C(3)C(4)	1.478	C(2)C(3)C(4)	116.76
C(2)H(2)	1.074[b]	TeC(2)H(2)	122.93
C(3)H(3)	1.079[b]	C(2)C(3)H(3)	121.59

[a] Data from Fringuelli and Taticchi.[3]
[b] Bond lengths are based on corresponding data of other congeners, assuming tellurophene to be planar.

Fanfani et al.[4] have determined the structure of tellurophene-2-carboxylic acid by X-ray diffraction (Table III). The structure consists of dimers, two crystallographically nonequivalent molecules being linked together by hydrogen bonds. The two molecules of the dimer are slightly twisted; the dihedral angle between the ring planes is 166.30°. Comparison with 2-furan, 2-thiophene, and 2-selenophene carboxylic acids[5,6] shows that modifications to the ring geometry occur that are similar to those observed in the unsubstituted rings. The values for the C(2)C(6) distance (1.414, 1.481, 1.438, and 1.423 Å for 2-furan, 2-thiophene, 2-

[2] A. D'Annibale, L. Lunazi, F. Fringuelli, and A. Taticchi, Mol. Phys. **27**, 257 (1974).
[3] F. Fringuelli and A. Taticchi, Gazz. Chim. Ital. **103**, 453 (1973).
[4] L. Fanfani, A. Nunzi, P. F. Zanazzi, A. R. Zanzari, and M. A. Pellinghelli, Cryst. Struct. Commun. **1**, 273 (1972).
[5] P. Hudson, Acta Crystallogr. **15**, 919 (1962).
[6] M. Nardelli, G. Fava, and G. Giraldi, Acta Crystallogr. **15**, 737 (1962).

TABLE III

Structural Parameters of Tellurophene-2-Carboxylic Acid[a,b]

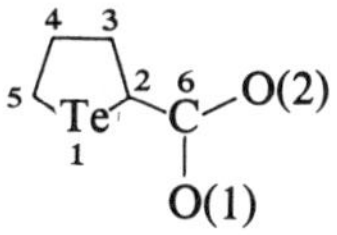

Bond length (Å)		Angle (degrees)	
TeC(5)	2.047	C(2)TeC(5)	81.5
TeC(2)	2.057	TeC(5)C(4)	111.7
C(5)C(4)	1.357	TeC(2)C(3)	111.7
C(2)C(3)	1.384	C(5)C(4)C(3)	118.8
C(3)C(4)	1.412	C(4)C(3)C(2)	116.3
C(2)C(6)	1.423	TeC(2)C(6)	123.4
C(6)O(1)	1.298	C(3)C(2)C(6)	124.8
C(6)O(2)	1.270	C(2)C(6)O(1)	116.7
		C(2)C(6)O(2)	121.7
		O(1)C(6)O(2)	121.6

[a] Data from Fanfani et al.[4]
[b] Bond lengths and angles are mean values of the two crystallographically independent molecules of dimer.

selenophene and 2-tellurophene carboxylic acid) indicate that conjugation of the heterocyclic ring with the carboxylic group decreases: furan > tellurophene > selenophene > thiophene, an order in reverse with respect to the order of ground-state aromaticities (see Section II,C). Evidently the greater the "internal" π-electron delocalization in the ring, the smaller the "external" π-electron delocalization to the substituent, and canonical forms **2–4** contribute less to the resonance hybrid the higher the aromaticity of the ring.

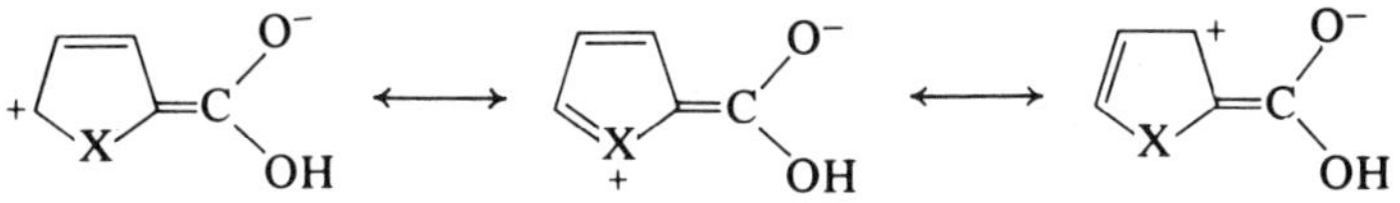

In the crystalline state the 2-thiophene, 2-selenophene, and 2-tellurophene carboxylic acids adopt a (X,O)-trans conformation (see Section II,D) whereas the 2-furoic acid is in the reversed (X,O)-cis conformation.

B. Physical Properties

1. General

Tellurophene is a light yellow malodorous liquid having values of boiling point (b.p.), melting point (m.p.), density (d_4), refractive index (n_D),

TABLE IV
Physical Constants of Tellurophene and Its Congeners

Property	Furan	Thiophene	Selenophene	Tellurophene
b.p. (°C, 760 mmHg)	31.36[a]	84.16[a]	110[b]	151[c]
m.p. (°C)	−85.65[a]	−38.25[a]	−38[b]	−36[d]
d_4^{20}	0.9514[a]	1.0649[a]	1.5251[b]	2.13[d]
n_D^{20}	1.4214[a]	1.5289[a]	1.5642[b]	1.6844[c]
MR_D (cm^3 mol^{-1})	18.27	24.32	27.74[b]	32.08[c]
χ_M (−10^6, cm^3 mol^{-1})	43.09	57.38	66.82	74.20[e]
μ (D, benzene 25°)	0.72[f]	0.54[g]	0.52[h]	0.46[i]
μ (D, gas phase)	0.66[j]	0.53[k]	0.39[l]	0.19[m]

[a] See "Handbook of Chemistry and Physics," 51st ed. Chemical Rubber Co., Cleveland, Ohio, 1970–1971.
[b] See Yu. K. Yur'ev, *J. Gen. Chem. (USSR)* **16**, 851 (1946).
[c] Data from Fringuelli and Taticchi.[14]
[d] Data from Mack.[48]
[e] Data from Fringuelli et al.[12]
[f] See A. L. McClellan, "Tables of Electric Dipole Moments," Freeman, San Francisco, 1963.
[g] See H. Lumbroso, D. M. Bertin, and P. Cagniant, *Bull. Soc. Chim. Fr.*, 1720 (1970).
[h] See H. Lumbroso and C. Carpanelli, *Bull. Soc. Chim. Fr.*, 3198 (1964).
[i] Data from H. Lumbroso et al.[8]
[j] See M. A. Sirvetz, *J. Chem. Phys.* **19**, 1609 (1951).
[k] See N. M. Pozdeev, L. I. Panikovskaya, L. N. Liskovskaya, and A. Kh. Mamleev, *Primen. Mol. Spektrosk. Khim. Sb. Dokl Sib. Soveshch.* **3**, 26 (1964).
[l] See footnote c of Table I.
[m] Data from Brown and Crofts.[1]

molar refraction (MR_D), and diamagnetic susceptibility (χ_M) greater than those of the congener rings (Table IV). All these physical properties increase linearly with increasing heteroatom atomic weight for thiophene, selenophene, and tellurophene, while furan exhibits values smaller than might be predicted from naive extrapolation of the data of its congeners. Similar correlations are observed between the physical properties and the electronegativity of the heteroatoms (again furan deviates from linearity). All four congeners give the same relative molar response with a flame ionization detector in gas chromatographic analysis.[7]

2. Dipole Moment

The dipole moment of tellurophene has been determined in gas phase[1] ($\mu = 0.19$ D), in benzene[8] (0.46 D), and in carbon tetrachloride solution[9]

[7] S. Clementi, S. Savelli, and M. Vergoni, *Chromatographia* **5**, 413 (1972).
[8] H. Lumbroso, D. M. Bertin, F. Fringuelli, and A. Taticchi, *Chem. Commun.*, 342 (1973).
[9] H. Lumbroso, D. M. Bertin, F. Fringuelli, and A. Taticchi, *C.R. Hebd. Seances Acad. Sci., Ser. C* **277**, 203 (1973).

(0.36 D). Passing from furan to tellurophene, the magnitude of the dipole moment regularly decreases.

The direction of the dipole moment has been determined by a study on the orientational influence of tellurophene on aromatic solvents, hexadeuterobenzene and hexafluorobenzene, as reflected in proton chemical shifts.[10] It was found that tellurophene behaves as do the other congeners, giving upfield solvent shifts in hexadeuterobenzene with respect to carbon tetrachloride. On the other hand, pyrrole gives downfield solvent shifts in aromatic solvents. This is an indication that the dipole moment of tellurophene is directed, as in the congener rings,[11] from the ring (positive pole) to the heteroatom (negative).

Other parameters besides the electronegativity of the heteroatom (tellurium is less electronegative than carbon) are responsible for the observed direction of the dipole moment of tellurophene: the geometry of the ring, the nonbonding electron pairs of the tellurium atom, etc.

The difference between the dipole moments of the aromatic compound and the corresponding tetrahydro derivative is called the mesomeric moment and represents a measure of the π-electron delocalization. The mesomeric dipole moment is directed, in all the congener systems, from the heteroatom toward the ring. The values obtained (furan 1.03 D, thiophene 1.35 D, selenophene 1.29 D, tellurophene 1.17 D) are in excellent quantitative agreement with other aromaticity indices based on structural and magnetic properties[12] (see discussion in Section II,C).

The dipole moments in benzene of a number of 2-substituted tellurophenes have been determined[13] and compared with the moments of the corresponding derivatives of furan, thiophene, and selenophene (Table V).

Some consideration has been given to the mesomeric moments of "regular" groups (Cl, Br, I, Me) and the conformational preference of the acyl, methylthio, and hydroxymethyl derivatives (see Section II,D).

3. *UV Spectrum*

Only a qualitative report on the UV spectrum of tellurophene is available.[14] The spectrum (in *n*-hexane) exhibits three bands at 209, 241,

---

[10] F. Fringuelli, S. Gronowitz, A.-B. Hörnfeldt, and A. Taticchi, *J. Heterocycl. Chem.* **11**, 827 (1974).

[11] G. Marino, *J. Heterocycl. Chem.* **9**, 817 (1972).

[12] F. Fringuelli, G. Marino, A. Taticchi, and G. Grandolini, *J. Chem. Soc., Perkin Trans. 2*, 332 (1974).

[13] H. Lumbroso, D. M. Bertin, F. Fringuelli, and A. Taticchi, *J. Chem. Soc., Perkin Trans. 2*, in press.

[14] F. Fringuelli and A. Taticchi, *J. Chem. Soc., Perkin Trans. 1*, 199 (1972).

TABLE V

DIPOLE MOMENTS (DEBYE UNITS) OF 2-SUBSTITUTED DERIVATIVES OF FURAN,
THIOPHENE, SELENOPHENE, AND TELLUROPHENE IN BENZENE

Substituent	Furan	Thiophene	Selenophene	Tellurophene
H	0.72[a]	0.54[b]	0.52[c]	0.46
Cl	1.60[a]	1.48[a]	—	1.43
Br	1.46[a]	1.35[a]	1.37[c]	1.44
I	1.14[a]	1.18[a]	—	—
Me	0.72[a]	0.67[a]	0.56	0.64
SMe	1.60	1.47	1.37	1.30
COMe	3.23	3.37[a]	3.18[e]	2.97
CHO	3.54[d]	3.48[b]	3.25[c]	3.18
COOMe	2.24	1.91	2.12	1.95
CONMe$_2$	3.58	3.95	3.77	3.60
CH$_2$OH	1.99	—	—	1.75

[a] See A. L. McClellan, "Tables of Experimental Dipole Moments," Vol. 2, Rahara Enterprises, El Cerrito, California, 1974, and footnote *f* of Table IV.
[b] See footnote *g* of Table IV.
[c] See H. Lumbroso, D. M. Bertin, J. Morel, and C. Paulmier, *Bull. Soc. Chim. Fr.*, 1921, 1924 (1973).
[d] See D. M. Bertin, C. Chatain-Cathaud, and M. C. Fournié-Zaluski, *C.R. Hebd. Seances Acad. Sci., Ser C* **274**, 1112 (1972).
[e] See S. V. Tsukerman, V. D. Orlov, and V. F. Lavrushin, *Zh. Strukt. Khim.* **10**, 263 (1969).

and 279 nm. Although there is some controversy[15–17] over the theoretical interpretation of the electronic spectra of five-membered heteroaromatic compounds, the band below 220 nm has been ascribed to diene absorption. On this ground tellurophene should exhibit a degree of "double bond fixation" similar to that of furan and greater than that of thiophene and selenophene. The hypothesis is confirmed by other studies.[12]

4. Infrared and Raman Spectra

The full vibrational assignment of the absorption bands of infrared (IR) and Raman (R) spectra of tellurophene and a set of deuteriated tellurophenes (2-d$_1$, 2,5-d$_2$, 3,4-d$_2$, 2,3,4-d$_3$, and 2,3,4,5-d$_4$) has recently been reported.[18]

[15] R. M. Silverstein and G. C. Bassler, "Spectrometric Identification of Organic Compounds," 2nd ed. Wiley, New York, 1968.
[16] H. H. Jaffé and N. Orchin, "Theory and Application of Ultraviolet Spectroscopy," Wiley, New York, 1962.
[17] C. N. R. Rao, "Ultraviolet and Visible Spectroscopy," 2nd ed. Butterworths, London, 1967.
[18] G. Paliani, R. Cataliotti, A. Poletti, F. Fringuelli, A. Taticchi, and M. G. Giorgini, *Spectrochim. Acta Part A* **32**, 1089 (1976).

The light molecule and isotopic species 2,5-d_2, 3,4-d_2, and 2,3,4,5-d_4 belong to the point group C_{2v} with 21 normal vibrations, classified as:

$$8A_1(IR, R) + 7B_1(IR, R) + 3A_2(R) + 3B_2(IR, R)$$

Vibrations of species A_1 and B_1 occur in the molecular plane.

The two deuteriated derivatives 2-d_1 and 2,3,4-d_3 belong to the point group C_s, and their fundamental vibrations may be classified as:

$$15A'(IR, R) + 6A''(IR, R)$$

TABLE VI

FUNDAMENTAL INFRARED FREQUENCIES OF LIQUID TELLUROPHENE AND SOME OF ITS DEUTERIATED DERIVATIVES

No.[a]	d_0	2-d_1	2,5-d_2	3,4-d_2	2,3,4-d_3	d_4	Approximate description[a]
	A_1	A'	A_1	A_1	A'	A_1	
ν_1	3084	3083	2301	3082	3080	2308	ν(C—H)
ν_2	3045	3043	3044	2274	2262	2262	ν(C—H)
ν_5	1432	1423	1415	1414	1410	1405	ν(ring)
ν_4	1316	1300	1272	1256	1234	1180	ν(ring)
ν_6	1079	1079	1045	984	924	824	δ(C—H)
ν_7	984	855	799	815	834	768	δ(C—H)
ν_3	687	660	638	668	644	630	ν(ring)
ν_8	380	379	375	374	372	370	δ(ring)
	A_2	A''	A_2	A_2	A''	A_2	
ν_9	912	910	905	778	744	741	γ(C—H)
ν_{10}	(690)[b]	680	544	656	616	548	γ(C—H)
ν_{11}	507	501	498	460	437	437	γ(ring)
	B_1	A'	B_1	B_1	A'	B_1	
ν_{12}	3084	2304	2301	3082	2310	2308	ν(C—H)
ν_{13}	3030	3030	3031	2243	2240	2240	ν(C—H)
ν_{14}	1516	1505	1498	1476	1464	1455	δ(ring)
ν_{15}	1246	1224	1213	1136	1074	995	δ(C—H)
ν_{16}	1079	1021	895	900	810	811	δ(C—H)
ν_{17}	797	748	716	789	726	704	$(\nu + \sigma)$(ring)
ν_{18}	552	541	533	536	530	522	δ(ring)
	B_2	A''	B_2	B_2	A''	B_2	
ν_{20}	884	856	834	810	790	696	γ(C—H)
ν_{19}	674	547	545	570	514	502	γ(C—H)
ν_{21}	354	345	334	343	334	326	γ(ring)

[a] Numbering and approximate description from Rico et al.[19]
[b] Nonobserved band.

[19] M. Rico, M. Barrachina, and J. M. Orza, *J. Mol. Spectrosc.* **24**, 133 (1967).

The in-plane vibrations (species A') give polarized Raman lines.

The vibrational assignment is based on many experimental and theoretical criteria, such as the intensity and polarization of the Raman lines, the contours and P–R separations in gas phase IR spectra, classical concepts such as group frequencies, and comparison with the spectra of related compounds; good correspondence was found for the isotopic product and sum rules. The fundamental frequencies of liquid tellurophene and its deuteriated derivatives are reported in Table VI.

The vibrational spectra of thiophene,[20] selenophene,[21] and tellurophene are very similar, but furan[19] has a somewhat different spectroscopic behavior (Table VII).

TABLE VII

Fundamental Frequencies of Liquid Furan, Thiophene, Selenophene, and Tellurophene

	Approximate description	C_{2v}	Furan[a]	Thiophene[b]	Selenophene[c]	Tellurophene[d]
ν_1	C—H stretch	A_1	3159	3110	3110	3084
ν_2	C—H stretch		3128	3086	3063	3045
ν_5	Ring stretch		1483	1408	1419	1432
ν_4	Ring stretch		1380	1360	1341	1316
ν_6	C—H def i.p.[e]		1140	1081	1080	1079
ν_7	C—H def i.p.		1061	1033	1010	984
ν_3	Ring stretch		986	833	758	687
ν_8	Ring def i.p.		873	606	456	380
ν_9	C—H def o.o.p.[e]	A_2	863	900	905	912
ν_{10}	C—H def o.o.p.		728	686	685	690
ν_{11}	Ring def o.o.p.		613	565	541	507
ν_{12}	C—H stretch	B_1	3148	3110	3100	3084
ν_{13}	C—H stretch		3120	3073	3054	3030
ν_{14}	Ring stretch		1556	1506	1515	1516
ν_{15}	C—H def i.p.		1270	1250	1243	1246
ν_{16}	C—H def i.p.		1171	1081	1080	1079
ν_{17}	Ring (def + stretch)		1040	871	820	797
ν_{18}	Ring def i.p.		873	750	623	552
ν_{20}	C—H def o.o.p.	B_2	839	864	870	884
ν_{19}	C—H def o.o.p.		745	712	700	674
ν_{21}	Ring def o.o.p.		601	453	394	354

[a] Data from Rico et al.[19]
[b] Data from Rico et al.[20]
[c] Data from Magdesieva.[21]
[d] Data from Paliani et al.[18]
[e] i.p., in plane; o.o.p., out of plane.

[20] M. Rico, J. M. Orza, and J. Morcillo, *Spectrochim. Acta* **21**, 689 (1965).
[21] N. N. Magdesieva, *Adv. Heterocycl. Chem.* **12**, 1 (1970).

In the sequence thiophene, selenophene, tellurophene some normal modes fall at virtually the same frequency (v_6, v_{10}, v_{15}, and v_{16}); others undergo a slight decrease in frequency (v_1, v_2, v_4, v_7, v_{11}, v_{12}, v_{13}, and v_{19}) or slight increase in frequency (v_5, v_9, v_{14}, and v_{20}), and the remaining modes exhibit a drastic drop in frequency (v_3, v_8, v_{17}, v_{18}, and v_{21}).

Many factors are responsible for the observed differences: the mass and the electronegativity of the heteroatom, the π-electron delocalization, the geometry of the ring, the vibrational couplings of the normal modes, etc. These factors often operate against one another, and it is therefore impossible completely to rationalize the differences observed; also accurate force fields are not available.

Only for some modes is it possible to interpret satisfactorily the different frequencies observed; the ring modes, in particular, are those that allow the best qualitative comparison. So, the increase in frequency of the v_5 mode (symmetric stretching of the double bonds) in the order thiophene, selenophene, tellurophene, furan may be related to a change, in the same order, toward a situation of lower aromaticity, with greater localization of the double bonds.[12] The great decrease in frequency of the v_3 and v_{17} modes (symmetric and antisymmetric C—X—C stretching, respectively) and the ring deformation modes v_8, v_{18}, and v_{21} has been attributed to mass and geometry effects, which influence these modes in the same direction.

Andrieu et al.[22] have studied the intensity and mutiplicity of the v(C=O) and v(C=C) bonds in 2-formyl- and 2-acetyltellurophene. The higher frequency and intensity of the v(C=O) band in the formyl compared with the acetyl derivative, as in the corresponding congener derivatives, has been related to inductive and steric effects. It is hypothesized that the doublet in the region of the v(C=O) vibration is due to a conformational equilibrium of (Te,O)-cis and (Te,O)-trans forms, but evidence to exclude Fermi resonance is not given.

5. Microwave Spectrum

Brown and Crofts[1] analyzed the microwave spectra of tellurophene for three isotopic constitutions of the molecule ($C_4H_4{}^{130}Te$, $C_4H_4{}^{128}Te$, $C_4H_4{}^{126}Te$). The line frequencies, computed from the rotational constants obtained from least-squares fitting of low J transitions agree well with those observed. The constancy of the A rotational constant for the three isotopic species indicates that the tellurium atom is on the a axis. This, with the very small inertial defects, is consistent with a planar molecular structure of C_{2v} symmetry. Stark shift measurements correspond to a dipole moment of 0.186 D.

[22] C. G. Andrieu, D. Debruyne, and Y. Mollier, *C.R. Hebd. Seances Acad. Sci., Ser. C* **280**, 977 (1975).

6. NMR Spectra

a. The ¹H-NMR Spectrum of Tellurophene. The ^{1}H resonance spectrum of tellurophene is composed of two chemically shifted multiplets, that at lower field is assigned to the α-hydrogens. The assignment of the chemical shifts has been made by analogy with thiophene and selenophene and on the basis of the expected substituent effects[14,23] and by analysis of the deuterio derivatives.[18] The spectral parameters at infinite dilution are reported in Table VIII, together with the relevant data of the congener systems. The signs of the coupling in tellurophene are all positive. The absolute sign was determined[2] from the spectrum of tellurophene partially oriented in the nematic phase of a liquid crystalline solvent (4-n-butyl-4′-methoxyazoxybenzene). The ^{1}H-NMR spectrum of tellurophene shows also satellite spectra due to both ^{13}C and ^{125}Te.[23]

TABLE VIII

¹H-NMR PARAMETERS FOR TELLUROPHENE AND CONGENER RINGS AT INFINITE DILUTION[a]

Compound	H-2	H-3	$J_{2,3}$	$J_{2,4}$	$J_{2,5}$	$J_{3,4}$
Furan[b]	437.57	374.42	1.75	0.85	1.40	3.30
Thiophene[b]	430.90	419.55	4.90	1.04	2.84	3.50
Selenophene[b]	473.05	433.57	5.40	1.46	2.34	3.74
Tellurophene[c]	530.94	463.56	6.58	1.12	1.82	3.76

[a] In hertz relative to internal TMS at 60 MHz.
[b] See J. M. Read, C. T. Mathis, and J. H. Goldstein, *Spectrochim. Acta* **21**, 85 (1965).
[c] Data from D'Annibale *et al.*[2]; F. Fringuelli and A. Taticchi, unpublished results.

Table VIII shows that the α-hydrogen resonances occur, in all four systems, at lower field than the corresponding β-ones. The β-proton chemical shifts increase regularly with decreasing heteroatom electronegativity, whereas the α-proton chemical shifts are not directly related to electronegativity and vary in an irregular way reflecting the paramagnetic contributions of shielding by the heteroatoms. Such contributions become more important with increasing availability of the d-orbitals and the consequent lowering of the excitation energy term involved in this effect.[24]

The vicinal coupling constants ($J_{2,3}$ and $J_{3,4}$) vary in the congener series systematically with the Pauling electronegativity of the heteroatoms and with the bond angles at C(2) and C(3) (Figure 1).

---

[23] F. Fringuelli, S. Gronowitz, A.-B. Hörnfeldt, I. Johnson, and A. Taticchi, *Acta Chem. Scand., Ser. B*, **28**, 175 (1974).
[24] J. D. Roberts, "Nuclear Magnetic Resonance," pp. 22–25, McGraw-Hill, New York, 1959.

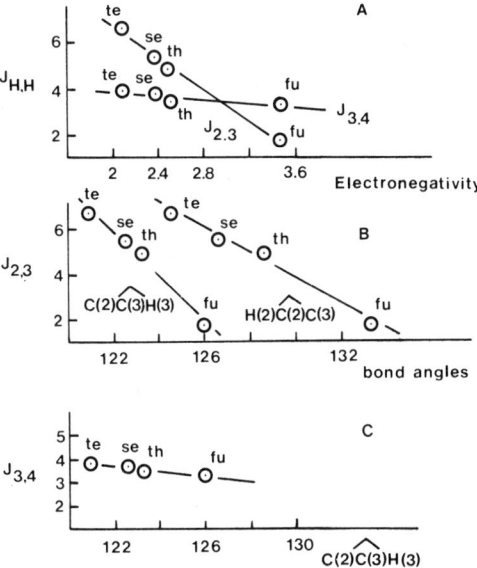

FIG. 1. Plots of vicinal $J_{H,H}$ coupling constants of furan (fu), thiophene (th), selenophene (se), and tellurophene (te) against the electronegativity of the heteroatom (plot A) and bond angles (plots B and C).

The proton chemical shifts of the α-hydrogens of tellurophene are strongly influenced by the polarity of solvent (Table IX), whereas only small differences are observed for the β-protons. Aromatic solvents (C_6D_6, C_6F_6) induce an upfield shift in both protons: this effect has been employed in the determination of the direction of the dipole moment[10] (see Section II,B,2).

b. *The ^{1}H-NMR Spectra of Substituted Tellurophenes.* A systematic investigation of the ^{1}H-NMR spectra of twelve 2-substituted tellurophenes has been carried out under homogeneous and comparable

TABLE IX

Solvent Dependence of H-2 and H-3 Proton Signals of Tellurophene[a]

Proton	$CDCl_3^b$	$CD_3COCD_3^c$	CCl_4^d	$C_6D_6^d$	$C_6F_6^d$	Pure liquid[e]
H-2	8.87	8.97	8.77	8.56	8.72	8.64
H-3	7.78	7.79	7.70	7.58	7.59	7.65

[a] δ-Values relative to TMS as internal standard.
[b] Data from Fringuelli and Taticchi.[14]
[c] Data from Fringuelli *et al.*[23]
[d] Data from Fringuelli *et al.*[10]
[e] F. Fringuelli and A. Taticchi, unpublished results.

conditions[23,25] (Table X). In all four systems, electron-withdrawing substituents deshield the ring protons, and electron-releasing groups produce the opposite effect. The coupling constants for the tellurophene derivatives fall in well-defined intervals in the same relative order as in thiophene and selenophene, but the magnitudes of the couplings are somewhat greater in tellurophene than in the other congeners.[26–28]

TABLE X

¹H-NMR PARAMETERS OF 2-SUBSTITUTED TELLUROPHENES[a]

Substituent	Chemical shifts (δ)			Coupling constants (Hz)		
	H-3	H-4	H-5	$J_{3,4}$	$J_{3,5}$	$J_{4,5}$
CHO	8.62	8.05	9.56	4.10	1.32	6.77
COMe	8.44	8.00	9.41	4.22	1.16	6.78
COOH	8.53	7.93	9.40	4.20	1.34	6.76
COOMe	8.49	7.92	9.38	4.11	1.33	6.79
SMe	7.42	7.55	8.81	4.03	1.28	6.93
CH$_2$OH	7.41	7.64	8.77	3.88	1.25	6.83
Cl	7.33	7.34	8.75	4.26	1.47	7.33
Br	7.72	7.41	8.91	4.27	1.49	7.28
I	8.11	7.32	9.13	4.06	1.54	7.10
Me	7.23	7.47	8.64	3.90	1.26	7.14
CH(OCOMe)Me	7.60	7.67	8.87	4.10	1.82	6.10
CONMe$_2$	7.94	7.87	9.19	4.10	1.95	6.00

[a] In deuterioacetone.[23,25]

The chemical shifts and the coupling constants are sensitive to a change in the heteroatom, and good linear correlations are obtained against the Pauling electronegativities for all twelve substituents.

The proton shifts relative to those of the protons of the parent compound show that the sensitivity of the 5-proton to substituent effects from the 2-position increases in the order furan < thiophene < selenophene < tellurophene.

The ^{125}Te–H coupling constants for the three ring hydrogens of some tellurophene derivatives have been also determined.[23] The values are in the same relative order $J_{Te-H_5} \gg J_{Te-H_4} > J_{Te-H_3}$ as observed previously for the selenophene derivatives.[29]

[25] F. Fringuelli, S. Gronowitz, A.-B. Hörnfeldt, I. Johnson, and A. Taticchi, *Acta Chem. Scand.* In press.
[26] S. Gronowitz, I. Johnson, and A.-B. Hörnfeldt, *Chem. Scripta* **7**, 76 (1975).
[27] S. Gronowitz, I. Johnson, and A.-B. Hörnfeldt, *Chem. Scripta* **7**, 111 (1975).
[28] S. Gronowitz, I. Johnson, and A.-B. Hörnfeldt, *Chem. Scripta* **7**, 211 (1975).
[29] S. Gronowitz, I. Johnson, and A.-B. Hörnfeldt, *Chem. Scripta* **3**, 94 (1973).

c. *The ^{13}C-NMR Spectrum of Tellurophene.* The carbon chemical shifts and coupling constants of tellurophene have been determined in deuterioacetone and compared with those of the congener systems under strictly homogeneous conditions.[23] The assignment of the α- and β-carbons was established from the undecoupled spectrum and was based on the magnitude of the direct carbon–hydrogen couplings. The smaller splitting was associated with the lower field absorption: thus, the carbon chemical shifts of tellurophene are in the reverse order with respect to the proton chemical shifts.

TABLE XI

^{13}C-NMR PARAMETERS OF TELLUROPHENE AND CONGENER RINGS[a]

	Chemical shifts (ppm)		Coupling constants (Hz)	
Compound	C-2	C-3	J_{C2-H2}	J_{C3-H3}
Furan	143.6	110.4	201	175
Thiophene	125.6	127.3	185	168
Selenophene	131.0	129.8	189	166
Tellurophene	127.3	138.0	183	159

[a] In deuterioacetone solution[23] using TMS as internal reference.

The carbon NMR parameters of the four congener systems are reported in Table XI. The β-carbon shifts systematically shift upfield with decreasing electronegativity of the heteroatom while the α-carbon shifts vary irregularly.

Good linear correlations were observed between the carbon chemical shifts and the total charge densities at the α- and β-carbons, calculated by a complete neglect of differential overlap (CNDO/2) treatment.[30]

The direct J_{C3-H3} couplings decrease regularly along the series from furan to tellurophene; in contrast, the J_{C2-H2} values show a peculiar behavior, that for selenophene being greater than that for thiophene. The possible reasons for this discrepancy have been discussed.[23]

d. *The ^{13}C-NMR Spectra of Substituted Tellurophenes.* The spectra of a number of 2-substituted tellurophenes have been determined[23,25] (Table XII). The signals were assigned using the following criteria: (i) the quaternary 2-carbon was identified by its lower intensity and the absence of direct coupling; (ii) the 5-carbon was identified by the largest direct coupling; (iii) the β-carbons were assigned on the basis of the relative order and magnitude of the long-range couplings and of the expected substituent effects.

[30] V. Galasso, private communication.

TABLE XII

^{13}C-NMR PARAMETERS OF 2-SUBSTITUTED TELLUROPHENES[a]

Substituent	Chemical shifts (δ)				Coupling constants (Hz)								
	C-2	C-3	C-4	C-5	J_{C3-H3}	J_{C4-H4}	J_{C5-H5}	J_{C3-H4}	J_{C3-H5}	J_{C4-H5}	J_{C4-H3}	J_{C5-H3}	J_{C5-H4}
CHO	151.5	148.1	139.4	138.7	161	163	184	5.6	11.0	2.5	4.5	11.2	4.9
COMe	153.5	143.4	139.8	137.8	161	163	182	5.3	10.7	2.5	4.7	10.8	4.6
COOH	137.6	144.7	138.6	138.1	159	161	183	5.6	11.0	2.4	4.8	10.8	5.0
COOMe	139.0	144.5	138.6	137.4	164	163	184	5.6	10.7	3.0	5.0	11.2	4.8
SMe	142.1	136.3	137.8	125.6	161	161	185	6.0	10.5	2.7	5.0	11.0	4.0
CH$_2$OH	155.3	132.2	137.4	124.9	158	160	183	5.8	10.8	3.0	4.9	10.6	5.0
Cl	136.4	139.1	136.0	128.7	165	164	185	6.0	10.0	2.2	4.2	9.8	4.8
Br	110.0	142.6	137.4	131.5	166	164	187	5.8	10.5	2.5	4.6	10.9	4.3
I	68.9	149.2	139.4	135.0	—	—	—	—	—	—	—	—	—
Me	144.6	137.5	136.8	124.9	—	—	—	—	—	—	—	—	—
CH(OCOMe)Me	152.7	134.8	137.4	127.1	160	164	185.6	5.7	10.7	2.0	5.0	10.4	4.6
CONMe$_2$	146.6	138.4	138.4	132.6	163	166	185.6	5.7	10.7	2.0	4.5	10.7	—

[a] In deuterioacetone at 25.14 MHz with TMS as internal standard.[23,25]

The long-range coupling constants over three bonds (^3J) are larger than those over two bonds (^2J) for thiophenes, selenophenes, and tellurophenes; substituted furans seem to be anomalous in this.[23] Linear correlations are obtained when the relative shifts of the furans, selenophenes, and tellurophenes are plotted against the relative shifts of the substituted thiophenes. A comparison of the slopes of these straight lines indicates that the ease of transmission of the substituent effects to the 5-carbon increases in the series furan < thiophene < selenophene < tellurophene.

Good correlations were also obtained between the 5-proton and 5-carbon shifts for all the four heteroaromatic systems. This is an indication that probably the same transmission mechanism is operating for both carbon and proton shifts.

7. *Photoelectron Spectra*

The photoelectron spectrum of tellurophene has been recorded and discussed.[31,32] Comparison with the photoelectron spectrum of the tetrahydro derivative allows the assignment[33] of the ionization energies of the uppermost occupied molecular orbitals of tellurophene as follows: 8.40 eV (π_2), 8.88 (π_3), 10.8 (σ_{Te-C2}), 11.5 (σ_{Te-C2}), 11.8 (π_1). This assignment is preferred to one previously proposed by analogy with the thiophene spectrum.[32]

The photoelectron data of the four congener systems are reported in Table XIII. A reversal is observed in the sequence of the two highest occupied molecular orbitals in tellurophene with respect to the other three

TABLE XIII

IONIZATION ENERGIES (eV) OF π-MOs OF TELLUROPHENE AND CONGENER RINGS

Molecular orbital	Furan[a]	Thiophene[a]	Selenophene[b]	Tellurophene[b]
π_3	8.89	8.87	8.92	8.88
π_2	10.32	9.49	9.18	8.40
π_1	14.4	12.1	12.0	11.8

[a] See D. W. Turner, A. Baker, A. D. Baker, and C. R. Brundle, in "Molecular Photoelectron Spectroscopy," p. 329. Wiley (Interscience), New York, 1970; P. J. Derrick, L. Asbrink, O. Edquist, B. O. Johnson, and E. Lindholm, *Int. J. Mass Spectrom. Ion Phys.* **6**, 177 (1971); J. H. D. Eland, *Ibid.*, **2**, 471 (1969).
[b] Data from Fringuelli *et al.*[33]

[31] G. Distefano, S. Pignataro, G. Innorta, F. Fringuelli, G. Marino, and A. Taticchi, *Chem. Phys. Lett.* **22**, 132 (1973).
[32] W. Schäfer, A. Shweig, S. Gronowitz, A. Taticchi, and F. Fringuelli, *Chem. Commun.*, 541 (1973).
[33] F. Fringuelli, G. Marino, A. Taticchi, G. Distefano, F. P. Colonna, and S. Pignataro, *J. Chem. Soc., Perkin Trans.* 2, 276 (1976).

five-membered rings. This effect is also shown in Fig. 2, where the ionization energies of the two upper orbitals are plotted against the Pauling electronegativities of the heteroatoms. The energy of the π_3-orbital (which extends exclusively over the carbon atoms) is approximately constant and independent of the heteroatom electronegativity, but the π_2-orbital energy depends markedly on the heteroatom, increasing as the electronegativity decreases. This leads to the observed "crossing" between selenophene and tellurophene.

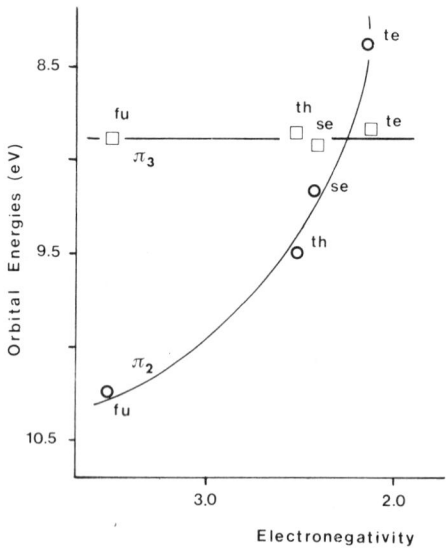

FIG. 2. Plots of the energies of the π_2 and π_3 orbitals of furan (fu), thiophene (th), selenophene (se), and tellurophene (te) vs. the electronegativity of the heteroatoms (Koopmans' theorem is assumed to be valid). From G. Distefano, P. Pignataro, G. Innorta, F. Fringuelli, G. Marino, and A. Taticchi, *Chem. Phys. Lett.* **22**, 132 (1973).

Support for the "reverse sequence hypothesis" is derived from a comparative study on the photoelectron spectra of numerous α-substituted derivatives of furan, thiophene, selenophene, and tellurophene.[33] The vertical ionization energies (I_1 and I_2) of the two highest MOs (π_2 and π_3) are reported in Table XIV. An examination of the correlations obtained between the ionization energies of molecular orbitals of substituted furans, selenophenes, and tellurophenes vs. those of substituted thiophenes makes clear the diverse nature of the upper orbitals in tellurophenes. Slopes close to 1, very high correlation coefficients and low standard deviations are obtained when I_1 and I_2 values for substituted furans and selenophenes are plotted against the corresponding values for substituted thiophenes. On the contrary, slopes differing

TABLE XIV

Vertical Ionization Energies (eV) of π_3 and π_2 MOs of 2-Substituted Five-Membered Heteroaromatic Congener Compounds[a]

Substituent	Furan		Thiophene		Selenophene		Tellurophene	
	$I_1(\pi_3)$	$I_2(\pi_2)$	$I_1(\pi_3)$	$I_2(\pi_2)$	$I_1(\pi_3)$	$I_2(\pi_2)$	$I_1(\pi_2)$	$I_2(\pi_3)$
Me	8.37	10.13	8.43	9.23	8.40	8.96	8.20	8.43
H	8.89	10.32	8.87	9.49	8.92	9.18	8.40	8.88
CONMe$_2$	8.86	10.41	8.84	9.40	8.85	9.10	8.39	8.89
Cl	—	—	8.89	9.63	8.83	9.34	8.68	8.89
Br	—	—	8.82	9.58	—	—	8.59	8.84
I	—	—	8.52	9.47	—	—	8.34	8.52
COOH	9.16	10.72	9.14	9.73	9.19	9.45	8.62	9.15
COOMe	9.00	10.56	8.98	9.61	9.05	9.26	8.51	9.00
NO$_2$	9.75	11.13	9.73	10.21	9.64	9.88	—	—
CH$_2$Cl	—	—	8.89	9.49	—	—	—	—
CHO	—	—	9.37	9.87	—	—	—	—
CN	9.47	10.99	—	—	—	—	—	—
SMe	8.58	10.32	8.63	9.37	—	—	—	—

[a] Data are from Fringuelli et al.[33] and references cited therein.

greatly from unity, with very poor correlation coefficients and high standard errors are observed when the energies of the first and second bands of substituted tellurophenes are plotted against the corresponding energies of thiophene derivatives. However, very good correlations with unit slope are again obtained if the *first* ionization energies of tellurophenes are plotted against the *second* ionization energies of thiophenes, and vice versa.

It has been observed[33] that the differences (Δ) between I(π_2) and I(π_1) parallel the "ground-state aromaticities" of the congener rings, as estimated by several different approaches,[12] in the sense that a higher "aromaticity" corresponds to a smaller energy difference: thiophene ($\Delta = 2.61$ eV) > selenophone ($\Delta = 2.85$ eV) > tellurophene ($\Delta = 3.40$ eV) > furan ($\Delta = 4.08$ eV).

The effect of the ring on the orbitals mainly localized on the substituents has also been briefly discussed.[33]

C. Ground-State Aromaticity

Tellurophene is a typical aromatic molecule: a planar cyclic unsaturated compound with six delocalized π-electrons, it exhibits an enhanced stability over simple olefinic compounds and tends (as the congeners do) to react by substitution rather than addition.

In this section the discussion is confined to a quantitative comparison of some "ground state" properties of tellurophene and its congener systems; the "chemical aromatic properties" are discussed in Section III,C.

The "aromaticity" of furan and thiophene has been compared by nearly all the criteria by which aromaticity may be assessed. All the available information, based either on thermodynamic, structural, or magnetic approaches, indicates that thiophene is more aromatic than furan and that both are less aromatic than benzene.[34] The aromaticity of selenophene has been estimated to be very similar to that of thiophene from studies on chemical, spectroscopic,[21] and magnetic properties.[35]

The quantitative comparison of some aromatic properties of all the four congener rings, under strictly homogeneous conditions, has recently been carried out by us.[12,36] Owing to experimental difficulties, it was not possible to determine the aromatic resonance energy directly from measurements of heats of combustion or hydrogenation.

TABLE XV

AROMATICITY INDICES FOR TELLUROPHENE AND CONGENER RINGS[a]

Compound	A^b	B^c	$\Sigma\Delta N^d$	J^e	μ_m^f
Furan	7.67	1.72	1.42	0.87	1.03
Thiophene	11.56	3.85	0.90	0.93	1.35
Selenophene	10.44	2.94	1.02	0.91	1.29
Tellurophene	8.50	1.85	1.30	0.88	1.17

[a] Data from Fringuelli et al.[12]
[b] The dilution shift parameter.
[c] From the NMR spectra of 2-methyl derivatives.
[d] Sum of the differences in bond orders of the three nonequivalent bonds.
[e] The Julg parameter.
[f] The mesomeric dipole moment (in D).

All the aromaticity indices that have been calculated are summarized in Table XV. The first two indices are derived from the NMR spectra: the A parameter is based on the "dilution shift method";[37,38] the B parameter is based on the uniformity of the methyl effects on aromatic proton chemical shifts.[39] The next two indices, $\Sigma\Delta N$ and Julg's (J)

[34] M. J. Cook, A. R. Katritzky, and P. Linda, *Adv. Heterocycl. Chem.* **17**, 255 (1974).
[35] W. Czieslik, D. Sutter, H. Dreizler, C. L. Norris, S. L. Rock, and W. H. Flyglare, *Z. Naturforsch. A* **27**, 1961 (1972).
[36] F. Fringuelli, G. Marino, and A. Taticchi, *Gazz. Chim. Ital.* **103**, 1041 (1973).
[37] C. R. Kanekar, G. Govil, C. L. Khetrapal, and M. M. Dhingra, *Proc. Indian Acad. Sci.* **64**, 315 (1966).
[38] M. Pasdeloup and J. P. Laurent, *J. Chim. Phys.* **69**, 1022 (1972).
[39] G. S. Reddy and J. A. Goldstein, *J. Am. Chem. Soc.* **83**, 5020 (1961).

parameter,[40] are based on structural criteria. As a consequence of the π-electron delocalization, the C—C bond lengths of an aromatic ring tend to be intermediate between those of pure single and pure double bonds. The more aromatic is the ring, the more similar are the bond orders of the various nonequivalent bonds. ΣΔN and J parameters are both obtained from an elaboration of the bond lengths. The last index is the mesomeric dipole moment (μ_m), which is also related to the π-electron delocalization and therefore has been proposed as a criterion of aromaticity.[41]

Although the criteria used have limited validity from a theoretical point of view and although some experimental data are not very accurate, nevertheless the agreement among the results is excellent. All the criteria except one agree in establishing the following order of decreasing ground-state aromaticity: benzene > thiophene > selenophene > tellurophene > furan.

The only exception is represented by the diamagnetic molar susceptibility exaltation (Λ), defined as the difference between the experimental molar susceptibility and the susceptibility estimated for the identical but not electron-delocalized structure. This method leads to a completely different and not credible order of aromaticity (selenophene > benzene). The procedure probably needs to be refined in order to give satisfactory results with heterocyclic compounds; moreover, the Pascal constants for selenium and tellurium are likely not to be sufficiently accurate.

The agreement among the other six empirical aromaticity parameters is not only qualitative, but even quantitative, as evidenced by the linearity of the plots of Fig. 3, in which the indices B, ΣΔN, J, and μ_m are diagrammed against the "dilution shift parameter," A.

The A values for benzene, thiophene, and furan give also an excellent linear correlation with the Pauling resonance energies. From this plot, empirical resonance energies for selenophene (29 kcal/mol) and tellurophene (25 kcal/mol) were estimated by interpolation.

The order of aromaticity observed (thiophene > selenophene > tellurophene > furan) may be rationalized by taking into account two properties of the heteroatoms: the electronegativity and the covalent radius.

The more electronegative the heteroatom is, the more "compact" is the p-orbital and the less efficient is the overlap with the adjacent p-orbitals of the carbon atoms. In terms of valence bond description, the structures in which a positive charge is localized on the heteroatom are

[40] A. Julg and P. Francois, *Theor. Chim. Acta* **8**, 249 (1967).
[41] E. D. Bergman and I. Agranat, in "Aromaticity, Pseudoaromaticity, Antiaromaticity" (E. D. Bergman and B. Pullman, eds.), Isr. Acad. Sci. Humanities, p. 15. Academic Press, New York, 1971.

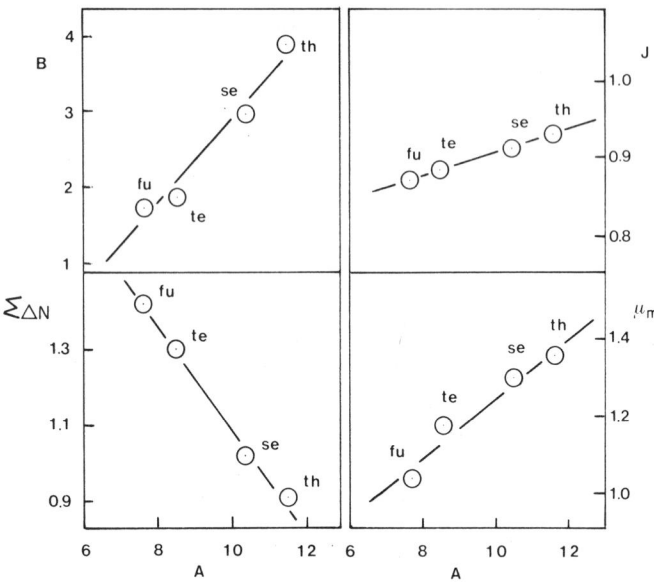

FIG. 3. Plots of A parameters vs. B, $\Sigma\Delta N$, J, and μ_m parameters for furan (fu), thiophene (th), selenophene (se), and tellurophene (te). From F. Fringuelli, G. Marino, A. Taticchi, and G. Grandolini, *J. Chem Soc., Perkin Trans. 2*, 332 (1974).

less important when its electronegativity is greater: the conjugation ability should hence decrease in the order Te > Se > S > O.

However, the overlap between the p-orbitals of the heteroatom and ring carbons depends also upon the C—X bond length and the disparity between the size of the orbitals. From this point of view, the overlap should be maximum in the case of oxygen, and the ability to conjugate should be in the order O > S > Se > Te. The observed order of ground-state aromaticity (with a maximum for the thiophene ring) is the result of the interplay of these opposing factors.

D. Conformational Aspects of Tellurophene Derivatives

The (X,O)-cis (**5**) and (X,O)-trans (**6**) conformational equilibrium of 2-formyl, 2-acetyl, and 2-*N*,*N*-dimethylcarboxamidotellurophene has

R = H; Me; NMe$_2$; OMe; OH

(X,O)-*cis*
(**5**)

(X,O)-*trans*
(**6**)

been investigated by dipole moment[13] and computer simulation of lanthanide-induced shift techniques.[42] The investigation has also been extended to the corresponding derivatives of furan, thiophene, and selenophene, so that internally consistent and directly comparable data are available.

The two methods agree in the conformational preference (Table XVI) in all the cases examined except three (2-formylfuran, 2-acetyltellurophene, 2-N,N-dimethylcarboxamidothiophene). These discrepancies have been ascribed to inexact theoretical assumptions and to specific solvent effects.

TABLE XVI

CONFORMATIONAL PREFERENCE OF 2-SUBSTITUTED FURANS, THIOPHENES, SELENOPHENES, AND TELLUROPHENES: PERCENT OF cis FORM

Substituent	Furan	Thiophene	Selenophene	Tellurophene	Method[a]
CHO	83	100	80	72	DM
	44	99	98	96	LIS
COMe	51	92	70	47	DM
	53	79	87	90	LIS
CONMe$_2$	11	56	33	11	DM
	5	2	5	—	LIS
SMe	48	51	59	65	DM
COOMe	55	33	59	42	DM
COOH	100	0	0	0	XR

[a] DM = dipole moment (benzene solution) (Lumbroso et al.[13]).
LIS = lanthanide-induced shifts (CDCl$_3$) (Caccamese et al.[42]).
XR = X-ray (Hudson[5]; Fanfani et al.[4]).

The trans-conformational preference observed in amides has been attributed to steric factors.

Infrared investigation[22] in the $\nu(C=O)$ region of 2-formyl- and 2-acetyltellurophenes seems to confirm the results obtained using the dipole moment method, but the results are dubious, since the possibility of Fermi resonance was not excluded.

The dipole moment method has also been used to investigate the conformational equilibria of the 2-methylthio and 2-methoxycarbonyl derivatives. These molecules do not show a conformational preference (Table XVI).

The dipole moments of 2-acetyl-5-methoxycarbonyl and 2,5-bismethoxycarbonyltellurophene were calculated assuming that these compounds exist in four planar conformers [(Te,O)-cis–cis, cis–trans, trans–cis, and trans–trans], the energy difference of which depends only

[42] S. Caccamese, G. Montaudo, A. Recca, F. Fringuelli, and A. Taticchi, *Tetrahedron* **30**, 4129 (1974).

on the factors governing the equilibrium of the monosubstituted derivatives. The agreement between calculated and observed values was excellent.[13]

A (Te,O)-cis chelated form with O—H···Te bridge has been hypothesized for 2-hydroxymethyltellurophene.[13] X-Ray investigation[4] shows that crystalline 2-tellurophenecarboxylic acid exists in a (Te,O)-trans conformation, as 2-thiophene- and 2-selenophenecarboxylic acids;[6] in contrast, 2-furoic acid is in the reversed (O,O)-cis conformation.[5]

III. Synthesis and Chemical Reactivity of Tellurophene and Its Derivatives

A. SYNTHESIS OF THE TELLUROPHENE RING

This section is confined to the synthesis of tellurophene and its derivatives through direct construction of the tellurophene ring; the numerous interconversions yielding derivatives from the preformed tellurophene nucleus are considered in the subsequent section.

There are two different basic methods for obtaining the five-membered heterocycles. In the first approach, the ring is constructed directly from aliphatic compounds; the other and less common approach makes use of congener rings or other heterocyclic systems as starting materials. All the syntheses reported for the tellurophene system have been based on the first principle.

After some previous unfruitful attempts,[43,44] the first synthesis of the tellurophene nucleus was finally accomplished in 1961 by Braye *et al.*,[45,46] who prepared tetraphenyltellurophene by two alternative procedures, using 1,4-dilithiotetraphenylbutadiene and tellurium tetrachloride (yield 56%) and from 1,4-diiodotetraphenylbutadiene and lithium telluride (82%).

A variation of the above scheme gave tetrachlorotellurophene by heating for 40 hours at 250° a mixture of hexachloro-1,3-butadiene and metallic tellurium.[47]

A method of broad scope for the preparation of 2,5-disubstituted tellurophene derivatives has been described by Mack,[48,49] based on the

[43] A. McMahon, T. G. Pearson, and P. L. Robinson, *J. Chem. Soc.*, 1644 (1933).
[44] W. G. Zoellner, *Diss. Abstr.* **19**, 3139 (1959); *C.A.* **53**, 17141i (1959).
[45] E. H. Braye, W. Hübel, and I. Caplier, *J. Am. Chem. Soc.* **83**, 4406 (1961).
[46] U.S. Patent 3,149,101; [*C.A.* **63**, 1819 (1965)]; U.S. Patent 3,151,140; [*C.A.* **61**, 16097 (1964)].
[47] W. Mack, *Angew. Chem.* **77**, 260 (1965); *Angew. Chem., Int. Ed. Engl.* **4**, 245 (1965).
[48] W. Mack, *Angew. Chem.* **78**, 940 (1966); *Angew. Chem., Int. Ed. Engl.* **5**, 896 (1966).
[49] British Patent 1,107,698; [*C.A.* **69**, 77110 (1968)].

reaction [Eq. (1)] of sodium telluride and diacetylenes in methanol at room temperature.

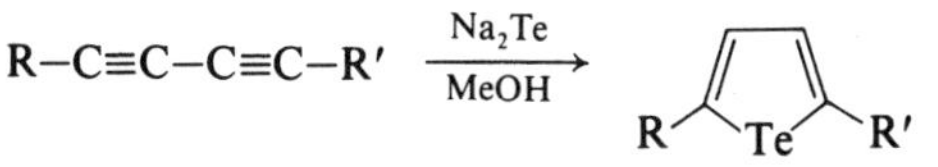

$$R-C\equiv C-C\equiv C-R' \xrightarrow[\text{MeOH}]{\text{Na}_2\text{Te}} \underset{R\quad Te\quad R'}{\text{[tellurophene]}} \quad (1)$$

R, R' = H, H; C(OH)Me$_2$, C(OH)Me$_2$; CH$_2$OH, CH$_2$OH; Ph, Ph; C$_4$H$_7$, C$_4$H$_7$; pyrrolidinylmethyl, pyrrolidinylmethyl; H, C(OH)Me$_2$

Unsubstituted tellurophene (1) was first prepared using this method; however, the report[48] did not include sufficient experimental details. A full description of this synthesis was reported later by Fringuelli and Taticchi.[14,50]

When monodeuteriomethanol is used instead of methanol as solvent, 2,3,4,5-tetradeuteriotellurophene is obtained.[18]

In the same year, Barton and Roth[51] prepared tellurophene using as starting material the 1,4-trimethylsilyl derivative of butadiyne, but the yields were smaller.

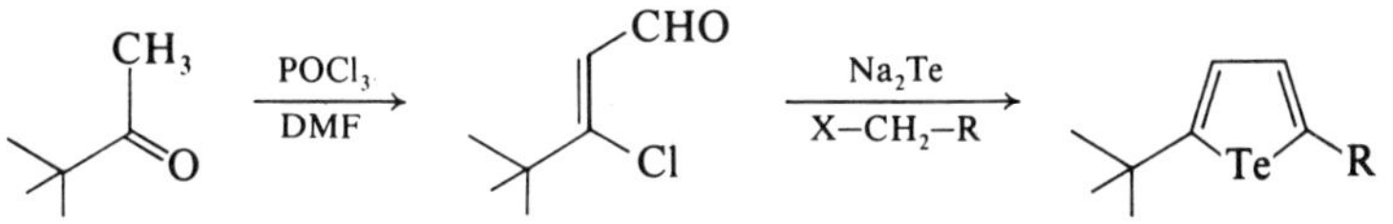

$$\quad (2)$$

R = CO$_2$Et; NO$_2$; CHO; COMe

Cagniant et al.[52] have extended to tellurophenes a method of synthesis, recently developed for thiophenes and selenophenes,[53,54] based on the condensation of a β-halovinyl aldehyde with sodium telluride and a halogeno derivative of general formula X—CH$_2$—R [Eq. (2)]. Some mono and polycyclic derivatives of tellurophene have been prepared using the above procedure.

Quinones containing the tellurophene ring were obtained by Müller et al.[55] by the reaction of an appropriate rhodium complex with amorphous tellurium (e.g., Section VI,D).

[50] F. Fringuelli and A. Taticchi, *Ann. Chim. (Rome)* **62**, 777 (1972).
[51] T. J. Barton and R. W. Roth, *J. Organomet. Chem.* C66, **39** (1972).
[52] P. Cagniant, R. Close, G. Kirsch, and D. Cagniant, *C.R. Hebd. Seances Acad. Sci., Ser. C* **281**, 187 (1975).
[53] P. Cagniant, P. Perin, G. Kirsch, and D. Cagniant, *C.R. Hebd. Seances Acad. Sci., Ser. C* **277**, 37 (1973).
[54] P. Cagniant and G. Kirsch, *C.R. Hebd. Seances Acad. Sci., Ser. C* **281**, 111 (1975).
[55] E. Müller, E. Luppold, and W. Winter, *Synthesis*, 265 (1975).

B. Chemical Properties of Tellurophene. Preparation of Tellurophene Derivatives

Tellurophene is fairly stable to air and light at room temperature; in the dark, and at 0° it can be stored for long periods.[50]

Tellurophene behaves as a typical five-membered heteroaromatic ring: it readily undergoes electrophilic substitution and is more reactive at the alpha than at the beta position. Numerous 2-monosubstituted and 2,5-disubstituted derivatives have been obtained by direct substitution or by transformation of functional groups using conventional procedures. Since tellurophene is decomposed by strong mineral acids, good yields require alkaline, neutral, or only moderately acid conditions. 2-Acetyl- (**7**), 2-trifluoroacetyl- (**8**), and 2-formyl (**9**) tellurophene were obtained (Scheme 1) by electrophilic substitution using the appropriate reagent.

A key intermediate for the synthesis of several derivatives is 2-lithiotellurophene, obtained by metallation of tellurophene with 2-butyllithium in ethereal solution. Treatment of 2-lithiotellurophene with dimethyl sulfate, N-methylformanilide, acetaldehyde, carbon dioxide, hexachloroethane, hexabromoethane, and dimethyl sulfide gives, respectively, 2-methyltellurophene (**10**), tellurophene-2-carboxaldehyde (**9**), 1(2-tellurienyl)ethanol (**11**), 2-tellurophenecarboxylic acid (**12**), 2-chloro- (**13**), 2-bromo- (**14**), and 2-methylthiotellurophene (**15**).

An attempt to prepare 2-fluorotellurophene by treating 2-lithiotellurophene with perchloryl fluoride was unsuccessful;[25] the procedure gives good results when applied to thiophene and selenophene.[29]

The reaction of 2-lithio derivative with trans-chlorovinyl iodosodichloride and sodium nitrite in dimethylformamide gives with furan, thiophene, and selenophene the corresponding 2-iodo and 2-nitro derivatives. The same reaction, when applied to tellurophene, gives 2-iodotellurophene (**16**) and the unforeseen compound di-2-tellurienyl telluride (**17**); the expected 2-nitrotellurophene was not obtained.[25] Other attempts to prepare 2-nitrotellurophene were unsuccessful.

Some of the above reactions, when used consecutively, yield 2,5-disubstituted products.

The main interconversions of tellurophene derivatives are summarized in Scheme 1.

2-d_1, 2,5-d_2, 3,4-d_2, and 2,3,4-d_3 tellurophenes were recently prepared[18] by hydrogen–deuterium exchange via lithium intermediate starting from tellurophene or its tetradeuterio derivative.

Owing to the very high $\alpha:\beta$ reactivity ratio, no 3-substituted derivatives of tellurophene (except two deuterated compounds) have been prepared so far.

Tellurophene and its derivatives react with halogens giving 1:1 addition products at the tellurium atom.[45,47–49]

a: $(CF_3CO)_2O$–75°; b: nBuLi–Me_2S; c: nBuLi–CO_2; d: nBuLi–HCONHPh; e: nBuLi–ClCH:CH.ICl_2–$NaNO_2$; f: nBuLi–Me_2SO_4; g: nBuLi–MeCHO; h: Ac_2O–$SnCl_4$; i: nBuLi–C_2Cl_6/C_2Br_6; k: HMPA; l: CH_2N_2; m: $LiAlH_4$; n: N_2H_4–KOH; o: I_2–Py–KOH; p: Ac_2O–Py.

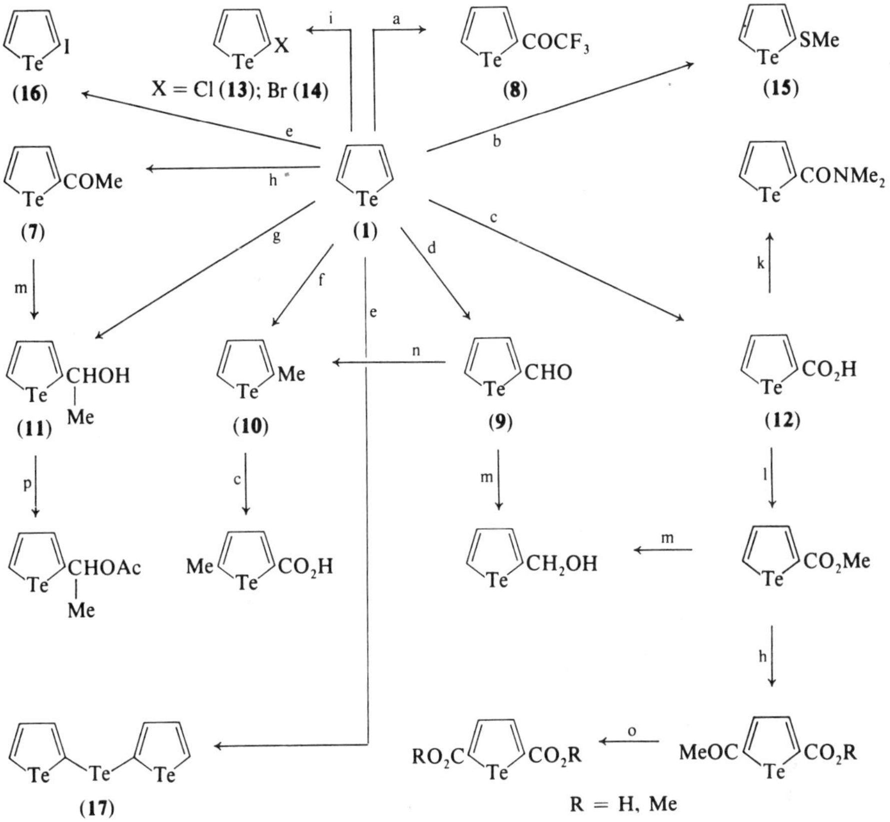

SCHEME 1

C. QUANTITATIVE STUDIES OF REACTIVITY

To date, the quantitative studies on tellurophene nucleus reactivity have come exclusively from our laboratory and are devoted to electrophilic substitution, to some side-chain solvolysis reactions, and to carboxylic acid ionization.

1. Electrophilic Substitutions

The reactivity of tellurophene toward electrophilic reagents has been quantitatively compared with that of the congener systems in three different substitution reactions: formylation by phosgene and dimethylformamide, acetylation by acetic anhydride and by tin tetrachloride, and trifluoroacetylation by trifluoroacetic anhydride.[56,57]

TABLE XVII

RELATIVE REACTIVITIES (k/k_{Th}) OF TELLUROPHENE AND CONGENER RINGS IN ELECTROPHILIC SUBSTITUTION[a]

Substrate	Acetylation (25°)	Trifluoroacetylation (75°)	Formylation (30°)
Furan	11.9	140	107
Thiophene	1	1	1
Selenophene	2.28	7.33	3.64
Tellurophene	7.55	46.4	36.8

[a] Reactivities relative to thiophene.[57]

In the first reaction, the relative rates were obtained by a kinetic approach and are the ratios of the second-order rate constants; in the last two reactions, the relative rates were determined using a competitive procedure. The relevant data are summarized in Table XVII. In all three reactions examined, the reactivity sequence is furan > tellurophene > selenophene > thiophene. The reactivity data are consonant, as evidenced by the linear plots obtained when log (k/k_{Th}) values for one reaction are plotted against the log (k/k_{Th}) values for another (Fig. 4).

As concerns the factors leading to the observed reactivity scale, it is our opinion that the relative ground-state energies play a more important role than the relative stabilities of the intermediate carbocations. From a qualitative point of view, it is observed that the more aromatic is the starting molecule, the smaller is the rate of substitution. The order of reactivity is in fact the reverse of the order of the ground-state aromaticities, as determined by several different approaches: thiophene > selenophene > tellurophene > furan (see Section II,C). A quantitative confirmation of this hypothesis has been obtained by analysis of the activation parameters for the formylation reaction.[57]

The activation entropies are constant within experimental error for all the members of the series, and the relative rates appear to be controlled by the activation enthalpies. A good linear correlation is obtained when the $\Delta H^{\ddagger}$ values are plotted against the resonance energies of the four rings (Fig. 5). The linearity of this plot confirms that the differences in ground-state energy play a fundamental role in determining the relative reactivities of formylation at the α-positions.

[56] F. Fringuelli, G. Marino, G. Savelli, and A. Taticchi, *Chem. Commun.*, 1441 (1971).
[57] S. Clementi, F. Fringuelli, P. Linda, G. Marino, G. Savelli, and A. Taticchi, *J. Chem. Soc., Perkin Trans. 2*, 2097 (1973).

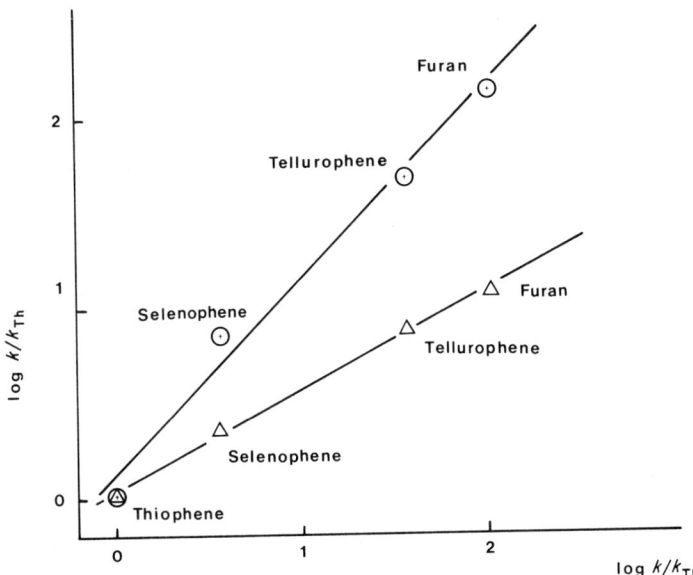

FIG. 4. Relative reactivities (log $k/k_{\text{thiophene}}$) for acetylation (Δ) and trifluoroacetylation (O), against the relative reactivities for formylation. From S. Clementi, F. Fringuelli, P. Linda, G. Marino, G. Savelli, and A. Taticchi, *J. Chem. Soc., Perkin Trans. 2*, 2097 (1973).

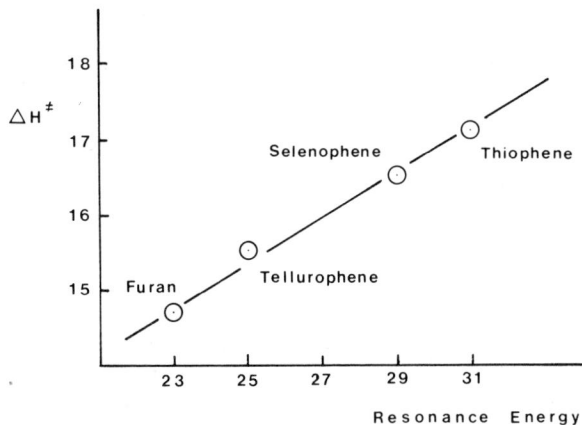

FIG. 5. Plot of activation enthalpies for formylation of furan, thiophene, selenophene, and tellurophene against their resonance energies. From S. Clementi, F. Fringuelli, P. Linda, G. Marino, G. Savelli, and A. Taticchi, *J. Chem. Soc., Perkin Trans. 2*, 2097 (1973).

The $\Delta S^{\ddagger}$ values bring a contribution to another vexed question, that of the position of the transition state along the reaction coordinate for the same substitution of various aromatic systems. The strict similarity of the entropy values may represent an indication that the disposition of the atoms around the reaction center is similar, or, in other words, that for all four compounds the transition states of the reaction lie in a similar position along the reaction coordinate. The high negative $\Delta S^{\ddagger}$ values (ca. -28 cal mol^{-1} deg^{-1}) are consistent with a high degree of formation of a new bond, i.e., the transition states resemble the σ complexes.

As concerns the sensitivity to substituent effects, the available data are limited, because of synthetic difficulties, to the effect of the methyl group in the α-position on the reactivity of position 5 in the formylation and trifluoroacetylation reaction[58] (Table XVIII).

TABLE XVIII

k_{Me}/k_H REACTIVITY RATIOS FOR ELECTROPHILIC SUBSTITUTION OF 2-SUBSTITUTED FIVE-MEMBERED HETEROAROMATIC CONGENERS[a]

Substrate	Formylation (20°)	Trifluoroacetylation (75°)
Furan	880	1700[b]
Thiophene	290	380[b]
Selenophene	300	280
Tellurophene	620	500

[a] Data from Clementi et al.[58]
[b] See S. Clementi and G. Marino, *Tetrahedron* **25**, 4599 (1969).

The order of sensitivity to substituent effects, as derived from the k_{Me}/k_H ratios, is the following: furan > tellurophene > selenophene ≃ thiophene.

2. Side-Chain Reactions

Quantitative data are available for the reaction of solvolysis of 1(2-tellurienyl)ethyl acetates in 30% ethanol.[58,59] The first-order rate constants for the reaction at 60° of all the four congener systems, the activation parameters, the rates relative to the thiophene derivative (k/k_{Th}), and the rates of the 5-methyl-substituted compounds relative to the corresponding unsubstituted derivatives (k_{Me}/k_H) are summarized in Table XIX.

[58] S. Clementi, F. Fringuelli, P. Linda, G. Marino, G. Savelli, A. Taticchi, and J. L. Piette, unpublished results.
[59] F. Fringuelli, G. Marino, and A. Taticchi, *Gazz. Chim. Ital.* **102**, 534 (1972).

TABLE XIX
Rate Constants, Activation Parameters, and Relative Rates for the Solvolysis of 1(2-Aryl)Ethyl Acetates in 30% Ethanol

Aryl group	$10^4 k$ (at 60°) (sec^{-1})	$\Delta H\ddagger$ (kcal mole^{-1})	$\Delta S\ddagger$ (cal mol^{-1} deg^{-1})	k_X/k_{Th}^a	k_{Me}/k_H^b
Furyl^c	19.79	18.39	−15.9	3.03	160
Thienyl^c	6.53	19.83	−13.7	1	70
Selenienyl^d	10.92	18.52	−16.7	1.67	23.3
Tellurienyl^d	36.79	17.34	−17.8	5.63	11.8

^{a} Rate relative to thiophene derivative at 60°.
^{b} Rate of the 5-methyl-substituted compound relative to the corresponding unsubstituted derivative at 25°.
^{c} See E. A. Hill, M. L. Gross, M. Stasiewicz, and M. Manion, *J. Am. Chem. Soc.* **91**, 7381 (1969).
^{d} Data from Clementi *et al.*[58]

Side-chain reactions in which a delocalizable positive charge is developed in the transition state are frequently considered together with electrophilic aromatic substitutions;[60] the analogy between these two classes of reactions has been confirmed also in heteroaromatic chemistry.[61] However, a comparison of the solvolysis data with those referring to electrophilic substitutions (Tables XVII–XIX) indicates that tellurophene behaves in a different way in the two types of reactions. As concerns the reactivity, tellurophene is less reactive than furan in electrophilic substitution but more reactive than furan in the solvolysis of 1(2-aryl)ethyl acetates; as concerns the k_{Me}/k_H reactivity ratios, tellurophene appears to be more sensitive to substituent effects than selenophene in electrophilic substitutions and less sensitive in side-chain reactions.

The reasons for these discrepancies have been discussed mainly in terms of different charge distributions in the transition states of these two classes of reactions.[58]

The heteroatom in a heterocyclic ring can be considered as a special kind of "endocyclic" substituent replacing a CH=CH moiety in the benzene ring. Therefore it is possible to attribute sigma constants to heteroatoms as a quantitative measure of their electronic effect.

The σ^+ value for α-tellurium (−0.92), as evaluated from electrophilic substitution and side-chain solvolyses[59] reactions in which a carbonium ion is formed in the transition state, is intermediate between the σ^+ values of α-selenium (−0.88) and α-oxygen (−0.93); the σ^+ value of sulfur (−0.79) is the lowest in the series.

[60] L. M. Stock and H. C. Brown, *Adv. Phys. Org. Chem.* **1**, 35 (1963).
[61] G. Marino, *Adv. Heterocycl. Chem.* **13**, 235 (1971).

3. Ionization of Carboxylic acids

Tellurophene-2-carboxylic acid is the weakest in the congener series (Table XX). Nevertheless it is still stronger than benzoic acid, in spite of the fact that tellurium atom is less electronegative than carbon and is electron-releasing by resonance. However, it must be recalled that the positive end of the dipole moment is on the hydrocarbon part of the ring (Section II,B,2). A linear dependence between the pK_a values and C=O stretching frequencies of monomer and dimer form has been observed.[62]

TABLE XX

pK_a VALUES OF CONGENER 2-CARBOXYLIC ACIDS[a]

Compound	pK_a	
	H_2O, 25°	H_2O–EtOH, 50%, 25°
2-Furoic acid	3.16[b]	4.54
2-Thiophenecarboxylic acid	3.53[c]	5.05
2-Selenophenecarboxylic acid	3.60[d]	5.14
2-Tellurophenecarboxylic acid	3.97	5.48

[a] Data from Fringuelli et al.[62,63]
[b] See P. O. Lumme, Suomen. Kemistil. **B33**, 87 (1960).
[c] See A. R. Butler, J. Chem. Soc. B, 867 (1970).
[d] See D. Spinelli, G. Guanti, and C. Dell'Erba, Ric. Sci. **38**, 1048 (1968).

The ionization constants of the tellurophene-2-carboxylic acid and a number of 5-substituted derivatives have been determined in water at 25° by using a potentiometric procedure.[63] The thermodynamic pK_a values are summarized in Table XXI.

TABLE XXI

IONIZATION CONSTANTS OF 5-SUBSTITUTED TELLUROPHENE-2-CARBOXYLIC ACIDS IN WATER AT 25°[a]

Substituent	H	Me	$COCH_3$	COOH	CO_2^-
pK_a^T	3.97	4.16	3.36	3.11	4.24

[a] Data from Fringuelli et al.[63]

The application of the Hammett equation gives an excellent linear correlation if σ_p constants are used. The ρ values and the statistical parameters for the ionization of the 2-carboxylic acids of all four congener rings are summarized in Table XXII. The furan ring appears to be

[62] F. Fringuelli and A. Taticchi, J. Heterocycl. Chem. **10**, 89 (1973).
[63] F. Fringuelli, G. Marino, and A. Taticchi, J. Chem. Soc. Perkin Trans. 2, 1738 (1972).

TABLE XXII
Transmission of Substituent Effects in the Ionization of 2-Carboxylic Acids of Congener Series

Carboxylic acid	ρ^a	r^b	s^c	n^d	References
Furan	1.41	0.988	0.03	6	e
Thiophene	1.23	0.996	0.02	7	f
Selenophene	1.23	0.998	0.02	7	g
Tellurophene	1.20	0.999	0.01	5	63

[a] Hammett reaction constant.
[b] Correlation coefficient.
[c] Standard deviation.
[d] Number of substituent.
[e] W. E. Catlin, *Iowa State Coll. J. Sci.* **10**, 65 (1935).
[f] Footnote c of Table XX.
[g] Footnote d of Table XX.

the most sensitive to substituent effects, whereas the other three rings exhibit very similar behavior.

The Hammett σ_α constants of heteroatoms considered as "endocyclic" substituents, calculated from the pK_a values, are all positive and decrease following the electronegativities of the heteroatoms.[63]

D. Molecular Complexes

Tellurophene and its derivatives give molecular complexes of a different kind with a variety of substances such as mercuric chloride, picric acid, 2,4,7-trinitrofluorenone, 1,3,5-trinitrobenzene, transition metal carbonyls and chlorides, and tetracyanoethylene. Some of these complexes have been used in the characterization of tellurophene derivatives and will be referred to in the Appendix.

Several coordination complexes of tellurophene with transition metals are known.[64] A complex having the formula $C_7H_4CrO_3Te$ and probably the structure **18** was obtained by heating tellurophene with $(CH_3CN)_3Cr(CO)_3$.

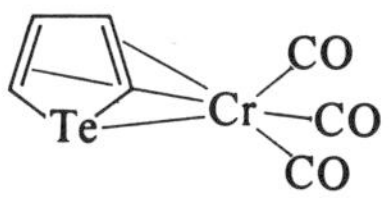

(18)

Reaction of tellurophene with Na_2PdCl_4 in methanol at room temperature gives a mixture of two complexes having the formulas

[64] K. Öfele and E. Dotzaucr, *J. Organometal. Chem.* **42**, C87 (1972).

($C_4H_4Te)_2PdCl_2$ and $(C_4H_4TePdCl_2)_2$, for which the structures **19** and **20** were proposed. The analogous reaction with tetrachlorotellurophene afforded $(C_4Cl_4Te)_2PdCl_2$ (**21**).

$$\begin{array}{ccc} L_1\diagdown\!\!\!\diagup Cl & L_1\diagdown\!\!\!\diagup Cl\diagdown\!\!\!\diagup Cl & L_2\diagdown\!\!\!\diagup Cl \\ Pd & PdPd & Pd \\ L_1\diagup\!\!\!\diagdown Cl & Cl\diagup\!\!\!\diagdown Cl\diagup\!\!\!\diagdown L_1 & Cl\diagup\!\!\!\diagdown L_2 \\ (19) & (20) & (21) \end{array}$$

L_1 = Tellurophene $\qquad L_2$ = Tetrachlorotellurophene

Reaction of tellurophene with $Fe_3(CO)_{12}$ in benzene yielded a mixture of compounds not containing the tellurophene nucleus.[64] Also the reaction of tetraphenyltellurophene with metal carbonyls is very difficult. In only one experiment with $Fe_3(CO)_{12}$ was a small quantity of a complex obtained.[45] On the basis of its IR spectrum, the structure $(C_{28}H_{20}Te)Fe(CO)_3$ was suggested.

Tellurophene, like the congener heterocycles, forms charge-transfer complexes with tetracyanoethylene.[65]

Two charge-transfer bands were observed for furan, thiophene, and tellurophene and only a single band for selenophene, because of the small difference in energy between the two upper molecular orbitals of this ring (see Section II,B,7). The spectral data, the stability-constant value, and the empirical calculation of ionization energies are in favor of a $\pi \to \pi^*$ nature of these complexes and indicate that the inner orbitals of the donors are involved in the charge-transfer interaction.

IV. Other Tellurium Heterocycles

A. TETRAHYDROTELLUROPHENE

1. Synthesis and Chemical Properties

The currently accepted name for compound **23** is tetrahydrotellurophene, although cyclotellurobutane has also been used.

Tetrahydrotellurophene has been prepared[66,67] by the sequence of reactions shown in Eq. (3).

$$\underset{X = Br;\ I}{\underset{(22)}{\begin{array}{c}\diagup\!\!\!\diagdown\\XX\end{array}}} \longrightarrow \underset{(22)}{\begin{array}{c}\diagup\!\!\!\diagdown\\ Te\\XX\end{array}} \longrightarrow \underset{(23)}{\begin{array}{c}\diagup\!\!\!\diagdown\\ Te\end{array}} \qquad (3)$$

[65] G. G. Aloisi, S. Santini, and G. Savelli, *J. Chem. Soc., Faraday Trans. 1*, **71**, 2045 (1975).
[66] G. T. Morgan and F. H. Burstall, *J. Chem. Soc.*, 180 (1931).
[67] W. V. Farrar and J. M. Gulland, *J. Chem. Soc.*, 11 (1945).

The first step was achieved by three alternative procedures: (i) the action of aluminum telluride on 1,4-dihalobutane at temperatures ranging from 125° to 175°; this method is very tedious and the yield is poor; (ii) the reaction between amorphous tellurium and 1,4-diiodobutane at 130°–140°, which gives a good yield; (iii) the reaction between sodium telluride made *in situ* and 1,4-dibromobutane; this is the most reliable method. The subsequent reduction of 1,1-dihalotetrahydrotellurophene (22) has been carried out using sulfur dioxide. By a similar procedure 3,3'-bistetrahydrotellurophene (24) has been prepared.[68,69]

(24)   (25)   (26)

Tetrahydrotellurophene is oxidized rapidly in air to tetrahydrotellurophene 1-oxide (25); it reacts with halogens and methyl iodide to give 1,1-dihalo (22) and 1-methyl-1-iodotetrahydrotellurophene (26), respectively, and gives molecular complexes with mercuric chloride: it reacts explosively with concentrated nitric acid and develops a red coloration with sulfuric acid.[66]

2. Physical Properties

Tetrahydrotellurophene is a pale yellow oil of penetrating and repulsive odor. The dipole moment (1.63 D in benzene at 25°) is lower[8] than those of the congener heterocycles tetrahydrofuran (1.75 D), tetrahydrothiophene (1.89 D), and tetrahydroselenophene (1.81 D). The positive pole of the dipole moment is on the ring and the negative pole is on the heteroatom, as in the other congener systems.[10,70] The values of the dipole moments of the tetrahydro derivatives and their directions indicate that there is no simple relationship with the electronegativity of the heteroatoms.[9]

Djerassi and coworkers[71] have analyzed the modes of fragmentation of all the congener tetrahydro derivatives upon electron impact. The ability of the different heteroatoms to stabilize the molecular ions increases in the order $O < S < Se < Te$.

[68] E. Buchta and K. Greiner, *Naturwissenschaften* **46**, 532 (1959) [*C.A.* **54**, 6688 (1960)].
[69] E. Buchta and K. Greiner, *Chem. Ber.* **94**, 1311 (1961).
[70] T. J. Barton, R. W. Roth, and J. G. Verkade, *J. Am. Chem. Soc.* **94**, 8854 (1972).
[71] A. M. Duffield, H. Budzikiewicz, and C. Djerassi, *J. Am. Chem. Soc.* **87**, 2920 (1965).

TABLE XXIII

^{1}H-NMR DATA OF TETRAHYDROTELLUROPHENE AND ITS CONGENERS[a]

Compound	CCl$_4$		C$_6$D$_6$		C$_6$F$_6$	
	H-2	H-3	H-2	H-3	H-2	H-3
Tetrahydrofuran[b]	3.61	1.79	3.57	1.43	3.51	1.84
Tetrahydrothiophene[b]	2.75	1.91	2.54	1.47	2.71	1.99
Tetrahydroselenophene[c]	2.79	1.96	2.65	1.64	2.83	2.04
Tetrahydrotellurophene[c]	3.10	2.03	2.91	1.69	3.18	2.10

[a] δ values relative to internal TMS.
[b] Data from Barton et al.[70]
[c] Data from Fringuelli et al.[10]

The ^{1}H-NMR spectrum of tetrahydrotellurophene has been recorded in several solvents,[10] and the chemical shifts, with those of the congener rings, are reported in Table XXIII. In all four systems the α-hydrogen resonances occur at lower field than the β-hydrogen resonances. The β-hydrogen shifts vary systematically with the electronegativity of the heteroatoms, whereas the α-hydrogen shifts do not. This behavior is similar to that of the corresponding aromatic systems. Photoelectron spectroscopy[72] has determined the first three ionization energies of the four congener tetrahydro derivatives (Tables XXIV). Mixing of the "nonbonding" electrons of the heteroatoms with the σ-system is apparent from the shape of the first PES band and from the comparison of the absolute values of the first ionization energy of these compounds with those of the corresponding compounds H$_2$X. Such mixing is highest for the oxygen derivative and gradually decreases down to the tellurium compound.

TABLE XXIV

IONIZATION ENERGIES (eV) OF n_π AND C$_2$X MOs OF
TETRAHYDROTELLUROPHENE AND CONGENER RINGS[a]

Compound	Ionization energies		
	n_π(b$_1$)	C$_2$X(a$_1$)	C$_2$X(b$_2$)
Tetrahydrofuran	9.53	11.4	13.0 ± 0.5
	9.65		
Tetrahydrothiophene	8.42	10.9	⩾11.9
Tetrahydroselenophene	8.14	10.5	⩾11.4
Tetrahydrotellurophene	7.73	10.0	10.7

[a] Data from Pignataro and Distefano.[72]

[72] S. Pignataro and G. Distefano, Chem. Phys. Lett. **26**, 356 (1974).

B. Benzo[b]tellurophene and Derivatives

(27)

1. Synthesis

Although the chemistry of congener benzo[b]heterocycles containing oxygen,[73] sulfur,[74] and selenium[75–78] has been the subject of many investigations, the chemistry of the last member of the series, benzo[b]-tellurophene (27), has received less attention.

The first synthesis of the benzo[b]tellurophene nucleus was claimed by Mazza and Melchionna,[79] who reported the preparation of 3-hydroxybenzo[b]tellurophene (telluroindoxyl) by cyclization of the phenyltelluroglycolic-o-carboxylic acid in the presence of sodium acetate and acetic anhydride. However, there is some doubt on this synthesis; it seems not to be reproducible.[80,81]

Piette et al.[81,82] developed the first synthesis of benzo[b]tellurophene (27) and a number of its derivatives by the sequence of reactions shown in Eq. (4) and Eq. (5).

$$\text{o-CHO-C}_6\text{H}_4\text{-TeMe} \xrightarrow{\text{BrCH}_2\text{R}, \Delta} \text{o-CHO-C}_6\text{H}_4\text{-TeCH}_2\text{R} \xrightarrow{\text{Ac}_2\text{O/Pyr.}} \text{benzo[b]tellurophene-2-R} \quad (4)$$

R = COMe (28); CO$_2$H (29)

Cyclization of appropriate o-(telluroether)benzaldehydes in presence of pyridine and acetic anhydride gives 2-acetylbenzo[b]tellurophene (28) and benzo[b]tellurophene-2-carboxylic acid (29). The acid is then decarboxylated to benzo[b]tellurophene (27) by heating with copper in quinoline.

---

[73] P. Cagniant and D. Cagniant, *Adv. Heterocycl. Chem.* **18**, 338 (1975).
[74] B. Iddon and R. M. Scrowston, *Adv. Heterocycl. Chem.* **11**, 178 (1975).
[75] T. Q. Minh, L. Christiaens, and M. Renson, *Bull. Soc. Chim. Fr.*, 2244 (1974).
[76] Q. T. Minh, F. Mantovani, P. Faller, L. Christiaens, and M. Renson, *Bull. Soc. Chim. Fr.*, 3955 (1972).
[77] N. M. Magdesieva and V. A. Vdovin, *Khim. Geterotsikl. Soedin.* **1**, 15 (1972).
[78] T. Q. Minh, L. Christiaens, and M. Renson, *Bull. Soc. Chim. Fr.*, 2239 (1974).
[79] F. P. Mazza and E. Melchionna, *Rend. Acad. Sci. Fis. Mat. Napoli* **34**, 54 (1928) [*C.A.* **23**, 2956 (1929)].
[80] W. V. Farrar, *Research Suppl.* **4**, 177 (1951).
[81] J.-L. Piette and M. Renson, *Bull. Soc. Chim. Belges* **80**, 521 (1971).
[82] J.-L. Piette, J.-M. Talbot, J.-C. Genard, and M. Renson, *Bull. Soc. Chim. Fr.* 2468 (1973).

Reaction of the *o*-(methyltelluro)benzaldehyde with phosphoranes and subsequent cyclization in the presence of pyridine[82] gives 2-acetyl- (**28**), 2-ethoxycarbonyl- (**30**), 2-cyano- (**31**), and 2-benzoylbenzo[*b*]-tellurophene (**32**).

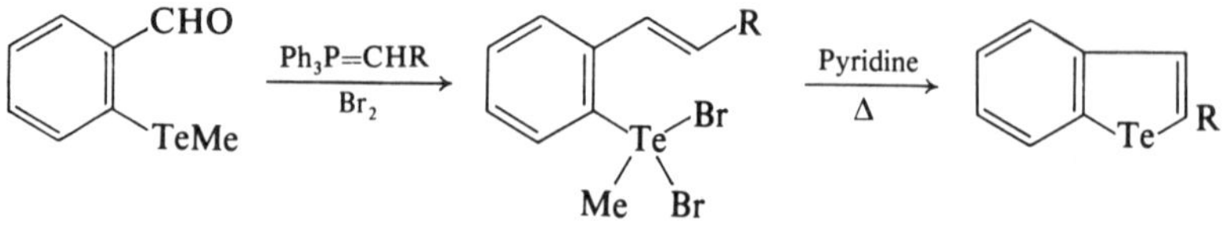

$$\text{(5)}$$

R = COMe (**28**); CO$_2$Et (**30**); CN (**31**); COPh (**32**)

Another synthesis of the benzo[*b*]tellurophene system has been effected by Sadekov and Minkin,[83] who prepared 2-phenyl-3-chloro-benzo[*b*]tellurophene (**34**) by treating diphenylacetylene with tellurium tetrachloride and reducing the 1,1-dichloro adduct (**33**) [Eq. (6)].

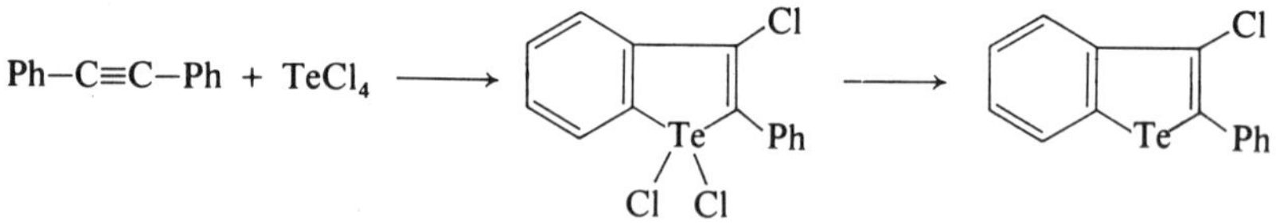

Ph—C≡C—Ph + TeCl$_4$ ⟶

(**33**)   (**34**)

$$\text{(6)}$$

Belgian workers[84,85] were able to synthesize a variety of 2-arylidene-2,3-dihydro-3-oxobenzo[*b*]tellurophene (telluroaurones) of general formula **35**.

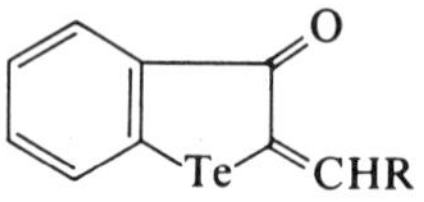

(**35**)

In a very recent work, Renson and co-workers[85] synthesized the telluroindoxyl by the reactions shown in Eq. (7). The telluroindoxyl exists not as 3-hydroxybenzo[*b*]tellurophene, but in the more stable

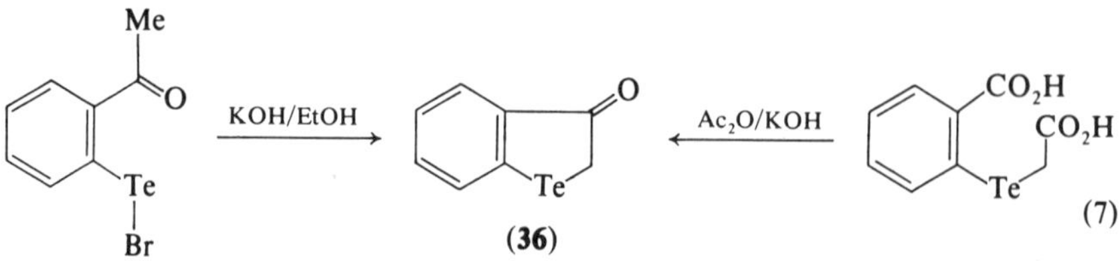

$$\xrightarrow{\text{KOH/EtOH}} \quad \xleftarrow{\text{Ac}_2\text{O/KOH}} \quad \text{(7)}$$

(**36**)

[83] I. D. Sadekov and I. V. Minkin, *Khim. Geterotsikl. Soedin.* **7**, 138 (1971) [*C.A.* **75**, 20102 (1971)].

[84] J. L. Piette, P. Andre, and M. Renson, *C.R. Hebd. Seances Acad. Sci., Ser. C* 1035 (1973).

[85] J. M. Talbot, J.-L. Piette, and M. Renson, *Bull. Soc. Chim. Fr.*, 249 (1976).

tautomeric form (36). This very interesting compound is reduced to benzo[b]tellurophene, oxidized to telluroindigo, and is the key compound to synthesize[86] 3-substituted benzo[b]tellurophenes (Scheme 2) by the Wittig reaction.

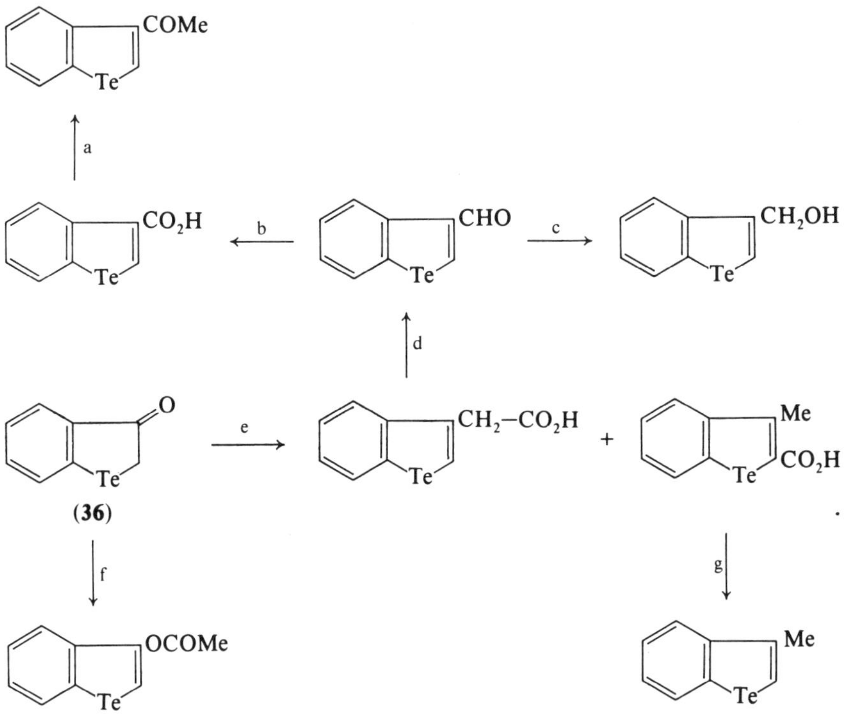

a: CH_3Li; b: Ag_2O; c: $NaBH_4$; d: $C_5H_5N \rightarrow O$; e: $Ph_3P=CH-R/NaOH/EtOH$; f: Ac_2O; g: $Cu_2Cr_2O_4$

SCHEME 2

2. Physical Properties

A complete assignment[87,88] of the rather complex NMR spectrum of benzo[b]tellurophene has been carried out. H-2 proton has been identified through its coupling with ^{125}Te nucleus; H-7 was separated from H-4 by the use of an europium complex with 2-acetylbenzo[b]tellurophene; H-5 and H-6 were identified from their meta couplings. All the proton–proton coupling constants (vicinal and long range) have been assigned with the exception of $J_{2,6}$ and $J_{3,4}$, which were not observed.

[86] J. M. Talbot, J.-L. Piette, and M. Renson, unpublished results.
[87] G. Llabres, M. Baiwir, J. Denoel, J.-L. Piette, and L. Christiaens, *Tetrahedron Lett.* 3177 (1972).
[88] P. Faller and J. Weber, *Bull. Soc. Chim. Fr.*, 3193 (1972).

The values of the chemical shifts and the coupling constants of all the congener benzo[b]heterocycles are reported in Table XXV. The spectroscopic behavior of the heterocyclic moiety of these bicyclic systems is similar to that exhibited by the monocyclic systems (see Section II,B,6): the H-2 chemical shifts occur at lower field than the corresponding H-3 shifts; the H-3 shifts are a function of the heteroatom Pauling electronegativity and increase regularly on going from benzo[b]furan to benzo[b]tellurophene; the H-2 chemical shifts vary irregularly, probably because of the anisotropic effects of the heteroatoms.

TABLE XXV

^{1}H-NMR Parameters of Congener Benzo[b]Heterocycles

δ and J	Benzo[b]-furan[b]	Benzo[b]-thiophene[c]	Benzo[b]-selenophene[d]	Benzo[b]-tellurophene[d]
H-2	7.52	7.27	7.79	8.55
H-3	6.66	7.19	7.42	7.84
H-4	7.49	7.70	7.69	7.71
H-5	7.13	7.25	7.24	7.26
H-6	7.19	7.22	7.14	7.03
H-7	7.42	7.77	7.77	7.82
J_{23}	2.19	5.57	5.57	6.95
J_{37}	0.87	0.86	0.65	0.43
J_{45}	7.89	8.09	7.97	7.97
J_{46}	1.28	1.16	1.02	1.07
J_{47}	0.80	0.73	0.48	0.51
J_{56}	7.27	7.22	7.22	7.24
J_{57}	0.92	1.17	1.17	1.08
J_{67}	8.43	8.06	8.27	8.00
J_{34}	—	—	0.14	—
J_{26}	—	0.30	—	—

[a] In CCl_4 relative to internal TMS.
[b] See P. J. Black and M. L. Hofferman, *Aust. J. Chem.* **18**, 353 (1965).
[c] See K. D. Bartle, D. W. Jones, and R. S. Matthesis, *Tetrahedron* **27**, 5177 (1971).
[d] Data from Llabres et al.[87]

Faller and Weber[88] analyzed the dependence of the $J_{2,3}$ coupling constants both with the Pauling electronegativity values of the heteroatoms and with the chemical shifts.

The IR and UV spectra of benzo[b]tellurophene have been studied only qualitatively.[81] They are very similar to those of the congener systems containing selenium and sulfur atoms.

An infrared study of the 2-carboxylic acids has shown that the C=O absorption frequency is a direct function of the electron-withdrawing power of the heteroatom.[62]

The behavior of benzo[b]tellurophene under electron impact has been analyzed by Buu-Hoi and co-workers.[89] The mass spectrum is similar to those of benzo[b]thiophene and benzo[b]selenophene. The base peak corresponds to a detellurated species, probably benzocyclobutadiene.

Although spectroscopic data (IR, UV, NMR) have been recorded for some individual derivatives, there is no systematic study of the spectra of substituted benzo[b]tellurophenes.

3. Chemical Properties

Benzo[b]tellurophenes react with halogens (Cl_2, Br_2, I_2) and methyl bromide, giving addition products to the tellurium atom, like the parent nonannelated compounds.[81,82,84]

Benzo[b]tellurophene is metalled in the 2-position using *n*-butyl-lithium,[82] to give the 2-lithio derivative (37), which is a useful intermediate for the preparation of 2-substituted benzo[b]tellurophene derivatives (see Scheme 3).

2-Acetylbenzo[b]tellurophene (28) can be prepared by direct electrophilic substitution using acetyl chloride and Friedel–Crafts

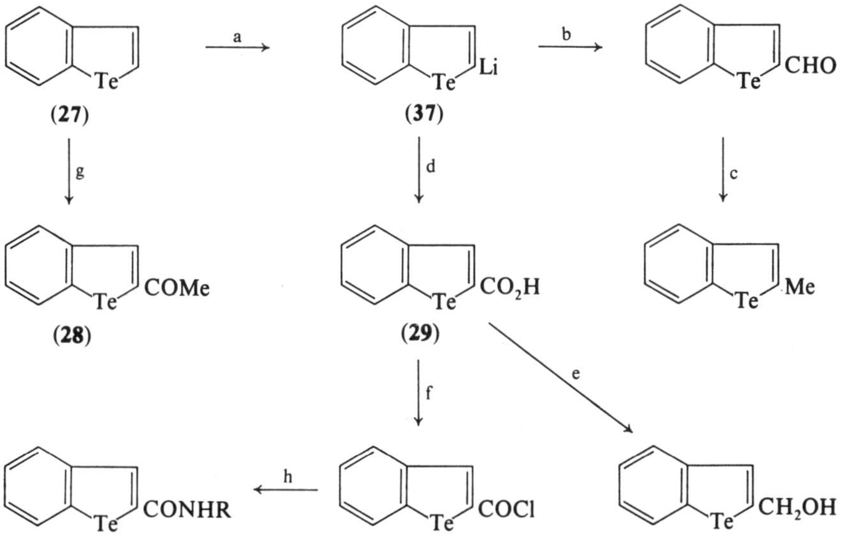

a: *n*BuLi; b: HCONMe$_2$; c: NH$_2$·NH$_2$/KOH; d: CO$_2$; e: LiAlH$_4$; f: dichloromethyl butyl ether; g: Ac$_2$O–BF$_3$; h: NH$_3$ or PhNH$_2$

SCHEME 3

[89] N. P. Buu-Hoi, M. Mangane, M. Renson, and J.-L. Piette, *J. Heterocycl. Chem.* **7**, 219 (1970).

catalysts; however, the substitution is complicated by side-chain reactions and the yields are poor. The 3-acetyl isomer was not found in the reaction mixture, but sure conclusions about the orientation of the substitution in this system cannot be drawn because of the low yields obtained.

4. Quantitative Studies of Reactivity

The ionization constants of the 2-carboxylic acids of benzo[b]-tellurophene and its congeners have been determined in an ethanol–water mixture;[62] the strength of the acids appears to be a direct function of the electronegativity of the heteroatoms. Comparison with the pK_a values of the corresponding nonannelated monocyclic compounds shows that the fused benzene ring increases the acidity in a constant way for all the congener heterocycles. Linear correlations between the pK_a values and the dimer and monomer C=O stretching frequencies have been observed.[62]

TABLE XXVI

RATE CONSTANTS AND REACTIVITY RATIOS IN THE SOLVOLYSIS OF 1-(2-ARYL)ETHYL ACETATES IN 30% ETOH AT 60°

Aryl group	$10^5\,k$ (sec^{-1})	k_X/k_{Th}^a	k_B/k_M^b
2-Benzo[b]furyl	0.945^{cd}	1.50	4.78 × 10^{-3}
2-Benzo[b]thienyl	0.629^c	1	9.62 × 10^{-3}
2-Benzo[b]selenienyl	1.044	1.66	9.56 × 10^{-3}
2-Benzo[b]tellurienyl	3.256	5.18	8.85 × 10^{-3}

[a] Reactivity ratios to 1-(2-benzo[b]thienyl)ethyl acetate.
[b] Reactivity ratios between bicyclic (B) and monocyclic (M) systems in the title reaction.
[c] See reference c of Table XIX.
[d] Extrapolated value.

Kinetic data are available for the reaction of solvolysis of the 1-(2-aryl)ethyl acetates in 30% ethanol.[58] The second-order rate constants (at 60°), the rates relative to the benzothiophene derivative, and the rates relative to the corresponding monocyclic derivatives are summarized in Table XXVI. The reactivity order in the congener series is the following: benzo[b]tellurophene > benzo[b]selenophene > benzo[b]furan > benzo[b]thiophene.

Benzo-fusion decreases the reactivity by a factor of about 10^2 in the case of the thiophene, selenophene, and tellurophene rings, and by about 2 × 10^2 in the case of furan.

C. Dibenzotellurophene and Derivatives

1. Synthesis and Chemical Reactivity

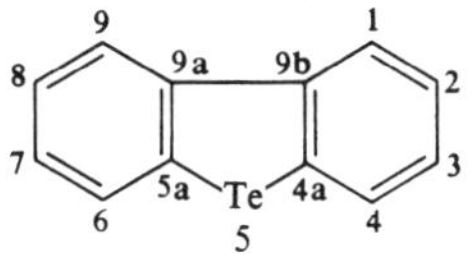

(38)

A variety of names has been used in the literature for this compound: biphenylenetelluride, diphenylenetelluride, biphenylylenetelluride, dibenzotellurophene (38). Although the first names have been used recently, we prefer the last one, by analogy with the names used for the congener systems dibenzofuran, dibenzothiophene, and dibenzoselenophene. The name dibenzotellurophene is now that adopted by *Chemical Abstracts*. Numbering systems different from that shown in 38 are occasionally met in the literature.

The first synthesis of dibenzotellurophene was accomplished by Courtot and Bastani[90] by treating diphenyl (39) with tellurium tetrabromide or tetrachloride and then reducing the adduct (40) with potassium hydrogen sulfite [Eq. (8)].

Dibenzotellurophene has been also obtained by treating dibenzothiophene-5-dioxide (41) with tellurium[91,92] [Eq. (8)].

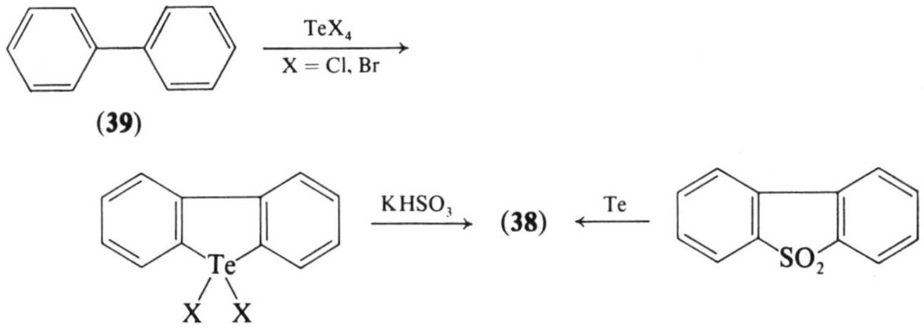

Three additional routes to dibenzotellurophene or its 3,7-dimethyl derivative have been reported by Hellwinkel and Fahrbach[93–95]: (i) by

[90] C. Courtot and M.-G. Bastani, *C.R. Hebd. Seances Acad. Sci.* **203**, 197 (1936). [*C.A.* **30**, 7107 (1936)].
[91] N. M. Cullinane, A. G. Rees, and C. A. J. Plummer, *J. Chem. Soc.*, 151 (1939).
[92] R. Passerini and G. Purrello, *Ann. Chim. (Rome)* **48**, 738 (1958).
[93] D. Hellwinkel and G. Fahrbach, *Tetrahedron Lett.*, 1823 (1965).
[94] D. Hellwinkel and G. Fahrbach, *Justus Liebigs Ann. Chem.* **712**, 1 (1968).
[95] D. Hellwinkel and G. Fahrbach, *Chem. Ber.* **101**, 574 (1968).

thermal decomposition of an appropriate diphenylylmercuric compound (**42**) in the presence of tellurium [Eq. (9)]; (ii) by treating the 2,2′-

(9)

(**42**) R = H, Me

dilithiodiphenyl (**45**) with tellurium tetrachloride or trimethyltelluronium iodide [Eq. (10)]; (iii) by thermal decomposition of the spiro compounds **43** and **44**.

(10)

R = H (**43**) R = Me (**44**)

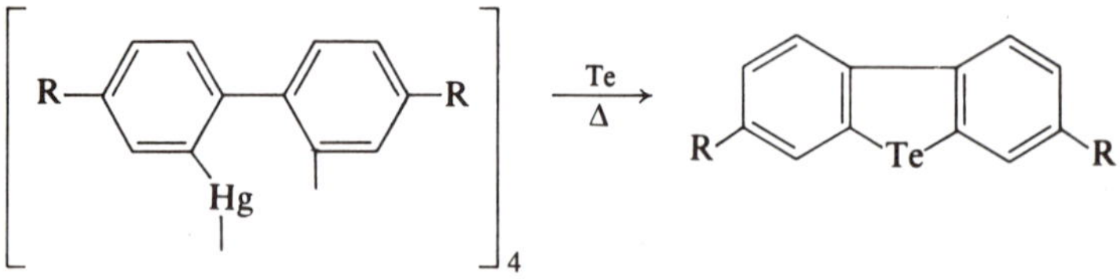

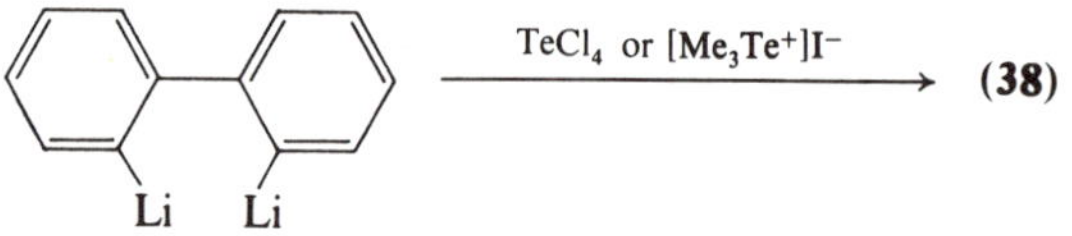

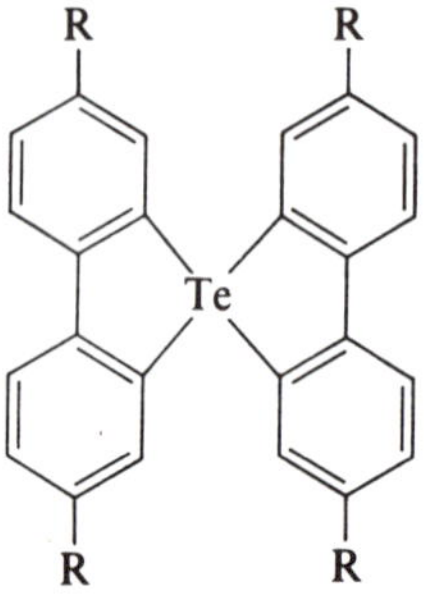

Cohen *et al.*[96,97] prepared octafluorodibenzotellurophene (**47**) by heating 2,2′-diiodooctafluorodiphenyl (**45**) with tellurium at 325° or 2,2′-octafluorodiphenyllithium (**46**) with tellurium tetrachloride [Eq. (11)].

(11)

R = I (**45**) R = Li (**46**) (**47**)

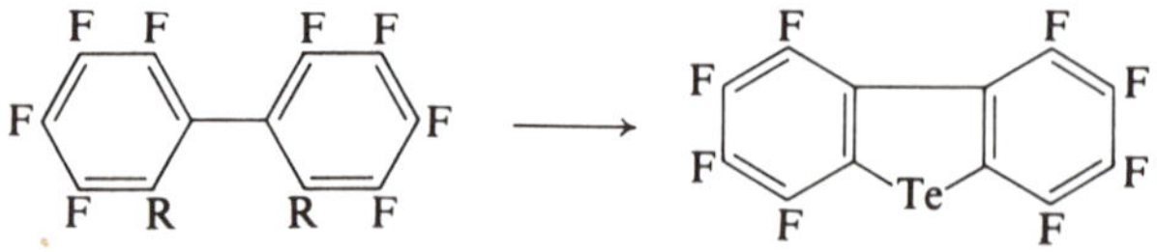

[96] S. C. Cohen, M. L. N. Reddy, and A. G. Massey, *Chem. Commun.*, 451 (1967).
[97] S. C. Cohen, M. L. N. Reddy, and A. G. Massey, *J. Organomet. Chem.* **11**, 563 (1968).

Dibenzotellurophene reacts with electrophiles giving substitution or addition products: with nitric acid in acetic acid solution the main product is the 2-nitro derivative[92]; with the halogens the main product is, as in the case of tellurophene and benzotellurophene, an 1:1 addition product at the tellurium atom.[93]

The reaction of dibenzotellurophene with butyllithium causes the opening of the tellurophene ring and gives 2,2'-diphenyllithium and dibutyl telluride; the kinetics of the reaction has been discussed.[95]

2. *Physical Properties*

The crystal and molecular structure of dibenzotellurophene and 5,5-diiododibenzotellurophene have been determined by using three-dimensional X-Ray diffraction techniques.[98] The two six-membered rings of the diphenyl system are twisted out of plane from each other. Each benzene ring, considered separately, is planar, but the central five-membered ring is not.

The comparison of the geometries of heterocyclic moiety of the dibenzo congeners (Table XXVII) shows that bond lengths and bond angles have the same trend along the series as in the monocyclic non-annelated compounds (see Section II,A).

TABLE XXVII

BOND DISTANCES (Å) AND ANGLES (DEGREES) WITHIN THE HETEROCYCLIC MOIETY OF DIBENZENE CONGENERS[a]

Compound	C(4a)X	C(4a)XC(5a)	XC(4a)C(9b)	C(4a)C(9b)C(9a)
Dibenzofuran[b]	1.404	104.1	112.3	105.3
Dibenzothiophene[c]	1.740	91.5	112.3	111.9
Dibenzoselenophene[d]	1.899	86.7	112.3	114.3
Dibenzotellurophene[e]	2.087	81.7	112.1	117.1

[a] X-ray analysis; for numbering system see compound **38**.
[b] Two X-ray studies are reported: See O. Dideberg, L. Dupont, and J. M. Andre, *Acta Crystallogr., Sect. B* **28**, 1002 (1972); A. Banerjee, *ibid.* **29**, 2070 (1973); the study of O. Dideberg *et al.* appears more accurate because it has a lower final R index and lower estimated standard deviations of bond lengths and bond angles.
[c] See R. M. Schaffrin and J. Trotter, *J. Chem. Soc.*, 1561 (1970).
[d] See H. Hope, C. Knobler, and J. D. McCullough, *Acta Crystallogr., Sect. B* **26**, 628 (1970).
[e] McCullough.[98]

The mass spectrum of the octafluorodibenzotellurophene (**47**) has been determined.[99] Comparison with the spectral data of the octafluoro derivatives of dibenzothiophene and dibenzoselenophene indicates that Te—C bonds are less stable under electron bombardment than those with Se—C or S—C bonds.

[98] J. D. McCullough, *Inorg. Chem.* **14**, 1142, 2639 (1975).
[99] S. C. Cohen, *Org. Mass. Spectrom.* **6**, 373 (1972).

D. Other Systems Containing the Tellurophene Nucleus

A variety of quinones containing the tellurophene nucleus of general formula **49** have been obtained by Müller et al.[55,100] by reaction between an appropriate rhodium complex (**48**) and amorphous tellurium as shown in Eq. (12).

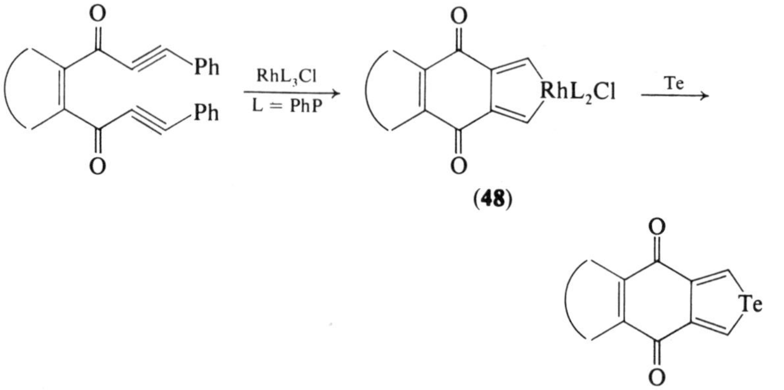

$$\text{(48)} \xrightarrow{\text{Te}} \text{(49)} \quad (12)$$

Cagniant et al.[52] have prepared several polycyclic compounds containing the tellurophene nucleus (**50–52**) using the general procedure described in Section III,A [Eq. (2)].

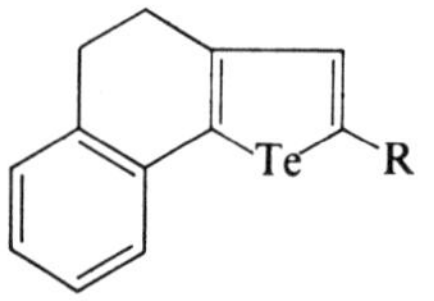

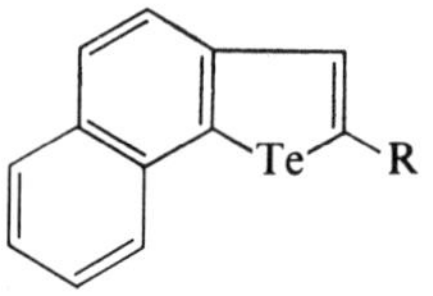

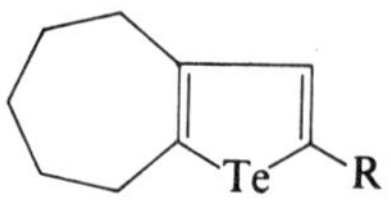

R = H; CO_2Et; CO_2H

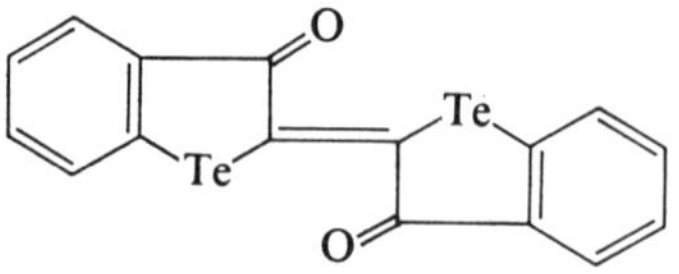

Telluroindigo (**53**) and the benzotelluropheno-indole derivative (**54**) have been prepared starting from telluroindoxyl (**36**) by oxidation and by reaction with phenylhydrazine in acetic acid, respectively.[85]

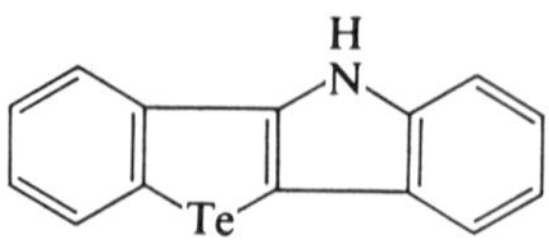

[100] E. Müller, E. Luppold, and W. Winter, Chem. Ber. **108**, 237 (1975).

Sec. IV.D]   TELLUROPHENE AND RELATED COMPOUNDS

Interesting spiran systems (**43** and **44**) with a tetravalent tellurium atom are known.[93,95,101] They can be prepared by treatment of 2,2'-dilithiodiphenyl with a variety of reagents such as $TeCl_4$, $Te(OMe)_6$, $Te(OMe)_4$, 5(*p*-toluenesulfonylimido)dibenzotellurophene and 5-methyldibenzotelluronium iodide. The 5,5'-spiro-bisdibenzotellurophene (**43**) is stable under normal conditions and reacts with electrophiles to form telluronium salts with ring opening. On thermolysis, **43** gives dibenzotellurophene, diphenylene, and traces of tetraphenylene. The ^{1}H-NMR spectrum of **44** shows only a single Me-proton signal down to $-55°$, indicating rapid exchange between the trigonal bipyramid and tetragonal pyramid structures. The spiro compound (**43**) with butyllithium undergoes opening of the tellurophene ring; the mechanism of this reaction has been discussed.[95] Some typical reactions are reported in Scheme 4.

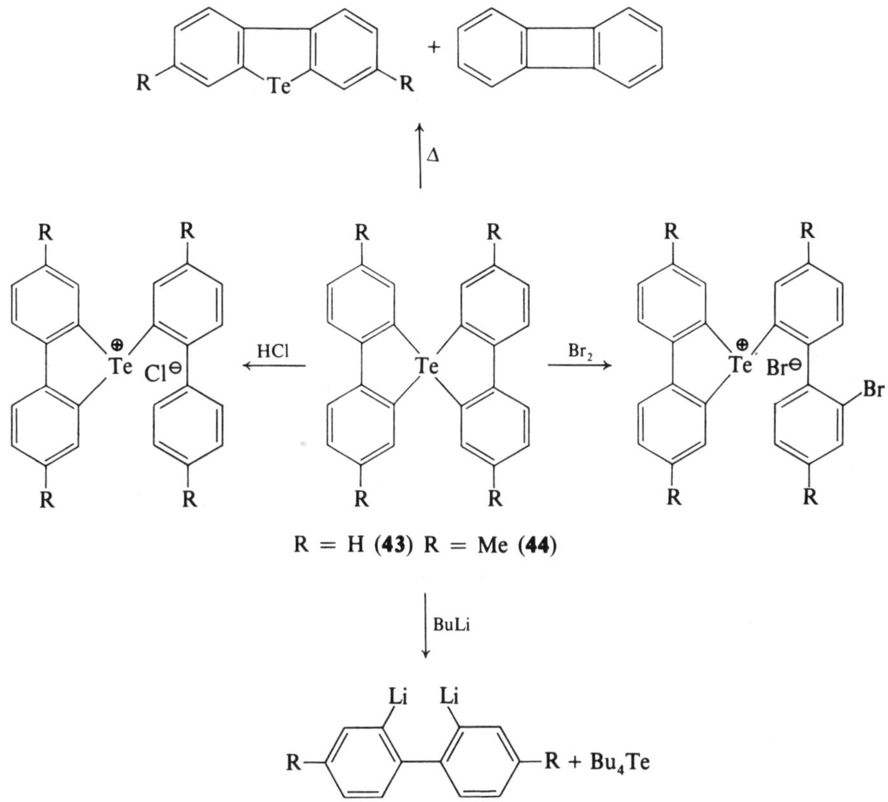

R = H (**43**)   R = Me (**44**)

SCHEME 4

[101] D. Hellwinkel, *Ann. N.Y. Acad. Sci.* **192**, 158 (1972) [*C.A.* **77**, 61725 (1972)].

V. Appendix. Tables of Tellurophene Derivatives and Related Compounds

All the known compounds having the tellurophene moiety and their available physical and chemical properties are listed in Tables XXVIII–XXXVI. The properties are symbolized as follows: m.p., melting point; b.p., boiling point; IR, infrared; ^{1}H, proton magnetic resonance; ^{13}C, carbon magnetic resonance; UV, ultraviolet; χ, diamagnetic susceptibility; μ, dipole moment; n, refractive index; pK, ionization constant; M, mass fragmentation; IE, ionization energy; dv, functional group derivatives; MR, molar refraction; d, density; R, Raman spectrum; k, kinetic measurements; G, geometrical parameters; C, molecular complex.

Appropriate references for each property are given in parentheses.

TABLE XXVIII

TELLUROPHENES

(1)

Compound	Physical and chemical properties (references)
Tellurophene	m.p.; b.p.; IR; ^{1}H; ^{13}C; UV; χ; μ; n; IE; MR; d; R; k; G; C (for references see text)
2-Me	b.p. (12,14); ^{1}H (12,14,25); ^{13}C (25); μ (8,9,13); IE (31,33); MR (13); k (58); C (65)
2-SMe	b.p. (31,13); ^{1}H (31,23); ^{13}C (23); μ (13); IE (31); MR (13)
2-CO$_2$Me	b.p. (14); IR (14); ^{1}H (14,23); ^{13}C (23); μ (8,9,13); n (14); IE (31,33); MR (13)
2-CO$_2$H	m.p. (14); IR (14,62); ^{1}H (14,23); ^{13}C (23); pK (62,63); IE (31,33); G (4)
2-CHO	b.p. (14); IR (14,22); ^{1}H (14,23,42); ^{13}C (23); μ (8,9,13); IE (31); MR (13)
2-COMe	b.p. (14); IR (14,22); ^{1}H (14,23,42); ^{13}C (23); UV (14); μ (8,9,13); n (14); IE (31); MR (13)
2-CH$_2$OH	b.p. (13); IR (14); ^{1}H (14,23); ^{13}C (23); μ (8,9,13); MR (13)
2-CMe$_3$	b.p. (52); n (52)
2-Cl	b.p. (25); ^{1}H (25); ^{13}C (25); μ (13); IE (33); MR (13)
2-Br	b.p. (25); ^{1}H (25); ^{13}C (25); μ (13); IE (33); MR (13)
2-I	^{1}H (25); ^{13}C (25); IE (33)
2-COCF$_3$	b.p. (57); IR (57); ^{1}H (57)
2-CH(OH)Me	b.p. (14); IR (14); ^{1}H (14); n (14)
2-CH(OAc)Me	b.p. (59); ^{1}H (25,59); ^{13}C (25); n (59); k (58,59); IR (59)
2-CONMe$_2$	m.p. (42); b.p. (42); IR (42); ^{1}H (25,42); ^{13}C (25); μ (13); IE (33); MR (13)
2-TeC$_4$H$_3$Te	m.p. (25); ^{1}H (25)

TABLE XXVIII—continued

Compound	Physical and chemical properties (references)
2-CMe_2OH	m.p. (49)
2-D	b.p. (18); IR (18); R (18)
1,1-Br_2	m.p. (48)
2,5-Ph_2	m.p. (49); b.p. (49)
2,5-$(C_4H_6NHMe)_2$	b.p. (49)
2,5-$(C_4H_7)_2$	b.p. (49)
2,5-$(CO_2Me)_2$	m.p. (63); IR (63); ^{1}H (63); μ (13); MR (13)
2-COMe; 5-CO_2Me	m.p. (63); IR (63); ^{1}H (63); μ (9,13); MR (13)
2-Me; 5-CO_2H	m.p. (14); ^{1}H (14); pK (63)
2,5-$(CO_2H)_2$	m.p. (63); pK (63)
2-COMe; 5-CO_2H	m.p. (14); IR (14); ^{1}H (14); pK (63)
2-Me; 5-CH(OH)Me	IR (58); ^{1}H (58); k (58)
2-Me; 5-CH(OAc)Me	b.p. (58); IR (58); ^{1}H (58); k (58)
2-Me; 5-CHO	^{1}H (58)
2-Me; 5-$COCF_3$	k (58)
2-CO_2Et; 5-CMe_3	b.p. (52); n (52)
2-CO_2H; 5-CMe_3	m.p. (52)
2-NO_2; 5-CMe_3	b.p. (52); m.p. (52)
2-CHO; 5-CMe_3	b.p. (52); n (52); dv (52)
2-COMe; 5-CMe_3	b.p. (52); m.p. (52); dv (52)
2,5-$(CH_2OH)_2$	m.p. (48,49); ^{1}H (48,49)
2,5-$[C(OH)Me_2]_2$	m.p. (48,49); ^{1}H (48,49)
2,5-D_2	b.p. (18); IR (18); ^{1}H (18); R (18)
3,4-D_2	b.p. (18); IR (18); ^{1}H (18); R (18)
2,3,4-D_3	b.p. (18); IR (18); ^{1}H (18); R (18)
1,1-Br_2; 2,5-Ph_2	m.p. (49)
1,1-Br_2; 2,5-$(C_4H_7)_2$	m.p. (49)
2,3,4,5-Cl_4	m.p. (47); b.p. (47); IR (47); UV (47)
2,3,4,5-Ph_4	m.p. (45,46); UV (45); C (45)
2,3,4,5-D_4	b.p. (18); IR (18); R (18)
1,1,2,3,4,5-Cl_6	m.p. (47)
1,1-Br_2; 2,3,4,5-Ph_4	m.p. (45)

TABLE XXIX
Tetrahydrotellurophenes

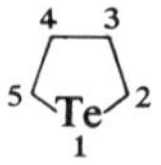

$$\underset{1}{\underset{5}{\overset{4\ \ \ 3}{\diagdown\text{Te}\diagup}}} 2$$

Compound	Physical and chemical properties (references)
Tetrahydrotellurophene	b.p. (66); ^{1}H (10); μ (9,13); n (66); M (71); IE (72); MR (13); C (66)
1-O	m.p. (66)
3-C$_4$H$_7$Te	m.p. (68,69); b.p. (68,69)
1,1-Br$_2$	m.p. (66,67)
1,1-Cl$_2$	m.p. (66)
1,1-I$_2$	m.p. (66)
1-I; 1-Me	m.p. (66)
1-Br; 1-(CH$_2$)$_4$Br	m.p. (66)
1-I; 1-(CH$_2$)$_4$Br	m.p. (66)
1-Br; 1-(CH$_2$)$_4$C$_4$H$_8$TeBr	m.p. (66)
1-I; 1-(CH$_2$)$_4$C$_4$H$_8$TeBr	m.p. (66)
1-Br; 1-OC$_4$H$_8$TeBr	m.p. (66)

TABLE XXX
Benzo[b]tellurophenes

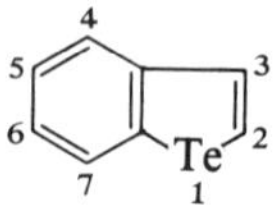

(27)

Compound	Physical and chemical properties (references)
Benzo[b]tellurophene	m.p. (81); ^{1}H (81,87,88); UV (81); M (89); C (81)
2-CO$_2$H	m.p. (81,82); IR (62,82); ^{1}H (82); pK (62)
2-COCl	m.p. (82); IR (82); ^{1}H (82)
2-CH$_2$OH	m.p. (82); ^{1}H (82)
2-CONH$_2$	m.p. (82); IR (82); ^{1}H (82)
2-CONHPh	m.p. (82); IR (82); ^{1}H (82)
2-COMe	m.p. (82); IR (82); ^{1}H (82)
2-Me	m.p. (82); ^{1}H (82)
2-CO$_2$Et	m.p. (82); IR (82); ^{1}H (82)
2-COPh	m.p. (82); IR (82); ^{1}H (82)
2-CN	m.p. (82); IR (82); ^{1}H (82)
2-CH(OAc)Me	m.p. (58); IR (58); ^{1}H (58); UV (58); k (58)
3-CH$_2$CO$_2$H	IR (86); ^{1}H (86)
3-CH$_2$CO$_2$Et	IR (86); ^{1}H (86)
3-CH$_2$CO$_2$Me	IR (86); ^{1}H (86)
3-CH$_2$CN	IR (86); ^{1}H (86)
3-CH$_2$COMe	IR (86); ^{1}H (86)
3-Me	^{1}H (86); C (86)
3-CHO	IR (86); ^{1}H (86); dv (86)

TABLE XXX—continued

Compound	Physical and chemical properties (references)
3-CH_2OH	^1H (86)
3-CO_2H	m.p. (86)
3-COMe	m.p. (86)
3-OCOMe	m.p. (86)
1,1-Cl_2	m.p. (81,82)
1,1-Br_2	m.p. (81,82)
1,1-I_2	m.p. (81,82)
1-Br; 1-Me	m.p. (81)
2-Ph; 3-Cl	(83)
2-CO_2H; 3-Me	IR (86); ^1H (86)
2-CO_2Et; 3-Me	IR (86); ^1H (86)
2-CO_2Me; 3-Me	IR (86); ^1H (86)
2-CN; 3-Me	IR (86); ^1H (86)
2-CH_2OH; 3-Me	IR (86); ^1H (86)
2-$CONH_2$; 3-Me	IR (86); ^1H (86)
2-COMe; 3-Me	IR (86); ^1H (86)
2-CHO; 3-Me	IR (86); ^1H (86)
1,1-Cl_2; 2-COCl	m.p. (82)
1,1-Br_2; 2-CO_2H	m.p. (82)
1,1-Br_2; 3-Me	m.p. (86)
1,1,3-Cl_3; 2-Ph	(83)

TABLE XXXI

TELLUROAURONES

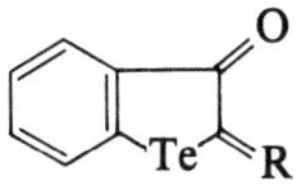

Compound	Physical and chemical properties (references)
Telluroindoxyl (R = H_2)	m.p. (85); IR (85); ^1H (85); M (85); dv (85)
R = H_2; 1,1-Cl_2	m.p. (85)
R = H_2; 1,1-Br_2	m.p. (85)
R = H_2; 1,1-I_2	m.p. (85)
R = H_2; 1-I; 1-Me	m.p. (85)
R = CH—NMe_2	m.p. (85); IR (85); ^1H (85)
R = CH—Ph	m.p. (84,85); IR (84,85); ^1H (84,85)
R = CH—C_6H_4—Me(o)	m.p. (85); IR (85); ^1H (85)
R = CH—C_6H_4—NO_2(m)	IR (85); ^1H (85)
R = CH—NH—C_6H_4—NMe_2	IR (85); ^1H (85)
R = CH—C_6H_4—I(p)	IR (84); ^1H (84)
R = CH—C_6H_4—Cl(p)	IR (84); ^1H (84)
R = CH—C_6H_4—NMe_2	IR (84); ^1H (84)
R = CH—C_6H_4—Cl(o)	IR (84); ^1H (84)
R = CH—$C_{10}H_7$(α)	IR (84); ^1H (84)
R = CH—C_6H_4—OMe(p)	IR (84); ^1H (84)
R = CH—C_6H_4—Me(p)	IR (84); ^1H (84)

TABLE XXXII
FUSED ALICYCLIC TELLUROPHENES

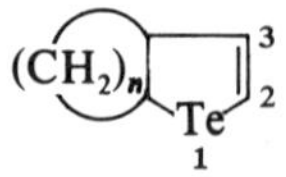

$(CH_2)_n \underset{Te}{\underset{1}{\boxed{}_2^3}}$

Compound	Physical and chemical properties (references)
$n = 4$	b.p. (52)
$n = 4$; 2-CO_2Et	b.p. (52); n (52)
$n = 4$; 2-CO_2H	m.p. (52)
$n = 4$; 2-NO_2	m.p. (52)
$n = 4$; 2-COMe	b.p. (52); dv (52)
$n = 5$	b.p. (52); n (52)
$n = 5$; 2-CO_2Et	b.p. (52); n (52)
$n = 5$; 2-CO_2H	m.p. (52)

TABLE XXXIII
DIBENZOTELLUROPHENES

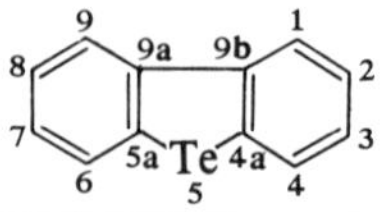

Compound	Physical and chemical properties (references)
Dibenzotellurophene	m.p. (90,91,92); UV (92); M (94); C (94); G (98)
2-NO_2	m.p. (92); UV (92)
5-O	m.p. (94)
5-N–SO_2–C_6H_4Me(p)	m.p. (94, 101)
3-Me, 7-Me	m.p. (94); C (94)
5,5-Cl_2	m.p. (90,93,94)
5,5-Br_2	m.p. (90)
5-,5-I_2	m.p. (93,94); G (98)
3,7-Me_2; 5,5-Cl_2	m.p. (94)
1,2,3,4,6,7,8,9-F_8	m.p. (96,97); M (99)

TABLE XXXIV
TELLUROPHENE QUINONES

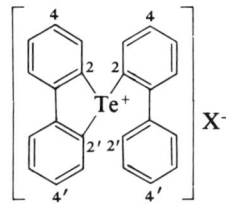

Compound	Physical and chemical properties (references)
A = C_6H_4; 2,5-Ph_2	m.p. (55); IR (55); UV (55); M (55)
A = C_6H_4; 2,5-$(C_6H_4$-p-Me$)_2$	m.p. (55); IR (55); UV (55); M (55)
A = $C_{10}H_6$; 2,5-Ph_2	m.p. (55); IR (55); UV (55); M (55)
A = $C_{10}H_6$; 2,5-$(C_6H_4$-p-Me$)_2$	m.p. (55); IR (55); UV (55); M (55)
A = C_4SMe_2; 2,5-Ph_2	m.p. (55); IR (55); UV (55); M (55)
A = C_8H_4S; 2,5-Ph_2	m.p. (55,100); IR (55,100); UV (55,100); M (55,100)

TABLE XXXV
DIPHENYLYLENE-DIPHENYLYL-TELLURONIUM SALTS

Compound	Physical and chemical properties (references)
X = OH	m.p. (93,94)
X = Cl	m.p. (93,94)
X = I	m.p. (93,94)
X = BPh_4	m.p. (94)
X = Br; 2'-Br	m.p. (94)
X = I; 2'-I	m.p. (94)
X = Cl; (4,4')Me_4	m.p. (94)
X = Br; 2'-Br; (4,4')Me_4	m.p. (94)

TABLE XXXVI
MISCELLANEOUS

Compound	Physical and chemical properties (references)
Telluroindigo [$C_{16}H_8O_2Te_2$]	m.p. (85)
10H-[1]Benzotelluropheno[3,2-b]indole-[$C_{14}H_9NTe$]	m.p. (85)
Bis-2,2'-biphenylylene-tellurium[$C_{24}H_{16}Te$]	m.p. (94,95); M (94); C(94)
Bis-(4,4'-dimethyl-2,2'-biphenylylene) tellurium [$C_{28}H_{24}Te$]	m.p. (94); ^{1}H (94)
Naphthotellurophene [$C_{12}H_8Te$]	m.p. (52); C (52)
4,5-Dihydronaphthotellurophene[$C_{12}H_{10}Te$]	m.p. (52); b.p. (52)
4,5-Dihydronaphthotellurophene-2-carboxylic acid [$C_{13}H_{10}O_2Te$]	m.p. (52)
Ethyl-4,5-Dihydronaphthotellurophene-2-carboxylate [$C_{15}H_{14}O_2Te$]	b.p. (52)

Note Added in Proof

Since the completion of this review, additional papers have been published and new data are available. The literature has been now covered to the end of 1975.

Two patents[102,103] report practical utilization of tellurophene and some of its derivatives. These are (1) increasing the fire resistance of hydraulic fluids and (2) causing higher autoignition temperatures.

The mass spectrum of tellurophene has been recorded and compared with those of other congeners.[104] The tellurophene spectrum shows the molecular ion as the base peak, the M-HTe and M-acetylene peaks, and an important ion of mass 52 corresponding to the loss of tellurium.

The photolysis of 2-phenyltellurophene and two of its deuterioderivatives has been investigated.[105] The photolysis of the undeuteriated compound gives phenylvinylacetylene and tellurium only, in contrast with the 2-substituted thiophenes and selenophenes which, in addition, give 3-substituted derivatives. The ^{125}Te-labeled NMR parameters of tellurophene and seven of its derivatives have been obtained.[106]. Large substituent-caused shifts have been observed and the ^{125}Te is more sensitive to substituent effects than ^{77}Se in corresponding 2-substituted selenophenes.

[102] U.S. Patent 3,730,889 (*CA* **79**, 44208).
[103] U.S. Patent 3,795,619 (*CA* **81**, 124071).
[104] F. Fringuelli and A. Taticchi, in preparation.
[105] T. J. Barton, C. R. Tully, and R. W. Roth, *J. Organomet. Chem.* **108**, 183 (1976).
[106] T. Drakenberg, F. Fringuelli, S. Gronowitz, A.-B. Hörnfeldt, I. Johnson, and A. Taticchi, *Chem. Scripta*, submitted for publication.

The chemistry of tellurophene, benzo[b]tellurophene, and dibenzotellurophene during the period 1972–1974 has been briefly reviewed in a general review[107] concerning the organic tellurium compounds.

The photoelectron spectrum of benzo[b]tellurophene has been recorded[108] with He(I) and He(II) radiation and compared with those of other benzo[b]congeners.

A new synthesis of dibenzotellurophene by way of 2-biphenyltellurophenonium trichloride has been reported.[109] The procedure gives a good yield. Preparation of dibenzotellurophene dibromide and dibenzotellurophene diiodide have been also described.[109]

---

[107] K. J. Irgolic, *J. Organomet. Chem.* **103**, 91 (1975).
[108] J. F. Müller, *Helv. Chim. Acta* **58**, 2646 (1975).
[109] J. D. McCullough, *Inorg. Chem.* **14**, 2285 (1975).

New Developments in the Chemistry of Oxazolones

ROBERT FILLER

Department of Chemistry, Illinois Institute of Technology, Chicago, Illinois

AND Y. SHYAMSUNDER RAO

Department of Chemistry, Kennedy-King College, Chicago, Illinois

I. Introduction	176
II. 2-Oxazolin-5-ones	176
A. Methods of Preparation	176
1. Use of Polyphosphoric Acid	177
2. From 5-Oxazolonium Perchlorates	178
3. From Other Heterocycles and Open-Chain Compounds	178
4. Application of Bergmann's Reaction	179
5. Miscellaneous Methods	180
B. Reactions	180
1. Behavior toward Enzymes	180
2. Reaction with Diazomethane	181
3. Conversion into Other Heterocycles with Amines and Hydrazines	182
4. Reaction with Oxygen	184
5. Cycloaddition Reactions of Mesoionic Oxazolones	185
6. Hydrolysis of Azlactones	186
7. Photochemistry and Thermolysis	187
C. Stereochemistry	188
1. Optically Active Oxazolones and Racemization	188
2. Geometrical Isomerism	189
D. Bis-oxazolones	190
III. 3-Oxazolin-5-ones	191
A. Methods of Preparation	191
B. Reactions	194
1. Behavior under Friedel–Crafts Conditions	194
2. Reaction with Grignard Reagents	194
3. Reaction with Hydrazoic Acid	195
4. Special Reactions	195
5. Photoreactions	196
IV. 2-Oxazolin-4-ones	198
A. Preparation	198
1. From Benzoyl Isocyanate and Diazoalkanes	198

2. Alkylation with Ethyl Bromoacetate	199
3. Cyclization of α-Haloimides	200
4. 2-Amino-2-oxazolin-4-ones	200
B. Reactions	200
V. 4-Oxazolin-2-ones	202
A. Preparation	202
1. From Cyclic Carbonates	202
2. From Amine Derivatives	202
3. From Benzoin Derivatives	203
4. Miscellaneous Methods	203
B. Reactions	203
VI. 3-Oxazolin-2-ones	205
A. Preparation	205
B. Reactions	206

I. Introduction

Major advances in the chemistry of oxazolones have been made since the author's last review of this subject.[1] More recently, 2-oxazolin-5-ones and 3-oxazolin-5-ones were surveyed.[2] The whole field of oxazolone chemistry has burgeoned during the past decade, and the purpose of this review is to present the highlights of these new developments.

The structures of the classes of oxazolones are shown below:

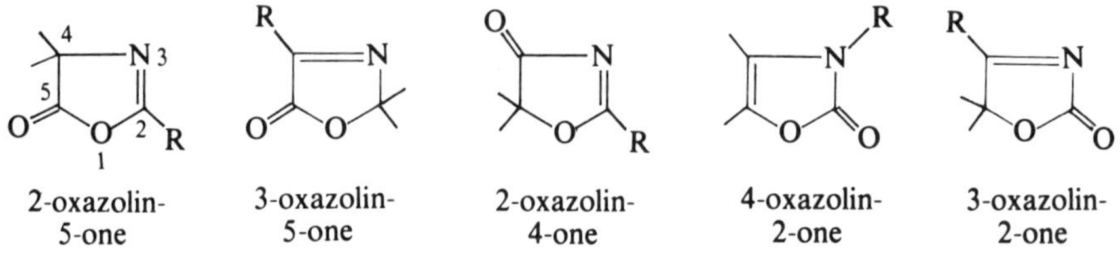

| 2-oxazolin-5-one | 3-oxazolin-5-one | 2-oxazolin-4-one | 4-oxazolin-2-one | 3-oxazolin-2-one |

This nomenclature is current usage in *Chemical Abstracts*.

II. 2-Oxazolin-5-ones

A. METHODS OF PREPARATION

While the classical Erlenmeyer synthesis of azlactones (condensation of aldehydes with hippuric or aceturic acid in the presence of acetic

[1] R. Filler, *Adv. Heterocycl. Chem.* **4**, 75 (1965).
[2] W. Steglich, *Fortschr. Chem. Forsch.* **12**, 77 (1969).

anhydride and sodium acetate) continues to be the most commonly used method of preparation, several recent variations deserve mention.

1. *Use of Polyphosphoric Acid*

Rao[3] recently reported that aromatic aldehydes condense with hippuric acid in polyphosphoric acid (PPA) to give the generally inaccessible lower-melting geometric isomers of 4-aryl-methylene-2-phenyl-2-oxazolin-5-ones, e.g., **1**, Eq. (1). The Erlenmeyer method and its minor variations, such as use of 4-chloromethylene or 4-ethoxymethylene azlactones, always afford the geometric isomer,[1,2] e.g., **2**. Compound **2** is quantitatively isomerized to **1** in PPA [Eq. (2)]. The

$$\text{PhCHO} + \text{PhCONHCH}_2\text{CO}_2\text{H} \xrightarrow{\text{PPA}} \quad (1)$$

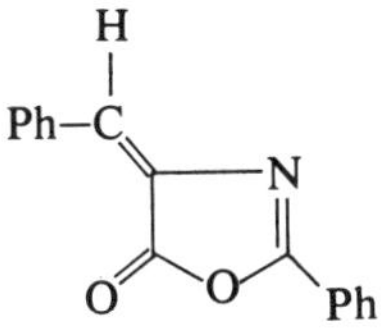

(1) E isomer

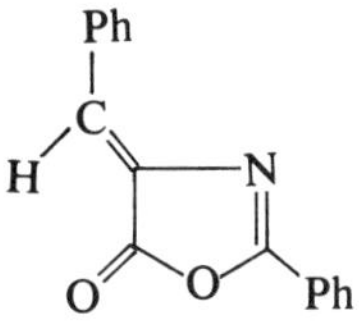

$$\mathbf{(2)} \xrightarrow{\text{PPA}} \mathbf{(1)} \quad (2)$$

(2) Z isomer

method has been employed for the preparation of the labile isomers derived from *p*-methyl-, chloro-, nitro-, and methoxybenzaldehydes, which have not previously been reported. A major advantage of this modification is that hydroxybenzylidene azlactones can be prepared, whereas the acetoxy compound is obtained when acetic anhydride is used. Moreover, acetophenone and its ring-substituted analogs react in the PPA system to give the hitherto unknown α-arylethylidene azlactones [**3**, Eq. (3)]. The usually unreactive phenylacetaldehyde, hydro-

$$\text{ArCOMe} + \text{PhCONHCH}_2\text{CO}_2\text{H} \xrightarrow{\text{PPA}} \quad (3)$$

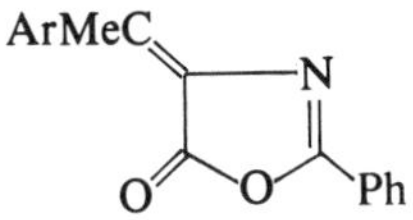

(3)

---

[3] Y. S. Rao, *J. Org. Chem.* **41**, 722 (1976).

cinnamaldehyde, and 2-methylcyclohexanone also condense under these conditions. The scope and limitations of this method, the reasons for its efficacy, and the mechanism of the isomerization process are under investigation.

2. From 5-Oxazolonium Perchlorates

Boyd and Wright[4-6] prepared 2-phenyloxazol-5-onium perchlorate and observed that this compound condenses with benzaldehyde to give the E isomer 1.

3. From Other Heterocycles and Open-Chain Compounds

a. From 2-Thiohydantoins. Zubenko[7,8] showed that heating 5-arylidene-2-thiohydantoins with a 2–4-fold excess of aroyl chlorides in pyridine gave 2-oxazolin-5-one derivatives [Eq. (4)]. The advantage of

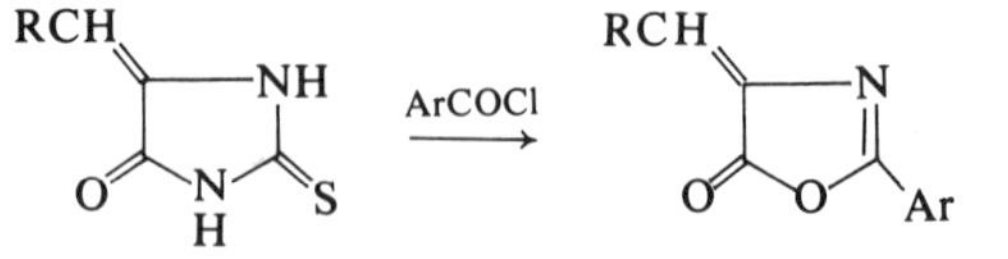

this method is that it provides a convenient route to azlactones bearing a substituted phenyl group in the 2-position by obviating the preparation of ring-substituted hippuric acids. Yields are high in the range of 79–93%.

b. Rearrangement of Acylamino Heterocycles. The E isomer of 2-methyl-4-phenylmethylene-2-oxazolin-5-one (**4**) was prepared by rearrangement,[9] as shown in Scheme 1. Compound **2** was one of several products obtained by rearrangement of 3-benzamido-1,4-diphenyl azetidin-2-one.[10]

[4] G. V. Boyd, *Chem. Commun.*, 1410 (1968).
[5] G. V. Boyd, *Jerusalem Symp. Quant. Chem. Biochem.* **3**, 166 (1971) [*CA* **81**, 91401 (1974)].
[6] G. V. Boyd and P. H. Wright, *J. Chem. Soc., Perkin Trans. 1,* 909, 914 (1972).
[7] V. G. Zubenko, *Dopov. Akad. Nauk Ukr. RSR, Ser. B* **30**, 547 (1968) [*CA* **69**, 106596 (1968)].
[8] V. G. Zubenko, USSR Patent 188,976 (1966) [*CA* **67**, P73601 (1967)].
[9] J. C. Howard, *J. Org. Chem.* **36**, 1073 (1971).
[10] C. W. Bird, *J. Chem. Soc. C,* 3155 (1971).

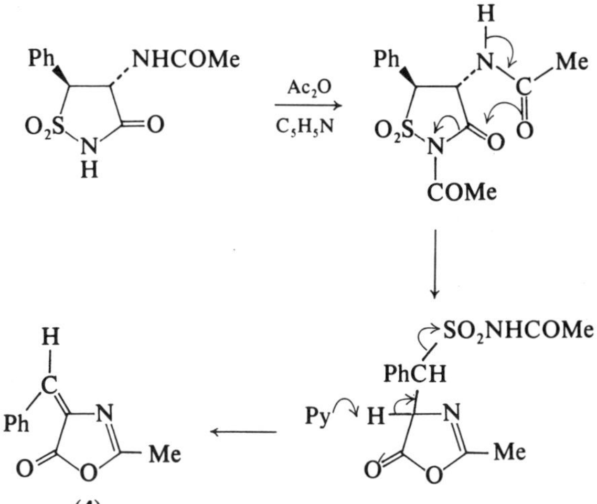

SCHEME 1

4. Application of Bergmann's Reaction

Riordan and Stammer[11,12] applied Bergmann's procedure for the syntheses of a series of azlactones. Amino acids, such as phenylalanine, tyrosine, valine, leucine, and isoleucine were converted to N-(α-methylcinnamoyl) derivatives. Azlactonization followed by oxidation using pyridine hydrobromide perbromide gave the corresponding unsaturated azlactone having the Z configuration around the newly formed double bond in the 4-position [Eq. (5)].

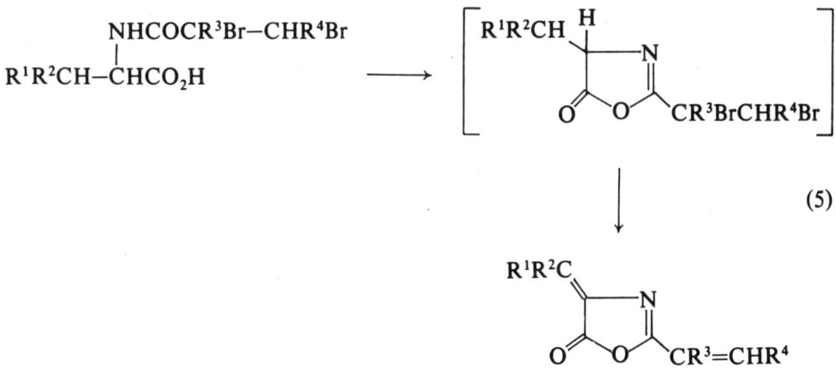

[11] J. M. Riordan and C. H. Stammer, *Tetrahedron Lett.*, 4969 (1971).
[12] J. M. Riordan and C. H. Stammer, *J. Org. Chem.* **39**, 654 (1974).

5. Miscellaneous Methods

DL-Amino acids react with N-phenylbenzimido chlorides in benzene containing triethylamine to give saturated azlactones[13] [Eq. (6)].

$$\text{PhCCl=NR} + \text{H}_2\text{N-CHEt-CO}_2\text{H} \longrightarrow \text{[azlactone]} \quad (6)$$

Neutral amino acids, e.g., alanine, valine, isovaline, and methionine, react with orthoacetate esters in dimethylacetamide to give mixtures of N-acetamido acid esters and saturated azlactones[14] [Eq. (7)].

$$\text{R}^1\text{CHCO}_2\text{H} + \text{R}^2\text{CH(OEt)}_3 \longrightarrow \text{[azlactone]} + \text{R}^1\text{CHCO}_2\text{Et} \quad (7)$$
$$\text{NH}_2 \qquad\qquad\qquad\qquad\qquad\qquad\qquad\qquad \text{NHCOR}^2$$

B. REACTIONS

1. Behavior toward Enzymes

The 2-oxazolin-5-ones are excellent acylating agents for a variety of enzymes. Thus, 2-phenyloxazolin-5-one and 4,4-dimethyl-2-phenyl-2-oxazolin-5-one react with α-chymotrypsin, trypsin, and papain to form stable acyl enzymes.[15–18] The azlactone from p-nitrobenzoylvaline reacted with α-chymotrypsin and trypsin, and it was observed that the enzymic activity of chymotrypsin decreased with increasing number of p-nitrobenzoylvaline residues.[19] The reaction of α-chymotrypsin with **2** has been studied extensively.[20–25] In this reaction, the oxazolone dis-

---

[13] E. I. Boksiner, *Zh. Org. Khim.* **8**, 604 (1972).
[14] S. V. Rogozhin, Y. A. Davidovich, and V. V. Kershak, *Izv. Akad. Nauk SSSR, Ser. Khim.*, 1593 (1971).
[15] J. de Jersey and B. Zerner, *Biochemistry* **8**, 1967 (1969).
[16] J. de Jersey, P. Willadsen, and B. Zerner, *Biochemistry* **8**, 1959 (1969).
[17] J. K. Stoops, D. J. Horgan, M. T. C. Runnegar, J. de Jersey, E. C. Webb, and B. Zerner, *Biochemistry* **8**, 2026 (1969).
[18] J. de Jersey, *Biochemistry* **9**, 1761 (1970).
[19] H. Siemeniewski and T. Baranowski, *Acta Biochem. Pol.* **16**, 243 (1969).
[20] K. Brocklehurst and K. Williamson, *Chem. Commun.*, 462 (1966).
[21] K. Brocklehurst and K. Williamson, *Chem. Commun.*, 666 (1967).
[22] K. Brocklehurst and K. Williamson, *Biochem. Biophys. Res. Commun.* **26**, 175 (1967).
[23] J. de Jersey and B. Zerner, *Biochem. Biophys. Res. Commun.* **28**, 173 (1967).
[24] J. de Jersey and B. Zerner, *Biochemistry* **8**, 1975 (1969).
[25] K. Brocklehurst and K. Williamson, *Tetrahedron* **30**, 351 (1974).

appears rapidly with the formation of two new species, depending on pH. At pH = 5, a species with λ_{max} 305 nm, postulated involving a histidine residue, was formed; and at pH = 8 (λ_{max} 285 nm), an acyl enzyme involving a serine moiety was proposed. The identity of the second species was questioned and subsequently shown to be structure

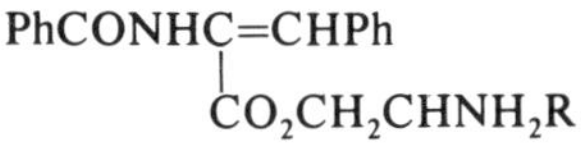

$$\begin{array}{c} \text{PhCONHC=CHPh} \\ | \\ \text{CO}_2\text{CH}_2\text{CHNH}_2\text{R} \end{array}$$

(5)

5.[23,24] The reaction of either the E isomer (1) or the Z isomer (2) leads to the acylenzyme, probably α-benzamidocinnamoyl-α-chymotrypsins

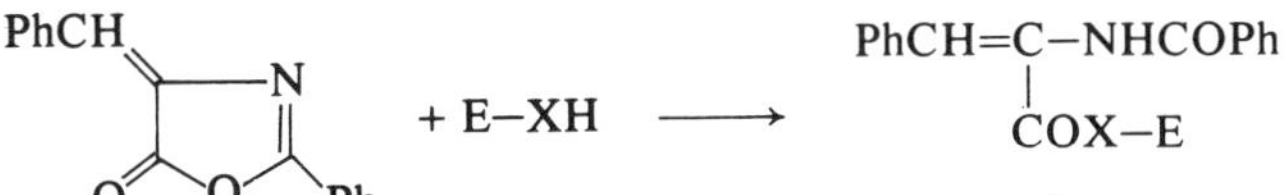

PhCH⟩=⟨oxazolone⟩-Ph + E—XH ⟶ PhCH=C—NHCOPh | COX—E   (8)

(1) or (2)   (6)
                    X = active site

(6). It was also observed that α-benzamido-*trans*-cinnamoyl-α-chymotrypsin (derived from 2) is hydrolyzed faster than the cis analog (from 1) in the pH range 7–10.[25]

2. Reaction with Diazomethane

4-Arylmethylene-2-oxazolin-5-ones react with diazomethane to afford cyclopropane derivatives,[26–33] such as 7.[28] The geometric isomer 8 was prepared indirectly [Eq. (9)].[28] Recently, Stammer and co-workers[32]

---

[26] W. I. Awad, A. K. Fateen, and M. A. Zayed, *Tetrahedron* **20**, 891 (1964).
[27] R. A. Pages and A. Burger, *J. Med. Chem.* **9**, 766 (1966).
[28] R. A. Pages and A. Burger, *J. Med. Chem.* **10**, 435 (1967).
[29] M. Bernabe, E. Fernandez-Alvarez, and S. Penades Ullate, *An. Quim.* **68**, 501 (1972).
[30] M. Bernabe, E. Fernandez-Alvarez, and S. Penades Ullate, *An. Quim.* **68**, 1005 (1972).
[31] M. A. F. Elkaschef, F. M. E. Abdel-Megeid, and S. M. A. Yassin, *J. Prakt. Chem.* **316**, 363 (1974).
[32] J. W. Hines, E. G. Breitholle, M. Sato, and C. H. Stammer, *J. Org. Chem.* **41**, 1466 (1976).
[33] A. Mustafa, W. Asker, A. H. Harhash, and A. M. Fleifel, *Tetrahedron* **21**, 2215 (1965).

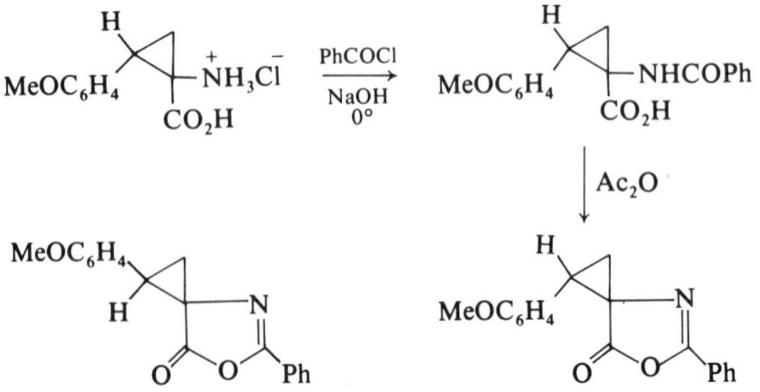

prepared the spiroazlactone (9) with the objective of hydrolyzing it to cyclopropylphenylalanine (10). The authors postulated that, under the strongly acidic conditions, the cyclopropyl ring in 10 is readily opened to give 11, probably via a benzyl carbocation. The reactions are summarized in Scheme 2. Similarly, attempts to prepare the cyclopropylog

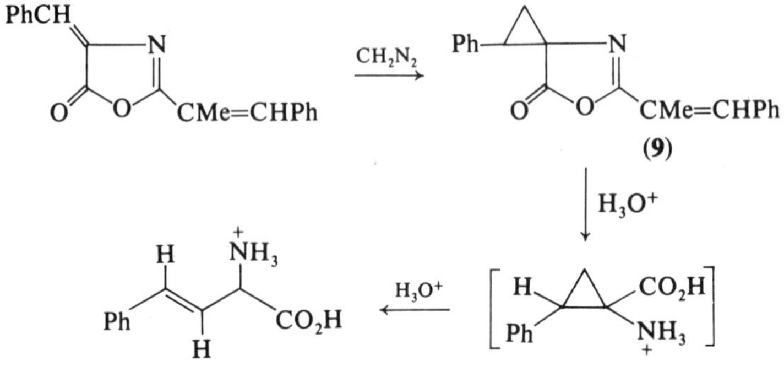

SCHEME 2

of DOPA via the corresponding spiroazlactone, led to destruction of the cyclopropane ring.

3. Conversion into Other Heterocycles with Amines and Hydrazines

2-Styryl-4-arylmethylene-2-oxazolin-5-ones react with aromatic amines to form anilides (12).[34] In the presence of acetic acid and sodium acetate, however, 2-aryl-4-arylmethylene-2-oxazolin-5-ones and aromatic amines give imidazolin-5-ones (13).[35,36] Under similar conditions,

[34] A. F. M. Fahmy and M. O. A. Orabi, *Ind. J. Chem.* **10**, 961 (1971).
[35] A. M. Islam, A. M. Khalil, and I. I. Abd El-Gawad, *Aust. J. Chem.* **26**, 827 (1973).
[36] A. M. Islam, A. M. Khalil, and M. L. El-Houseni, *Aust. J. Chem.* **26**, 1701 (1973).

2-aryl-4-arylazo-2-oxazolin-5-ones react with amines to afford triazole-carboxamides,[37,38] by initial ring opening and alternate ring closure.

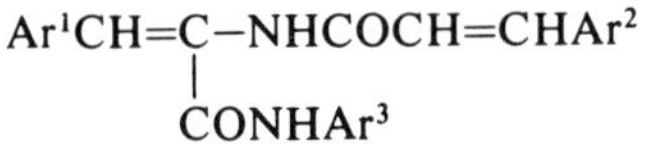

N-Phenylhydroxylamine reacts with 5-oxazolones to give 1,2,5-oxadiazin-3-ones (**14**).[39] When heated with hydrazine in dilute alkali, triazines are formed [Eq. (10)].[40–42] Hydrolysis of azlactones, followed by treatment with thiosemicarbazide, yields 6-azauracils (**15**).[42] The reaction of **2** with substituted hydrazines in boiling ethanol yields isomeric pyrazolidin-3-ones (**16**, **17**) in addition to the open-chain hydrazide. The formation of these heterocycles results from Michael addition reactions, as shown in Scheme 3.[43] Compounds **16** and **17** were isolated from the alcoholic mother liquor, whereas the hydrazide **18** crystallized.

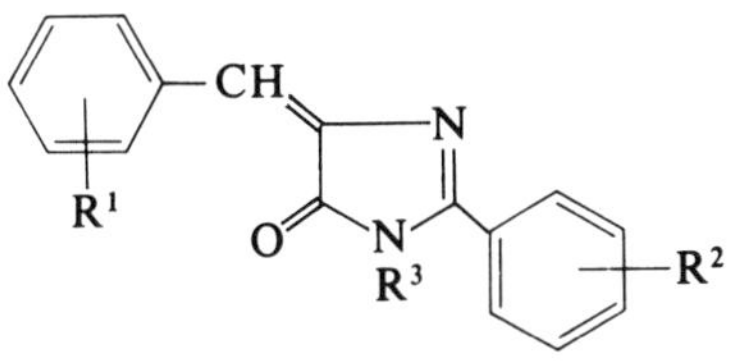

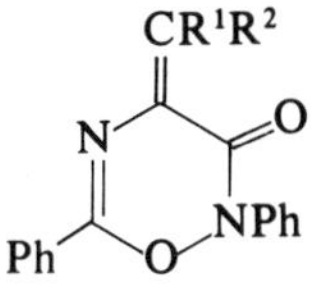

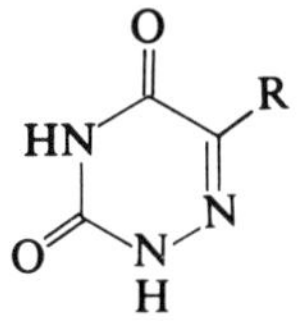

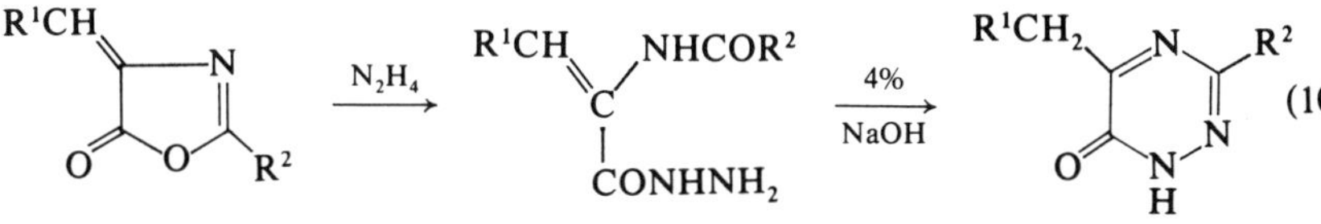

[37] A. M. Khalil, I. I. Abd El-Gawad, and H. M. Hassan, *Aust. J. Chem.* **27**, 2509 (1974).
[38] Filler,[1] p. 92.
[39] A. Mustafa, W. Asker, A. H. Harhash, T. M. S. Abdin, and E. M. Zayed, *Justus Liebigs Ann. Chem.* **714**, 146 (1968).
[40] K. Nalepa, V. Bekarek, and J. Slouka, *J. Prakt. Chem.* **714**, 851 (1972).
[41] K. Nalepa, *Monatsh. Chem.* **98**, 1230 (1967).
[42] P. Pec, J. Slouka, and K. Nalepa, *Acta Univ. Palacki. Olomuc., Fac. Recumnatur.* **33**, 401 (1971).
[43] D. S. Iyengar, K. K. Prasad, and R. V. Venkataratnam, *Aust. J. Chem.* **27**, 2439 (1974).

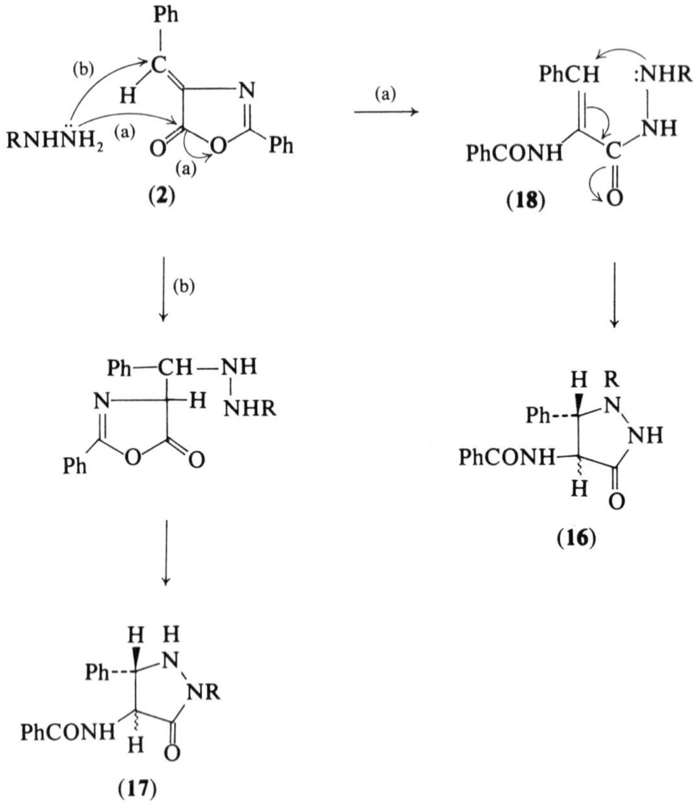

SCHEME 3

Other heterocyclic systems prepared from these oxazolones include oxazolo[5,4-b]quinolines,[44] pyran-2-ones,[45] pseudoxazolones,[46] and indole derivatives.[47]

4. *Reaction with Oxygen*

Warnhoff and coworkers[48] oxygenated 4-isobutylidene-2-phenyl-2-oxazolin-5-one (**19**) in the presence of triethylamine. The resulting product undergoes decarboxylation to form the imide [**20**, Eq. (11)].

[44] T. Kametani, T. Yamanaka, and K. Ogasawara, *J. Chem. Soc. C*, 385 (1969).
[45] T. Hiraoka and Y. Kishida, *Chem. Pharm. Bull.* **16**, 1576 (1968).
[46] W. Steglich, P. Gruber, G. Höfle, and W. König, *Angew. Chem., Int. Ed. Engl.* **10**, 653 (1971).
[47] A. DeAntoni, G. Allegri, and C. Costa, *Gazz. Chim. Ital.* **100**, 1056 (1970).
[48] R. Bisson, R. B. Yeats, and E. W. Warnhoff, *Can. J. Chem.* **50**, 2851 (1972).

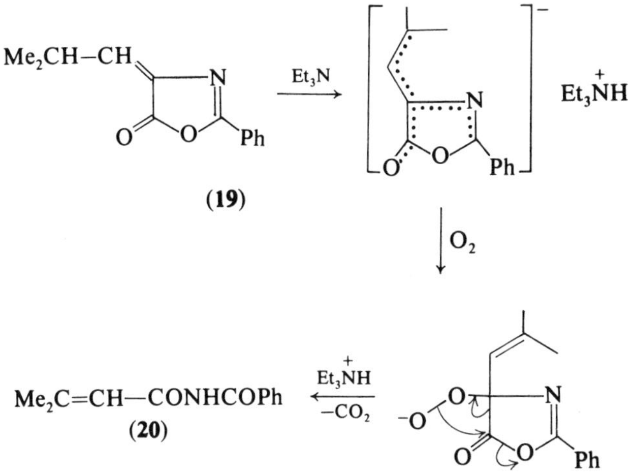

5. Cycloaddition Reactions of Mesoionic Oxazolones

Saturated azlactones, such as the isomeric **21** and **22**, possess mesoionic character and behave as dipolar species in 1,3-dipolar cycloaddition reactions with dipolarophiles,[49-51] e.g., acetylene dicarboxylates [Eq. (12)]. Decarboxylation of the adduct from either **21** or **22** gives the same pyrrole derivative (**23**).

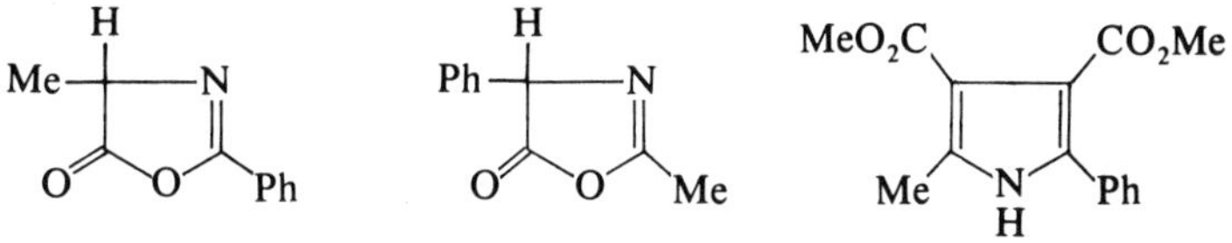

---

[49] H. Gotthardt, R. Huisgen, and H. O. Bayer, *J. Am. Chem. Soc.* **92**, 4340 (1970).
[50] H. O. Bayer, H. Gotthardt, and R. Huisgen, *Chem. Ber.* **103**, 2356 (1970).
[51] R. Huisgen, H. Gotthardt, and H. O. Bayer, *Chem. Ber.* **103**, 2368 (1970).

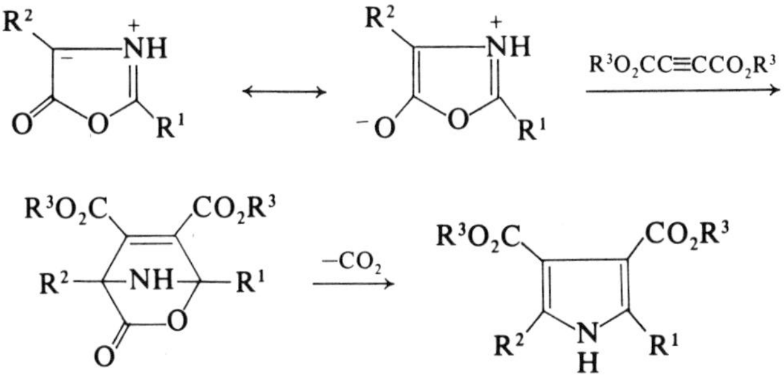

With dimethyl fumarate, methyl acrylate, and methyl methacrylate as dipolarophiles, substituted pyrrolines are obtained.[52-54] In contrast, the reaction of 2-phenyl-2-oxazolin-5-onium perchlorate with benzylideneaniline follows a different course to give the imidazolin-4-one (24).[55]

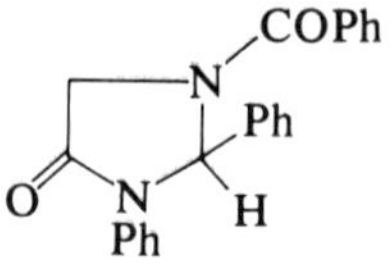

(24)

6. Hydrolysis of Azlactones

The acidic hydrolysis of oxazolones, such as 2, could proceed by the addition of water to either the C=O or C=N linkages, as shown in Scheme 4. In an elegant series of experiments using H_2O^{18} and mass spectrometry, Steglich et al.[56] demonstrated that attack occurs, in fact,

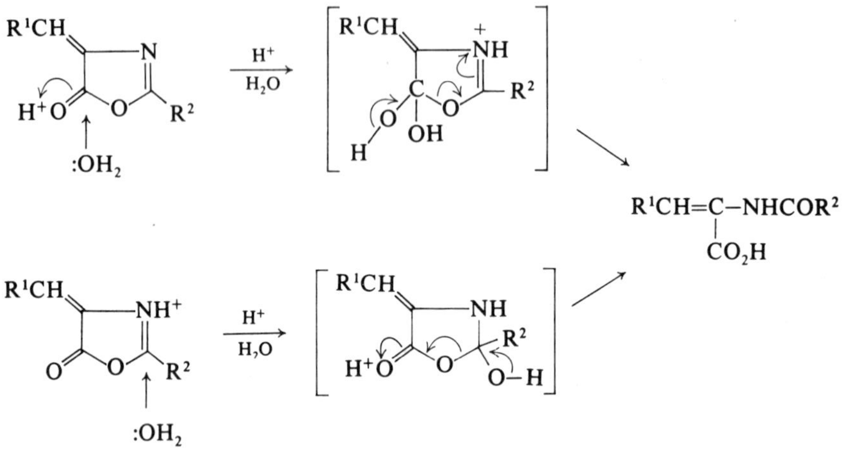

SCHEME 4

---

[52] A. Padwa and J. Smolanoff, J. Am. Chem. Soc. **93**, 548 (1971).
[53] A. Padwa, M. Dharan, J. Smolanoff, and S. I. Wetmore, J. Am. Chem. Soc. **95**, 1945 (1973).
[54] N. S. Narasimhan, H. Heimgartner, H. J. Hansen, and H. Schmid, Helv. Chim. Acta **56**, 1351 (1973).
[55] A. R. Knowles, A. Lawson, G. V. Boyd, and R. A. Newberry, Tetrahedron Lett., 485 (1971).
[56] W. Steglich, V. Austel, and A. Prox, Angew. Chem., Int. Ed. Engl. **7**, 726 (1968).

at the C=N bond. The acetic acid-catalyzed ring opening of oxazolone rings has also been investigated.[57,58]

7. Photochemistry and Thermolysis

When the Z isomer of 2-phenyl-4-phenylmethylene-2-oxazolin-5-one (2) is irradiated in 2-propanol with 365-nm light, Z→E (1) isomerization occurs. With 253.7-nm light, however, solvent participation leads to formation of 25, after ring cleavage and alternate cyclization,[59–62]

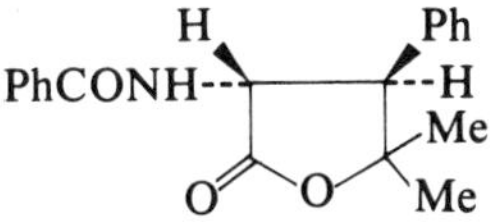

(25)

Equilibration of the Z–E isomers also occurs in acetonitrile.[25] With 2-acetoxyphenylmethylene oxazolones, isomerization to the E isomer was observed.[63] Isomerizations to benzamidocoumarin derivatives under photochemical conditions have also been reported.[64,65]

While the 2-oxazolin-5-ones are thermally stable, the introduction of a 4-acyl or aroyl group, a potential site for further reaction, markedly decreases the stability. Thus, when 2,4-dialkyl-4-aroyl-2-oxazolin-5-ones are heated to 180°, decarboxylation occurs readily with the formation of trisubstituted oxazoles in yields of 71–95%[66] [Eq. (13)].

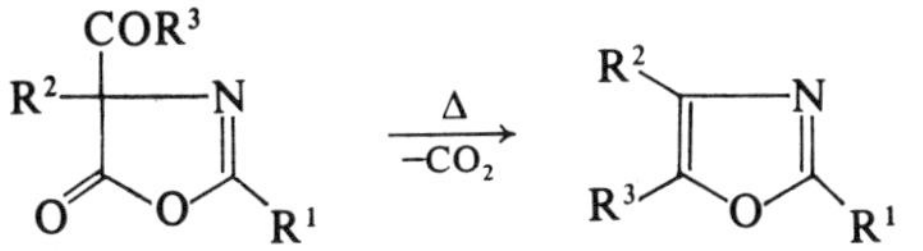

$$\xrightarrow{\Delta}_{-CO_2}$$ (13)

[57] C. Chauqui, S. Atala, A. Marquez, and H. Rodriguez, *Tetrahedron* **29**, 1197 (1973).
[58] H. Rodriguez, C. Chauqui, S. Atala, and A. Marquez, *Tetrahedron* **27**, 2425 (1971).
[59] N. Baumann, M. Sung, and E. F. Ullman, *J. Am. Chem. Soc.* **90**, 4157 (1968).
[60] E. F. Ullman and N. Baumann, *J. Am. Chem. Soc.* **90**, 4158 (1968).
[61] E. F. Ullman and N. Baumann, *J. Am. Chem. Soc.* **92**, 5892 (1970).
[62] N. Baumann, *Chimia* **27**, 471 (1973).
[63] G. W. Kirby and J. Michael, *J. Chem. Soc., Perkin Trans. 1*, 115 (1973).
[64] R. Walter, T. Purcell, and H. Zimmer, *J. Heterocycl. Chem.* **3**, 235 (1966).
[65] E. F. Ullman, U.S. Patent 3,689,391 (1972) [CA **78**, 3585 (1973)].
[66] G. Höfle and W. Steglich, *Chem. Ber.* **104**, 1408 (1971).

C. Stereochemistry

1. Optically Active Oxazolones and Racemization

While several optically active oxazolones have been prepared, these intermediates are likely to undergo racemization in peptide synthesis.[67,68] Thus, 2-phenyl-L-4-benzyl-2-oxazolin-5-one (**26**) was synthesized, and its rate of racemization with nucleophiles such as *p*-nitrophenoxide ion and phenylalanine methyl ester was studied. The rates of

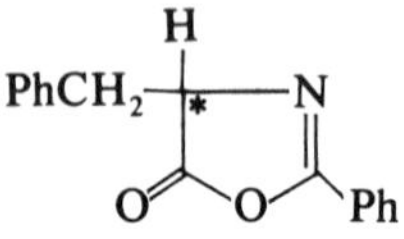

(**26**)

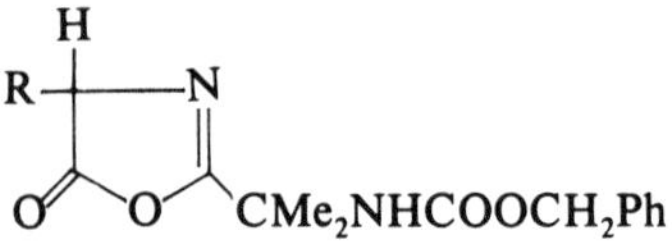

R = Me, PhCH$_2$ (**27**)

ring opening were in the order *p*-nitrophenoxide > phenylalanine > pH buffer 8 > water, while the rates of racemization of optically active peptide oxazolones, e.g., **27**, with several nucleophiles were also found to follow the same order.[69,70] The competing processes are shown in Eq. (14).[71,72] Optical activity decreases with increasing polarity (and

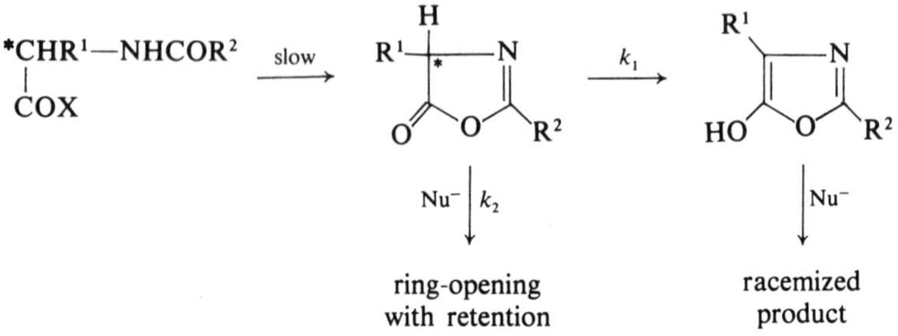

$$*CHR^1-NHCOR^2 \xrightarrow{slow} \text{(oxazolone)} \xrightarrow{k_1} \text{(hydroxy form)} \quad (14)$$
$$|$$
$$COX$$

ring-opening with retention / racemized product

basicity) of the solvent. The most important factor in controlling the racemization process is the nucleophilicity: basicity ratio of the attacking nucleophile, with ethyl glycinate having the best ratio.[73,74] Goodman and Glaser[75,76] observed that amino acid oxazolones are formed far more

---

[67] M. Goodman and K. C. Steuben, *J. Org. Chem.* **27**, 3409 (1962).
[68] M. Goodman and L. Levine, *J. Org. Chem.* **29**, 2918 (1964).
[69] M. Goodman and W. J. McGahren, *J. Am. Chem. Soc.* **87**, 3028 (1965).
[70] M. Goodman and W. J. McGahren, *J. Am. Chem. Soc.* **88**, 3887 (1966).
[71] D. S. Kemp and I. Rebek, *J. Am. Chem. Soc.* **92**, 5792 (1970).
[72] D. S. Kemp and S. W. Chien, *J. Am. Chem. Soc.* **89**, 2745 (1967).
[73] W. J. McGahren and M. Goodman, *Tetrahedron* **23**, 2017 (1967).
[74] M. Goodman and W. J. McGahren, *Tetrahedron* **23**, 2031 (1967).
[75] M. Goodman and C. B. Glaser, *Tetrahedron Lett.*, 3473 (1969).
[76] M. Goodman and C. B. Glaser, *J. Org. Chem.* **25**, 1954 (1970).

readily than peptide oxazolones. With basic reagents, oxazolones racemize due to proton abstraction at the acidic asymmetric center. Racemization of a peptide oxazolone occurs twelve times faster than ring opening, while amino acid oxazolones racemize 200 times faster than ring cleavage. With α-nucleophiles (containing two adjacent nucleophilic centers), which exhibit enhanced nucleophilicity relative to basicity, there is the possibility of both nucleophilic and electrophilic ("biphylic") interaction with the C=O group of the oxazolones. The competing reactions, ring opening vs. racemization, are also governed by temperature, solvent, and concentration. Siemion and co-workers[77-79] showed that azlactones of N-acylamino acids racemize 1000 times faster than the amino acid methyl esters in the presence of triethylamine, while t-butyl esters of amino acids are more stable toward racemization.

$$2 \quad \text{R}^1 \underset{O}{\overset{H}{\underset{\parallel}{\bigtriangleup}}} \text{R}^2 \rightleftharpoons \text{R}^1 \underset{O}{\overset{H}{\underset{\parallel}{\bigtriangleup}}}\text{NH}^+ \text{R}^2 + \text{R}^1 \underset{O}{\overset{}{\underset{\parallel}{\bigtriangleup}}} \text{R}^2 \quad (15)$$

In some cases, oxazolones may undergo autoracemization [Eq. (15)]. The presence of hydrogen at the 4-position of the oxazolone and α to the attacking atom in the nucleophile may be important for retention or loss of optical activity in these reactions. Similar studies were carried out by Grahl-Nielsen.[80]

2. Geometrical Isomerism

In the previous review on oxazolones in this series,[1] the unresolved controversy regarding the relative geometrical configurations of 1 and 2 was discussed in detail.[81] The problem has since been attacked with vigor by several groups of investigators, using NMR[82-84] and X-ray crystallographic[85,25] techniques. The structures are now clearly es-

[77] I. Z. Siemion and L. Wilschowitz, *Z. Naturforsch. B* **26**, 762 (1971).
[78] I. Z. Siemion and A. Dzugaj, *Rocz. Chem.* **40**, 1699 (1966).
[79] I. Z. Siemion, *Rocz. Chem.* **42**, 237 (1968).
[80] O. Grahl-Nielsen, *Chem. Commun.*, 1588 (1971).
[81] Filler,[1] p. 95.
[82] K. Brocklehurst, H. S. Price, and K. Williamson, *Chem. Commun.*, 884 (1968).
[83] A. P. Morgenstern, C. Schuitj, and W. T. Nauta, *Chem. Commun.*, 321 (1969).
[84] A. Maquestiau, Y. Van Haverbeke and R. N. Muller, *Bull. Soc. Chem. Belg.* **83**, 259 (1974).
[85] K. Brocklehurst, R. P. Bywater, R. A. Palmer, and R. Patrick, *Chem. Commun.*, 632 (1971).

tablished, with **1** having the E, and **2** the Z, configuration, as indicated at the beginning of this chapter. Mention was made earlier concerning the use of polyphosphoric acid to effect Z → E isomerizations.[3] The entire subject of geometrical isomerism in 2-aryl(aralkyl)-4-arylmethylene-(alkylmethylene)-2-oxazolin-5-ones is treated in detail in a very recent review.[86]

D. BIS-OXAZOLONES

A number of bis-oxazolones[87] have been prepared in recent years. A representative example of the 4,4' (unsaturated) type is compound **28**, prepared as an intermediate in the synthesis of DOPA dimer.[88,89] A series of 2,2' (saturated) bis-oxazolones (**29**) were prepared by Chilean chemists,[90–95] who studied their reactions with ethyl esters of amino

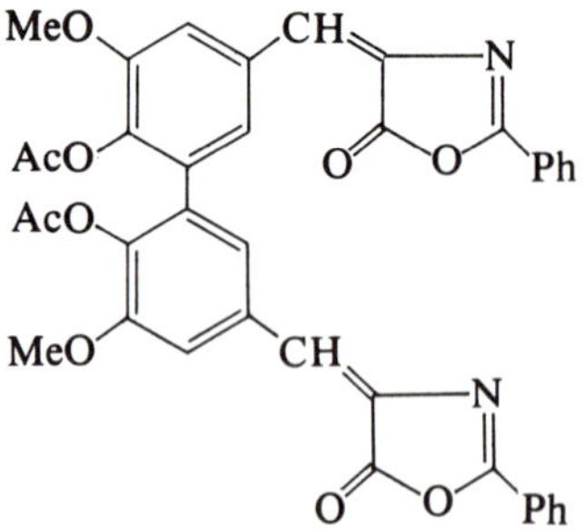

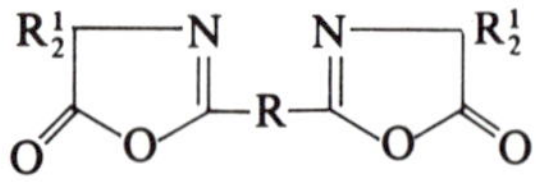

---

[86] Y. S. Rao and R. Filler, *Synthesis*, 749 (1975).
[87] Filler,[1] p. 80.
[88] Y. Omote, Y. Fujinuma, and N. Sugiyama, *Chem. Commun.*, 190 (1968).
[89] Y. Omote, Y. Fujinuma, and N. Sugiyama, *Bull. Chem. Soc. Jpn.* **42**, 1752 (1969).
[90] L. D. Azan, *An. Fac. Quim. Farm., Univ. Chile* **20**, 43 (1968) [*CA* **74**, 142328 (1971)].
[91] S. Diaz Palominos, *An. Fac. Quim. Farm., Univ. Chile* **20**, 47 (1968) [*CA* **74**, 142327 (1971)].
[92] R. Torres Gaona, *An. Fac. Quim. Farm., Univ. Chile* **19**, 124 (1967) [*CA* **70**, 115531 (1969)].
[93] T. Pinto, *An. Fac. Quim. Farm., Univ. Chile* **21**, 53 (1969) [*CA* **77**, 19995 (1972)].
[94] G. F. Mella, *An. Fac. Quim. Farm., Univ. Chile* **20**, 77 (1968) [*CA* **74**, 125594 (1971)].
[95] M. E. Aquilera, *An. Fac. Quim. Farm., Univ. Chile* **19**, 9 (1967) [*CA* **71**, 22314 (1969)].

Sec. III.A] THE CHEMISTRY OF OXAZOLONES 191

acids to yield dipeptides. Terephthaloyldiglycine was employed in the preparation of bisoxazolone dyes (**30**).⁹⁶ A bis-oxazolone in which the

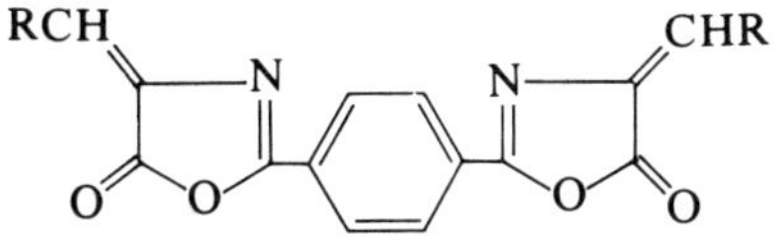

(**30**) 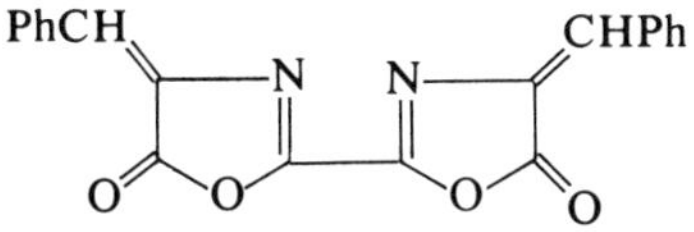   (**31**)

two rings are directly linked at the 2 and 2' positions (**31**) was prepared from oxalyl diglycine and benzaldehyde.⁹⁵ Compound **32** has also been obtained by the mixed anhydride method.⁹⁷

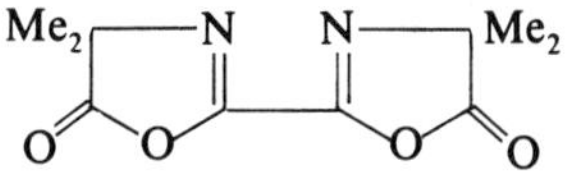

(**32**)

III. 3-Oxazolin-5-ones

3-Oxazolin-5-ones are frequently referred to as pseudoxazolones. The chemistry of these compounds was reviewed in the late 1960s.²,⁹⁸

A. Methods of Preparation

The 2-arylidenepseudoxazolones are generally prepared by cyclization of N-(α-halophenylacetyl)-α-amino acids.⁹⁹,¹⁰⁰ Thus, 2-phenylmethylene-4-phenyl-3-oxazolin-5-one (**35**) was obtained by treating N-(α-chlorophenylacetyl)-α-aminophenylacetic acid (**34**) with acetic anhydride⁹⁹ or from the more accessible N-mandeloyl-α-aminophenylacetic acid (**33**) and acetyl chloride¹⁰⁰ [Eq. (16)], presumably via the unstable chlorooxazolone, which rapidly loses hydrogen chloride.

---

⁹⁶ R. Schickfluss and W. Steckelberg, Ger. Offen. 2,225,546 (1973) [*CA* **80**, 97348 (1974)].
⁹⁷ L. D. Taylor, T. E. Platt, and M. H. Mach, *J. Polym. Sci., Part B* **8**, 537 (1970).
⁹⁸ Y. Iwakura, *Yuki Gosei Kagaku Shi* **26**, 479 (1968) [*CA* **69**, 67253 (1968)].
⁹⁹ G. Lucente, C. Gallina, and A. Romeo, *Ann. Chim. (Rome)* **56**, 1192 (1966).
¹⁰⁰ G. Adembri, *Ann. Chim. (Rome)* **50**, 374 (1960).

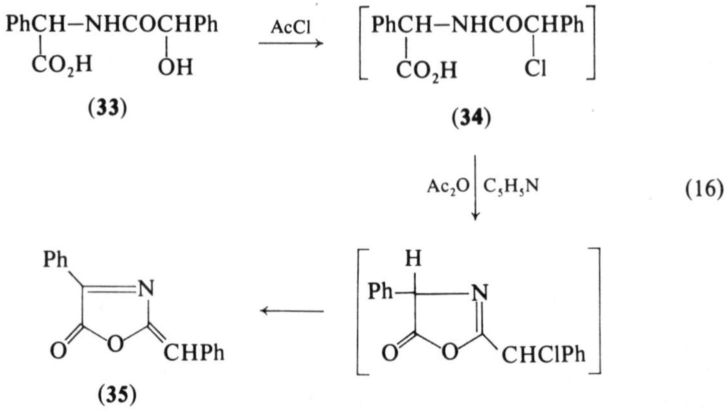

$$\text{PhCH-NHCOCHPh} \atop {|| \atop CO_2HOH}} \quad \xrightarrow{\text{AcCl}} \quad \left[{\text{PhCH-NHCOCHPh} \atop {|| \atop CO_2HCl}} \right]$$

(33) (34)

$$\downarrow Ac_2O \,|\, C_5H_5N \qquad (16)$$

(35) ← [pseudoxazolone intermediate with CHClPh]

Iwakura and co-workers[101,102] reported that cyclodehydration of N-acryloyl-α-amino acids in the presence of acetic anhydride and pyridine gives pseudoxazolones [Eq. (17)], whereas in the absence of pyridine, only 2-oxazolin-5-ones are obtained.

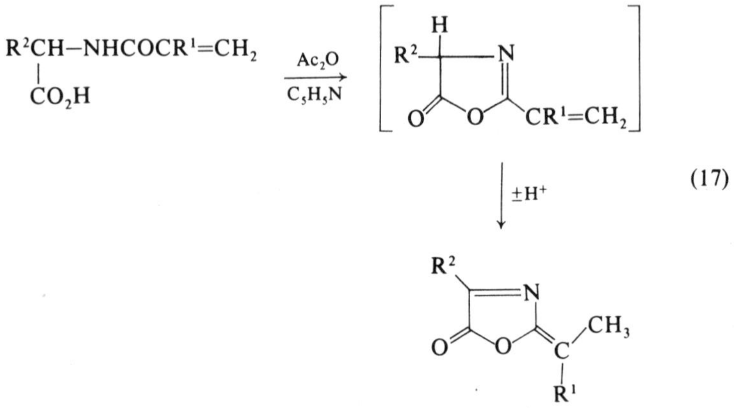

$$R^2CH-NHCOCR^1=CH_2 \atop {| \atop CO_2H}} \quad \xrightarrow[C_5H_5N]{Ac_2O} \quad \left[\text{intermediate with } CR^1=CH_2 \right]$$

$$\downarrow \pm H^+ \qquad (17)$$

[product oxazolone with =C(CH$_3$)R^1 side chain]

2-Perfluoroalkyl pseudoxazolones,[103,104] e.g., **36**, are useful intermediates for separation of enantiomers of the parent amino acids.

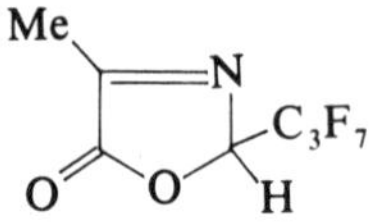

(36) [structure with Me, N, C$_3$F$_7$, H]

[101] Y. Iwakura, F. Toda, and Y. Torii, *Tetrahedron Lett.*, 4427 (1966).
[102] Y. Iwakura, F. Toda, and Y. Torii, *Tetrahedron* **23**, 3363 (1967).
[103] W. Steglich, E. Frauendorfer, and F. Weygand, *Chem. Ber.* **104**, 687 (1971).
[104] F. Weygand, W. Steglich, and W. Oettmeier, *Chem. Ber.* **103**, 818 (1970).

Sec. III.A] THE CHEMISTRY OF OXAZOLONES 193

Several rearrangement reactions lead to pseudoxazolones. Thus, 5-acyloxyoxazoles, in the presence of nitrogen bases, such as 4-dimethylaminopyridine, rearrange to a mixture of 2-acyl-3-oxazolin-5-ones and 4-acyl-2-oxazolin-5-ones[105,106] [Eq. (18)]. Recently, the racemate of the

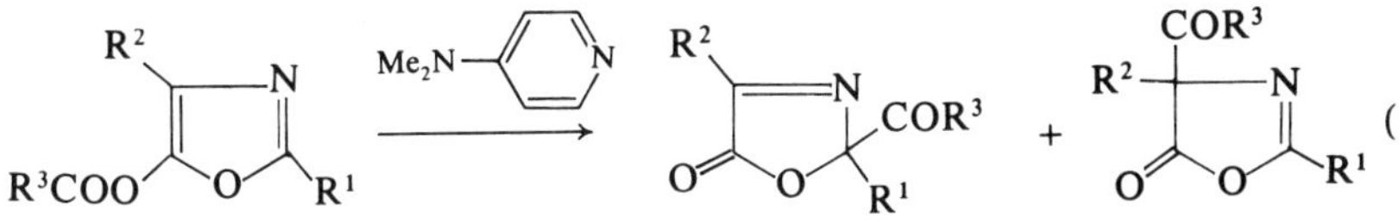

hydroxylactone sulfoxide (**37**), was converted into the 3-oxazolin-5-one (**40**) on treatment with acetyl chloride[107] (Scheme 5). The authors

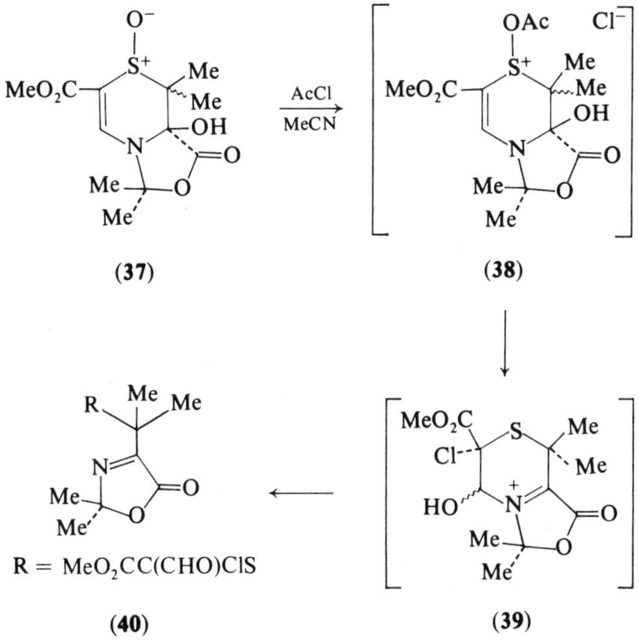

SCHEME 5

suggest that this transformation probably proceeds via species **39**, formed from the acetoxysulfonium salt (**38**) by a Pummerer-type reaction and a 1,3-hydroxyl shift.

[105] W. Steglich and G. Höfle, *Tetrahedron Lett.*, 4727 (1970).
[106] W. Steglich and G. Höfle, *Chem. Ber.* **102**, 899 (1969).
[107] R. J. Stoodley and R. B. Wilkins, *Chem. Commun.*, 796 (1974).

B. REACTIONS

1. Behavior under Friedel–Crafts Conditions

Compound **35** reacts with benzene in the presence of anhydrous aluminium chloride to give N-(diphenylacetyl)phenylglycine[108] [Eq. (19)]. Similar results were obtained with 2-isopropylidene pseudoxa-

$$35 \xrightarrow{C_6H_6}_{AlCl_3} \left[\begin{array}{c} \text{Ph—CH—N} \\ | \quad \quad \| \\ O=C \quad C-CHPh_2 \\ \diagdown O \diagup \end{array} \right] \xrightarrow{H_3O^+} \begin{array}{c} PhCHCO_2H \\ | \\ NHCOCHPh_2 \end{array} \quad (19)$$

zolones.[109,110] 2-Dichloromethylene-4-alkyl-3-oxazolin-5-ones react with toluene to give **41**.[111]

$$\begin{array}{c} R \\ \diagdown \\ \quad \quad N \\ \| \\ O=C \quad C=C(C_6H_4Me)_2 \\ \diagdown O \diagup \end{array}$$

(**41**)

2. Reaction with Grignard Reagents

Adembri and co-workers[112,113] reported that excess methylmagnesium iodide reacts with **35** to give two of the four possible pairs of enantiomers of **42**, according to the mechanism proposed in Eq. (20). However, they could not duplicate the results of Mustafa and co-workers,[108] who claimed that the reaction proceeds without ring closure to give the amide **43**. Reaction of **35** with phenylmagnesium bromide takes a different course, with the formation of the oxazolidone **44**, by addition to the C=N linkage. 2-Isopropylidene-4-alkylpseudoxazolones react with the phenyl Grignard reagent to afford pyrrolidine-2,4-diones.[114]

---

[108] A. Mustafa, A. E. Sammour, M. M. Noureldeen, T. Salama, and M. K. Hilmy, *Justus Liebigs Ann. Chem.* **689**, 189 (1965).
[109] Y. Iwakura, F. Toda, and Y. Torii, Japanese Patent 70 04,489 (1970) [*CA* **73**, 14509 (1970)].
[110] Y. Iwakura, F. Toda, and Y. Torii, *J. Org. Chem.* **32**, 3202 (1967).
[111] Y. Iwakura, F. Toda, M. Kosugi, and Y. Torii, *J. Org. Chem.* **36**, 3990 (1971).
[112] G. Adembri, M. Scotton, and P. Tedeschi, *Chim. Ind.* (Rome) **48**, 1346 (1966).
[113] G. Adembri, M. Scotton, G. Speroni, and P. Tedeschi, *Gazz. Chim. Ital.* **103**, 3 (1973).
[114] Y. Iwakura, F. Toda, and Y. Torii, Japanese Patent 71 25,380 (1971) [*CA* **75**, 140677 (1971)].

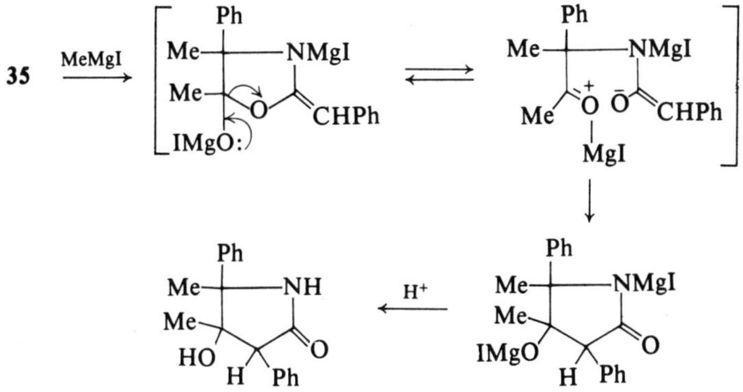

Sec. III.B] THE CHEMISTRY OF OXAZOLONES 195

$$35 \xrightarrow{MeMgI} \left[\text{structure with Ph, Me, NMgI, IMgO, CHPh} \right] \rightleftharpoons \left[\text{structure with Ph, Me, NMgI, MgI, CHPh} \right] \quad (20)$$

(42) ← H⁺ ← (IMgO intermediate)

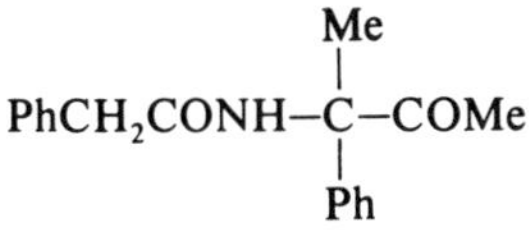

(43) PhCH₂CONH—C(Me)(Ph)—COMe

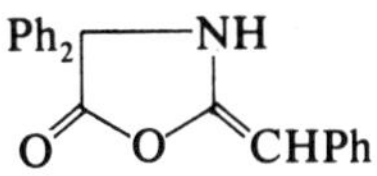

(44) Ph₂-NH / O-CHPh oxazoline

3. Reaction with Hydrazoic Acid

2-Isopropylidene-4-alkyl-3-oxazolin-5-ones react with hydrazoic acid to yield ureas[115] [Eq. (21)].

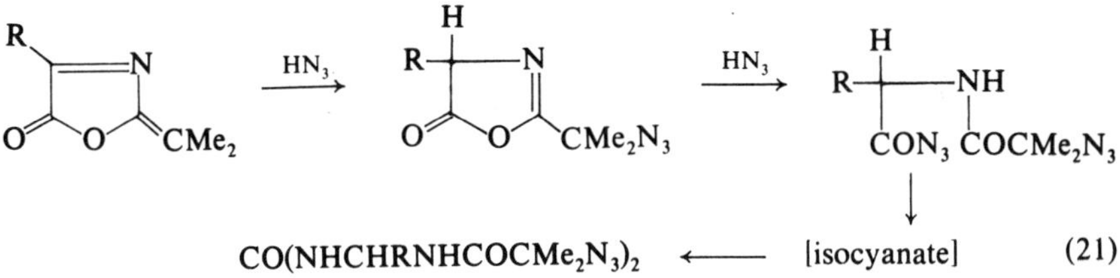

$$\text{oxazolinone} \xrightarrow{HN_3} \text{R-N-CMe}_2\text{N}_3 \xrightarrow{HN_3} \text{R-NH, CON}_3, \text{COCMe}_2\text{N}_3$$

$$\text{CO(NHCHRNHCOCMe}_2\text{N}_3)_2 \leftarrow [\text{isocyanate}] \quad (21)$$

4. Special Reactions

2-Trifluoromethyl-4-isopropyl-3-oxazolin-5-one (**45**) reacts with aryl-thioesters of arylsulfonic acids to yield derivatives which, on heating, give benzothiazines[116] [Eq. (22)]. With arylsulfenyl chlorides, 2-

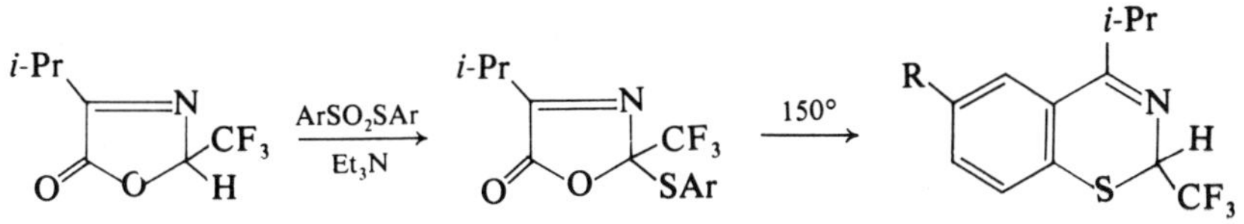

$$(\mathbf{45}) \xrightarrow[\text{Et}_3\text{N}]{\text{ArSO}_2\text{SAr}} \text{intermediate} \xrightarrow{150°} \text{benzothiazine} \quad (22)$$

[115] Y. Iwakura, F. Toda, and Y. Torii, *J. Org. Chem.* **33**, 2541 (1968).
[116] P. Gruber, L. Muller, and W. Steglich, *Chem. Ber.* **106**, 2863 (1973).

trifluoromethyl-4-isopropyl-4-thioaryl-2-oxazolin-5-ones are obtained, and these compounds yield alkene derivatives on heating.

Pyrolysis of the product isolated from the reaction of **45** and 2,4-dinitrofluorobenzene gives an acyloxyindazole[117] [Eq. (23)].

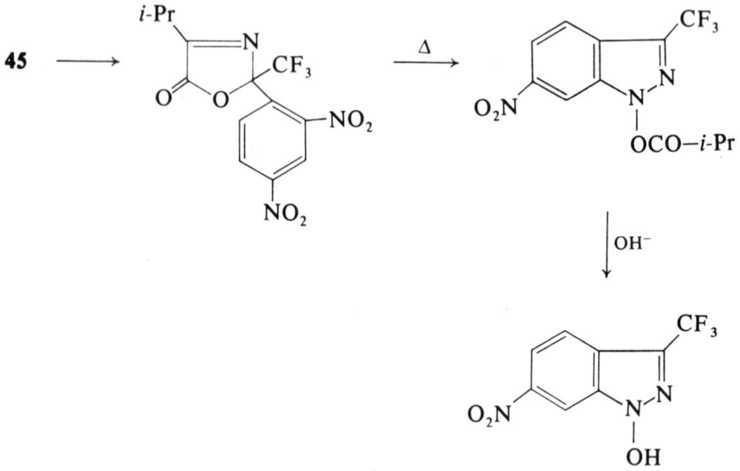

(23)

5. *Photoreactions*

The previously proposed structure for the photodimer (**46**) of 2-phenylmethylene-4-methyl-3-oxazolin-5-one[118] has been confirmed by X-ray crystallographic analysis,[119] which indicated the centrosymmetric structure originally postulated. The 4-phenyl analog (**35**) also dimerizes to the cyclobutane structure when irradiated in benzene.[120]

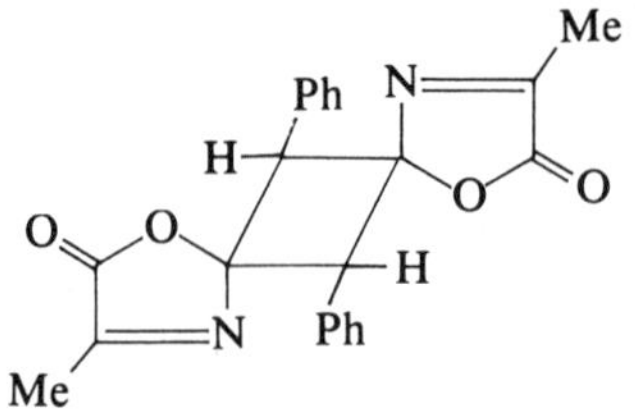

(**46**)

[117] W. Steglich, B. Kübel, and P. Gruber, *Chem. Ber.* **106**, 2870 (1973).
[118] R. Filler and E. J. Piasek, *J. Org. Chem.* **28**, 221 (1963).
[119] R. Filler, B. S. Green, and L. Heller, unpublished results.
[120] G. Adembri, F. M. Carlini, P. Sarti-Fantoni, and M. Scotton, *Tetrahedron Lett.*, 3347 (1972).

Irradiation of phenyl-$2H$-azirines in the presence of carbon dioxide leads to the formation of the 3-oxazoline-5-one system[121-123] and, in some cases, to the isomeric 2-oxazolin-5-one[122] [Eq. (24)]. The azirines serve as incipient nitrile ylides, whose 1,3-dipolar structure permits cycloaddition to the dipolarophile CO_2,[123] [Eq. (25)]. The reverse reaction, photolytic extrusion of CO_2 from pseudoxazolones, is synthetically useful, since the dipolar nitrile ylide thus formed can be trapped with a variety of dipolarophiles. Thus, 2,2,4-triphenyl-3-oxazolin-5-one (48) is readily converted into the stabilized ylide (49)[124] [Eq. (26)], and the use of methyl acrylate,[122] acrylonitrile,[122] and dimethylacetylene dicarboxy-

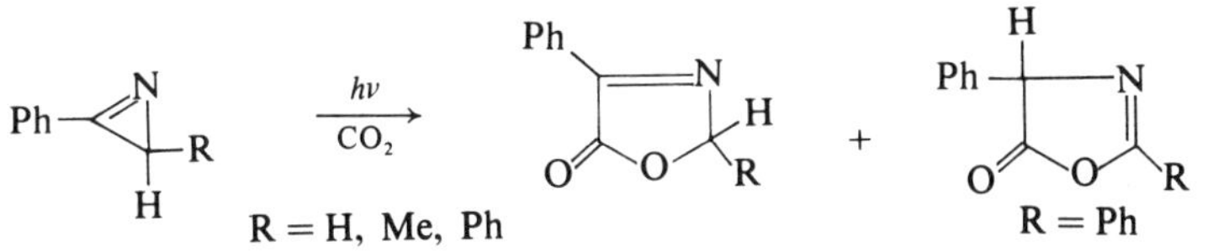

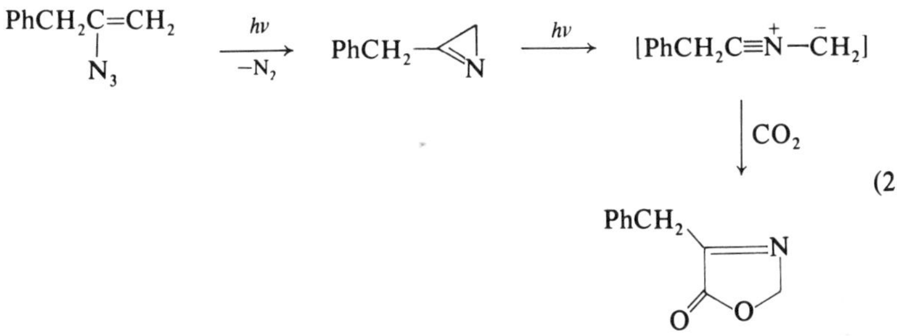

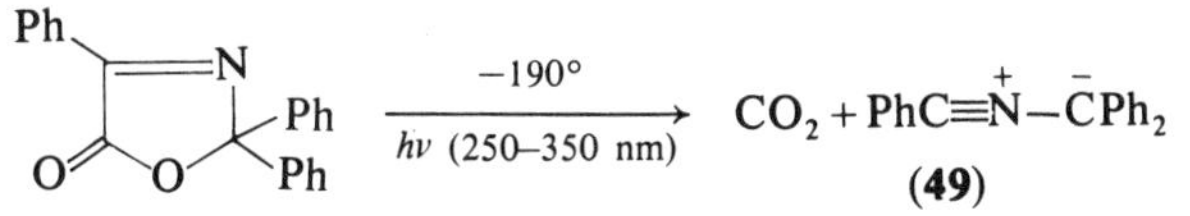

late[125] with other pseudoxazolones leads to pyrrole and pyrroline derivatives.

---

[121] H. Giezendanner, M. Märky, B. Jackson, H. J. Hansen, and H. Schmid, *Helv. Chim Acta* **55**, 745 (1972).

[122] A. Padwa and S. I. Wetmore, *J. Am. Chem. Soc.* **96**, 2414 (1974).

[123] A. Orahovats, B. Jackson, H. Heimgartner, and H. Schmid, *Helv. Chim. Acta* **56**, 2007 (1973).

[124] W. Sieber, P. Gilgen, S. Chaloupka, H. J. Hansen, and H. Schmid, *Helv. Chim. Acta* **56**, 1679 (1973).

[125] W. Steglich, P. Gruber, H. U. Heininger, and F. Kneidl, *Chem. Ber.* **104**, 3816 (1971).

IV. 2-Oxazolin-4-ones

Quite a number of compounds belonging to this class have been reported recently. Their importance has been enhanced because the 2-amino derivatives have found use as tranquilizers, antidepressants, memory aids, and hunger suppressants. The antibiotic indolmycin (**50**) contains this ring.[126]

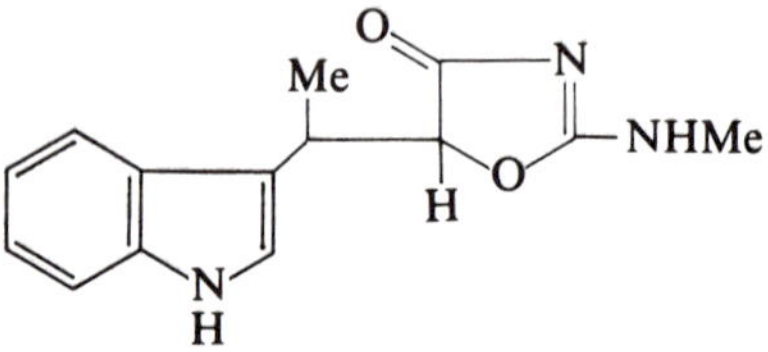

(**50**)

A. Preparation

1. *From Benzoyl Isocyanate and Diazoalkanes*

Sheehan and Izzo[127] reported that benzoyl isocyanate reacts with diazomethane to form 2-phenyl-2-oxazolin-4-one (**51**), [Eq. (27)]. This reaction has been employed by Neidlein and Bottler[128] in the synthesis of **52**. Phenyl- and diphenyldiazomethane react with benzoyl isocyanate.[129,130] In this fashion, 2,5,5-triphenyl-2-oxazolin-4-one was isolated.[130] The preparation of this compound by another route had been

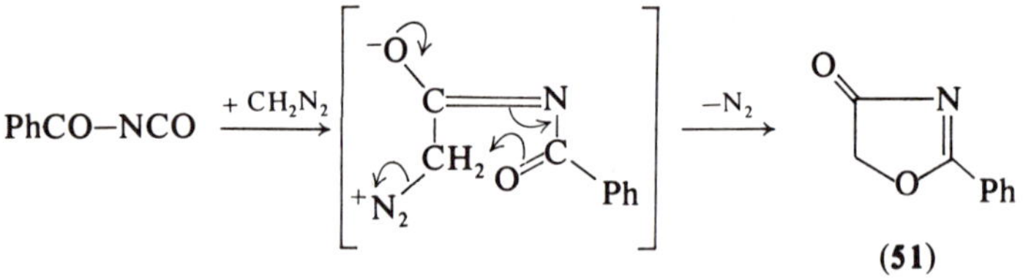

(**51**)   (27)

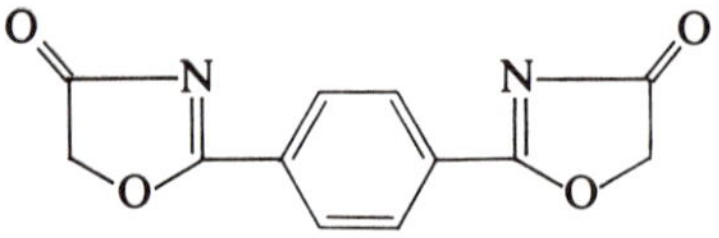

(**52**)

---

[126] U. Hornemann, L. H. Hurley, M. K. Speedie, and H. G. Floss, *J. Am. Chem. Soc.* **93**, 3028 (1971).
[127] J. C. Sheehan and P. T. Izzo, *J. Am. Chem. Soc.* **71**, 4059 (1949).
[128] R. Neidlein and R. Bottler, *Chem. Ber.* **100**, 698 (1967).
[129] E. J. Browne, E. E. Nunn, and J. B. Polya, *J. Chem. Soc. C*, 1515 (1970).
[130] Y. S. Rao and R. Filler, *Chem. Commun.*, 1622 (1970).

reported previously.[131] Benzoyl isocyanate and 2,2,2-trialkoxy-2,2-dihydro-1,3,5-dioxaphosphalenes form 5-acyl-2-oxazolin-4-ones[132-134] [Eq. (28)]. Substitution of carboalkoxy and carbamyl isocyanates af-

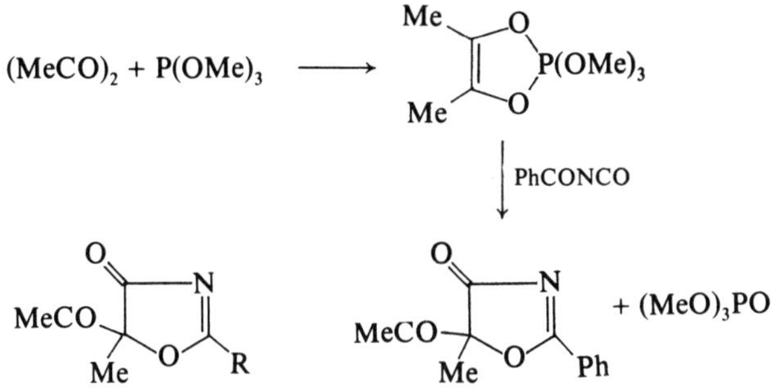

$$(MeCO)_2 + P(OMe)_3 \longrightarrow \begin{array}{c} Me \\ Me \end{array}\! P(OMe)_3$$

↓ PhCONCO   (28)

(MeCO / Me) oxazoline with R; R = OR′, NPh$_2$ (**53**) and MeCO / Me / Ph oxazoline + (MeO)$_3$PO

fords the 2-alkoxy or amino oxazolone (**53**).[134] Finally, benzoyl isocyanate and dimethyloxosulfonium methylide form products that, on pyrolysis at 140°, yield **51**[135] [Eq. (29)].

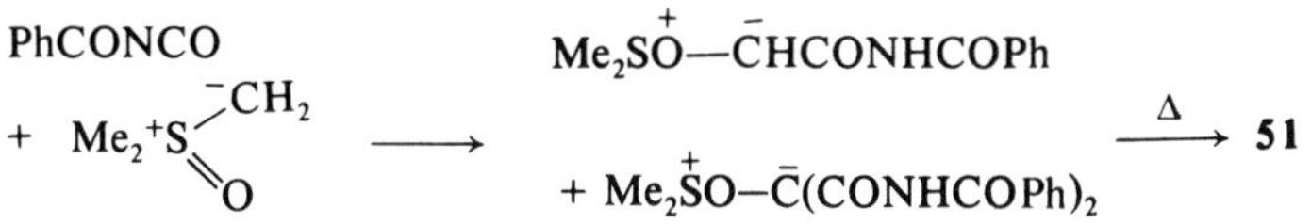

$$\begin{array}{c} PhCONCO \\ + Me_2{}^+S{-}^-CH_2\!\!=\!\!O \end{array} \longrightarrow \begin{array}{c} Me_2\overset{+}{S}O{-}\bar{C}HCONHCOPh \\ + Me_2\overset{+}{S}O{-}\bar{C}(CONHCOPh)_2 \end{array} \xrightarrow{\Delta} 51 \quad (29)$$

2. Alkylation with Ethyl Bromoacetate

The principal product obtained by the reaction of 1-methyl-1,4,5,6-tetrahydronicotinamide and ethyl bromoacetate has been identified by Troxler[136,137] as the substituted-2-oxazolin-4-one (**54**).

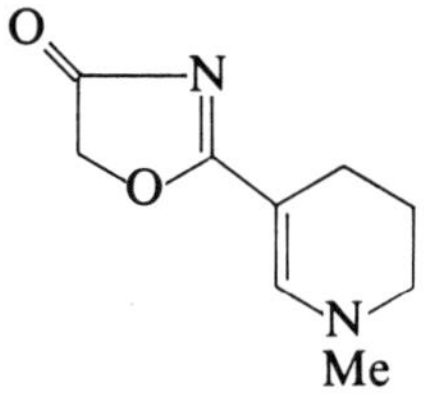

(**54**)

[131] K. Hohenlohe-Oehringen, *Monatsh. Chem.* **89**, 588 (1958).
[132] F. Ramirez and C. D. Telefus, *J. Org. Chem.* **34**, 376 (1969).
[133] F. Ramirez, *Synthesis*, 103 (1974).
[134] F. Ramirez, C. D. Telefus, and V. A. V. Prasad, *Tetrahedron* **31**, 2007 (1975).
[135] O. Tsuge, K. Sakai, and M. Tashiro, *Tetrahedron* **29**, 1984 (1973).
[136] F. Troxler, *Helv. Chim Acta* **56**, 374 (1973).
[137] F. Troxler, *Helv. Chim. Acta* **56**, 1815 (1973).

3. Cyclization of α-Haloimides

Rao and Filler[130] prepared a number of 2-phenyl-5,5-disubstituted 2-oxazolin-4-ones by base-catalyzed cyclization of halogenated bis-amides, such as α-chloro-α,α-diphenylacetylbenzamides [Eq. (30)]. When chloroacetamide reacts with oxalyl chloride, the bis-oxazolone (**55**) is obtained.[138]

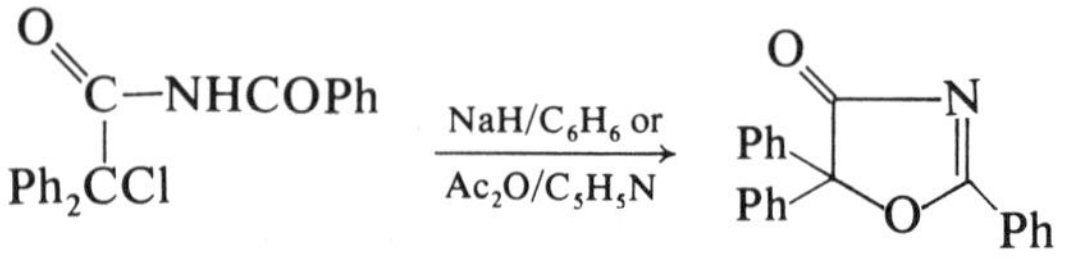

$$\underset{Ph_2CCl}{\overset{O}{\underset{\|}{C}}-NHCOPh} \xrightarrow[Ac_2O/C_5H_5N]{NaH/C_6H_6 \text{ or}} \text{Ph-oxazolone} \qquad (30)$$

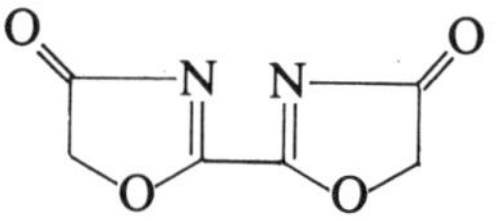

(**55**)

4. 2-Amino-2-oxazolin-4-ones

These are prepared by base-catalyzed cyclization of bromoacylureas. The patent literature abounds in these compounds because of their medicinal value. Howell and co-workers[139,140] condensed ethyl mandelate with dialkylcyanamides to prepare 2-dialkylamino-2-oxazolin-4-ones [Eq. (31)].

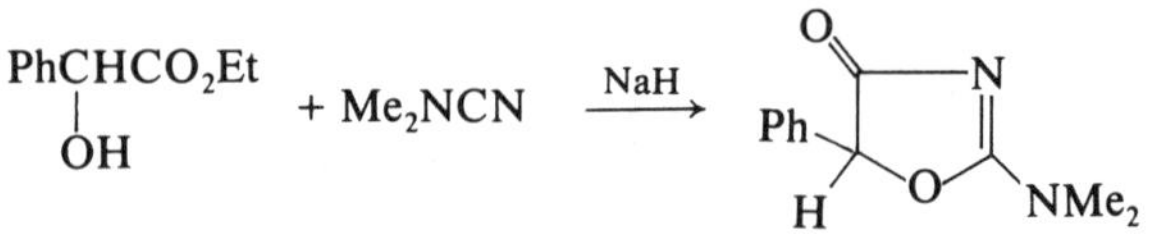

$$\underset{OH}{\overset{PhCHCO_2Et}{|}} + Me_2NCN \xrightarrow{NaH} \text{product} \qquad (31)$$

B. REACTIONS

The 2-oxazolin-4-one ring is readily opened under mild hydrolytic conditions to yield open-chain amides.[127, 132] The alkoxy and phenoxy groups in **53** are cleaved rapidly by water in organic solvents, whereas the amino derivative (**53**, R = NPh$_2$) requires mild acid catalysis.

---

[138] I. V. Smolanka, S. M. Khripak, and V. I. Staninents, *Ukr. Khim. Zh.* **32**, 202 (1966) [*CA* **64**, 15863 (1966)].

[139] See, e.g., C F. Howell, R. A. Hardy, and N. Q. Quinones, U.S. Patent 3,321,470 (1967) [*CA* **67**, 64382 (1967)].

[140] R. A. Hardy, C. F. Howell, and N. Q. Quinones, U.S. Patent 3,313,688 (1967) [*CA* **67**, 90791 (1967)].

However, the ring remains intact and oxazolidinediones are isolated[134] [Eq. (32)]. When Y is phenoxy, the tetrahedral intermediate is relatively

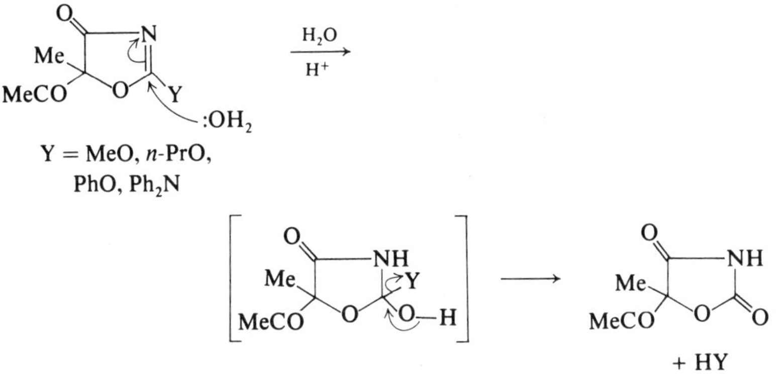

Y = MeO, n-PrO, PhO, Ph$_2$N

(32)

+ HY

stable and isolable. 1,3 O → N-alkyl migration occurs on heating the 2-alkoxy compounds.[134] Methyl migration is much more facile than that of the propyl group. The products are the corresponding 3-alkyl-5-acetyl-5-methyloxazolidine-2,4-diones. The migration is related to the Chapman rearrangement of O-alkylimino carboxylate esters. The relative rates, Me > n-Pr, suggest that steric factors are more important than enhanced carbocation character.

Irradiation of a solution of **51** in benzene causes an interesting isomerization to 4-phenyl-4-oxazolin-2-one (**56**), via an epoxy isocyanate[141] [Eq. (33)]. In the presence of 1,1-dimethoxyethylene, a 2 + 2 cycloaddition occurs in competition with α-cleavage to give **57**. Compound **51** is also reported to react with dimethylacetylene dicarboxylate

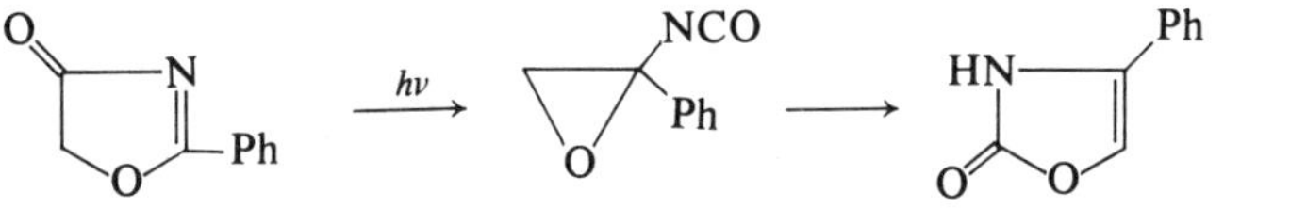

(33)

(51)   (56)

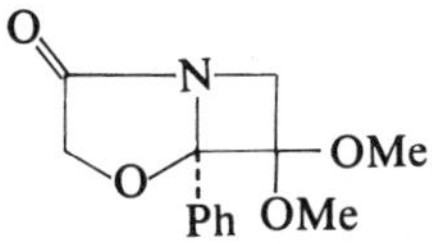

(57)

and dibenzoylacetylene to give substituted furans.[142]

Finally, **51** reacts with hydrazine hydrate to yield 1,2,4-triazoles.[129]

[141] T. H. Koch and R. M. Rodehorst, *Tetrahedron Lett.*, 4039 (1972).
[142] K. T. Potts and J. Marshall, *Chem. Commun.*, 1000 (1972).

V. 4-Oxazolin-2-ones

A. Preparation

1. From Cyclic Carbonates

The method of preparation is depicted in Eq. (34).[143-145]

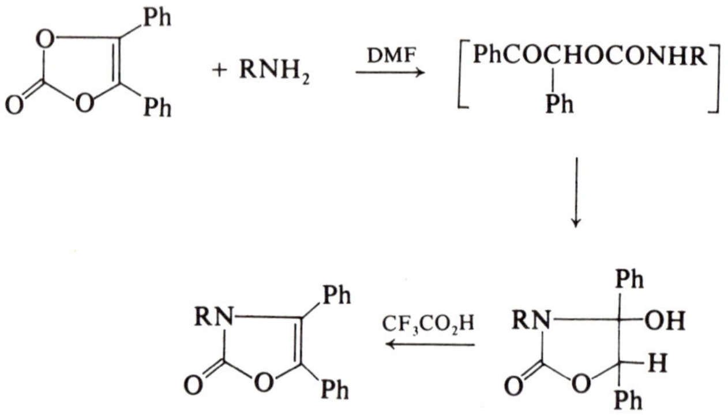

$$\text{(34)}$$

Intermediate: [PhCOCHOCONHR with Ph substituent], leading via CF$_3$CO$_2$H from the hydroxy intermediate (RN—C(Ph)(OH)—O—CH(Ph)—, with H) to the 3-RN-4,5-diphenyl-4-oxazolin-2-one.

2. From Amine Derivatives

N-Phenylphenacylamine, on heating with ethyl chloroformate, affords 3,5-diphenyl-4-oxazolin-2-one[146] which is obtained also from N-phenyl-N-phenacylglycine ethyl ester and sodium methoxide [Eq. (35)].

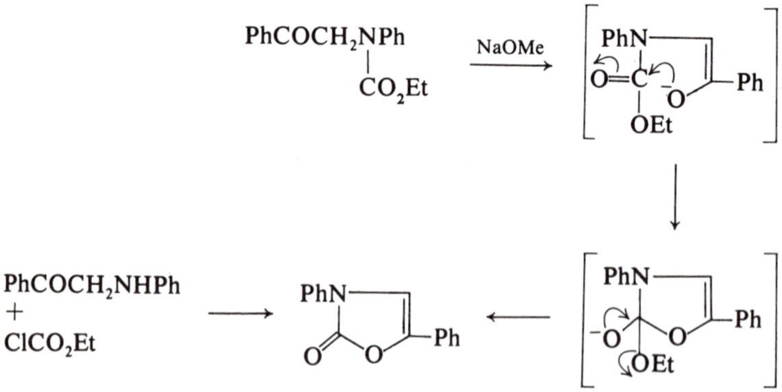

$$\text{(35)}$$

---

[143] J. C. Sheehan and F. S. Guziec, *J. Am. Chem. Soc.* **94**, 6561 (1972).
[144] J. C. Sheehan and F. S. Guziec, *J. Org. Chem.* **38**, 3034 (1973).
[145] E. Eimers and H. Rudolph, Ger. Offen, 2,304,589 (1974) [*CA* **82**, 58461 (1975)].
[146] H. Brachwitz, *J. Prakt. Chem.* **313**, 667 (1971).

3. From Benzoin Derivatives

The reaction of benzoin with potassium isocyanate in DMSO or pyridine gives carbamates, which undergo cyclodehydration to form 4-oxazolin-2-ones. Instead of benzoin, α-bromocarbonyl compounds may be used[147-149] [Eq. (36)].

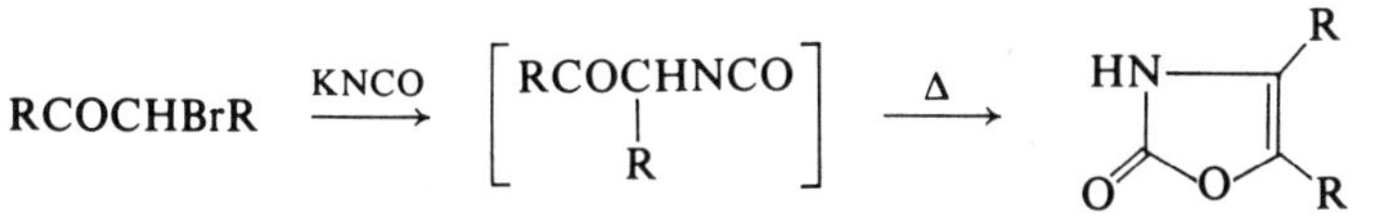

$$\text{RCOCHBrR} \xrightarrow{\text{KNCO}} \left[\begin{array}{c} \text{RCOCHNCO} \\ | \\ \text{R} \end{array} \right] \xrightarrow{\Delta} \text{HN-}\underset{\text{O}}{\overset{\text{R}}{\diagdown}}\text{R} \quad (36)$$

4. Miscellaneous Methods

A number of other approaches to 4-oxazolin-2-ones have been described, including hydrolysis of 2-amino-4,5-diphenyl-5-cyano-oxazole,[150] photolysis of 3-hydroxyisoxazoles,[151] and thermal rearrangement of an isoxazolin-3-one[152] [Eq. (37)].

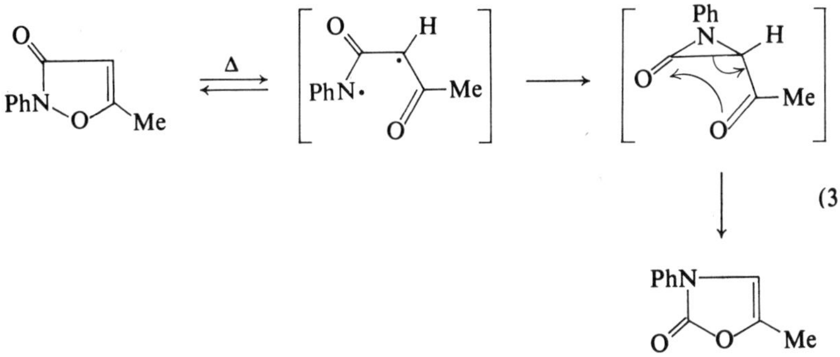

B. REACTIONS

Although they were originally assumed to be unreactive, these oxazolones exhibit considerable reactivity. Thus, compound **58** reacts with

---

[147] V. M. Dziomko and A. V. Ivashchenko, *Zh. Org. Khim.* **9**, 2191 (1973).
[148] V. M. Dziomko and A. V. Ivashchenko, USSR Patent 414,262 (1974) [*CA* **80**, 133412 (1974)].
[149] V. M. Dziomko and A. V. Ivashchenko, USSR Patent 427,015 (1974) [*CA* **81**, 77899 (1974)].
[150] K. Gewald, H. Schaeffer, and B. Plumbohm, *J. Prakt. Chem.* **315**, 44 (1973).
[151] M. Nakagawa, T. Nakamura, and K. Tomita, *Agr. Biol. Chem.* **38**, 2205 (1974).
[152] A. R. Gagneux and R. Göschke, *Tetrahedron Lett.*, 5451 (1966).

POCl₃, followed by ethylenediamine, to form an imidazoimidazole[147] [Eq. (38)]. 4-Oxazolin-2-ones also react with hydrazine hydrate to yield

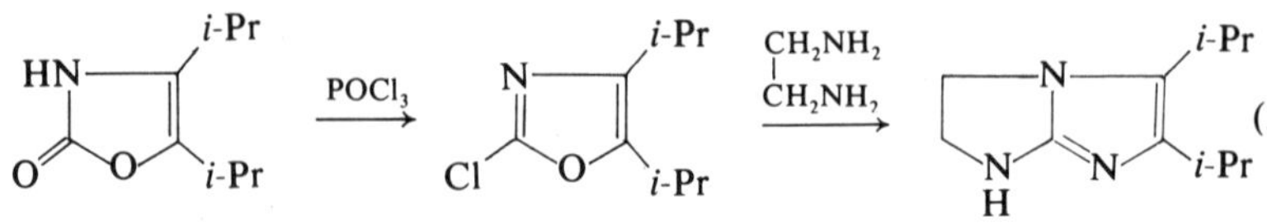

[Structure diagrams for Eq. (38): compound (58) → via POCl₃ → chloro-oxazoline → via CH₂NH₂/CH₂NH₂ → imidazoimidazole] (38)

1,2,4-triazin-3-ones,[153] and with primary[154] and secondary amines.[155] Compound 59 is converted into an N-hydroxyhydantoin in a two-step process by benzyl nitrite[156] [Eq. (39)]. The use of 4,5-diphenyl-4-

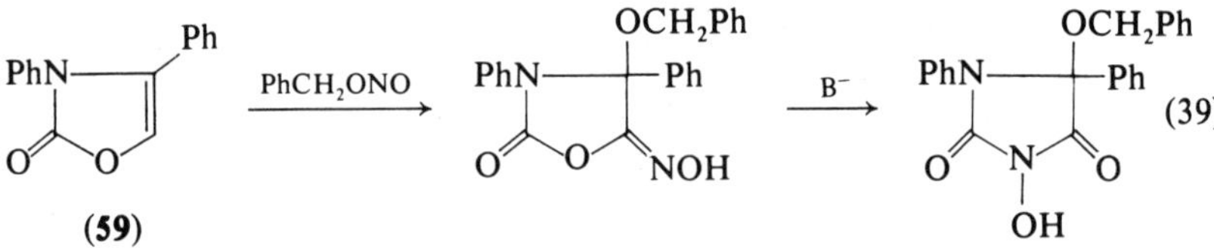

[Structure diagrams for Eq. (39): compound (59) → via PhCH₂ONO → oxime intermediate → via B⁻ → N-hydroxyhydantoin] (39)

oxazolin-2-one derivatives as a protective group for α-amino acids[143, 144] is illustrated in Eq. (40). Oxidation with m-chloroperbenzoic acid in

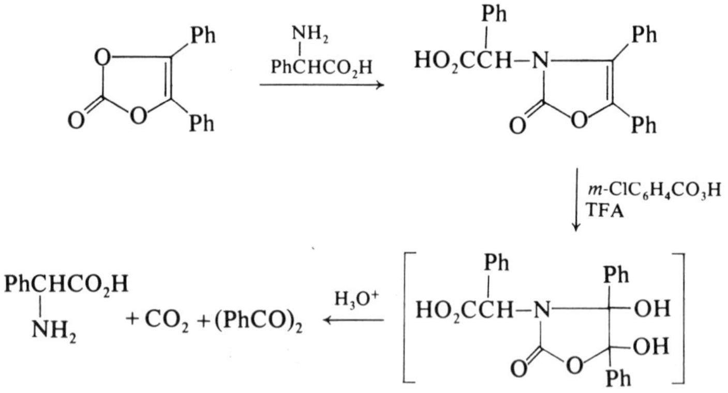

[Structure diagrams for Eq. (40): 4,5-diphenyl-4-oxazolin-2-one + PhCHCO₂H/NH₂ → N-substituted oxazolinone → via m-ClC₆H₄CO₃H/TFA → diol intermediate → via H₃O⁺ → PhCHCO₂H/NH₂ + CO₂ + (PhCO)₂] (40)

CF₃CO₂H gives the vic-diol, which undergoes ring cleavage to regenerate the amino acid. This method has been used for the synthesis of peptides.

[153] O. P. Shvaika and G. P. Klimisha, *Dopov. Akad. Nauk Ukr RSR, Ser B* **32**, 350 (1970) [*CA* **73**, 35433 (1970)].
[154] M. F. Saettone, V. Nuti, and A. Dasettimo, *Gazz. Chim. Ital.* **96**, 1615 (1966).
[155] M. F. Saettone and A. Marsili, *Tetrahedron Lett.*, 6009 (1966).
[156] A. Marsili, M. F. Saettone, and E. Bucci, *J. Org. Chem.* **33**, 2884 (1968).

VI. 3-Oxazolin-2-ones

A. Preparation

Notwithstanding a statement in the earlier review,[157] these compounds have been known since 1944. Spielman[158] reported that treatment of the silver salt of 5,5-dimethyloxazolidine-2,4-dione with ethyl iodide gave the 3-ethyl derivative. However, the compound was later shown[159] to be 5,5-dimethyl-4-ethoxy-3-oxazolin-2-one **(60)**, the product of O-

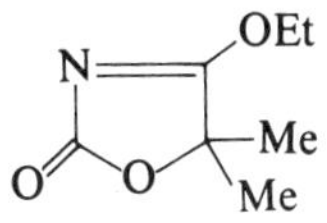

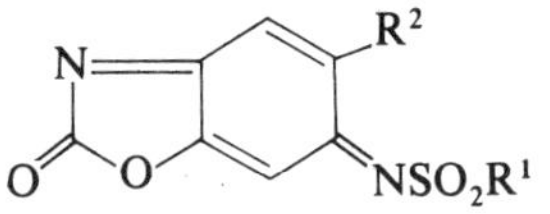

alkylation. Fused 3-oxazolin-2-ones, such as **61–63**, have also been described.[160–162] The reaction of the α-cyanoisopropyl ester of chloroformic acid **(64)** with aniline yields the 3-oxazolin-2-one **(66)**,[163] rather than the intermediate carbamate **(65)**, which was claimed earlier.[164] The reaction may proceed as indicated in Eq. (41). The 4-chloro analog **(67)**, obtained by a similar reaction from acetone cyanohydrin and phosgene, is converted into **66** on treatment with aniline.

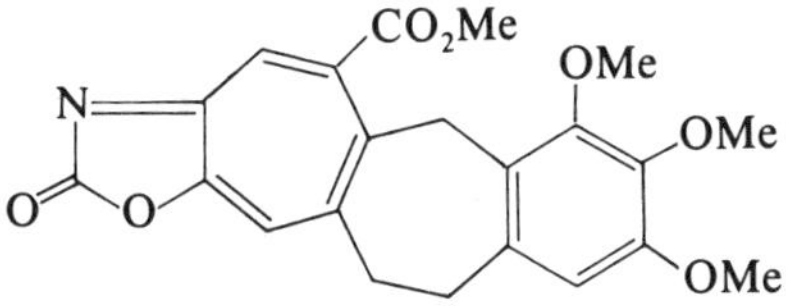

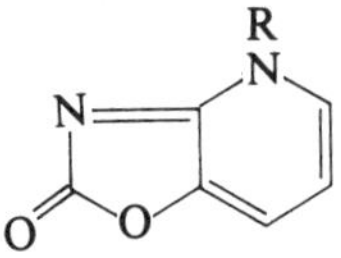

[157] Filler,[1] p. 76.
[158] M. A. Spielman, *J. Am. Chem. Soc.* **66**, 1244 (1944).
[159] J. S. H. Davies and W. H. Hook, *J. Chem. Soc.*, 30 (1950).
[160] I. Baxter, D. W. Cameron, and M. R. Thoseby, *J. Chem. Soc. C*, 850 (1970).
[161] J. Schreiber, W. Leimgruber, M. Pesaro, P. Schudel, T. Threlfall, and A. Eschenmoser, *Helv. Chim. Acta* **44**, 581 (1961).
[162] E. Renk, F. Ostermayer, and R. Denss, Swiss Patent 461,489 (1968) [*CA* **70**, 68344 (1969)].
[163] V. V. Dovlatyan and E. N. Ambartsumyan, *Arm. Khim Zh.* **23**, 49 (1970) [*CA* **73**, 14411 (1970)].
[164] I. G. Khaskin, V. I. Kondratenko, and V. T. Vdovichenko, USSR Patent 183,733 (1966) [*CA* **66**, P 37619 (1967)].

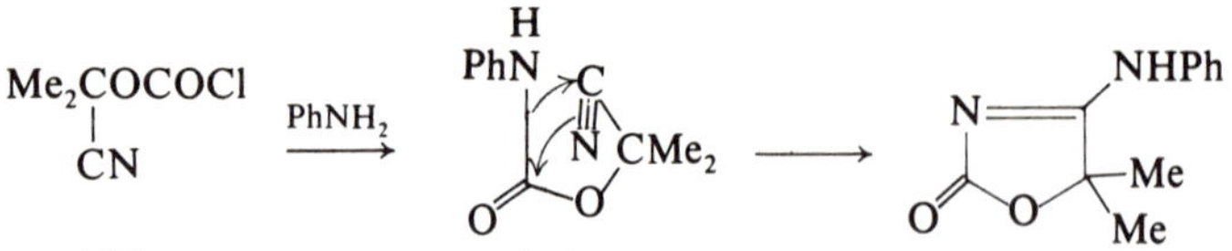

$$\text{Me}_2\text{COCOCl} \atop \text{CN} \quad \xrightarrow{\text{PhNH}_2} \quad (65) \quad \longrightarrow \quad (66) \qquad (41)$$

(64)

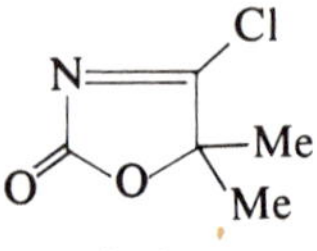

(67)

Recently, Hofmann and co-workers[165] prepared a series of 3-oxazolin-2-ones by treating tertiary α-hydroxyketones with chlorosulfonyl isocyanate, followed by hydrolysis of the intermediate to give carbamates, which cyclize to the oxazolones (**68**) on heating [Eq. (42)].

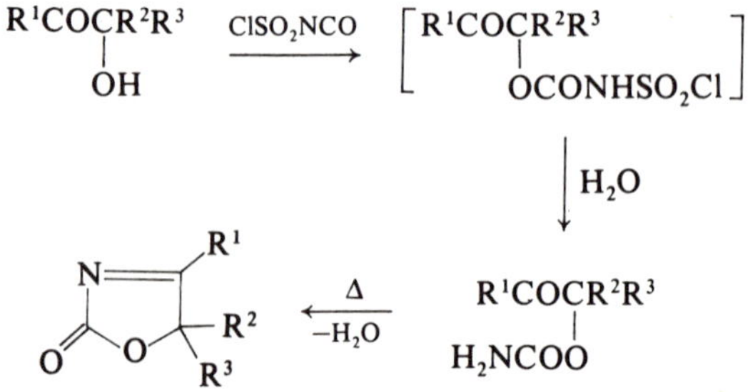

$$\text{R}^1\text{COCR}^2\text{R}^3 \atop \text{OH} \quad \xrightarrow{\text{ClSO}_2\text{NCO}} \quad \left[\text{R}^1\text{COCR}^2\text{R}^3 \atop \text{OCONHSO}_2\text{Cl} \right] \quad \bigg\downarrow \text{H}_2\text{O} \qquad (42)$$

$$(68) \xleftarrow{\Delta \atop -\text{H}_2\text{O}} \text{R}^1\text{COCR}^2\text{R}^3 \atop \text{H}_2\text{NCOO}$$

B. REACTIONS

Reduction of **68** with lithium aluminum hydride occurs at the C=N linkage to give the corresponding oxazolidones.[165] Alkaline hydrolysis of **68** produces the starting hydroxyketones.[165] The enol ether (**60**) undergoes O → N-alkyl migration on heating to 180° to afford the oxazolidine-2,4-dione. This observation probably accounts for the discrepancy in results mentioned earlier.[158,159]

---

[165] H. Hofmann, R. Wagner, and J. Uhl, *Chem. Ber.* **104**, 2135 (1971).

The Chemistry of Isoxazolidines

YOSHITO TAKEUCHI

Department of Chemistry, College of General Education, The University of Tokyo, Komaba, Meguro-ku, Tokyo, Japan

AND FUMIO FURUSAKI

Central Research Laboratory, Mitsubishi Chemical Industries Ltd., Kamoshida-cho, Midori-ku, Yokohama, Japan

I. Introduction	208
A. History	208
B. Nomenclature	209
II. Synthetic Methods	209
A. Isoxazolidines from Intermolecular Cycloaddition between Nitrones and Olefins	210
1. Regio- and Stereoselectivity of the Reaction	210
2. Cycloaddition between Open-Chain Nitrones and Olefins	213
3. Isoxazolidines from Cyclic Nitrones and Olefins	222
4. Isoxazolidines from Oximes and Olefins	224
5. Dimerization of Nitrones	224
6. Isoxazolidines from Nitronic Esters and Olefins	224
B. Isoxazolidines by Intramolecular Cyclization of Nitrones	229
C. Miscellaneous Methods	232
III. Physical Properties	234
A. Basicity	234
B. Dipole Moments	234
C. X-Ray Analysis	235
D. Inversion at Nitrogen	236
IV. Spectroscopic Properties	237
A. IR Spectra	237
B. UV Spectra	238
C. NMR Spectra	238
D. Mass Spectra	239
V. Chemical Properties	241
A. Thermal Stability	241
B. Photolysis	242
C. Reduction (Hydrogenolysis)	243

D. Oxidation 244
E. Decomposition by Acid 244
F. Decomposition by Base 246
G. Substitution 246
H. Quaternization 247
VI. Uses of Isoxazolidines 248
   A. Physiological Activity 248
   B. Polymeric Isoxazolidines 250

I. Introduction

A. History

The preparation of isoxazolidine derivatives was first reported by Bodforss[1] in 1918, and then by Kohler[2-4] in 1924. In 1942, King[5] succeeded in synthesizing the parent isoxazolidine (1); since then, however, little was added to this field for more than 10 years. During 1959–1960, a few novel routes to isoxazolidines were proposed, which included the intramolecular cyclization of olefinic nitrones of LeBel[6] and the intermolecular cycloaddition reaction of nitrones and olefins of Huisgen,[7] Brown,[8,9] and Delpierre.[10] Meanwhile Huisgen[11-13] systematized 1,3-dipolar cycloaddition as a versatile synthetic method for heterocycles, and the nitrone–olefin cyclization turned out to be a typical example.

There is no comprehensive review on the chemistry of isoxazolidines except for the specific aspect of the nitrone–olefin cycloaddition (1964)[14] and on the cycloaddition of nitronic esters with olefins (1969).[15] This review will cover the chemistry of isoxazolidines in the literature up to May 1974.

[1] S. Bodforss, *Ber. Deut. Chem. Ges.* **51**, 192 (1918).
[2] E. P. Kohler, *J. Am. Chem. Soc.* **46**, 503 (1924).
[3] E. P. Kohler and G. R. Barrett, *J. Am. Chem. Soc.* **46**, 2105 (1924).
[4] E. P. Kohler and G. R. Barrett, *J. Am. Chem. Soc.* **48**, 1770 (1926).
[5] H. King, *J. Chem. Soc.*, 432 (1942).
[6] N. A. LeBel and J. J. Whang, *J. Am. Chem. Soc.* **81**, 6334 (1959).
[7] R. Grashey, R. Huisgen, and L. Leitermann, *Tetrahedron Lett.* 9 (1960).
[8] C. W. Brown, K. Marsden, M. A. T. Rogers, C. M. B. Tylor, and R. Wright, *Proc. Chem. Soc.*, 254 (1960).
[9] C. W. Brown and M. A. T. Rogers, British Patent 850,418 [*CA* **55**, 6498 (1961)].
[10] G. R. Delpierre and M. Lamchen, *Proc. Chem. Soc.*, 386 (1960).
[11] R. Huisgen, *Proc. Chem. Soc.*, 357 (1961).
[12] R. Huisgen, *Angew. Chem.* **75**, 604 (1963).
[13] R. Huisgen, *Angew. Chem.* **75**, 742 (1963).
[14] J. Hamer and A. Macaluso, *Chem. Rev.* **64**, 473 (1964).
[15] A. T. Nielsen, in "The Chemistry of the Nitro and Nitroso Groups" (H. Feuer, ed.), Part 1, p. 453. Wiley, New York, 1969.

B. Nomenclature

The parent isoxazolidine (**1**) can be named as 1,2-oxazolidine according to the extension of the Hantzsch–Widman system. However, since the corresponding heterocycle (**2**) with the greatest number of noncumulative double bonds is conventionally called isoxazole, **1** is most frequently referred to as isoxazolidine, which is used in *Chemical Abstracts* (*CA*). The hydro-isoxazole or -isoxazoline nomenclature is used solely in connection with the condensed ring; for instance, compound **3** may be called either octahydro-1,2-benzisoxazole (*CA* nomenclature) or 4,5-tetramethyleneisoxazolidine (as a derivative of the isoxazolidine), or 7-oxa-8-azabicyclo[4.3.0]nonane (**3'**) (Stelzner method). A bis-isoxazolidine is called tetrahydro-2*H*-isoxazolo[2,3-*b*]-isoxazole (**4**) (*CA* method), or conventionally isoxazolizidine (**4'**).

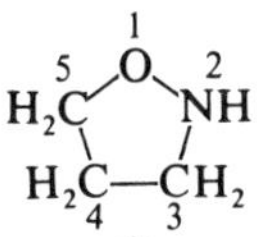

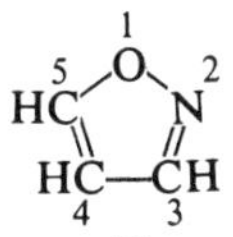

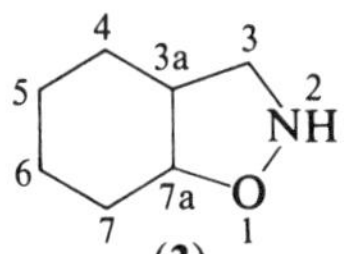

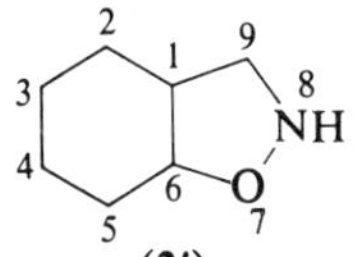

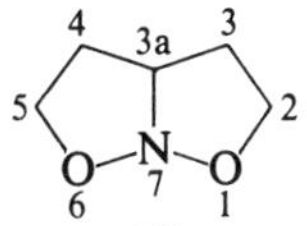

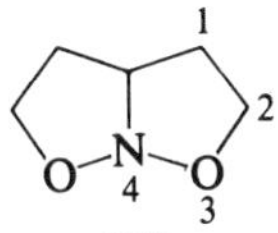

II. Synthetic Methods

Of the possible methods of isoxazolidine ring formation (Scheme 1), the most important involves simultaneous formation of the 1:5 and 3:4 bonds (**a**). A limited number of reactions involve formation of the 1:5 and 2:3 bonds (**b**). This is also a standard synthesis of the isoxazole ring.

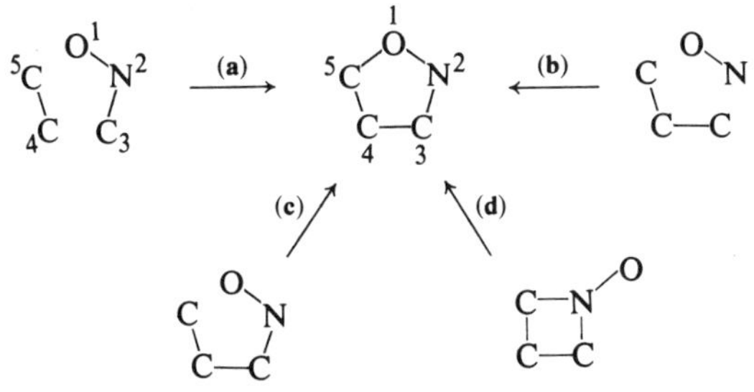

SCHEME 1

As described later, however, attempted reduction of an isoxazole to an isoxazolidine in most cases causes ring opening. Cyclization of open-chain systems at the 1:5 bond is often a good method (**c**), but for the isoxazolidine ring this method is not particularly convenient. A few examples of ring enlargement of azetidine *N*-oxides are reported (**d**). Almost all the isoxazolidines so far reported have been prepared by process **a**.

A. Isoxazolidines from Intermolecular Cycloaddition between Nitrones and Olefins

1. *Regio- and Stereoselectivity of the Reaction*

The cycloaddition between nitrones and olefins **5–9** (Scheme 2) has been thoroughly investigated.[11–16] Huisgen[12,13,17] formulates the reaction as a concerted 1,3-dipolar cycloaddition, but Firestone[18–20] formulates it as a stepwise biradical mechanism. A compromise process has been suggested;[21] the accumulated evidence, however, seems to favor the concerted mechanism.[22]

The most interesting aspect of the 1,3-dipolar cycloaddition is the high regioselectivity. Monosubstituted olefins (**5**) almost always give 5-substituted isoxazolidines (I) with no detectable amount of 4-isomers (II). Similarly, formation of 5,5-disubstituted isoxazolidines (III) is also predominant when 1,1-disubstituted olefins (**6**) are employed.

Occasionally, the reverse orientation is obtained. The reaction of a trans-disubstituted nitrone, α-phenyl-*N*-methylnitrone (**10a**), with nitroethylene gives the 4-substituted isoxazolidine (II).[23] With *N-t*-butylnitrone (**11a**), however, the same olefin gives the 5-substituted isomer (I).[23] Each reaction is regiospecific in that, of the two possible orientations, only one is produced in the reaction. The reaction of phenyl vinyl sulfone with nitrones is one of the rare examples where regiospecificity is lost. Thus, both nitrones **10a** and **11a** give a mixture of adducts I and II (31:68), and (70:30), respectively.[23] Recently, the regioselectivity of 1,3-dipolar cycloadditions have been interpreted by

---

[16] G. R. Delpierre and M. Lamchen, *Q. Rev., Chem. Soc.* **19**, 329 (1965).
[17] R. Huisgen, *J. Org. Chem.* **33**, 2291 (1968).
[18] R. A. Firestone, *J. Org. Chem.* **33**, 2285 (1968).
[19] R. A. Firestone, *J. Chem. Soc. A*, 1570 (1970).
[20] F. A. Firestone, *J. Org. Chem.* **37**, 2181 (1972).
[21] R. D. Harcourt, *J. Mol. Struct.* **12**, 351 (1972).
[22] B. M. Benjamin and C. J. Collins, *J. Am. Chem. Soc.* **95**, 6145 (1973).
[23] J. Sims and K. N. Houk, *J. Am. Chem. Soc.* **95**, 5798 (1973).

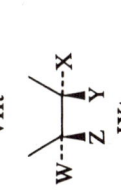

SCHEME 2. Cycloaddition reaction of nitrones and olefins.

the FMO (frontier molecular orbital) treatment by Houk,[23-26] Sustmann,[27,28] Bastide,[29] and Grée.[30]

$$\begin{array}{c} H \\ R^1 \end{array} C = \overset{+}{N} \begin{array}{c} R^N \\ O^- \end{array}$$

(10)

trans-Disubstituted nitrones

	R^1	R^N		R^1	R^N		R^1	R^N
a	Ph	Me	h	n-Pr	Ph	o	Et$_2$NCO	Ph
b	Pr	cyclohexyl	i	PhCH(Me)	C$_6$F$_5$	p	Ar	OMe
c	Pr	Pr	j	Ph	Ph	q	COOMe	OMe
d	Me	Et	k	Ar	Ar	r	COOEt	OMe
e	i-Pr	Me$_2$CH(CN)	l	PhCO	Ph	s	CN	OMe
f	COOMe	Me	m	styryl	Ph			
g	Me	Ph	n	ArNHCO	Ph			

$$\begin{array}{c} H \\ H \end{array} C = \overset{+}{N} \begin{array}{c} R^N \\ O^- \end{array}$$

(11)

N-Substituted nitrones

a: $R^N = t$-Bu
b: $R^N = 1$-ethylcyclohexyl

The stereochemistry of the reaction is also interesting. It is stereoselective in that a variable amount of two possible stereoisomers (per regioisomer) are almost always formed. In principle any reaction between a nitrone and an olefin could give four products, i.e., two regioisomers each with two stereoisomers. Depending on the structure of the nitrone or of the olefin, some of the isomers may be "degenerate"; thus no regio- and stereoisomers exist for adducts obtained from ethylene or tetracyanoethylene. Olefins are classified into five groups: monosubstituted (5), 1,1-disubstituted (6), cis-disubstituted (7), trans-disubstituted (8), and tri- or tetrasubstituted (9). For each group, the four isomeric products are classified, according to the position of the substituent X (regioisomer), and second, according to its configuration (relative to R^1

[24] K. N. Houk, J. Sims, R. E. Duke, R. W. Strozier, and J. K. George, *J. Am. Chem. Soc.* **95**, 7287 (1973).
[25] K. N. Houk, J. Sims, C. R. Watts, and L. J. Luskus, *J. Am. Chem. Soc.* **95**, 7301 (1973).
[26] K. N. Houk, *J. Am. Chem. Soc.* **94**, 8953 (1972).
[27] R. Sustmann, *Tetrahedron Lett.* 2717 (1971).
[28] R. Sustmann and H. Trill, *Angew. Chem., Int. Ed. Engl.* **11**, 838 (1972).
[29] J. Bastide, N. El Ghandour, and H. Rousseau, *Tetrahedron Lett.*, 4225 (1972).
[30] R. Grée, F. Tonnard, and R. Carrié, *Tetrahedron Lett.*, 135 (1974).

at C3) (stereioisomers). For disubstituted olefins (**6–8**), the most important substituent is coded as X. The priority decreases in the order COOR > COR > CN > Ph > NO_2 > alkyl > others > H. For tri-, or tetrasubstituted olefins (**9**), the coding is indicated for individual compounds.) In Scheme 2, where twenty possible isomers are tabulated and numbered, I and II are regioisomeric, and It and Ic, etc., are stereoisomeric with each other.

The stereochemistry is not always clearly established. Such cycloadducts are called simply type I, etc. When a mixture of two stereoisomers is of unknown stereochemistry, they are denoted as type Itc, etc.

If there are two (or more) identical substituents in the olefin or in the nitrone degeneracy results. Thus, if X = W for olefins (**7**), or X = Z for olefins (**8**), cycloadducts Vt and VIt (hence Vc and VIc), or VIIt and VIIIc (hence VIIc and VIIIt), become identical (degeneracy in regiochemistry). If this type of degeneracy is involved, the type symbol will be placed in brackets, e.g., [Vt]. There is possibly another type of degeneracy; thus, if $R^1 = R^2$, cycloadducts It and Ic, for instance, become identical (degeneracy in stereochemistry). The same situation is involved when X = Y for olefins (**6**). In such cases the suffix t or c will be deleted and instead, the type symbol will be placed in parentheses, e.g., (I). Much the same principle will, if appropriate, be applied for isoxazolidines obtained from olefins (**9**).

Generally, the inversion at nitrogen is rapid and invertomer pairs are not isolated. In a few rare cases, however, a high nitrogen inversion barrier renders the invertomers separable.[31] This point will be discussed later in more detail. The kinetic aspects of the nitrone–olefin cycloaddition have been extensively investigated by Huisgen[32] and by Boyle.[33]

2. *Cycloaddition between Open-Chain Nitrones and Olefins*

a. N-Alkylnitrones. Cycloaddition reactions between *N*-alkylnitrones and olefins **5**, such as styrene regiospecifically, afford isoxazolidines of type I.[34,35] The nitrone **11b** with cyclohexene,[8,33] norborene,[33] *trans*-cyclooctene,[33] bicyclo[4.2.0]oct-7-ene,[33] and maleic anhydride[34,35] gives cycloadducts V.

b. α,N-Dialkylnitrones. Cycloadditions of α,*N*-dialkyl nitrones include those of **10b** and **10c** with ethyl acrylate, acrylonitrile, and styrene

---

[31] K. Müller and A. Eschenmoser, *Helv. Chim. Acta* **52**, 1823 (1969).
[32] R. Huisgen, H. Seidl, and I. Brüning, *Chem. Ber.* **102**, 1102 (1969).
[33] L. W. Boyle, M. J. Peagram, and G. H. Whitham, *J. Chem. Soc. B*, 1728 (1971).
[34] J. E. Baldwin, A. K. Qureshi, and B. Sklarz, *Chem. Commun.*, 373 (1968).
[35] J. E. Baldwin, A. K. Qureshi, and B. Sklarz, *J. Chem. Soc. C*, 1073 (1969).

to give isoxazolidines of type I.[36-38] Again, nitrone **10b** gives isoxazolidine III with methyl methacrylate[7,36] and isoxazolidines V with norbornene or norbornadiene.[39]

Reactions of α-methyl-N-ethylnitrone (**10d**) are interesting since some olefins form reverse-oriented cycloadducts. For example, **10d** and perfluoropropene give a mixture of adducts of type Xt and Xc in a 3:2 ratio.[40] Reverse-oriented cycloadducts IVt and IVc are also obtained from **10d** and methyl methacrylate.[41,42]

The reaction of α-isopropyl-N-(1-cyanoisobutyl)nitrone (**10e**) has been extensively investigated.[43] The orientation varies, and depends on the steric as well as electronic environment of the olefin. With olefins **5**, such as acrylonitrile, **10e** gives isoxazolidines I regiospecifically. With methyl methacrylate or methacrylonitrile, isoxazolidines of type III are obtained. However, the cycloadduct from crotonate ester or crotonitrile has "reverse" orientation (type VIIIt). The reactions with diethyl maleate or fumarate give adducts of types V and VII, respectively.

Only one reaction of α-methoxycarbonyl-N-methylnitrone (**10f**) was reported; with dimethyl acetylenedicarboxylate a 2:1 addition takes place to give a perhydroisoxazolo[4,5-d]isoxazole.[44]

c. *α-Alkyl-N-arylnitrones.* Much work has been done on the reactions of nitrones with an aryl substituent at the carbon or the nitrogen; these nitrones are more stable and easier to prepare than their dialkyl analogs.

The reaction of α-methyl-N-phenyl nitrone (**10g**)[45] and α-(n-propyl)-N-phenylnitrone (**10h**)[7,45] with styrene gives isoxazolidines of type I. The reaction of nitrone **10h** with olefins **7**, such as N-phenylmaleimide[46,47] and acenaphthylene,[48] gives fused isoxazolidines V. An unstable nitrone (**10i**) can be trapped by N-phenylmaleimide to give a cycloadduct V.[49]

[36] R. Huisgen, H. Hauck, R. Grashey, and H. Seidl, *Chem. Ber.* **101**, 2568 (1968).
[37] A. D. Nikolaeva and V. S. Perekhod'ko, *Zh. Org. Khim.* **8**, 2297 (1972) [*CA* **78**, 58287 (1973)].
[38] G. Kamai, A. D., Nikolaeva, and V. S. Perekhod'ko, *Zh. Org. Khim.* **5**, 244 (1969) [*CA* **70**, 105901 (1969)].
[39] R. Huisgen, R. Grashey, H. Hauck, and H. Seidl, *Chem. Ber.* **101**, 2043 (1968).
[40] I. L. Knunyants, E. G. Bykhovskaya, V. N. Forsin, I. V. Galakhov and L. I. Regulin, *Zh. Vses. Khim. Obshch.* **17**, 356 (1972) [*CA* **77**, 101432 (1972)].
[41] H. E. De la Mare and G. M. Coppinger, *J. Org. Chem.* **28**, 1068 (1963).
[42] K. Nakagawa, H. Onoue, and K. Minami, *Chem. Pharm. Bull.* **17**, 835 (1969).
[43] M. Masui, K. Suda, M. Yamauchi, and C. Yijima, *Chem. Pharm. Bull.* **21**, 1605 (1973).
[44] E. Winterfeldt, W. Krohn, and H. Stracke, *Chem. Ber.* **102**, 2346 (1969).
[45] R. Huisgen, R. Grashey, H. Hauck, and H. Seidl, *Chem. Ber.* **101**, 2548 (1968).
[46] I. Brüning, R. Grashey, H. Hauck, R. Huisgen, and H. Seidl, *Org. Synth.* **46**, 96 (1966).
[47] R. Huisgen, H. Hauck, R. Grashey, and H. Seidl, *Chem. Ber.* **103**, 736 (1969).
[48] R. Huisgen, R. Grashey, H. Seidl, and H. Hauck, *Chem. Ber.* **101**, 2559 (1968).
[49] R. E. Banks, R. N. Haszeldine, and P. J. Miller, *Tetrahedron Lett.* 4417 (1970).

Sec. II.A]   THE CHEMISTRY OF ISOXAZOLIDINES   215

d. α-Aryl-N-alkylnitrone. Of the open-chain nitrones, the nitrone **10a** has been most extensively investigated. The reactions of this nitrone with various olefins are regio- and stereoselective and provide us with information on the nature of the orientation and stereochemistry of the reaction. The reaction with such olefins **5** as alkylethylene,[39,50] butyl vinyl ether,[48] allyl alcohol,[39] acrylonitrile,[36] styrene,[45] and ethyl acrylate,[32] affords isoxazolidines of type I, but with phenyl vinyl sulfone a mixture of 4- and 5-substituted isoxazolidines (Itc and IIt), and with nitroethylene, 4-substituted isoxazolidines (type IItc) are formed.[23]

The reaction with olefins **6**, such as methyl methacrylate,[21,36] 1-methyl-1-phenylethylene,[48] and 1,1-diphenylethylene,[48] gives isoxazolidines of type III. With dimethyl maleate and *cis*-1,2-dibenzoylethylene (**7**), addition is stereospecific to afford isoxazolidine [Vc]. The reaction with methyl *cis*-cinnamate affords a cycloadduct VItc.[51] With olefins **8**, such as dimethyl fumarate and *trans*-1,2-dibenzoylethylene, the addition is stereoselective to give a mixture of [VIIt] and [VIIc].[48] Reactions with crotonate[36] or *trans*-cinnamate[51] esters are also reverse-oriented to give adducts [VIIIt] and [VIIIc].

Cycloaddition of the nitrone **10a** with dimethyl citraconate[47] and mesaconate,[47] methyl β,β-dimethyl acrylate,[36] and mesityl oxide[36] (for all of them Y = H) gives cycloadducts X. Some of the results are summarized in Table I.

The reactions of nitrone **10a** with carvone or eucarvone are also reported.[36] Reaction with various cyclic olefins, such as cyclopentene,[39]

TABLE I
ISOXAZOLIDINES FROM α-PHENYL-N-METHYLNITRONE (**10a**) WITH OLEFINS **6–9**

	Olefin				Isoxazolidine type	
	W	X	Y	Z	(ratio %)	References
6	H	COOMe	Me	H	IIItc (60:40)	36
	H	Ph	Me	H	IIItc (55:45)	48
	H	Ph	Ph	H	III	48
7	COOMe	COOMe	H	H	[Vc]	47
	COPh	COPh	H	H	[Vc]	47
	Ph	Ph	H	H	VItc (72:18)	51
8	H	COOMe	H	COOMe	[VIIt] (55) + [VIIc] (45)	47
	H	COPh	H	COPh	VIItc	47
	H	CN	H	CN	[VIIt]	47
	H	COOMe	H	Me	VIII	36
	H	COOMe	H	Ph	VIIItc (73:17)	51
9	Me	COOMe	H	Me	X	36
	Me	COOMe	H	COOMe	X	47
	Me	COMe	H	Me	X	36
	COOMe	COOMe	H	Me	X	47

[50] I. Ikeda, G. Takemoto, and S. Komori, *Kogyo Kagaku Zasshi* **74**, 220 (1971) [*CA* **75**, 5766 (1971)].

[51] Y. Iwakura, K. Uno, Y. Kihara, M. Setsu, and S. Yamamoto, *Nippon Kagaku Kaishi* 1448 (1972) [*CA* **77**, 139864 (1972)].

cyclohexene,[33,39] norbornene,[33,39,52] *trans*-cyclooctene,[33] norbornadiene,[39] dimethyl bicyclo[2.2.1]hept-2-ene-5,6-dicarboxylate,[39] bicyclo-[4.2.0]oct-7-ene,[33] 1,2-dihydronaphthalene,[48] indene,[48] and acenaphthylene,[48] gives cis-fused isoxazolidines of type V, in most cases stereospecifically. Reaction with dimethyl tricyclo[4.2.2.0^{2,5}]deca-3,7,9-triene-7,8-dicarboxylate is stereoselective, and a mixture of fused adducts **12a** and **12b** (type Vt and Vc) are obtained.[53]

	R^1	R^2	R^3
a:	Me	H	Ph
b:	Me	Ph	H
c:	Ph	H	Ph
d:	Ph	Ph	H

(12)

A reaction between heteroaromatic nitrone, α-(5-nitro-2-imidazolyl)-*N*-methylnitrone and cyclooctene was reported.[54]

e. *α,N-Diarylnitrones.* α,*N*-Diphenylnitrone (**10j**) and its ring-substituted derivatives (**10k**) have been thoroughly investigated. 1-Hexene,[39] 1-hexadecene,[55] 10-undecenoate, oleate, and linoleate esters,[55] allyl alcohol,[7,39] vinyl ethyl ether,[56] vinyl acetate,[57] acrylonitrile,[36] methyl acrylate,[36] and styrene[7,45,57,58] give 5-substituted isoxazolidines. 1,1-Disubstituted olefins (**6**) afford isoxazolidine of type III: e.g., 1,1-diphenylethylene with α,*N*-diphenylnitrone.[48,57] The reaction with 1-methyl-1-phenylethylene is stereoselective, and adducts IIIt and IIIc are obtained in the ratio of ca. 4:1.[48] Reaction of nitrones **10j** and **10k** with ketene diethylacetal[48,59] or 1-phenyl-1-pyrrolidinoethylene[60] affords isoxazolidines of type III.

The reaction with *cis*-olefin (**7**) is stereoselective to give epimeric mixtures of Vt and Vc, or of VIt and VIc. With *trans*-olefins (**8**), the products are also epimeric isoxazolidine mixtures. Some reactions are

---

[52] R. R. Fraser and Y. S. Lin, *Can. J. Chem.* **46**, 801 (1968).
[53] G. Bianchi, A. Gamba, and R. Gandolfi, *Tetrahedron* **28**, 1601 (1972).
[54] P. Kulsa and C. S. Rooney, German Offen. 2,100,242 (1971) [*CA* **75**, 110312 (1971)].
[55] H. Basu and H. Schlenk, *Chem. Phys. Lipids* **6**, 266 (1971) [*CA* **75**, 35292 (1971)].
[56] R. Poul and S. Tchelitcheff, *Bull. Soc. Chim. Fr.*, 4179 (1967).
[57] G. Cum, M. C. Aversa, and N. Uccella, *Gazz. Chim. Ital.* **98**, 782 (1968).
[58] I. Brüning, R. Grashey, H. Hauck, R. Huisgen, and H. Seidl, *Org. Synth.* **46**, 127 (1966).
[59] R. Scarpati, D. Sica, and C. Santacroce, *Gazz. Chim. Ital.* **96**, 375 (1966).
[60] O. Tsuge, M. Tashiro, and Y. Nishihara, *Tetrahedron Lett.*, 3796 (1967).

reverse-oriented. The reaction of nitrone **10j** with chalcones or substituted chalcones gives epimeric mixtures of Vt and Vc.[51] The reactions of **10j** with trisubstituted olefins are also regiospecific and stereoselective. For instance, **10j** with methyl β,β-dimethylacrylate[36] and with methyl ethylidenecyanoacetate[64] gives a mixture of Xt and Xc, respectively. Reaction of **10j** with methyl ethylidenemalonate or methyl benzylidenecyanoacetate also give an epimeric mixture of Xt and Xc.[64] With methyl benzylidenemalonate, adduct Xc is obtained.[64] Some results are summarized in Table II.

When vinylacetylenes,[65,66] methylthioethynylethylene,[67] and

TABLE II
ISOXAZOLIDINES FROM α,N-DIPHENYLNITRONE (**10j**) WITH OLEFINS **6–9**

Type	Olefin				Isoxazolidine type (ratio %)	References
	W	X	Y	Z		
6	H	Ph	Ph	H	(III)	48
	H	Ph	Me	H	IIIt(79) + IIIc(21)	48
	H	OEt	OEt	H	(III)	48, 59
	H	Ph	Pyrrolidino	H	III	60
7	COOMe	COOMe	H	H	[Vt](90) + [Vc](10)	61, 62
	Ph	COOMe	H	H	VIt(50) + VIc(50)	51, 61, 62
	Ph	CN	H	H	VIt(37) + VIc(63)	61
	Ph	$SO_2C_6H_4Me$	H	H	V	63
	Ph	SO_2Ph	H	H	V	63
8	H	COOMe	H	COOMe	[VIIt](20) + [*VIIc*](80)	61, 62
	H	COOMe	H	Ph	VIIIt + VIIIc (trace)	61, 62
	H	COOMe	H	Me	VIIItc	36
	H	CN	H	CN	[VIIc]	61
	H	CN	H	Ph	VIIIt + VIIIc (trace)	61, 62
	H	COPh	H	COPh	[VIIt]	47
	H	COOEt	H	p-$NO_2C_6H_4$	VIII	36
	H	NO_2	H	Ph	VIIIt[85] + VIIIc(15)	61
	H	COPh	H	p-$NO_2C_6H_4$	VIII	51
	H	COPh	H	p-ClC_6H_4	VIIItc(90:10)	51
	H	COPh	H	Ph	VIIItc(97:3)	51
	H	COPh	H	p-MeC_6H_4	VIIItc(96:4)	51
	H	COPh	H	p-$MeOC_6H_4$	VIIItc(94:6)	51
9	Me	COOMe	H	Me	Xtc	36
	H	COOMe	COOMe	Me	Xt(5) + Xc(95)	64
	H	COOMe	COOMe	Ph	Xc	64
	H	COOMe	CN	Me	Xt(75) + Xc(25)	64
	H	COOMe	CN	Ph	Xt(80) + Xc(20)	64
	H	OEt	OEt	Et	X	59

[61] M. Joucla, D. Grée, and J. Hamelin, *Tetrahedron* **29**, 2315 (1973).
[62] M. Joucla and J. Hamelin, *C.R. Hebd. Seances Acad. Sci., Ser. C*, **273** 769 (1971).
[63] S. M. Yarnal and V. V. Badiger, *J. Indian Chem. Soc.* **48**, 453 (1971).
[64] M. Joucla, J. Hamelin, and R. Carrié, *Bull. Soc. Chim. Fr.*, 3116 (1973).
[65] V. N. Chistokletov and A. A. Petrov, *Zh. Obshch. Khim.* **32**, 2385 (1962) [*CA* **58**, 9040 (1963)].
[66] V. N. Chistokletov, L. Kvagina, and A. A. Petrov, *Zh. Org. Khim.* **1**, 369 (1965) [*CA* **62**, 16092 (1965)].
[67] S. I. Radchenko, V. N. Chistokletov, and A. A. Petrov, *Zh. Obshch. Khim.* **35**, 1735 (1965) [*CA* **64**, 3514 (1966)].

trimethylsilylethynylethylene[68] are used as the dipolarophile, nitrone **10j** reacts with the double bond exclusively to give isoxazolidines of type I. Nitrone **10j** and 1-methyl-1-(trimethylsilylethynyl)ethylene gives III; with 1,3-pentadiene, **10j** gives adduct I; and with (1-pentyn-4-enyl)-ethylene, a 1:2 adduct was obtained.[66]

The cycloaddition of nitrones **10j** or **10k** with various cyclic olefins is also stereoselective, and the ratio of *exo* (Vc)- and *endo* (Vt)-adducts depends on the structure of olefins, and to some extent on the nature of the ring-substituent, althouth generalization is difficult.[69] The reaction of nitrone **10j** with dimethyl tricyclo[4.2.2.0^{2,5}]deca-3,7,9-triene-7,8-dicarboxylate is also stereoselective. The *endo*-(**12c**)/*exo*(**12d**) adduct ratio is 0.5.[53] The cycloaddition of nitrone **10j** with some complex α,β-unsaturated ketones is reported.[70] The reaction with lumisantonin is regioselective to give a mixture of VIc (24%) and Vc (8%)[71] (the carbonyl side chain is taken as X). The reaction with dihydrofuran is regio- and stereospecific.[56]

Cycloaddition of nitrones with cyclanone enamines are extensively investigated. The reaction is regio- and stereospecific, and always gives one product. Thus, nitrones **10k**, when caused to react with 1-pyrrolidino-1-cyclohexene, give the 5-pyrrolidinoisoxazolidine (**13a**), and with 1-pyrrolidino-1-cyclopentene, the isoxazolidine (**13b**).[60,72–74]

	n	R^1	R^2
a:	$n=4$,	$-(CH_2)_4-$	
b:	$n=3$,	$-(CH_2)_4-$	
c:	$n=4$,	Ph	OH
d:	$n=3$,	Ph	OH
e:	$n=3$,	$-(CH_2)_2-O-(CH_2)_2-$	

(**13**)

Some reactions with morpholinoenamines are anomalous; thus with nitrone **10j** and 1-morpholino-1-cyclohexene, a 5-hydroxyamino-isoxazolidine (**13c**) is formed;[60] this, in turn, can be prepared from the same nitrone (**10j**) and *N*-phenylhydroxyaminocyclohexene. With 1-morpholino-1-cyclopentene, a mixture of adducts **13d** (12.4%) and

---

[68] I. G. Kolokol'tseva, V. N. Chistokletov, M. D. Stadnichuk, and A. A. Petrov, *Zh. Obshch. Khim.* **38**, 1820 (1968) [*CA* **70**, 78067 (1969)].
[69] Y. Iwakura, K. Uno, S. Hong, and T. Hongu, *Bull. Chem. Soc. Jpn.* **45**, 192 (1972).
[70] O. Tsuge and I. Shinkai, *Tetrahedron Lett.*, 3847 (1970).
[71] T. Sasaki and S. Eguchi, *J. Org. Chem.* **33**, 4389 (1968).
[72] Y. Nomura, F. Furusaki, and Y. Takeuchi, *Bull. Chem. Soc. Jpn.* **40**, 1740 (1967).
[73] Y. Nomura, F. Furusaki, and Y. Takeuchi, *Bull. Chem. Soc. Jpn.* **43**, 3002 (1970).
[74] O. Tsuge, M. Tashiro, and Y. Nishihara, *Nippon Kagaku Zasshi* **92**, 72 (1971).

13e (4.2%) are obtained.[73] With a sterically overcrowded enamine, 1-pyrrolidino-3,5-dimethylcyclohexene, a reverse-oriented cycloadduct (**14**) was obtained.[75] However, the reaction of nitrone **10j** with a mixture of 1-pyrrolidino-3- and 5-methylcyclohexene gives exclusively the 5-pyrrolidino-isoxazolidines of type IX.[76]

(**14**)

f. α-Benzoyl-N-aryl Nitrones. α-Benzoyl-*N*-phenylnitrone (**10l**) and its ring-substituted derivatives are versatile precursors for isoxazolidines. As compared with nitrones previously described, this nitrone is more reactive, and the cycloaddition is more regioselective, or almost regiospecific. Thus, the reaction of benzoylnitrone (**10l**) and ethylene proceeds smoothly at room temperature.[77] Other nitrones, such as **10a** or **10j**, do not react with ethylene even at 100°.[39]

The reaction of nitrone **10l** with such olefins **5** as propylene, *n*-hexene, allyl alcohol, styrene, acrylic acid, methyl acrylate, acrylonitrile, and butadiene are all regiospecific to give 5-substituted isoxazolidines of type I.[77] The reaction with olefins **6** seems to be regio- and stereospecific; thus, nitrone **10l** and methyl methacrylate give cycloadduct IIIt as the sole product.[77] Similarly, **10l** and 1-methyl-1-phenylethylene gives adduct IIIt.[77] In the reactions with olefins **7** or **8**, the stereospecificity is more evident.[77,78] The isoxazolidine obtained is always one of the two possible epimers. The reaction of nitrone **10l** with trisubstituted olefins is highly stereoselective.[64] Some results are summarized in Table III.

The reaction of nitrone **10l** with cyclic olefins, such as maleic anhydride,[61] *N*-phenylmaleimide,[77] cyclopentene,[77] norbornene,[77] indene,[77] and acenaphthylene,[77] is normal and gives isoxazolidine of type V. Some reactions of α-(6-uracilyl)-*N*-phenylnitrone are also reported; those with allyl alcohol, methyl acrylate, acrylamide, and styrenes, all give 5-substituted 3-uracilylisoxazolidines (type I).[79,80] With methyl methacrylate a type III cycloadduct was formed.[79,80]

[75] Y. Nomura, F. Furusaki, and Y. Takeuchi, *Bull. Chem. Soc. Jpn.* **43**, 1913 (1970).
[76] Y. Nomura, F. Furusaki, and Y. Takeuchi, unpublished work, 1970.
[77] R. Huisgen, H. Hauck, H. Seidl, and M. Burger, *Chem. Ber.* **102**, 1117 (1969).
[78] J. W. Lown and B. E. Landberg, *Can. J. Chem.* **52**, 798 (1974).
[79] T. Sasaki and M. Ando, *Bull. Chem. Soc. Jpn.* **41**, 2960 (1968).
[80] T. Sasaki, Japanese Patent 70 34586 (1970) [*CA* **74**, 76440 (1971)].

TABLE III
ISOXAZOLIDINES FROM α-BENZOYL-N-PHENYLNITRONES (101) WITH OLEFINS 6–9

Type	Olefin				Isoxazolidine type (ratio %)	References
	W	X	Y	Z		
6	H	COOMe	Me	H	IIIt	77
	H	Ph	Me	H	III	77
7	COOMe	COOMe	H	H	[Vt]	61, 62, 77
	Ph	COOMe	H	H	VIt	61, 62
	Ph	CN	H	H	VIt	61, 62
8	H	COOMe	H	COOMe	[VIIIt]	61, 62, 77
	H	COPh	H	COPh	VII	77, 78
	H	CN	H	CN	[VIIIt]	61
	H	COOMe	H	Me	VIII	62, 77
	H	COOMe	H	Ph	VIIIt	61, 62
	H	CN	H	Ph	VIIIt	61, 62
	H	NO$_2$	H	Ph	VIIIt	61
9	H	COOMe	COOMe	Me	Xt(5) + Xc(95)	64
	H	COOMe	CN	Me	Xt(70) + Xc(30)	64
	H	COOMe	COOMe	Ph	Xc	64
	H	COOMe	CN	Ph	Xt(82) + Xc(18)	64

Reaction of some α-heteroaromatic-N-arylnitrones [α-(5-substituted 2-furyl)-,[56,81,82] α-(2-thienyl)-,[82] α-(2-benzimidazolyl)-,[83] and α-(2-, 3-, and 4-pyridyl)-N-phenylnitrone[56]] are also reported. The reactivity of these nitrones is much the same as with diphenylnitrone (10j), and the course of the reaction is mostly regiospecific and stereoselective.

A conjugated nitrone, α-styryl-N-phenylnitrone (10m), also reacts with olefins to give the normal cycloadducts; e.g., with styrene and allyl chloride, isoxazolidines of type I are formed.[84] A reaction with 1-morpholino-1-cyclohexene is described.[85]

(15)

The nitrone 15, which is a 2:1 adduct between nitrosobenzene and 2,3-dimethylbutadiene, gives an adduct of type III with methyl methacrylate.[86] Reactions of a few of α-arylcarbamyl-N-phenylnitrone

[81] T. Sasaki, T. Yoshioka, and I. Izure, *Bull. Chem. Soc. Jpn.* **41**, 2964 (1968).
[82] Y. Iwakura, K. Uno, and T. Hongu, *Bull. Chem. Soc. Jpn.* **42**, 2882 (1969).
[83] T. Sasaki and T. Ohishi, *Bull. Chem. Soc. Jpn.* **41**, 3012 (1968).
[84] N. Singh and S. Mohan, *Chem. Commun.*, 787 (1968).
[85] N. Singh and K. Krishan, *Indian J. Chem.* **11**, 1076 (1973).
[86] E. Oikawa and S. Tsubaki, *Bull. Chem. Soc. Jpn.* **46**, 1819 (1973).

Sec. II.A] THE CHEMISTRY OF ISOXAZOLIDINES 221

(10n) with acrylonitrile are also reported.[87,88] The reaction of α-(diethylcarbamoyl)-N-phenylnitrone (10o) and maleic anhydride gives cycloadduct of type V.[89]

g. Trisubstituted Nitrones. Cycloadditions of α,α-diphenyl-N-methyl- or -N-phenylnitrone (16a or 16b) with olefins are also known. The nitrone (16a) with but-1-en-3-one,[90] and nitrone (16b) with styrene[45] give isoxazolidines of type I. The reaction with dimethyl maleate smoothly gives isoxazolidine of type V; however, no corresponding isoxazolidine is formed with dimethyl fumarate probably due to a steric effect.[47] No reaction takes place with styrene.[91] The reaction of 16b with dimethyl tricyclo[4.2.2.0^{2,5}]deca-3,7,9-triene-7,8-dicarboxylate is also described.[53] The nitrone 16a gives an adduct of type I with methyl acrylate.[91]

$$R^2 \underset{R^1}{\overset{}{>}}C=\overset{+}{N}\underset{O^-}{\overset{R^N}{<}}$$

(16)

Trisubstituted nitrones

	R^1	R^2	R^N		R^1	R^2	R^N
a:	Ph	Ph	Me	d:	COOEt	COOEt	MeO
b:	Ph	Ph	Ph	e:	COOMe	COOMe	MeO
c:	CH$_2$COOMe	COOMe	Ph	f:	NO$_2$	NO$_2$	MeO

$$\text{Ph}\underset{-O}{\overset{+}{>}}N=CH-\!\!\!\left\langle\right\rangle\!\!\!-CH=\overset{+}{N}\underset{O^-}{\overset{Ph}{<}}$$

(17)

The bisnitrone (17) with methyl methacrylate, styrene, and N-phenylmaleimide gives normal bisisoxazolidines.[92–94] The meta-isomer of 17 and a few of its cycloadditions are also reported.[94] The thiophene analog of 17 also reacts normally with methyl acrylate, allyl alcohol, methyl methacrylate, and N-phenylmaleimide.[95]

[87] Yu, V. Svetkin, N. A. Akmanova, and G. I. Plotnikova, *Zh. Org. Khim.* **8**, 2429 (1972) [*CA* **78**, 58288 (1973)].
[88] Yu. V. Svetkin, N. A. Akmanova, and G. I. Karataeva, *Zh. Org. Khim.* **8**, 2431 (1972) [*CA* **78**, 136143 (1973)].
[89] K. W. Ratts and A. N. Yao, *J. Org. Chem.* **31**, 1689 (1966).
[90] A. Lablache-Combier and M. L. Villaume, *Tetrahedron* **24**, 6951 (1968).
[91] D. A. Kerr and D. A. Wilson, *J. Chem. Soc. C*, 1718 (1970).
[92] Y. Iwakura, M. Akiyama, and S. Shiraishi, *Bull. Chem. Soc. Jpn.* **38**, 513 (1965).
[93] Mitsubishi Petrochemical Co. Ltd., British Patent 1,129,976 (1968) [*CA* **70**, 20525 (1969)].
[94] G. Manecke and J. Klawitter, *Macromol. Chem.* **108**, 292 (1967).
[95] T. Hongu, Y. Iwakura, Japanese Patent 71 25 (1971) [*CA* **75**, 6600 (1971)].

3. Isoxazolidines from Cyclic Nitrones and Olefins

a. *Alicyclic Nitrones.* The reaction between 5,5-dimethyl-1-pyrroline-1-oxide (**18a**) and ethyl acrylate at room temperature gives cycloadduct of type I, but the same reaction at 100° gave the regioisomer II. The initial product is I, which at the higher temperature isomerizes to the more stable II.[10,96] The reactions of oxide **18a** with acrylamide, *N-t*-butyl acrylamide, and polysubstituted olefins are also described.[97] 4,5,5-Trimethyl-1-pyrroline-1-oxide (**18b**) with cyclohexene gives V.[8] Some reaction of 1-piperideine-1-oxide (**19a**),[8] 2,3-dihydro-1,4-oxazine-*N*-oxide (**19b**),[98] the oxaziridine-nitrone (**20**),[99] and -dinitrone (**21**)[99] have been reported.

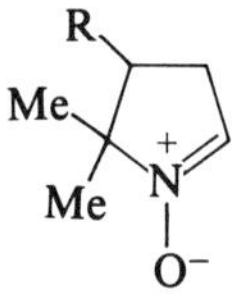

(18) a: R = H
     b: R = Me

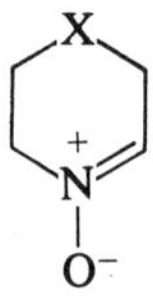

(19) a: X = CH$_2$
     b: X = O

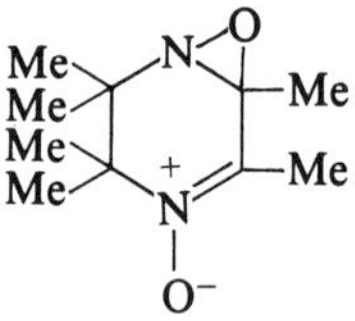

(20)

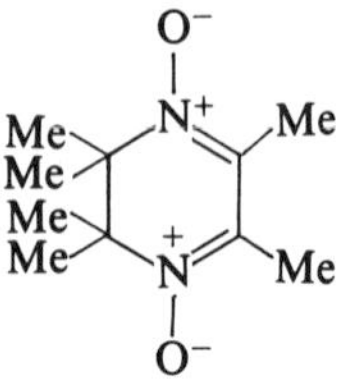

(21) 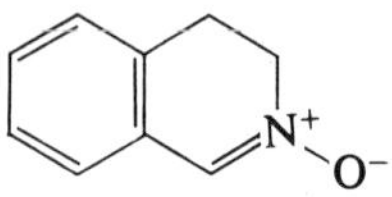 (22)

The reaction of 3,4-dihydroisoquinoline-*N*-oxide (**22**) with olefins such as allyl alcohol,[39] styrene,[45] methyl acrylate,[36,100] and ethyl acrylate[36] gives stereospecifically isoxazolidine of type It.

Some reactions of the oxide **22** with polysubstituted olefins are given in Table IV.

The reactions with various cyclic olefins are also described. The products have structures similar to those from other nitrones. With maleic anhydride,[47] *N*-phenylmaleimide,[47] norbornene,[7,39] norbornadiene,[39] indene,[48] dimethyl bicyclo[2.2.1]hept-2-ene-5,6-

[96] G. R. Delpierre and M. Lamchen, *J. Chem. Soc.*, 4693 (1963).
[97] B. G. Murray and A. F. Turner, *J. Chem. Soc. C*, 1338 (1966).
[98] J. F. Elsworth and M. Lamchen, *J. Chem. Soc. C*, 2423 (1968).
[99] M. Lamchen and T. W. Mittag, *J. Chem. Soc. C*, 1917 (1968).
[100] R. Huisgen and H. Seidl, *Tetrahedron Lett.*, 2019 (1963).

TABLE IV
Isoxazolidines from Dihydroisoquinoline-N-oxide (22) and Olefins 6–9

Type	Olefin				Isoxazolidine	References
	W	X	Y	Z		
6	H	COOMe	Me	H	IIIt + IIIc	36
	H	Ph	Ph	H	(III)	48
	H	OEt	OEt	H	(III)	48
7	COOMe	COOMe	H	H	Vc	47
8	H	COOMe	H	COOMe	[VIIt]	47, 100
	H	COOMe	H	Me	VIIIc	36
9	Me	COOEt	H	Me	X	36

dicarboxylate,[39] and cyclooctatetraene,[53] cis-fused isoxazolidines of type V are obtained. The reactions of 2-phenylisatogen (23) with various

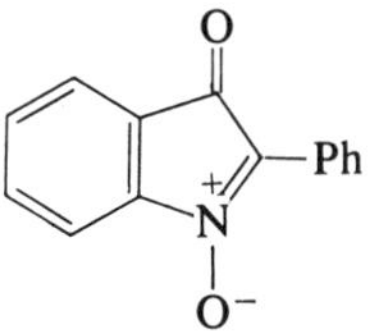

(23)

olefins are also described; with butyl vinyl ether,[77] styrene,[77] acrylonitrile,[101] ethyl acrylate,[77] and nitroethylene,[101] the reaction is regiospecific to give isoxazolidines of type I. The reaction of the nitrone 23 with ethyl fumarate, cyclopentene, and norbornene gives the expected cycloadducts.[7]

b. *Aromatic N-Oxides.* The reaction between quinoline-N-oxide and dihydronaphthalene-1,4-oxide gives an isoxazolidine (24).[102] It seems that this is a unique case in which an aromatic N-oxide gives an isoxazolidine. In other cases, ring opening of the isoxazolidines takes place to give aromatic compounds.[32] Thus, the reaction of pyridine N-oxide with perfluoropropene gives α-(1,2,2,2-tetrafluoroethyl)pyridine via an isoxazolidine of type II, followed by loss of COF_2.[103]

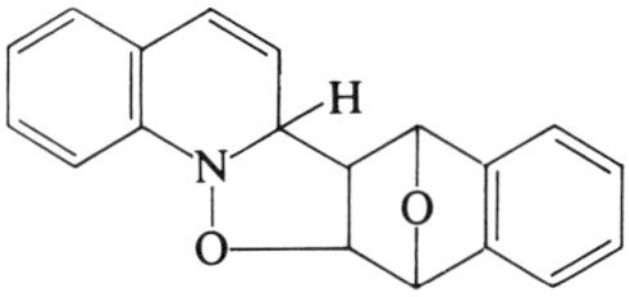

(24)

[101] W. E. Noland and D. A. Jones, *Chem. Ind. (London)*, 363 (1962).
[102] G. Wittig and G. Steinhoff, *Justus Liebig's Ann. Chem.* **676**, 21 (1964).
[103] E. A. Mailey and L. R. Ocone, *J. Org. Chem.* **33**, 3343 (1968).

4. Isoxazolidines from Oximes and Olefins

Cyclization of oximes with olefins is another route to isoxazolidines. Only a few oximes are reactive, however, and so this method is not very versatile. Reaction of formaldoxime with acrylonitrile[104-106] or methyl acrylate[104,105] gives 5-substituted isoxazolidines together with N-alkylated species. Benzophenone oxime reacts with but-1-en-3-one to give a similar mixture.[90,107] However, oximes such as acetaldehyde or cyclohexanone oxime do not react with acrylonitrile or acrylic ester,[106] although they do react with dimethyl acetylenedicarboxylate to give a 1:2-adduct.[108]

5. Dimerization of Nitrones

Some reactive nitrones dimerize to isoxazolidines immediately on formation. Thus, a nitrone **16c**, which is prepared from N-phenylhydroxylamine and dimethyl acetylenedicarboxylate, dimerized to isoxazolidine (**25**).[44] N-Phenylhydroxylamine and n-butyraldehyde gives α-(n-propyl)-N-phenylnitrone, which dimerizes in a similar manner.[109]

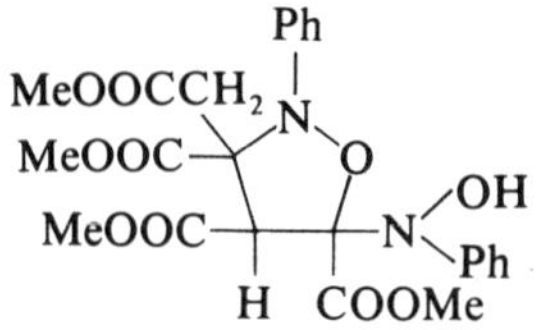

(25)

6. Isoxazolidines from Nitronic Esters and Olefins

Nitronic esters are simply N-alkoxynitrones and can easily be prepared, in particular the methyl esters, by the reaction of diazoalkanes with nitro compounds. Of methyl nitronates, α-aryl-N-methoxynitrones (**10p**) are most widely investigated. Reactions of the ester **10p** with mono- or disubstituted olefins, such as methyl

---

[104] M. Ochiai, M. Obayashi, and K. Morita, *Tetrahedron* **23**, 2641 (1967).
[105] K. Morita, M. Ochiai, and M. Obayashi, Japanese Patent, 70 18,294 (1970) [*CA* **73**, 77232 (1970)].
[106] H. Hjeds, B. Jerslev, and K. J. Ross-Petersen, *Dansk. Tidsskr. Farm.* **46**, 97 (1972) [*CA* **77**, 88375 (1972)].
[107] A. Lablache-Combier, M. Villaume and R. Jacquesy, *Tetrahedron Lett.*, 4959 (1967).
[108] E. Winterfeldt and W. Krohn, *Angew. Chem.* **79**, 722 (1967).
[109] W. Kliegel, *Tetrahedron Lett.*, 2627 (1969).

acrylate,[110,111] acrylonitrile,[110–112] but-1-en-3-one,[111] and dimethyl methylidenemalonate,[111] regiospecifically give isoxazolidines of type I or III. Reactions of α-alkoxycarbonyl-N-methoxynitrones are described. With styrene, allyl chloride, but-1-en-3-one, acrylonitrile, or methyl acrylate, α-methoxycarbonyl-N-methoxynitrone (**10q**) gives isoxazolidines of type I.[113] The reaction of the ethoxycarbonyl analog (**10r**) with vinylacetylene proceeds across the double bond to give the 5-ethynylisoxazolidine.[114] The stereospecificity is revealed by comparison

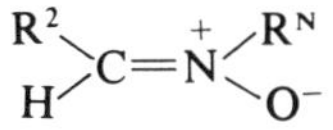

$$\underset{H}{\overset{R^2}{>}}C=\overset{+}{N}\underset{O^-}{\overset{R^N}{<}}$$

(**26**)

cis-Disubstituted nitrones

	R^2	R^N
a:	COOMe	OMe
b:	CN	OMe

with the reaction of corresponding *cis*-nitrones (**26**). With maleic anhydride or N-methyl maleimide, *trans*-nitrone (**10q**) gives adduct **27a** while *cis*-nitrone (**26a**) gives adduct **27b**.[115,116] Reaction of **10q** and **26a** with dimethyl fumarate or fumaronitrile affords adducts of types VIIc

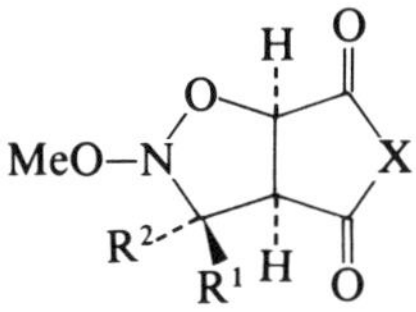

(**27**)

	R^1	R^2
a:	H	COOMe
b:	COOMe	H

X = O or NMe

---

[110] V. A. Tartakovskii, S. S. Smagin, I. E. Chlenov, and S. S. Novikov, *Izv. Akad. Nauk SSSR, Ser. Khim.*, 552 (1965) [*CA* **63**, 594 (1965)].

[111] V. A. Tartakovskii, Z. Ya. Lapshina, I. A. Savost'yanova, and S. S. Novikov, *Zh. Org. Khim.* **4**, 236 (1968) [*CA* **68**, 95734 (1968)].

[112] V. A. Tartakovskii, I. E. Chlenov, S. S. Smagin, and S. S. Novikov, *Izv. Akad. Nauk SSSR, Ser. Khim.*, 583 (1964) [*CA* **60**, 15852 (1964)].

[113] V. A. Tartakovskii, I. E. Chlenov, S. L. Ioffe, G. V. Lagodzinskaya, and S. S. Novikov, *Zh. Org. Khim.* **2**, 1593 (1966) [*CA* **66**, 64799 (1967)].

[114] V. A. Tartakovskii, O. A. Luk'yanov, N. I. Shlykova, and S. S. Novikov, *Zh. Org. Khim.* **3**, 980 (1967) [*CA* **67**, 100039 (1967)].

[115] R. Grée and R. Carrié, *Tetrahedron Lett.*, 4117 (1971).

[116] R. Grée and R. Carrié, *Tetrahedron Lett.*, 2987 (1972).

and VIIt, respectively.[116] The stereochemistry of the reaction of *trans*-
(**10s**) and *cis*- (**26b**) α-cyano-*N*-methoxynitrone is more interesting. The
trans-nitrone (**10s**) and acrylonitrile, methyl acrylate or but-1-en-3-one
gives a pair of epimers: **28c** (X = CN, COOMe, or Ac) (type Ic) and **28t**
(type It). The *cis*-nitrone (**26b**) and acrylonitrile or but-1-en-3-one gives
an epimeric mixture of **29t** (X = CN or Ac) (type It) and **29c** (X = CN
or Ac) (type Ic) while with methyl acrylate only adduct **29t**
(X = COOMe) is obtained. The interconversion of the epimers is also
described.[117]

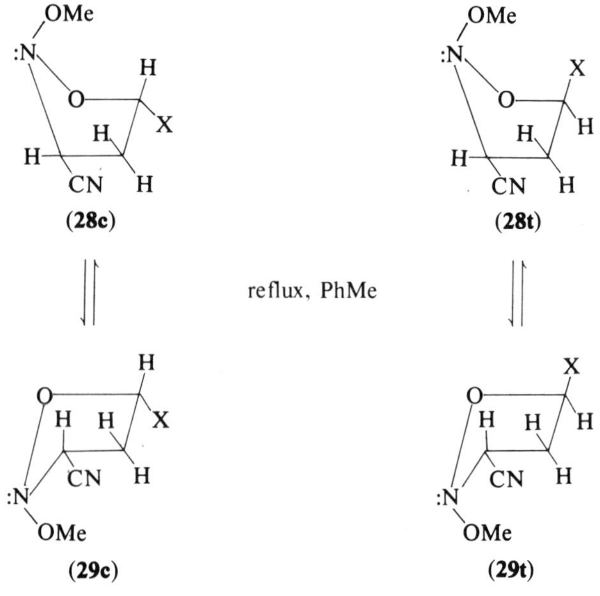

(**28c**)   (**28t**)

reflux, PhMe

(**29c**)   (**29t**)

X = CN, COOMe, or Ac

α,α-Diethoxycarbonyl-*N*-methoxynitrone (**16d**) also gives 5-substituted isoxazolidines with ethylene, propylene, and allyl chloride.[118,119] α,α-Dimethoxycarbonyl-*N*-methoxynitrone (**16e**) and acrylonitrile gives a 5-cyanoisoxazolidine, which can be separated into two invertomers: **30** and **31**.[31] The reaction of nitrone **16d** with substituted dienes takes place across the less hindered double bond to give adducts of type I.[114]

[117] R. Grée, F. Tonnard and R. Carrié, *Tetrahedron Lett.*, 453 (1973).
[118] V. A. Tartakovskii, I. A. Savost'yanova, and S. S. Novikov, *Zh. Org. Khim.* **4**, 240 (1968) [*CA* **68**, 95738 (1968)].
[119] G. A. Shvekhgeimer, N. I. Sobtsova, and A. Baranski, *Rocz. Chem.* **46**, 1543 (1972) [*CA* **78**, 84476 (1973)].

Sec. II.A]   THE CHEMISTRY OF ISOXAZOLIDINES

(30) and (31): structures with OMe, MeOOC groups, N-O ring, H, CN substituents.

Polynitroalkanes, when treated with diazomethane, give α-nitronitrones. Thus, from dinitromethane and trinitromethane, α-nitro-N-methoxynitrone[120] and α,α-dinitro-N-methoxynitrone (**16f**)[121] are formed. Both nitrones give adducts of type I when treated with olefins (**5**).[120,121] Notably, the nitronic ester **16f** is so reactive that it cyclizes with ethylene.[114] The reaction of **16f** with cyclopentene[122] gives adduct of type V, and with cyclopentadiene[116] a mixture of adducts of types V and VI is obtained. The nitronic ester **16f** with butadiene (equimolar) affords 5-vinylisoxazolidine; if the nitrone is in excess, further cyclization to the bisisoxazolidine takes place.[123] Cyclization of **16f** with 1- and 2-substituted butadienes is reported.[123]

The reaction of a trinitromethyl halide with diazomethane gives a nitrone intermediate, which can be trapped by a suitable olefin.[124–127]

Tetranitromethane and an olefin form a nitrone, which reacts further with another molecule of the olefin to give an isoxazolidine.[128]

	R^1	R^2		R^1	R^2
a:	H	H	f:	OH	H
b:	H	Ph	g:	H	COOMe
c:	H	Me	h:	OEt	H
d:	Me	H	i:	OMe	H
e:	Ph	H	j:	OMe	COOMe

[32]

---

[120] V. A. Tartakovskii, I. E. Chlenov, N. S. Morozova, and S. S. Novikov, *Izv. Akad. Nauk SSSR, Khim.*, 370 (1966) [*CA* **64**, 17567 (1966)].

[121] V. A. Tartakovskii, I. E. Chlenov, G. A. Langodzinskaya, and S. S. Novikov, *Dokl. Akad. Nauk SSSR* **161**, 136 (1965) [*CA* **62**, 14646 (1965)].

[122] A. A. Onishchenko and V. A. Tartakovskii, *Izv. Akad. Nauk SSSR, Ser. Khim.*, 948 (1970) [*CA* **73**, 34718 (1970)].

[123] V. A. Tartakovskii, O. A. Kuk'yanov, N. I. Shlykova, and S. S. Novikov, *Zh. Org. Khim.* **4**, 231 (1968) [*CA* **68**, 95740 (1968)].

[124] A. A. Onishchenko, I. E. Chlenov, L. M. Makarenkova, and V. A. Tartakovskii, *Izv. Akad. Nauk SSSR, Ser. Khim.*, 1560 (1971) [*CA* **75**, 98480 (1971)].

[125] K. V. Altukhov, E. V. Ratsino, and V. V. Perekalin, *Zh. Org. Khim.* **9**, 269 (1973) [*CA* **78**, 111180 (1973)].

[126] V. A. Tartakovskii, L. A. Nikonova, and S. S. Novikov, *Izv. Akad. Nauk SSSR, Ser. Khim.*, 1290 (1966) [*CA* **65**, 16809 (1966)].

[127] V. A. Tartakovskii, A. A. Fainzil'berg, V. I. Gulevskaya, and S. S. Novikov, *Izv. Akad. Nauk SSSR, Ser. Khim.*, 621 (1968) [*CA* **69**, 106598 (1968)].

[128] L. M. Andreeva, K. V. Altukhov, and V. V. Perekalin, *Zh. Org. Khim.* **8**, 1419 (1972) [*CA* **77**, 126479 (1972)] and references therein.

Reactions of cyclic nitronic esters are also described. 3-Nitroisoxazoline-*N*-oxide (**32a**) is itself a cyclic nitrone, and forms isoxazolizidines (**33a**) when allowed to react with olefins (**5**) such as ethylene,[112,129,130] vinyl acetate,[129] propylene,[129] and styrene.[130] With tetramethylethylene, cyclopentene, and cyclohexene, the corresponding isoxazolizidines are also obtained.[131]

X = H or residue of **5**

	R¹	R²			R¹	R²
a:	H	H	d:		Me	H
b:	H	Ph	e:		Ph	H
c:	H	Me	f:		OH	H

From the nitrone **32b** and ethylene or styrene, a mixture of the two epimers (**33b**) (configuration at C7) is formed.[130,132]

(**34**)

a: R = Ph
b: R = COOMe

In a similar manner, nitrones **32c**,[133] **32d**,[129] and **32e**[112] with various monosubstituted olefins afford the corresponding isoxazolizidines (**33c–e**). 4-Hydroxy-3-nitroisoxazoline-*N*-oxide (**32f**) is thoroughly investigated. With olefins **5**, such as propylene, styrene, vinyl acetate, methyl acrylate and acrylonitrile, adducts **33f** are formed.[134] Reactions of alkoxy or alkoxycarbonyl derivatives of **32** are also known; these include **32g**,[133] **32h**,[135] **32i**,[135] and **32j**[135]; **32j** is reactive enough to cyclize

[129] V. A. Tartakovskii, A. A. Onishchenko, I. E. Chlenov, and S. S. Novikov, *Dokl. Akad. Nauk SSSR* **164**, 1081 (1965) [*CA* **64**, 2079 (1966)].
[130] V. M. Shitkin, V. A. Korenevskii, V. G. Osinov, M. V. Kashutina, S. L. Ioffe, I. E. Chlenov, and V. A. Tartakovskii, *Zh. Org. Khim.* **8**, 864 (1972) [*CA* **77**, 33840 (1972)].
[131] V. A. Tartakovskii, A. A. Onishchenko, and S. S. Novikov, *Zh. Org. Khim.* **3**, 588 (1967) [*CA* **67**, 32625 (1967)].
[132] I. E. Chlenov, M. V. Kashutina, S. L. Ioffe, S. S. Novikov, and V. A. Tartakovskii, *Izv. Akad Nauk SSSR, Ser. Khim.*, 2085 (1969) [*CA* **72**, 12627 (1970)].
[133] V. A. Tartakovskii, A. A. Onishchenko, V. A. Smirnyagin, and S. S. Novikov, *Zh. Org. Khim.* **2**, 2225 (1966) [*CA* **66**, 75940 (1967)].
[134] V. A. Tartakovskii, A. A. Onishchenko, I. E. Chlenov, and S. S. Novikov, *Dokl. Akad. Nauk SSSR* **167**, 844 (1966) [*CA* **65**, 3852 (1966)].
[135] V. A. Tartakovskii, A. A. Onishchenko, and S. S. Novikov, *Zh. Org. Khim.* **3**, 1079 (1967) [*CA* **67**, 100042 (1967)].

with ethylene. 3-Phenylisoxazoline-N-oxide (**34a**) is another example of a reactive nitrone; thus it reacts with ethylene, styrene, and methyl acrylate to give isoxazolizidines.[136] 3-Methoxycarbonylisoxazoline-N-oxide (**34b**) shows much the same reactivity toward styrene and methyl acrylate.[136]

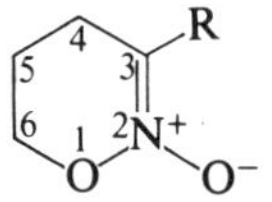

(**35**)

R = H or Me

Six-membered cyclic nitronic esters, the 4H-5,6-dihydrooxazine-N-oxides (**35**), exhibit much the same reactivity as their five-membered analogs. Reactions with styrene, methyl acrylate, and allyl alcohol give isoxazolidines of type I.[137]

B. Isoxazolidines by Intramolecular Cyclization of Nitrones

Besides the cycloaddition reaction between nitrones and olefins, the most widely used method of preparing isoxazolidines is the closely related intramolecular cyclization of unsaturated nitrones.

LeBel and Whang[6] found that N-methylhydroxylamine and 5-hexenal gives an isoxazolidine (**37a**) via the nitrone (**36**), which can also be prepared by oxidizing the corresponding hydroxylamine with HgO.

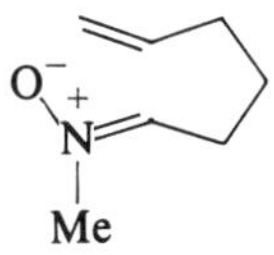

	R^1	R^2
a:	H	H
b:	H	Me
c:	Me	H

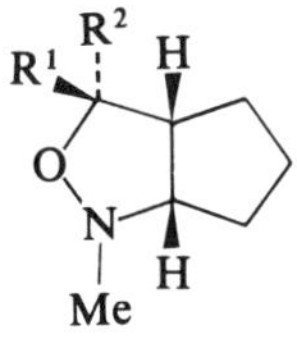

(**36**) (**37**)

Generally the oxygen atom of the nitrone moiety attacks the olefinic carbon atom which lies farther from the nitrone. The stereochemistry of cyclization is not specific; sometimes cis-ring fusion takes place exclusively, while in other cases trans-ring fusion accompanies cis-fusion. For instance, the reaction of N-methylhydroxylamine with *trans-* or *cis-*5-heptenal is stereospecific; thus from the former the *cis-*isoxazolidine

---

[136] I. E. Chlenov, V. I. Khudak, V. A. Tartakovskii, and S. S. Novikov, *Izv. Akad. Nauk SSSR, Ser. Khim.*, 2266 (1969) [*CA* **72**, 31663 (1970)].

[137] I. E. Chlenov, N. S. Morozova, V. I. Khudak, and V. A. Tartakovskii, *Izv. Akad. Nauk SSSR, Ser. Khim.*, 2641 (1970) [*CA* **74**, 141662 (1971)].

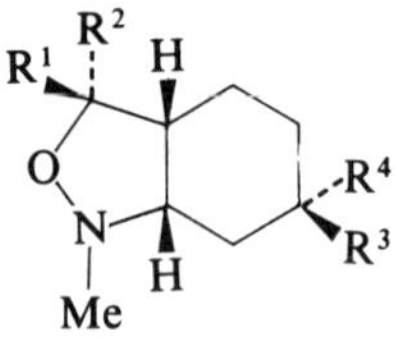

(38)

	R¹	R²	R³	R⁴
a:	H	H	H	H
b:	Me	Me	Me	H
c:	H	H	Me	H
d:	H	H	H	Me
e:	Me	Me	H	H
f:	Me	Me	H	Me

(**37b**) is obtained, and from the latter the trans isomer (**37c**) is the product.[138] A similar reaction with 6-hexenal, however, affords a mixture of cis- and trans-fused isomers (**38a** and **39a**) in a 2:1 ratio in addition to the bridged isoxazolidine (**40a**).[139] The ratio of cis- and trans-fused isomer seems to depend on the presence of alkyl groups on the terminal carbon of the double bond. The reaction of (+)-citronellal with N-methylhydroxylamine gives predominantly trans isomer **39b** over cis isomer **38b** [cis:trans = 3:97 (25°); 13:87 (138°)].[138]

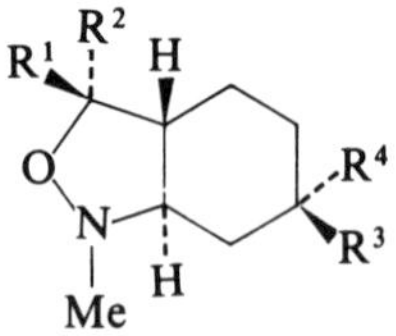

(39)

	R¹	R²	R³	R⁴
a:	H	H	H	H
b:	Me	Me	Me	H
c:	H	H	Me	H
d:	Me	Me	H	H

The reaction of 3-methyl-6-heptenal with N-methylhydroxylamine is more complex: it gives two cis-isomers (**38c** and **38d**), one trans isomer (**39c**), and one bridged isoxazolidine (**40b**), in a 19.4:15.4:60.1:5.0 ratio.[139]

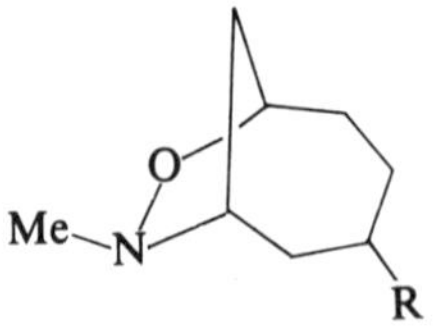

(40)

a:	R = H
b:	R = Me

[138] N. A. LeBel, M. E. Post, and J. J. Whang, *J. Am. Chem. Soc.* **86**, 3759 (1964).
[139] N. A. LeBel and E. G. Banucci, *J. Org. Chem.* **36**, 2440 (1971).

Reaction of various cyclic unsaturated aldehydes with N-methylhydroxylamine can afford complex polycyclic isoxazolidines;[140-142] e.g., from norbornenylacetaldehyde, adduct **41** is formed.[140]

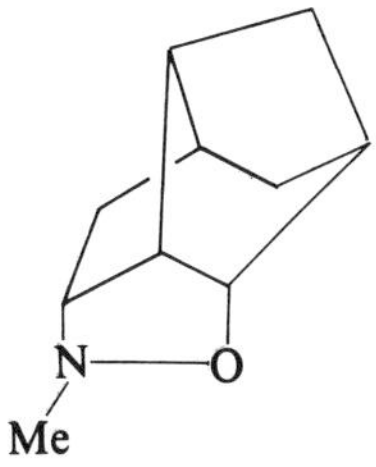

(41)

o-Allyloxybenzaldehydes possess the necessary structural requirement for cyclization. Thus, with an N-alkylhydroxylamine *cis*-isoxazolidines (**42**) and occasionally bridged isoxazolidines (**43**) are obtained.[143-145] As expected, *o*-allyloxybenzaldoxime itself cyclized to isoxazolidine (**42**; R = R^1 = H).[143].

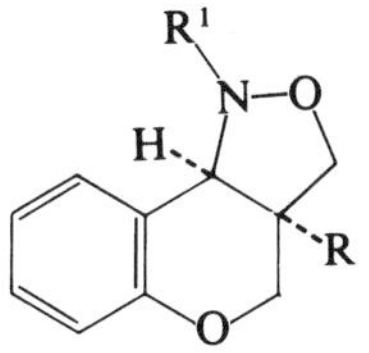

(42) 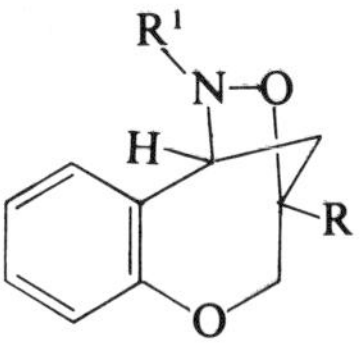   (43)

δε-Unsaturated ketones also form isoxazolidines with hydroxylamines. 6-Hepten-2-one and N-methylhydroxylamine gives cis-fused isoxazolidine (**44a**), and from 7-octen-2-one a mixture of cis- and trans-fused isoxazolidines (1:2) are obtained.[138,146] A *trans*-secoketosteroid also cyclized with N-methylhydroxylamine.[147] The reaction of 6,7-octadien-2-one with N-methylhydroxylamine in ethanol affords three isoxazolidines **44b**, **44c**, and **45**.[148]

[140] N. A. LeBel, *Trans. N.Y. Acad. Sci.* **27**, 858 (1965).
[141] N. A. LeBel, G. M. J. Slusarczuk, and L. A. Spurlock, *J. Am. Chem. Soc.* **84**, 4360 (1962).
[142] N. A. LeBel, N. D. Ojha, J. R. Menke, and R. J. Newland, *J. Org. Chem.* **37**, 2896 (1972).
[143] W. Oppolzer and K. Keller, *Tetrahedron Lett.*, 1117 (1970).
[144] W. Oppolzer and H. P. Weber, *Tetrahedron Lett.*, 1121 (1970).
[145] W. Oppolzer and K. Keller, *Tetrahedron Lett.*, 4313 (1970).
[146] M. Raban, F. B. Jones, E. H. Carlson, E. Banucci, and N. A. LeBel, *J. Org. Chem.* **35**, 1496 (1970).
[147] M. L. Mihailović, L. Lorenc, and Z. Maksimović, *Tetrahedron* **29**, 2683 (1973).
[148] N. A. LeBel and E. Banucci, *J. Am. Chem. Soc.* **92**, 5278 (1970).

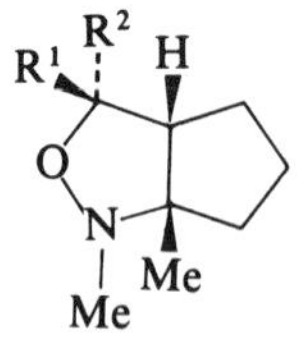

(44)

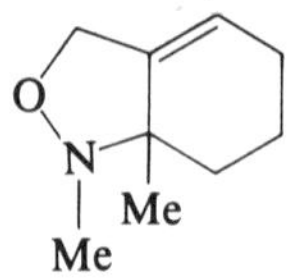

	R¹	R²
a:	H	H
b:	Me	EtO
c:	EtO	Me

(45)

C. Miscellaneous Methods

King prepared N-ethoxycarbonylisoxazolidine by the base-catalyzed condensation of 1,3-dibromopropane with N-hydroxyurethane, which was decarbethoxylated by hydrochloric acid to give the parent isoxazolidine (1).[5] If 1,3-dibromopropane is treated with potassium benzohydroxamate, N-benzoylisoxazolidine is obtained.[149]

Base-catalyzed condensation of α,β-unsaturated ketones and hydroxylamine gives hydroxyisoxazolidines; thus from mesityl oxide, adduct 46 is obtained.[150] Compound 46 is also formed from hydroxylamine and the condensation product of mesityl oxide and benzaldoxime.[151] The reaction of N,N-dimethyl-α-bromomethylacrylamide with N-methylhydroxylamine gives 3-(N,N-diethylcarbamoyl)-2-methylisoxazolidine.[152]

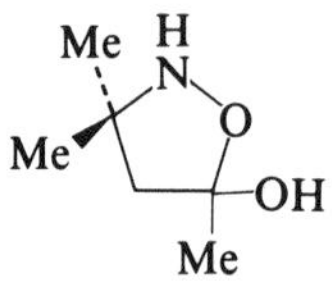

(46)

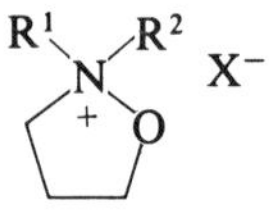

(47)

	R¹	R²	X
a:	$(CH_2)_2Cl$	$(CH_2)_2Cl$	Cl
b:	$(CH_2)_2Cl$	$(CH_2)_2Cl$	I
c:	$(CH_2)_2Cl$	Me	I
d:	Et	Et	Cl

---

[149] J. E. Johnson, J. R. Springfield, J. S. Hwang, L. J. Hayes, W. C. Cunningham, and D. L. McClagherty, *J. Org. Chem.* **36**, 284 (1971).

[150] A. Belly, R. Jacquier, F. Petrus, and J. Verducci, *Bull. Soc. Chim. Fr.*, 330 (1972).

[151] A. Ya. Tikhonov and L. B. Volodarskii, *Izv. Akad. Nauk SSSR, Ser. Khim.*, 2372 (1973) [*CA* **80**, 47888 (1974)].

[152] J. R. Smythies and D. Leaver, German Offen. 2,165,980 (1972) [*CA* **77**, 114415 (1972)].

Reaction of bis(β-chloroethyl)-γ-chloropropylamine with H_2O_2-acetic anhydride gives the isoxazolidinium chloride **47a**.[153,154] Various isoxazolidinium salts of this type have been prepared.[153-157] Dehydrohalogenation of N-(γ-chloroalkyl)hydroxylamine is an obvious method of preparing isoxazolidines. Thus, a hydroxy-4H-1,3-benzoxazine (**48**) gives the isoxazolidine **49** with sodium methoxide.[158,159]

(**48**)   (**49**)

Zinc reduction of isoxazoline-N-oxides is a general route to N-hydroxyisoxazolidines.[3-4] Unfortunately, attempted reduction of isoxazole,[160-161] 2-isoxazoline,[162] or 4-isoxazoline[163] to isoxazolidine always causes cleavage of N—O bonds to open-chain products.

(**50**)

1,6-Addition of cycloheptatriene and a nitrosobenzene affords the isoxazolidines (**50**).[164-167] The effect of the substituent X of nitrosobenzene upon the rate of reaction is NO_2 > Cl > H > Me > NMe_2 in line with the Hammett equation.[167]

---

[153] M. Ishidate, Y. Sakurai, and M. Torigoe, *Chem. Pharm. Bull.* **9**, 485 (1961).
[154] M. Ishidate and N. Sakurai, Japanese Patent 2,975 (1957) [*CA* **52**, 11950 (1958)].
[155] M. Ishidate and N. Sakurai, Japanese Patent 4,529 (1962) [*CA* **58**, 10205 (1963)].
[156] Y. Kuwada, *Chem. Pharm. Bull.* **8**, 807 (1960).
[157] M. Ishidate and N. Sakurai, Japanese Patent 4,790 (1962) [*CA* **58**, 13963 (1963)].
[158] D. B. Reisner and B. J. Ludwig, *J. Heterocycl. Chem.* **6**, 953 (1969).
[159] D. B. Reisner, B. J. Ludwig, H. M. Bates, and F. M. Berger, German Offen. 2,010,418 (1970) [*CA* **73**, 120644 (1970)].
[160] R. A. Barnes, in "Heterocyclic Compounds" (R. C. Elderfield, ed.), Vol. 5, p. 476. Wiley, New York, 1952.
[161] N. K. Kochetkov and S. D. Sokolov, *Adv. Heterocycl. Chem.* **2**, 412 (1963).
[162] N. K. Kochetkov and S. D. Sokolov, *Adv. Heterocycl. Chem.* **2**, 417 (1963).
[163] H. Seidl, R. Huisgen, and R. Knorr, *Chem. Ber.* **102**, 904 (1969).
[164] G. Kresze and C. Schulz, *Tetrahedron* **12**, 7 (1961).
[165] J. Hutton and W. A. Waters, *Chem. Commun.*, 634 (1966).
[166] P. Burns and W. A. Waters, *J. Chem. Soc. C*, 27 (1969).
[167] S. Ito, S. Narita, and K. Endo, *Bull. Chem. Soc. Jpn.* **46**, 3517 (1973).

Some azetidine-*N*-oxides are known to form isoxazolidines by thermal ring-enlargement.[168]

III. Physical Properties

A. Basicity

The parent isoxazolidine (1) itself is a strongly basic oil,[5] b.p. 70°–80°/50 mmHg,[169] with pK_a 5.05.[170] Substituted isoxazolidines are slightly weaker bases than 1. pK_a values (5.83–5.90) of some fused isoxazolidines similar to bicyclic nitrones **37**, **38**, and **39** are reported.[138,140]

Polarographic half-wave potential of the salts **47b–d**, $E_{1/2}$ vs. SCE (pH 3.5), are −0.488 −0.460, and −0.811, respectively.[153]

B. Dipole Moments

Lumbroso measured the dipole moments of various isoxazolidines.[171] The dipole moment of the parent isoxazolidine (1) is 2.88 D in C_6H_6 at

TABLE V
Dipole Moment of Isoxazolidines

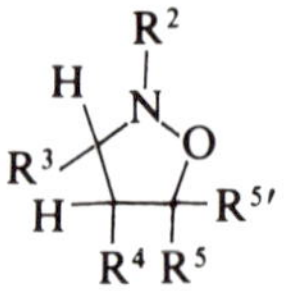

R^2	R^3	R^4	R^5	$R^{5\prime}$	Solvent		Reference
					Benzene	Dioxane	
H	H	H	H	H	2.88D	3.10D	171
Ph	Ph	H	Ph	H	2.47	—	171
Ph	Ph	H	Ph	Ph	2.19	—	171
Ph	Ph	Ph	Ph	H	2.54	—	171
Ph	p-NO$_2$C$_6$H$_4$	H	Ph	H	—	5.15	171
Ph	p-NO$_2$C$_6$H$_4$	H	Ph	Ph	—	4.35	171
Ph	H	H	OAc	H	2.95	—	171
COOEt	Ph	H	H	H	2.97	—	171
Me	Ph	H	COOEt	H	2.48	—	32

---

[168] Y. Suzuki, T. Watanabe, K. Tsukamoto, and Y. Hasegawa, German Offen. 2,317,980 (1973) [*CA* **80**, 37092 (1973)].

[169] R. O. C. Norman, R. Purchase, and C. B. Thomas, *J. Chem. Soc., Perkin Trans. 1*, 1701 (1972).

[170] H. J. Brass, J. O. Edwards, and N. J. Fina, *J. Chem. Soc., Perkin Trans. 2*, 726 (1972).

[171] H. Lumbroso, D. M. Bertin, and G. Cum, *C.R. Hebd. Seances Acad. Sci., Ser. C* **269**, 5 (1969).

Sec. III.C] THE CHEMISTRY OF ISOXAZOLIDINES 235

25°; since the (N—O) bond dipole moment is 1.0 D, the (N—Ph) bond dipole moment in N-phenylisoxazolidines seems to be smaller than that in N-phenylpyrrolidine. Some examples are given in Table V.

C. X-Ray Analysis

X-Ray analyses of some isoxazolidines are described. The structure of the isoxazolidine obtained from N-phenylhydroxylamine and acetone was determined with the aid of X-ray analysis.[172]

The stereochemistry of the two invertomers, **30** and **31**, has been determined by X-ray analysis,[173] the results of which aid the interpretation of the variable-temperature NMR study of these isomers.[31]

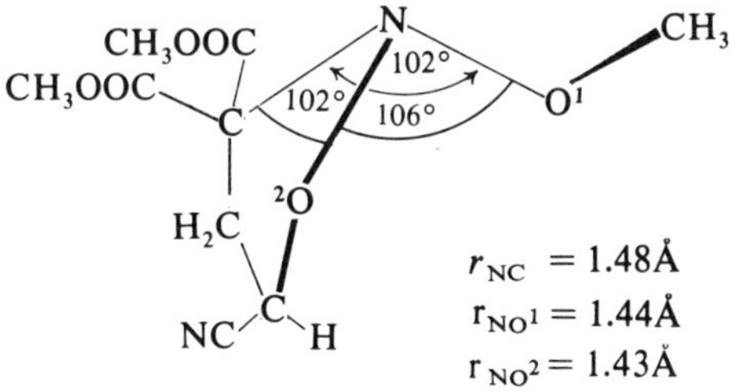

$r_{NC} = 1.48\text{Å}$
$r_{NO^1} = 1.44\text{Å}$
$r_{NO^2} = 1.43\text{Å}$

(51)

It is also shown that the isoxazolidine ring has an "envelope" conformation (**51**).[173] Ginzburg et al.[174] found that the two isoxazolidine rings, each in an "envelope" conformation, make a dihedral angle of 115° with each other in an isoxazolizidine. X-ray analysis of the compound **52** has also been described.[144]

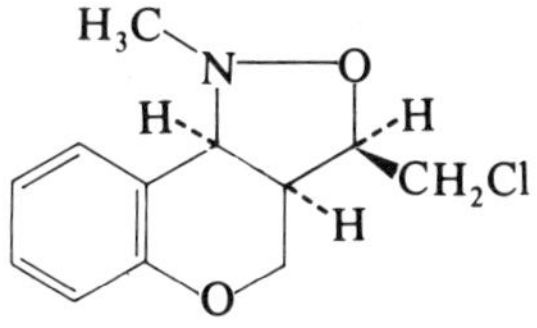

(52)

---

[172] R. Foster, J. Iball, and R. Nash, *Chem. Commun.*, 1414 (1968).
[173] M. Dobler, J. D. Dunitz and D. M. Hawley, *Helv. Chim. Acta* **52**, 1831 (1969): K. Müller, *Helv. Chim. Acta* **53**, 1112 (1970).
[174] S. L. Ginzburg, M. G. Neiganz, L. A. Novakovskaya, S. S. Novikov, V. A. Tartakovskii, I. E. Chlenov, Z. A. Akopyan, A. I. Gusev, and Y. T. Struchkov, *Zh. Strukt. Khim.* **10**, 877 (1969) [*CA* **72**, 36724 (1970)].

D. Inversion at Nitrogen

As a cyclic saturated amine, the inversion at nitrogen as well as that of the ring is in principle possible for isoxazolidines: in rings such as piperidine, morpholine, and pyrrolidine, usually the inversion is very rapid on the NMR time scale even at reasonably low temperature unless a very electronegative atom (or group) like chlorine is substituted on nitrogen.[175] Recently, however, it has been found that the presence of a heteroatom adjacent to nitrogen, as in a tetrahydro-1,2-oxazine, considerably reduces the rate of nitrogen inversion, probably owing to the lone-pair interaction.[176] It is expected that nitrogen inversion in isoxazolidines is slow on the NMR time scale.

A few pairs of invertomers have in fact been separated[31,116,117] or at least the presence of invertomers proved by variable temperature NMR.[146,177–179] Riddell[177] measured the ^{1}H NMR or N-methylisoxazolidine, and found that the $-CH_2-N$ methylene protons are magnetically nonequivalent at low temperature in $CDCl_3$. This nonequivalence was ascribed to slow nitrogen inversion based on the following evidence. The coalescence temperature (T) rises from 42° to 62°, and the energy barrier increases from 15.6 kcal to 16.9 kcal in D_2O. Such an increase in the barrier is unlikely if ring inversion is involved, and it is usually associated with the energy required for breakage of the hydrogen bond formed between the nitrogen lone pair and the solvent. The dielectric constant of the solvent seems to have little if any effect on the barrier to nitrogen inversion.[180]

Griffith followed the NMR signals of $MeOCH_2N-$ of N-methoxymethylisoxazolidine or $CH(CH_3)_2$ of N-isopropylisoxazolidine and found that the size of the substituent greatly influences the exchange rate.[178]

Raban et al.[146] examined the low-temperature ^{1}H NMR of **44a** and observed two invertomers. The *endo*-Me:*exo*-Me ratio depends on the solvent: 0.4 in toluene-d_8, CS_2, CH_2Cl_2, and acetone-d_6, 0.5 in $CDCl_3$, 0.6 as neat liquid, 1.0 in methanol–water (4:1), 1.4 in acetic acid–acetone. In nonaqueous solvents the exo isomer, which cannot hydrogen-bond, is favored, whereas in aqueous media or the like, the endo isomer, which can hydrogen-bond, increases.

Lagod-Zinskaya measured the nitrogen-inversion of eight N-methoxy-3,3-dinitroisoxazolidines, and suggested that the larger barrier

---

[175] J. B. Lambert, W. C. Oliver, and B. S. Packard, *J. Am. Chem. Soc.* **93**, 933 (1971).
[176] F. G. Riddell, D. A. R. Williams, C. Hootelé, and N. Reid, *J. Chem. Soc. B*, 1739 (1970); J. S. Splitter and M. Calvin, *Tetrahedron Lett.*, 4111 (1973).
[177] F. G. Riddell, *Chem. Commun.*, 1403 (1968).
[178] D. L. Griffith and B. L. Olson, *Chem. Commun.*, 1682 (1968).
[179] T. V. Lagod-Zinskaya, *Zh. Strukt. Khim.* **11**, 31 (1970) [*CA* **73**, 9212 (1970)].
[180] D. L. Griffith and J. D. Roberts, *J. Am. Chem. Soc.* **87**, 4089 (1965).

Sec. IV.A] THE CHEMISTRY OF ISOXAZOLIDINES 237

(15 kcal) is an indication that, in addition to nitrogen inversion, ring inversion is involved in the rate process observed.[179]

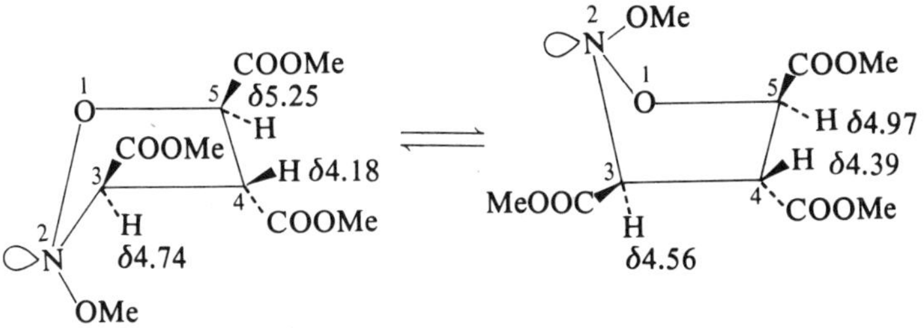

$J_{3,4} = 2.2$ Hz
$J_{4,5} = 5.4$
(53)

$J_{3,4} = 9.5$ Hz
$J_{4,5} = 5.0$
(54)

The isoxazolidine **53**, when heated to reflux in toluene, forms an invertomer equilibrium **53** ⇌ **54**. The activation energy $\Delta G^{\ddagger}$ for this inversion is 26.4 kcal/mole (110°C). The dihedral angle H_3–C–C–H_4 changes from ca. 100° to 140°, as indicated by the coupling constants.[116] The invertomer equilibria **28c** ⇌ **29c** and **28t** ⇌ **29t** have been mentioned. The activation energy $\Delta G^{\ddagger}$ for the latter process is 25.4 kcal/mole (110°C).[117]

IV. Spectroscopic Properties

A. IR Spectra

There seems to be no characteristic band for the isoxazolidine ring, hence the IR spectrum has little value in the diagnosis of the isoxazolidine structure. There have, however, been some assignments made in connection with ν(C–O) or ν(N–O). 5-Cyanoisoxazolidine shows ν(C–N) at 1035 cm^{-1}.[104] For isoxazolidine **55**, peaks at 1078 and 918 cm^{-1} are assigned ν(C–N) and ν(N–O), respectively.[38] LeBel et al.[138] pointed out that the isoxazolidine **38a** and its analog have no absorption in the O–H, N–H, C=O, or C=C regions, and the peak at 950 cm^{-1} was assigned to ν(N–O). ν(N–O) in isoxazolizidine **33** is found in the range 1010–1060 cm^{-1}.[129,134,181]

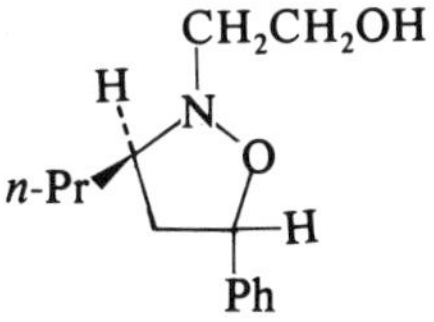

(55)

[181] A. I. Ivanov, V. I. Slovetskii, V. A. Tartakovskii, and S. S. Novikov, *Khim. Geterotsikl. Soedin. Akad. Nauk Latv. SSR*, 197 (1966) [*CA* **65**, 4848 (1966)].

B. UV Spectra

The isoxazolidine ring is UV-transparent since it is saturated, and any absorption recorded for isoxazolidine derivatives is due to the substituents.[182] Interestingly, the two invertomers **30** and **31** have different intensities, although the peak position is identical.[31]

C. NMR Spectra

The proton chemical shifts of isoxazolidine **1**[169] and *N*-benzoylisoxazolidine[149] will give the standard δ values for H_2–H_5 of the isoxazolidine ring.

Generally, the N–H peak appears at $\delta 5.0$ as a broad singlet[150]; irradiation of the N–H proton does not affect the other signals.[183] Inversion of the nitrogen causes a change in ring proton chemical shifts as shown in the equilibrium **53** ⇌ **54**.[31,116,117]

The chemical shift difference between geminal protons in isoxazolidines is quite sensitive to the molecular structure; thus in the *cis*-isoxazolidine (**56a**) $\Delta\delta(H_{4gem}) = 0.69$ ppm, and for the trans isomer (**56b**) $\Delta\delta(H_{4gem}) \leq 0.15$.[45] A similar difference is recorded for $\Delta\delta(Me_{3gem})$ of *gem*-dimethylisoxazolidines such as **38e** and **39d**.[139]

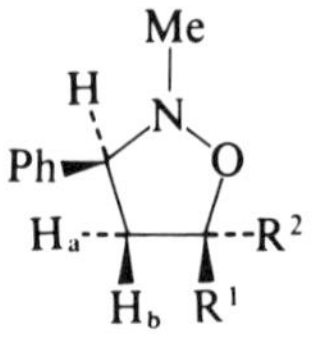

<pre>
           Me
       H |
        \\ N
    Ph—⟨   \\O
       Hₐ--   ⟩--R²
           |
         H_b R¹

         (56)
              R¹   R²
        a: Ph   H
        b: H   Ph
</pre>

The stereochemistry of the ring fusion can be determined by the shift of the bridgehead proton. Comparison of δH_a of isomers **57** and **58** reveals that the proton in the former is equatorial to the cyclohexane ring.[139,184] Vicinal proton–proton coupling constants of isoxazolidines are also sensitive to the stereochemistry ($J_{cis} > J_{trans}$).[36,39,47,52] Representative examples **59** and **60** are given.

---

[182] N. A. LeBel and J. J. Whang, *J. Am. Chem. Soc.* **89**, 3076 (1967).
[183] G. Cum, M. C. Aversa, N. Uccella, and M. Gattuso, *Atti Soc. Peloritana Sci. Fis. Mat. Nat.* **14**, 413 (1968) [*CA* **73**, 130370 (1970)].
[184] N. A. LeBel and T. A. Lajiness, *Tetrahedron Lett.* 2173 (1966).

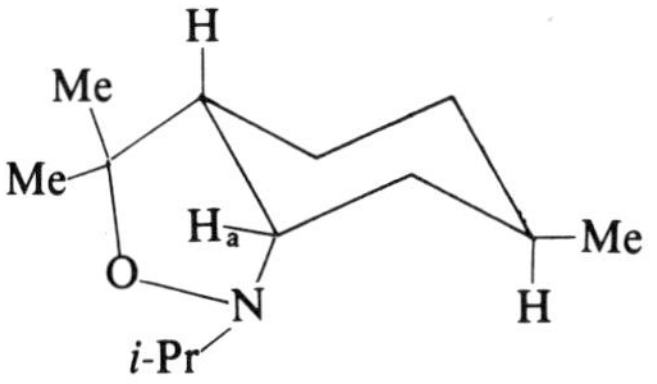

Sec. IV.D]   THE CHEMISTRY OF ISOXAZOLIDINES   239

H$_a$ δ2.78
(57)

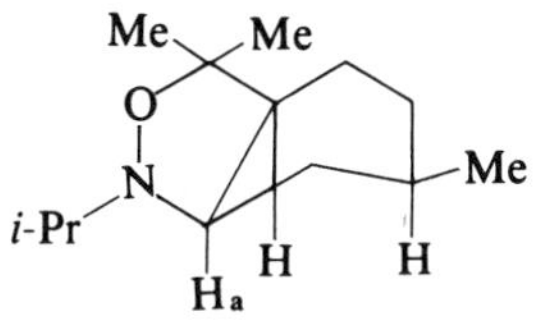

H$_a$ δ2.05
(58)

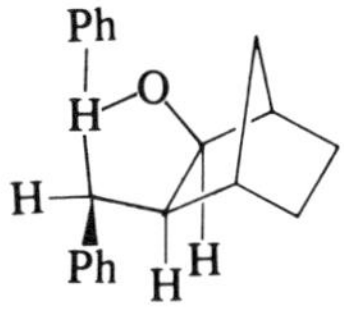

cis $J_{3,4}$ = 9.3 Hz

(59)

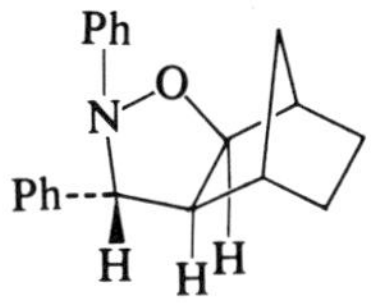

trans $J_{3,4}$ = 7.0 Hz

(60)

In Table VI, proton chemical shifts and coupling constants for some monocyclic isoxazolidines are tabulated.

NMR data for nuclei other than protons are rather scarce. ^{19}F data are available for two isomeric fluoroisoxazolidines.[40]

D. Mass Spectra

The mass spectrum of the isoxazolidine ring (61) shows a 4-fold cleavage; loss of a phenyl residue from the molecular ion also occurs.[107]

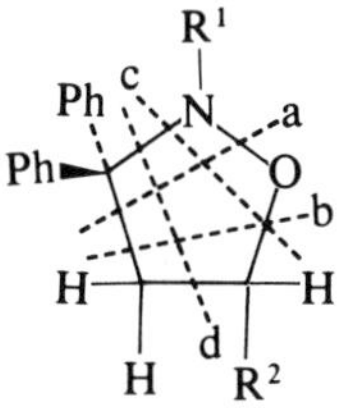

(61) 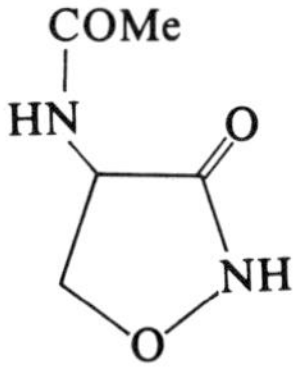   (62) 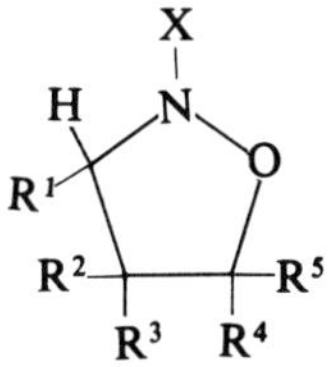   (63)

Isoxazolidine 1 itself exhibits strong peaks due to the ethylene and formaldoxime ion radicals.[104] The mass spectrum of 5-methoxy-carbonyl-5-methylisoxazolidine has [R—C≡O$^+$] peak typical of carboxylic acid derivatives.[104] For cycloserine (62), the initial fragmentation

TABLE VI
NMR Spectra of Simple Isoxazolidines

$$R^{3'} \underset{R^{3}}{\overset{R^{2}}{\underset{R^{4'}}{\bigg|}}} \underset{R^{4} \; R^{5}}{\overset{O}{\underset{}{\bigg|}}} R^{5}$$

								δ													
R^2	R^3	$R^{3'}$	R^4	$R^{4'}$	R^5	$R^{5'}$	H_2	H_3	$H_{3'}$	H_4	$H_{4'}$	H_5	$H_{5'}$	$J_{3,4}'$	$J_{3,4}$	$J_{4,5}'$	$J_{4,5}$	$J_{4',5}$	$J_{4,5'}$	$J_{4,4'}$	$J_{5,5'}$	Reference
H	H	H	H	H	H	H	4.75	3.0		2.1			3.7									169
PhCO	H	H	H	H	H	H		3.82		2.17			3.77					7				149
H	Me	Me	H	H	Me	OH	5.3			1.96										13		150
H	Me	Me	H	H	Me	OMe	5.2			1.93												150
t-Bu	H	H	H	H	H(NO$_2$)	NO$_2$(H)		(2.6–3.2)		(2.6–3.2)		(5.4–5.6)										23
Me	Ph	H	NO$_2$	H	H	H		3.87			5.46	4.62	4.34		8.0							23
Me	Ph	H	H	NO$_2$	H	H		3.95		4.95		4.15	4.44	6.0		6.7	3.0	8.0			10.5	23
Me	Ph	H	D	SO$_2$Ph	H	H		3.87				4.25	4.47								10.5	23
Me	Ph	H	H	H	Ph	H		3.76		2.36	3.05	5.22		6.8	9.1		7.4	7.5	11.4	9.5		45
Ph	Ph	H	H	H	Ph	H		4.79		2.36	2.98	5.04		7.7	7.5		9.6	5.5	11.6			45
Ph	Ph	H	H	H	H	Ph		4.60		2.52	2.68	5.27		7.5	5.5	7.4		7.4	12.0			45
Ph	Ph	H	H	H	Me	Ph		4.43		2.60	3.10			9.0	7.8				12.5			48
Ph	Ph	H	H	H	Ph	Me		4.67		2.66	2.94			8.5	8.5				12.0			48
Ph	Ph	H	H	H	Ph	Ph		4.54		3.06	3.40			8.0	8.0				12.0			48
Me	Ph	H	H	H	Ph	Ph		3.81		3.11	3.31			10.0	6.8				12.0			48
Ph	Ph	H	H	H	CN	H		4.28		2.48	2.97		4.70	6.2	8.2		4.0	8.2				36
Ph	Ph	H	H	H	OAc	H		4.38		2.49	3.13		6.55	6.8	9.3		2.0	6.1	13.7			57
Ph	Ph	H	H	H	COOMe	H		5.05			4.05		4.80		8.7			7.1				61
Ph	Ph	H	COOMe	H	H	COOMe		4.99		3.79		5.04		8.1		7.8						61
Ph	Ph	H	H	COOMe	COOMe	H		4.77		4.00			5.10	6.4			4.8					61
Ph	Ph	H	COOMe	H	H	COOMe		5.07			4.05		5.25		8.6			7.8				61

is a loss of ONH_2 which was proved to originate by migration of the 4-H to the oxygen.[185]

Caruso et al.[186] found that the loss of (X–NHO) from isoxazolidine (63) gives the base peak. The mechanism of this cleavage was discussed.

Isoxazolidines 13a and 13b, which are formed from cycloaddition between α,N-diphenylnitrone and enamines, cleave into enamine and nitrone moieties (retro-1,3-dipolar cycloaddition upon electron impact).[187]

Isoxazolidines 33 lose NO_2 in the first stage, and ring opening follows.[188]

V. Chemical Properties

A. THERMAL STABILITY

a. *Thermal Retrofission and Thermal Isomerization.* When heated, some of the isoxazolidines formed from olefins and nitrones cleave into the components (thermal retro-1,3-dipolar cycloaddition);[8,10,36,43,64,96,99] e.g., isoxazolidines formed from the nitrone 18a and an $\alpha\beta$-unsaturated ester decompose into their components when heated to 100° under reduced pressures,[8,10] and vacuum distillation of isoxazolidines from the nitrone 10j re-form the nitrone (19%).[36]

When heated, some isoxazolidines isomerize to the more stable isomers. 5-Ethoxycarbonylisoxazolidine obtained from the nitrone 18a and ethyl acrylate gives the more stable 4-substituted isomer.[96] An extensive study of the thermal isomerization of fused isoxazolidines was carried out by LeBel et al..[139,141,184] Thus, the trans-fused isoxazolidine 39b isomerizes into a mixture of two cis-fused isomers, 38b and 38f. Starting from one of the last two compounds, the same equilibrium mixture is obtained.[139,184] Thermal equilibration of *trans*-isoxazolidine (39d)[139] with the cis-fused isomer (38e) is also described, and the isomerization of the bridged isoxazolidine (40a) to the fused isoxazolidine (38a).[139,184] The time required for attaining equilibrium depends on the temperature. At 180° the trans-isomer 39c disappears after about 200 hours; at 200° and 235°, it takes only 2 hours and 45 minutes, respectively. In almost all cases the cis-fused isomer is more stable and predominates in the mixture.

---

[185] A. Tatematsu, S. Sugiura, S. Inoue, and T. Goto, *Org. Mass. Spectrosc.* 1, 205 (1968).

[186] F. Caruso, G. Cum, and N. Uccella, *Tetrahedron Lett.,* 3711 (1971).

[187] Y. Nomura, F. Furusaki, and Y. Takeuchi, *J. Org. Chem.* 37, 502 (1972).

[188] Yu. V. Rozynov, N. S. Vul'fson, V. A. Puchkov, V. A. Tartakovskii, A. A. Onishchenko, and S. S. Novikov, *Khim. Geterotsikl. Soedin.,* 36 (1969) [*CA* 70, 105759 (1969)].

b. *Thermal Decomposition.* Some isoxazolidines are thermally unstable, and decomposition takes place if excess heat is applied. When isoxazolidine **64** is heated above its melting point, it decomposes, with evolution of gas, into methanol, water, benzonitrile, and a residue which is probably nitrated polystyrene.[2]

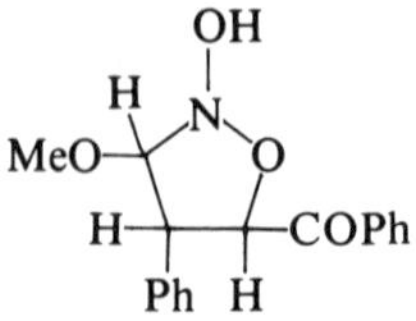

(64)

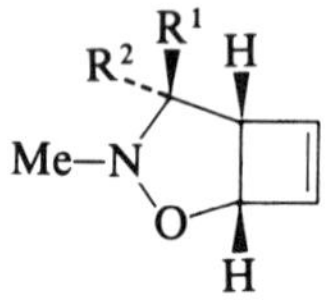

(65)

	R^1	R^2
a:	Ph	H
b:	H	Ph

When isoxazolidine **25** (the 2:2 adduct of phenylhydroxylamine and dimethyl acetylenedicarboxylate) is heated in refluxing benzene, elimination of phenylhydroxylamine takes place.[44]

The fused isoxazolidines **12a** and **12b** thermally decompose into isoxazolidine **65a** and **65b**, and dimethyl phthalate (retro Diels–Alder reaction).[53].

B. Photolysis

LeBel and Whang[182] obtained a tetrahydro-1,3-oxazine (**66**) and an open-chain amide by irradiating (2537 Å) the fused isoxazolidine **39b**. The yield of **66** is increased if the reaction is photosensitized with benzophenone or fluorenone.

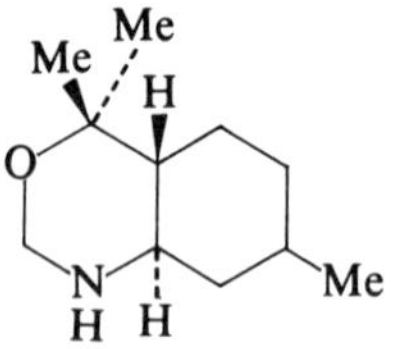

(66)

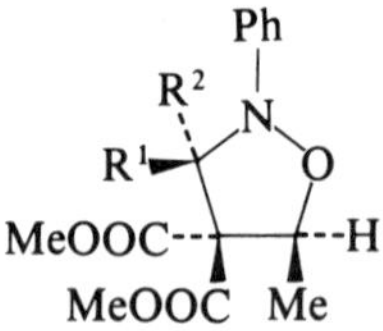

(67)

	R^1	R^2
a:	H	PhCO
b:	PhCO	H
c:	H	Ph

Irradiation of the isoxazolidine **67a** in methanol causes isomerization to the epimer (**67b**). No reaction, however, takes place when the phenyl analog (**67c**) is irradiated under the same conditions.[64]

C. Reduction (Hydrogenolysis)

When isoxazolidines are hydrogenated, N—O bond cleavage usually take place to give 3-amino alcohols. Hydrogenation of monocyclic isoxazolidines with Ni,[36,39,45,48,189] Pd–C,[90,150] Pd–CaCO$_3$,[65,66] platinum oxide,[190] aluminum amalgam,[45] Na–BuOH,[45] or Zn–acetic acid[45] all afford 3-amino alcohols. Since the configuration is retained during the reaction, hydrogenolysis is frequently used as a means to establish the stereochemistry of isoxazolidines. Thus, the diastereomeric isoxazolidines **56a** and **56b** give the corresponding *erythro*- (**68a**) and *threo*- (**68b**) alcohols.[45] Lithium aluminum hydride also effects N—O cleavage.[96]

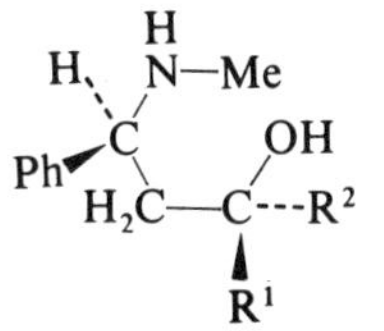

	R^1	R^2
a:	Ph	H
b:	H	Ph

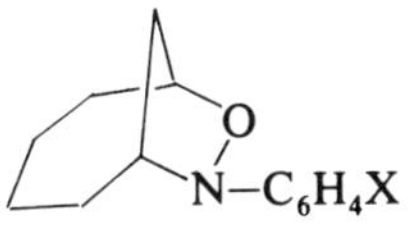

It is possible to hydrogenate a double bond in an isoxazolidine without effecting N—O cleavage if milder conditions are applied.[8,184]

When Pd–C[166] or PtO$_2$[167] is used as the catalyst, reduction of the double bonds in the side chain precedes the hydrogenolysis. Thus the unsaturated isoxazolidine **50** can be reduced to the saturated compound **69** (Pd–C or PtO$_2$), which is then cleaved by Zn–AcOH to an aminocycloheptanol.[167] Quaternary isoxazolidinium salts are also reduced to *N,N*-dialkylaminoalcohols (e.g., **70a → 70b**).[52,142,153,191]

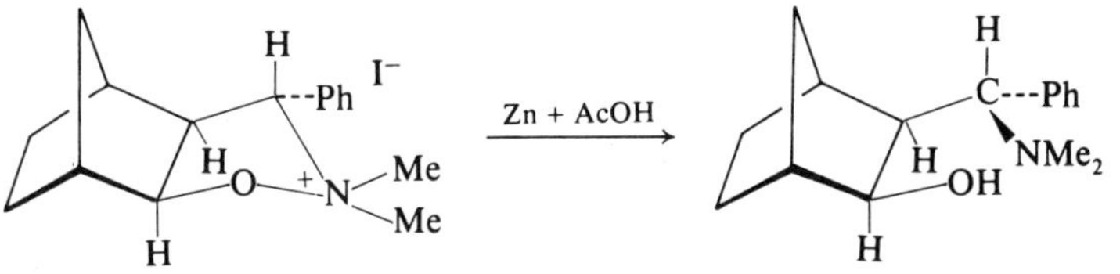

---

[189] I. Ikeda, G. Takemoto, and S. Komori, *Kogyo Kagaku Zasshi* **74**, 419 (1971).
[190] F. Hoffmann, Netherland Appl. 6,512,329 (1966) [*CA* **65**, 7156 (1966)].
[191] J. J. Tufariello and E. J. Trybulski, *Chem. Commun.*, 720 (1973).

Some 5-methoxycarbonylisoxazolidines give γ-lactams on hydrogenation.[47] Initially, 3-amino alcohols are formed; however, lactam formation immediately occurs with retention of configuration. With 4-ethoxycarbonylisoxazolidines, lactam formation is not favorable, and only the 3-amino alcohol is obtained.[36]

Generally, the hydrogenolysis with Raney–Ni gives 3-amino alcohols; with Pd–C[7,36,39,45,48,90] or Pt–C[66] as the catalyst, deamination sometimes accompanies the reaction. Thus, reduction of 2,3,5-triphenylisoxazolidine with Ni(H$_2$) affords the expected 3-amino alcohol, but with Pd–C, 1,3-di-phenylpropanol and aniline are formed.[7,45]

Sometimes loss of a side chain accompanies the N–O cleavage. In particular, hydroxyl groups are often removed.[59,145,192] On attempted reduction of N-methoxyisoxazolidines with LiAlH$_4$, loss of methanol takes place to give 2-isoxazolines.[110]

D. Oxidation

Oxidation of the isoxazolidine **64** with permanganate causes complete decomposition of the isoxazolidine ring. Two moles of benzoic acid are obtained.[2] Oxidation of the chloro compound **50**, X = Cl, with periodic acid affords only an intractable purple resin; with permanganate p,p'-dichloroazoxybenzene is obtained.[166]

LeBel[140] investigated the oxidation of isoxazolidine **38f** with H$_2$O$_2$ and of compound **44a** with a peracid. Oxidation of isoxazolidine **1** with Pb(OAc)$_4$ gives 2-isoxazoline.[169]

E. Decomposition by Acid

When isoxazolidines are treated with an acid, N–O cleavage and further reactions take place. With N-hydroxy- or N-alkoxyisoxazolidines, sometimes water or alcohol is eliminated to give 2-isoxazolines.

When N-hydroxy-3,4,5-triphenylisoxazolidine is treated with conc. H$_2$SO$_4$, a mixture of 3,4,5-triphenylisoxazolidine, 3,4,5-triphenyl-2-isoxazoline, and 3,4,5-triphenylisoxazole was obtained.[3] The reaction of hydroxy compound **64** is more complex. Dilution of a conc. H$_2$SO$_4$ solution of the compound **64** affords 3-hydroxyisoxazolidine. When the

---

[192] A. D. Baker, J. E. Baldwin, S. P. Kelly, and J. DeBernardis, *Chem. Commun.*, 344 (1969).

H_2SO_4 solution is kept standing for 15 days, an isoxazoline-N-oxide (**71**) is obtained.[2]

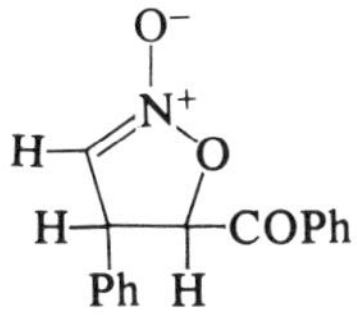

(**71**)

It is reported that if N-methoxyisoxazolidine is heated with gaseous HCl, either (a) formation of 2-isoxazoline with loss of methanol, or (b) ring opening to leave β-imino alcohol, takes place[111,113,120]; thus the compound **72a** gives a mixture of isoxazoline **73** and imino alcohol **74**.[111] A similar reaction can be performed with conc. H_2SO_4.[113] Other acids (dil. H_2SO_4, N_2O_4, or BF_3 etherate) are effective to give isoxazoline.[114,118] Sometimes further hydrolysis of imino alcohols takes place.[120,193] For instance, the N-methoxyisoxazolidine **72b** gives malic acid.[120]

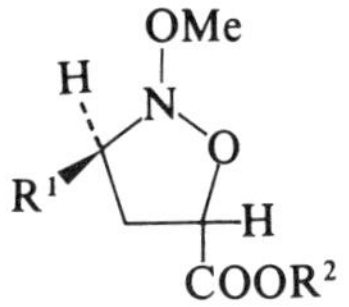

(**72**) 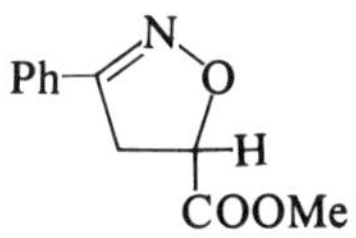   (**73**)

	R^1	R^2
a:	Ph	Me
b:	NO_2	Et

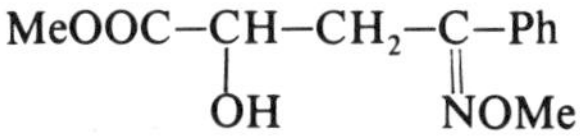

MeOOC—CH—CH$_2$—C—Ph
         | ||
         OH   NOMe

(**74**)

When 2,3-diphenyl-5,5-diethoxyisoxazolidine was treated with hydrochloric acid, cleavage of the N—O bond takes place to give ethyl β-phenyl-β-(o- or p-chloroanilino)propionate (through Orton rearrangement).[59]

When 5-pyrrolidinoisoxazolidine (**13a**) was treated with glacial acetic acid, 2-6-dibenzylidenecyclohexanone was isolated.[72] A similar decomposition of **13a** can be effected by dil. HCl.[74,194] When methanolic HCl

---

[193] V. I. Erashko, S. A. Shevelev, and A. A. Fainzil'berg, *Izv. Akad. Nauk SSSR, Ser. Khim.*, 2117 (1968) [*CA* **70**, 87638 (1969)].

[194] F. Furusaki, Ph.D. Thesis, University of Tokyo, Japan, 1969.

was employed, cyclohexanone and o- and p-chloroaniline were detected in addition to dibenzylidenecyclohexanone.[194] With glacial acetic acid–water (9:1), however, no N—O cleavage takes place, and instead substitution of the pyrrolidine moiety by a hydroxyl group occurs.[74]

When an isoxazolizidine is treated with an acid, either complete decomposition takes place, or cleavage of one N—O bond occurs to give an isoxazoline. Thus, hydrolysis of the isoxazolizidine **33a** with dil. H_2SO_4 gave 1,5-dihydroxy-3-pentanone.[129] From the same compound 3-(2-hydroxyethyl)-2-isoxazoline was obtained when treated with gaseous HCl in benzene.[195]

F. Decomposition by Base

Under very mild conditions, the isoxazolidine ring is not cleaved by base. However, under severe conditions, cleavage of N—O bonds and further fragmentation takes place.

Alkaline (NaOH or KOH) treatment of N-methoxyisoxazolidines causes loss of methanol to give 2-isoxazolines.[113] Howevever, in most cases N—O cleavage is brought about to give propanol derivatives. Sometimes further fragmentation takes place.[118]

If structural requirements are fulfilled, recyclization follows the N—O cleavage. Thus, from the nitronic ester (**72a**), 3-phenyl-5,6-dioxo-1,2-oxazine is formed.[111] Treatment of compound **39b** with t-BuOK gives an 1,3-oxazine (**66**).[140,182]

Basic ring opening of isoxazolidinium salts (methohalides) might be followed by Hofmann degradation; thus, the salt **75** gives compounds **76** and **77**.[140] Base treatment of isoxazolizidines is also described.[134]

G. Substitution

The isoxazolidine NH group is expected to possess properties common to aliphatic amines, and in fact the nitrogen atom is easily alkylated and acylated. Thus, the hydrochloride of compound **1** is acylated by acyl chloride in the presence of a tertiary amine.[5,196–200] Reaction of **1**

---

[195] V. A. Tartakovskii, A. A. Onishchenko, G. V. Lagodzinskaya and S. S. Novikov, *Zh. Org. Khim.* **3**, 765 (1967) [*CA* **67**, 53206 (1967)].

[196] G. Pifferi, German Offen. 2,019,659 (1969) [*CA* **74**, 22819 (1971)].

[197] G. Pifferi, P. Consonni, A. Diena, and B. R. D. Turco, *J. Med. Chem.* **15**, 851 (1972).

[198] C. H. Boehringer, French Patent 1,573,667 (1969) [*CA* **72**, 90479 (1970)].

[199] G. Pifferi, U.S. Patent 3,696,096 (1972) [*CA* **78**, 16166 (1973)].

[200] C. H. Boehringer, French Patent 1,560,344 (1969) [*CA* **72**, 79018 (1970)].

with trityl chloride affords 2-tritylisoxazolidine.[201] With a phosphonate it reacts by an S_N2 mechanism to give a phosphonamide.[170] Methylation can be effected either by methyl iodide[90] or methyl tosylate.[147]

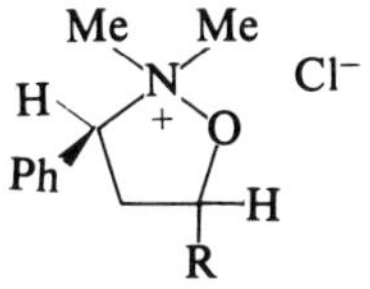

(75)

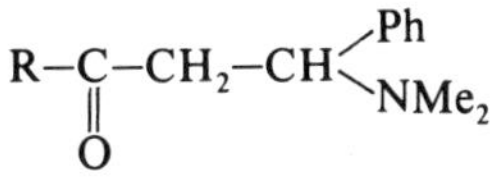

(76)

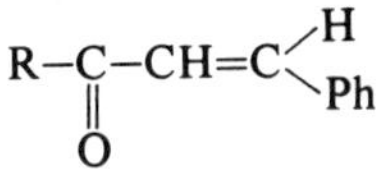

(77)

The reactions of N-hydroxyisoxazolidines are well documented. With cupric acetate, a copper compound is formed[2,4]; with acetyl chloride,[3] benzoyl chloride,[2,3] and acetic anhydride[4] it gives N-acyl derivatives.

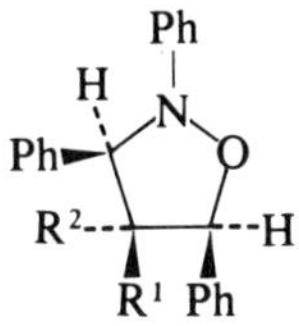

(78)

(79)

	R^1	R^2
a:	COOMe	H
b:	D	COOMe

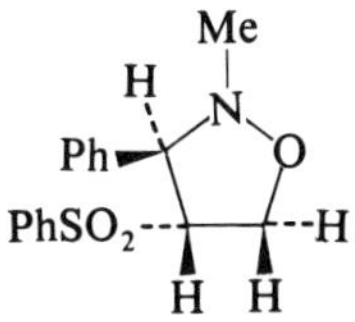

If activated, a hydrogen at the 4-position is exchangeable by deuterium in the presence of base. Thus, when compound **78a** was treated with MeOD–MeONa, 4-deuterioisoxazolidine (**78b**) is formed, with the inversion of configuration.[61] With KOH-D_2O in THF, compound **79** gives the 4-deuterio derivative with retention of configuration. No deuterium exchange took place in the 5-$PhSO_2$ analog.[23]

H. Quaternization

Isoxazolidine **1**[5] and N-substituted isoxazolidines[50,52,97,138,141,142,190,191] are easily methylated to form quaternary salts. It is noteworthy that although compound **38** is quaternized, its five-membered analog (**37**) is not.[138]

[201] Shell Research Ltd., Belgian Patent 625,441 (1963) [CA **61**, 3120 (1964)].

VI. Uses of Isoxazolidines

So far no large-scale demand has been found for isoxazolidines. However, some isoxazolidines and their quaternary salts are physiologically active; furthermore, some polymeric isoxazolidines are known, and it seems that the practical use of isoxazolidines deserves more thorough investigation.

A. Physiological Activity

2-(Substituted anilinocarbonyl)isoxazolidines (**80**) are effective as the herbicides for a variety of weeds, such as wild oats (*Avena fatua*), charlock (*Sinapis arvensis*), red shanks (*Polygonum persicaria*), stinging nettle (*Urtica urens*), and chickweed (*Stellaria media*).[200]

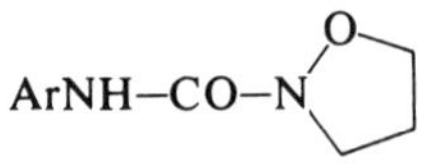

$$ArNH-CO-N\begin{pmatrix}O\\ \end{pmatrix}$$

(**80**)

N-Aroylisoxazolidines (**81**) have central nervous depressant activity without muscle relaxant effect. Compounds **81a** and **81b** are particularly harmless and active toward the inhibition of conditioned avoidance responses of rats. Compounds **81c** and **81d** are practically ineffective.[196,197,199] The thioamide analog of the compound **81a** also possesses depressant properties.[202]

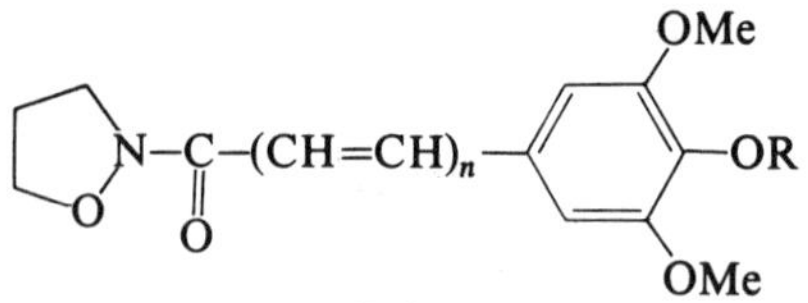

(**81**)

	R	n
a:	Me	$n=0$
b:	Ac	$n=0$
c:	Me	$n=1$
d:	Ac	$n=1$

The terephthaloylisoxazolidine (**82**) is antiphlogistic, analgesic, and effective as an antipyretic agent.[198] Similarly, compounds **83** are antiinflammatory, antipyretic, and diuretic.[159]

---

[202] G. Pifferi, German Offen. 2,102,246 (1972) [*CA* **76**, 99690 (1972)].

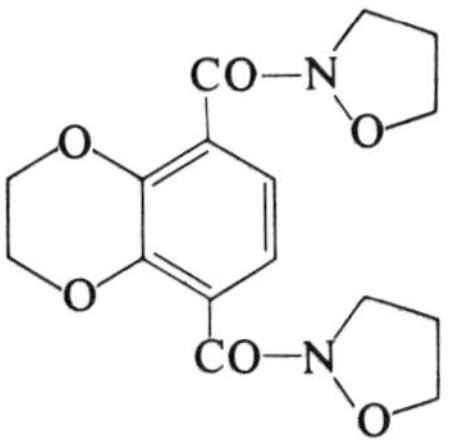

(82) 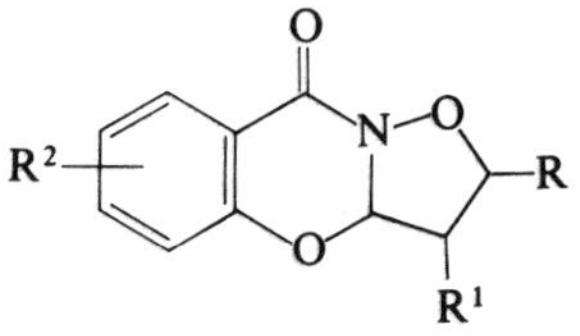   (83)

The antitrichomonal and antiamebic properties of compound **84** have been investigated, but no significant effect was observed.[203] The imidazolylnitrone adducts **85** exhibit antibacterial and antiprotozoological properties.[54] Isoxazolizidines have a mutagenic action on *Actinomyces olivaceus*, a kind of ray fungus. This activity is particularly effective in the isoxazolizidine **33f**.[204]

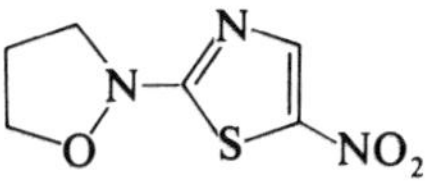

(84) 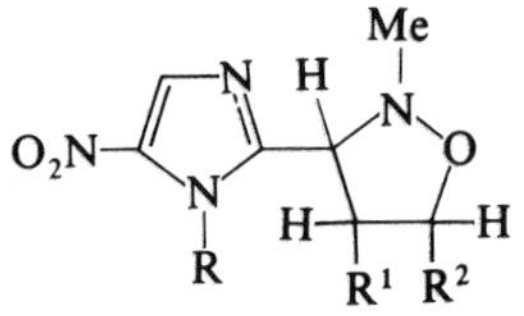   (85)

Ishidate *et al.*[153,205-208] found that isoxazolidinium salts **47a** and the like are effective in the therapy of malignant tumors and have strong biological activity toward Yoshida sarcoma. It appears that the salt **47d** is inactive to Yoshida sarcoma.[153,156,209]

The relation between the size of the alkyl substituent in the salt **75** and its antibacterial action was investigated. The bacteriostatic action is maximal with $R = C_{12}H_{25}$, which is still active toward *Staphylococcus aureus* 209P with the maximum dilution of 400,000. For use as a disinfectant or a bacteriocide the salt of **75** is possible.[50]

[203] L. M. Werbel, E. F. Elslager, A. A. Phillips, D. F. Worth, P. J. Islip, and M. C. Neville, *J. Med. Chem.* **12**, 521 (1969).
[204] L. L. Gumanov, *Spetsifichnost Khim. Mutageneza, Mater. Vses. Simp., 1967*, p. 65 (1968) [*CA* **71**, 58047 (1969)].
[205] M. Ishidate, Y. Sakurai, H. Imamura, and A. Moriwaki, *Chem. Pharm. Bull.* **8**, 444 (1960).
[206] M. Ishidate, Y. Sakurai, and E. Matsui, *Chem. Pharm. Bull.* **8**, 89 (1960).
[207] H. Imamura, *Chem. Pharm. Bull.* **8**, 449 (1960).
[208] Y. Sakurai and M. Torigoe, *Acta Unio Int. Contra Cancrum* **16**, 757 (1960) [*CA* **54**, 23051d (1960)].
[209] T. Yoshida, H. Sato, T. Tashiro, H. Imamura, Y. Sakurai, and M. Ishidate, *Gann, Suppl.* **50**, 4 (1960) [*CA* **62**, 7561 (1964)].

B. Polymeric Isoxazolidines

Reaction between α,α'-(p-phenylene)-bis(N-phenylnitrone) and m-phenylene-bismaleimide gives an isoxazolidine polymer (**86**) that does not melt above 300°.[92,93] The polymer **87** obtained from the same nitrone and ethylene dimethacrylate has a lower melting point

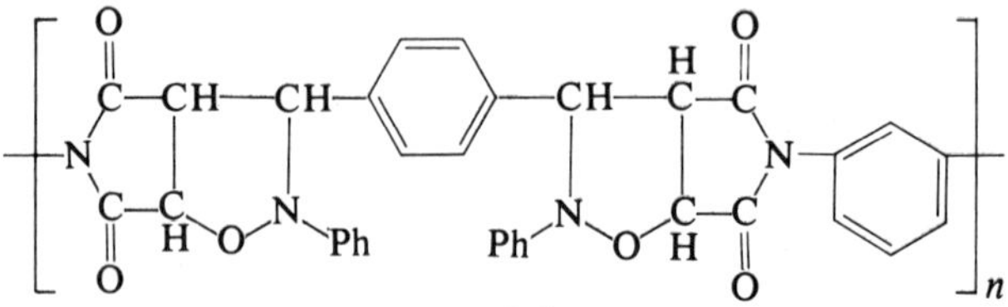

(**86**)

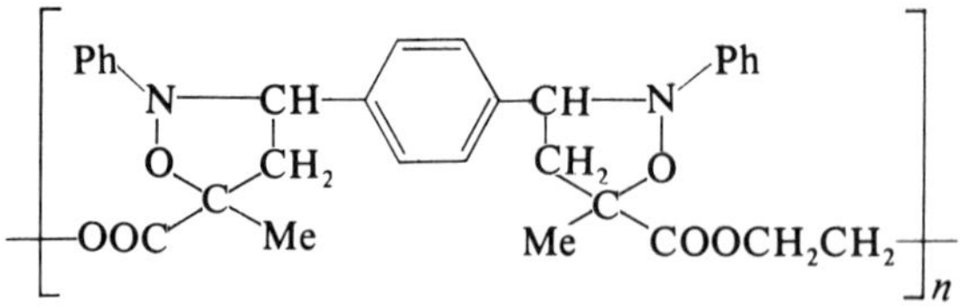

(**87**)

(100°–107°).[92,93] From the same nitrone, or glyoxal-bis(N-phenyl)-nitrone and m- or p-phenylene-bismaleimide, polymers, **88–91** are obtained.[94] Some physical properties of these polymers are summarized in Table VII.

If polybutadiene is caused to react with nitrones, the isoxazolidine ring is incorporated in the butadiene rubber chain. The carbon black-

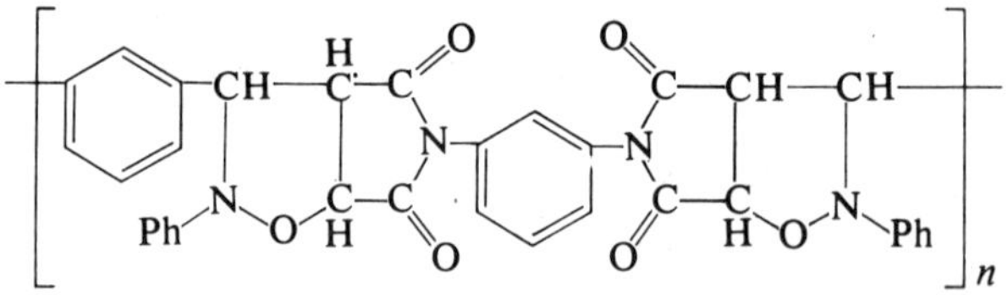

(**88**)

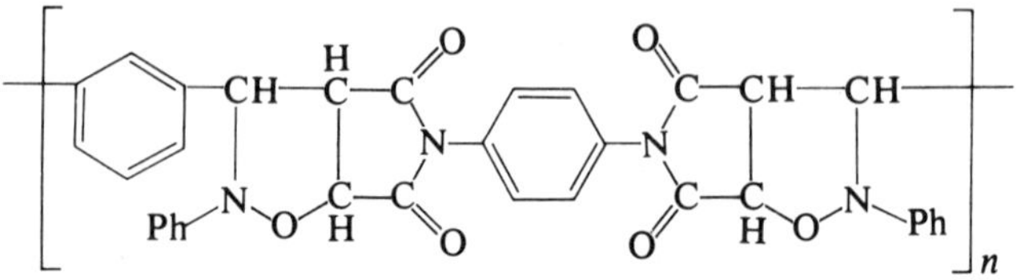

(**89**)

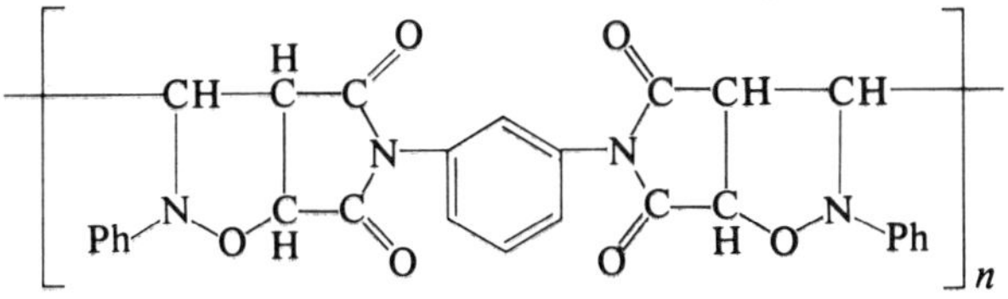

(90)

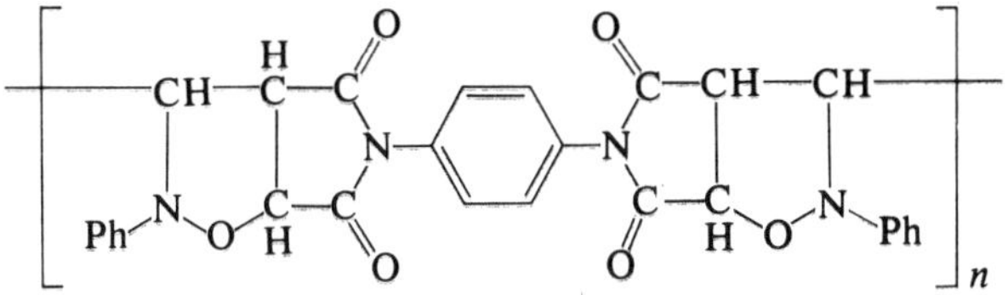

(91)

TABLE VII
Physical Properties of Polymeric Isoxazolidines

Polymer	η_{inh}^a	$\bar{M}^{2b}$	References
86	0.19^c	—	92, 93
87	0.09^c	—	92, 93
88	0.15^d	3640	94
89	0.14^d	2460	94
90	0.18^d	2540	94
91	0.24^d	3390	94

^{a} Inherent viscosity.
^{b} Mean molecular weight.
^{c} In $(Me_2N)_3PO$ at 30°; $c = 0.5$ gm/100 ml.
^{d} In dimethylformamide at 25°; $c = 0.5$ gm/100 ml.

filled vulcanizate of the polybutadiene modified with isoxazolidine exhibits improved M_{100} and tensile strength over the original butadiene rubber. This increase is common to all modified butadiene rubbers except the one modified with α-ethyl-N-phenylnitrone. Furthermore M_{100} increases as the isoxazolidine content increases. The M_{100} of the unfilled vulcanizate of the modified polybutadiene is much the same as the M_{100} of the unfilled vulcanizate of the original polybutadiene, and does not depend on the isoxazolidine content. The M_{100} and T_B (tensile strength) differences between filled and unfilled vulcanizates depend on the nature of the incorporated isoxazolidines.[210]

[210] K. Tada, Y. Numata, and T. Katsumura, *J. Appl. Polym. Sci.* **15**, 117 (1971).

(2 + 2)-Cycloaddition and (2 + 2)-Cycloreversion Reactions of Heterocyclic Compounds

D. N. REINHOUDT

Koninklijke-Shell-Laboratorium, Amsterdam, The Netherlands

I. Introduction	254
II. Thermal (2 + 2)-Cycloadditions	257
A. Mechanism	257
B. Reactions with Compounds Having Olefinic or Other Double Bonds	260
C. Reactions with Acetylenes	264
1. Nonaromatic Heterocycles	264
2. Heteroaromatics	270
D. Reactions with Heterocumulenes	276
1. Ketenes	276
2. Other Heterocumulenes	283
III. Photochemical (2 + 2)-Cycloadditions	285
A. Mechanism	285
B. Intermolecular Reactions of Nonaromatic Heterocycles	286
1. Dimerization	286
2. Reactions with Carbonyl Compounds	289
3. Reactions with Olefins	289
4. Reactions with Singlet Oxygen	294
5. Reactions with Acetylenes	294
C. Intermolecular Reactions of Heteroaromatics	295
1. Dimerization	295
2. Reactions with Carbonyl Compounds	296
3. Reactions with Olefins	297
4. Reactions with Acetylenes	299
D. Intramolecular Reactions	300
1. Nonconjugated Dienes	300
2. Conjugated Dienes	303
IV. (2 + 2)-Cycloreversion Reactions	310
A. Mechanism	311
B. Thermal (2 + 2)-Cycloreversion Reactions	312
1. Intermolecular Reactions	312
2. Intramolecular Reactions	313
C. Photochemical (2 + 2)-Cycloreversion Reactions	319
1. Intermolecular Reactions	319
2. Intramolecular Reactions	320
Appendix	321

I. Introduction

In the past few decades cycloaddition reactions have amply demonstrated their usefulness as direct and elegant routes for the synthesis of heterocyclic compounds.[1,2] Extensive reviews have been given of their applications,[3,4] including the synthesis of four-membered heterocycles by (2 + 2)-cycloaddition. Muller and Hamer[5] presented a survey of all the work published in this field prior to 1965. More recent reviews covering the synthesis of oxetans,[6] β-lactams,[7] and thietes[8] include the relevant (2 + 2)-cycloadditions. In the vast majority of the reactions described in these reviews, noncyclic reactants combine to give heterocyclic molecules. However, in the past decade many (2 + 2)-cycloaddition reactions have been reported in which at least one reactant is a heterocyclic compound that enters into reaction via its nucleus. It is these (2 + 2)-cycloadditions of heterocyclic compounds that are surveyed here. To complete the picture, the retrograde reactions, viz. (2 + 2)-cycloreversions, have been included as well. For the sake of clearness it is emphasized that the present review does not cover (2 + 2)-cycloadditions of heterocycles reacting via the side chain.

The work reported here relates to (i) intermolecular reactions, and (ii) intramolecular reactions, where the two reacting π-electron systems constitute part of one heterocyclic molecule. In the first type of reaction two σ-bonds are formed [Eqs. (1) and (2)], whereas in the second type either

---

[1] Various systems have been used for the classification of cycloaddition reactions. In this review the author has adopted Huisgen's proposal to classify them according to the number of atoms that each of the reactants contributes to the resulting cyclic reaction product.[2] In conformity with this definition the reactions described here are (2 + 2)-cycloadditions. This system of classification is less ambiguous than others, such as a classification referring to, e.g., 1,2-cycloadditions, or a classification based on the number of electrons, [2 + 2], involved in the reaction. In the former system, the designation 1,2- refers to the termini of the longest π-electron system taking part in the reaction. Consequently, the term 1,2-cycloaddition covers—apart from reactions resulting in the formation of a four-membered ring—the processes in which a three-membered ring is formed, such as reactions of olefins with carbenes or nitrenes. In the latter system the term [2 +2]-cycloaddition covers not only [2π + 2π] reactions yielding a four-membered ring, but also [2π + 2σ] processes giving a five-membered ring.

[2] R. Huisgen, R. Grashey, and J. Sauer, in "Chemistry of the Alkenes" (S. Patai, ed.), p. 739. Wiley (Interscience), New York, 1964.

[3] J. Hamer, "1,4-Cycloaddition Reactions." Academic Press, New York, 1967.

[4] R. Huisgen, Angew. Chem. **75**, 604 (1963).

[5] L. L. Muller and J. Hamer, "1,2-Cycloaddition Reactions." Wiley (Interscience), New York, 1967.

[6] D. R. Arnold, Adv. Photochem. **6**, 301 (1968).

[7] A. K. Mukerjee and R. C. Srivastava, Synthesis **5**, 327 (1973).

[8] J. F. King, Acc. Chem. Res. **8**, 10 (1975).

one σ-bond and a new π-bond are created via electrocyclization of a conjugated diene system [Eq. (3)] or two σ-bonds are formed from two isolated π-bonds [Eq. (4)]. Intermolecular reactions have been reported

$$\begin{matrix} a \\ \| \\ b \end{matrix} + \begin{matrix} c \\ \| \\ d \end{matrix} \rightleftarrows \begin{matrix} a-c \\ | \ | \\ b-d \end{matrix} \qquad (1)$$

$$\begin{matrix} a \\ \| \\ b \end{matrix} + \begin{matrix} c \\ \|\| \\ d \end{matrix} \rightleftarrows \begin{matrix} a-c \\ | \ \| \\ b-d \end{matrix} \qquad (2)$$

$$\begin{matrix} a=b \\ | \\ c=d \end{matrix} \rightleftarrows \begin{matrix} a-b \\ | \ \| \\ c-d \end{matrix} \qquad (3)$$

$$\begin{matrix} a=b \\ c=d \end{matrix} \rightleftarrows \begin{matrix} a-b \\ | \ | \\ c-d \end{matrix} \qquad (4)$$

to occur with various π-electron systems, e.g., olefins, acetylenes, (hetero)cumulenes, and carbonyl compounds. They have been effected both thermally and by direct or sensitized irradiation with UV light. In the literature on this subject the emphasis is more on the scope and application of (2 + 2)-cycloadditions for the synthesis of heterocycles than on the actual mechanism of these reactions. Therefore, in most cases it remains unclear whether the four-membered ring is formed via a concerted or via a stepwise process.[9,10] The work related to the mechanism of these reactions and the relevant theoretical considerations have been summarized in a separate part of each section in this review.

The impact of (2 + 2)-cycloaddition and (2 + 2)-cycloreversion reactions of heterocyclic compounds on organic chemistry over the last 10 years is clearly illustrated by several examples. Various members of the important β-lactam antibiotics, penicillin and cephalosporin C, as well as structurally related heterobicyclic compounds have been obtained by (2 + 2)-cycloaddition of heterocycles with ketenes (Section II,D,1).[11] Intramolecular photochemical (2 + 2)-cycloadditions of 2-pyrones yield 2-oxabicyclo[2.2.0]hex-5-en-3-ones, which upon further irradiation afford cyclobutadienes (Section III,D,2).[12] Intermolecular (2 + 2)-cycloadditions of vinylene carbonates with olefins and with acetylenes offer a simple route to cyclobutanes and cyclobutenes, respectively (Sections III,B,3 and 5).[13] (2 + 2)-Cycloaddition and (2 + 2)-cycloreversion reactions have contributed substantially to the development of the chemistry

[9] R. Huisgen, *Angew. Chem.* **82**, 783 (1970).
[10] R. Gompper, *Angew. Chem.* **81**, 348 (1969).
[11] M. S. Manhas and A. K. Bose, "Synthesis of Penicillin, Cephalosporin C and Analogs." Dekker, New York, 1969.
[12] C. Y. Lin and A. Krantz, *Chem. Commun.*, 1111 (1972).
[13] H.-D. Scharf, *Angew. Chem.* **86**, 567 (1974).

of medium-sized heterocycles. Prinzbach et al.[14] were the first to use an intramolecular (2 + 2)-cycloaddition of 7-oxabicyclo[2.2.1]hepta-2,5-dienes as the key step in the synthesis of oxepins; the same reaction scheme was successfully employed for the preparation of azepines (Section III,D,1).[14] Ring enlargement of five- and six-membered heterocycles by two carbon atoms has been achieved by (2 + 2)-cycloaddition of acetylenes with heterocycles, followed by valence bond isomerization of the intermediates.[15,16] Thus, heteroaromatic compounds such as 1*H*-pyrroles and thiophenes have been converted into azepines and thiepins (Section IV,B,2).[17,18] Many dimerizations of heterocycles proceed via intermolecular (2 + 2)-cycloaddition, and it was found that the deactivation of deoxyribonucleic acids upon irradiation with UV light is partly due to intermolecular (2 + 2)-cycloaddition of the pyrimidine bases.[19]

The author has attempted to give an exhaustive review of the literature on (2 + 2)-cycloadditions and (2 + 2)-cycloreversions published up to the end of 1974; however, some relevant work published in early 1975 has been incorporated as well. A difficulty encountered in searching the literature was that it was not until recently that (2 + 2)-cycloaddition reactions have been recognized as such by *Chemical Abstracts*. Another problem was that many reactions, in particular the intramolecular (2 + 2)-cycloadditions, have been classified as rearrangements, isomerizations, etc. It is hoped that the present review may be of some assistance in this respect and that it may prove to be a useful source of information.

The available material has been divided into three main sections, covering the thermal, photochemical, and retrograde (2 + 2)-cycloadditions. Each main section comprises a number of subsections, starting with a brief summary of the mechanism of the particular type of reaction. Further subsections differ from section to section, depending on the subject matter to be discussed. Generally, a separation has been made between intermolecular and intramolecular reactions, with other parameters, such as type of reactant, degree of unsaturation of the heterocyclic nucleus, and ring size determining the order of presentation.

---

[14] H. Prinzbach, *Pure Appl. Chem.* **16**, 17 (1968).
[15] R. M. Acheson, G. Paglietti, and P. A. Tasker, *J. Chem. Soc., Perkin Trans. 1*, 2496 (1974).
[16] D. N. Reinhoudt and C. G. Kouwenhoven, *Rec. Trav. Chim.* **92**, 865 (1973).
[17] R. P. Gandhi and V. K. Chadha, *Indian J. Chem.* **9**, 305 (1971).
[18] D. N. Reinhoudt and C. G. Kouwenhoven, *Chem. Commun.*, 1233 (1972).
[19] E. Fahr, *Angew. Chem.* **81**, 581 (1969).

II. Thermal (2 + 2)-Cycloadditions

Thermal (2 + 2)-cycloaddition reactions have never been reviewed so far, although occasionally a few reactions have been discussed in other review articles.[20,21] The literature on this subject is summarized here in four subsections. First the mechanistic aspects of thermal (2 + 2)-cycloaddition reactions are dealt with and subsequently a review is given of (2 + 2)-cycloadditions of heterocycles with olefins and compounds having other double bonds, with acetylenes, and with heterocumulenes. The reactions with acetylenes are discussed under two separate headings, covering (1) reactions with nonaromatic heterocycles and (2) reactions with heteroaromatics. The reactions included are exclusively *inter*molecular (2 + 2)-cycloadditions. No examples are known of intramolecular thermal (2 + 2)-cycloadditions of two isolated π-electron systems or of thermal electrocyclizations of conjugated 4π-electron systems of heterocyclic compounds (Appendix).

A. Mechanism

The most intriguing question with regard to the formation of a (hetero)cyclobutane or (hetero)cyclobutene ring from two isolated π-electron systems is whether the two new σ-bonds are formed in a concerted reaction, i.e., simultaneously, or in a two-step process via an intermediate, and, if so, what kind of intermediate. As to the concerted reaction, two modes of addition can be envisaged. In view of the requirement that the orbital symmetry should be preserved during the bond formation, a concerted thermal $[_{\pi}2_s + _{\pi}2_s]$ cycloaddition reaction is predicted to be a nonallowed high-energy process.[22] On the other hand, in the allowed $[_{\pi}2_s + _{\pi}2_a]$ process the two reactants have to approach each other at right angles, resulting in steric interaction between the relevant groups; moreover, a perpendicular approach of the two π-systems involves a very inefficient orbital overlap. Consequently, one would expect the (2 + 2)-cycloadditions described in this section to proceed not in a concerted manner, but via two or more steps.

Experimental evidence in support of this has been presented by

---

[20] M. Baumgarth, *Chem. Ztg* **96**, 361 (1972).
[21] R. Fuks and H. G. Viehe, *in* "Chemistry of Acetylenes" (H. G. Viehe, ed.), p. 425. Dekker, New York, 1969.
[22] R. B. Woodward and R. Hoffmann, "The Conservation of Orbital Symmetry." Verlag Chemie, Weinheim, 1970.

Huisgen and co-workers,[23,24] who studied the reactions of tetracyanoethylene with enol ethers, e.g., 3,4-dihydro-2H-pyran. They observed a pronounced solvent effect on the rate of cycloaddition, indicating that a dipolar intermediate was involved. This intermediate was captured upon 1,4-dipolar addition of ketones or nitriles [Eq. (5)].[24] In other cases it

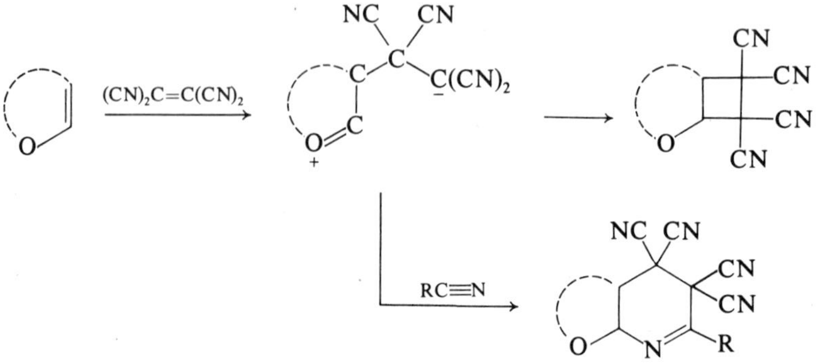

(5)

was the formation of by-products that pointed to the occurrence of dipolar intermediates, such as in the reaction of 7-oxa-2,3-benzobicyclo-[2.2.1]hepta-2,5-diene with chlorocyanoacetylene [Eq. (6)].[25] Apart from the expected (2 + 2)-cycloadduct an isomeric species was isolated,

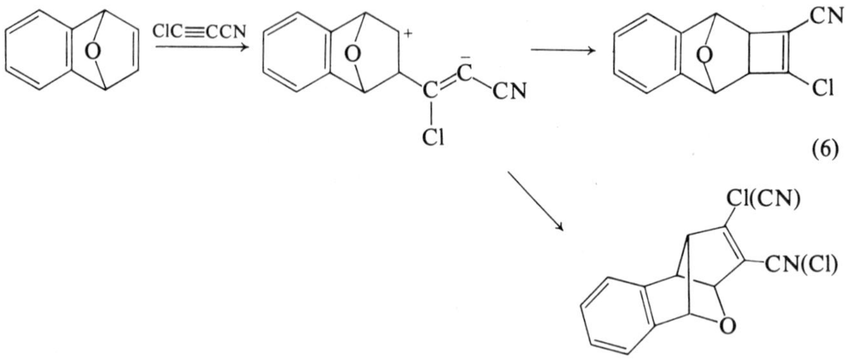

(6)

which might have been formed by a Wagner–Meerwein-type rearrangement via a dipolar intermediate. A similar reaction with 4-phenyl-4H-1,2,4-triazole-3,5-dione yielded exclusively such a type of rearranged product.[25] In many (2 + 2)-cycloaddition reactions linear Michael adducts were formed as the result of a hydrogen transfer in the dipolar in-

[23] R. Huisgen and R. Schug, *Chem. Commun.*, 59 (1975).
[24] R. Schug and R. Huisgen, *Chem. Commun.*, 60 (1975).
[25] T. Sasaki, K. Kanematzu, and M. Uchide, *Tetrahedron Lett.*, 4855 (1971).

termediate.[26,27] The fact that the ratio between (2 + 2)-cycloadducts and Michael adducts decreased with increasing solvent polarity points to the occurrence of a common intermediate, but definite proof is still lacking.

More recent views on the theory of (2 + 2)-cycloaddition, in particular with respect to the question whether the two novel σ-bonds are formed via a concerted or a stepwise mechanism, have been presented by Epiotis.[28] He predicts that, if in the reaction of two π-electron systems one of the reactants has an electron-donating and the other an electron-accepting character, the activation energy of the concerted non-allowed $[_{\pi}2_s +\, _{\pi}2_s]$-cycloaddition will be lowered, so that such a reaction may occur in a concerted manner under relatively mild conditions. As this condition is satisfied in most of the reported thermal (2 + 2)-cycloaddition reactions of heterocyclic compounds, care must be taken in drawing any conclusions as regards the reaction pathways followed.

A mechanistically interesting class of (2 + 2)-cycloadditions is constituted by the reactions of heterocycles with heterocumulenes. Huisgen and co-workers[29] made a special study of the mechanism of reactions of olefins with ketenes. Their conclusion that generally these reactions proceed in a concerted manner with some charge separation in the transition state is based on experimental evidence, e.g., a high stereospecificity, a low activation enthalpy, a large negative activation entropy, and a weak solvent effect. Brady and O'Neal[30] arrived at the same conclusion on the basis of the kinetics of the (2 + 2)-cycloaddition of diphenylketene and 3,4-dihydro-2H-pyran. According to Woodward and Hoffmann,[22] these (2 + 2)-cycloaddition reactions of heterocumulenes may be concerted as the transition state of the allowed $[_{\pi}2_s +\, _{\pi}2_a]$ mode of addition is stabilized by a favorable interaction between the highest occupied olefin orbital and the vacant orthogonal π^* antibonding orbital of the heterocumulene. Wagner and Gompper, however, have described a model according to which these cycloadditions proceed via a dipolar intermediate.[31] Evidence has been provided[32] in support both of concerted and of stepwise reactions, depending on the nature of the olefinic reactant and the polarity of the solvent. The general conclusion is that a concerted $[_{\pi}2_s +\, _{\pi}2_a]$ pathway is

[26] H. Plieninger and D. Wild, *Chem. Ber.* **99**, 3070 (1966).
[27] R. M. Acheson, J. N. Bridson, and T. S. Cameron, *J. Chem. Soc., Perkin Trans. 1*, 968 (1972).
[28] N. D. Epiotis, *Angew. Chem.* **86**, 825 (1974).
[29] R. Huisgen, L. A. Feiler, and P. Otto, *Chem. Ber.* **102**, 3405, 3444 (1969); R. Huisgen, L. A. Feiler, and G. Bisch, *ibid.*, 3460.
[30] W. T. Brady and H. R. O'Neal, *J. Org. Chem.* **32**, 612 (1967).
[31] H. U. Wagner and R. Gompper, *Tetrahedron Lett.*, 2819 (1970).
[32] T. L. Gilchrist and R. C. Storr, "Organic Reactions and Orbital Symmetry," p. 158. Cambridge Univ. Press, London and New York, 1972.

feasible for reactions with cumulenes (in particular ketenes) having a vacant orthogonal π^* antibonding orbital of low energy. A stepwise process will be the preferred reaction pathway when the dipolar intermediates are effectively stabilized, as is the case in, e.g., the (2 + 2)-cycloaddition reactions of ketenes with enamines.

B. Reactions with Compounds Having Olefinic or Other Double Bonds

The reactions described in this subsection can be divided into two types. In the first an electron-rich double bond of the heterocyclic nucleus reacts with an electron-deficient olefin, and in the second an electron-deficient double bond of the heterocyclic nucleus reacts with an electron-rich olefin.

Diarylthiirene 1,1-dioxides (**1**) have been proposed to react via (2 + 2)-cycloaddition reaction with the carbon–carbon double bond of enamines.[33] The initially formed cycloadducts (**2**) cannot be isolated as they rearrange by (2 + 2)-cycloreversion of the cyclobutane ring to give divinyl sulfones (**3**). If the enamine double bond forms part of a (hetero)cyclic system, the ultimate result is ring enlargement by a three-atom moiety, as illustrated by the conversion of **4** into **5**, which is effected in 37% yield. It is interesting to compare this reaction with that of diphenylcyclopropenone—a carbocyclic analog of **1**—with enamines, which proceeds via a different type of cycloaddition, with participation of the free electron pair of the nitrogen atom of the enamine group.[34-35] Thiete 1,1-dioxides react with enamines to give the corresponding (2 + 2)-cycloadducts.[36]

A number of (2 + 2)-cycloadditions of N-substituted maleimides with enamines occurred at 80°; the reaction products (**7**) were formed in 40–90% yield.[37] Similarly, 4-phenyl-4H-1,2,4-triazole-3,5-dione entered into a (2 + 2)-cycloaddition reaction via its electron-deficient nitrogen–nitrogen double bond. The reaction with enamines to give **8**,[37] or with 1,4-dioxene to give **9**,[38] took place at −60°. With less electron-rich olefins, such as indene[38] or *trans*-2,3-dimethylmethylenecyclopropane,[39]

---

[33] M. H. Rosen and G. Bonet, *J. Org. Chem.* **39**, 3805 (1974).
[34] M. A. Steinfels and A. S. Dreiding, *Helv. Chim. Acta* **55**, 702 (1972).
[34b] T. Eicher and S. Böhm, *Chem. Ber.* **107**, 2186, 2215 (1974).
[35] D. N. Reinhoudt and C. G. Kouwenhoven, *Tetrahedron Lett.*, 3751 (1973).
[36] L. A. Paquette, R. W. Houser, and M. Rosen, *J. Org. Chem.* **35**, 905 (1970).
[37] R. H. Rynbrandt, *J. Heterocycl. Chem.* **11**, 787 (1974).
[38] E. Koerner von Gustorf, D. V. White, B. Kim, D. Hess, and J. Leitich, *J. Org. Chem.* **35**, 1155 (1970).
[39] D. J. Pasto and A. Fu-Tai Chen, *Tetrahedron Lett.*, 2995 (1972).

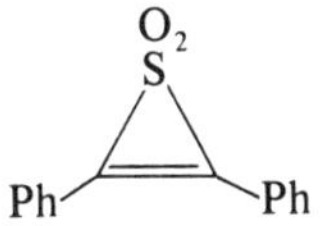

the corresponding (2 + 2)-cycloadducts (**10**) were formed at room temperature. Interesting examples are known of (2 + 2)-cycloadditions of 4-phenyl-4H-1,2,4-triazole-3,5-dione with vinyl esters.[40] At 60° mixtures of the cycloadducts (**11**) and 1,2-substituted-4-phenyl-1,2,4-triazolidine-3,5-diones (**12**) were obtained. Probably, both products were formed by rearrangement of a common 1,4-dipolar intermediate [Eq. (7)].[40] Other heterocycles that react via an electron-deficient nitrogen–nitrogen double bond with olefins are phthalazine-1,4-dione[41] and

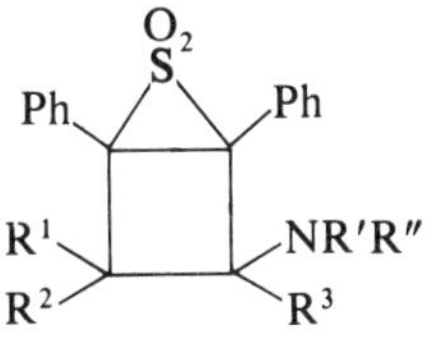

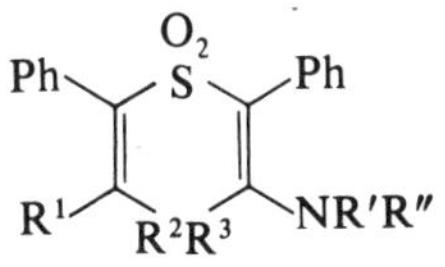

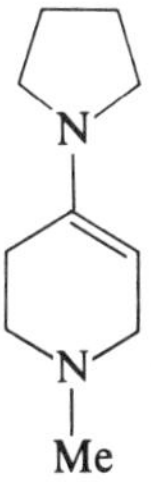

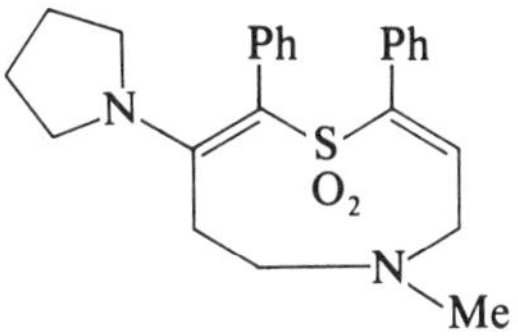

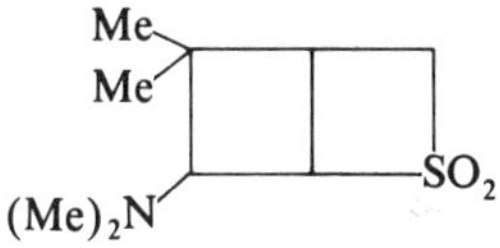

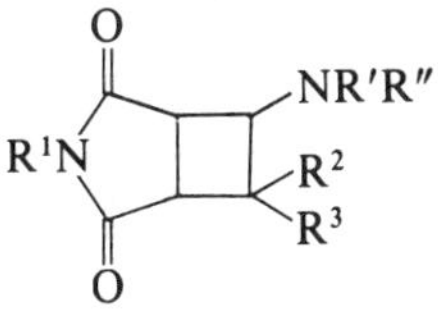

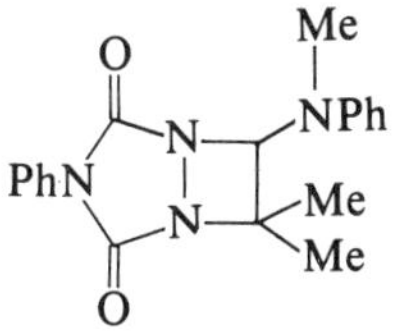

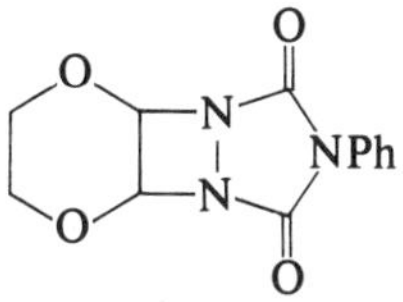

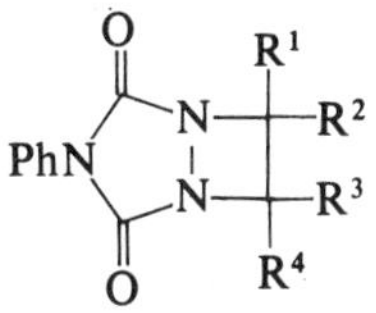

[40] K. B. Wagener, S. R. Turner, and G. B. Butler, *J. Org. Chem.* **37**, 1454 (1972).
[41] O. L. Chapman and S. J. Dominianni, *J. Org. Chem.* **31**, 3862 (1966).

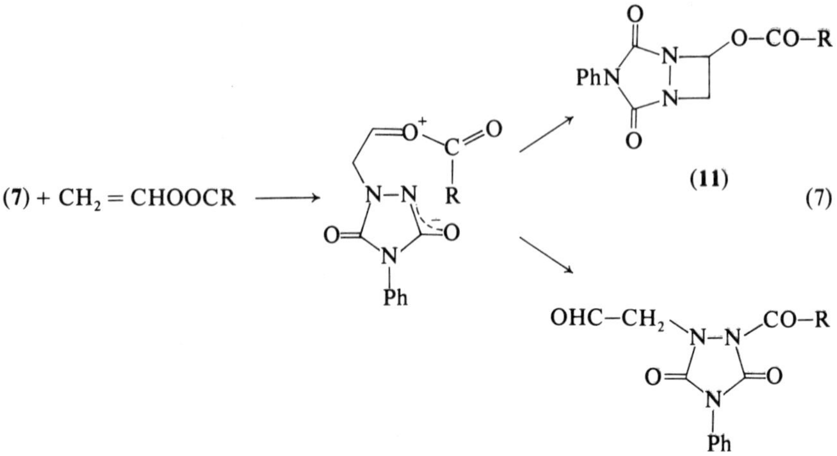

(7) + CH₂=CHOOCR ⟶ [intermediate] ⟶ (11) ⟶ (7)

↘

(12)

pyridazine-3,6-dione,[42] which were prepared *in situ* by oxidation of the corresponding hydrazides. They reacted with indene to give **13** and with styrene to give **14**. In all the reactions of this type the heterocyclic nucleus enters into the (2 + 2)-cycloaddition via an electron-deficient double bond.

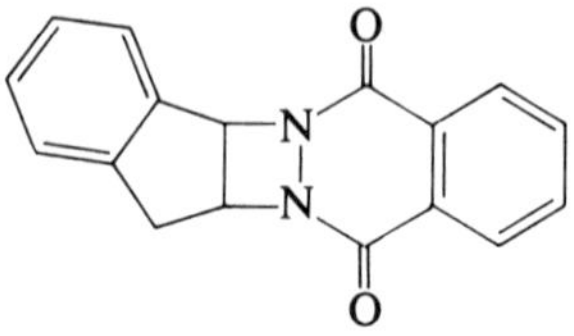

(13) 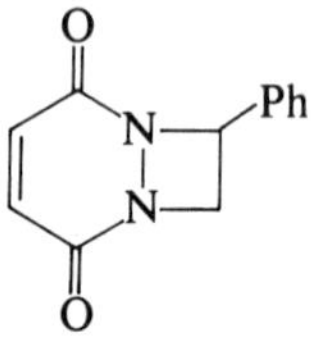   (14)

The opposite type of reaction has also been reported, *viz.* one in which the heterocyclic molecule reacts via an electron-rich double bond with electron-poor olefins, in particular with tetracyanoethylene. Tanny and Fowler[43] found that 2-azabicyclo[3.1.0]hex-3-enes reacted with tetracyanoethylene via a (2 + 2)-cycloaddition of the enamine double bond to give **15**. Other electron-deficient reactants, such as *N*-phenylmaleimide, reacted differently, yielding an 8-azabicyclo[3.2.1]oct-2-ene (**16**). This type of reaction possibly occurs via a concerted [$_\pi 2 +_\sigma 2 +_\pi 2$]-cycloaddition.[43] At room temperature tetracyanoethylene also readily formed (2 + 2)-cycloadducts with heterocycles that contained a vinyl ether group; for instance, 3,4-dihydro-2*H*-pyran, 2,3-dihydrofuran, and 2,2-dimethyl-1,3-dioxole afforded the adducts **17–19** in yields of

[42] M. Lora-Tamayo, P. Navarro, and J. L. Soto, *Ann. Chem.* **748**, 96 (1971).
[43] S. R. Tanny and F. W. Fowler, *J. Org. Chem.* **39**, 2715 (1974).

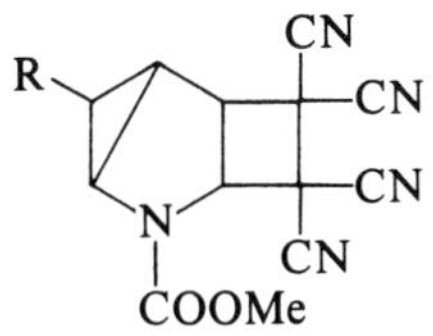

Sec. II.B]   (2 + 2)-CYCLOADDITION AND CYCLOREVERSION

(15), 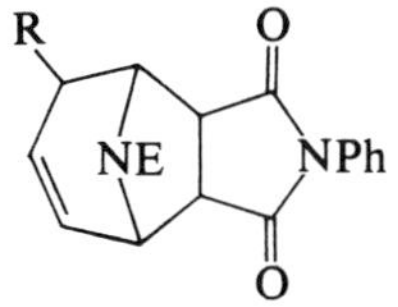 (16) E = COOMe, (17), 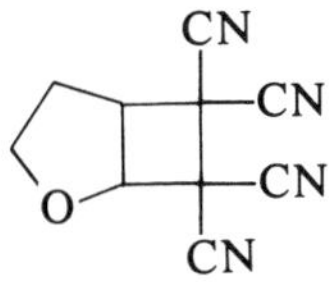 (18), 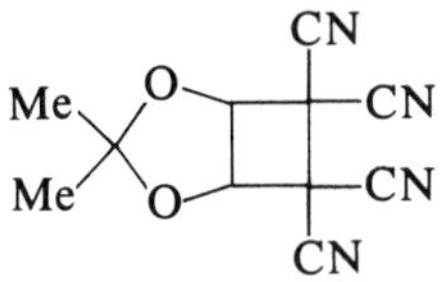 (19), 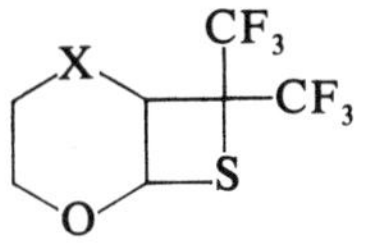 (20)

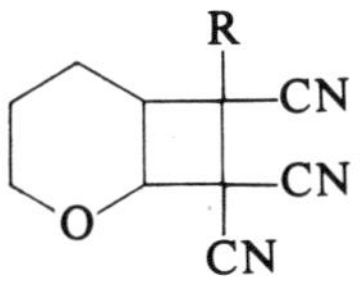

a: X = O
b: X = CH$_2$

85–100%[44,45] The mechanism of these reactions, which proceed via a dipolar intermediate, was discussed in Section II,A. It is not yet clear whether or not such reactions take place through an initially formed charge-transfer complex. In general, such a complex disappeared and a (2 + 2)-cycloadduct was obtained, but in the case of 1,4-dioxene the formation of a charge-transfer complex was observed without subsequent (2 + 2)-cycloaddition.[45,46] Huisgen and co-workers have measured the relative rate constants of cycloadditions with various enol ethers.[45] They reported that 2,3-dihydrofuran reacts three hundred times faster than 3,4-dihydro-2H-pyran, which they attributed to relief of the larger angle and conformational strain in the five-membered ring. The yield of the cycloadduct 17, which amounted to only 45% with chlorotricyanoethylene (R = Cl), was almost quantitative with tetracyanoethylene (R = CN) owing to the lower electron density in the latter compound.[44]

1,4-Dioxene and 3,4-dihydro-2H-pyran have been reported to afford (2 + 2)-cycloadducts (20) via addition to the carbon–sulfur double bond of hexafluorothioacetone at −78°.[47] Although no conclusive evidence has been given, the 2-oxa-8-thiabicyclo[4.2.0]octane structure (20b) was preferred to others on the basis of mass spectrometric data.[47]

[44] J. K. Williams, D. W. Wiley, and B. C. McKusick, *J. Am. Chem. Soc.* **84**, 2210, 2216 (1962).
[45] R. Huisgen and G. Steiner, *Tetrahedron Lett.*, 3763 (1973).
[46] P. D. Bartlett, *Quart. Rev.* **24**, 473 (1970).
[47] W. J. Middleton, *J. Org. Chem.* **30**, 1395 (1965).

An interesting thermal (2 + 2)-cycloaddition of 1,2-dihydropyridines has been reported by Liberatore *et al.* [Eq. (8)]. Low-temperature reduc-

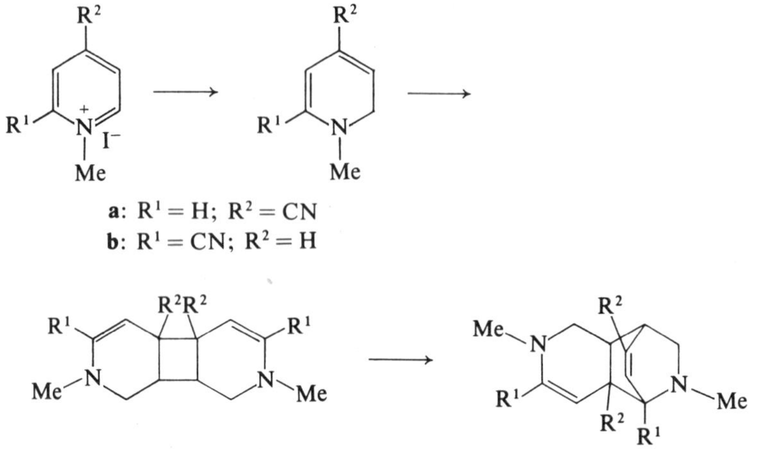

a: $R^1 = H$; $R^2 = CN$
b: $R^1 = CN$; $R^2 = H$

(21)　　(22)　　(8)

tion of 2- or 4-cyano-1-methylpyridinium iodide with sodium borohydride yielded dimers of structure **21**.[48,49] At higher temperatures these subsequently isomerized via a 1,2-divinylcyclobutane rearrangement into **22**. The authors argued that the dimeric cycloadducts **21** had most probably been formed via a two-step biradical dimerization mechanism.[49]

(2 + 2)-Cycloadditions of *heteroaromatics* with compounds having olefinic or other double bonds have not been reported so far.

C. Reactions with Acetylenes

1. *Nonaromatic Heterocycles*

A useful method for achieving annulation of a heterocyclic ring with a cyclobutene ring consists in effecting (2 + 2)-cycloaddition reactions between electron-deficient acetylenes, e.g., dimethyl acetylenedicarboxylate, and heterocyclic compounds having an electron-rich carbon–carbon double bond in the nucleus. Similarly, electron-rich acetylenes react with heterocycles containing an electron-deficient

---

[48] F. Liberatore, A. Casini, V. Carelli, A. Arnone, and R. Mondelli, *Tetrahedron Lett.*, 3829 (1971).

[49] F. Liberatore, A. Casini, V. Carelli, A. Arnone, and R. Mondelli, *J. Org. Chem.* **40**, 559 (1975).

carbon–carbon or a carbon–nitrogen double bond. Both types of reaction have also been described for carbocyclic compounds.[50,51]

The electron density of a carbon–carbon double bond in a heterocyclic nucleus is influenced not only by the nature of the other ring atoms, e.g., various unsaturated heterocycles with nitrogen as the heteroatom contain an "enamine" type of double bond, but also by the presence of any electron-donating (NR'R" or SR) or electron withdrawing (NO_2, COOR) substituents.

Five- and six-membered heterocycles substituted with an amino group (23–25) have been reported to enter into (2 + 2)-cycloaddition reactions with dimethyl acetylenedicarboxylate.[16,36,52–54] In some cases these

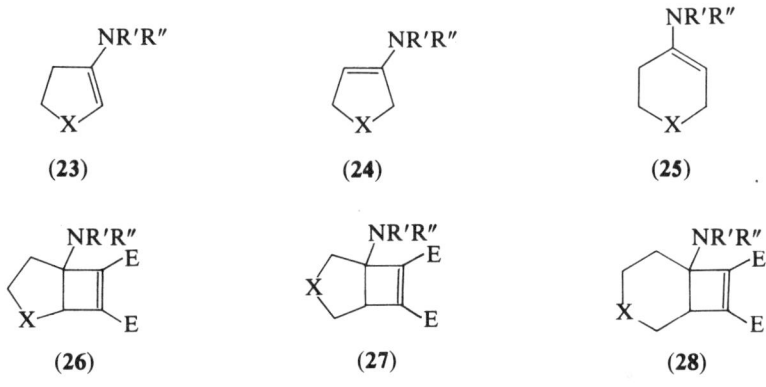

X = S, O, NMe or POPh; E = COOMe

products (26–28) were actually isolated, whereas others were identified on the basis of NMR spectroscopy (Table I).[16,36] In other cases they were only proposed as intermediates in the conversion of five- and six-membered heterocycles into the corresponding seven- and eight-membered heterocycles, because the (2 + 2)-cycloaddition was followed by a rapid intramolecular (2 + 2)-cycloreversion of the annulated cyclobutene ring such as in 27 or 28 (Section IV,B,2).[53,54] The reaction rate proved to depend on the nature of the amino substituent [dimethylamino < 1-(piperidinyl) ≤ 4-(morpholinyl) < 1-(pyrrolidinyl)], the position of the enamine double bond in the ring relative to the heteroatom,

[50] Å. G. Cook, in "Enamines: Synthesis, Structure and Reactions" (A. G. Cook, ed.), p. 211. Dekker, New York, 1969.
[51] H. G. Viehe, in "Chemistry of Acetylenes" (H. G. Viehe, ed.), p. 861. Dekker, New York, 1969.
[52] D. N. Reinhoudt and C. G. Leliveld, *Tetrahedron Lett.*, 3119 (1972).
[53] K. C. Brannock, R. D. Burpitt, V. Wilson Goodlett, and J. G. Thweatt, *J. Org. Chem.* **28**, 1464 (1963).
[54] G. Märkl and H. Baier, *Tetrahedron Lett.*, 4439 (1972).

TABLE I

REACTIONS OF **23–25** WITH DIMETHYL ACETYLENEDICARBOXYLATE IN DIETHYL ETHER AT 25°

Compound	X	NR'R"	Substituent in **23–25**	Yield of **26–28** (%)	References
23	S	NC_4H_8	—	93	16
23	S	NC_4H_8O	—	98	16
23	S	NC_5H_{10}	—	95	16
23	S	NC_4H_8	4-Me	24	16
23	S	NC_4H_8O	4-Me	98	16
23	S	NC_4H_8	4,4-Me_2	92	16
23	S	NC_5H_{10}	4,4-Me_2	93	16
24	O	NC_4H_8	—	17	53
24	S	NC_4H_8	—	30	16
24	S	NC_4H_8O	—	38	16
24	S	NC_5H_{10}	—	34	16
24	S	NC_4H_8	2-Me	43	16
24	S	NC_4H_8O	2-Me	32	16
25	NMe	NC_4H_8	—	50	36
25	S	NC_5H_{10}	—	60	36
25	$POPh^a$	NC_4H_8O	—	8	54

^{a} Reaction in refluxing toluene; yield calculated from the corresponding phosphacyclooctadiene.

and the electron density in the acetylene triple bond.[16] For instance, methyl propiolate is capable of entering into (2 + 2)-cycloaddition reactions only with the most reactive 1-(pyrrolidinyl)-substituted heterocycles. These reactions have been found to give exclusively one type of adduct (**29**).[16]

Most heterocycles that contain a 1,3-dienamine moiety react as a diene and give Diels–Alder adducts. However, 3-dimethylamino-7,8-dihydro-2H-oxocin (**30**) has been claimed to give a (2 + 2)-cycloadduct (**31**) with dimethyl acetylenedicarboxylate (Section IV,B,2).[55]

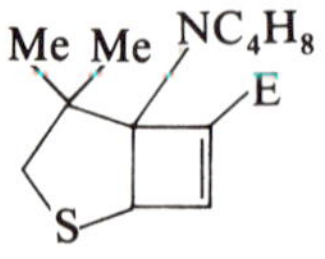

(29) 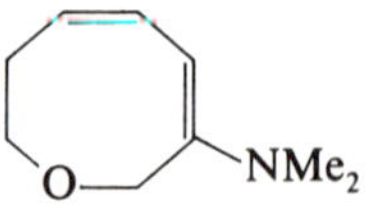   (30) 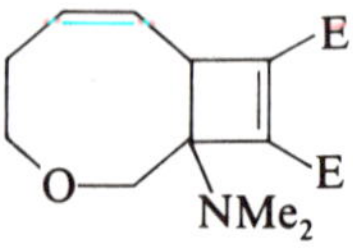   (31)

E = $COOCH_3$

[55] L. A. Paquette and R. W. Begland, *J. Am. Chem. Soc.* **88**, 4685 (1966).

Sec. II.C] (2 + 2)-CYCLOADDITION AND CYCLOREVERSION

A variety of reactions of nitrogen-containing heterocycles with acetylenic esters have been reported and reviewed by Acheson.[56] In several instances electrophilic attack at nitrogen took place, and ultimately 1:2 cycloadducts were formed. Recently some (2 + 2)-cycloadditions of these heterocycles with electron-deficient acetylenes have been disclosed. The formation of a (2 + 2)-cycloadduct was proposed to be the first step in the conversion of 1-methyl-5-methylmercapto-2,3-dihydro-1H-pyrrole (32) with dimethyl acetylenedicarboxylate into 1-methyl-7-methylmercapto-5,6-bis(methoxycarbonyl)-2,3-dihydro-1H-azepine (34) in 14% yield, but the cycloadduct (33) was not detected.[57] Relatively stable (2 + 2)-cycloadducts have been derived from 1,3-disubstituted-1,4-dihydropyridines (35) and dimethyl acetylenedicarboxylate.[58,59] Reaction invariably took place at the more electron-rich C(5)–C(6) double bond and the yields of 36 varied from 32% (R^1 = CONEtPh) to 52% (R^1 = CN). When the 1,4-dihydropyridine ring in 35 was substituted with a carboxylate or an amide group (R^1 = COOH, CONH$_2$, or CONHCH$_2$CH$_2$Ph) the cyclobutene formation was accompanied by a displacement of this group to give 38 in 1–14% yield. This reaction was assumed to involve a hydrogen transfer in the intermediate

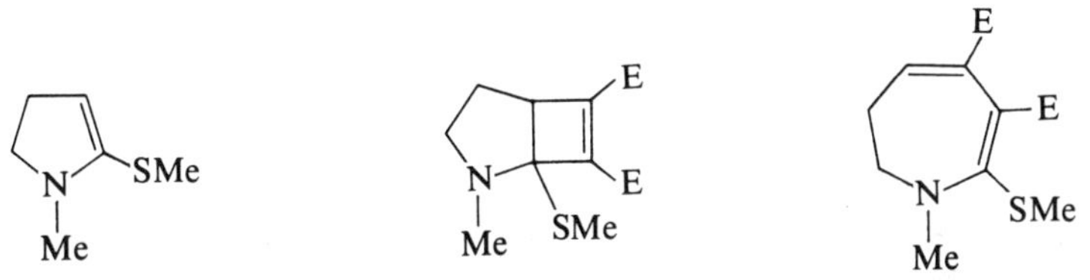

(32)   (33)   (34)

E = COOMe

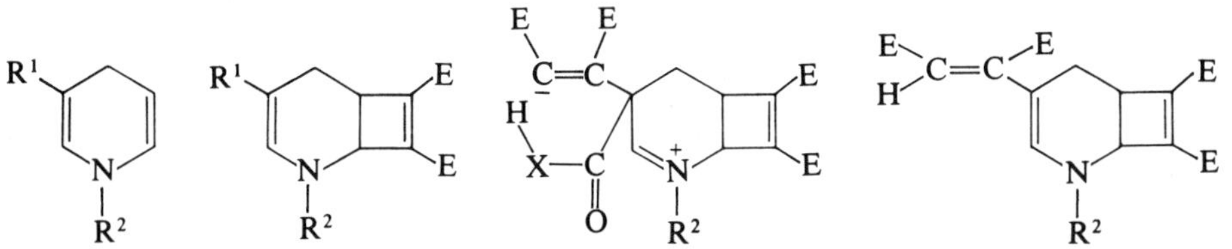

(35)   (36)   (37)   (38)

R^2 = CH$_2$Ph; E = COOMe

[56] R. M. Acheson, *Adv. Heterocycl. Chem.* **1**, 125 (1963).
[57] T. Oishi, S. Murakami, and Y. Ban, *Chem. Pharm. Bull. Jpn.* **20**, 1740 (1972).
[58] R. M. Acheson and N. D. Wright, *Chem. Commun.*, 1421 (1971).
[59] R. M. Acheson, N. D. Wright, and P. A. Tasker, *J. Chem. Soc., Perkin Trans. 1*, 2918 (1972).

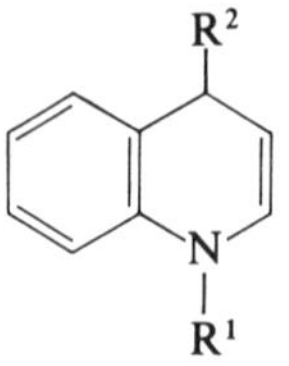

(39)
a: $R^1 = Me$; $R^2 = H$
b: $R^1 = CH_2Ph$; $R^2 = CN$

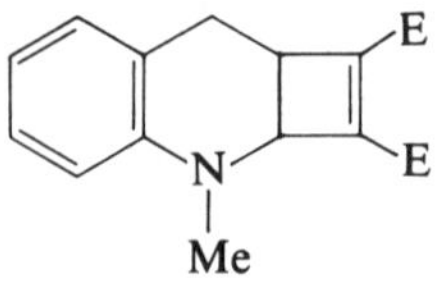

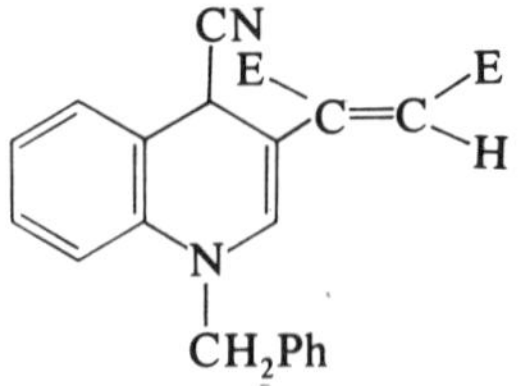

(40) E = COOMe

(41)

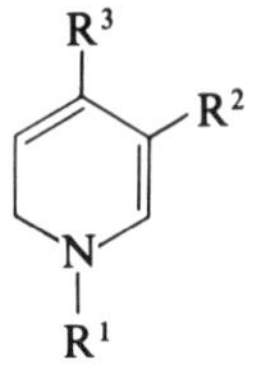

(42) 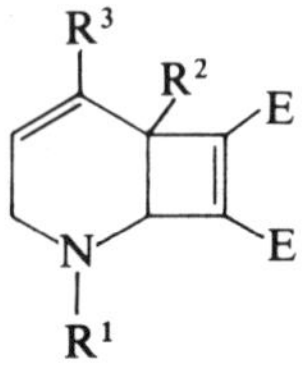   **(43)** 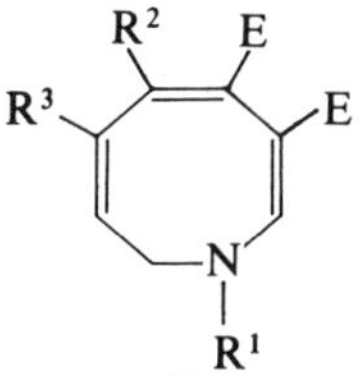   **(44)**

E = COOMe

(37).[59] Heterocycles with a similar 1,4-dihydropyridine ring, such as *N*-substituted 1,4-dihydroquinolines (**39**), have also been allowed to react with dimethyl acetylenedicarboxylate. Depending on the substituent at the ring, a (2 + 2)-cycloadduct (**40**)[60] or a linear Michael adduct (**41**)[59] was formed. The (2 + 2)-cycloadducts (**43**) of 1,2-dihydropyridines (**42**) with dimethyl acetylenedicarboxylate are far less stable. Only NMR spectroscopy at −10° to 0° has provided evidence for the formation of **43**. At room temperature the (2 + 2)-cycloaddition was followed by isomerization to the corresponding 1,2-dihydroazocine (**44**).[15,61] The reaction took a different course when other dienophiles were employed; for instance, with *N*-phenylmaleimide or maleic anhydride, Diels–Alder-type adducts were formed. Reaction of a 1,2-dihydropyrazine (**45**) with dimethyl acetylenedicarboxylate yielded a bicyclic compound, which was shown to be not the expected (2 + 2)-cycloadduct **46**, but the isomeric 2,7-diazabicyclo[4.2.0]octa-2,4-diene (**47**). This compound was claimed to result from initial (2 + 2)-cycloaddition, ring opening, and subsequent *intra*molecular (2 + 2)-cycloaddition [Eq. (9)].[62]

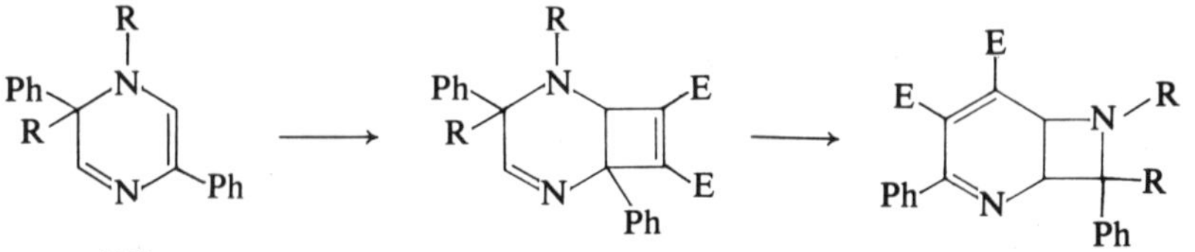

(45) → **(46)** → **(47)**   (9)

[60] P. G. Lehman, *Tetrahedron Lett.* 4863 (1972).
[61] R. M. Acheson and G. Paglietti, *Chem. Commun.*, 665 (1973).
[62] J. W. Lown and M. H. Akhtar, *Tetrahedron Lett.*, 3727 (1973).

In another group of (2 + 2)-cycloaddition reactions, the heterocyclic nucleus reacts via an electron-deficient carbon–carbon or carbon–nitrogen double bond with electron-rich aminoacetylenes (ynamines). For instance, thiete 1,1-dioxides, N-benzylmaleimide, and 2,3-bis(methoxycarbonyl)-7-oxabicyclo[2.2.1]hepta-2,5-diene reacted with 1-diethylamino-1-propyne and with 1-phenyl-2-(1-pyrrolidinyl)-acetylene to give the (2 + 2)-cycloadducts **48**, **49**, and **50**, respectively.[36,37,63] The latter product was thermally rather unstable, and its structure was identified on the basis of its conversion with 2,4,6-trimethylbenzonitrile oxide into **51**.[63] (2 + 2)-Cycloaddition via a carbon–nitrogen double bond has been reported to take place in the reactions of 3,3-dimethyl-3H-indoles and 3,4-dihydroisoquinoline with ynamines, e.g., 1-dimethylamino-2-phenylacetylene, in the presence of boron trifluoride.[64] The (2 + 2)-cycloadducts **52** and **53** were not isolated, but

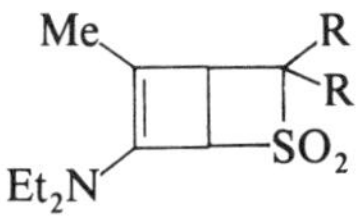

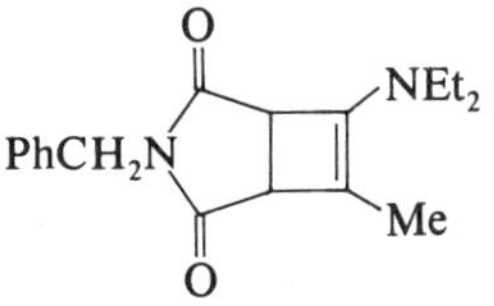

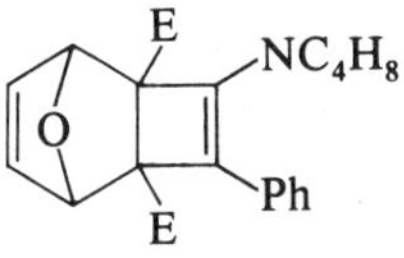

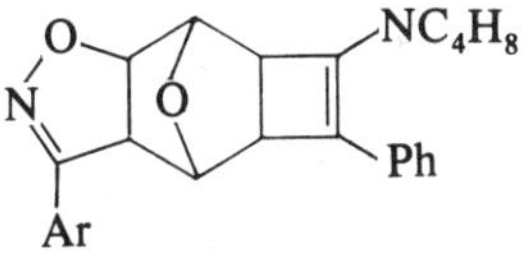

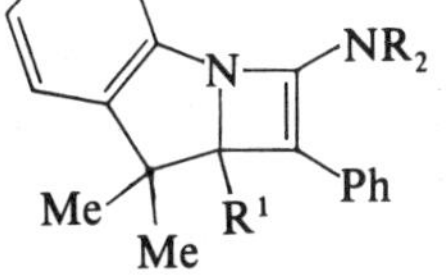

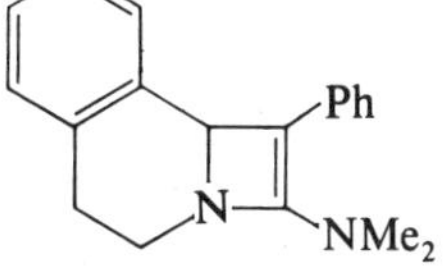

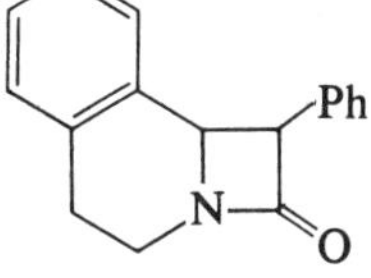

their formation was inferred from the generation of β-lactam (**54**) upon hydrolysis of the enamine of **53** in the presence of traces of water. In the absence of water, isomers of **52** and **53**, viz. 5H-benzo[b]azepines and 2,3-dihydrobenzo[d]azocines, were isolated in 65–78% yield.

[63] D. N. Reinhoudt and C. G. Kouwenhoven, *Rec. Trav. Chim.* **95**, 67 (1976).
[64] R. Fuks and H. G. Viehe, *Chem. Ber.* **103**, 573 (1970).

(55)   (56)

The reaction of 7-oxa-2,3-benzobicyclo[2.2.1]hepta-2,5-diene (**55**) with chlorocyanoacetylene at 90–100° has been reported to give **56** in 50% yield.[25]

2. *Heteroaromatics*

A useful reaction for the synthesis of unsaturated seven-membered heterocycles is the (2 + 2)-cycloaddition of heteroaromatic compounds, e.g., 1H-pyrrole, furan, or thiophene derivatives, with acetylenes. In combination with a subsequent intramolecular (2 + 2)-cycloreversion (Section IV,B,2) of the annulated cyclobutene moiety, ring enlargement with two carbon atoms can be achieved. 1-Heterocycloheptatrienes, such as benzo[b]azepines,[26,27,65,66] benzo[b]oxepins,[67,68] benzo[b]-thiepins,[69,70] and thiepins,[18,71] have been successfully prepared in this way; other routes are either nonexistent or laborious.[72] In these compounds the reacting carbon–carbon double bond constitutes part of a $(4n + 2)\pi$-electron system and in the (2 + 2)-cycloaddition the resonance energy of the aromatic nucleus is lost. Just like the nonaromatic heterocycles, heteroaromatic compounds have been reported to undergo (2 + 2)-cycloaddition reactions both with electron-deficient and with electron-rich acetylenes.

Reactions with dimethyl acetylenedicarboxylate have been found to occur with 1H-indoles and also with heteroaromatics substituted with an electron-donating amino substituent, in particular a 1-pyrrolidinyl group. The first (2 + 2)-cycloaddition reaction of an indole was described by Plieninger and Wild.[26] They found that 2-ethoxy-1H-indole (**57**, R = H) reacts with dimethyl acetylenedicarboxylate in refluxing dioxane to give a mixture of three products, identified as cis and trans

[65] F. Fried, J. B. Taylor, and R. Westwood, *Chem. Commun.*, 1226 (1971).
[66] M.-S. Lin and V. Snieckus, *J. Org. Chem.* **36**, 645 (1971).
[67] D. N. Reinhoudt and C. G. Kouwenhoven, *Rec. Trav. Chim.* **93**, 129 (1974).
[68] D. N. Reinhoudt and C. G. Kouwenhoven, *Tetrahedron Lett.*, 5203 (1972).
[69] D. N. Reinhoudt and C. G. Kouwenhoven, *Chem. Commun.*, 1232 (1972).
[70] D. N. Reinhoudt and C. G. Kouwenhoven, *Tetrahedron* **30**, 2431 (1974).
[71] D. N. Reinhoudt and C. G. Kouwenhoven, *Tetrahedron* **30**, 2093 (1974).
[72] L. A. Paquette, in "Non-Benzenoid Aromatics I" (J. P. Snyder, ed.), p. 249. Academic Press, New York, 1969.

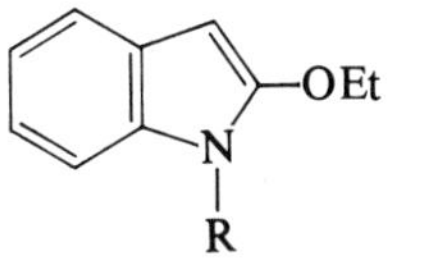

Sec. II.C] (2 + 2)-CYCLOADDITION AND CYCLOREVERSION 271

(57) 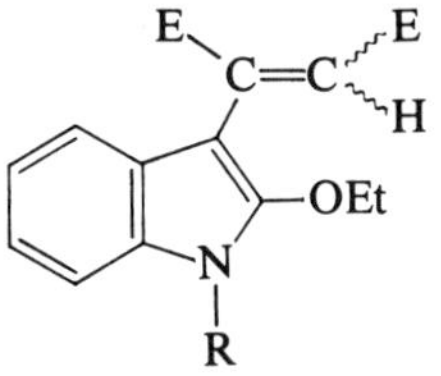 (58) 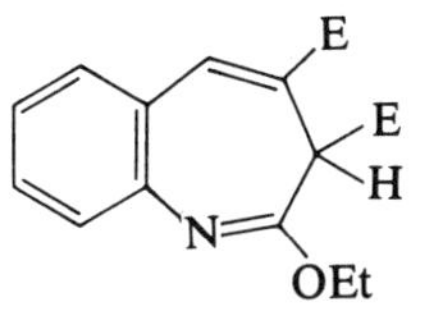 (59)

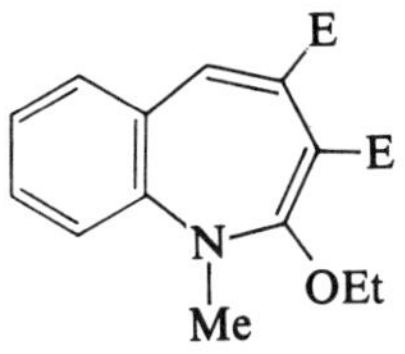

(60) 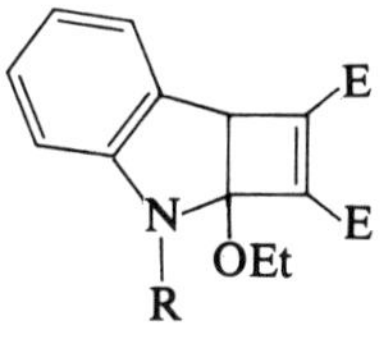 (61)

E = COOMe

Michael adducts **58** and 2-ethoxy-3,4-bis(methoxycarbonyl-3*H*-benzo-[*b*]azepine (**59**). The latter product was obtained in 20% yield. 2-Ethoxy-1-methyl-1*H*-indole (**57**, R = CH$_3$) yielded exclusively the corresponding 1*H*-benzo[*b*]azepine (**60**), in a much higher yield (72%). The (2 + 2)-cycloadduct (**61**) was proposed to be the unstable intermediate in both reactions. 1,3-Dimethyl-1*H*-indole (**62a**) was reported to react with dimethyl acetylenedicarboxylate only in the presence of boron trifluoride diethyl etherate as a catalyst.[65] In the reaction mixture, two isomeric Michael adducts and a thermally unstable product were

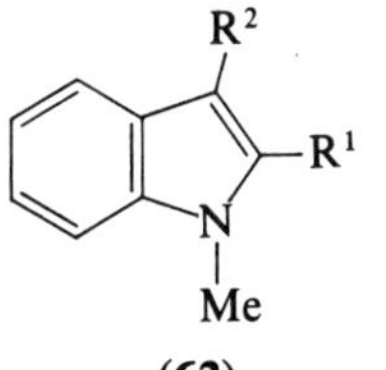

(62) 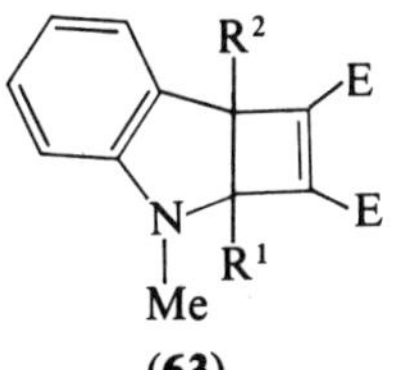 (63)

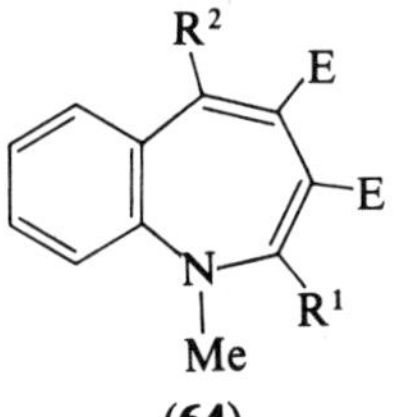

(64) 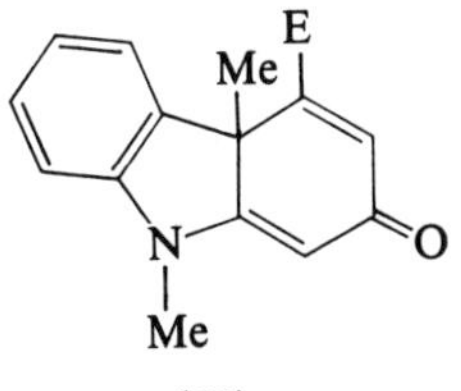 (65)

a: R^1 = H, R^2 = Me
b: R^1 = R^2 = Me; E = COOMe
c: R^1 = R^2 = H

detected. The latter disappeared rapidly upon heating and a 1H-benzo-[b]azepine (**64**) was isolated in 10% yield. Consequently, the unstable intermediate was postulated to be an annulated cyclobutene derivative (**63**), formed by (2 + 2)-cycloaddition. Blocking of the 2-position, which renders the Michael addition impossible, as in 1,2,3-trimethyl-1H-indole (**62b**), resulted in an increase in yield of 1H-benzo[b]azepine (**64b**) to 30%.[65] The fact that this yield is still relatively low was due to the occurrence of a competitive process, the elimination of methanol, which resulted in the formation of **65**. The noncatalyzed reaction of 1-methyl-1H-indole (**62c**) with dimethyl acetylenedicarboxylate required 6 days' refluxing in acetonitrile.[27] 1-Methyl-3,4-bis(methoxycarbonyl)-1H-benzo[b]azepine was isolated in 20% yield together with various other products, including the Michael adducts **58** (R = Me, H replaces OEt).

Related to these reactions are those of thiazoles, which yielded 2:1 cycloaddition products with dimethyl acetylenedicarboxylate. Originally the products were assumed to have either structure **66** or the corresponding monocyclic structure.[73] A conceivable route from thiazole to **66** would involve a (2 + 2)-cycloaddition with the thiazole carbon–nitrogen double bond, followed by ring enlargement and another (2 + 2)-cycloaddition with the carbon–nitrogen double bond of the 1,4-thiazepine. Structures of type **67** have also been proposed,[74] on the analogy of similar reactions of pyridine.[56] Recently, this problem has been solved by X-ray crystallography of the adduct formed with 2-methylthiazole, which revealed a structure of type **68**.[75] The generation of **68** is explained in terms of a [1,5] sigmatropic shift in **67** or by assuming the formation of an intermediate vinyl sulfide. Therefore it would seem that (2 + 2)-cycloaddition reactions of nitrogen-containing heteroaromatics with electron-deficient acetylenes are limited to 1H-indoles.

(**66**)   (**67**)   (**68**)

E = COOMe

[73] D. H. Reid, F. S. Skelton, and W. Bonthrone, *Tetrahedron Lett.*, 1797 (1964).
[74] R. M. Acheson, M. W. Foxton, and G. R. Miller, *J. Chem. Soc.*, 3200 (1965).
[75] P. J. Abbott, R. M. Acheson, U. Eisner, D. J. Watkin, and J. R. Carruthers, *Chem. Commun.*, 155 (1975).

Another group of (2 + 2)-cycloadditions of heteroaromatics with dimethyl acetylenedicarboxylate, which has a wider scope, comprises the reactions of amino-substituted heteroaromatics with structure **69**. These reactions proceed at temperatures varying from −30° to 100°. The thermal stability of the cycloadduct **70** is greatly dependent on the nature of the heteroatom. For instance, the cycloadduct (**70a**) obtained from 3-(1-pyrrolidinyl)benzo[b]furan (**69a**) was sufficiently stable at room temperature and could be isolated in 58% yield,[67,68] whereas the cycloadduct of **69b**, 6,7-bis(methoxycarbonyl)-5-(1-pyrrolidinyl)-2-thia-3,4-benzobicyclo[3.2.0]hepta-3,6-diene (**70b**), could only be characterized by NMR and IR spectroscopy at −20°.[69,70] The reaction with 1-acetyl-3-(1-piperidinyl)-1H-indole (**69d**) was carried out in refluxing dioxane, and under these conditions the cycloadduct (**70d**) isomerized to the corresponding 1H-benzo[b]azepine.[66] The cycloadduct (**71**) formed from **69d** and methyl propiolate was isolated in a yield of 26% from a similar reaction after 2 days. At higher temperatures the cycloadducts **70** undergo intramolecular (2 + 2)-cycloreversion to afford the corresponding benzo[b]oxepins, benzo[b]thiepins, and 1H-benzo[b]azepines (Section IV,B,2). The rate of the cycloaddition reaction depends on the "enamine" character of the C(2)–C(3) double bond. Reinhoudt and Kouwenhoven[70] found that 3-(1-pyrrolidinyl)benzo[b]thiophene reacted at −30°, whereas 3-(1-piperidinyl)benzo[b]thiophene could be converted with dimethyl acetylenedicarboxylate only at 130°. Similarly, a temperature of 100° was needed for the conversion of 1-acetyl-3-(1-piperidinyl)-1H-indole (**69d**) whereas 3-(1-pyrrolidinyl)benzo[b]furan gave a smooth reaction at room temperature.

(**69**)   (**70**)   E = COOMe   (**71**)

a: X = O; NR'R" = NC$_4$H$_8$
b: X = S; NR'R" = NC$_4$H$_8$
c: X = S; NR'R" = NC$_5$H$_{10}$
d: X = NCOMe; NR'R" = NC$_5$H$_{10}$

In compounds **69** the reacting heteroaromatic ring is fused to a benzene ring, but nonannulated thiophenes substituted with a 1-pyrrolidinyl group yielded the same type of (2 + 2)-cycloadducts.[18,71] The reactions of 3-(1-pyrrolidinyl)thiophenes (**72a,b**) with dimethyl acetylenedicarboxylate occurred at −30°, and the (2 + 2)-cycloadducts **73** were identified by low-temperature NMR and IR spectroscopy. Under the prevailing reaction conditions, the adducts decomposed into sulfur and the corresponding benzene derivatives via a thiepin inter-

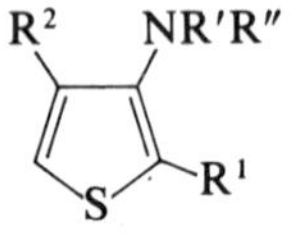

(72)

(73)

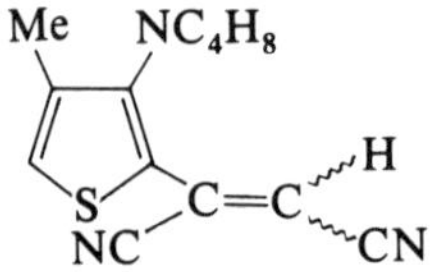

(74)

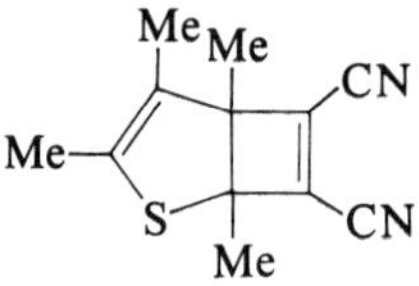

(75)

a: $R^1 = R^2 = H$; $NR'R'' = NC_4H_8$
b: $R^1 = H$, $R^2 = Me$; $NR'R'' = NC_4H_8$
c: $R^1 = Me$, $R^2 = H$; $NR'R'' = NC_4H_8$
d: $R^1 = H$, $R^2 = Me$; $NR'R'' = NC_5H_{10}$
E = COOMe

mediate (Section IV,B,2). The rate of cycloaddition was lowered by the presence of a methyl substituent in the 2-position (**72c**), and 3-(1-piperidinyl)thiophene, which has a less reactive "enamine" double bond, failed to give any (2 + 2)-cycloaddition reaction at all under similar conditions. The reaction of **72b** with dicyanoacetylene instead of dimethyl acetylenedicarboxylate yielded, at low temperature (−70°), a Michael adduct (**74**), which polymerized at higher temperatures.[71]

The presence of an amino substituent in the thiophene facilitates these (2 + 2)-cycloaddition reactions. Thiophenes lacking such a substituent require temperatures of 100–200° to react with dimethyl acetylenedicarboxylate or dicyanoacetylene.[76,77] However, it is most likely that a Diels–Alder addition rather than a (2 + 2)-cycloaddition is the first step in the formation of the resulting benzene derivatives.[78] A similar reaction of tetramethylthiophene with dicyanoacetylene in the presence of aluminum trichloride occurred at room temperature.[79] The (2 + 2)-cycloadduct, 6,7-dicyano-1,3,4,5-tetramethyl-2-thiabicyclo[3.2.0]hepta-

[76] R. Helder and H. Wynberg, *Tetrahedron Lett.*, 605 (1972).
[77] H. J. Kuhn and K. Gollnick, *Tetrahedron Lett.*, 1909 (1972).
[78] H. J. Kuhn and K. Gollnick, *Chem. Ber.* **106**, 674 (1973).
[79] H. Wynberg and R. Helder, *Tetrahedron Lett.*, 3647 (1972).

3,6-diene (**75**), rather than the corresponding thiepin, was formed in 57% yield (Section IV,B,2).[80]

Reactions of electron-*rich* acetylenes with heteroaromatic compounds are rare. Such (2 + 2)-cycloadditions have been proposed as the first step in the boron-trifluoride-catalyzed reaction of isoquinoline with 1-dimethylamino-2-phenylacetylene[64] and in the reaction of 1,2,4-triazines with 1-diethylamino-1-propyne,[81] but no experimental

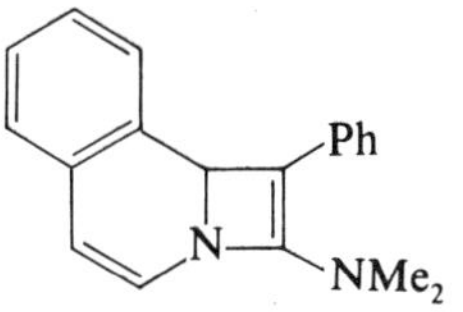

(**76**) 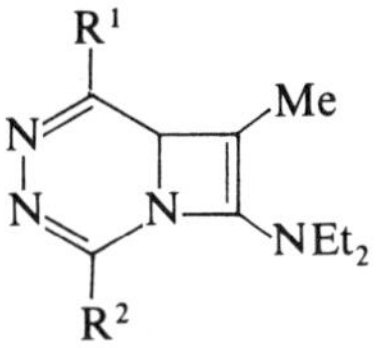   (**77**) 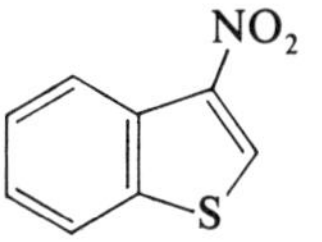   (**78**) 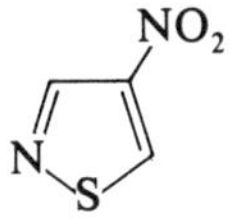   (**79**)

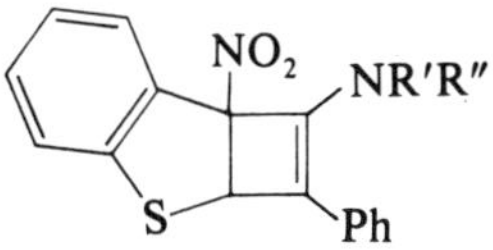

(**80**) 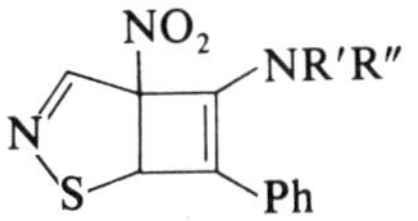   (**81**) 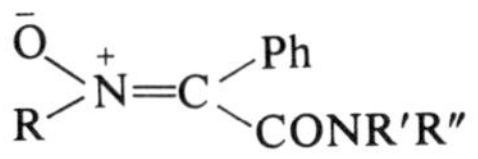   (**82**)

a: R = 3-benzo[*b*]thienyl
b: R = 4-isothiazolyl

evidence has been provided for the formation of intermediates **76** and **77**. So far only two such reactions have been actually observed. Reinhoudt and Kouwenhoven[63,82] described the reactions of 3-nitrobenzo[*b*]thiophene (**78**) and of 4-nitroisothiazole (**79**) with ynamines in diethyl ether. The corresponding (2 + 2)-cycloadducts **80** and **81** were obtained in yields of 43 and 14%, respectively. This (2 + 2)-cycloaddition was accompanied by a competitive reaction, *viz.* the formation of a nitrone (**82**), which was suggested to result from a (2 + 2)-cycloaddition of a nitrogen–oxygen double bond of the nitro group with the ynamine, followed by rearrangement. The ratio of (2 + 2)-cycloadduct to nitrone decreased with increasing solvent polarity.

[80] D. N. Reinhoudt, H. C. Volger, C. G. Kouwenhoven, H. Wynberg, and R. Helder, *Tetrahedron Lett.*, 5269 (1972).
[81] H. Neunhoeffer and H.-W. Frühauf, *Tetrahedron Lett.*, 3355 (1970).
[82] D. N. Reinhoudt and C. G. Kouwenhoven, *Tetrahedron Lett.*, 2503 (1974).

D. Reactions with Heterocumulenes

1. Ketenes

(2 + 2)-Cycloadditions of heterocycles with ketenes have been reported to occur via endocyclic carbon–carbon, carbon–nitrogen, or nitrogen–nitrogen double bonds and via endocyclic carbon–carbon triple bonds. The reactions proceeding via the carbon–nitrogen double bond have received a great deal of attention since they yield heterobicyclic compounds related to the penicillin or the cephalosporin C antibiotics.[7]

Annulated cyclobutanones (**83–86**) were obtained in high yields upon reaction of ketenes with a number of heterocycles containing a vinyl ether moiety (Table II).[29,30,83–85] The reported cycloadducts **83** and **84** were all formed in a stereospecific manner, in agreement with the polarization of the two π-electron systems.[30,83] The reactions were fast except when the reacting endocyclic carbon–carbon double bond was not substituted with an oxygen atom, such as the formation of **87** from 2,5-dihydrofuran (Table II). Furthermore, the various ketenes showed

TABLE II

ANNULATED CYCLOBUTANONES FROM KETENES AND HETEROCYCLES

Cycloadduct	R^1	R^2	Temp. (°C)	Yield (%)	References
83	Ph	Ph	25	100	29
84	Ph	Ph	25	90	29, 30, 83
84	Me	Me	25	73	84
84	H	H	100	12	83
85	Ph	Ph	25	52	29
86	Ph	Ph	25	80	29
87	Ph	Ph	100	51	85
87	Et	i-Bu	180	30	84

the usual differences in reactivity. It was found that 2,3-dihydrofuran reacted more rapidly with diphenylketene than 3,4-dihydro-2H-pyran.[29] Similarly, the rate of formation of **85** was faster than that of **86**.[29] This, together with the reported difference between the rates of (2 + 2)-cycloaddition of 2,3-dihydrofuran and of 3,4-dihydro-2H-pyran with tetracyanoethylene,[45] suggests that this difference in rate is quite a general phenomenon in (2 + 2)-cycloaddition reactions of related five- and six-membered heterocycles (Section II,B).

[83] C. D. Hurd and R. D. Kimbrough, *J. Am. Chem. Soc.* **82**, 1373 (1960).
[84] R. H. Hasek, P. G. Gott, and J. C. Martin, *J. Org. Chem.* **29**, 1239 (1964).
[85] R. D. Kimbrough and R. W. Askins, *J. Org. Chem.* **32**, 3683 (1967).

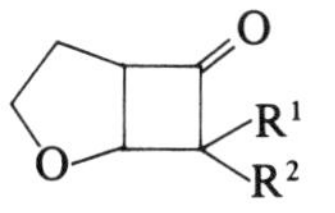

Sec. II.D] (2 + 2)-CYCLOADDITION AND CYCLOREVERSION

(83) 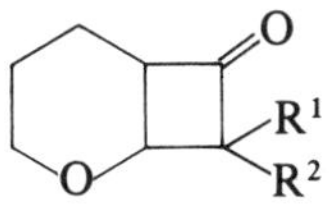 (84) 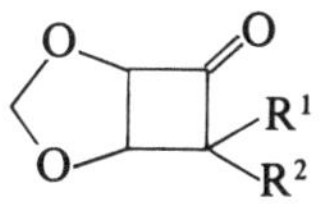 (85) 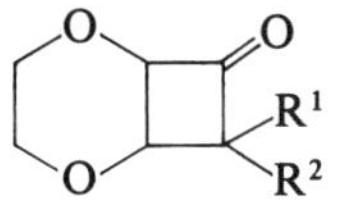 (86) 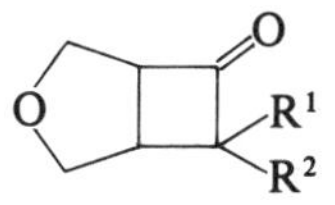 (87) 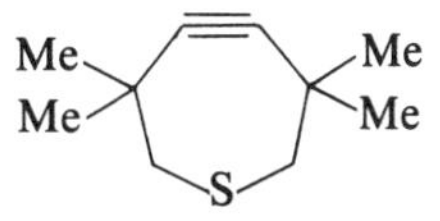 (88) 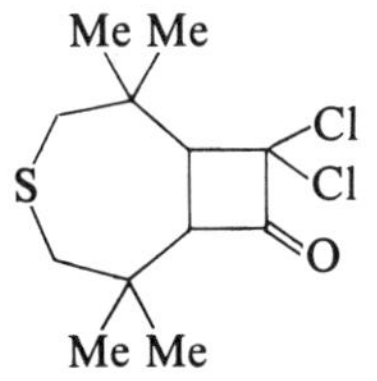 (89) 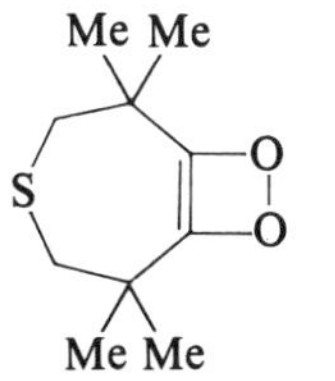 (90) 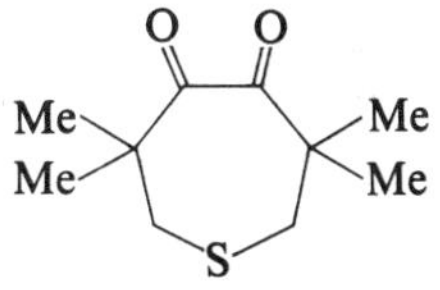 (91)

Krebs and Kimling[86] reported the unique (2 + 2)-cycloaddition of a heterocyclic compound, 3,3,7,7-tetramethyl-5-thiacyclohept-1-yne (**88**), via an endocyclic carbon–carbon triple bond with *in situ* prepared dichloroketene. The reaction proceeded much faster with **88** than with other acetylenes, because of the relief of ring strain upon conversion to the annulated cyclobutenone (**89**). The oxidation of **88** in the dark was assumed to proceed via a similar (2 + 2)-cycloaddition with oxygen to give the corresponding annulated 1,2-dioxete (**90**), with subsequent rearrangement to 3,3,6,6-tetramethylhexahydrothiepin-4,5-dione (**91**).

The (2 + 2)-cycloaddition reactions of ketenes with heterocycles via an endocyclic imino group have contributed substantially to the synthesis of penicillin (**92a**) and cephalosporin C (**93a**) antibiotics and structurally related heterobicyclic β-lactams.[7,11] Reactions with relatively stable ketenes, e.g., diphenylketene, have been reported as well as reactions of heterocycles with mixtures of acid chlorides and tertiary amines.

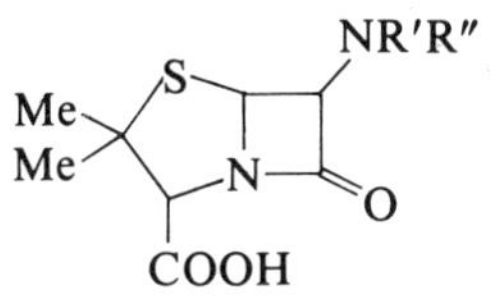

(92a) 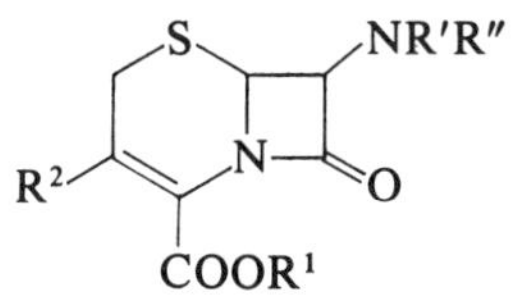 (93a)

[86] A. Krebs and H. Kimling, *Ann. Chem.*, 2074 (1974).

TABLE III
Penicillin and Cephalosporin Derivatives from Reactions of Heterocycles with Ketenes

Compound	Structure	Substituents	Yield (%)	References
92b	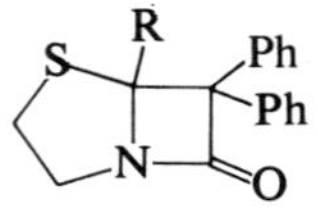	$R = Ph$	63	89
		$R = SCONHPh_2$	68	94
		$R = NHCOCHPh_2$	44	94
		$R = NHCOMe$	54	94
92c	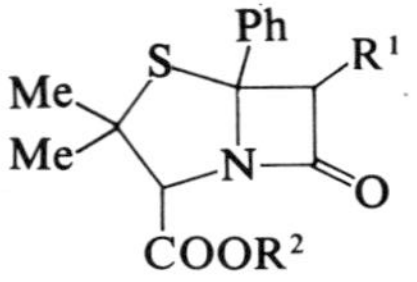	$R^1 = N$-phthaloylamino, $R^2 = Me$	58	90
		$R^1 = MeO, R^2 = CHPh_2$	90	95
92d	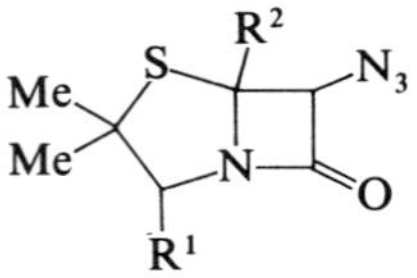	$R^1 = H, R^2 = COOt\text{-}C_4H_9$	Good	96
		$R^1 = H, R^2 = Ph$	87	95
		$R^1 = COOMe, R^2 = H$	8	91
		$R^1 = COOCH_2Ph, R^2 = H$	13	92
92e	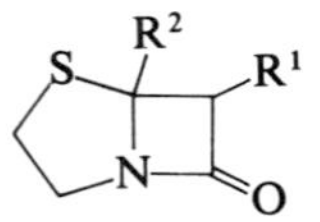	$R^1 = MeO, R^2 = SMe$	73	97
		$R^1 = N_3, R^2 = Ph$	70	95
		$R^1 = PhO, R^2 = Ph$	70	95, 98
		$R^1 = MeO, R^2 = Ph$	90	95
		$R^1 = MeO, R^2 = p\text{-}C_6H_4COOCH_2Ph$	70	95
		$R^1 = PhO, R^2 = p\text{-}C_6H_4COOCH_2Ph$	63	95
		$R^1 = PhO, R^2 = 2\text{-furyl}$	70	95

92f	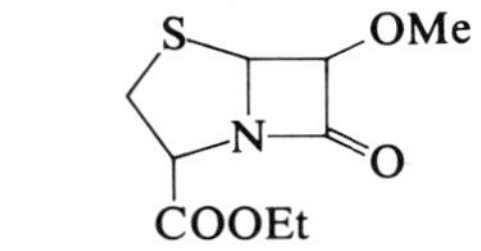 (structure: S, OMe, N, COOEt, O)	—	11	95
92g	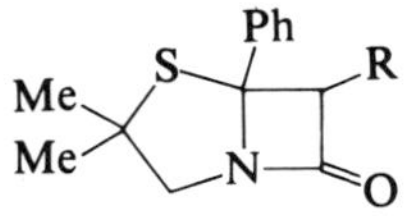 (structure: S, Me, Me, Ph, R, N, O)	R = 3-(5-phenyl-2,4-diketo-oxazolidyl)	28	99
93b	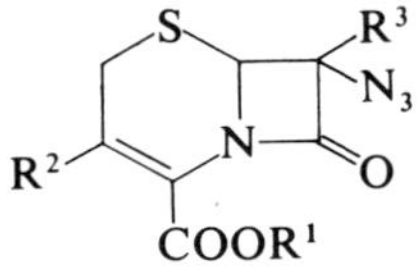 (structure: S, R², R³, N₃, N, COOR¹, O)	$R^1 = R^2 = Me, R^3 = H$	52	93
		$R^1 = CH_2C_6H_4OMe, R^2 = CH_2OCOMe, R^3 = H$	56	93
		$R^1 = R^2 = Me, R^3 = OMe$	Low	93
94	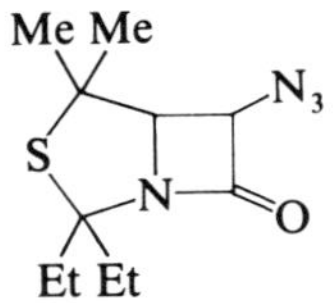 (structure: Me Me, S, Et Et, N₃, N, O)	—	18	100
95	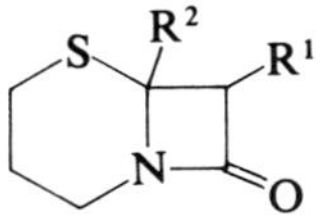 (structure: S, R², R¹, N, O)	$R^1 = MeO, R^2 = Ph$	63	95, 98
		$R^1 = PhO, R^2 = Ph$	81	95, 98
		$R^1 = N_3, R^2 = Ph$	23	101
		$R^1 = N_3, R^2 = p\text{-}NO_2C_6H_4$	55	101
		$R^1 = N\text{-phthaloylamino}, R^2 = Ph$	43	102

(continued)

TABLE III—continued

Compound	Structure	Substituents	Yield (%)	References
96	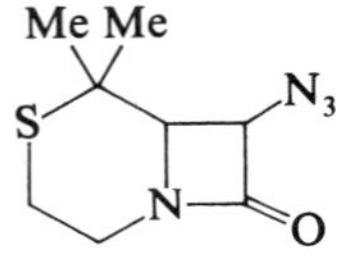 Me Me, S, N₃, N, O	—	30	101
97	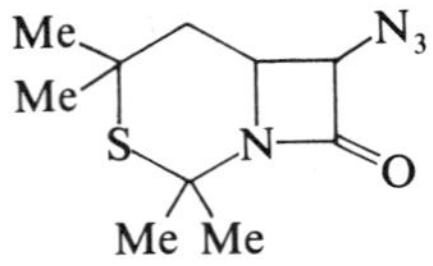 Me Me, Me Me, S, N₃, N, O	—	Trace	101

[93] R. W. Ratcliffe and B. G. Christensen, *Tetrahedron Lett.*, 4649, 4653 (1973).
[94] R. Pfleger and A. Jäger, *Chem. Ber.* **90**, 2460 (1957).
[95] A. K. Bose, M. S. Manhas, J. S. Chib, H. P. S. Chawla, and B. Dayal, *J. Org. Chem.* **39**, 2877 (1974).
[96] A. K. Bose, G. Spiegelman, and M. S. Manhas, *Chem. Commun.*, 321 (1968).
[97] A. K. Bose, J. L. Fahey, and M. S. Manhas, *J. Heterocycl. Chem.* **10**, 791 (1973).
[98] A. K. Bose, B. Dayal, H. P. S. Chawla, and M. S. Manhas, *Tetrahedron Lett.*, 2823 (1972).
[99] J. C. Sheehan and G. D. Laubach, *J. Am. Chem. Soc.* **73**, 4752 (1951).
[100] A. K. Bose, G. Spiegelman, and M. S. Manhas, *J. Chem. Soc. C*, 188 (1971).
[101] A. K. Bose, V. Sudarsanam, B. Anjaneyulu, and M. S. Manhas, *Tetrahedron* **25**, 1191 (1969).
[102] S. M. Desphande and A. K. Mukerjee, *J. Chem. Soc. C*, 1241 (1966).

The latter reactions might proceed via acylation of the imino nitrogen atom with subsequent ring closure or via (2 + 2)-cycloaddition of an *in situ* generated ketene with the imino group.[87,88] Therefore, these reactions have been included in this section although in most cases evidence for the formation of the intermediate ketene is lacking. The reactions were carried out under high-dilution conditions, in order to avoid side reactions of the ketene, at temperatures between 25° and 80°. Depending on the structures of the ketene and heterocycle either 1:1 or 2:1 cycloadducts were formed, together with Michael-type adducts.[7] The (2 + 2)-cycloadduct (**92b**; R = Ph) of diphenylketene and 2-phenyl-4,5-dihydro-1,3-thiazole served as a model compound in IR studies of the structure of the penicillin skeleton,[89] and later this method of β-lactam formation was applied to the synthesis of both penicillins and cephalosporins. *In situ* generated functionalized ketenes, such as azido- or *N*-phthaloylaminoketenes, have been allowed to react with 2-thiazolines[90-92] and 6*H*-1,3-thiazines to give **92** and **93**, respectively.[93] The results of such reactions with other ketenes have been summarized in Table III. The relative positions of the heteroatoms in the heterocyclic nucleus have also been varied in order to establish any relationship between the structure of these heterobicyclic lactams (**94–97**) and their biological activity. The best yields were obtained with heterocycles in which the endocyclic imino group carries an aryl substituent. The stereochemistry of the resulting β-lactams, in which the substituent atoms may be either cis or trans, depended on the type of substituent. In most cases the stereochemistry around the β-lactam ring was trans, but in some instances cis isomers resulted from the (2 + 2)-cycloadditions.

Other heterocycles have also been allowed to react with ketenes via (2 + 2)-cycloaddition of an imino group. Heterobicyclic compounds related to the penicillins or the cephalosporins in which the sulfur atom is replaced by oxygen (*O*-penam and *O*-cepham)[103] were obtained

---

[87] A. K. Bose, B. Anjaneyulu, S. K. Bhattacharya, and M. S. Manhas, *Tetrahedron* **23**, 4769 (1967).
[88] J. P. Luttringer and J. Streith, *Tetrahedron Lett.*, 4163 (1973).
[89] S. A. Ballard, D. S. Melstrom, and C. W. Smith, *in* "The Chemistry of Penicillin" (H. T. Clarke, J. R. Johnson, and R. Robinson, eds.), p. 973. Princeton Univ. Press, Princeton, New Jersey, 1949.
[90] J. C. Sheehan, H. Wayne Hill, and E. L. Buhle, *J. Am. Chem. Soc.* **73**, 4373 (1951).
[91] A. K. Bose, G. Spiegelman, and M. S. Manhas, *J. Am. Chem. Soc.* **90**, 4506 (1968).
[92] R. A. Firestone, N. S. Maciejewicz, R. W. Ratcliffe, and B. G. Christensen, *J. Org. Chem.* **39**, 437 (1974).
[103] Nomenclature according to Manhas and Bose.[11]

from 2-methyl-4,5-dihydrooxazole[102] and 2-aryl-5,6-dihydro-4*H*-1,3-oxazines[104] in yields of 26% (**98**) and 25–65% (**99**). Similarly, an analog of penicillin in which the sulfur atom is incorporated in a substituent (**100**) was prepared by (2 + 2)-cycloaddition.[105] Various heterobicyclic lactams in which the β-lactam ring is fused to a nitrogen-containing six-membered ring (**101–103**) have been reported by Bose *et al.*[7,87,106,107] Several seven-membered heterocycles were found to react with ketenes via an endocyclic imino group, e.g., 4,7-dihydro-1,3-benzo[*e*]-

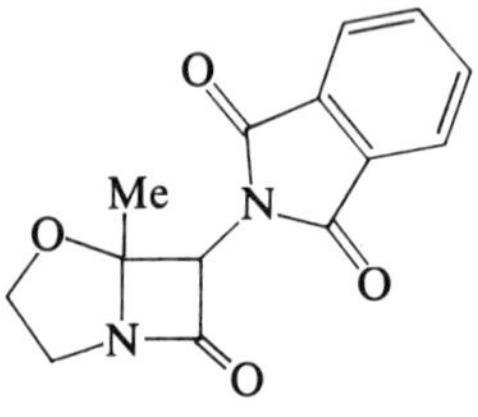

(**98**) 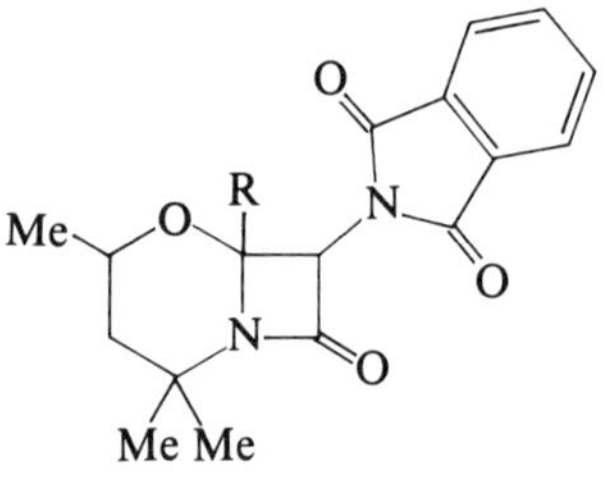   (**99**) 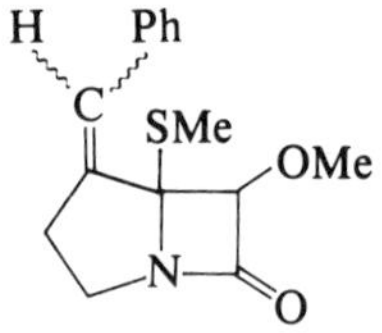   (**100**)

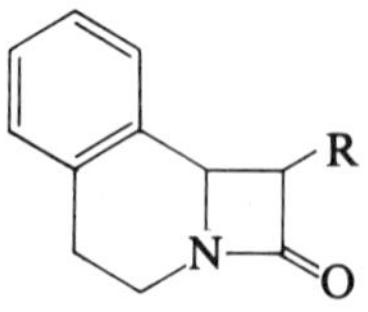

(**101**) 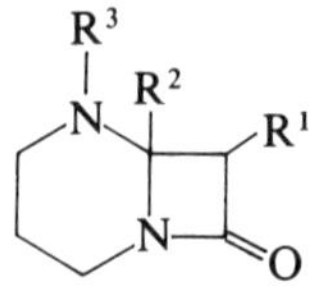   (**102**) 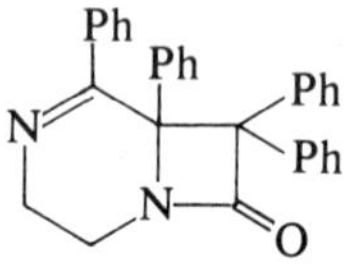   (**103**)

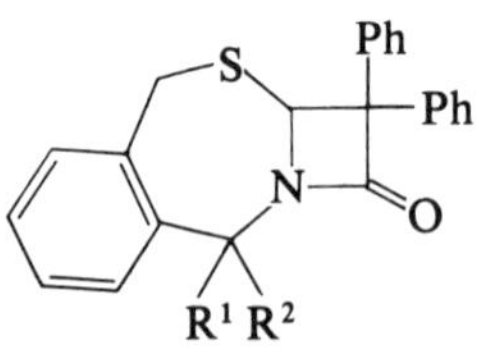

(**104**) 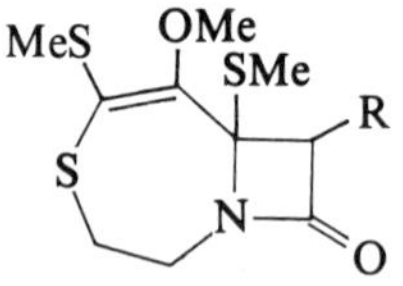   (**105**) 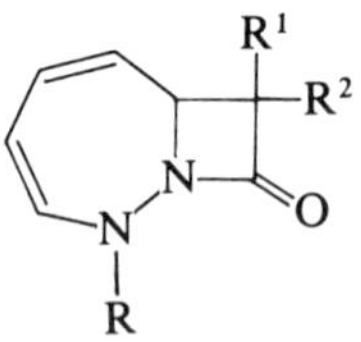   (**106**) 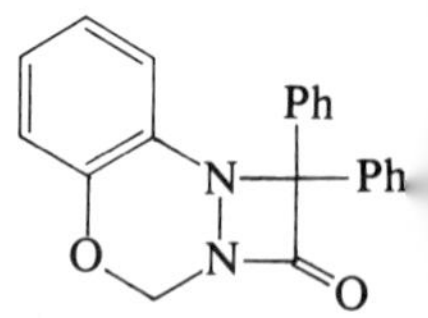   (**107**)

---

[104] J. C. Sheehan and M. Dadîc, *J. Heterocycl. Chem.* **5**, 779 (1968).
[105] A. K. Bose and J. L. Fahey, *J. Org. Chem.* **39**, 115 (1974).
[106] A. K. Bose, J. C. Kapur, J. L. Fahey, and M. S. Manhas, *J. Org. Chem.* **37**, 3437 (1972).
[107] A. K. Bose, J. C. Kapur, and M. S. Manhas, *Synthesis* **6**, 891 (1974).

thiazepines,[108] 2,3-dihydro-1,4-thiazepines,[97] and 1H-1,2-diazepines[88] yielded the corresponding (2 + 2)-cycloadducts **104–106**.

Diphenylketene gave a fast reaction with the cis-diaza group in 2H-benzo[e]-1,3,4-oxadiazines; the (2 + 2)-cycloadducts (**107**) were formed in yields ranging from 33 to 90%, depending on the nature of the substituent R.[109]

2. Other Heterocumulenes

(2 + 2)-Cycloaddition reactions of heterocycles with heterocumulenes other than ketenes are relatively rare. An interesting example is the thermal reaction of diphenylketene-N-(p-tolyl)imine with dibenzo[c,f]diazepine.[110] The cycloadduct (**108**) was formed in 68% yield in the absence of light. It was concluded that the actual (2 + 2)-cycloaddition of ketenimines with symmetrically substituted azobenzenes is not a photochemical process. The light is required only for the conversion of *trans*-azobenzene into its thermodynamically less stable cis isomer, since the trans isomer itself does not react with the ketenimine. In the dibenzo[c,f]diazepine the reacting azo group is incorporated in the seven-membered ring and is consequently in a cis configuration.

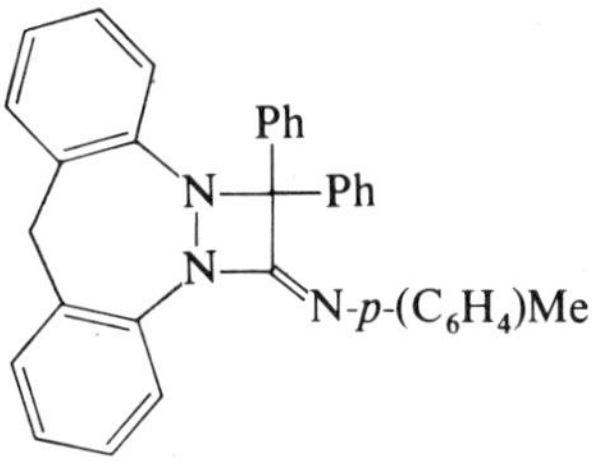

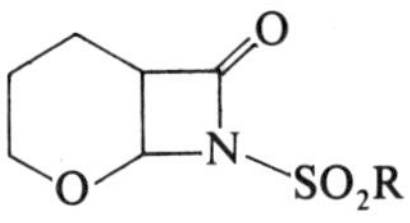

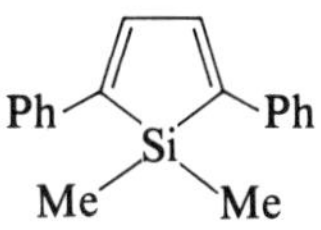

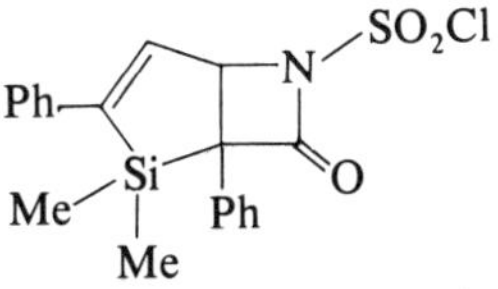

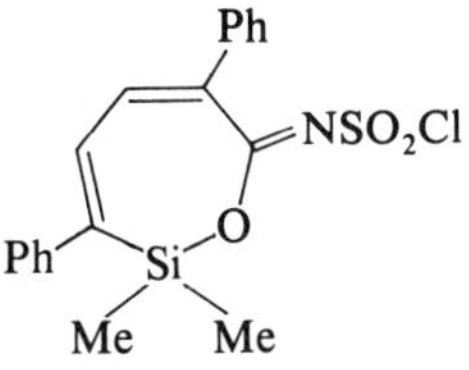

[108] D. N. Reinhoudt, *Rec. Trav. Chim.* **92**, 20 (1973).
[109] W. Ried and E. Kahr, *Chem. Ber.* **103**, 331 (1970).
[110] M. W. Barker and R. H. Jones, *J. Heterocycl. Chem.* **9**, 555 (1972).

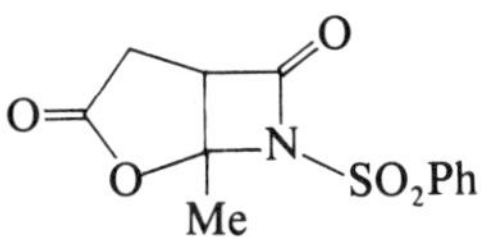

Sulfonyl isocyanates have been found to react with 5-methyl-2,3-dihydrofuran-2-one[111] and with 3,4-dihydro-2H-pyran[112] at room temperature to give (2 + 2)-adducts in nearly quantitative yields. At higher temperatures only the corresponding sulfonamides were formed, by rearrangement of the initially produced (2 + 2)-cycloadducts **109** and **110**. Recently the (2 + 2)-cycloaddition of 1,1-dimethyl-2,5-diphenyl-1-silacyclopenta-2,4-diene (**111**) with N-chlorosulfonyl isocyanate has been reported to occur in chloroform at 0°. At higher temperatures the cycloadduct (**112**) rearranged to the 1-oxa-2-silacyclohepta-3,5-diene (**113**).[113]

The reaction of azirines (**114**) with benzoyl isothiocyanate has been discussed in terms of a (2 + 2)-cycloaddition of the azirine with the carbon-sulfur double bond, followed by rearrangement to the 2-benzamidothiazole (**116**).[114] Experimental evidence for the formation of the (2 + 2)-cycloadduct (**115**) has not been provided. Reactions of azirines with other heterocumulenes such as thiobenzoyl isocyanate and

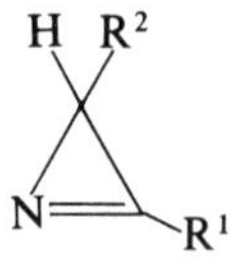

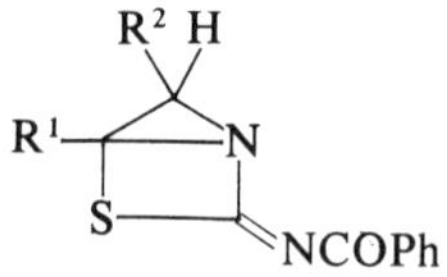

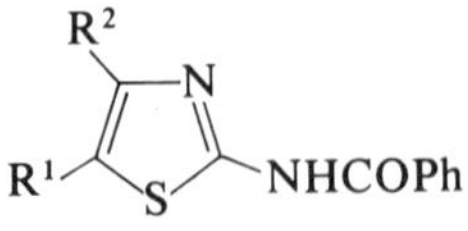

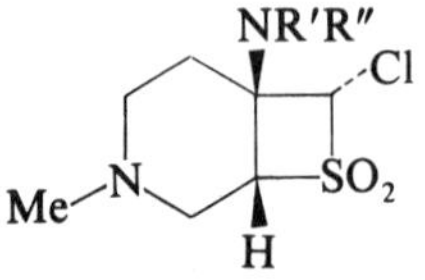

benzoyl isocyanate yielded (4 + 2)-cycloadducts or rearranged (4 + 2)-cycloadducts.

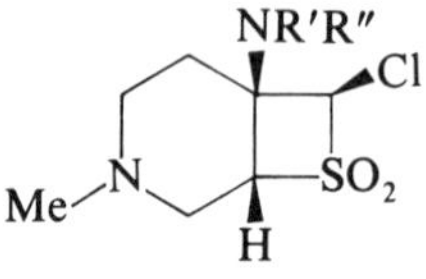

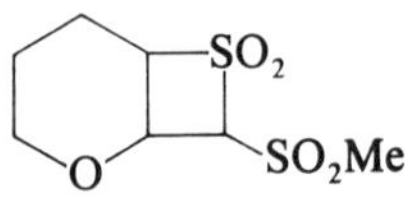

Chlorosulfene, generated *in situ* from chloromethanesulfonyl chloride and triethylamine, reacted with 1-methyl-4-(4-morpholinyl)-1,2,5,6-tetrahydropyridine to form **117a** and **117b**, the two possible stereoisomers of 7-chloro-8,8-dioxo-3-methyl-6-(4-morpholinyl)-8-thia-3-azabicyclo[4.2.0]octane, in a ratio of 56 to 44.[115] The reaction of the

[111] G. Westphal, *Tetrahedron* **25**, 5199 (1969).
[112] F. Effenberger and R. Gleiter, *Chem. Ber.* **97**, 1576 (1964).
[113] T. J. Barton and R. J. Rogido, *J. Org. Chem.* **40**, 582 (1975).
[114] V. Nair and K. H. Kim, *J. Org. Chem.* **39**, 3763 (1974).
[115] L. A. Paquette, *J. Org. Chem.* **29**, 2854 (1964).

corresponding carbocyclic compounds afforded only one isomer. A single stereoisomer (**118**) was isolated from the (2 + 2)-cycloaddition of methylsulfonylsulfene and 3,4-dihydro-2H-pyran.[116]

III. Photochemical (2 + 2)-Cycloadditions

Both intermolecular and intramolecular (2 + 2)-cycloaddition reactions of heterocyclic compounds have been reported to occur upon irradiation. The intermolecular reactions of nonaromatic heterocycles and heteroaromatics will be dealt with successively to cover their reactions with the same type of compound (dimerization), with carbonyl compounds, with alkenes, with singlet oxygen, and with acetylenes. The intramolecular (2 + 2)-cycloaddition reactions, which are in fact isomerizations, have been divided into two groups. The first group comprises the reactions in which a four-membered ring is formed from two nonconjugated π-electron systems, and the second comprises similar reactions of conjugated 1,3-(hetero)diene systems. The latter reactions have been summarized according to the ring size of the starting heterocycle. The mechanistic aspects of the photochemical reactions are dealt with separately in Section III,A.

Some of the reactions included have been discussed in other review articles on the photochemistry of heterocyclic compounds,[117] dimerization reactions,[19,118,119] photochemical oxetane synthesis,[6] and photochemical addition reactions.[120]

Experimental details, such as type of light source, filter or sensitizer used, have been omitted; for such information the reader is referred to the original papers. Only those details that have a direct bearing upon the products formed are given in this review.

A. Mechanism

The mechanisms of photochemical (2 + 2)-cycloaddition reactions have received a great deal of attention in the literature.[6,120–122] Some general remarks on these mechanisms are summarized here. Two types of excited states are important in photochemical (2 + 2)-cycloadditions

[116] G. Opitz, K. Rieth, and G. Walz, *Tetrahedron Lett.*, 5269 (1966).
[117] S. T. Reid, *Adv. Heterocycl. Chem.* **11**, 1 (1970).
[118] A. Mustafa, *Adv. Photochem.* **2**, 63 (1964).
[119] L. J. Kricka and A. Ledwith, *Synthesis* **6**, 539 (1974).
[120] R. Steinmetz, *Fortschr. Chem. Forsch.* **7**, 445 (1967).
[121] W. L. Dilling, *Chem. Rev.* **67**, 373 (1967).
[122] W. G. Dauben, L. Salem, and N. J. Turro, *Acc. Chem. Res.* **8**, 41 (1975).

of heterocyclic compounds. The first type is formed by a $\pi \to \pi^*$ transition in which an electron is transferred from a bonding to an antibonding π-orbital. The second type is obtained from an $n \to \pi^*$ transition in which an electron is transferred from a nonbonding atomic orbital to an antibonding π^*-orbital. The latter type of transition is important for compounds that contain a heteroatom or a carbonyl group. In general $n \to \pi^*$ transitions occur at lower energies than $\pi \to \pi^*$ transitions. The excited species may react in its excited singlet state (lifetime $\sim 10^{-9}$ sec), but in most cases intersystem crossing to the corresponding triplet state (lifetime $\sim 10^{-4}$ sec) is fast compared to the rate of the chemical reaction. Consequently, *intra*molecular (2 + 2)-cycloadditions are more likely to proceed via the excited singlet state than *inter*molecular reactions.

In terms of conservation of orbital symmetry, the concerted photochemical $[_\pi 2_s + _\pi 2_s]$ cycloaddition is an allowed process, but on the basis of the available experimental evidence it is most likely that the majority of the (2 + 2)-cycloadditions summarized in this section are non-concerted reactions.

In addition to direct excitation, photochemical (2 + 2)-cycloadditions have been performed in the presence of a sensitizer. In such reactions the electronic excitation energy is transferred from a sensitizing molecule to the reactant by molecular collision. Studies of sensitizer molecules with different triplet energies have provided useful information on the mechanism of these reactions.

The mechanisms of two particular classes of photochemical (2 + 2)-cycloaddition reactions have been studied extensively, *viz.* (i) of reactions in which either the heterocyclic or the substrate contains an enone moiety,[123,124] which are believed to occur via the excited triplet state of the enone, and (ii) of oxetane formation, which is thought to proceed via $n \to \pi^*$ excitation of the carbonyl group, followed by intersystem crossing to the triplet state and addition to the carbon–carbon double bond in its ground state[6]; in some cases, however, an oxetane has been reported to be formed from a ground-state ketone and an excited-state olefin (Section III,C,2).

B. Intermolecular Reactions of Nonaromatic Heterocycles

1. *Dimerization*

Formation of a dimer by photochemical (2 + 2)-cycloaddition is a quite general reaction of heterocycles.[117–119] The dimerization of benzo-

---

[123] P. de Mayo, *Acc. Chem. Res.* **4**, 41 (1971).
[124] P. G. Bauslaugh, *Synthesis* **2**, 287 (1970).

Sec. III.B] (2 + 2)-CYCLOADDITION AND CYCLOREVERSION

[b]thiophene 1,1-dioxide, which was reported years ago,[118] has recently been reinvestigated.[125,126] It was found that the anti head-to-head (**119**) and anti head-to-tail (**120**) isomers are formed in a ratio of 2.7. This dimerization was proposed to occur via attack of a molecule in the excited triplet state ($E_T \sim 50$ kcal/mol) on another molecule in the ground state. The preferential formation of isomer **119** was attributed to a better stabilization of the more polar transition state in the formation of **119** relative to **120** in view of the observed effect of the polarity of the solvent on the ratio of **119** to **120**. Similar dimerizations have been reported for 2-nitrobenzo-1,4-dithiin (**121**)[127] and 1,1-dimethyl-2,5-diphenyl-1-silacyclopenta-2,4-diene (**111**).[128] Other heterocycles that dimerize via (2 + 2)-cycloaddition, either upon direct or upon sensitized irradiation, are 1H-dibenzo[b,f]-azepine derivatives (**122**),[129] vinylene carbonates (**123**),[13,130] and 1,3-substituted imidazolin-2-ones (**124**).[131,132]

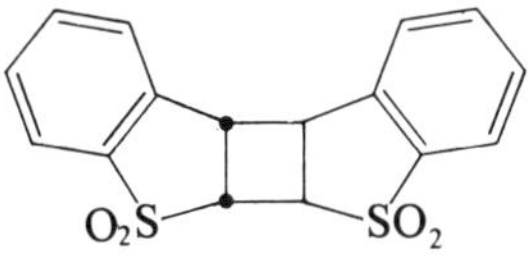

(**119**) 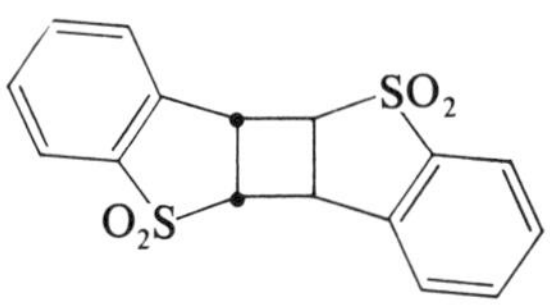 (**120**)

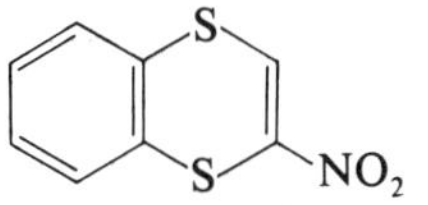

(**121**) 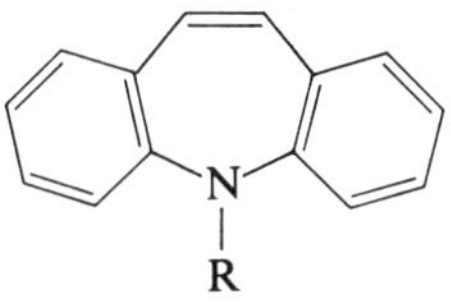 (**122**) 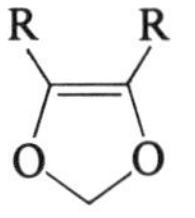 (**123**) 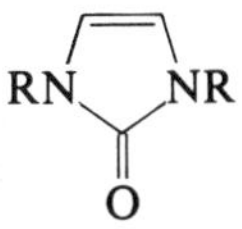 (**124**)

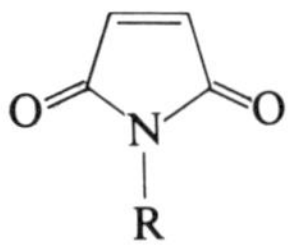

(**125**) 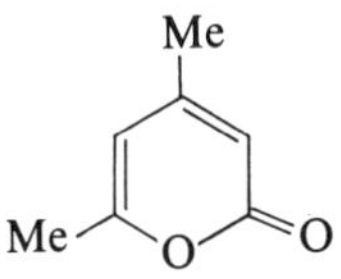 (**126**) 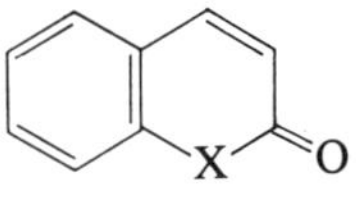 (**127**) X = O (**129**)
           (**128**) X = NR

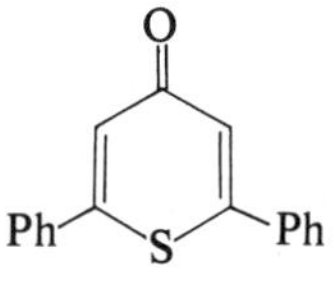

[125] D. N. Harpp and C. Heitner, *J. Org. Chem.* **35**, 3256 (1970).
[126] D. N. Harpp and C. Heitner, *J. Amer. Chem. Soc.* **94**, 8179 (1972).
[127] W. E. Parham, P. L. Stright, and W. R. Hasek, *J. Org. Chem.* **24**, 262 (1959).
[128] Y. Nakadaira and H. Sakurai, *Tetrahedron Lett.*, 1183 (1971).
[129] L. J. Kricka, M. C. Lambert, and A. Ledwith, *Chem. Commun.*, 224 (1973).
[130] W. Hartmann and R. Steinmetz, *Chem. Ber.* **100**, 217 (1967).
[131] G. Steffan and G. O. Schenck, *Chem. Ber.* **100**, 3961 (1967).
[132] G. Steffan, *Chem. Ber.* **101**, 3688 (1968).

Various heterocycles with an enone system dimerize via (2 + 2)-cycloaddition, e.g., maleimides (**125**),[119] 4,6-dimethyl-2-pyrone (**126**),[133] coumarin (**127**),[134] carbostyril (**128**),[135] 2,6-diphenyl-4H-thiopyran-4-one (**129**),[136] and several naturally occurring furocoumarins.[117,137]

Reactions that are not actually dimerizations but are closely related to them have been reported for molecules in which two heterocyclic rings are linked by a polymethylene chain of variable length, e.g., N,N'-polymethylenedimaleimides (**130**),[138] 7,7'-polymethylenedioxycoumarins (**131**),[139] and 6,6'-polymethylene-bis(4-methyl-2-pyrone) (**132**).[133] Depending on the length of the polymethylene chain, intramolecular (2 + 2)-cycloaddition of the heterocyclic moieties with formation of "dimers" linked by an additional polymethylene bridge or intermolecular reaction occurs.

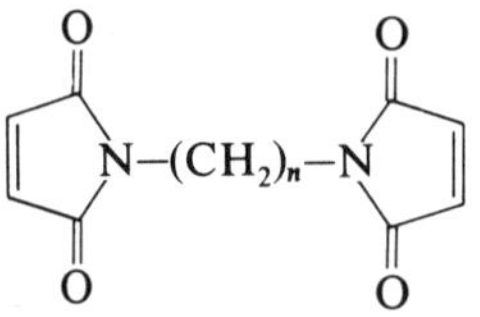

(**130**) 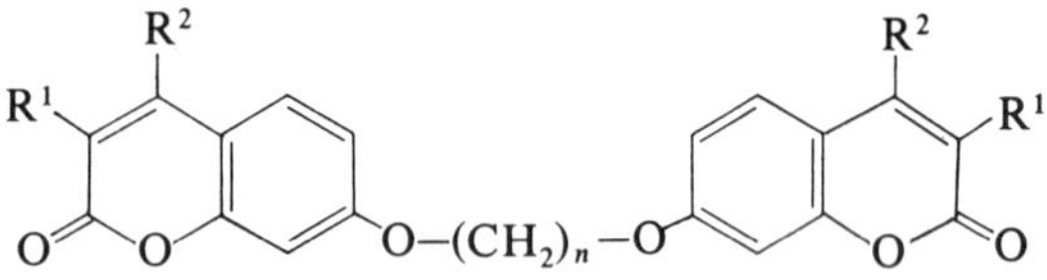   (**131**)

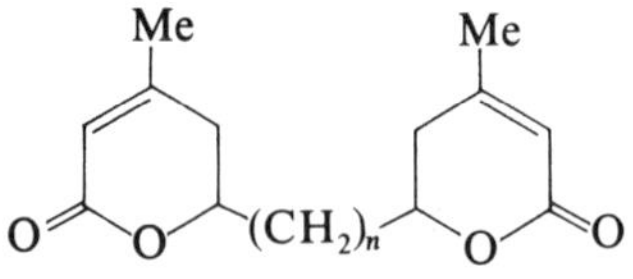

(**132**) 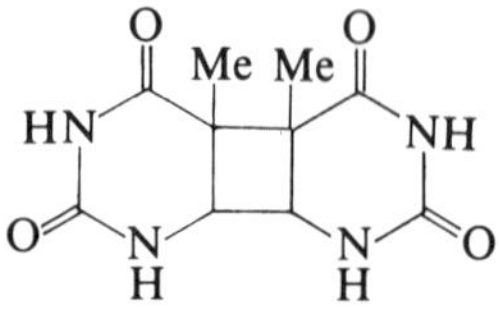   (**133**)

A biologically significant dimerization of the same type is the (2 + 2)-cycloaddition of pyrimidine bases in deoxyribonucleic acids, which is considered to be at least partly responsible for their observed deactivation upon irradiation with UV light.[19] It has been reported that irradiation of thymine yields a (2 + 2)-cycloadduct (**133**).[140] Other thymine and also uracil derivatives behaved similarly.[141]

[133] M. van Meerbeck, S. Toppet, and F. C. de Schryver, *Tetrahedron Lett.*, 2247 (1972).
[134] C. H. Krauch, S. Farid, and G. O. Schenck, *Chem. Ber.* **99**, 625 (1966).
[135] O. Buchardt, *Acta Chem. Scand.* **18**, 1389 (1964).
[136] N. Sugiyama, Y. Sato, H. Kataoka, C. Kashima, and K. Yamada, *Bull. Chem. Soc. Jpn.* **42**, 3005 (1969).
[137] C. H. Krauch and S. Farid, *Chem. Ber.* **100**, 1685 (1967).
[138] J. Put and F. C. de Schryver, *J. Am. Chem. Soc.* **95**, 137 (1973).
[139] L. H. Leenders, E. Schouteden, and F. C. de Schryver, *J. Org. Chem.* **38**, 957 (1973).
[140] R. Beukers and W. Berends, *Biochim. Biophys. Acta* **41**, 550 (1960).
[141] J. G. Burr, *Adv. Photochem.* **6**, 193 (1968).

2. Reactions with Carbonyl Compounds

The formation of oxetanes by photochemical (2 + 2)-cycloaddition of carbonyl compounds, such as aldehydes, ketones, and quinones, with carbon–carbon double bonds has been reported for various heterocyclic compounds. Maleic anhydride,[142] isocoumarin (and its derivatives),[143,144] benzo[b]thiophene 1,1-dioxide,[144] 1,3-dihydroimidazol-2-ones,[131,132] vinylene carbonate,[145] and 1,4-dioxene[146] yielded the corresponding annulated oxetanes **134–139**.

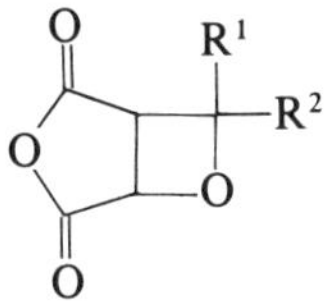

(**134**) 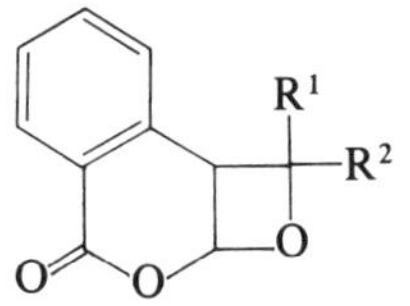 (**135**) 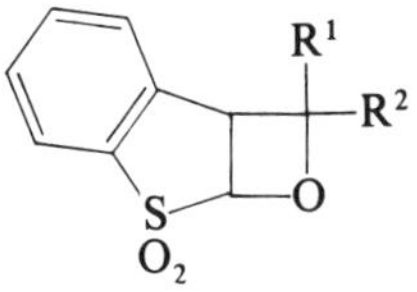 (**136**)

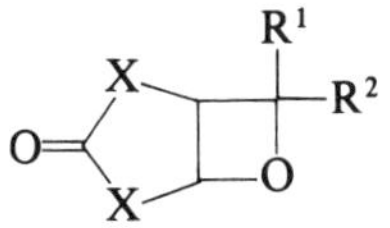

(**137**) X = NR
(**138**) X = O

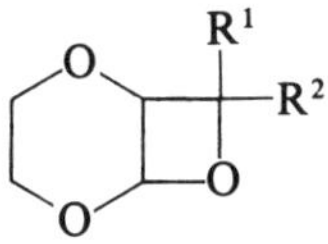

(**139**)

3. Reactions with Olefins

The intermolecular (2 + 2)-cycloaddition reactions of heterocyclic compounds can be divided into two groups, depending on whether the heterocycle contains an enone moiety or not.
5-Methyl-2,3-dihydrofuran,[147] 3,4-dihydro-2H-pyran,[148] and 1,4-dioxene[146] yielded the corresponding (2 + 2)-cycloadducts (**140–142**)

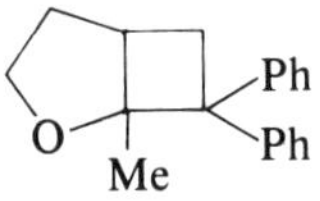

(**140a**) 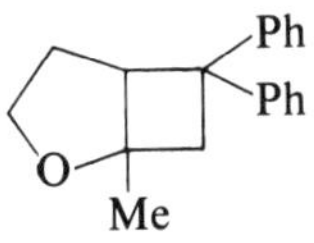 (**140b**)

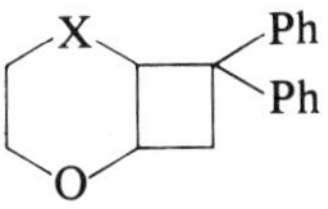

(**141**) X = CH$_2$
(**142**) X = O

---

[142] N. J. Turro, P. Wriede, J. C. Dalton, D. Arnold, and A. Glick, *J. Am. Chem. Soc.* **89**, 3950 (1967).
[143] C. H. Krauch and S. Farid, *Tetrahedron Lett.*, 4783 (1966).
[144] C. H. Krauch, S. Farid, and G. O. Schenck, *Chem. Ber.* **98**, 3102 (1965).
[145] S. Farid, D. Hess, and C. H. Krauch, *Chem. Ber.* **100**, 3266 (1967).
[146] N. R. Lazear and J. H. Schauble, *J. Org. Chem.* **39**, 2069 (1974).
[147] H. M. Rosenberg and M. P. Servé, *J. Org. Chem.* **36**, 3015 (1971).
[148] P. Servé, H. M. Rosenberg, and R. Rondeau, *Can. J. Chem.* **47**, 4295 (1969).

with 1,1-diphenylethylene or with *trans*-stilbene. With 5-methyl-2,3-dihydrofuran two isomers (**140a** and **140b**) were formed in a ratio of 5 to 1, whereas 3,4-dihydro-2*H*-pyran yielded only one isomer (**141**). The reactions occurred via an excited triplet state of 1,1-diphenylethylene or via an excited singlet state of *trans*-stilbene, as was concluded from sensitization and quenching experiments.[147] Servé et al.[148] have postulated that the formation of **141** is a nonconcerted process. The specificity of the reaction has been attributed to polarization of the carbon–carbon double bonds, which directs the initial attack. A number of substituted vinylene carbonates also yield (2 + 2)-cycloadducts (**143**) with olefins; subsequent hydrolysis offers a simple route to 1,2-*cis*-dihydroxycyclobutanes.[149,150] Dichlorovinylene carbonate has been found to enter into a (2 + 2)-cycloaddition reaction with benzene upon irradiation. The (2 + 2)-cycloadduct (**144**) is photochemically unstable and rearranges to the (4 + 2)-adduct (**145**).[13,151,152] Naphthalene reacted similarly with dichlorovinylene carbonate.[13] Barton et al.[153] prepared, via a (2 + 2)-

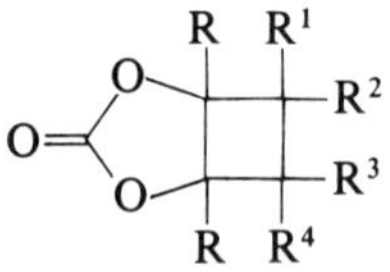

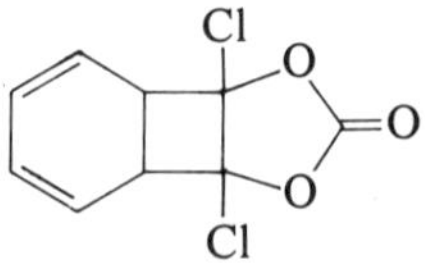

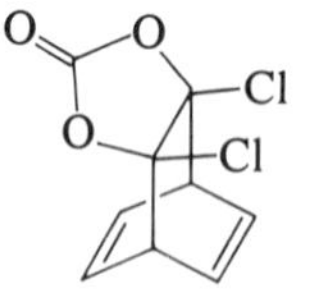

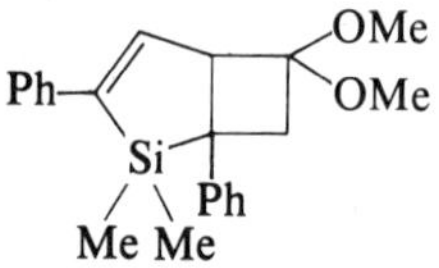

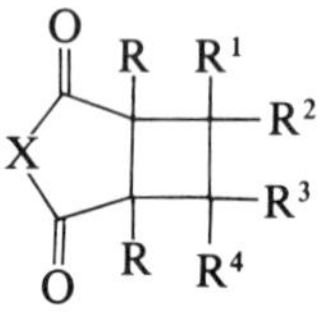

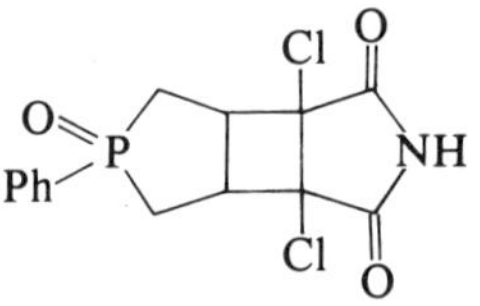

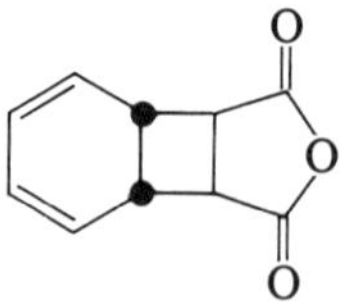

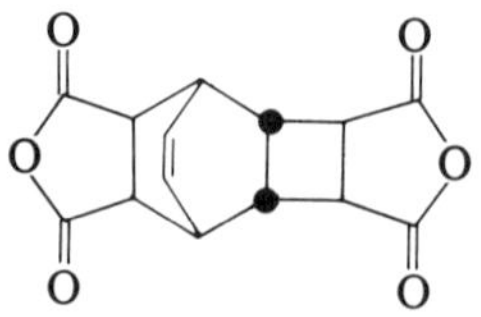

[149] W. Hartmann, *Chem. Ber.* **101**, 1643 (1968).
[150] H.-M. Fischler, H.-G. Heine, and W. Hartmann, *Tetrahedron Lett.*, 1701 (1972).
[151] H.-D. Scharf and R. Klar, *Tetrahedron Lett.*, 517 (1971).
[152] P. Lechtken and G. Hesse, *Ann. Chem.* **754**, 1 (1971).
[153] T. J. Barton, R. C. Kippenhan, and A. J. Nelson, *J. Am. Chem. Soc.* **96**, 2272 (1974).

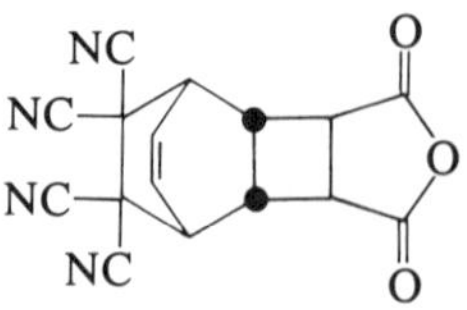

Sec. III.B] (2 + 2)-CYCLOADDITION AND CYCLOREVERSION

cycloaddition of 1,1-dimethyl-2,5-diphenyl-1-silacyclopenta-2,4-diene (**111**) with 1,1-dimethoxyethene, adduct **146** as starting material for the synthesis of a nonannulated silepin. They obtained the cycloadduct **146** in 70% yield.

Reactions of some heterocycles with cyclic enones have also been reported[154] as well as (2 + 2)-cycloaddition reactions of heterocycles having an enone moiety with olefins. In general it is assumed that the reactive species in these reactions are excited triplet states ($E_T \sim 70$ kcal/mol),[123] but in some instances excited singlet states have also been proposed.[155]

Maleic anhydrides and maleimides gave (2 + 2)-cycloadducts upon irradiation in the presence of a variety of compounds with an olefinic group such as (halogenated) olefins,[156,157] 2,5-dimethoxy-2,5-dihydrofuran,[158] vinylene carbonate,[130] 1,3-diphenyl- or 1,3-diacetyl-1,3-dihydroimidazol-2-one,[131,132] and 2,5-dihydrothiophene 1,1-dioxide.[157,159] The yields of the cycloadducts with the general structure **147** varied from 10 to 90%. Märkl and Schubert[160] applied this type of reaction to the synthesis of a phosphepin 1-oxide starting from the cycloadduct **148** (Section IV,B,2). Maleic anhydride also formed a (2 + 2)-cycloadduct with benzene, as was recently shown by Hartmann et al.[161] and Bryce-Smith et al.[162] In the original work the (2 + 2)-cycloadduct **149** was proposed as the primary photoproduct, which subsequently reacted to give the 2:1 adduct **150**.[163] Since all attempts to intercept **149** failed, the intermediacy of this (2 + 2)-cycloadduct was rejected in favor of a zwitterionic species.[164,165] Recently, however, **149** has been trapped with tetracyanoethylene[161,162] to give **151**, which definitely proves its structure. Other heterocyclic enones reacted with olefins, e.g., 4-butenolide with cycloalkenes,[166] 3,3-dimethyl-2,3-dihydropyran-2,4-dione with 1,4-dioxene,[167,168] and 2,2-dimethyl-3,4-

---

[154] T. S. Cantrell, *J. Org. Chem.* **39**, 3063 (1974).
[155] P. P. Wells and H. Morrison, *J. Am. Chem. Soc.* **97**, 154 (1975).
[156] R. Steinmetz, W. Hartmann, and G. O. Schenck, *Chem. Ber.* **98**, 3854 (1965).
[157] H.-D. Scharf and F. Korte, *Angew. Chem.* **77**, 452 (1965).
[158] J. C. Hinshaw, *J. Org. Chem.* **39**, 3951 (1974).
[159] V. Sh. Shaikhrazieva, R. S. Enikeev, and G. A. Tolstikov, *J. Org. Chem. USSR* **7**, 1831 (1971).
[160] G. Märkl and H. Schubert, *Tetrahedron Lett.*, 1273 (1970).
[161] W. Hartmann, H.-G. Heine, and L. Schrader, *Tetrahedron Lett.*, 3101 (1974).
[162] D. Bryce-Smith, R. R. Deshpande, and A. Gilbert, *Tetrahedron Lett., 1627* (1975).
[163] H. J. F. Angus and D. Bryce-Smith, *J. Chem. Soc.*, 4791 (1960).
[164] D. Bryce-Smith, *Pure Appl. Chem.* **16**, 47 (1968).
[165] D. Bryce-Smith, R. Deshpande, A. Gilbert, and J. Grzonka, *Chem. Commun.*, 561 (1970).
[166] M. Tada, T. Kokubo, and T. Sato, *Tetrahedron* **28**, 2121 (1972).
[167] P. Margaretha, *Tetrahedron* **27**, 6209 (1971).
[168] P. Margaretha, *Tetrahedron* **29**, 1317 (1973).

dihydro-2H-pyran-4-one with olefins.[169] It was shown that 2,2-dimethyl-3,4-dihydro-2H-pyran-4-one afforded mixtures of cis- and trans-annulated (2 + 2)-cycloadducts (**152a** and **b**).[169] Such (2 + 2)-cycloaddition reactions have been applied to the synthesis of grandisol, one of the four components of the boll weevil sex pheromone, starting from **153**, which was obtained by photochemical reaction of ethylene with 4-methyl-5,6-dihydro-2H-pyran-2-one in 56% yield.[170] 2,6-Diphenyl-4H-thiopyran-4-one 1,1-dioxide gave a (2 + 2)-cycloadduct with cyclohexene.[171] Both coumarins[137,155,172] and chromones[173,174] entered into (2 + 2)-cycloaddition reactions with olefins to give **154** and **156**, respectively. Hanifin and Cohen[174] found that the reaction with 1,1-dimethoxyethylene yielded only one (2 + 2)-cycloadduct (**156**, $R^1 = R^2 = H$, $R^3 = R^4 = OMe$). This specificity has been explained in terms of the formation of a stabilized 1,4-diradical intermediate by electrophilic attack of C_α of the n → π* chromone triplet on the most nucleophilic carbon atom of the olefin. There is no conclusive evidence to indicate that a π-complex between the reactants is involved.

Similar photoadditions of olefins to the nitrogen analogs of coumarins, carbostyril, and isocarbostyril are also known to give, e.g.,

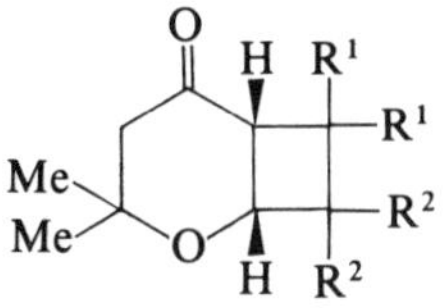

(**152a**) 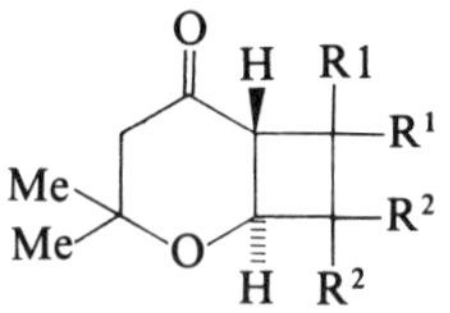   (**152b**) 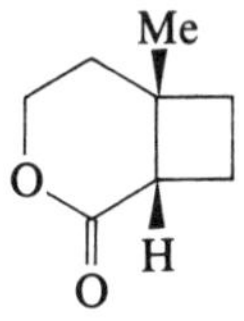   (**153**)

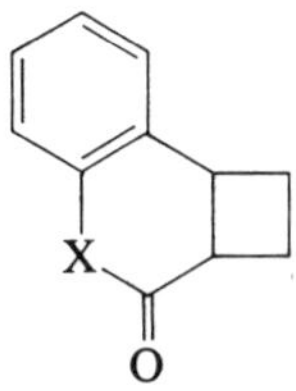

(**154**) X = O   (**156**) 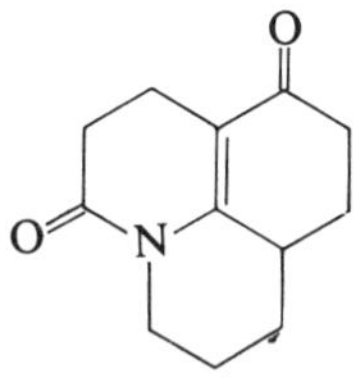   (**157**) 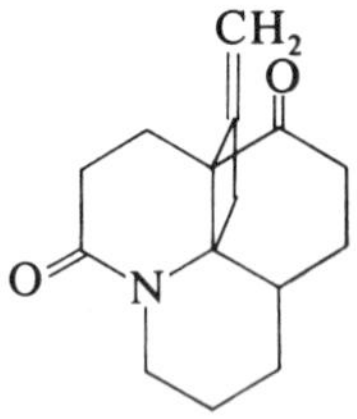   (**158**)
(**155**) X = NH

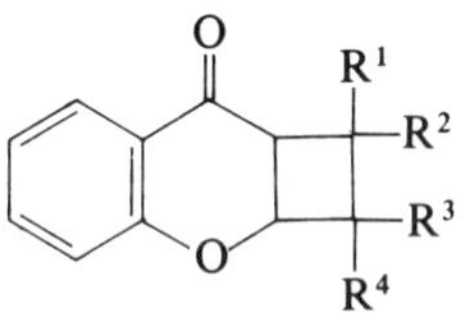

[169] P. Margaretha, *Ann. Chem.*, 727 (1973).
[170] R. C. Gueldner, A. C. Thompson, and P. A. Hedin, *J. Org. Chem.* **37**, 1854 (1972).
[171] N. Ishibe, K. Hashimoto, and M. Sunami, *J. Org. Chem.* **39**, 103 (1974).
[172] C. H. Krauch and W. Metzner, *Chem. Ber.* **99**, 88 (1966).
[173] J. W. Hanifin and E. Cohen, *Tetrahedron Lett.*, 5421 (1966).

155,[175-177] Nitrogen-containing heterocyclic enone systems also reacted with allene to give (2 + 2)-cycloadducts, as was shown in alkaloid synthesis.[178] As the keystep in the synthesis of an annotinine derivative, allene was added to **157** and the adduct **158** was obtained in quantitative yield. Various uracils have been modified by photochemical (2 + 2)-cycloaddition with olefins, e.g., vinylene carbonate,[179,180] vinyl ethers, vinyl acetates, and ketene acetals yielded **159**.[181] Very recently the photochemical addition of cyanoethylenes to 2-pyridones has been observed to yield mixtures of tetrahydroazocin-2-ones (**160**) and (2 + 2)-cycloadducts (**161**).[182]

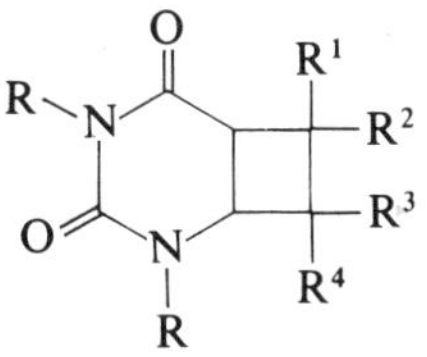

(159) 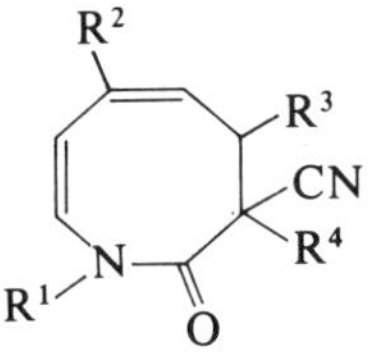   (160) 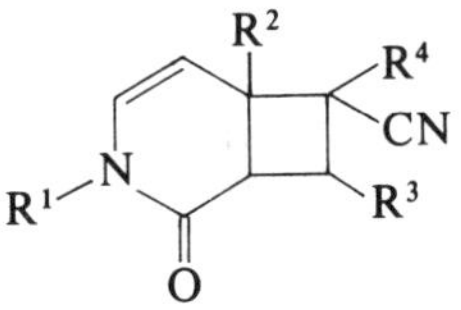   (161)

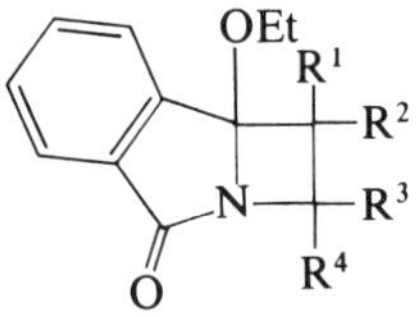

(162) 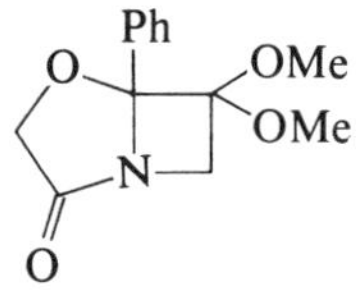   (163) 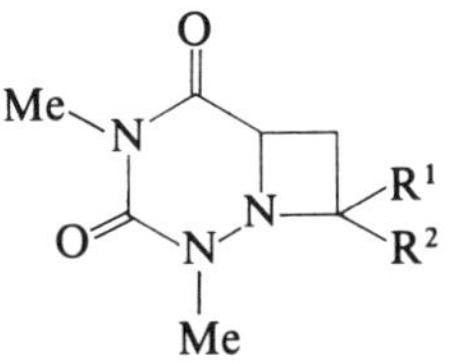   (164)

In all the reactions discussed so far in this section, the heterocyclic nucleus reacts via an endocyclic carbon–carbon double bond. However, recently Koch et al.[183,184] reported the (2 + 2)-cycloaddition via a carbon–nitrogen double bond. 3-Ethoxyisoindol-1-one yielded (2 + 2)-cycloadducts (**162**) with both electron-rich and electron-neutral olefins, but not with electron-deficient olefins.[183] 2-Phenyl-4,5-dihydrooxazol-4-

[174] J. W. Hanifin and E. Cohen, *J. Am. Chem. Soc.* **91**, 4494 (1969).
[175] G. R. Evanega and D. L. Fabiny, *Tetrahedron Lett.*, 2241 (1968).
[176] G. R. Evanega and D. L. Fabiny, *J. Org. Chem.* **35**, 1757 (1970).
[177] G. R. Evanega and D. L. Fabiny, *Tetrahedron Lett.*, 1749 (1971).
[178] K. Wiesner, I. Jirkovský, M. Fishman, and C. A. J. Williams, *Tetrahedron Lett.*, 1523 (1967).
[179] D. E. Bergstrom and W. C. Agosta, *Tetrahedron Lett.*, 1087 (1974).
[180] R. Beugelmans, J. L. Fourrey, S.-D. Gero, M.-T. LeGoff, D. Mercier, and V. Ratovelomanana, *C.R. Hebd. Seances Acad. Sci., Ser. C* **274**, 882 (1972).
[181] J. A. Hyatt and J. S. Swenton, *J. Am. Chem. Soc.* **94**, 7605 (1972).
[182] K. Somekawa, T. Shimou, K. Tanaka, and S. Kumamoto, *Chem. Lett.*, 45 (1975).
[183] T. H. Koch and K. H. Howard, *Tetrahedron Lett.*, 4035 (1972).

one reacted only with 1,1-dimethoxyethylene to give **163**.[184] Similarly, the carbon–nitrogen double bond in 6-azauracils entered into a (2 + 2)-cycloaddition reaction with vinyl ethers.[185,186] This reaction offers a simple route to 5-substituted 6-azauracils by hydrolysis of the carbon–nitrogen bond in the adduct **164**.

4. Reactions with Singlet Oxygen

Schaap[187] isolated stable bicyclic 1,2-dioxetanes (**165**) upon photochemical addition of singlet molecular oxygen[188] to 1,3-dioxole or 1,4-

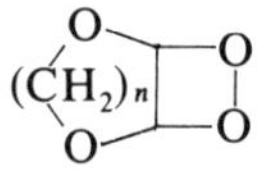

(**165**)
a: $n = 1$
b: $n = 2$

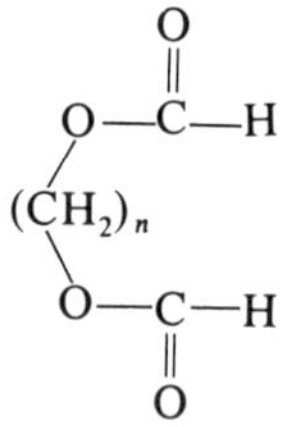

(**166**)
a: $n = 1$
b: $n = 2$

dioxene at $-78°$. The adducts rearranged thermally at $60°$ to methylene or ethylene diformate (**166**).

5. Reactions with Acetylenes

Annulation of heterocycles with a cyclobutene ring has been achieved by photochemical (2 + 2)-cycloaddition with acetylenes. Both maleic anhydride and N-substituted maleimides yielded 3-oxa- or 3-azabicyclo[3.2.0]-hept-6-ene-2,4-diones (**167**).[189,190] Vinylene carbonates also entered into a cycloaddition reaction with acetylenes to afford **168**, which has been employed as starting material for the synthesis of cyclobutadiene(tricarbonyl)iron or cyclobutenedione.[191–193] 3,4-Dihydro-2H-pyran and 5-methyl-2,3-dihydrofuran reacted with diphenylacetylene to

[184] T. H. Koch and R. M. Rodehorst, *Tetrahedron Lett.*, 4039 (1972).
[185] J. A. Hyatt and J. S. Swenton, *Chem. Commun.*, 1144 (1972).
[186] J. S. Swenton and R. J. Balchunis, *J. Heterocycl. Chem.* **11**, 917 (1974).
[187] A. P. Schaap, *Tetrahedron Lett.*, 1757 (1971).
[188] D. R. Kearns, *Chem. Rev.* **71**, 395 (1971).
[189] W. Hartmann, *Chem. Ber.* **102**, 3974 (1969).
[190] R. F. Childs and A. W. Johnson, *J. Chem. Soc. C*, 874 (1967).
[191] R. H. Grubbs, *J. Am. Chem. Soc.* **92**, 6693 (1970).
[192] J. Tancrede and M. Rosenblum, *Synthesis* **3**, 219 (1971).
[193] J. C. Hinshaw, *Chem. Commun.* 630 (1971).

give the (2 + 2)-cycloadducts **169** and **170**.[194,195] On the basis of sensitization and quenching experiments it was assumed that both reactions proceed via the first excited triplet state of diphenylacetylene. 3H-

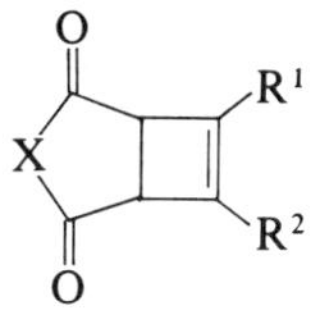

(**167**) X = O, NR

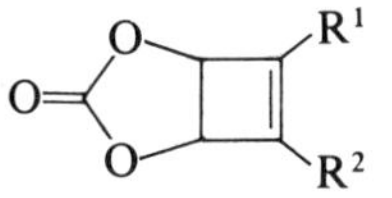

(**168**)

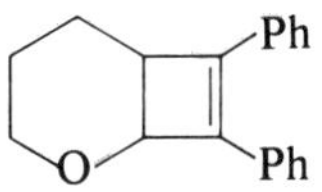

(**169**)

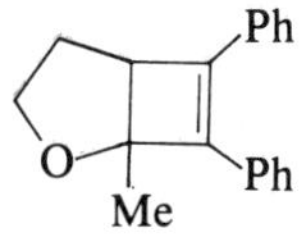

(**170**)

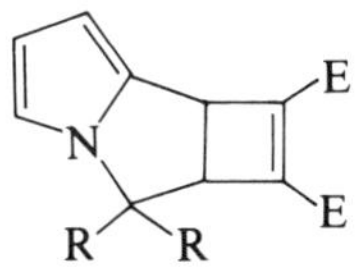

(**171**) E = COOMe

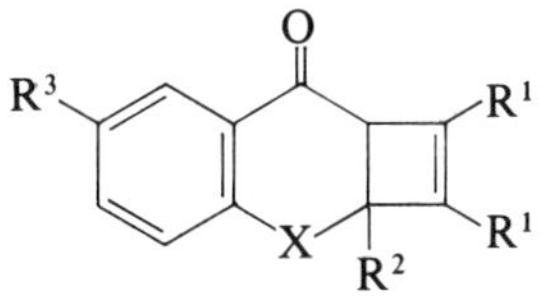

(**172**) X = O, S

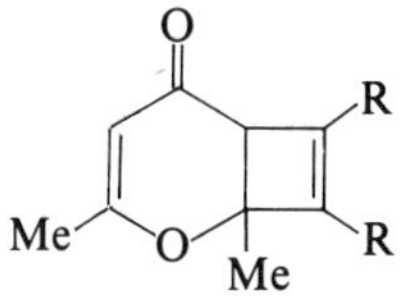

(**173**)

pyrrolizine and 3,3-dimethyl-3H-pyrrolizine afforded low yields of (2 + 2)-cycloadducts with dimethyl acetylenedicarboxylate (**171**).[196] Various heterocyclic enones, e.g., chromones, flavone, 1-thiaflavone, and 2,6-dimethyl-4-pyrone, were added to acetylenes to give cycloadducts of types **172–173**.[174,197,198] 2-Pyridones reacted similarly.[199] The yields of these reactions were generally high (60–100%).

C. Intermolecular Reactions of Heteroaromatics

1. Dimerization

The dimerization of benzo[b]furan was observed by Schenck and co-workers upon irradiation in the presence of sensitizers with high triplet energy (e.g., propiophenone or acetophenone[200,201]). Both the anti and the syn dimers (**174a** and **174b**) were formed, in a ratio of 3:1.[200] In the

---

[194] H. M. Rosenberg and P. Servé, *J. Org. Chem.* **33**, 1653 (1968).
[195] M. P. Servé and H. M. Rosenberg, *J. Org. Chem.* **35**, 1237 (1970).
[196] D. Johnson and G. Jones, *J. Chem. Soc., Perkin Trans 1*, 2517 (1972).
[197] A. Schönberg and G. D. Khandelwal, *Chem. Ber.* **103**, 2780 (1970).
[198] J. W. Hanifin and E. Cohen, *J. Org. Chem.* **36**, 910 (1971).
[199] A. I. Meyers and P. Singh, *Tetrahedron Lett.*, 4073 (1968).
[200] C. H. Krauch, W. Metzner, and G. O. Schenck, *Chem. Ber.* **99**, 1723 (1966).
[201] S. Farid, S. E. Hartman, and C. D. de Boer, *J. Am. Chem. Soc.* **97**, 808 (1975).

presence of lower triplet-energy carbonyls (e.g., benzaldehyde or benzophenone) oxetanes were obtained (Section III,C,2).

(174a) (174b)

2. Reactions with Carbonyl Compounds

Furan has been found to form oxetanes with a variety of carbonyl compounds, e.g., ketones,[202–205] aldehydes,[206] and ethyl cyanoformate.[207] In most reactions the (2 + 2)-cycloaddition occurred specifically to give a 2,7-dioxabicyclo[3.2.0]hept-3-ene (175) rather than the 2,6-isomer (176). Only the addition of ethyl cyanoformate yielded mixtures of 175 and 176 (R^1 = OEt and R^2 = CN), in a ratio of 2:1.[207] Two subsequent (2 + 2)-cycloadditions of benzophenone and furans have been reported to give two isomeric products, 177 and 178.[205] Substituted furans yielded similar oxetanes.[203] Benzo[b]furans, furocoumarins, and furochromones also proved to undergo (2 + 2)-cycloaddition reactions with carbonyl compounds such as ketones, aldehydes, and quinones. Invariably one type of oxetane was formed (179).[137,143,144,200–202,208,209] In the case of 2-methoxycarbonylbenzo[b]furan, evidence has been provided that the oxetane was produced by addition of the excited triplet state of the "olefinic" reactant to the ground state of the ketone.[208]

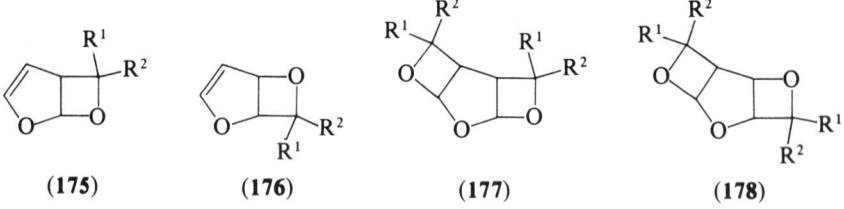

(175)   (176)   (177)   (178)

[202] G. O. Schenck, W. Hartmann, and R. Steinmetz, *Chem. Ber.* **96**, 498 (1963).
[203] C. Rivas and E. Payo, *J. Org. Chem.* **32**, 2918 (1967).
[204] E. B. Whipple and G. R. Evanega, *Tetrahedron* **24**, 1299 (1968).
[205] M. Ogata, H. Watanabe, and H. Kano, *Tetrahedron Lett.*, 533 (1967).
[206] K. Shima and H. Sakurai, *Bull. Chem. Soc. Jpn.* **39**, 1806 (1966).
[207] Y. Odaira, T. Shimodaira, and S. Tsutsumi, *Chem. Commun.*, 757 (1967).
[208] C. de Boer, *Tetrahedron Lett.* 4977 (1971).
[209] Y. Kawase, S. Yamaguchi, H. Ochiai, and H. Horita, *Bull. Chem. Soc. Jpn.* **47**, 2660 (1974).

(179)   (180)   (181)   (182)

Only one thiophene has been reported to undergo (2 + 2)-cycloaddition with carbonyl compounds, *viz.*, 2,5-dimethylthiophene.[210] In this reaction, too, only one type of isomer (**180**) was obtained, in 50–60% yield. *N*-Benzoyl-1*H*-pyrrole reacted with two molecules of ketone to give cycloadduct **181**.[211] Similarly, *N*-acyl-1*H*-indoles reacted with certain ketones to produce adducts **182**.[212] The lack of reactivity of other heterocycles, such as 1*H*-pyrroles, oxazoles, and isoxazoles, has been attributed to a quenching effect on the excited ketone of the nonbonded electrons on the heteroatom.[212]

3. Reactions with Olefins

2-Methyl furoate was found to react with 2,3-dimethylbutene-2 upon irradiation to give **183**.[213] Furan reacted with 2,5-diphenyl-1,3,4-oxadiazole upon irradiation, with or without a sensitizer, to yield cycloadduct **184**.[214]

Various photochemical (2 + 2)-cycloadditions of heteroaromatic compounds have been reported in which an enone moiety is incorporated either into the olefinic reagent or into the heteroaromatic compound. Both furan and thiophene have been found to give cycloaddition reactions with maleic anhydride derivatives in the presence of a sensitizer.[202,215] The cycloadducts (**185** and **186**) were formed in high yield, but in the case of 2,5-dimethylthiophene, cyclobutane formation was the minor pathway, as oxetane formation predominated.[210] Cyclic enones, such as 2-cyclopenten-1-one and 2-cyclohexen-1-one reacted with furan to afford mixtures of (2 + 2)-cycloadducts (**187a**, R = H) and (**188**),

[210] C. Rivas and R. A. Bolivar, *J. Heterocycl. Chem.* **10**, 967 (1973); C. Rivas, M. Velez, and O. Cresente, *Chem. Commun.*, 1474 (1970).
[211] C. Rivas, M. Velez, M. Cucarella, R. A. Bolivar, and S. E. Flores, *Acta Cient. Venez.* **22**, 145 (1971).
[212] D. R. Julian and G. D. Tringham, *Chem. Commun.*, 13 (1973).
[213] T. S. Cantrell, *J. Org. Chem.* **40**, 1447 (1975).
[214] O. Tsuge, K. Oe, and M. Tashiro, *Tetrahedron* **29**, 41 (1973).
[215] G. O. Schenck, W. Hartmann, S.-P. Mannsfeld, W. Metzner, and C. H. Krauch, *Chem. Ber.* **95**, 1642 (1962).

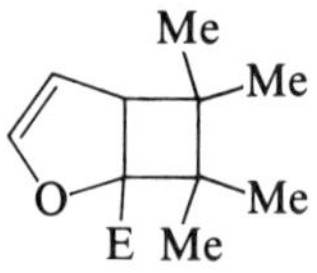

(183) E = COOMe

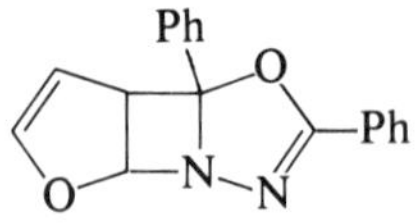

(184)

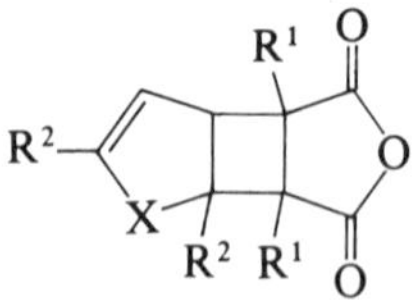

(185) X = O
(186) X = S

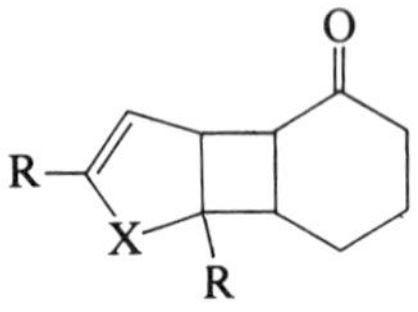

(187)
a: X = O
b: X = S
c: X = NMe

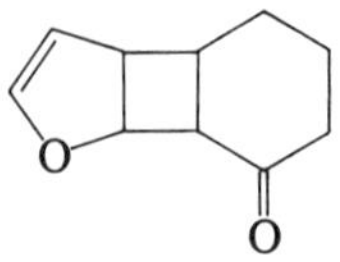

(188)

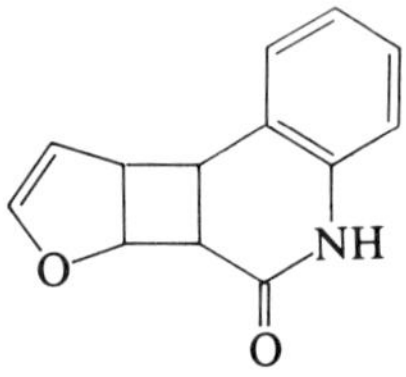

(189)

together with some (4 + 2)-cycloadduct.[154] Both thiophene and 2,5-dimethylthiophene yielded (2 + 2)-cycloadducts (**187b**), but N-methyl-1H-pyrrole gave a mixture of unstable adducts.[154] Upon hydrolysis, signals characteristic of aldehydes were found in the IR and NMR spectra. This might indicate the presence of an enamine in the crude reaction mixture, as is expected in the (2 + 2)-cycloadduct **187c**. Irradiation of furan in the presence of carbonstyril, another cyclic enone system, yielded a single isomer (**189**).[154] Cantrell[216,217] reported the (2 + 2)-cycloaddition of furans and thiophenes carrying an acetyl or benzoyl substituent in the 2-position with some olefins, viz., tetramethylethylene and isobutene. Apart from (2 + 2)-cycloadducts (**190**), they obtained oxetanes and (4 + 2)-cycloadducts.

Benzo[b]furans yielded (2 + 2)-cycloadducts (**191**) with dimethylmaleic anhydride, via excitation of an initially formed charge-transfer complex, together with oxetans.[218] Similar (2 + 2)-cycloadducts (**192**) were obtained from quinones and benzo[b]furan and some of its derivatives.[143,200] Irradiation of N-acyl-1H-indoles in the presence of olefins, either with or without acetophenone as a sensitizer, produced a series of (2 + 2)-cycloadducts **193**, usually as a mixture of exo- and endoisomers.[219] The olefins that give a (2 + 2)-cycloadduct are either electron-rich olefins, such as ethyl vinyl ether and vinyl acetate, or

[216] T. S. Cantrell, *Chem. Commun.*, 155 (1972).
[217] T. S. Cantrell, *J. Org. Chem.* **39**, 2242 (1974).
[218] S. Farid and S. E. Shealer, *Chem. Commun.*, 296 (1973).
[219] D. R. Julian and R. Foster, *Chem. Commun.*, 311 (1973).

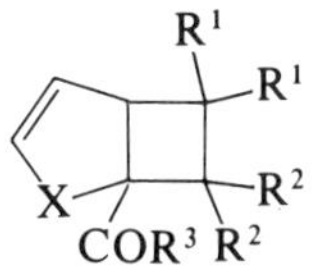

Sec. III.C] (2 + 2)-CYCLOADDITION AND CYCLOREVERSION

(190)
a: X = O
b: X = S
c: X = NMe

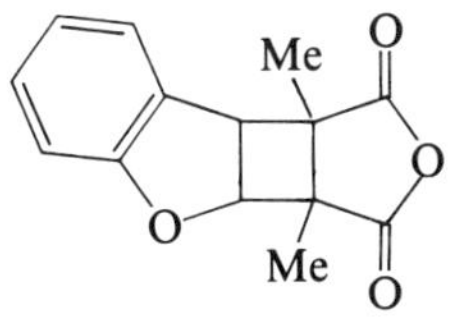

(191)

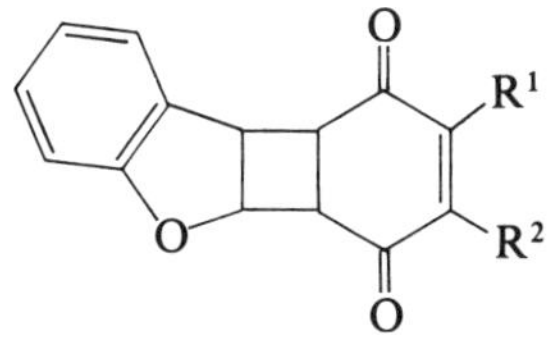

(192)

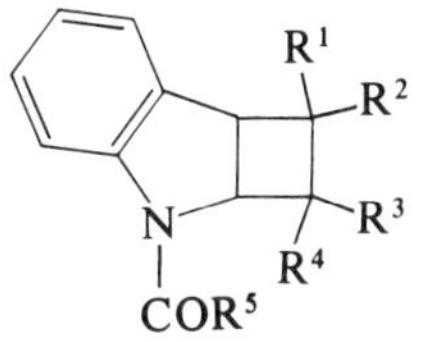

(193)

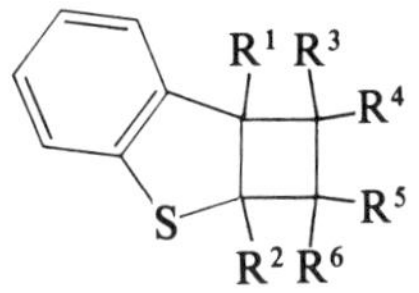

(194)

electron-deficient olefins, such as methyl acrylate and acrylonitrile. Competition experiments have shown that the cycloaddition is more efficient with electron-deficient than with electron-rich olefins. Evidence that a triplet excited 1H-indole is involved was obtained from quenching experiments. Benzo[b]thiophene has been reported to react with haloolefins upon irradiation, either direct or sensitized, to give (2 + 2)-cycloadducts (194).[220,221] The yields of these reactions were as high as 90%. These photoadditions have been proposed to occur via a benzo[b]thiophene triplet state. The observed selectivity has been attributed to equilibration of the diradical intermediates.

4. Reactions with Acetylenes

The results of the photochemical reactions of dimethyl acetylenedicarboxylate with furan,[222] thiophene,[78] and 1 H-pyrrole[17] were found to differ widely. Furan yielded either a 1:1 or a 2:1 adduct, and thiophene yielded a sulfur-free reaction product, viz., dimethyl phthalate. These products were formed by (4 + 2)-cycloadditions, which—in the case of thiophene—was followed by extrusion of sulfur from the initial Diels–Alder adduct. 1H-pyrrole, however, reacted in a different way because a 1H-azepine (195) was isolated. This product was most probably generated by (2 + 2)-cycloaddition of the acetylene to the

[220] D. C. Neckers, J. H. Dopper, and H. Wynberg, J. Org. Chem. 35, 1582 (1970).
[221] I. Murata, T. Tatsuoka, and Y. Sugihara, Tetrahedron Lett., 199 (1974).
[222] R. P. Gandhi and V. K. Chadha, Chem. Commun., 552 (1968).

C(2)–C(3) double bond of the pyrrole nucleus with subsequent (2 + 2)-cycloreversion of the resultant adduct (**196**).

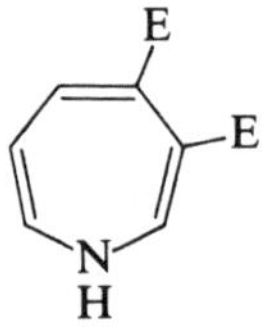

(**195**) 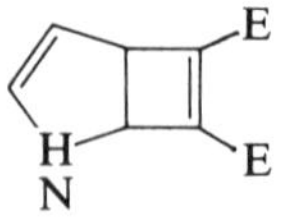   (**196**) 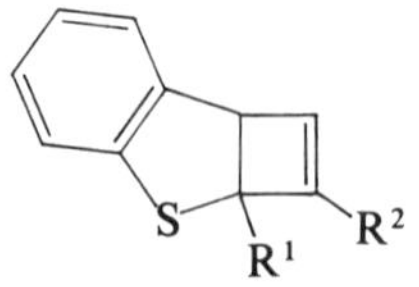   (**197**) 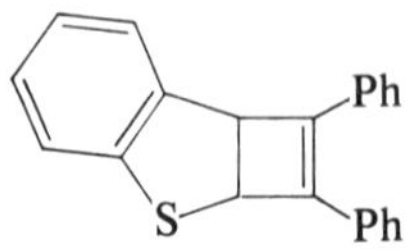   (**198**)

E = COOMe

Benzo[b]thiophenes readily formed (2 + 2)-cycloadducts with dimethyl acetylenedicarboxylate, methyl propiolate, and methyl phenylpropiolate, either upon direct or upon photosensitized irradiation. Quite unexpectedly, the isolated (2 + 2)-cycloadducts (**197**) were not the ones originating from a simple (2 + 2)-cycloaddition.[223,224] Their formation has been discussed in terms of a rearrangement of the original (2 + 2)-cycloadduct, e.g., **198**, but other possible routes have not been ruled out. Diphenylacetylene gave the unrearranged (2 + 2)-cycloadduct **198**, although this compound accounted for only 10% of the total reaction products. The major product also had structure **197** ($R^1 = R^2 = Ph$).[225] The additions of methyl phenylpropiolate and methyl propiolate were highly stereoselective. This has been interpreted as the result of a highly polarized excited state of benzo[b]thiophene.[223]

D. Intramolecular Reactions

1. Nonconjugated Dienes

Upon irradiation with UV light ($\lambda > 200$ nm) the two carbon–carbon double bonds of pentakis(pentafluoroethyl)-1-azabicyclo[2.2.0]hexa-2,5-diene (**199**; Section III,D,2) entered into an intramolecular (2 + 2)-cycloaddition reaction to give the corresponding 1-azaprismane (**200**).[226] This highly strained molecule was thermally stable up to 160°, at which temperature it decomposed into a mixture of **199** and pentakis(pentafluoroethyl)pyridine. In terms of conservation of orbital symmetry this high stability can be well understood. Just as in prismane

[223] J. H. Dopper and D. C. Neckers, *J. Org. Chem.* **36**, 3755 (1971).
[224] D. C. Neckers, J. H. Dopper, and H. Wynberg, *Tetrahedron Lett.*, 2913 (1969).
[225] W. H. F. Sasse, P. J. Collin, and D. B. Roberts, *Tetrahedron Lett.*, 4791 (1969).
[226] M. G. Barlow, J. G. Dingwall, and R. N. Haszeldine, *Chem. Commun.*, 1580 (1970).

isomerization of **200** to the thermodynamically most stable isomer, the pyridine, via the symmetry-allowed [$_\pi 2_s + {}_\pi 2_a + {}_\pi 2_a$] cycloreversion would afford a six-membered ring in which the π-components form an antibonding orbital.[227]

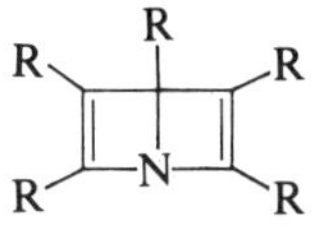

(199) 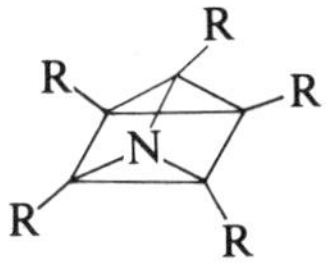 (200)

R = C$_2$F$_5$

(201) 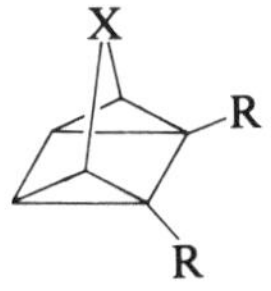 (202) 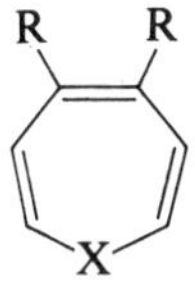 (203)

X = O, NR'

2,3-Bis(methoxycarbonyl)-7-oxa- and 7-azabicyclo[2.2.1]hepta-2,5-diene (**201**; R = COOMe) isomerized upon direct or sensitized irradiation to give the corresponding 7-oxa- or 7-azaquadricyclenes (**202**; R = COOMe) in yields of 60–90%.[14,228–234] These intramolecular photochemical (2 + 2)-cycloadditions have up to now been regarded as some of the few examples of *concerted* (2 + 2)-cycloadditions.[235] However, recently Kaupp has provided evidence for the occurrence of a two-step process.[236] The 7-oxa- or 7-azaquadricyclenes (**202**) were thermally far less stable than the 1-azaprismane (**200**) and the quadricyclene. They isomerized at 100° (X = O, R = COOMe)[228,231] or even at room

[227] Woodward and Hoffman,[22] p. 107.
[228] H. Prinzbach, M. Arguëlles, and E. Druckrey, *Angew. Chem.* **78**, 1057 (1966).
[229] H. Prinzbach and P. Vogel, *Helv. Chim. Acta* **52**, 396 (1969).
[230] E. Payo, L. Cortés, J. Mantecón, C. Rivas, and G. de Pinto, *Tetrahedron Lett.*, 2415 (1967).
[231] W. Eberbach, M. Perroud-Arguëlles, H. Achenbach, E. Druckrey, and H. Prinzbach, *Helv. Chim. Acta* **54**, 2579 (1971).
[232] H. Prinzbach, R. Fuchs, and R. Kitzing, *Angew. Chem.* **80**, 78 (1968).
[233] H. Prinzbach, G. Kaupp, R. Fuchs, M. Joyeux, R. Kitzing, and J. Markert, *Chem. Ber.* **106**, 3824 (1973).
[234] H. Günther, J. B. Pawliczek, B. D. Tunggal, H. Prinzbach, and R. H. Levin, *Chem. Ber.* **106**, 984 (1973).
[235] Gilchrist and Storr,[32] p. 184.
[236] G. Kaupp, *Ann. Chem.*, 844 (1973).

temperature to give the corresponding oxepin or azepine derivatives (**203**).[232,233] This rapid reaction was attributed to the fact that the concerted $[_\pi 2_s + _\pi 2_a + _\pi 2_a]$ cycloreversion, which gives rise to an antibonding orbital level in the quadricyclene–cycloheptatriene conversion, now results in the highest occupied ground-state level of the oxepin or azepine, which might be weakly bonding.[237] Recently, however, this reaction type has also been discussed in terms of a multistep process.[238] Nevertheless, regardless of the mechanism, these intramolecular photochemical (2 + 2)-cycloadditions of 7-heteroquadricyclenes have opened up a simple route for the synthesis of both oxepin and azepine derivatives (**203**).

Irradiation of annulated 7-oxabicyclo[2.2.1]hepta-2,5-dienes (**204**) yielded the corresponding intramolecular (2 + 2)-cycloadducts (**205**). These products were thermally rather unstable, and even at room temperature they rapidly isomerized to the corresponding oxepin derivatives.[239] Annulation to a benzene ring lowered the stability of the 7-oxa- or 7-azaquadricyclene to such an extent that irradiation of compounds **206** resulted in the formation of a benzo[*d*]oxepin or -azepine (**207**).[239–243] Attempts to detect the corresponding 7-oxa- or azaquadricyclenes at −40° to −50° or to intercept these by the addition of dimethyl acetylenedicarboxylate failed.[239,241,243]

(**204**)   (**205**)   (**206**)   (**207**)
                          a: X = O   a: X = O
                          b: X = NR   b: X = NR

Various cage compounds have been obtained by photochemical intramolecular (2 + 2)-cycloadditions of heterocyclic compounds. The 2,7-dioxatricyclo[6.2.0.0³,⁶]deca-4,9-diene derivative **208** (R = Me) underwent intramolecular (2 + 2)-cycloaddition upon irradiation.[244] The di-

[237] Woodward and Hoffman,[22] p. 111.
[238] E. Haselbach and H.-D. Martin, *Helv. Chim. Acta* **57**, 472 (1974).
[239] H. Prinzbach, P. Würsch, P. Vogel, W. Tochtermann, and C. Franke, *Helv. Chim. Acta* **51**, 911 (1968).
[240] G. R. Ziegler and G. S. Hammond, *J. Am. Chem. Soc.* **90**, 513 (1968).
[241] G. R. Ziegler, *J. Am. Chem. Soc.* **91**, 446 (1969).
[242] P. D. Rosso, J. Oberdier, and J. S. Swenton, *Tetrahedron Lett.*, 3947 (1971).
[243] G. Kaupp, J. Perreten, R. Leute, and H. Prinzbach, *Chem. Ber.* **103**, 2288 (1970).
[244] R. Criegee and R. Rucktäschel, *Chem. Ber.* **103**, 50 (1970).

oxabishomocubane (**209**) was formed in 74% yield only from the syn isomer. A similar oxahomocubane (**211**) was obtained by irradiation of **210**.[245] Analogous reactions have been reported for the 7,8-diazatricyclo[4.2.2.0^{2,5}]deca-3,9-dienes **212**, which yielded the corresponding diazabasketanes (**213**).[246,247]

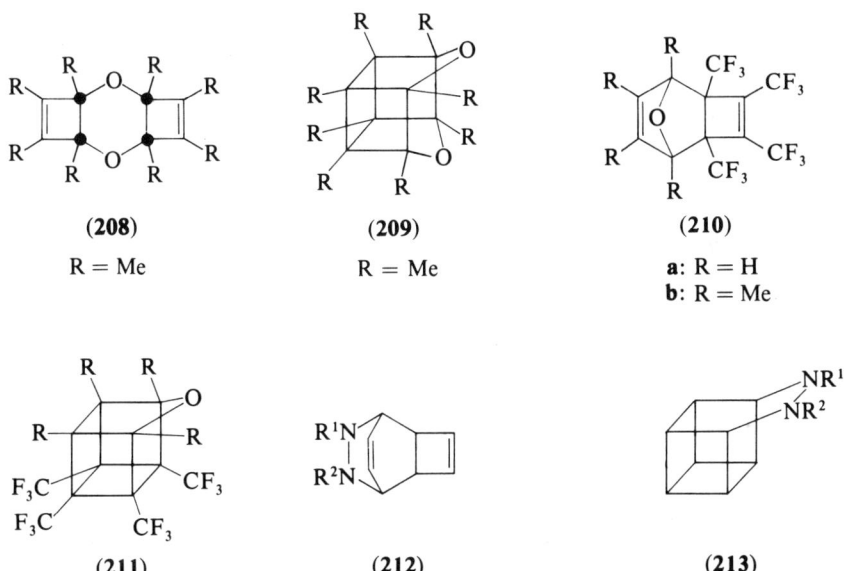

(**208**)
R = Me

(**209**)
R = Me

(**210**)
a: R = H
b: R = Me

(**211**)

(**212**)

(**213**)

2. Conjugated Dienes

a. Five-Membered Rings. Irradiation of tetrakis(trifluoromethyl)-thiophene has recently been shown to result in an intramolecular (2 + 2)-cycloaddition.[248] The structure of the reaction product, 1,2,3,4-tetrakis-(trifluoromethyl)-5-thiabicyclo[2.1.0]pent-2-ene (**214**), was confirmed by the fact that reaction with furans gave the Diels–Alder adduct (**215**).[245] The isolation of **214** as the product of irradiation of a thiophene supports earlier suggestions that photochemical isomerizations of thiophenes and other heteroaromatics might proceed, at least partly, via bicyclic intermediates of type **214**.[249–251] A similar bicyclic furan has

---

[245] Y. Kobayashi, J. Kumadaki, A. Ohsawa, and Y. Sekine, *Tetrahedron Lett.*, 2841 (1974).
[246] R. Askani, *Chem. Ber.* **102**, 3304 (1969).
[247] L. A. Paquette, *J. Am. Chem. Soc.* **92**, 5765 (1970).
[248] H. A. Wiebe, S. Braslavsky, and J. Heicklen, *Can. J. Chem.* **50**, 2721 (1972).
[249] R. M. Kellogg, *Tetrahedron Lett.* 1429 (1972).
[250] E. E. van Tamelen and T. H. Whitesides, *J. Am. Chem. Soc.* **93**, 6129 (1971).
[251] P. Beak and W. Messer, *Tetrahedron* **25**, 3287 (1969).

been proposed as an intermediate in various photochemical reactions of furan.[214] Such intramolecular (2 + 2)-cycloaddition reactions have also been observed with nonaromatic heterocycles. Low-temperature (−78°) NMR spectroscopic studies of the photochemical isomerization of 4,4-diethyl-3,5-dimethyl-4H-pyrazole 1-oxide to 5,5-diethyl-3,4-dimethyl-5H-pyrazole 1-oxide revealed the presence of a bicyclic intermediate (**216**), which underwent a thermal rearrangement at about −20° to give either the starting material or the final product.[252,253] The same type of intermediate (**217**) has been detected at −50° in the photochemical isomerizations of fully alkylated 3H-pyrazoles. It is only in these reactions that the formation of the 1,2-diazabicyclo[2.1.0]pent-2-enes can compete with cyclopropene formation by nitrogen elimination.[254]

(**214**)　　(**215**)　　(**216**)　　(**217**)

b. Six-Membered Rings. Pyridines have been reported to undergo intramolecular (2 + 2)-cycloaddition upon irradiation in two different ways. The parent pyridine was found to be converted into 2-azabicyclo[2.2.0]hexa-2,5-diene (**218**).[255] This compound was thermally unstable, and the reverse reaction giving pyridine proceeded with a half-life of 2.5 minutes at 25°. Irradiation of pyridine in the presence of aqueous sodium borohydride yielded the thermally stable 2-azabicyclo[2.2.0]-hex-5-ene (**219**), which was shown to be formed by reduction of **218**. Picolines and lutidines reacted similarly, but owing to the thermal instability of the (2 + 2)-cycloadducts they could be identified only as the partially reduced bicyclic isomers of type **219**. Irradiation ($\lambda >$ 270 nm) of pentakis(pentafluoroethyl)pyridine afforded pentakis(pentafluoroethyl)-1-azabicyclo[2.2.0]hexa-2,5-diene in quantitative yield (**199**, Section III,D,1).[226] This pyridine isomer was further converted into a 1-azaprismane (**200**) upon irradiation with light of shorter wavelength ($\lambda >$ 200 nm).

[252] W. R. Dolbier and W. M. Williams, *Chem. Commun.*, 289 (1970).
[253] W. M. Williams and W. R. Dolbier, *J. Am. Chem. Soc.* **94**, 3955 (1972).
[254] G. L. Closs, W. A. Böll, H. Heyn, and V. Dev, *J. Am. Chem. Soc.* **90**, 173 (1968).
[255] K. E. Wilzbach and D. J. Rausch, *J. Am. Chem. Soc.* **92**, 2178 (1970).

Sec. III.D] (2 + 2)-CYCLOADDITION AND CYCLOREVERSION

Similar intramolecular (2 + 2)-cycloaddition reactions have been reported for 1,2-dihydropyridines[256] and 1,2-dihydropyridazines.[257] The corresponding 2-aza- and 2,3-diazabicyclo[2.2.0]hex-5-enes (**220** and **221**) were isolated in good yields.

(**218**)

(**219**)

(**199**)
R = C_2F_5

(**220**) X = CH_2
(**221**) X = NCOOMe

(**222**) X = O
(**223**) X = NR

(**224**) X = O
(**225**) X = NR

(**226**)

(**227**)

Of special interest are the photochemical intramolecular (2 + 2)-cycloadditions of 2-pyrones and 2-pyridones because the resulting 2-oxa- or 2-azabicyclo[2.2.0]hex-5-en-3-ones are precursors of cyclobutadiene and stable complexes of cyclobutadiene.[12,258–267] In 1964 Corey and Streith[258] reported the photochemical conversion of 2-pyrone (**222**) and N-methyl-2-pyridone (**223**) into the corresponding bicyclic isomers (**224** and **225**) at −20°. More recent studies performed at a temperature as low as 8°K with 2-pyrones in a matrix of argon or nitrogen showed that the intramolecular (2 + 2)-cycloaddition is only a minor photochemical pathway.[12,260–262] The major reaction proved to be

[256] F. W. Fowler, *J. Org. Chem.* **37**, 1321 (1972).
[257] L. J. Altman, M. F. Semmelhack, R. B. Hornby, and J. C. Vederas, *Chem. Commun.*, 686 (1968).
[258] E. J. Corey and J. Streith, *J. Am. Chem. Soc.* **86**, 950 (1964).
[259] W. H. Pirkle and L. H. McKendry, *J. Am. Chem. Soc.* **91**, 1179 (1969).
[260] O. L. Chapman, C. L. McIntosh, and J. Pacansky, *J. Am. Chem. Soc.* **95**, 244 (1973).
[261] C. L. McIntosh and O. L. Chapman, *J. Am. Chem. Soc.* **95**, 247 (1973).
[262] R. G. S. Pong and J. S. Shirk, *J. Am. Chem. Soc.* **95**, 248 (1973).
[263] M. Rosenblum and C. Gatsonis, *J. Am. Chem. Soc.* **89**, 5074 (1967).
[264] J. Agar, F. Kaplan, and B. W. Roberts, *J. Org. Chem.* **39**, 3451 (1974).
[265] H. Furrer, *Chem. Ber.* **105**, 2780 (1972).
[266] R. C. de Selms and W. R. Schleigh, *Tetrahedron Lett.*, 3563 (1972).
[267] E. Ager, G. E. Chivers, and H. Suschitzky, *J. Chem. Soc., Perkin Trans. 1*, 1125 (1973).

the formation of an aldoketene (**226**), which, however, rapidly reverted to 2-pyrone at about 20°. Consequently, at room temperature only the intramolecular (2 + 2)-cycloaddition reaction is actually observed. The ketene is most probably an intermediate in various photochemical reactions of 2-pyrones.[259] Irradiation of tetramethylpyrazinone yielded an unstable bicyclic isomer (**227**), which upon *in situ* hydrogenation could be trapped in the form of the corresponding 2,5-diazabicyclo[2.2.0]-hexan-3-one.[265]

c. *Seven-Membered Rings.* The photochemical intramolecular (2 + 2)-cycloaddition of heterocycloheptadienes or -trienes is a quite general reaction and provides a very useful method for the preparation of the otherwise difficultly accessible heterobicyclo[3.2.0]heptenes or heterobicyclo[3.2.1]heptadienes.

The reaction of oxepin (**228**; R = H) was complicated by the simultaneous photochemical reaction of benzene oxide, the valence isomer of oxepin.[268] The results varied with solvent, temperature, and wavelength.[269] The reaction proceeded with high selectivity to 2-oxabicyclo[3.2.0]hepta-3,6-diene (**229**; R = H) upon irradiation ($\lambda > 310$ nm) at room temperature. In most other cases the reaction was attended with the formation of phenol, probably from benzene oxide via Dewar benzene oxide, as this compound is known to isomerize photo-

(**228**) (**229**) (**230**) (**231**) (**232**)

a: R = H   a: R = H   a: R = H
b: R = Me b: R = Me b: R = Cl

chemically to phenol.[270] Irradiation of 2,7-dimethyloxepin (**228**; R = Me) yielded the rather unstable 1,3-dimethyl-2-oxabicyclo[3.2.0]-hepta-3,6-diene (**229**; R = Me) together with some unidentified reaction products.[271] Similar intramolecular (2 + 2)-cycloaddition reactions have been observed for 2,3-dihydrooxepins,[272] 2,7-dihydrooxepin-2,7-diones,[273] and benzo[*b*]oxepins,[67,274] giving **230**–**232** in good yields.

[268] E. Vogel and H. Günther, *Angew. Chem.* **79**, 429 (1967).
[269] J. M. Holovka and P. D. Gardner, *J. Am. Chem. Soc.* **89**, 6390 (1967).
[270] E. E. van Tamelen and D. Carty, *J. Am. Chem. Soc.* **93**, 6102 (1971).
[271] L. A. Paquette and J. H. Barrett, *J. Am. Chem. Soc.* **88**, 1718 (1966).
[272] L. A. Paquette, J. H. Barrett, R. P. Spitz, and R. Pitcher, *J. Am. Chem. Soc.* **87**, 3417 (1965).
[273] G. J. Fonken, *Chem. Ind. (London),* 1575 (1961),
[274] H. Hofmann and P. Hofmann, *Tetrahedron Lett.,* 4055 (1971).

Sec. III.D] (2 + 2)-CYCLOADDITION AND CYCLOREVERSION

Benzo[b]thiepins (**233**) have been found to undergo the same type of intramolecular (2 + 2)-cycloaddition upon irradiation, yielding 3,4-benzo-2-thiabicyclo[3.2.0]hepta-3,6-dienes (**234**).[70,275] In one particular case, viz., with compound **233** (R^1 = H, R^2 = R^3 = COOMe, R^4 = OH), this cycloaddition reaction was followed by a photorearrangement of **234** (R^1 = H, R^2 = R^3 = COOMe, R^4 = OH) into **235**.[70] This observation supports earlier suggestions that the ultimate products of photochemical reactions of benzo[b]thiophene and acetylenes are formed via (2 + 2)-cycloaddition and subsequent rearrangement (Section III,C,4).

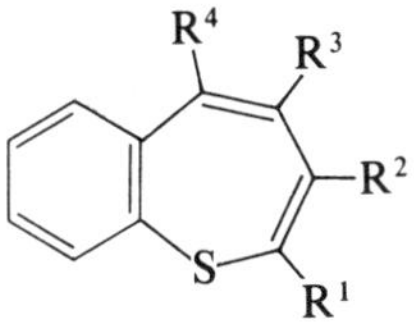

(**233**) 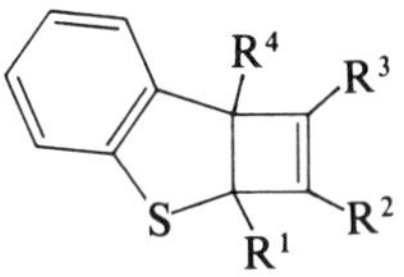   (**234**) 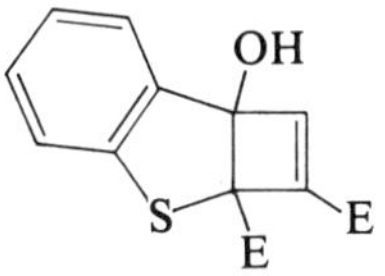   (**235**)

E = COOMe

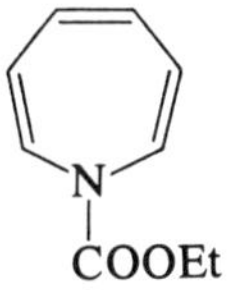

(**236**) COOEt   (**237**) COOMe   (**238**) 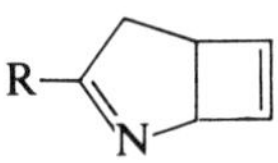   (**239**)

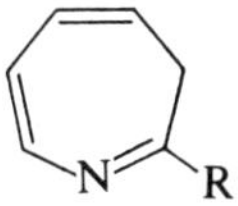

a: R = NMe_2
b: R = NH_2
c: R = OEt

A large number of azepines have been reported to undergo intramolecular (2 + 2)-cycloaddition upon irradiation. N-Ethoxycarbonyl-1H-azepine (**236**) isomerized to give 2-ethoxycarbonyl-2-azabicyclo[3.2.0]hepta-3,6-diene (**237**) in 43% yield.[271] The corresponding 2-, 3-, and 4-methyl-1-methoxycarbonyl-1H-azepines afforded a mixture of the two possible photoisomers. The ratios of these isomers depended on the position of the methyl substituent.[276] The observed selectivity has been related to the steric requirements of the methyl groups in the bicyclic isomers. The intramolecular (2 + 2)-cycloaddition reaction in 2-substituted-3H-azepines (**238**) proceeded with high selectivity to the corresponding 2-azabicyclo-[3.2.0]hepta-2,6-dienes (**239**). The other possible isomers, i.e., the 6-azabicyclo-[3.2.0]hepta-2,6-dienes, were not formed at all.[277] It has been suggested

[275] H. Hofmann and B. Meyer, *Tetrahedron Lett.*, 4597 (1972).
[276] L. A. Paquette and D. E. Kuhla, *J. Org. Chem.* **34**, 2885 (1969).
[277] R. A. Odum and B. Schmall, *Chem. Commun.*, 1299 (1969).

that this selectivity is due to the fact that the formation of the 6-aza isomers is hampered by the loss of resonance energy in the amidine or imidate moieties. On the other hand, it seems to be a general phenomenon that upon irradiation annulated cyclobutene rings are formed more easily than annulated azetine rings. The formation of such an azetine, 5-dimethylamino-6-azabicyclo[3.2.0]hept-6-ene (**240**), has been proposed as an intermediate step in the photolysis of 2-dimethylamino-4,5-dihydro-3*H*-azepine (Section IV,B,1).[278] Experimental evidence for the occurrence of such an annulated 3,4-dihydroazete

(**240**)

(**241**)

(**242**)

(**243**)

(**244**)

(**245**)

(**246**)

(**247**)

(**248**)

(**249**)

(**250**)

was provided by Koch and Brown.[279] Upon irradiation of 4,5-dihydro-2-ethoxy-4,4,6-trimethyl-3*H*-azepine (**241**) in the presence of sodium methoxide and benzophenone as a photosensitizer, they isolated 5-ethoxy-7-methoxy-1,3,3-trimethyl-6-azabicyclo[3.2.0]heptane (**243**). This compound was most probably formed by solvent addition to **242**. Irradiation of 2,3-dihydro-1*H*-azepines (**244**) yielded stable isomers (**245**),[280] and similar intramolecular (2 + 2)-cycloaddition

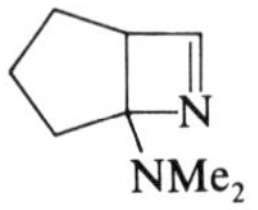

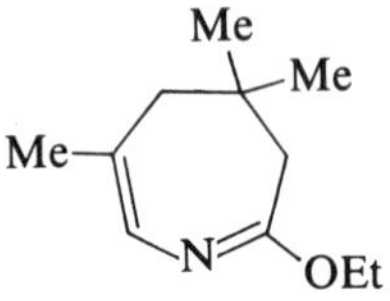

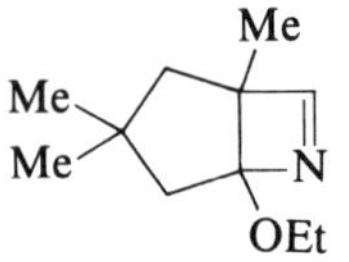

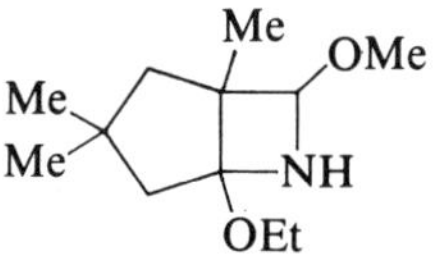

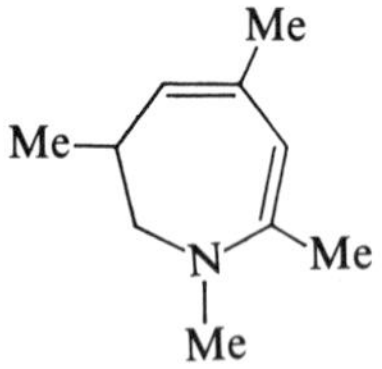

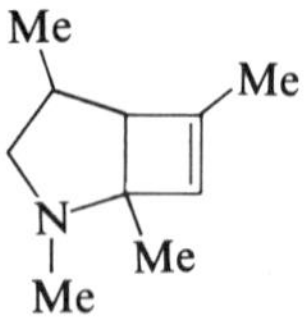

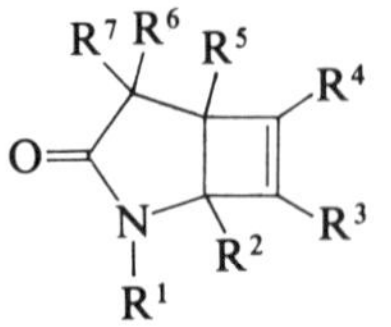

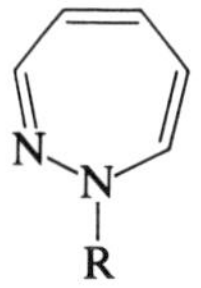

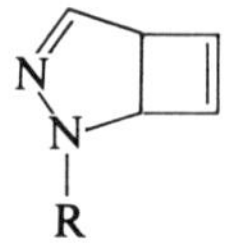

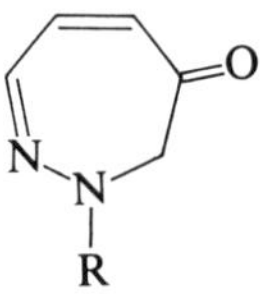

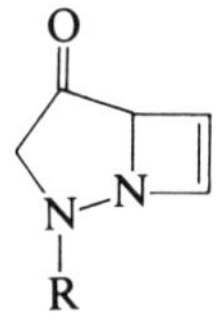

[278] E. Lerner, R. A. Odum, and B. Schmall, *Chem. Commun.*, 327 (1973).
[279] T. H. Koch and D. A. Brown, *J. Org. Chem.* **36**, 1934 (1971).
[280] L. A. Paquette, *J. Am. Chem. Soc.* **86**, 4092 (1964).

reactions have been reported for 2,3-dihydro-1*H*-azepin-2-ones, giving compounds **246**.[281-285]

Photochemical intramolecular (2 + 2)-cycloaddition within the framework of a heterocyclic seven-membered ring has further been observed in 1*H*-1,2-diazepines, 1,3-oxazepines, and 1,2-phosphazepines.

The teams of Streith[286,287] and Snieckus[288,289] studied the reactions of 1*H*-1,2-diazepines (**247**) and found that, just as in the reactions of 3*H*-azepines,[277] intramolecular (2 + 2)-cycloaddition proceeds to give a cyclobutene rather than a 3,4-dihydroazete. In most cases the corresponding bicyclic isomers (**248**) were isolated, whereas in other cases decomposition took place. Annulated 1,2-dihydroazetes (**250**) were

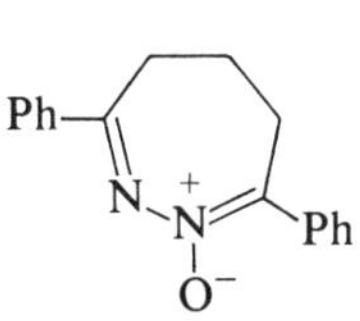

(251) 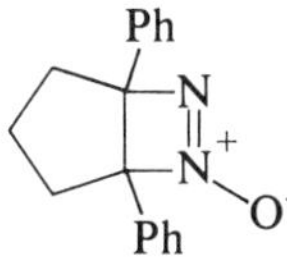   (252) 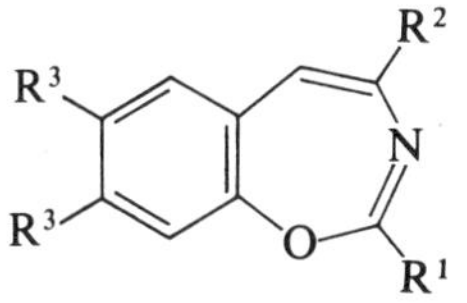   (253) 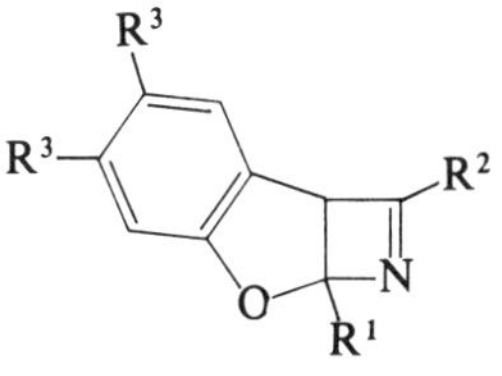   (254)

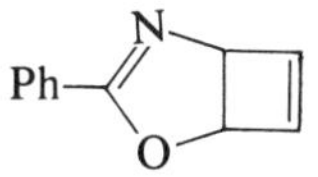

(255) 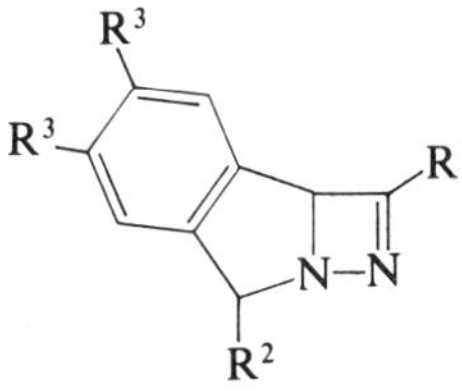   (256) 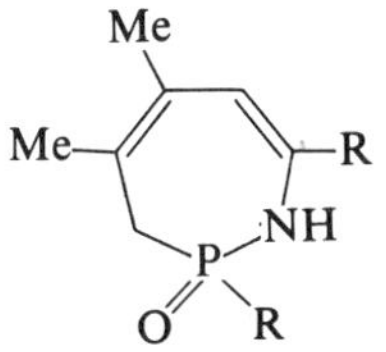   (257) 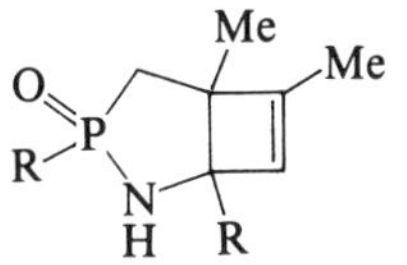   (258)

R = Ph

formed upon irradiation of 6,7-dihydro-1*H*-1,2-diazepin-6-ones (**249**); the corresponding alcohols reacted similarly.[290-292] 3,7-Diphenyl-5,6-dihydro-4*H*-1,2-diazepine 1-oxide (**251**) was reported to decompose

[281] O. L. Chapman and E. D. Hoganson, *J. Am. Chem. Soc.* **86**, 498 (1964).
[282] L. A. Paquette, *J. Am. Chem. Soc.* **86**, 500 (1964).
[283] E. Vogel, R. Erb, G. Lenz, and A. A. Bothner-By, *Ann. Chem.* **682**, 1 (1965).
[284] L. A. Paquette and W. C. Farley, *J. Org. Chem.* **32**, 2725 (1967).
[285] F. R. Atherton and R. W. Lambert, *J. Chem. Soc., Perkin Trans. 1*, 1079 (1973).
[286] J. Streith, J. P. Luttringer, and M. Nastasi, *J. Org. Chem.* **36**, 2962 (1971).
[287] J. P. Luttringer, N. Pérol, and J. Streith, *Tetrahedron*, in press.
[288] G. Kan, M. T. Thomas, and V. Snieckus, *Chem. Commun.*, 1022 (1971).
[289] T. Tsuchiya and V. Snieckus, *Can. J. Chem.* **53**, 519 (1975).
[290] W. J. Theuer and J. A. Moore, *Chem. Commun.*, 468 (1965).
[291] J.-L. Derocque, W. J. Theuer, and J. A. Moore, *J. Org. Chem.* **33**, 4381 (1968).
[292] J. A. Moore, E. J. Volker, and C. M. Kopay, *J. Org. Chem.* **36**, 2676 (1971).

partly upon irradiation, but the bicyclic isomer (**252**) was detected as well.[253] Another intramolecular (2 + 2)-cycloaddition of a heterocyclic azabutadiene moiety was reported for benzo[*f*]-1,3-oxazepines (**253**), yielding **254**,[293,294] although a corresponding nonfused heterocycle, 2-phenyl-1,3-oxazepine, gave 3-phenyl-2-oxa-4-azabicyclo[3.2.0]hepta-3,6-diene (**255**).[295] 1*H*-Benzo[*e*]-1,2-diazepines yielded, upon irradiation, the tricyclic isomers (**256**).[296] Finally irradiation of the 1,3-dihydro-1,2-phosphazepine **257** was shown to give the corresponding isomer **258**.[297]

d. Eight-Membered Rings. Irradiation of 6,7-bis(methoxycarbonyl)-1-phenyl-1,2-dihydroazocine (**44**; R^1 = Ph, R^2 = R^3 = H) yielded a 3-

(**259**)
E = COOMe

(**260**)

a: X = O
b: X = S
c: X = NMe

(**261**)

azabicyclo[4.2.0]octa-4,7-diene (**259**).[15] Three 14π-electron heterocycles (**260**) have been reported to give an intramolecular (2 + 2)-cycloaddition reaction upon irradiation. The stereochemistry of the isomers (**261**) has been shown to be cis annulation.[298-300]

IV. (2 + 2)-Cycloreversion Reactions

As far as the mechanistic aspects are concerned, there is no difference between cycloaddition reactions and cycloreversion reactions, and therefore the latter have been included in this review. The question of whether cycloaddition or cycloreversion takes place depends on the relative free energies of reactants and products. Although most (2 + 2)-cycloreversions are achieved thermally, several examples are known of

[293] C. Lohse, *Tetrahedron Lett.*, 5625 (1968).
[294] J. B. Bremner and P. Wiriyachitra, *Aust. J. Chem.* **26**, 437 (1973).
[295] T. Tezuka, O. Seshimoto, and T. Mukai, *Chem. Commun.*, 373 (1974).
[296] A. A. Reid, J. T. Sharp, and S. J. Murray, *Chem. Commun.*, 827 (1972).
[297] J.-P. Lampin and F. Mathey, *Tetrahedron* **28**, 5367 (1972).
[298] W. Schroth and B. Werner, *Angew. Chem.* **79**, 684 (1967).
[299] D. L. Coffen, Y. C. Poon, and M. L. Lee, *J. Am. Chem. Soc.* **93**, 4627 (1971).
[300] H.-J. Shue and F. W. Fowler, *Tetrahedron Lett.*, 2437 (1971).

photochemical (2 + 2)-cycloreversions of heterocycles. In this review both the thermal and the photochemical reactions have been separated into intermolecular and intramolecular cycloreversions.

A. Mechanism

Four different types of (2 + 2)-cycloaddition and -cycloreversion reactions of heterocyclic compounds are known: intermolecular and intramolecular reactions, both thermal and photochemical. Three of these have already been discussed in the previous sections on (2 + 2)-cyclo*additions*, and as far as the mechanism is concerned both the forward and the reverse reaction suffer from the same ambiguity: Do they proceed via a concerted or a nonconcerted mechanism; do they involve an ionic or a diradical intermediate; are they symmetry-allowed or forbidden? So far only one reaction type is known to be limited to the reverse reaction, *viz.*, the thermal intramolecular reaction [Eq. (10)], in which one σ-bond is broken and a conjugated 4π-electron system is

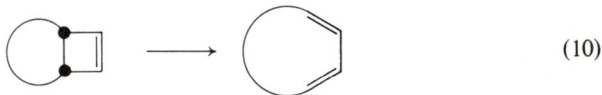 (10)

formed, either in a concerted or in a stepwise process. From the point of view of orbital symmetry conservation, such an electrocyclic process is allowed for a conrotatory ring opening. For cyclobutadienes cis fused to heterocycles, this consequently leads to cyclodienes containing a sterically highly unfavorable trans double bond. The heterocyclic systems reviewed here are cis fused bicyclic compounds, and consequently the "classical" orbital symmetry treatment predicts these (2 + 2)-cycloreversions to be stepwise processes. However, a more refined theory, in which contributions of polar structures have been taken into account,[28] predicts that substitution of the cyclobutene ring with polar groups will lower the activation energy for both the conrotatory and the disrotatory way of ring opening. Hence, when steric constraints prevent the conrotatory process, these reactions may take place via a concerted disrotatory process. The intramolecular (2 + 2)-cycloreversions discussed in this section support the predictions of Epiotis very nicely. Although in most cases the emphasis has been put on the synthetic rather than the kinetic aspects of the reactions, there are some kinetic data available on intramolecular (2 + 2)-cycloreversions of heterocyclic compounds that point to the occurrence of a concerted reaction with some charge separation in the transition state.[67]

B. Thermal (2 + 2)-Cycloreversion Reactions

1. Intermolecular Reactions

A (2 + 2)-cycloreversion was observed when the dimers (**174a** and **174b**) of benzo[b]furan were heated at 250°.[200] Benzo[b]furan was isolated as the only product. Similarly, the dimer of 2,6-diphenyl-4H-thiopyran-4-one (**129**) underwent (2 + 2)-cycloreversion upon refluxing in benzene.[136] The cycloadducts of flavone and diphenylacetylene (**262**)[197] and 3,4-benzo-2,5-dioxabicyclo[4.2.0]octa-3,7-diene (**261a**)[298] eliminated diphenylacetylene and acetylene, respectively, when heated at 250° to 300°.

Of particular interest is the thermal decomposition of azepines, diazepines, and triazepines. 3,4-Benzo-1,7-diazabicyclo[3.2.0]hepta-3,6-dienes (**256**) were found to decompose at 180° with loss of a nitrile or hydrogen cyanide. The other products of the (2 + 2)-cycloreversion were highly unstable isoindoles and could not be isolated.[296] A similar type of reaction has been proposed for the decomposition of 4,5-dihydro-3H-azepines[278] and N-methylazepines,[301] but the corresponding bicyclic intermediates **240** and **263** have not been detected. The question whether these cycloreversions are thermal or photochemical reactions remains unresolved.

There exists some confusion about the reported (2 + 2)-cycloreversion of 2,3,4,5,6-pentaphenyl-1,7-diazabicyclo[3.2.0]hepta-2,6-diene (**264**), a compound that has been proposed as a thermal isomer of

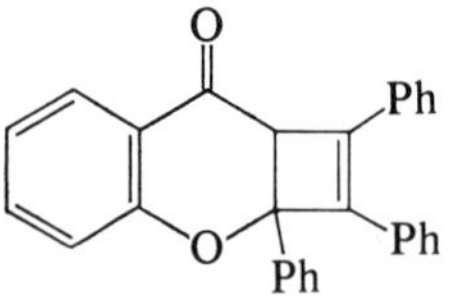

(**262**) 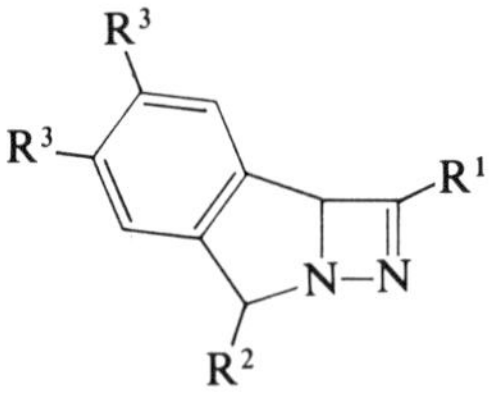   (**256**) 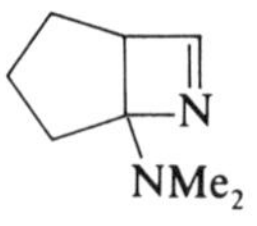   (**240**) 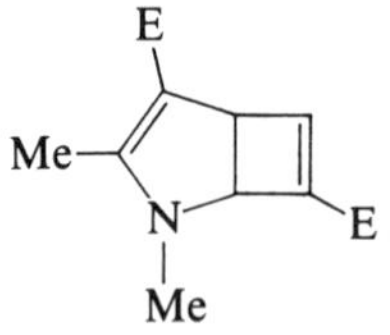   (**263**)

E = COOMe

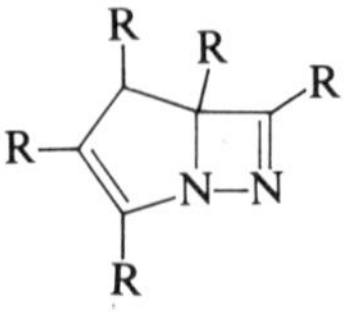

(**264**)   (**265**)
R = Ph

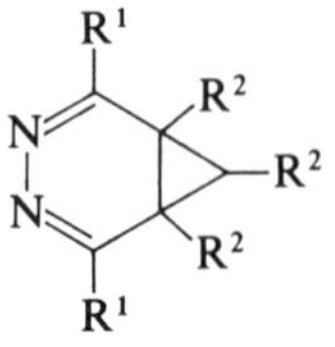

[301] R. F. Childs and A. W. Johnson, *J. Chem. Soc. C.*, 1950 (1966).

Sec. IV.B] (2 + 2)-CYCLOADDITION AND CYCLOREVERSION 313

pentaphenyl-5H-1,2-diazepine or (the corresponding) 3,4-diazanorcaradiene (**265**; $R^1 = R^2 = Ph$).[302] Upon heating at 235°–245° this isomer afforded 2,3,4,5-tetraphenyl-1H-pyrrole and benzonitrile, but Sauer et al.[303] showed that a similar decomposition of a 3,4-diazanorcaradiene **265** ($R^1 \neq R^2$) yielded a pyrrole that could not have been formed via an intermediate with structure **264**. Therefore, the question whether or not the elimination of nitriles from 3,4-diazanorcaradienes (or 3,4,7-triazanorcaradienes[304]) occurs via **264** remains to be answered. Other intermolecular (2 + 2)-cycloreversions have been involved in order to explain the reactions of isoquinolines with ynamines[64] and the thermal decomposition of hexaphenyl-1,5-diazocine into benzonitrile and pentaphenylpyridine.[305]

2. Intramolecular Reactions

The majority of the intramolecular (2 + 2)-cycloreversions of heterobicyclic compounds proceed via ring opening of an annulated cyclobutene or azetine ring. Only in one instance has the same reaction of a cyclobutane ring been proposed as an intermediate step, viz., in the reaction of a thiirene 1,1-dioxide with enamines (Section II,B).[33]

a. Five- and Six-Membered Heterobicyclic Compounds. Recently the (2 + 2)-cycloreversion of 1,2,3,4-tetrakis(trifluoromethyl)-5-thiabicyclo-[2.1.0]pent-2-ene (**214**) has been reported to occur at 160° with a half-life time of 5.1 hours; this reaction occurred smoothly at 25° in the presence of phosphines.[306] Similar thermal (2 + 2)-cycloreversions of diazabicyclo[2.1.0]pent-2-enes (**216** and **217**) were intermediate steps in the photochemical rearrangement of 3H- and 4H-pyrazole derivatives.[252-254]

(**214**)   (**218**)   (**225**)   (**266**)

[302] M. A. Battiste and T. J. Barton, *Tetrahedron Lett.*, 1227 (1967).
[303] G. Heinrichs, H. Krapf, B. Schröder, A. Steigel, T. Troll, and J. Sauer, *Tetrahedron Lett.*, 1617 (1970).
[304] H. W. Heine and J. Irving, *Tetrahedron Lett.*, 4767 (1967).
[305] E. M. Burgess and J. P. Sanchez, *J. Org. Chem.* **39**, 940 (1974).
[306] Y. Kobayashi, I. Kumadaki, A. Ohsawa, and Y. Sekine, *Tetrahedron Lett.*, 1639 (1975).

Dewar pyridine, 2-azabicyclo[2.2.0]hexa-2,5-diene (**218**), thermally reverted to pyridine at room temperature with a half-life time of 2.5 minutes (E_a = 16 kcal/mol).[255] Far more stable were 2-azabicyclo-[2.2.0]hex-5-en-3-ones (**225**).[265-267] The kinetics of the thermal (2 + 2)-cycloreversion of **225** (R = Me) in the temperature range of 130° to 160° have been reported ($\Delta H^{\ddagger}$ = 33.2 kcal mol^{-1}; $\Delta S^{\ddagger}$ = + 2.7 cal mol^{-1} deg^{-1}).[266] An interesting difference in rate was observed between **225** (R = H) and its methyl homolog **225** (R = Me). At 130° the former reverted ten times as rapidly to 2-pyridone as the latter did to 1-methyl-2-pyridone; this difference has been related to the intermediacy of the lactim tautomer of **225** (R = H) in the former reaction. Dewar benzene oxide, 2,3-epoxybicyclo[2.2.0]hex-5-ene (**266**), isomerized to an equilibrium mixture of benzene oxide/oxepin at 115° with a half-life time of 18 minutes.[270] The relatively high thermal stability of such strained bicyclic heterocycles has been attributed to the fact that the symmetry-allowed conrotatory process would give rise to an unfavored cis,trans heterocyclic diene.[265]

b. Seven- and Eight-Membered Heterobicyclic Compounds. A number of bicyclic seven- and eight-membered heterocycles show (2 + 2)-cycloreversion on heating. The reaction conditions required vary widely with the structure and in particular with the nature of the endo- or exocyclic substituents linked to the σ-bond that is cleaved. High-temperature (2 + 2)-cycloreversions have been reported for 6,7 bis(methoxycarbonyl)-3-phenyl-3-phosphabicyclo[3.2.0]hept-6-en-3-one (**267**),[160] 3,4-benzo-2-thiabicyclo[3.2.0]hepta-3,6-dienes (**197**),[233] and 3,4-benzo-2,5-dioxa, 2,5-dithia-, and 2,5-diaza-bicyclo[4.2.0]octa-3,7-dienes (**261**).[298-300]

Various azabicyclo[3.2.0]heptenes and -heptadienes as well as azabicyclo[4.2.0]octenes and -octadienes have been found to undergo (2 + 2)-cycloreversion. 2-Azabicyclo[3.2.0]hepta-3,6-dienes (**268**) were converted into the corresponding 1*H*-azepines (**269**) at temperatures ranging from 20° to 125°.[17,271,301] A significant difference in thermal stability was observed between compounds **268a, b** and **c**. Whereas **268a** and **268b** have been proposed as intermediates in reactions that occur at room temperature,[17,301] compound **268c**, 1-*ethoxycarbonyl*-2-azabicyclo[3.2.0]hepta-3,6-diene, required a temperature as high as 125° to be converted into the corresponding azepine (**269c**). Similarly, 2-*ethoxycarbonyl*-2,3-diazabicyclo[3.2.0]heptadiene (**248**; R = COOEt) required a temperature of 120° in order to give the corresponding 1*H*-1,2-diazepine (**247**).[287] (2 + 2)-Cycloreversions under mild conditions have also been reported for 3,4-benzo-2-azabicyclo[3.2.0]hepta-3,6-dienes (**270**), the primary products in the reactions of indoles with

Sec. IV.B] (2 + 2)-CYCLOADDITION AND CYCLOREVERSION

dimethyl acetylenedicarboxylate to give benzo[b]azepins (**271**) (Section II,C,2).[26,27,65] Likewise, a fast (2 + 2)-cycloreversion of 6,7-bis(methoxycarbonyl-2-methyl-1-mercaptomethyl 2-azabicyclo[3.2.0]-hept-6-ene (**33**) has been suggested as the second step in the reaction of 1-methyl-5-methylmercapto-2,3-dihydro-1H-pyrrole with dimethyl acetylenedicarboxylate (Section II,C,1).[57] Thermally far more stable are

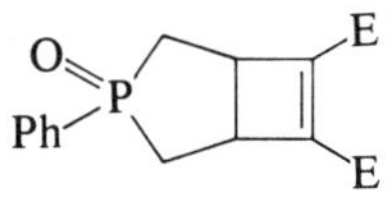

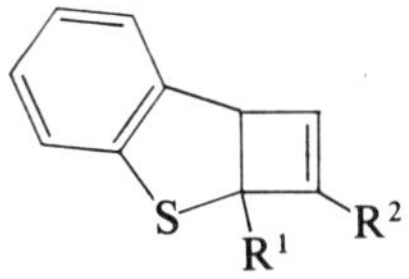

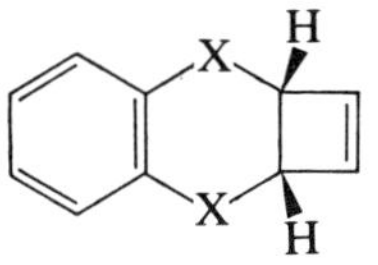

(**267**)   (**197**)   (**261**)

E = COOMe

a: X = O
b: X = S

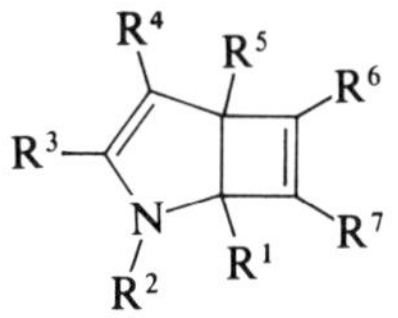

(**268**) 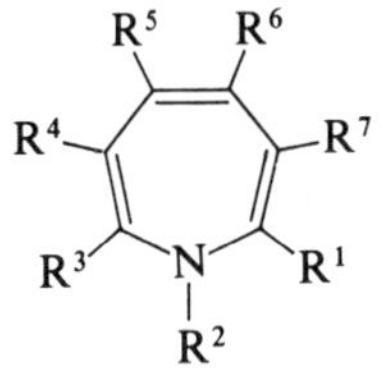   (**269**)   (**270**)   (**271**)

a: $R^1 = R^2 = R^3$ = Me; $R^4 = R^7$ = E;
   $R^5 = R^6$ = H
b: R^1–R^5 = H; $R^6 = R^7$ = E
c: R^1 = COOEt; R^2–R^7 = H

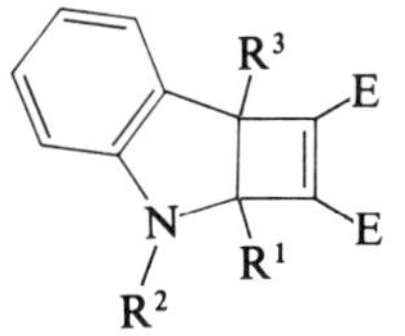

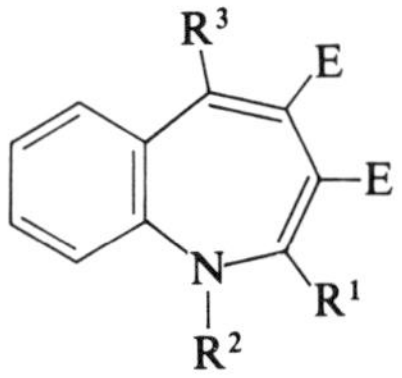

a: R^1 = H; $R^2 = R^3$ = Me
b: $R^1 = R^3$ = H; R^2 = Me
c: R^1 = OEt; $R^2 = R^3$ = H
d: R^1 = OEt; R^2 = Me, R^3 = H

E = COOMe

3-oxo-2-azabicyclo[3.2.0]hept-6-enes (**272**), which display an intramolecular (2 + 2)-cycloreversion only at 430° to 500° yielding 2,3-dihydro-1H-azepin-2-ones (**273**).[281,283] Rapid (2 + 2)-cycloreversions have been observed for 2-azabicyclo[4.2.0]-octa-4,7-diene derivatives (**43**), at room temperature, and for a 3,4-benzo-2-azabicyclo[4.2.0]-octa-3,7-diene (**40**), at 80°.[15,60,61] A similar cycloreversion has been proposed in the reactions of 1,2-dihydropyrazines with dimethyl acetylenedicarboxylate (Section II,B).[62] In contrast to these compounds, 2-azabicyclo[4.2.0]octa-3,7-dienes (**36** and **38**) proved to be "stable"; no experimental data were presented (Section II,C,1).[59]

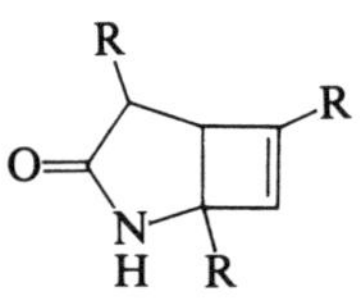

(272) 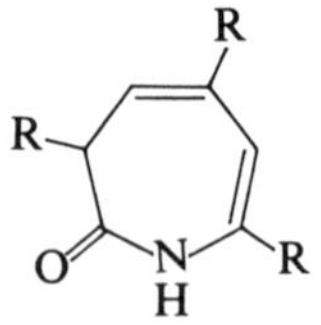 (273) 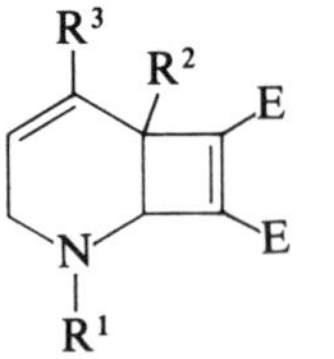 (43) 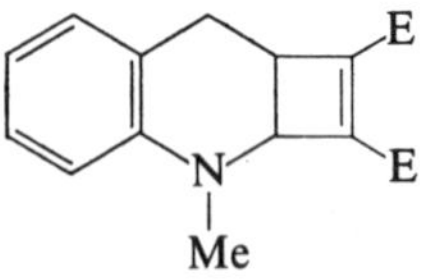 (40)

a: R = H
b: R = Me

E = COOMe

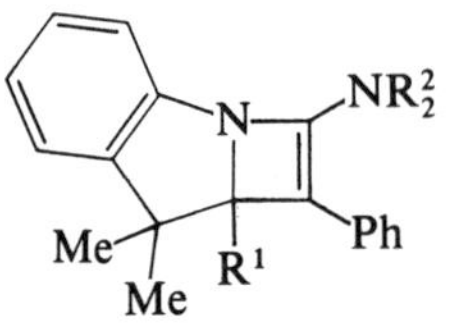

(52) 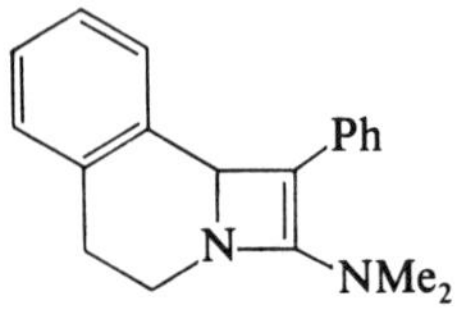 (53) 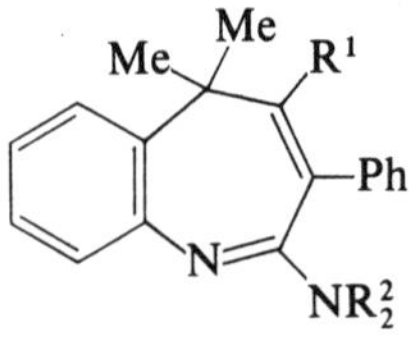 (274)

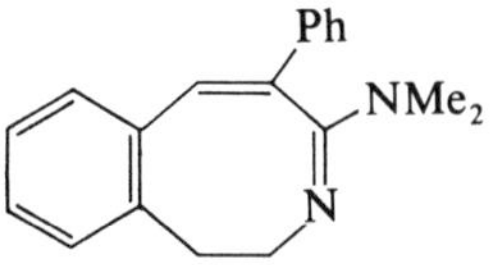

(275) 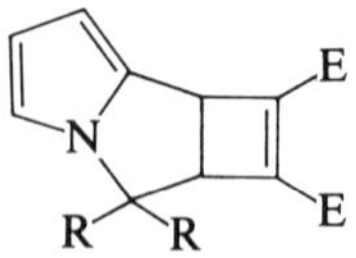 (171) 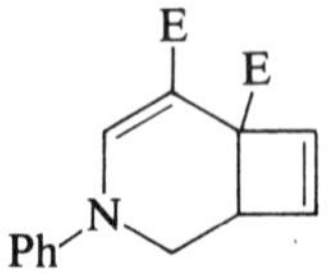 (259) 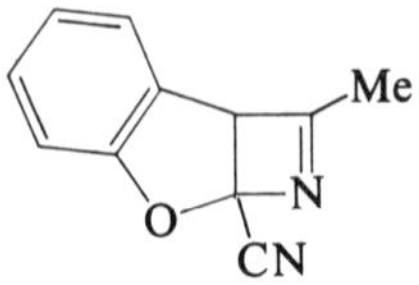 (276)

E = COOMe

In all these heterobicyclic compounds the nitrogen atom is in the 2-position, but (2 + 2)-cycloreversions of heterobicyclic compounds with the nitrogen atom in other positions are also known. 4,5-Benzo-1-azabicyclo[4.2.0]octa-4,7-dienes (**53**) and 2,3-benzo-1-azabicyclo-[3.2.0]hepta-2,6-dienes (**52**) isomerized to **274** and **275**, respectively, at the temperature at which they were formed (37°), but Fuks and Viehe[64] provided chemical evidence for the intermediate formation of **53**. The 3-azabicyclo[3.2.0]hept-6-ene ring system and the 3-azabicyclo[4.2.0]-octa-4,7-diene ring system are far more stable. Compound **259** was found to be thermally stable[15]; **171** (R = Me) gave a (2 + 2)-cycloreversion upon heating at 180–200°.[196] Less stable is the 3,4-benzo-2-oxa-7-azabicyclo[3.2.0]hepta-3,6-diene (**276**), which isomerized in refluxing toluene to give the corresponding benzo[f]1,3-oxazepine (**253**; R^1 = CN, R^2 = Me, R^3 = H).[293]

A comparison of the results for the various types of azaheterobicyclic compounds clearly indicates that the rate of the (2 + 2)-cycloreversion is greatly dependent on the position of the nitrogen atom.

In particular, when the σ-bond that is cleaved in the reaction is linked via an endocyclic bond to a nitrogen atom, a rapid cycloreversion of the cis-fused cyclobutene ring is observed. These results might be explained in terms of an effective stabilization of the intermediate in a two-step process by the nitrogen atom, but they also fit in nicely with the concepts of (2 + 2)-cycloreversion developed by Epiotis.[28] The polar contribution of the electron-donating endocyclic amino group might well result in a sufficient lowering of the activation energy of the concerted "nonallowed" disrotatory (2 + 2)-cycloreversion. The effect of the nitrogen atom is less pronounced when the endocyclic amino group is substituted with an electron-withdrawing ethoxycarbonyl group or even negligible when it constitutes part of an endocyclic amido or imino group. A nitrogen atom in position 1 or 7 also lowers the activation energy of the cycloreversion and in these cases the ring opening might occur via a symmetry-allowed conrotatory process with simultaneous isomerization of the *trans*-azabutadiene moiety that is formed. The presence of a nitrogen atom in the 3-position has no significant effect on the rate of (2 + 2)-cycloreversions.

The accelerating effect of a nitrogen atom attached to the σ-bond that is broken in the (2 + 2)-cycloreversion is not restricted to endocyclic nitrogen atoms. Heterobicyclic compounds substituted with an exocyclic amino group at one of the bridgehead carbon atoms also undergo (2 + 2)-cycloreversion at relatively low temperatures. These reactions are widely used in the synthesis of unsaturated and partially saturated medium-sized heterocycles. In combination with a (2 + 2)-cycloaddition of five- or six-membered heterocycles with acetylenes this reaction provides a quite generally applicable method for the ring enlargement of heterocycles with a two-carbon-atom moiety [Eq. (11)].

The syntheses of 4,5-bis(methoxycarbonyl)-3(1-pyrrolidinyl)-2,7-dihydrooxepin (**278a**)[53] and 1-phenyl-1-oxo-4,5-bis(methoxycarbonyl)-6-(4-morpholinyl)-1-phosphacycloocta-3,5-diene (**279c**)[54] have been proposed to involve such a (2 + 2)-cycloreversion of a bicyclic intermediate (**27a** and **28c**) at the reaction temperatures adopted (25° and 110°, respectively). Other heterobicyclic compounds were more stable, e.g., several 2- and 3-thiabicyclo[3.2.0]hept-6-enes (**26** and **27b**)[16] only isomerized to the corresponding dihydrothiepins (**277** and **278b**) at 80°–100°. A kinetic study of the (2 + 2)-cycloreversion of these compounds at 100° actually demonstrated that the rate of the (2 + 2)-cycloreversion depends on the electron-donating capacity of the amino group (1-pyrrolidinyl > 4-morpholinyl ⩾ 1-piperidinyl).[16] Similarly, 6-amino-3-heterobicyclo[4.2.0]oct-7-enes (**28**) isomerized at temperatures of 100°–120° to give the corresponding thiocin or azocine derivatives in quantitative yields.[36] Paquette and Begland[55] have reported the synthesis

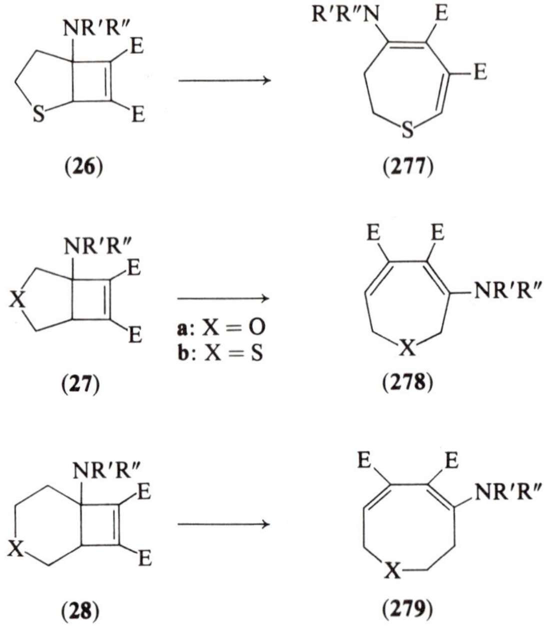

(26) → (277)

(27) → (278) (11)
a: X = O
b: X = S

(28) → (279)
a: X = S
b: X = NMe
c: X = POPh

E = COOMe

of 3-dimethylamino-4,5-bis(methoxycarbonyl)-9,10-dihydro-2H-oxecin (**280**) via a (2 + 2)-cycloreversion of the (2 + 2)-cycloadduct (**31**) of dimethyl acetylenedicarboxylate and 3-dimethylamino-7,8-dihydro-2H-oxocin (**30**) in 51% yield.

Important for the synthesis of 1-heterocycloheptatriene derivatives (**281**) are the thermal (2 + 2)-cycloreversions of 3,4-benzo-2-heterobicyclo[3.2.0]cyclohepta-3,6-dienes (**70**).[66,67,70] The rate of the (2 + 2)-cycloreversion depends substantially on the nature of the heteroatom (X). For instance, the isomerization of **70d** proceeded only after 2 days' refluxing in dioxane,[66] whereas the isomerization of **70a** took place at 100° ($k = 0.9 \times 10^{-3}$ sec^{-1}) and that of **70b** even at room temperature ($k = 10^{-3}$ sec^{-1}).[67,70] Reinhoudt and Kouwenhoven[67] observed a relatively small solvent effect in the (2 + 2)-cycloreversion of **70**, which points to a near-concerted reaction with some charge separation in the transition state.

The rate of the (2 + 2)-cycloreversion of 6,7-bis(methoxycarbonyl)-5-(1-pyrrolidinyl)-2-thiabicyclo[3.2.0]hepta-3,6-dienes (**73**) was even higher. The reaction was followed at −30° by NMR spectroscopy and

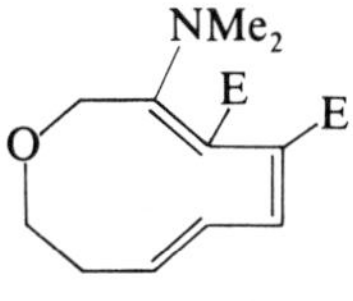

Sec. IV.C] (2 + 2)-CYCLOADDITION AND CYCLOREVERSION 319

(280)

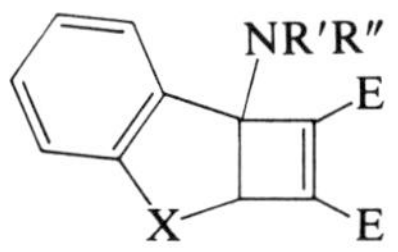

(70)

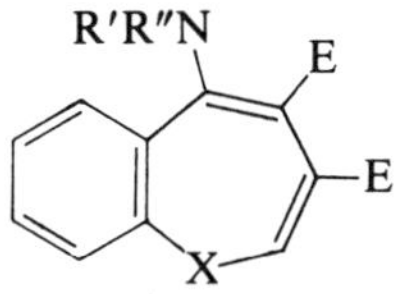

(281)

E = COOMe

a: X = O; NR'R" = NC$_4$H$_8$
b: X = S; NR'R" = NC$_4$H$_8$
c: X = S; NR'R" = NC$_5$H$_{10}$
d: X = NCOMe; NR'R" = NC$_5$H$_{10}$

(73)

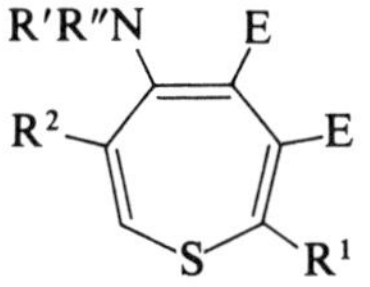

(282)

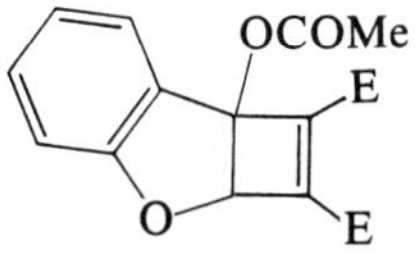

(283)

E = COOMe

found to afford the corresponding thiepins (282).[18,71] This (2 + 2)-cycloreversion of 73 is the only route available for the synthesis of the thermally unstable, antiaromatic thiepins. The rapid (2 + 2)-cycloreversions of these compounds are due to the presence of an amino substituent at one of the bridgehead carbon atoms. Analogous heterocycles with the same ring structure but lacking the amino group at C(5), e.g., 75, 197, and 283, are stable up to 250°.[67,80,223] Other electron-donating substituents, such as a methoxy group, have a similar effect, as can be concluded from the rapid (2 + 2)-cycloreversion of 5-methoxy-2-methyl-2-azabicyclo[3.2.0]hept-6-en-3-one to the corresponding 2,3-dihydro-1H-azepin-2-one.[285]

C. PHOTOCHEMICAL (2 + 2)-CYCLOREVERSION REACTIONS

1. *Intermolecular Reactions*

Photochemical (2 + 2)-cycloreversions of heterocyclic compounds are rare. The dimer of xanthotoxin, a naturally occurring furocoumarin, was reported to undergo a photochemical (2 + 2)-cycloreversion upon irradiation ($\lambda > 250$ nm).[137] Irradiation of 2-oxabicyclo[2.2.0]hex-5-ene-3-ones (224) at temperatures of 8° to 20°K resulted in the elimination of carbon dioxide [Eq. (12)].[12,307] Similarly, pyridine was found to give

(224) → ☐☐ + CO_2

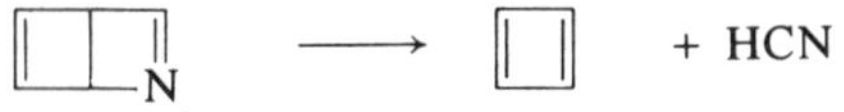

(218) → ☐☐ + HCN

(12)

cyclobutadiene and hydrogen cyanide upon irradiation in an argon matrix at 8°K.[307] This reaction was assumed by Chapman et al.[307] to involve a (2 + 2)-cycloreversion of 2-azabicyclo[2.2.0]hepta-2,5-diene (218).[255] Cyclobutadiene was also formed by photochemical (2 + 2)-cycloreversion of 284.[308] Irradiation of 1-methyl-3,5,7-triphenyl-1H-1,2-diazepine yielded 1-methyl-3,5-diphenyl-1H-pyrazole and phenylacetylene. This reaction has been interpreted as an intramolecular photochemical (2 + 2)-cycloaddition followed by (2 + 2)-cycloreversion of the intermediate formed (285).[288] Dimers of coumarin are split into the monomer upon irradiation with light having a wavelength of 277 nm.[309]

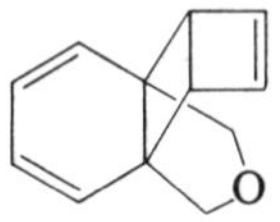

(284) 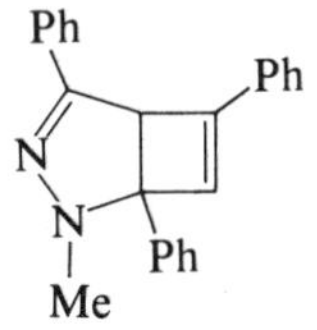   (285)

2. Intramolecular Reactions

Mukai et al.[295] reported the intramolecular (2 + 2)-cycloreversion of 3-phenyl-2-oxa-4-azabicyclo[3.2.0]hepta-3,6-diene (255) upon irradiation with light of 250 nm. Irradiation (λ = 254 nm) of 2-methyl-2-azabicyclo[2.2.0]hex-5-en-3-one (225; R = Me) yielded the corresponding 2-pyridone (223; R = Me).[266]

[307] O. L. Chapman, C. L. McIntosh, and J. Pacansky, J. Am. Chem. Soc. 95, 614 (1973).
[308] S. Masamune, M. Suda, H. Ona, and L. M. Leichter, Chem. Commun., 1268 (1972).
[309] M. Hasegawa, Y. Suzuki, and N. Kita, Chem. Lett., 317 (1972).

Appendix

RE SECTION II

After completion of this review, a thermal intramolecular (2 + 2)-cycloaddition of a conjugated 4π-electron system of a heterocyclic compound was disclosed. Padwa *et al.*[310] reported the conversion of 2,3,5,7-tetraphenyl-7*H*-1,4-diazepine **(286)** into 2,4,5,7-tetraphenyl-3,6-diazabicyclo[3.2.0]hepta-3,6-diene **(287)**, which occurs upon refluxing in benzene in 68% yield.

(286)   **(287)**

R = Ph

However, so far the assignment of structure **287** is entirely based on the chemical shifts of the two protons H(1) and H(2) and the coupling constant.

A similar type of reaction had previously been proposed as an intermediate step in the pyrolysis of 3,4,5,6,7-pentaphenyl-5*H*-1,2-diazepine, 7-benzyl-2,5-diphenyl-3,4,7-triazanorcaradiene, and hexaphenyl-1,5-diazocine, without any experimental support being given (Section IV,B,1).[302–305]

---

[310] A. Padwa, L. Gehrlein, and R. B. Kinnel, *J. Org. Chem.* **40**, 1683 (1975).

Recent Advances in Tetrazole Chemistry

R. N. BUTLER

Chemistry Department, University College, Galway, Ireland

I. Introduction	324
II. Physicochemical Studies	325
A. Nuclear Magnetic Resonance (NMR) Spectroscopy and Theoretical Calculations	325
1. Electronic Distributions, Protonation, and General NMR Studies	325
2. Chemical Shifts and Structural Assignments	327
3. Tautomerism	332
B. Mass Fragmentation	338
1. 2,5-Disubstituted Tetrazoles	338
2. 1,5-Disubstituted Tetrazoles	338
C. Tetrazole Complexes and Crystal Structures	340
1. Molecular Complexes	340
2. Cyclopolymethylenetetrazole Complexes with Metals	342
3. Complexes of Substituted Tetrazoles with Metals	343
4. General	346
D. Photolysis and Tetrazole Radicals	347
1. 5-Monosubstituted Tetrazoles	348
2. 2,5-Disubstituted Tetrazoles	349
3. 1,5-Disubstituted Tetrazoles	350
4. Tetrazolinyl Radicals	352
III. Studies of Synthesis and Mechanism	354
A. 5-Monosubstituted Tetrazoles	354
1. 5-Vinyltetrazoles	354
2. Biologically Active Tetrazoles ($-CN_4H$ versus $-CO_2H$)	355
3. General Syntheses	360
4. Reactions	361
B. 1-Monosubstituted and 1,5-Disubstituted Tetrazoles	370
1. Synthesis and Reactions of 1-Substituted Tetrazoles	370
2. Synthesis of 1,5-Disubstituted Tetrazoles	372
3. Reactions of 1,5-Disubstituted Tetrazoles	384
C. 2-Substituted and 2,5-Disubstituted Tetrazoles	392
1. Synthesis	392
2. Reactions	396
D. Mesoionic Tetrazoles	398
IV. Azidoazomethine Tetrazole Isomerism—Fused Tetrazoloheterocycles	402
A. General Characteristics. Monocyclic Systems	402
B. Bicyclic Fused Tetrazoloheterocycles	404
1. Tetrazoles Fused with Five-Membered Rings	404
2. Tetrazoles Fused with Six-Membered Rings	408

     C. Tricyclic and Higher Fused Tetrazoloheterocycles 416
        1. Tetrazolopyridazines Fused with Five-Membered Rings . . . 416
        2. Tetrazolopyridazines Fused with Six-Membered Rings . . . 418
        3. Tetrazolopyrimidines Fused with Five- and Six-Membered Rings . 420
        4. General Discussion of Fused Tetrazoloheterocycles 422
  V. Conclusion 424
     Notes Added in Proof 425

I. Introduction

The development of the chemistry of the tetrazoles up to the end of 1965 has been reviewed in two papers by Benson.[1,2] Shorter reviews of specific aspects of tetrazole chemistry include alkylation reactions of tetrazoles,[3] synthesis of tetrazoles from aminoguanidines,[4] influence of solvents on tetrazole synthesis from 1,3-cycloadditions of azides,[5] general aspects of tetrazole chemistry,[6] azidoazomethine tetrazole isomerism,[7,8] tetrazolinyl radicals,[9] and complexes of 1,5-disubstituted tetrazoles.[10]

The present review covers the development of tetrazole chemistry from 1965 to about September, 1975. References prior to 1965 are, however, included in a few cases where they are necessary for completion or comparison purposes. During the past 10 years the field has grown rapidly, probably owing in part to the industrial interest which the tetrazoles command. Areas that stand out, owing to the volume of work reported, include the applications of modern physical techniques to tetrazole derivatives and studies of pharmacologically active tetrazoles, particularly replacement of the carboxylic acid group in known active molecules by its analog the tetrazole ring ($-CN_4H$). The present review is considered to be complementary to the existing literature, and in many cases the individual sections cannot be put in proper perspective unless the previous excellent reviews by Benson[1,2] are also consulted.

[1] F. R. Benson, *in* "Heterocyclic Compounds" (R. C. Elderfield, ed.), Vol. 8, p. 1, Wiley, New York, 1967.
[2] F. R. Benson, *Chem. Rev.* **41**, 1 (1947).
[3] R. N. Butler, *Leicester Chem. Rev.* **10**, 12 (1969).
[4] F. Kurzer, *Angew. Chem., Int. Ed. Engl.* **2**, 468 (1963).
[5] P. K. Kadaba, *Synthesis*, 71 (1973).
[6] A. Albert, "Heterocyclic Chemistry," pp. 190–197, 206, 228, 234, 250. Athlone Press, London, 1968; M. H. Palmer, "Structure and Reactions of Heterocyclic Compounds," pp. 399–403. Edward Arnold, London, 1967.
[7] R. N. Butler, *Chem. Ind. (London)*, 371 (1973).
[8] M. Tisler, *Synthesis*, 123 (1973).
[9] F. A. Neugebauer, *Angew. Chem., Int. Ed. Engl.* **12**, 455 (1973).
[10] A. I. Popov, *Coord. Chem. Rev.* **4**, 463 (1969).

Tetrazole itself is an aromatic azapyrrole nucleus, which may exist in the tautomeric forms **1** and **2** (see Section II,A,3). The numbering system indicated in **1** and **2** is that used throughout the review. The structural unit, $-C(N_3)=N-$, is described either as an azidoazomethine, an iminoazide, or an imidoylazide, all of which terms are used widely in the literature.

II. Physicochemical Studies

A. NUCLEAR MAGNETIC RESONANCE (NMR) SPECTROSCOPY AND THEORETICAL CALCULATIONS

1. *Electronic Distributions, Protonation, and General NMR Studies*

The chemical shifts of carbon atoms in nitrogen heterocycles are generally governed by the number and location of nearby nitrogen atoms. Tetrazole, however, proved to be anomalous among azoles when an empirical equation that described the ^{13}C shifts in a series of heterocycles failed for tetrazole, the predicted shift from cyclopentadienide anion being 51 ppm as against the 41 ppm observed[11] (however, cf. Ref. 57). Correlations have been sought between $2p_z$ atomic orbital populations from extended Hückel calculations and ^{13}C chemical shifts in azoles including tetrazole.[12] A linear relationship was found between ^{13}C and ^1H chemical shifts in azoles and molecular orbital (MO) π-electron densities, suggesting that the electron densities may be the main factors in determining chemical shifts.[12] Tetrazole exhibited abnormal ^{13}C shielding but showed ^1H shifts in agreement with the π-electron densities.[12]

Theoretical calculations have been carried out on tetrazoles by a number of workers.[13-18] Ring geometries of the annular tautomers **1** and

---

[11] F. J. Weigert and J. D. Roberts, *J. Am. Chem. Soc.* **90**, 3543 (1968).
[12] B. M. Lynch, *Chem. Commun.*, 1337 (1968).
[13] M. Roche and L. Pujal, *Bull. Soc. Chim. Fr.*, 1097 (1969).
[14] M. J. S. Dewar and G. J. Gleicher, *J. Chem. Phys.* **44**, 759 (1966).
[15] W. Waznicki and B. Zurawski, *Acta Physl Pol.* **36**, 95 (1967) [*CA* **67**, 90327 (1967)].
[16] B. Zurawski, *Acta Phys. Pol. A.* **39**, 567 (1971) [*CA* **75**, 53036 (1971)].
[17] M. H. Palmer, R. H. Findlay, and A. J. Gaskell, *J. Chem. Soc., Perkin Trans. 2*, 420 (1974).
[18] M. A. Schroeder, R. C. Makino, and W. M. Tolles, *Tetrahedron* **29**, 3463 (1973).

2 have been calculated,[13,14] and energies, which are quite similar for both forms, have been predicted.[13,14] Dipole moments and ionization potentials have also been calculated for both forms of tetrazole.[15-17]

$$H_2N-\underset{\underset{H}{N-N}}{\overset{N-N}{\diagdown\mkern-6mu\diagup}} \longleftrightarrow H_2N^+=\underset{\underset{H}{N-N^-}}{\overset{N=N}{\diagdown\mkern-6mu\diagup}} \longleftrightarrow H_2N^+=\underset{\underset{H}{N-N}}{\overset{N=N}{\diagdown\mkern-6mu\diagup}}$$

(3)　　　　　　(4)　　　　　　(5)

These are discussed further in Section II,A,3. Calculations on 5-aminotetrazole and 2-methyl-5-aminotetrazole have suggested structures similar to those expected from classical resonance concepts.[18] For example, CNDO/2 calculations on compound 3 showed high negative charges at the 2- and 4-positions, as might be expected from resonance contributions such as 4 and 5.[18] The advantages and disadvantages of representing the molecular geometry as a regular pentagon or of using experimentally determined molecular geometries for such calculations have also been commented on.[18] These calculations also suggested[18] that resonance electron withdrawal from the 5-amino group was lower in 2-methyl-5-aminotetrazole than in compound 3, a result that is of interest in view of the resonance interactions detected by NMR in 2-alkyl-5-aryltetrazoles (see below). MO calculations have also been carried out on the molecular and electronic structure of the salt of guanidine with 5-aminotetrazole.[19]

The quadrupole resonance of ^{14}N has been correlated with electron density for a series of azoles.[20] The spectrum of tetrazole was uninformative. The molecule was piezoelectric, and only two resonances could be observed.[20] Nitrogen-14 NMR spectra of a number of 1,5-disubstituted tetrazoles have also been reported.[21] With trimethylenetetrazole four nitrogen signals were observed. The signal of the sp^3 nitrogen was higher upfield and sharper than those of the sp^2 nitrogen atoms.[21] An additivity rule for ^{14}N shifts in a wide range of azoles, including tetrazoles, has been formulated by Witanowski et al.[22,23] The shifts correlated well with calculated π-charge densities and the associated structural units and were, therefore, useful for dis-

---

[19] L. Paoloni, G. La Manna, and G. Camilletti, J. Mol. Struct. **20**, 135 (1974) [CA **80**, 107754 (1974)].
[20] L. Guibe and E. A. C. Lucken, Mol. Phys. **14**, 73 (1968).
[21] E. B. Baker and A. I. Popov, J. Phys. Chem. **76**, 2403 (1972).
[22] M. Witanowski, L. Stefaniak, H. Januszewski, and G. A. Webb, Bull. Acad. Pol. Sci., Ser. Sci. Chim. **21**, 71 (1973) [CA **78**, 104121 (1973)]; cf. also Witanowski et al.[59]
[23] M. Witanowski, L. Stefaniak, H. Januszewski, and J. Elguero, J. Chim. Phys. **70**, 697 (1973).

tinguishing between isomeric *N*-alkyl derivatives. For example, 1- and 2-methyltetrazole gave ^{14}N-Me shifts of +150 ppm and +103 ppm, respectively, from internal nitromethane.

Proton magnetic resonance (PMR) spectra have been used to locate protonation sites in azoles by comparing spectra of the neutral molecule in $CDCl_3$ with those of the cation in CF_3COOH.[24] For 1-methyltetrazole the shifts of the 5-CH proton and the 1-methyl group were 1.02τ and 5.73τ, respectively, in the neutral molecule and 0.12τ and 5.46τ, respectively, in the protonated form.[24] Although Barlin and Batterham[24] did not suggest a protonation site for the tetrazoles, the deshielding shift of the 5-CH proton (Δτ 0.9 ppm) may be indicative of a mixture of protonation at both the 4-N and 3-N positions. (For exclusive protonation of either the 4-N or 3-N atoms, deshielding shifts in the 5-CH proton of 1.25–1.41 ppm and 0.5–0.8 ppm, respectively, would be expected from a comparison with similar shifts in other azoles.[24]) Protonation of 2-methyltetrazole resulted in a deshielding of the 5-CH proton from 1.40τ to 0.88τ, and the 2-*N*-methyl group from 5.54τ to 5.40τ.[24] Raman spectra of protonated and nonprotonated tetrazole in 17.2–37.0% aqueous H_2SO_4 have also been reported, and protonation at the 4-N site has been suggested.[25] The weak basicities of substituted 5-phenyltetrazoles[26] and substituted 1-phenyl-5-methyltetrazoles[27] have also been examined and found to correlate well with the Hammett acidity function and the σ values of substituents on the phenyl ring.[26,27] A basic pK_a value of −3.0 has been quoted for tetrazole in aqueous H_2SO_4.[25] The weak basicities ($pK_b \sim 2$) of 1,5-disubstituted tetrazoles in formic acid solution have also been commented on by Popov[10] (see also Section II,C,1).

2. *Chemical Shifts and Structural Assignments*

Proton NMR spectra of a wide range of 5-substituted-1- and 2-alkyltetrazoles have been useful for distinguishing between the isomers since the chemical shift of the alkyl substituent was dependent on its location in the ring.[28-30] Examples of such tetrazoles that have been

---

[24] G. B. Barlin and T. J. Batterham, *J. Chem. Soc., B*, 516 (1967).
[25] M. M. Sokolova, V. A. Ostrovskii, G. I. Goldobskii, V. V. Mel'nikov, and B. V. Gidaspov, *Zh. Org. Khim.* **10**, 1085 (1974) [*CA* **81**, 49108 (1974)].
[26] V. N. Strel'tsova, N. P. Shirokova, G. I. Koldobskii, and B. V. Gidaspov, *Zh. Org. Khim.* **10**, 1081 (1974) [*CA* **81**, 62902 (1974)[.
[27] N. D. Agibalov, S. A. Enin, G. I. Koldobskii, G. I. Gidaspov, and B. V. Timofeeva, *Zh. Org. Khim.* **8**, 2414 (1972) [*CA* **78**, 96981 (1973)].
[28] F. L. Scott, R. N. Butler, and J. Feeney, *J. Chem. Soc., B*, 919 (1967).
[29] R. N. Butler, *Can. J. Chem.* **51**, 2315 (1973).
[30] R. N. Butler, *Can. J. Chem.* **50**, 1786 (1972).

TABLE I

CHEMICAL SHIFTS OF N-ALKYL PROTONS IN TETRAZOLES

$$R-C\underset{\underset{R^1}{N-N}}{\overset{N-N}{\|}}$$

Compound	R	R^1(1-isomer) τ (B unit)[a]	R^1(2-isomer) τ (C unit)[a]	References
6, 7 (R^1 = Me)[b]	H	5.73	5.54	24
8, 9 (R^1 = Me)	Br	5.87	5.60	24
10, 11 (R^1 = Me)	Me	5.90	5.70	31
12, 13 (R^1 = Me)	NH_2	6.18[e]	5.84	28
14, 15 (R^1 = Ph—CH_2)	NH_2	4.66	4.46	32
16, 17 (R^1 = $PhCH_2$)	NH_2	4.62	4.46	32
18, 19 (R^1 = $PhCH_2$)	$PhCH_2$	4.64	4.35	33
20, 21 (R^1 = Me)	Ph	5.84	5.75	34, 37
22, 23 (R^1 = CH_2COOEt)[c]	2-Thienyl	4.22[f]	4.15[f]	35
24, 25 (R^1 = CH_2COOEt)[c]	2-Furyl	4.25[f]	4.09[f]	35
26, 27 (R^1 = CH_2COOEt)[c]	CH_3OCH_2	4.50[f]	4.25[f]	35
28, 29 (R^1 = CH_2COOEt)[c]	Ph	4.30[f]	4.07[f]	35
30, 31 (R^1 = CH_2COOEt)[c]	NH_2	4.90[f]	4.57[f]	36
32, 33 (R^1 = CH_2COOEt)[c]	CH_3S	4.95[c]	4.67[g]	36
34[d]	p-XC_6H_4-C, N-N, N-C, NMe, N=N	5.80 (NMe)	—	30
35[d]	p-XC_6H_4-C, N-N, N-C, N, MeN-N	—	5.40 (NMe)	30

[a] Chemical shifts, from internal TMS, are for solutions in $CDCl_3$ at normal probe temperatures (ca. 35°) unless otherwise stated.
[b] The lower number of each pair refers to the 1-substituted isomer.
[c] A similar difference between the 1- and 2-alkyl signal was observed with the corresponding acids.
[d] X = CH_3, H, Br, Cl.
[e] Solvent D_2O. Shift from internal sodium 3-(trimethylsilyl)propane 1-sulfonate.
[f] Solvent DMSO.
[g] Solvent CCl_4.

[31] J. M. Markgraf, W. T. Bachmann, and D. P. Hollis, *J. Org. Chem.* **30**, 2387 (1965).
[32] F. L. Scott and J. C. Tobin, *J. Chem. Soc., C*, 703 (1971).
[33] L. Huff and R. A. Henry, *J. Med. Chem.* **13**, 777 (1970).
[34] R. R. Fraser and K. E. Haque, *Can. J. Chem.* **46**, 2855 (1968).
[35] A. K. Sorensen and N. A. Klitgaard, *Acta Chem. Scand.* **26**, 541 (1972).
[36] R. Raap and J. Howard, *Can. J. Chem.* **47**, 813 (1969).
[37] L. A. Lee and J. W. Wheeler, *J. Org. Chem.* **37**, 348 (1972).

$$=\overset{|}{\text{C}}-\text{N}-\overset{|}{\underset{\text{R}}{\text{C}}}= \quad > \quad =\text{N}-\text{N}-\overset{|}{\underset{\text{R}}{\text{C}}}= \quad > \quad =\text{N}-\underset{\text{R}}{\overset{|}{\text{N}}}-\text{N}=$$

(A) (B) (C)

prepared in recent years and on which NMR spectra have been reported are summarized in Table I. In each case the chemical shift of the 1-alkyl isomer is ca. 0.15–0.35 ppm upfield from that of the 2-isomer. The data represent an example of the more general chemical shift correlation in simple azoles where the alkyl chemical shift depends on the location of the alkyl group and decreases in the order A > B > C for the structural units shown.[29] The 1- and 2-alkyltetrazoles represent (B) and (C) units, respectively. The origin of the trend is a direct relationship between the electron density of the nitrogen atom and the chemical shift of the bonded alkyl group, the electron densities being influenced by the higher electron-withdrawing power of the —N= moiety over the —C= moiety. Similar correlations have been reported between electron densities and (i) ^{13}C chemical shifts[12,38,39] and (ii) amino proton chemical shifts in azoles.[40-42] There is a paucity of chemical shift data for amino groups bonded to the 1- and 2-ring nitrogen atoms of tetrazoles, but, in the few 5-substituted 1- and 2-aminotetrazoles that have been reported,[43] the 1-N-amino protons were considerably more shielded than those of the 2-N-amino group, $\Delta\tau$ 0.8–1.0 ppm. The correlation for the alkyltetrazoles breaks down when substituent effects outweigh structural effects in governing the electron density distribution in the ring.[28,29] It is interesting that a similar correlation between the structural units A, B, C and the ^{13}C chemical shifts of bonded N-methyl groups also seems to be emerging from the limited number of ^{13}C spectral data available (cf. Table II).

Interesting interannular conjugation effects in substituted phenyltetrazoles have been detected from NMR spectra.[34] Thus, for example, the phenyl protons of compound **36** appear as a singlet at 2.36τ while with the 2-methyl isomer **37** these protons appear as a multiplet with H ortho at 1.87τ and H meta/para at 2.59τ.[34] This is a general phenomenon among aryl azoles but the shifts are larger with the tetrazoles.[29] The shielding of the meta/para protons in compound **37** relative to **36** is considered to be due to an electron-donating resonance interaction between the rings[34] which is absent or greatly diminished

---

[38] B. M. Lynch and H. M. Dow, *Tetrahedron Lett.*, 2627 (1968).
[39] J. Elguero, E. Gonzalez, and R. Jacquier, *Bull. Soc. Chim. Fr.*, 2998 (1967).
[40] M. A. Khan and B. M. Lynch, *Can. J. Chem.* **49**, 3566 (1971).
[41] B. M. Lynch, B. C. McDonald, and J. G. K. Webb, *Tetrahedron* **24**, 3595 (1968).
[42] B. M. Lynch, *Tetrahedron Lett.*, 1537 (1969).
[43] R. Raap, *Can. J. Chem.* **47**, 3677 (1969).

TABLE II
^{13}C CHEMICAL SHIFTS OF ISOMERIC N-METHYL GROUPS IN AZOLES

Compound	δ-NMe (from TMS)	Structure unit	References
Tetrazole derivative			
1-Methyl-5-phenyl-	35.0^a	B	46
2-Methyl-5-phenyl-	39.4^a	C	46
1-Methyl-	33.7^b	B	57
2-Methyl-	38.8^b	C	57
Triazole derivative			
4-Methyl-1,2,4-	30.7^b	A	57
1-Methyl-1,2,4-	36.0^b	B	57
2-Methyl-1,2,3-	41.5^b	C	57
1-Methyl-1,2,3-	35.7^b	B	57
Diazole derivative			
1-Methylimidazole	32.6^b	A	57
1-Methylpyrazole	38.2^b	B	57

^{a}Solvent, CDCl$_3$.
^{b}Solvent, dioxane.

owing to steric loss of coplanarity in the 1-alkyl isomer (**36**). Comparisons of the shifts for a series of azoles have suggested that the deshielding of the ortho protons relative to benzene in compounds such as **37** is made up of components due to (i) an inductive effect of the heterocycle, (ii) a ring current effect in the heterocyclic ring, and (iii) nitrogen anisotropic effects when nitrogen atoms occupy positions β to the bonding site of the phenyl ring, as is the case with tetrazoles.[29] Similar effects have been observed for a series of p-nitrophenylazoles, including tetrazoles, where the phenyl ring is bonded to a nitrogen atom.[44,45] ^{13}C NMR spectra have also been useful for studying this phenomenon.[46,47] With compound **37** the chemical shift of C_2' was 126.4 ppm from TMS, and $\delta C_3' - \delta C_2'$ was 2.1 ppm while for compound **36** $\delta C_2'$ was 128.1 ppm and $\delta C_3' - \delta C_2'$ was 0.7 ppm.[46] It has been suggested[46] that interannular conjugation is extensive in C-phenylazoles if $\delta C_2' \sim 124.5-126.4$ ppm and $\delta C_3' - \delta C_2' \sim 3.2-3.8$ ppm, while if such conjugation is impeded $\delta C_2' \sim 128-128.5$ ppm and $\delta C_3' - \delta C_2' \sim 0-0.7$ ppm.[46] The UV spectra of the compounds **36** and **37**, λ_{max} 232 nm and 240 nm, respectively,[48] demonstrate rather mundanely that there are different extents of conjugation in both cases, but NMR spectroscopy is obviously a superior technique for studying this type of phenomenon.

[44] J. Elguero, R. Jacquier, and S. Mondon, *Bull. Soc. Chim. Fr.,* 1346 (1970).
[45] J. Elguero, R. Jacquier, and G. Tarrago, *Bull. Soc. Chim. Fr.,* 1345 (1970).
[46] M. Begtrup, *Acta Chem. Scand.* **27**, 3101 (1973).
[47] M. Begtrup, *Acta Chem. Scand.* **B28**, 61 (1974).
[48] B. Elpern and F. C. Nachod, *J. Am. Chem. Soc.* **72**, 3379 (1950).

NMR studies of the conformations of acetyl groups directly bonded to ring-nitrogen atoms of azoles including tetrazoles have also been reported.[49]

(36) (37)

Other recent work on tetrazole systems includes ^{23}Na and ^{7}Li NMR studies of the complexes formed between Na$^+$ and Li$^+$ ions with pentamethylenetetrazole.[50,51] In both cases weak 1:1 complexes were formed, and these were stronger for the lithium ion. The Na$^+$–PMT complex was stronger in nitromethane than in water, and as a result it has been suggested that Na$^+$–PMT interactions in biological systems occur in the lipid phase of membranes rather than in the aqueous phase.[50] Chemical shift differences between neutral tetrazoles and tetrazole anions have also been reported.[24,52] Higher shielding observed in the anions was mainly due to increased electron densities rather than changes in ring currents or anisotropic effects.[52] It is worth mentioning here that one of the earliest uses of the NMR technique on tetrazoles, which apparently is not widely known, was an interesting attempt to follow the thermal isomerization of 5-alkylaminotetrazoles (38) to 1-alkyl-5-aminotetrazoles (39).[53] The disappearance of the alkyl signals of compounds (38) and their gradual replacement by new lower-field alkyl signals of compounds (39) were clearly followed but intermediates in the process, which involves ring opening followed by closure on the (NHR) amino site,[54] could not be detected.

(38) (39)

---

[49] M. M. L. Pappalardo, J. Elguero, and C. Marzin, *C.R. Hebd. Seances Acad. Sci., Ser. C.* **277**, 1163 (1973) [*CA* **80**, 70186 (1974)].

[50] R. L. Bodner, M. S. Greenberg, and A. I. Popov, *Spectrosc. Lett* **5**, 489 (1972) [*CA* **78**, 76509 (1973)].

[51] E. T. Roach, P. R. Handy, and A. I. Popov, *Inorg. Nucl. Chem. Lett.* **9**, 359 (1973).

[52] S. Bradamante, G. Pagani, and A. Marchesini, *J. Chem. Soc., Perkin Trans. 2*, 568 (1973).

[53] A. G. Whittaker, D. W. Moore, J. N. Shoolery, and R. Jones, *J. Chem. Phys.* **25**, 366 (1956).

[54] R. A. Henry, W. G. Finnegan, and E. Lieber, *J. Am. Chem. Soc.* **77**, 3977 (1955).

3. Tautomerism

a. Annular. The possible annular tautomerism **40** ⇌ **41** in 5-substituted tetrazoles has aroused considerable interest and, not surprisingly, controversy since the balance of the tautomerism might be expected to be particularly sensitive to the environment. The main methods by which the problem has been tackled are NMR spectroscopy, dipole moment measurements, and theoretical calculations. Using proton NMR, Moore and Whittaker[55] compared the chemical shift of the 5-CH proton of tetrazole (measured in DMF) with that of 1-alkyl- and 2-alkyl-substituted tetrazoles (measured in the liquid state), and, from the similarity between the chemical shift of the 5-CH proton of the parent molecule with that of the 1-alkyl isomers, they concluded that form **40**,

R = H, was preferred.[55] A similar conclusion was arrived at from PMR spectra of tetrazole measured as a function of temperature down to −90° in deuterioacetone when separate 5-CH signals were observed (besides others attributed to slow, reversible, addition of acetone to the tetrazole N), and assigned on the basis of comparisons with *N*-methyl derivatives.[56] It was reported that form **40**, R = H, was present in 85% and form **41**, R = H, in 15%. An interesting ^{13}C study also indicated a preference for the 1-H form.[57] Thus, the C-5 chemical shift of tetrazole in DMF was 143.9 ppm, similar to that of 1-methyltetrazole (144.9 ppm) but considerably different from that of 2-methyltetrazole (153.5 ppm).[57] Similar results were obtained in dioxane, water, DMSO, and acetone.[57] Attempts to establish the tautomeric composition of 5-alkyltetrazoles (**40**, R = Me) by similar PMR procedures were unsuccessful since the 5-alkyl chemical shift differences between the 1- and 2-isomers were insignificant.[31] Also an attempt to distinguish between the different tautomers of 5-phenyltetrazole by using ^{15}N-labeled forms to detect the ^{15}N—^{1}H coupling was unsuccessful.[58] The 5-alkyltetrazoles

---

[55] D. W. Moore and A. G. Whittaker, *J. Am. Chem. Soc.* **82**, 5007 (1960).
[56] M. L. Roumestant, P. Viallefont, J. Elguero, and R. Jacquier, *Tetrahedron Lett.*, 495 (1969).
[57] J. Elguero, C. Marzin, and J. D. Roberts, *J. Org. Chem.* **39**, 357 (1974).
[58] P. Scheiner and J. F. Dinda, *Tetrahedron* **26**, 2619 (1970).

in SO_2,[31] in the concentration range 0.1–0.4 M, and the 5-phenyltetrazoles in DMSO,[58] in the concentration range ca. 1–2 M, were found to exist in hydrogen-bonded aggregates in solution. Comparisons of the two ^{14}N signals observed for tetrazole at 106 and ca. 20 ppm (in acetone and methanol) with the ^{14}N signals of 1-methyltetrazole (NMe, 150 ppm in chloroform) and 2-methyltetrazole (NMe, 101 and 103 ppm in CCl_4 and methanol, respectively) have led to suggestions that tetrazole exists primarily as the 2-H-tautomer **41**, R = H.[59] The ^{14}N data seem to contradict both the ^{13}C and ^1H data but the complexities that may arise from relaxation effects with the ^{14}N nuclei in tetrazoles may render these data less reliable for studies of tautomerism in these compounds.

Dipole moment measurements in solution suggest that tetrazole exists in the 1-H form. Thus the dipole moment of tetrazole in dioxane was 5.11 D,[60] while the dipole moments of 1-ethyltetrazole and 2-ethyltetrazole were 5.46 D and 2.65 D, respectively.[61] Numerous theoretical calculations[61-64] of the dipole moments of both forms of tetrazole agree with these results giving μ_D values for the 1-H form (**40**, R = H) of 4.8–5.23 D and the 2-H form (**41**, R = H) of 1.63–2.35 D. Calculations[13-18] of the relative energies of both forms suggested little difference between them, and in some cases[13] the 2-H form was considered to be slightly the more stable. Tautomerism in the gas phase was studied by using microwave spectra to determine dipole moments.[65] The values of μ_D arrived at for tetrazole, N-deuteriotetrazole, and C-deuteriotetrazole were 2.19, 2.14, and 5.30 D, respectively; accordingly, it was suggested that both tetrazole and N-deuteriotetrazole exist mainly as 2-H tautomers (**41**, R = H or D) in the gas phase while C-deuteriotetrazole exists as the 1-H form (**40**, R = D).[65] Mass spectral fragmentations of tetrazole in the gas phase also suggest the presence of both tautomers by comparison with the fragmentation patterns (Section II,B) of 1- and 2-substituted isomers.[66] Using 4- and 5-substituted 1-methylimidazolinium ions as models for the ionization of 1-H tetrazoles

---

[59] M. Witanowski, L. Stefaniak, H. Januszewski, Z. Grabowski, and G. A. Webb, *Tetrahedron* **28**, 637 (1972).

[60] K. A. Jensen and A. Friediger, *Kgl. Danske Videnskab. Selskab. Mat.-Fys. Medd.* **20**, 1 (1943) [*CA* **38**, 3629 (1944)].

[61] M. H. Kaufman, F. M. Ernsberger, and W. S. McEwan, *J. Am. Chem. Soc.* **78**, 4197 (1956).

[62] A. J. Owen, *Tetrahedron* **14**, 237 (1961).

[63] J. E. Bloor and D. L. Breen, *J. Am. Chem. Soc.* **89**, 6835 (1967).

[64] J. B. Lounsbury, *J. Phys. Chem.* **67**, 721 (1963).

[65] W. D. Krugh and L. P. Gold, *J. Mol. Spectrosc.* **49**, 423 (1974).

[66] D. M. Forkey and W. M. Carpenter, *Org. Mass Spectrosc.* **3**, 433 (1969) [*CA* **71**, 49069 (1969)].

(40) and 2-H tetrazoles (41), respectively, Charton[67] calculated macroscopic ionization constants for the tetrazoles, and from a comparison of these with observed values suggested that 5-substituted tetrazoles exist primarily as 2-H tautomers (41) in water.[67] It was also interestingly predicted that, if the 5-substituent R were strongly electron withdrawing, the 2-H form (41) would predominate, whereas if R were electron donating, the 1-H form (40) would be preferred. The predictions for tetrazole itself contradict the ^{13}C NMR data for aqueous solutions. No systematic experimental study of the effects of 5-substituents on the tautomerism has been reported to date.

b. Amino–Imino Tautomerism. The possibility of an amino–imino tautomerism in 5-aminotetrazoles (42) has also aroused considerable interest, particularly since 5-hydroxy- and 5-thioltetrazoles exist in the tetrazolinone forms.[68-70] It is generally accepted that tautomerism to an imino form occurs when there is a stabilizing interaction between a substituent on the amino group and the ring proton as, for example, with the intramolecular hydrogen bond in compound 44. This is suggested from anomalously low acidities[67,71,72] for compounds such as 44 and an insensitivity of the amino proton resonance in deuteriodimethyl sulfoxide solutions to dilution with nonpolar solvents, which is characteristic of intramolecular hydrogen bonding[73] rather than the strong intermolecular hydrogen bonding observed when the acetyl group is absent.[73]

Whether an amino–imino tautomerism of the type 42 ⇌ 43 occurs for 5-aminotetrazoles, which do not allow for intramolecular stabilizations

[67] M. Charton, *J. Chem. Soc., B,* 1240 (1969).
[68] E. Lieber, C. N. R. Rao, R. Pillai, J. Ramachandran, and R. D. Hites, *Can. J. Chem.* 36, 801 (1958).
[69] J. P. Horwitz, B. E. Fisher, and A. J. Tomasewski, *J. Am. Chem. Soc.* 81, 3076 (1959).
[70] J. C. Kauer and W. A. Sheppard, *J. Org. Chem.* 32, 3580 (1967).
[71] R. M. Herbst and W. L. Garbrecht, *J. Org. Chem.* 18, 1283 (1953).
[72] E. Lieber and E. Oftedahl, *Trans. Illinois State Acad. Sci.* 51, 41 (1958) [*CA* 56, 4151 (1962)].
[73] R. N. Butler, *J. Chem. Soc., B,* 680 (1969).

as in **44**, has proved to be problematic. In discussing this, it is necessary to draw a clear-cut distinction between compounds in the solid state and in solution, since failure to do this in the past has been one cause of ambiguity. Most solid-state studies have employed infrared (IR) spectra and, in these spectra with model compounds, the exocyclic C=N band[32,74] of 5-iminotetrazoles generally appears in the region 1640–1670 cm^{-1} whereas the cyclic C=N band of the tetrazole ring generally appears in the region 1600–1590 cm^{-1}.[32,74,75] With a number of compounds in which both forms are possible, the C=N absorption has fallen in or close to the region 1640–1670 cm^{-1}. This may suggest the presence of an imino form in the solid state. Such data have also been interpreted[76] to result from resonance canonical forms such as **46**. However, while such an explanation is valid, if it were the full explanation, the phenomenon should be more marked with dialkylaminotetrazoles, e.g., **48**, since the positive nitrogen should be more stabilized by

(45)
(47) R = Me; R^1 = Me

(46)
(48) R = Me, R^1 = Me

the alkyl groups. However, with compound **47** the C=N absorption[74] is at 1600 cm^{-1}, i.e., in the cyclic C=N range. This suggests that no resonance imino form is contributing significantly in this case and the normal form **47** is the dominant form. Recently, Nelson and Baglin[77] have claimed not only the detection but also the isolation of both amino and imino forms of hydrated 5-aminotetrazole, which were distinguished by laser-Raman absorptions. The amino form was monohydrated and showed an intense band at 1074 cm^{-1}, which was assigned to aromatic ring breathing. The imino form contained "$n \cdot H_2O$" molecules and showed no band at 1074 cm^{-1} but two weak bands at 1055 cm^{-1} and 1097 cm^{-1}, which were assigned to the separate double bonds of the nonaromatic imino form. Nelson and Baglin[77] concluded that the amino and imino forms of 5-aminotetrazole "are definitely interconvertible in

---

[74] D. B. Murphy and J. P. Picard, *J. Org. Chem.* **19**, 1807 (1954).
[75] R. N. Butler and T. M. McEvoy, unpublished work.
[76] G. Bianchi, A. J. Boulton, I. J. Fletcher, and A. R. Katritzky, *J. Chem. Soc., B*, 2355 (1971).
[77] J. H. Nelson and F. G. Baglin, *Spectrosc. Lett.* **5**, 101 (1972) [*CA* **77** 47707 (1972)].

the solid state" and that excess water favored the imino form, which was thermodynamically unstable with respect to the amino form. Schipanov and Postovskii[78] have concluded from IR studies of 1- and 2-methyl 5-amino- and 5-amidotetrazoles that the tautomeric equilibrium was dependent on the phase, the nature of the acyl group, and the position of ring methylation.

In a letter to the author, dated December 10, 1971, Professor Postovskii stated: "I am of the opinion that an amino–imino tautomerism obviously takes place in this case" (i.e., **42** ⇌ **43**). "According to the general views of Kabachnik[78a] the first form (**42**) is more available in solutions because the acidity of form (**43**) is probably stronger than that of (**42**). Only form (**42**) is probable in the solid state. I believe that the equilibrium (**42**) ⇌ (**43**) depends on the polarity of a solvent with the increasing of form (**43**) in more polar solvents."

Studies of possible tautomerism in solution have centered mainly around UV and NMR data. The UV spectra of 1-substituted 5-aminotetrazoles in 95% ethanol show only end absorption, λ_{max} 218–232 nm, characteristic of the amino form irrespective of whether the 5-amino substituent is primary, secondary, or tertiary.[74] These data suggest that only the amino form is present. With 2-methyl-5-methylaminotetrazole (**49**), the UV absorption is at λ_{max} 254 nm and is similar to that of the hydrochloride of 1,3-dimethyl-5-iminotetrazole (**50**) measured in an aqueous solution containing one equivalent of NaOH to neutralize the hydrochloride to **50a**.[79] The similarity in the UV spectra of compounds **49** and **50a** have led to suggestions that compound **49** exists in the imino form (**49a**).[79,80] (This assumes optical transparency for the methyl groups and that the increased ionic strength of the solution of **50a**, due to the NaCl, does not greatly alter the absorption maximum). However, Katritzky and co-workers[76] have pointed out that PMR spectra clearly suggest that 1-substituted 5-methylaminotetrazoles exist in the amino form in solution. Three signals observed for the 5-methylamino group in 1-substituted 5-methylaminotetrazoles in deuterodimethyl sufoxide[32,80] were found by these workers to be due to the amino form, which was about one-third deuteriated arising from the presence of deuterium oxide in the solvent,[76] not to an imino form as we had supposed.[80] At the same time, other workers[81] also warned of the presence of D_2O in DMSO-d_6. We have since confirmed the correctness

---

[78] V. P. Schipanov and I. Ya. Postovskii, *Spectrosk. Tr. Sib. Soveshch, 4th, 1965* 144 (1969) [*CA* **74**, 41715 (1971)].

[78a] M. I. Kabachnik, *Dokl. Akad. Nauk SSSR* **83**, 407 (1952) [*CA* **46**, 8499f (1952)].

[79] R. A. Henry, W. G. Finnegan, and E. Lieber, *J. Am. Chem. Soc.* **76**, 2894 (1954).

[80] R. N. Butler, *J. Chem. Soc., B,* 138 (1970).

[81] G. Barker, G. P. Ellis, and D. A. Wilson, *Chem. Ind. (London),* 656 (1970).

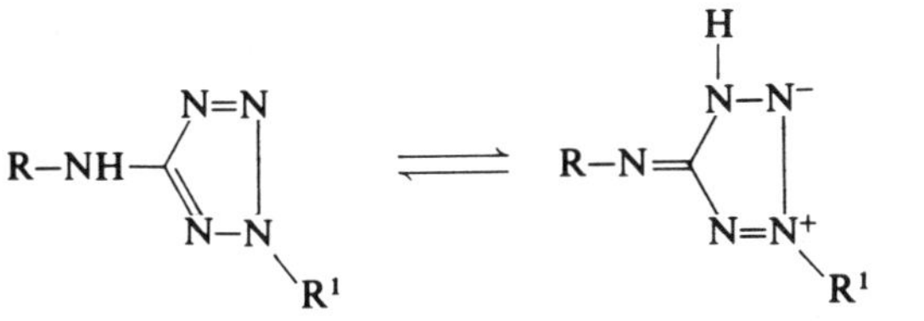

Sec. II.A] RECENT ADVANCES IN TETRAZOLE CHEMISTRY 337

$$R-NH-\underset{\underset{R^1}{N-N}}{\overset{N=N}{\diagup}} \rightleftharpoons R-N=\underset{\underset{R^1}{N=N^+}}{\overset{\overset{H}{|}}{\underset{}{N-N^-}}}$$

(49) R = R¹ = Me   (49a) R = R¹ = Me
(51) R = PhCH₂, R¹ = Me

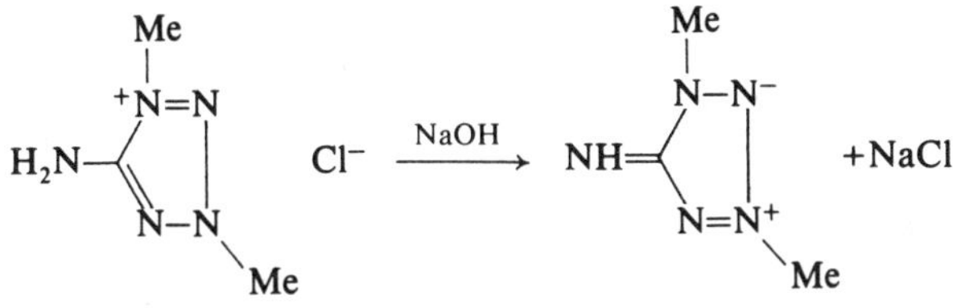

$$H_2N-\underset{\underset{Me}{N-N}}{\overset{\overset{Me}{|}}{\overset{+N=N}{\diagup}}} Cl^- \xrightarrow{NaOH} NH=\underset{\underset{Me}{N=N^+}}{\overset{\overset{Me}{|}}{\underset{}{N-N^-}}} + NaCl$$

(50)   (50a)

of Katritzky's NMR data and also examined[75] compound **51**, 5-benzylamino-2-methyltetrazole. The IR spectrum of this compound (solid state) showed a $>C=N-$ band at 1620 cm⁻¹ intermediate between the ranges mentioned above. The UV spectrum in methanol showed λ_{max} 257 nm similar to the other 2-alkyl-5-alkylaminotetrazoles. NMR spectra

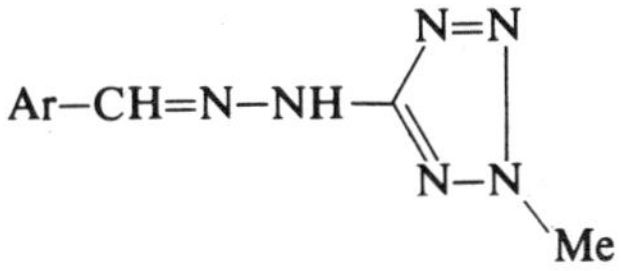

$$Ar-CH=N-NH-\underset{\underset{Me}{N-N}}{\overset{N=N}{\diagup}}$$

(52)

showed the benzyl CH₂ protons as a doublet (J, 6 Hz) at 5.52τ in deuteriochloroform and deuterioacetone and 5.66τ in deuteriodimethylsulfoxide. In each case the 2-NMe signal was a sharp singlet at 5.96τ. The NMR data are conclusive and suggest a strong preference for the amino form in solution. The UV data on these 2-alkyl-5-alkylaminotetrazoles are enigmatic and may possibly be due to special conjugation between the 2-substituent and the 5-amino group, the similarity in absorption with that of form **50a** being coincidental. This would be consistent with the suggestions, from theoretical calculations,[17] of a special resonance interaction in 5-amino-2-methyltetrazole and the resonance interactions detected by NMR in 2-alkyl-5-aryltetrazoles.[34] However, the presence of the alkyl group on the 5-amino substituent seems to play a significant role, since a series of the 2-methyl-5-substituted aminotetrazoles (**52**) showed UV spectra that were identical with their 1-methyl isomers[82] and in which the aminotetrazole end absorption appeared at λ_{max} 221–226 nm.[82]

B. Mass Fragmentation

1. *2,5-Disubstituted Tetrazoles*

The fragmentation pattern of compounds of type **53** was relatively simple and reasonably consistent. Fraser and Haque[34] reported a path involving loss of N_2 followed by loss of a hydrogen atom, leaving a fragment which then lost HCN (Scheme 1). This was confirmed by Forkey and Carpenter,[66] who found that the loss of H· and HCN may also occur in one step and that an additional fragmentation path involving

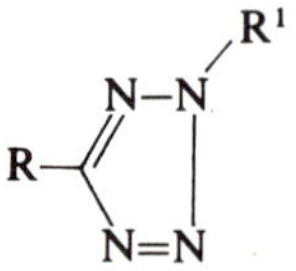

(**53**) R = H, alkyl, aryl
R¹ = alkyl

loss of RCN_2^- also occurred. Hence the overall fragmentation pattern for compounds of type **53** is as in Scheme 1. The compounds **53** (R' = Me) showed a unique M + 1 ion and displayed no molecular ion while the 1,5-disubstituted isomers showed molecular ion peaks and no abnormal M + 1 peaks, hence allowing a simple distinction between the isomers.[66]

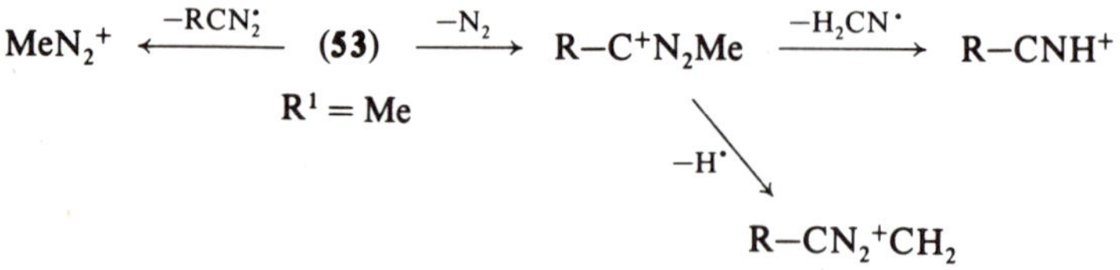

$$MeN_2^+ \xleftarrow{-RCN_2^-} (53) \xrightarrow{-N_2} R-C^+N_2Me \xrightarrow{-H_2CN\cdot} R-CNH^+$$
$$R^1 = Me \qquad \downarrow -H\cdot$$
$$R-CN_2^+CH_2$$

SCHEME 1

2. *1,5-Disubstituted Tetrazoles*

A more complex fragmentation pattern was exhibited by compounds such as **54**. For these materials, where R = aryl and R' = methyl, Fraser and Haque[34] reported a fragmentation pattern involving initial ejection

---

[82] R. N. Butler and F. L. Scott, *J. Chem. Soc., C*, 1711 (1968).

of $CH_2N_2^+$ or N_3^+ followed by loss of $CH_2N^\cdot$ or N_2. A precise mass measurement[66] showed that both $CH_2N_2^+$ and N_3^+ were ejected initially in a ratio of 2:1. However, for compound **54** (R = R' = Me) N_3^+ was lost rather than $CH_2N_2^+$.[66] In general for 1,5-disubstituted tetrazoles (**54**, R = H, or alkyl; R^1 = methyl) a number of different fragmentation patterns were found to be operating (Scheme 2).[66,83] The most significant was process (a) involving initial loss of $NH_2^\cdot$ followed by ejection of HCN. A composite of these two steps also occurred when $CH_2N_3^+$ was ejected (process b). Loss of a fragment R-$CN_2^\cdot$ also occurred (process c) giving rise to a strong $CH_3N_2^+$ peak. Other minor processes consisted of loss of N_2 and $N_3^\cdot$.[66] It is of interest that the mass spectra of compounds

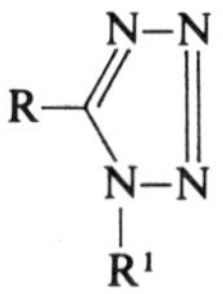

(**54**) R = H, alkyl, aryl
  R^1 = alkyl
(**55**) R = R^1 = COOMe
(**56**) R = R^1 = Ph
(**57**) R = PhO; R^1 = Ph

55 and **56** are reported[84] to display an intense M-28 peak, which appears to indicate an important loss of N_2. It seems clear that the fragmentation patterns observed for 1,5-disubstituted tetrazoles may vary considerably with the nature of the tetrazole ring substituents.

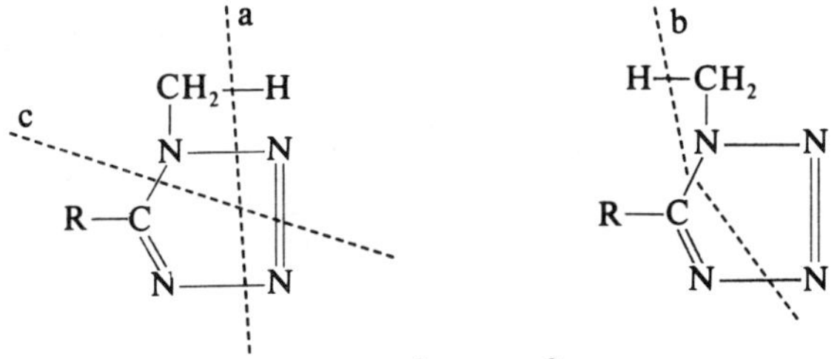

SCHEME 2

[83] C. Ainsworth, *J. Heterocycl. Chem.* **3**, 470 (1966).
[84] R. M. Moriarty, J. M. Kliegman, and C. Shovlin, *J. Am. Chem. Soc.* **89**, 5958 (1967).

The mass spectrum of 5-phenoxy-1-phenyltetrazole (**57**) displayed a fragmentation pattern[85] involving initial loss of N_2 followed by loss of H· and CO (Scheme 3). This fragmentation pattern provides evidence for an interesting 1,3-O–N phenyl migration in the material **57** since CO is ejected with retention of both phenyl rings.[85] Five different mass spectral fragmentation patterns have been described for 1-phenyl-5-mercaptotetrazole.[86] The fragmentations were sensitive to substituents

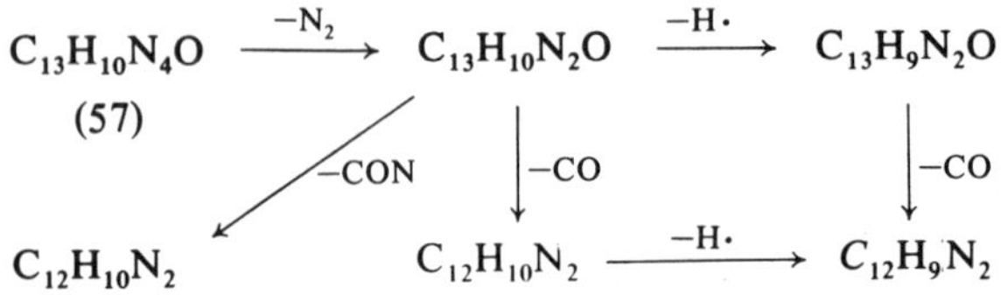

$$C_{13}H_{10}N_4O \xrightarrow{-N_2} C_{13}H_{10}N_2O \xrightarrow{-H\cdot} C_{13}H_9N_2O$$
(**57**)
$$\searrow{-CON} \quad \downarrow{-CO} \quad \downarrow{-CO}$$
$$C_{12}H_{10}N_2 \qquad C_{12}H_{10}N_2 \xrightarrow{-H\cdot} C_{12}H_9N_2$$

SCHEME 3

on the 1-phenyl ring, and the patterns for a variety of substituted phenyl derivatives have been reported.[87] Two mass spectral fragmentation patterns have been observed for 5-aminotetrazole.[88] These are (a), $NH_2CN_4H \rightarrow HN_3 \rightarrow HN_2$ and (b) $NH_2CN_4H \rightarrow CH_3N_3 \rightarrow HN_2$, H_2N_2, CH_2N. Interestingly, no fewer than seven metastable transitions were observed in the fragmentation of this nine-atom molecule.[88]

C. Tetrazole Complexes and Crystal Structures

A review of the complexes of 1,5-disubstituted tetrazoles, which covers the earlier literature and contains tables of comparative physical constants for the various complexes, has been published.[10] Recent studies have been concerned mainly with determining the nature of the bonding and the structure in tetrazole complexes and with assessing the π-donor ability of the tetrazole ring.

1. Molecular Complexes

Because of the weak protophilic properties of 1,5-disubstituted tetrazoles, it was not possible to obtain basicity constants for these compounds in water. In formic acid, however, pK_b values of ca. 2* were ob-

---

* Editors' note: Using the Authors'[91] value for the ionic product of formic acid, this corresponds to a pK_a value of ca. +4.6, and their figure for tetrazole itself to a pK_a of 4.8. Clearly, these values are not directly comparable with those determined in aqueous sulfuric acid (see Section II,A,1).

[85] F. L. Bach, J. Karliner, and G. E. Van Lear, *Chem. Commun.*, 1110 (1969).
[86] A. Antonowa, R. Borsdorf, R. Herzschuh, G. Fischer, and H. Engelmann, *J. Prakt. Chem.* **315**, 313 (1973) [*CA* **78**, 146978 (1973)].
[87] E. Lippmann, H. Loster, and A. Antonowa, *J. Signalaufzeichnungsmaterialien* **2**, 209 (1974) [*CA* **82**, 30496 (1975)].
[88] L. E. Brady, *J. Heterocycl. Chem.* **7**, 1223 (1970).

tained for a variety of tetrazoles, the leveling effect of solvent making it difficult to distinguish between individual compounds.[89-91] The π-donor ability of the tetrazole ring leads to the formation of charge-transfer complexes with π acids, such as tetracyanoethylene, tetracyanoquinodimethane, chloranil, and 1,3,5-trinitrobenzene.[92] In general, the π-donor ability of the tetrazoles was low and the complexes were weak and easily dissociated.[92] A correlation between the inductive effect of substituents on the tetrazole ring and the stability of the complexes was noted.[92]

Charge-transfer complexes between a range of 1,5-cyclopolymethylenetetrazoles (**58**) with iodine have been detected.[93] The donor properties of the tetrazoles were weak, and there was no simple correlation between the length of the hydrocarbon chain and the stability of the tetrazole–iodine complexes.[93] The compound pentamethylenetetrazole (**58**, $n = 5$) (PMT) was found to be relatively strong electron donor toward Lewis acids, but showed no basic properties in water.[94] In formic acid, however, the compounds **58** behaved as

(**58**)   (**59**)

fairly strong monoprotonic bases.[95] Whereas complexes of the type PMT·I$_2$ and PMT·IBr have been observed in solution,[96] a stable solid complex of formula PMT·ICl (**59**) has been isolated[96] and its crystal structure established.[97] In the complex the PMT molecule acts as a unidentate ligand with the iodine atom of the iodine monochloride bound to the 4-N atom of the tetrazole ring. The 4-N . . . I-Cl moiety is linear and coplanar with the tetrazole ring, which is itself planar. The seven-membered ring of the PMT molecule is in the chair conformation as in structure (**59**).[97]

---

[89] A. I. Popov and J. C. Marshall, *J. Inorg. Nucl. Chem.* **19**, 340 (1961).
[90] A. I. Popov and J. C. Marshall, *J. Inorg. Nucl. Chem.* **24**, 1667 (1962).
[91] T. C. Wehman and A. I. Popov, *J. Phys. Chem.* **72** 4031 (1968).
[92] T. C. Wehman and A. I. Popov, *J. Phys. Chem.* **70**, 3688 (1966).
[93] F. M. D'Itri and A. I. Popov, *J. Am. Chem. Soc.* **90**, 6476 (1968).
[94] A. I. Popov and R. D. Holm, *J. Am. Chem. Soc.* **81**, 3250 (1959).
[95] R. H. Erlich and A. I. Popov, *J. Phys. Chem.* **74**, 338 (1970).
[96] A. I. Popov, C. C. Bisi, and M. Craft, *J. Am. Chem. Soc.* **80**, 6513 (1958).
[97] N. C. Baenziger, A. D. Nelson, A. Tulinsky, J. M. Bloor, and A. I. Popov, *J Am. Chem. Soc.* **89**, 6463 (1967).

2. Cyclopolymethylenetetrazole Complexes with Metals

Numerous complexes of 1,5-cyclopolymethylenetetrazoles (**58**) with metals have been prepared, mainly using PMT. With first-row transition metal perchlorates, PMT forms complexes of the general type $M^{II}(PMT)_6(ClO_4)_2$.[98-101] The metals used were Fe, Mn, Co, Ni, Zn, and Cu. The complexes are soluble in water and polar nonaqueous solvents. Magnetic susceptibility measurements showed them to be of high spin type.[98,99] Complexes of type $M^{II}(PMT)_4(ClO_4)_2$ were also observed.[98,99] With Cu as metal, using acetic acid or 2,2-dimethoxypropane as solvent, the range of materials $Cu(PMT)_2ClO_4$, $Cu(PMT)_4(ClO_4)_2$ and $Cu(PMT)_6(ClO_4)_2$ was obtained.[100] The IR spectra of the complexes led to suggestions that in the hexakis (PMT)–transition metal complexes the PMT ligands are arranged around the central metal ion in such a way as to prevent direct coordination between the perchlorate group and the metal.[101] ESR spectra showing Cu hyperfine splitting were observed with samples of $Cu(PMT)_6(ClO_4)_2$ and $Cu(PMT)_4(ClO_4)_2$.[98] Spectrophotometric evidence has also been obtained for the formation of complexes between PMT and Cu(II), Ni(II) and Cr(III) when PMT is added to solutions of the perchlorates of these cations in formic acid.[102]

$$M^{II}(PMT)X_2 \qquad\qquad M^{II}(PMT)_2X_2$$
$$(60) \qquad\qquad\qquad (61)$$

When the anion was changed to a halogen, surprisingly different complexes were obtained. With chlorides and bromides of the metals Cr, Mn, Fe, Co, Ni, Cu, and Zn, PMT formed two types of complexes of general formulas **60** and **61**.[103] Both types of complexes were quite insoluble, **60** being the more stable. The metal ions in the compounds **60** are considered to be in octahedral environments, and these materials appeared to be polymeric.[103] The complexes **61** were considered to be tetrahedrally coordinated and to be monomeric.[103] The compounds **58** ($n = 3, 4, 5, 6, 7$) also form complexes with Ag^+ in which the general stoichiometry is Ag^+ (tetrazole)$_2NO_3^-$,[104] and whose stability is independent of the length of the hydrocarbon chain.[94,104] The crystal structure of one of these, nitratobis(pentamethylenetetrazole) silver (I), i.e. $[Ag(PMT)_2NO_3]_2$ has been determined.[105] The geometry about the silver

---

[98] H. A. Kuska, F. M. D'Itri, and A. I. Popov, *Inorg. Chem.* **5**, 1272 (1966).
[99] F. M. D'Itri and A. I. Popov, *Inorg. Chem.* **5**, 1670 (1966).
[100] F. M. D'Itri and A. I. Popov, *Inorg. Chem.* **6**, 597 (1967).
[101] F. M. D'Itri and I. A. Popov, *Inorganic Chem.* **6**, 1591 (1967).
[102] T. C. Wehman and A. I. Popov, *J. Inorg. Nucl. Chem.* **31**, 2951 (1969).
[103] D. M. Bowers and A. I. Popov, *Inorg. Chem.* **7**, 1594 (1968).
[104] F. M. D'Itri and A. I. Popov, *J. Inorg. Nucl. Chem.* **31**, 1069 (1969).
[105] R. L. Bodner and A. I. Popov, *Inorg. Chem.* **11**, 1410 (1972).

atom is a distorted tetrahedron consisting of bonded nitrate and both monodentate and bridging tetrazoles.[105] The monodentate tetrazole was coordinated to the silver at the 4-N atom, and the bridging tetrazoles were linked to the silver atom at the 3-N and 4-N atoms as in structure **62** (some bond lengths shown).

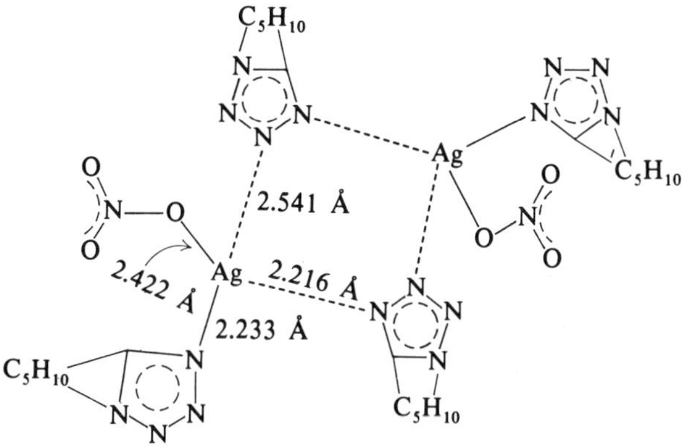

(**62**)

NMR studies of the interactions between Na^+ and Li^+ ions with PMT have been referred to in Section II,A,2.

3. Complexes of Substituted Tetrazoles with Metals

a. Complexes Containing Metal–Nitrogen Bonds. Cycloaddition of nitriles and the azide groups in *trans*-$Pd(PPh_3)_2(N_3)_2$ leads to trans tetrazole complexes of type **63**.[106,107] When these complexes were treated with HCl in alcoholic solution, the neutral tetrazole molecule was cleaved from the complex.[107] The cis form, **64**, of these complexes has also been reported from the reaction of the zero valent complexes $M[P(C_6H_5)_3]_4$ (M = Pd,Pt) with 5-substituted tetrazoles.[108] Compound **65** was also obtained by treating 5-cyclopropyl tetrazole with *cis*-$Cl_2Pt(PPh_3)_2$ and hydrazine.[108] This last reaction, when carried out with 5-bromo, 5-chloro, or 5-phenyltetrazole, gave trans monotetrazolatohydride complexes of type **66**.[108] In these complexes coordination of both the 1-N and 2-N tetrazole ring positions was observed.[108,109] Recently, the cis–trans isomerism **63** ⇌ **64** has been

[106] W. Beck and W. P. Felhammer, *Angew. Chem., Int. Ed. Engl.* **6**, 169 (1967).
[107] W. Beck, W. P. Felhammer, H. Bock, and M. Bauder, *Chem. Ber.* **102**, 3637 (1969).
[108] J. H. Nelson, D. L. Schmitt, R. A. Henry, D. W. Moore, and H. B. Jonassen, *Inorg. Chem.* **9**, 2678 (1970).
[109] D. A. Redfield, J. H. Nelson, R. A. Henry, D. W. Moore, and H. B. Jonassen, *J. Am. Chem. Soc.* **96**, 6298 (1974).

observed in solution and its mechanism investigated.[109] An X-ray structure analysis of *cis*-(Me$_2$PhP)$_2$ Pd^{II}(5-MeCN$_4$)$_2$ was carried out by Ansell,[110] and the cis configuration was confirmed. Both tetrazole rings are planar and bonded at the 1-N atom to the Pd, around which the configuration is a distorted square plane.[110] Complexes similar to **63**, in which the Pd is replaced by Ru and Cu, have also been obtained from the corresponding azides.[107]

[Ph$_3$P]$_2$ Pd^{II} [R—CN$_4$]$_2$

(**63**) *trans*
(**64**) *cis*
(**65**) *cis* (R = cyclopropyl)

(**66**)

Dichlorobis(1-methyltetrazole) zinc(II) has been reported and its structure determined.[111] The geometry around the zinc atom was that of a distorted tetrahedron. The tetrazoles were monodentate and again coordinated by a charge-transfer σ-bond between the 4-N atom and the Zn which was coplanar with the ring.[111] Cycloaddition of trialkyltin azides to nitriles or treatment of 5-substituted tetrazoles with bis(tri-*n*-butyltin) oxide or trimethyltin hydroxide gave 2-*N*-trialkyltin-5-substituted tetrazoles (**67**).[112,113] The (tri-*n*-butylstannyl)derivatives (**68**)

R—C≡N + R$^1_3$SnN$_3$ ⟶

(**67**)
(**68**) R^1 = *n*-C$_4$H$_9$

were polymeric in concentrated solutions.[112,113] Complexes of unsubstituted tetrazole with Fe(II), Ni(II), Zn(II), and Cd(II) have also been prepared,[114] and bis(tetrazolato) Cu(II) has been reported.[115] This compound is polymeric and contains one unpaired electron. Infrared

[110] G. B. Ansell, *J. Chem. Soc., Dalton Trans.*, 371 (1973).
[111] N. C. Baenziger and R. J. Schultz, *Inorg. Chem.* **10**, 661 (1971).
[112] K. Sisido, K. Nabika, T. Isida, and S. Kozima, *J. Organomet. Chem* **33**, 337 (1971); see also S. Kozima, T. Hitomi, T. Akiyama, and T. Isida, *J. Organomet. Chem.* **32**, 303 (1975).
[113] S. Kozima, T. Itano, N. Mihara, K. Sisido, and T. Isida, *J. Organomet. Chem.* **44**, 117 (1972).
[114] R. D. Holm and P. L. Donnelly, *J. Inorg. Nucl. Chem.* **28**, 1887 (1966).
[115] L. L. Garber, L. B. Sims, and C. H. Brubaker, *J. Am. Chem. Soc.* **90**, 2518 (1968).

bands at 328 cm^{-1} and 315 cm^{-1} have been attributed to Cu–N bonds.[115] Bonding of the tetrazole 2-N atom to copper has been reported in bis[2-(5-perfluoromethyltetrazolato)]-μ-1,2-bis(diphenylphosphino)ethane-bis[1,2-bis(diphenylphosphino)ethane] dicopper,[116] $Cu_2(CF_3CN_4)_2[(CH_2PPh_2)_2]_3$, the crystal structure of which has been determined.[116] In this molecule the geometry around the copper atom is a distorted tetrahedron and combination of 2-N bonding to copper, and the presence of the perfluoromethyl group at the tetrazole-5-position is reported to result in an unsymmetrical pattern of bond distances within the tetrazole ring.[116] Hexacarbonyls of molybdenum and tungsten, when treated with 5-substituted tetrazole anions, yield monomeric anionic pentacarbonyl metalates [M(CO)$_5$ tetrazole]$^-$, in which the tetrazoles behave as monodentate ligands bound at one of the ring N atoms.[117]

b. Complexes Containing Metal–Carbon Bonds. The compounds **69**, in which the tetrazole ligands are bonded to the central Au(III) atom at the 5-C atom, were prepared by treating (AsPh$_4$)[Au(N$_3$)$_4$] with substituted isocyanides.[106,118] The crystal structure of the tetrakis (1-isopropyltetrazol-5-ato) aurate(III) anion (**69**, R = isopropyl) has been determined.[119] The metal–carbon σ-bond was confirmed and the four tetrazole rings were found to be coordinated in a square-planar arrangement about the central Au(III) atom with the rings oriented so that the four rings lie in two mutually perpendicular planes with each pair of trans rings coplanar.[119] Treatment of bis(triphenylphosphine)bis(methylisocyanide) Pt(II) complexes with azide ion has led to complexes that are isomeric with structures **63** and **64**, but in which the alkyl group (R = Me) is at the 1-*N* tetrazole position and the metal is presumably bonded to the 5-C atom.[120] Evidence for a metal–carbon bond has also been reported for bis(1-methyltetrazol-5-yl) Ni(II).[121] The

(**69**)   (**70**)

[116] A. P. Gaughan, K. S. Bownan, and Z. Dori, *Inorg. Chem.* **11**, 601 (1972).
[117] J. C. Weiss and W. Beck, *Chem. Ber.* **105**, 3202 (1972).
[118] W. Beck, K. Burger, and W. P. Fehlhammer, *Chem. Ber.* **104**, 1816 (1970).
[119] W. P. Fehlhammer and L. F. Dahl, *J. Am. Chem. Soc.* **94**, 3370 (1972).
[120] P. M. Treichel, W. J. Knebel, and R. W. Hess, *J. Am. Chem. Soc.* **93**, 5424 (1971).
[121] L. L. Garber and C. H. Brubaker, *J. Am. Chem. Soc.* **88**, 4266 (1966).

compound was obtained by treating 1-methyltetrazol-5-yl lithium with dichlorobis(triethylphosphine) Ni(II) in THF at −50°. The analogous 1-cyclohexyl derivative has also been reported.[122] Both materials appear to be polymeric, to contain both Ni–C and Ni–N bonds, and to possess two unpaired electrons (magnetic moments, 2.90 for the methyl derivative and 2.98 μ_B for the cyclohexyl derivative).[121,122] The interesting 1,4-dimethyltetrazolin-5-ylidene iron carbonyls (**70**) have also been prepared from precursory (1,4-dimethyltetrazolium) carbonylferrates.[123] Ferrocenyltetrazole has been prepared from cyanoferrocene, but the tetrazole ring plays a substituent rather than a complexing role in this case.[124]

4. General

The complexing ability of substituted tetrazoline-5-thiones has rendered these materials useful for analytical determinations of a number of metals.[125–128] For example, *m*-phenylene-(1-tetrazoline-5-thione) (**71**) reacted quantitatively with Cu^{2+}, Ag^+, and Cd^{2+} ions, yielding precipitates of the insoluble metal complexes, which could be used for determining quantities of metal in the 0.5–50 mg range with errors of ca 0.5%.[125,126] Gravimetric determinations of transition metals, e.g., Cu^+, Ag^+, Au^+, Hg^+, can also be carried out using 1-substituted 5-mercaptotetrazole.[128,129] Treatment of a range of 2-alkyl, 2,5-disubstituted, and 1,5-disubstituted tetrazoles with aluminum hydride

(**71**)

---

[122] L. L. Garber and C. H. Brubaker, *J. Am. Chem. Soc.* **90**, 309 (1968).
[123] K. Öfele and C. G. Kreiter, *Chem. Ber.* **105**, 529 (1972).
[124] S. S. Washburne and W. R. Peterson, *J. Organomet. Chem.* **21**, 427 (1970).
[125] U. Agarwala and G. S. Johar, *Curr. Sci.* **48**, 139, 592 (1969) [*CA* **71**, 9399 (1969)].
[126] G. S. Johar and U. Agarwala, *Talanta* **17**, 355 (1970) [*CA* **73**, 10415 (1970)].
[127] G. S. Johar, *Labdev, Part A* **7**, 85 (1969) [*CA* **71**, 77023 (1969)].
[128] U. Agarwala and B. Singh, *Indian J. Chem.* **7**, 726 (1969) [*CA* **71**, 76861 (1969)].
[129] G. S. Johar and U. Agarwala, *Sci. Cult.* **35**, 479 (1969) [*CA* **73**, 56039 (1970)].

etherate at low temperatures yielded explosive AlH_3·tetrazole complexes.[130]

Recently, and X-ray crystal structure[131] of 5-bromotetrazole has been carried out, and the first single crystal Raman spectrum[132] for a tetrazole was measured on the same molecule. A table, that summarizes earlier literature on X-ray crystal structure determinations and contains a comparison of the bond distances in 5-aminotetrazole monohydrate, hydrazine salt of 5-aminotetrazole, sodium tetrazolate monohydrate, 2-methyl-5-aminotetrazole, 1,3-dimethyl-5-iminotetrazole, and pentamethylenetetrazole can be found in Baenziger and Schultz.[111] The tetrazole ring is clearly a planar resonance hybrid, and small differences in bond lengths of the order 0.02–0.04 Å, which have been noted,[116] may not be significant since the standard deviations are 0.02 Å. A crystal structure determination has also been carried out on 2,3-diphenyl-2H-tetrazolium-5-thiolate ("dehydrodithizone") (72).[133,134] In this molelcule the tetrazole ring atoms and the sulfur atom are coplanar, and both phenyl rings are twisted at 45° to this plane. The bond lengths within the tetrazole ring and the 5-C–S bond are intermediate between single and double bonds, and the tetrazole-5-thiolate unit is an aromatic resonance hybrid. There is little conjugation between the phenyl and tetrazole rings: the inter-ring C–N bond distance is 1.443 Å, the normal single C–N bond length.[133,134]

(72)

D. Photolysis and Tetrazole Radicals

Recently, the photolysis reactions of a range of tetrazoles have been investigated. Photolysis of tetrazoles in general involves ring fragmentation with evolution of nitrogen, but the specific reaction varies with the composition of the tetrazole species.

[130] N. R. Fetter, B. K. Bartocha, and K. Bodo, U.S. Patent 3,396,170 (1968) [*CA* **69**, 87170 (1968)].
[131] G. B. Ansell, *J. Chem. Soc., Perkin Trans. 2*, 2036 (1973).
[132] F. G. Baglin, *J. Chem. Phys.* **58**, 5534 (1973).
[133] Y. Kushi and Q. Fernando, *Chem. Commun.*, 1240 (1969).
[134] Y. Kushi and Q. Fernando, *J. Am. Chem. Soc.* **92**, 1965 (1970).

1. 5-Monosubstituted Tetrazoles

Irradiation of 5-phenyltetrazole (**73**) at 254 nm resulted in evolution of one molecule of N_2 and the formation of compound **74** along with lesser yields of the other products indicated in Eq. (1).[135,136] Further photodecomposition of compound **74** was the origin of the benzonitrile and the 3,5-diphenyl-1,2,4-triazole in the mixture.[135,136] The reaction is considered to involve loss of N_2 from the 3- and 4-positions of the ring[58,137] to yield a nitrilimine intermediate, $Ph-C\equiv N^+-NH^-$, which dimerizes to the dihydrotetrazine **74**, some of which is subsequently oxidized to the tetrazine form.

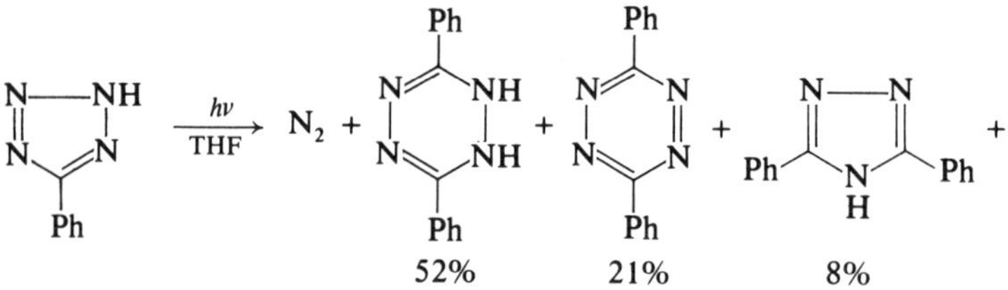

$$\text{(73)} \xrightarrow[\text{THF}]{h\nu} N_2 + \text{(74, 52\%)} + \text{(21\%)} + \text{(8\%)} + PhCN \quad (1)$$

The photolysis of 5-phenyltetrazolide, the anion of **73**, proved to be strikingly different.[135,138,139] Two molecules of nitrogen were evolved, and phenylcarbene was generated, probably via phenyldiazomethane. The reaction (Scheme 4) involved an initial photoexcitation of the anion to a long-lived triplet state, which proved to be an effective photosensitizer in the presence of suitable acceptors, e.g., conjugated dienes or oxygen. These quenched the evolution of nitrogen

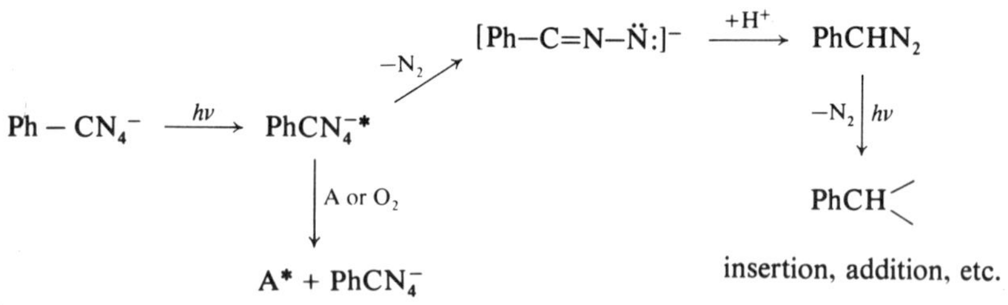

$$Ph-CN_4^- \xrightarrow{h\nu} PhCN_4^{-*} \xrightarrow{-N_2} [Ph-C=N-\ddot{N}:]^- \xrightarrow{+H^+} PhCHN_2 \xrightarrow[h\nu]{-N_2} PhCH{<}$$

$$\downarrow A \text{ or } O_2$$

$$A^* + PhCN_4^-$$

insertion, addition, etc.

SCHEME 4

[135] P. Scheiner, *J. Org. Chem.* **34**, 199 (1969).
[136] Y. B. Chae, K. S. Chang, and S. S. Kim, *Daehan Hwahak Hwoejee* **11**, 85 (1967) [*CA* **70**, 20031 (1969)].
[137] P. Scheiner and W. M. Litchman, *Chem. Commun.* 781 (1972); cf. also J. H. Boyer and P. J. A,. Frints, *J. Heterocycl. Chem.* **7**, 59, 71 (1970).
[138] P. Scheiner, *Tetrahedron Lett.*, 4863 (1969).
[139] P. Scheiner, *Tetrahedron Lett.*, 4489 (1971).

and halted the photolysis reaction.[138] Attempts to generate dicarbenes by photolysis of di(tetrazol-5-yl)benzenes resulted in consecutive reactions each involving a monocarbene pathway.[140] Aromatic carbenes have also been generated by thermolysis of 5-tolyltetrazoles which cleave initially to give diazomethane intermediates.[140a]

2. *2,5-Disubstituted Tetrazoles*

Both UV photolysis[141] and thermolysis[142] of 2,5-diphenyltetrazole (**75**) have been reported to yield a nitrilimine intermediate after evolution of one molecule of nitrogen. The same compounds were obtained from cycloaddition reactions with the products of tetrazole photolysis as from nitrilimines generated by the standard procedure of treating hydrazonyl halides (**76**) (Scheme 5) with base.[143] Thus photolysis of 2,5-diphenyltetrazole in the presence of cyanoacetylene yielded 1,3-diphenyl-5-cyanopyrazole,[144] and, in general, the expected 1,3-cycloaddition reactions of *C,N*-diphenylnitrilimine were observed in

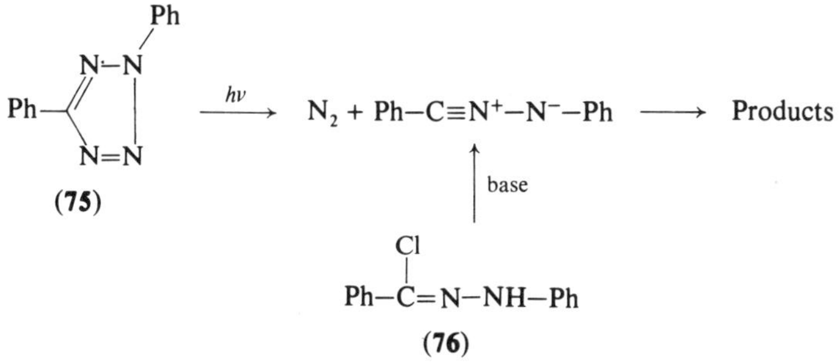

$$\text{(75)} \xrightarrow{h\nu} N_2 + Ph-C\equiv N^+-N^--Ph \longrightarrow \text{Products}$$

$$Ph-\underset{Cl}{C}=N-NH-Ph \xrightarrow{base}$$

(**76**)

SCHEME 5

photolysis reactions of 2,5-diphenyltetrazole.[145] When the photolysis was carried out in the absence of added dipolarophiles, the products were benzilosazone, 2,4,5-triphenyl-1,2,3-triazole, and benzanilide.[145] Photolysis of 2-methyl-5-phenyltetrazole (**21**) (Table I) does not appear to involve a free nitrilimine intermediate.[146] The main products are those shown in Eq. (2). A product directly analogous to compound **77** was

[140] P. Scheiner, E. Stockel, D. Cruset, and R. Noto, *J. Org. Chem.* **37**, 4207 (1972).
[140a] R. Gleiter, W. Rettig, and C. Wentrup, *Helv. Chim. Acta* **57**, 2111 (1974).
[141] J. S. Clovis, A. Eckell, R. Huisgen, and R. Sustmann, *Chem. Ber.* **100**, 60 (1967).
[142] R. Huisgen, J. Sauer, and M. Seidel, *Chem. Ber.* **94**, 2503 (1961).
[143] For reviews, see R. N. Butler and F. L. Scott, *Chem. Ind. (London)*, 1216 (1970); R. Huisgen, *Angew. Chem., Int. Ed. Engl.* **2**, 633, 565 (1963); **7**, 321 (1968).
[144] T. Sasaki and K. Kanematsu, *J. Chem. Soc., C*, 2147 (1971).
[145] C. S. Angadiyavar and M. V. George, *J. Org. Chem.* **36**, 1587 (1971).
[146] R. R. Fraser, G. Gurudata, and K. E. Haque, *J. Org. Chem.* **34**, 4118 (1969).

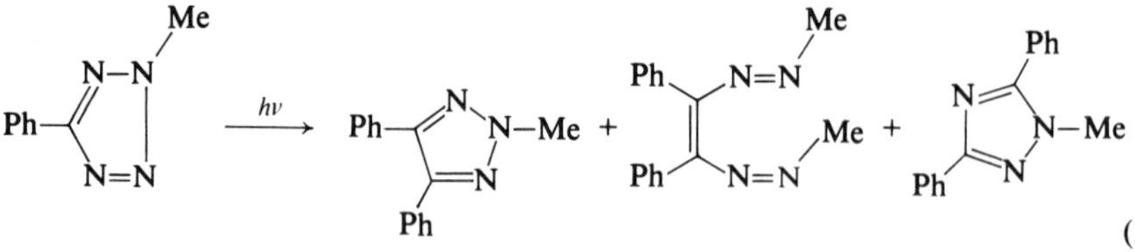

(21) → (77)   (2)

also obtained from the photolysis of 2,5-diphenyltetrazole in THF or benzene.[58] This is considered to arise either from a head-to-head dimerization of a nitrilimine or from decomposition of an intermediate formed in an initial tetrazole–tetrazole photodimerization, e.g., **78**.[58,146]

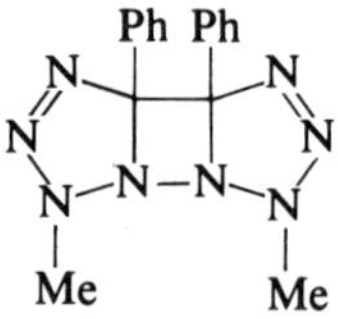

(78)

3. 1,5-Disubstituted Tetrazoles

UV photolysis of 1,5-disubstituted tetrazoles also results in evolution of N_2 and appears to involve an iminonitrene intermediate (Scheme 6).[147-149] Thus photolysis of 1,5-diphenyltetrazole (**56**) gave 2-

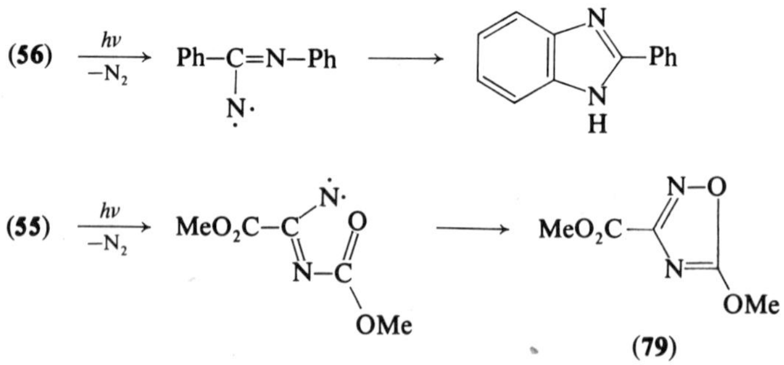

SCHEME 6

[147] W. Kirmse, *Angew. Chem.* **71**, 537 (1959).
[148] R. M. Moriarty and J. M. Kliegman, *J. Am. Chem. Soc.* **89**, 5959 (1967).
[149] J. Sauer and K. K. Mayer, *Tetrahedron Lett.*, 325 (1968).

phenylbenzimidazole,[147-149] and dimethyl tetrazole-1,5-dicarboxylate (**55**) gave the oxadiazole **79**.[84,150] Comparison of thermolysis and photolysis of compound **55** has led to suggestions that a triplet nitrene is involved in the photolysis and a singlet nitrene in the thermolysis.[150] Photolysis of 1-phenyl-5-aryloxytetrazoles also gave 2-aryloxybenzimidazoles from an imino-nitrene pathway.[85,151] With these compounds, a second pathway involving homolysis of the aryl–oxygen bond and ultimately giving rise to biphenyls has also been observed.[151]

$$
\begin{array}{c}
R^2 \\
| \\
R^1-N \\
\diagdown \\
N-N \\
\diagup\diagup \\
N-N \\
| \\
R^3
\end{array}
\qquad
\begin{array}{c}
R-CH_3 \\
\swarrow e^- \quad \searrow \gamma \\
R-CH_3^{\bullet -} \qquad RCH_3^{\bullet +} + 1e^- \\
\downarrow \qquad\qquad \downarrow -H^+ \\
R^- + CH_3^{\bullet} \qquad R-CH_2^{\bullet}
\end{array}
$$

(**80**)

R = 5-substituted tetrazole

SCHEME 7

Gamma irradiation of 1,5-disubstituted tetrazoles, in the solid state, resulted mainly in damage at the substituent sites.[152,153] For a series of tetrazoles (**80**), where R^3 was a methyl group, methyl radicals were formed owing to a cleavage of the 1-N–Me bond. When R^1 was an NO_2 group, cleavage of the exocyclic $N-NO_2$ bond yielding $NO_2^{\bullet}$ radicals also occurred.[153] Radicals arising from loss of an H atom at the alkyl side chain were also observed. The pattern of the reaction is outlined in Scheme 7, and it may be explained by an initial ejection of an electron from the system followed by loss of a proton from the resulting charged species. Simultaneously, electron capture by an undamaged parent molecule gives an anion that loses a Me$^{\bullet}$ radical by a process described as dissociative electron capture.

Photolysis of 1,4-dimethyltetrazolinone has been reported to involve evolution of nitrogen and gives *sym*-dimethylurea- and isourea-type products arising from solvent incorporation.[154] Photolysis of the Δ^2-

---

[150] R. M. Moriarty and P. Serridge, *J. Am. Chem. Soc.* **93**, 1534 (1971).
[151] P. D. Hobbs and P. D. Magnus, *J. Chem. Soc., Perkin Trans. 1*, 469 (1973).
[152] R. N. Butler, R. C. Catton, and M. C. R. Symons, *J. Chem. Soc., B* 378 (1970).
[153] R. C. Catton and R. N. Butler, *Can. J. Chem* **52**, 1248 (1974).
[154] W. S. Wadsworth, *J. Org. Chem.* **34**, 2994 (1969).

tetrazolines (**81**) also involves loss of N_2 and provides a new synthetic route to diaziridines in yields of preparative value.[155]

(**81**)

4. Tetrazolinyl Radicals

Radicals with spin located in a tetrazole ring[9] have been prepared by (i) oxidation of formazans,[156-159] (ii) reduction of tetrazolium salts,[156-160] and (iii) disproportionation of mixtures of formazans and tetrazolium salts in basic medium.[156] ESR spectra[156-161] of the radicals have nine-line hyperfine structure and clearly indicate a cyclic structure (**82**) as against a possible open-chain structure (**83**). Notable features of the ESR spectra were as follows, (i) appreciable spin density on all four ring N atoms (structure **83** would require high spin on the terminal N atoms only); (ii) higher splitting for $a^N 2,3$ (6.1–8.65 G) than for $a^N 1,4$ (5.4–5.8 G); (iii) intensitivity of the splitting constant $a^N 1,4$ to the nature of the substituent R^5; (iv) variation of $a^N 2,3$ with the substituents on N-2 and N-3. The value of $a^N 2,3$ varied from 8.65 G with $R^2 = R^3 = p$-anisyl to 8.0 G with $R^2 = R^3 = $ Ph, 7.50 with $R^2 = R^3 = $ 3,5-dichlorophenyl, 7.25 for $R^2 = R^3 = p$-carbethoxyphenyl and 6.65 with $R^2 = R^3 = p$-nitrophenyl, and in general a correlation was displayed between spin density and the Hammett σ constant.[161] The ESR spectrum of compound **82** ($R^2 = R^3 = $ Ph; $R^5 = -$N=N$-$Ph) derived from N,N'-diphenyl-C-phenylazoformazan, showed hyperfine splitting from only four N atoms, hence providing strong evidence in favor of the cyclic ring structure for the radicals.[157,158]

---

[155] T. Akiyama, T. Kitamura, T. Isida, and M. Kawansi, *Chem. Lett.*, 185 (1974) [*CA* **80**, 94932 (1974)]; see also H. Quast and L. Bieber, *Angew. Chem., Int. Ed. Engl.* **14**, 428 (1975).
[156] O. W. Maender and G. A,. Russell, *J. Org. Chem.* **31**, 442 (1966).
[157] F. A. Neugebauer, *Tetrahedron Lett.*, 2129 (1968).
[158] F. A. Neugebauer and G. A. Russell, *J. Org. Chem.* **33**, 2744 (1968).
[159] N. Azuma, K. Mukai, and K. Ishizu, *Bull. Chem. Soc. J.* **43**, 3960 (1970).
[160] Y. Deguchi and Y. Takagi, *Tetrahedron Lett.*, 3179 (1967).
[161] H. Le Maire, Y. Marechal, R. Ramasseul, and A. Rassat, *Bull. Soc. Chim. Fr.*, 372 (1965).

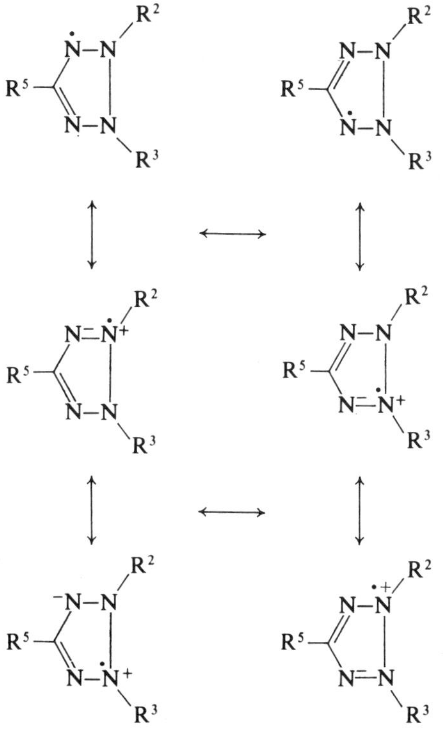

(82)

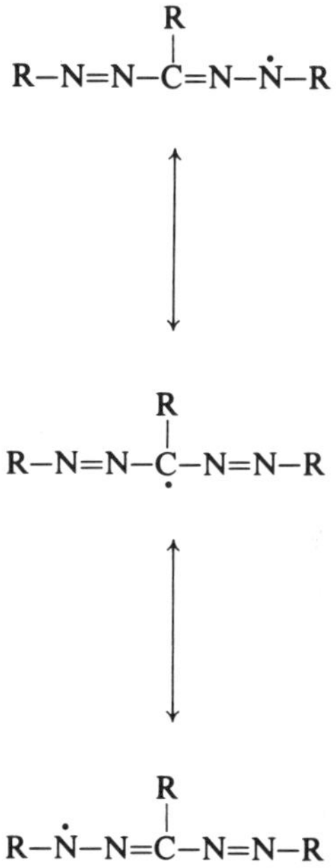

(83)

reported. These compounds generally involve a range of substituents where R^2 and R^3 are aromatic or highly conjugated and R^5 is generally an alkyl group or a carbethoxy substituent. The stability of the radicals

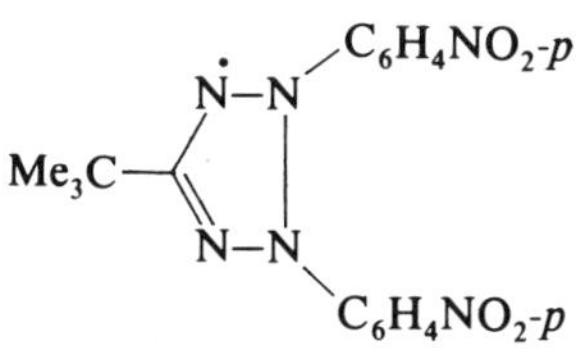

(84)

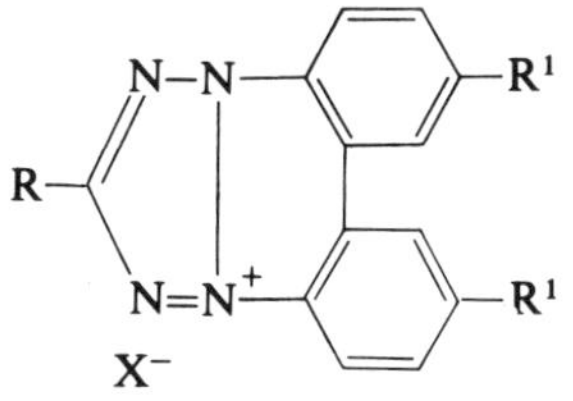

(85)

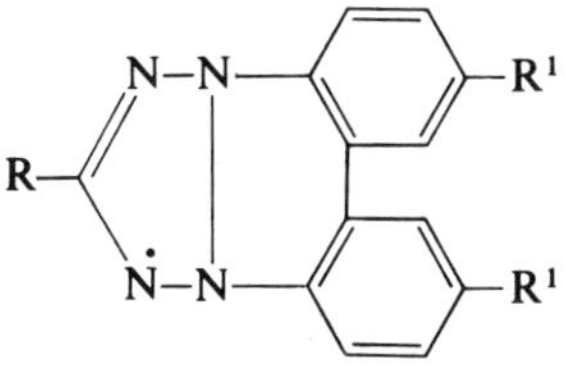

(86)

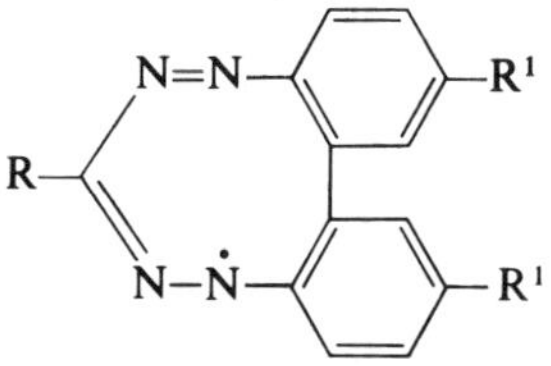

(86a)

increases with the electron withdrawing power of the substituents.[162,163] For example, the radical **84** is a stable crystalline solid[162,163] which decomposes at 126°–127°. Reduction of the tetrazolium salts (**85**) leads to the so-called Kuhn–Jerchel radicals.[164-167] ESR spectra[157,158,168] of these species have shown them to be tetrazolinyls (**86**) rather than the alternative valence isomers (**86a**).

III. Studies of Synthesis and Mechanism

A. 5-Monosubstituted Tetrazoles

1. 5-Vinyltetrazoles

An easy synthesis of 5-vinyltetrazoles (**87**) from substituted acrylonitriles and $Al(N_3)_3$ has been developed,[169] which contrasts with difficulties encountered in earlier attempts[170] to prepare such materials. The reaction failed when NH_4N_3, LiN_3, or NaN_3 was used in place of $Al(N_3)_3$. Treatment of tetracyanoethylene with azide ion in DMF at $-20°$ yielded the cyanovinyltetrazoles (**88**), which were isolated as tetramethylammonium salts.[171] Methylation of these materials with dimethyl sulfate occurred at the 2-*N*-position and gave the products **89**. A range of substituted 5-cyanovinyltetrazole salts (**90**) has been reported.[172] The materials are explosive and can be used as rocket propellants.[172] The vinyltetrazole (**91**) has been reported by Postovskii

$$H_2C=\overset{R}{C}-CN \xrightarrow[\text{THF}]{Al(N_3)_3} H_2C=\overset{R}{C}\underset{N\diagdown N\diagup\diagdown}{\overset{}{\overset{\displaystyle}{|}}}\!\!\!\!\!\!NH$$

(**87**)

$$NC-\overset{Y}{\underset{}{C}}=\overset{X}{\underset{}{C}}-\!\!\!\!\langle\overset{N-N}{\underset{N-N}{\ominus}}\!\rangle\; M^+ \xrightarrow{Me_2SO_4} (NC)_2C=\overset{CN}{\underset{}{C}}-\!\!\!\!\!\overset{N-N\diagup Me}{\underset{N=N}{|}}$$

(**88**) X = Y = CN   (**89**)
(**90**)

---

[162] F. A. Naugebauer, *Angew. Chem., Int. Ed. Engl.* **8**, 520 (1969).
[163] F. A. Naugebauer, *Tetrahedron* **26**, 4843 (1970).
[164] I. Hauser, D. Jerchel, and R. Kuhn, *Chem. Ber.* **82**, 195 (1949).
[165] R. Kuhn and D. Jerchel, *Justus Liebigs Ann. Chem.* **578**, 1 (1952).
[166] D. Jerchel and H. Fischer, *Justus Liebigs Ann. Chem.* **590**, 216 (1954).
[167] D. Jerchel and H. Fischer, *Chem. Ber.* **89**, 563 (1956).
[168] F. A. Naugebauer, *Chem. Ber.* **102**, 1339 (1969).
[169] C. Arnold and D. N. Thatcher, *J. Org. Chem.* **34**, 1141 (1969).
[170] R. A. Henry, U.S. Patent 3,096,312 (1963) [*CA* **59**, 14132 (1963)].
[171] M. Brown and R. E. Benson, *J. Org. Chem.* **31**, 3849 (1966).
[172] M. Brown, U.S. Patent 3,338,915 (1969) [*CA* **68**, 87299 (1968).].

and Smirnova.[173] Deformylation gave the acetonyltetrazole (**92**) via an enamine-imino tautomerism and hydrolysis.[173]

$$\underset{(91)}{\underset{|}{\text{Me}-\text{C}=\text{CH}-\text{CN}_4\text{H}}} \longrightarrow \underset{(92)}{\underset{\|}{\text{Me}-\text{C}-\text{CH}_2-\text{CN}_4\text{H}}}$$

2. Biologically Active Tetrazoles ($-CN_4H$ versus $-CO_2H$)

Studies of the biological activity of 5-substituted tetrazoles have been prompted by two properties of these materials: (i) a close similarity between the acidity of the tetrazole group ($-CN_4H$) and the carboxylic acid group ($-CO_2H$),[174] and (ii) the fact that the tetrazole function appears to be metabolically the more stable of the two.[175] Hence, biological activity observed in various substituted carboxylic acids might also be expected in the corresponding tetrazole compounds with longer duration of activity. Retention of activity by tetrazole analogs of known carboxylic acid counterparts has been observed in the antiinflammatory,[176,177] hypocholesterolemic,[178,179] antiinfective[180] and antiallergic[181] areas. In contrast tetrazole analogs of active estrogenic acids[182] and antiarrhythmic agents[183] displayed loss or considerable decrease of biological activity.

These tetrazoles are generally synthesized by treating the corresponding nitrile with azide ion in an appropriate solvent.[184,185] For example, a series of 5-(-3-pyridyltetrazoles) (**93**) was prepared in this way and compared with nicotinic acids.[184,185] The parent compound (**93**, R = H) was similar to nicotinic acid in lowering plasma free fatty acids (FFA) and cholesterol levels and had a considerably longer period of activity. Alkylation of the tetrazole ring at either the 1- or 2-positions resulted in loss of activity. The 5-aminoalkyltetrazoles (**94**, R' = H) represent

[173] I. Ya. Postovskii and N. B. Smirnova, *Dokl. Akad. Nauk. SSSR* **170**, 604 (1966) [*CA* **66**, 55452 (1967)].
[174] R. M. Herbst, *in* "Essays in Biochemistry" (S. Groff, ed.), pp. 141–155. Wiley, New York, 1956.
[175] D. W. Esplin and D. M. Woodbury, *J. Pharmacol. Exp. Therap.* **118**, 129 (1956) [*CA* **51**, 3825d (1957)].
[176] P. F. Juby, T. W. Hudyma, and M. Brown, *J. Med. Chem.* **11**, 111 (1968).
[177] P. F. Juby and T. W. Hudyma, *J. Med. Chem.* **12**, 396 (1969).
[178] R. L. Buchanan, V. Sprancmanis, and R. A. Partyka, *J. Med. Chem.* **12**, 1001 (1969).
[179] R. L. Buchanan and V. Sprancmanis, *J. Med. Chem.* **16**, 174 (1973).
[180] J. M. Essery, *J. Med. Chem.* **12**, 703 (1969).
[181] G. P. Ellis and D. Shaw, *J. Med. Chem.* **15**, 865 (1972).
[182] R. R. Grenshaw, G. M. Luke, and G. Bialy, *J. Med. Chem.* **15**, 1179 (1972).
[183] G. W. Adelstein, *J. Med. Chem.* **16**, 309 (1973).
[184] G. F. Holland, and J. N. Pereira, *J. Med. Chem.* **10**, 149 (1967).
[185] G. F. Holland, U.S. Patent 3,448,107 (1969) [*CA* **72**, 132780 (1970)].

(93)

(94) R¹ = H, alkyl

another range of tetrazoles prepared in this general way.[186-188] Alkylation of these compounds occurred preferentially at the 2-position. Varying degrees of antihypertensive activity were displayed by these compounds.[186-188] In general, a wide range of tetrazole derivatives in which the ring N atoms are both substituted and unsubstituted has been prepared and tested for biological activity with varying degrees of success. Because of the considerable interest shown in this area in recent years, it is desirable to generally summarize the main areas of study and the new types of tetrazoles synthesized. A summary is given in Table III.

Tetrazole analogs of amino acids have pK_a values that agree closely with those of the corresponding amino acids.[189] Close similarities between the spatial requirements of the tetrazolate and carboxylate ions have also been noted.[190] Incorporation of these tetrazole analogs of amino acids into peptides has been achieved,[191,192] and it was found that optically active tetrazole analogs of amino acids may be synthesized by normal procedures and incorporated into peptides, without protection of the tetrazole unit, by using the same procedures used for the synthesis of peptides with unprotected carboxy residues.[191] For example, the tetrapeptide (95) has been prepared and found to be equally as active as the carboxylic acid form (96) for stimulating gastric acid secretion.[191] A wide range and variety of tetrazole analogs of amino acids and peptides has been prepared by Grzonka and co-workers.[193-198] Infrared spectra

[186] S. Hayao, H. J. Havera, W. G. Strycker, T. J. Leipzig, and R. Rodrigues, *J. Med. Chem.* **10**, 400 (1967); B. V. Golomolzin and I. Ya. Postovskii, *Khim. Geterotsikl. Soedin.* **7**, 133 (1971) [*CA* **75**, 35932 (1971)].

[187] S. Hayao and W. G. Strycker, U.S. Patent 3,536,715 (1970) [*CA* **74**, 22886 (1971)].

[188] R. N. Schut and H. J. Havera, U.S. Patent 3,499,900 (1970) [*CA* **72**, 111509 (1970)].

[189] J. M. McManus and R. M. Herbst, *J. Org. Chem.* **24**, 1643 (1959).

[190] J. S. Morley, *Fed. Proc. Fed. Am. Soc. Exp. Biol.* **27**, 1314 (1968) [*CA* **70**, 17306 (1969)].

[191] J. S. Morley, *J. Chem. Soc., C* 809 (1969).

[192] Z. Grzonka, E. Rekowska, and B. Liberek, *Tetrahedron* **27**, 2317 (1971).

[193] Z. Grzonka and B. Liberek, *Tetrahedron* **27**, 1783 (1971).

[194] Z. Grzonka, *Rocz. Chem.* **47**, 2071 (1973) [*CA* **80**, 96331 (1974)].

[195] Z. Grzonka, *Rocz. Chem.* **47**, 1401 (1973) [*CA* **80**, 60176 (1974)].

[196] Z. Grzonka, *Rocz. Chem.* **46**, 1265 (1972) [*CA* **77**, 140507 (1972)].

[197] Z. Grzonka, *J. Chromatogr.* **51**, 310 (1970) [*CA* **74**, 9394 (1971)].

[198] Z. Grzonka and B. Liberek, *Rocz. Chem.* **45**, 967 (1971) [*CA* **75**, 118271 (1971)].

CH$_2$R
|
Z—Trp—Met—NH—CH—CO—Phe—NH$_2$

(95) R = CN$_4$H
(96) R = CO$_2$H

of these compounds show that they exist in the zwitterionic form similar to the carboxylic analogs.[199,200]

TABLE III
BIOLOGICALLY ACTIVE TETRAZOLES

Tetrazole type or derivative	Activity	References
5-Phenylsulfonylmethyl-	Hypoglycemic	201
5-Amino-1-(β-hydroxy-β-aryl)ethyl-	Antiinflammatory, analgesic	202
5-Dialkylarylmethyl-	Antiinflammatory, analgesic	203, 204
5-(2-Arylamino-pyrid-3-yl)	Antiinflammatory, analgesic	205, 206
5-Substituted (indazol-3-yl)oxymethyl	Antiinflammatory, analgesic	207
5-Aryloxy(or thio) alkyl-	Hypocholesterolemic	178, 208–212 255–257
5-Biphenyloxyalkyl	Hypolipemic	213
5-(2-Amino-4-chloro-5-sulfamoylphenyl)-	Diuretic, saluretic	214
5-Acylamino-	Antiinflammatory	215
1-Aryltetrazoline-5-thiones	Radioprotectives (X-irradiation)	216
5-Aryltetrazol-2-yl alkanoic acids	Antiinflammatory	217, 218, 258, 259
Tetrazolo[1,5-c]quinazolines	Fibrinolytic	219, 220
16,16-Dimethylprostaglandin	Bronchodilators, antihypertensive	221
2-(5-Tetrazolyl)benzoxazoles	Antiinflammatory	222
5-Heteroarylcarboxamido-	Antihistamine, antiallergic	223, 224,

(*continued*)

[199] Z. Grzonka, B. Liberek, and Z. Palacz, *Zesz. Nauk. Univ. Gdansk. [Chemia]* **1**, 201 (1971) [*CA* **78**, 110038 (1973)].

[200] Z. Grzonka, B. Liberek, and Z. Palacz, *Zesz. Nauk. Wydz. Mat., Fiz. Chem., Univ. Gdanski, Chem.* **2**, 914 (1972) [*CA* **81**, 120998 (1974)].

[201] J. McManus, *J. Med. Chem.* **12**, 550 (1969).

[202] A. Mugnaini, L. Polo Friz, E. Provinciali, P. Rugarli, A. Olivi, E. Zefelippo, L. Almirante, and W. Murmann, *Boll. Chim. Pharm.* **105**, 596 (1966) [*CA* **66**, 75958 (1967)].

[203] R. Aries, French Patent 1,580,970 (1969) [*CA* **72**, 132737 (1970)].

[204] T. Y. Shen and P. C. Dorn, Ger. Offen. 1,902,304 (1969) [*CA* **72**, 66946 (1970)].

[205] A. Dordilly, Ger. Offen. 1,934,551 (1970) [*CA* **72**, 90476 (1970)].

[206] R. Aries, French Patent 1,584,388 (1970) [*CA* **74**, 13157 (1971)].

[207] Cutter Laboratories Inc., British Patent 1,214,109 (1970) [*CA* **75**, 5910 (1971)].

[208] R. L. Buchanan and R. A. Partyka, U.S. Patent 3,517,024 (1970) [*CA* **73**, 66588 (1970)].

[209] Bristol-Meyer, French Patent M. 6926 (1969) [*CA* **74**, 87997 (1971)].

[210] R. A. Buchanan and R. A. Partyka, U.S. Patent 3,498,999 (1970) [*CA* **72**, 111485 (1970)].

TABLE III (*continued*)

Tetrazole type or derivative	Activity	References
Natural prostaglandins	—	260–264 225
Mono-, di-tetrazol-5-ylchromones	Antiallergic agents	226–235
Amincephallosporanic acids	Antibacterial	236–240, 275, 277
Tetrazolo[1,5-*a*]quinolines	Fungicides	241, 242
Tetrazolo[1,5-*b*]pyridazines	Hypotensive	243
Tetrazolo[5,1-*a*]isoindoles	Depressants, anticonvulsants	244
5-Polyacetoxyalkyl-	Sugars	245, 276, 279
5-Aryl, (or heteroaryl)-2 alkanoic acid derivative	Plant growth inhibitor	246
5-Mercaptopyridyl-	Hypoglycemic	247
5-Substituted pyridyl-	Antiinflammatory, antiedema	248–251
Benzodiazepines	Sedative, anticonvulsant	252–254
1-(1*H*-tetrazole)acetic acids	Plant growth regulators	265, 266, 274
Substituted indanyl-	Antiinflammatory, antipyretic	267, 268
5-Heteroarylmethyl-	Antilipemic, diabetes mellitus, dopamine β-hydroxylase	185, 269–271
5-(2-Benzo[*f*]quinolines)-	Bactericides	272
5-Arylamino-	Measles virus	273
1-Substituted 5-amino-	Arthritis, muscle relaxant	278

[211] D. T. Nash, L. Gross, W. Haw, and K. Agre, *J. Clin. Pharmacol.* **8**, 377 (1968) [*CA* **70**, 2390 (1969)].

[212] Bristol Meyer, French Patent 1,534,419 (1968) [*CA* **72**, 12731 (1970)].

[213] O. Wacker and C. Morel, Ger. Offen. 1,928,438 (1969) [*CA* **72**, 55463 (1970)].

[214] A. Popelak, A. Lerch, S. Kurt, E. Roesch, and K. Hardebeek, Ger. Offen. 1,815,922 (1970) [*CA* **73**, 45519 (1970)].

[215] G. Renier, R. Canevari, and J. C. Le Douarec, Ger. Offen. 2,228,216 (1973) [*CA* **78**, 84419 (1973)].

[216] P. N. Kulyabko, V. A. Kuz'menko, V. A. Fedorchenko, R. G. Dubenko, P. S. Pel'kis, I. M. Bazavova, and V. M. Andrianov, *Fiziol. Akt. Veshchestva* **5**, 114 (1973) [*CA* **81**, 86057 (1974)].

[217] R. T. Buckler, *J. Med. Chem.* **15**, 578 (1972).

[218] R. T. Buckler, Ger. Offen. 2,063,173 (1971) [*CA* **75**, 76804 (1971); cf. also *CA* **75**, 36048 (1971)].

[219] E R. Wagner, U.S. Patent 3,835,137 (1974) [*CA* **81**, 152258 (1974)].

[220] E. R. Wagner, U.S. Patent 3,838,126 (1974) [*CA* **82**, 4298 (1975)].

[221] E. E. Galantay, Ger. Offen. 2,405,255 (1974) [*CA* **81**, 152236 (1974)].

[222] H. Moeller and C. Gloxhuber, Ger. Offen. 2,314,238 (1974) [*CA* **82**, 4263 (1975)].

[223] G. P. Ellis, I. Collins, D. M. Waters, and D. E. Bays, Ger. Offen 2,415,767 (1974) [*CA* **82**, 16843 (1975)].

[224] G. P. Ellis, H. M. Ridgway, and D. E. Bays, Ger. Offen. 2,415,763 (1974) [*CA* **82**, 16844 (1975)].

References continued

[225] H. J. E. Hess, T. K. Schaof, and L. J. Czuba, Ger. Offen. 2,409,601 (1974) [*CA* **82**, 43427 (1975)].
[226] G. P. Ellis and D. Shaw, *J. Chem. Soc., Perkin Trans. 1*, 779 (1972).
[227] A. Nohara, *Tetrahedron Lett.*, 1187 (1974).
[228] B. T. Warren and J. W. Spicer, Ger. Offen, 2,360,355 (1974) [*CA* **81**, 91538 (1974)].
[229] S. Umio, Y. Sato, and M. Kobayashi, Japan. Patent 74 05974 (1974) [*CA* **81**, 63635 (1974)].
[230] A. Nohara, T. Umetani, and Y. Sanno, Ger. Offen. 2,317,899 (1973) [*CA* **80**, 14932 (1974)].
[231] J. Augstein, H. Cairns, D. Hunter, and J. King, Ger. Offen. 2,142,527 (1972) [*CA* **77**, 5488 (1972)].
[232] M. E. Peel and A. W. Oxford, Ger. Offen. 2,157,299 (1972) [*CA* **77**, 88509 (1972)].
[233] H. F. Hodson, J. F. Batchelor, and J. H. Gorvin, Ger. Offen. 2,344,814 (1974) [*CA* **80**, 146170 (1974)].
[234] H. Cairns, P. B. Johnson, and R. Minshull, Ger. Offen. 2235572 (1973) [*CA* **78**, 136301 (1973)].
[235] J. Augstein, H. Cairns, D. Hunter, and J. King, Ger. Offen. 2,142,556 (1972) [*CA* **77**, 5475 (1972)].
[236] H. Bickel and J. Mueller, Ger. Offen. 2,217,563 (1972) [*CA* **78**, 29792 (1973)].
[237] T. Takano and S. Horibe, Japan. Patent 71 24713 (1971) [*CA* **75**, 129822 (1971)].
[238] H. Bickel and J. Mueller, Ger. Offen. 2,120,504 (1971) [*CA* **76**, 85829 (1972)].
[239] T. Takano, Y. Kurita, H. Nikaido, M. Yonera, N. Konishi, and R. Okui, Japan. Patent 71 35751 (1971) [*CA* **76**, 14563 (1972)].
[240] G. L. Dunn and J. R. E. Hoover, S. African Patent 7,302,218 (1973) [*CA* **81**, 105544 (1974)].
[241] B. A. Dreikorn, Ger. Offen. 2,134,146 (1972) [*CA* **76**, 99676 (1972)].
[242] B. A. Dreikorn, Ger. Offen. 2,166,398 (1974) [*CA* **80**, 95967 (1974)].
[243] P. L. Anderson, W. J. Houlihan and R. E. Manning, *U.S. Patent* 3,637,690 (1972) [*CA* **76**, 11328 (1972)].
[244] W. J. Houlihan and M. K. Eberle, U.S. Patent 3,642,814 (1972) [*CA* **76**, 140839 (1972)].
[245] A. M. Seldes, E. G. Gros, I. M. E. Thiel, and J. O. Deferrari, *Carbohydr. Res.* **39**, 11, 47 (1975); J. O. Deferrari, A. M. Seldes, O. G. Maroza, and I. M. E. Thiel, *Ibid.* **17**, 237 (1971).
[246] E. F. George and W. D. Riddell, Ger. Offen. 2,310,049 (1973) [*CA* **80**, 23539 (1974)].
[247] B. Blank, N. W. DiTullis, C. K. Mias, F. F. Owings, I. G. Gleason, S. T. Ross, C. E. Berkoff, H. L. Saunders, J. Delarge, and C. L. Lapiere, *J. Med. Chem.* **17**, 1065 (1974).
[248] D. Jovanovic, French Patent 2,085,649 (1972) [*CA* **77**, 126639 (1972)].
[249] D. Jovanovic, French Patent 2,085,650 (1972) [*CA* **77**, 126645 (1972)].
[250] L. Chausse, French Patent 2,068,411 (1971) [*CA* **77**, 34513 (1972)].
[251] K. Sugiura, R. Ushijima, and K. Shimizu, Japan. Patent 7,242,770 (1972) [*CA* **78**, 72158 (1973)].
[252] A. Sallmann and R. Pfister, Swiss Patent 513,894 (1971) [*CA* **76**, 113221 (1972)].
[253] R. Y. Ning, L. H. Sternbach, W. Pool, and L. O. Randall, *J. Med Chem.* **16**, 879 (1973).
[254] H. B. Hester, U.S. Patent 3,717,653 (1973) [*CA* **78**, 136348 (1973)].
[255] W. A. Bolhofer, U.S. Patent 3,816,446 (1974) [*CA* **81**, 120606 (1974)].
[256] R. Aries, French Patent 2,173,778 (1973) [*CA* **80**, 95963 (1974)].

3. General Syntheses

The most general route to 5-substituted tetrazoles involves condensation of the appropriate nitrile with azide ion. The variation in the yields of tetrazoles obtained in different solvent systems has been summarized recently for much of the earlier work.[5] Although inorganic azides are usually used, 1,1,3,3-tetramethylguanidinium azide has proved to be a particularly useful source of azide ion, which gives high yields of

References continued

[257] R. L. Buchanan, U.S. Patent 3,634,444 (1972) [*CA* **76**, 113222 (1972)].
[258] R. T. Buckler and W. G. Strycker, Ger. Offen. 2,155,406 (1972) [*CA* **77**, 101621 (1972); cf. also *CA* **72**, 43683 (1970)].
[259] R. T. Buckler, S. Hayao, O. J. Lorenzetti, L. F. Sancilio, H. E. Hartzler, and W. G. Strycker, *J. Med. Chem.* **13**, 725 (1970).
[260] I. Collins and G. P. Ellix, Ger. Offen. 2,410,852 (1974) [*CA* **82**, 31352 (1975)].
[261] J. Augstein, H. Cairns, N. H. Rodgers, and R. C. Brown, Ger. Offen. 2,360,331 (1974) [*CA* **81**, 63639 (1974)].
[262] D. E. Bays, M. E. Peel, D. M. Waters, and G. P. Ellis, Ger. Offen. 2,332,731 (1974) [*CA* **80**, 146174(1974)].
[263] J. Augstein, H. Cairns, and N. H. Rodgers, Ger. Offen. 2,361,490 (1974) [*CA* **81**, 120633 (1974)].
[264] H. M. Ridgway, D. M. Waters, M. E. Peel, and G. P. Ellis, Ger. Offen. 2,407,744 (1974) [*CA* **81**, 169547 (1974)].
[265] T. Kamitani, S. Tsubouchi, and N. Yamamoto, Japan. Patent 71 35742 (1971) [*CA* **76**, 14550 (1972)].
[266] T. Kamitani and Y. Saito, Japan. Patent 72 47031 (1972) [*CA* **78**, 111331 (1973)].
[267] S. Noguchi, S. Kishimoto, M. Obayashi, I. Minamida, and K. Kawai, Japan, Patent 71 36619 (1971) [*CA* **76**, 14552 (1972)].
[268] T.-Y. Shen and C. P. Dorn, U.S. Patent 3,631,167 (1971) [*CA* **76**, 126994 (1972)].
[269] K. Shimizu, R. Ushijima, and K. Sugiura, Ger. Offen. 2,217,084 (1973) [*CA* **78**, 29778 (1973)].
[270] M. Fukumura, K. Shimaji, and S. Okano, Japan. Patent 72 07378 (1972) [*CA* **77**, 5489 (1972)].
[271] K. P. Boejesor and N. A. Klitgaard, *Acta Chem. Scand.* **25**, 1889 (1971).
[272] M. Murakami and Y. Nagano, Japan. Patent 71 34709 (1971) [*CA* **76**, 14548 (1972)].
[273] W. B. Scanlon and W. L. Garbrecht, S. African Patent 6804307 (1970) [*CA* **73**, 56100 (1970)].
[274] T. Kamitani and T. Teraji, Japan. Patents 70 21823/4 (1970) [*CA* **73**, 77256 and 77257 (1970)].
[275] R. Raap and U. R. Lemieux, U.S. Patent 3,468,874 (1969) [*CA* **72**, 55479 (1970)].
[276] U. Askani and R. Neidlein, *Dtsch. Apoth. Ztg.* **110**, 1502 (1970) [*CA* **74**, 42589 (1970)].
[277] Y. Kurita, T. Teraji, and N. Konishi, Japan. Patent 71 05150 (1971) [*CA* **75**, 20420 (1971); see also *CA* **75**, 36076 (1971)].
[278] Armour Pharmaceutical Co., British Patent 1,209,880 (1970) [*CA* **74**, 141818 (also 87995c) (1971)].
[279] J. J. Baker, A. M. Mian, and J. R. Tittensor, *Tetrahedron* **30**, 2939 (1974); J. J. Baker, P. Mellish, C. Riddle, A. R. Somerville, and J. R. Tittensor, *J. Med. Chem.* **17**, 764 (1974).

Sec. III.A] RECENT ADVANCES IN TETRAZOLE CHEMISTRY 361

tetrazoles when heated with nitriles without a solvent.[280,281] The presence of $AlCl_3$ has also been found to enhance the reaction when inorganic azides such as NaN_3 are used.[282] New tetrazoles, which have been prepared recently in this general way, include 5-aryloxy and 5-alkoxy tetrazoles (**97**),[283,284] 5-fluoroaminoalkyltetrazoles (**98**),[285] and the *N*-tetrazolylphosphinimines (**99**).[286] A series of 5-substituted aminotetrazoles, $RNH-CN_4H$ (R = acyl, aroyl) has also been reported.[287]

$$RO-CN_4H \qquad R^1-\underset{\underset{NF_2}{|}}{\overset{\overset{R^2}{|}}{C}}-CN_4H \qquad R_3P=N-CN_4H$$

(**97**)   (**98**)   (**99**)

Treatment of unsubstituted amidrazones with nitrous acid also provides a general route to 5-substituted tetrazoles.[1,288] This reaction has recently been used to obtain the ditetrazolylalkanes (**100**).[289] Other general synthetic routes to the 5-substituted tetrazole system, which have been known for many years, have been thoroughly reviewed by Benson.[1,2]

$$\underset{NH_2}{\overset{NH_2-N}{\diagdown}}C-(CH_2)_n-C\overset{\diagup N-NH_2}{\underset{\diagdown NH_2}{}} \xrightarrow{2HONO} \underset{N-NH}{\overset{N-N}{\|}}-(CH_2)_n-\underset{HN-N}{\overset{N-N}{\|}}$$

(**100**)

4. Reactions

a. Ring Substitution: Alkylation and Acylation. The reactions of 5-substituted tetrazoles may be divided between those in which the

---

[280] A. J. Papa, *J. Org. Chem.* **31**, 1426 (1966).
[281] A. J. Papa, U.S. Patent 3,429,879 (1969) [*CA* **70**, 87819 (1969)].
[282] Borg-Warner Corp., British Patent 1,163,355 (1969) [*CA* **71**, 112938 (1969)].
[283] D. Martin and A. Weise, *Chem. Ber.* **99**, 317 (1966).
[284] M. Hedayatulah, H. Iida, and L. Denivelle, *C.R. Hebd. Seances Acad. Sci., Ser. C* **271**, 146 (1970) [*CA* **73**, 98913 (1970)].
[285] R. J. Kosher, U.S. Patent 3,394,142 (1968) [*CA* **69**, 106714 (1968)].
[286] R. A. Mitsch, *J. Am. Chem. Soc.* **89**, 6297 (1967).
[287] V. P. Schipanov, S. L. Portnova, V. A. Krasnova, Yu. N. Sheinker, and I. Ya. Postovskii, *Zh. Org. Khim.* **1**, 2236 (1965) [*CA* **64**, 11056 (1966)].
[288] D. G. Neilson, R. Roger, J. W. M. Heatlie, and L. R. Newlands, *Chem. Rev.* **70**, 163 (1970).
[289] T. Kauffmann and L. Ban, *Chem. Ber.* **99**, 2600 (1966).

tetrazole ring is retained and those that involve cleavage of the ring with evolution of nitrogen. The former lead to N-substituted tetrazoles, the most important examples being alkylation reactions. Much work on the alkylation of tetrazoles was carried out in earlier years and is summarized in previous reviews.[1,2] However, interesting developments have also occurred recently. In these reactions mixtures of 1- and 2-alkyl isomers are usually obtained [Eq. (3)] and, in general, it appears that electron-donating substituents at the 5-position tend slightly to favor alkylation at the 1-N position of tetrazole anions while electron-withdrawing 5-substituents slightly favor the 2-position.[1,3,36] When the alkylation is carried out with diazomethane under neutral conditions, the 2-position is sometimes slightly favored.[3] The orientation of the products is also influenced by steric factors and is sensitive to the steric

$$R-\underset{\underset{H}{N-N}}{\overset{N-N}{\diagdown\!\!\!\diagup}} \xrightarrow[\text{base}]{R^1X} R-\underset{\underset{R^1}{N-N}}{\overset{N-N}{\diagdown\!\!\!\diagup}} + R-\underset{\underset{R^1}{N-N}}{\overset{N=N}{\diagdown\!\!\!\diagup}} \quad (3)$$

requirements of the alkylating agent.[36] In general, however, the product trends are quite precarious in these reactions, and it is difficult to predict which site is likely to be the most active for a particular tetrazole derivative. For example, benzylation of 5-aminotetrazole with para substituted benzyl chlorides under basic conditions occurs at both the 1- and 2-positions, the 2-isomer slightly predominating,[32] and the isomer distribution is insensitive to the nature of the para substituent of the benzyl chloride. The tetrazol-5-ylhydrazones (**101**) are alkylated preferentially at the 1-position, and only minor quantities of the 2-alkylated and exocyclic 5-N-alkylated products are formed.[290,291] Other recent alkylation studies include 5-cyanovinyltetrazoles,[171] 5-heterocyclic substituted tetrazoles,[184] 5-aryltetrazoles,[186] 5-aryloxyalkyltetrazoles,[178] and 5-arylthioalkyltetrazoles,[178] all of which are alkylated preferentially at the 2-position. Mixtures of the compounds **30–33** (Table I) were obtained from alkylations with ethylbromo- and ethylchloroacetates.[36] Basic alkylations of 5-R-substituted tetrazoles (R = Ph, PhCH$_2$,

$$\text{ArCH=N-NH}-\underset{\underset{H}{N-N}}{\overset{N-N}{\diagdown\!\!\!\diagup}} \xrightarrow[\text{NaOH}]{\text{MeI}} \text{Ar-CH=N-NH}-\underset{\underset{Me}{N-N}}{\overset{N-N}{\diagdown\!\!\!\diagup}}$$

(**101**)

[290] R. N. Butler and F. L. Scott, *J. Org. Chem.* **31**, 3182 (1966).
[291] R. N. Butler and F. L. Scott, *J. Org. Chem.* **32**, 1224 (1967).

Sec. III.A]   RECENT ADVANCES IN TETRAZOLE CHEMISTRY   363

piperazinylethyl) gave various mixtures of both isomers.[292] Alkylations of 5-substituted tetrazoles in basic media with α-chlorocarbonyl compounds (Cl—CH$_2$—CO—R) occurred preferentially at the 1-position with yields of the 2-isomer in the range 0–20%.[293] Alkylation of tetrazole analogs of benzilic acid esters giving mixtures of compounds **102** and **103**, [Eq. (4)], have also been reported.[294,295] Further examples of mixtures of other alkyltetrazoles, which have been prepared recently, are in Table I.

$$\underset{\underset{Na^+}{}}{Ph-\underset{\underset{R}{|}}{\overset{\overset{OH}{|}}{C}}-\underset{N-N}{\overset{N-N}{\diagup}}(-)} \xrightarrow{R^1Cl} Ph-\underset{\underset{R}{|}}{\overset{\overset{OH}{|}}{C}}-\underset{\underset{R^1}{|}}{\overset{N-N}{\diagup}}_{N-N} + Ph-\underset{\underset{R}{|}}{\overset{\overset{OH}{|}}{C}}-\overset{N=N}{\diagup}_{N-N_{R^1}} \qquad (4)$$

$$\qquad\qquad\qquad\qquad\qquad (\mathbf{102}) \qquad\qquad (\mathbf{103})$$

Selective 1-alkylation has been obtained by blocking the 2-position with a tri-*n*-butylstannyl group, e.g., compound **68**. In this case up to 90% alkylation at the 1-position could be obtained with normal reagents, such as methyl iodide or dimethyl sulfate, and the blocking group was removed in the course of the reaction.[296] An interesting conversion of 1-alkyl-5-phenyltetrazoles (**104**) into the corresponding 2-alkyl isomers (**105**) by heating with an alkyl iodide has also been observed. The reaction appears to involve 1,3-dialkyl-5-phenyltetrazolium salt formation followed by elimination of a molelcule of alkyl iodide to give the thermodynamically more stable isomer (Scheme 8).[297,298] The ditetrazoles **106** and **107** were obtained by treating the potassium salt of 5-phenyltetrazole with 1,2-dibromoethane [Eq. (5)]. Compound **106** was the major product.[299] These ditetrazoles were useful as cross-linking agents for curing unsaturated elastomers in a reaction that involved thermal elimination of nitrogen to form dinitrilimine 1,3-dipoles.[299]

---

[292] L. Huff and R. A. Henry, *J. Med. Chem.* **13**, 777 (1970).
[293] F. Einberg, *J. Org. Chem.* **35**, 3978 (1970).
[294] E. Balieu and N. A. Klitgaard, *Acta Chem. Scand.* **26**, 2951 (1972).
[295] E. Balieu and E. Bjarnov, *Acta Chem. Scand.* **27**, 1233 (1973).
[296] T. Isida, T. Akiyama, K. Nabika, K. Sisido, and S. Kozima, *Bull. Chem. Soc. Jpn.* **46**, 2176 (1973).
[297] T. Isida, S. Kozima, K. Nabika, and K. Sisido, *J. Org. Chem.* **36**, 3807 (1971).
[298] T. Isida, S. Kozima, S.-I. Fujimori, and K. Sisido, *Bull. Chem. Soc. Jpn.* **45**, 1471 (1972); for an interesting comparison of the alkylation reactions of compounds (**104**) with similar reactions of 5-substituted 1,2,3,4-thiatriazoles, see A. Holm, K. Schaumberge, N. Dahlberg, C. Christophersen, and J. P. Snyder, *J. Org. Chem.* **40**, 431 (1975).
[299] N. V. Schwartz, *J. Appl. Polym. Sci.* **16**, 2715 (1972) [*CA* **78**, 17288 (1973)].

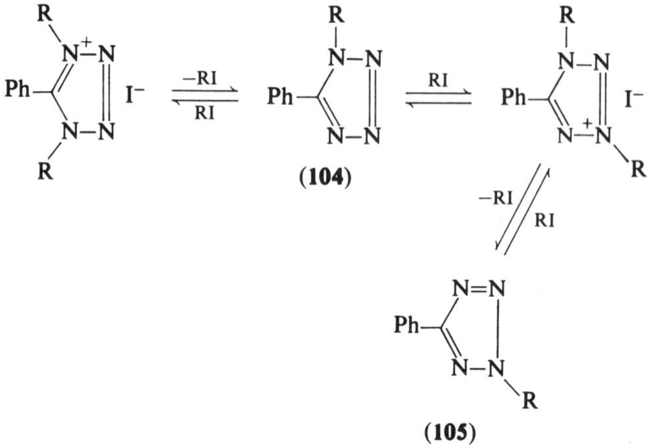

[SCHEME 8 diagram showing interconversion of structures (104) and (105) with RI/−RI]

SCHEME 8

Some further alkylation reactions leading to mesoionic tetrazoles are in Section III,D.

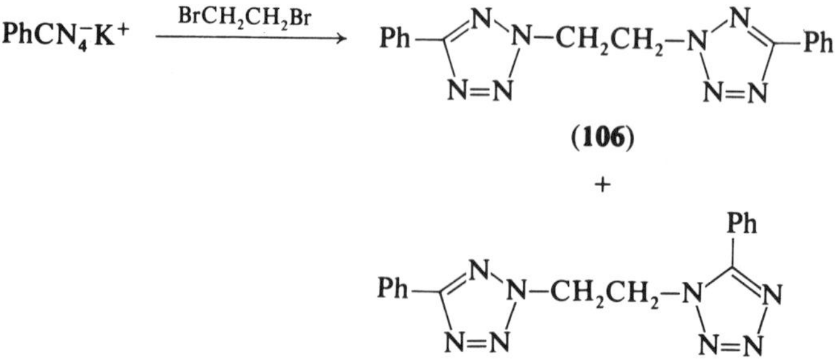

$$PhCN_4^- K^+ \xrightarrow{BrCH_2CH_2Br} \text{(106)} + \text{(107)} \quad (5)$$

[Structure (106): Ph-tetrazole-N-CH₂CH₂-N-tetrazole-Ph, both 2,5-disubstituted]

[Structure (107): isomeric bis-tetrazole with one 1,5- and one 2,5-substitution]

The acylation reactions of 5-substituted tetrazoles provide examples that bridge both of the processes of ring substitution and ring fragmentation. Acylation of 5-aryloxytetrazoles occurs at the 2-position yielding the acyltetrazoles (108) [Eq. (6)].[283] These compounds are low-melting solids. When they are heated gently in solution, nitrogen is evolved and the oxadiazoles (109) are formed.[283] The cyano-substituted tetrazoles (110) were obtained when 5-aryloxytetrazoles were treated with

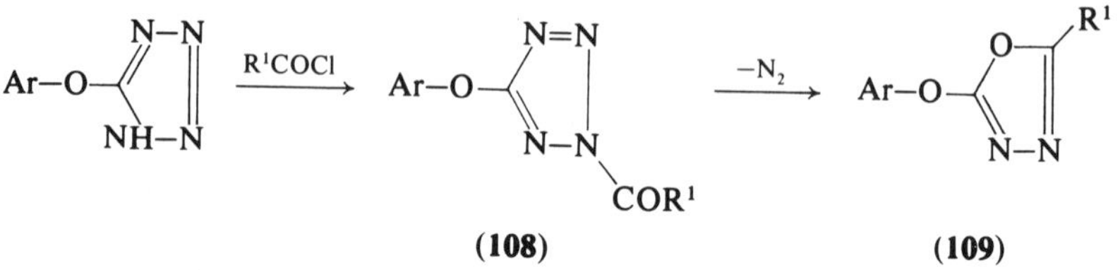

$$Ar-O-\underset{NH-N}{\overset{N-N}{\text{tetrazole}}} \xrightarrow{R^1COCl} Ar-O-\underset{\underset{COR^1}{N-N}}{\overset{N=N}{\text{tetrazole}}} \xrightarrow{-N_2} Ar-O-\underset{N-N}{\overset{O-R^1}{\text{oxadiazole}}} \quad (6)$$

(108)   (109)

cyanogenbromide [Eq. (7)].[283] A number of acylated 5-aminotetrazoles were obtained by Einberg[300] by terminating acylation reactions of 5-aminotetrazole early to prevent oxadiazole formation. 5-Formamido, 5-benzamido, and 5-acetamidotetrazoles were obtained in this work by treating 5-aminotetrazole with N,N-dimethylformamide, N,N-dimethylbenzamide, and N,N-dimethylacetamide, respectively, in basic solution.[300] Treatment of 5-dimethylaminotetrazole with p-toluenesulfonyl chloride in pyridine at 10° yields the unstable sulfonyltetrazole **111**.[301] This compound slowly lost nitrogen at room

$$\text{Ar-O} \begin{array}{c} \text{N-N} \\ \| \\ \text{HN-N} \end{array} \xrightarrow[\text{BrCN}]{\text{Et}_3\text{N}} \text{Ar-O} \begin{array}{c} \text{N=N} \\ \| \\ \text{N-N} \\ \text{CN} \end{array} \tag{7}$$

(110)

temperature both in the solid state and in solution. With warm water it yielded the semicarbazide **112**.[301] When the reaction between 5-dimethylaminotetrazole and p-toluenesulfonylchloride in pyridine was carried out at normal temperatures, compound **111** decomposed *in situ*, yielding the interesting pyridinium chloride **113**.[301]

$$\text{Me}_2\text{N} \begin{array}{c} \text{SO}_2\text{R} \\ \text{N-N} \\ \\ \text{N=N} \end{array} \qquad \text{Me}_2\text{N}-\overset{\overset{\displaystyle O}{\|}}{\text{C}}-\text{NHNH}-\text{SO}_2\text{R} \qquad \text{Me}_2\text{N}-\overset{}{\text{C}}=\text{NNHSO}_2\text{R}$$

(**111**) R = p-tolyl   (**112**)   (**113**)

b. Ring Fragmentation. In general, the reactions of 5-substituted tetrazoles with acyl halides have been mainly used for the synthesis of 1,3,4-oxa- and thiadiazoles without isolation of the intermediates (Scheme 9).[1,302] The reaction, which is now referrred to as the Huisgen reacton,[303] has been applied to new 5-vinyltetrazoles,[169] 5-aryltetrazoles,[303,304] and 5-perfluoroalkyltetrazoles.[305] When the reac-

[300] F. Einberg, *J. Org. Chem.* **32**, 3687 (1967).
[301] K. A. Jensen, A. Holm, and S. Rachlin, *Acta Chem. Scand.* **20**, 2795 (1966).
[302] A. Hetzheim and K. Mockel, *Adv. Heterocycl. Chem.* **7**, 183 (1966).
[303] N. Dahle, U.S. Patent 3,790,588 (1974) [*CA* **80**, 82993 (1974)].
[304] R. Huisgen, C. Axen, and H. Seidl, *Chem. Ber.* **98**, 2966 (1965).
[305] H. C. Brown and R. J. Kassal, *J. Org. Chem.* **32**, 1871 (1967).

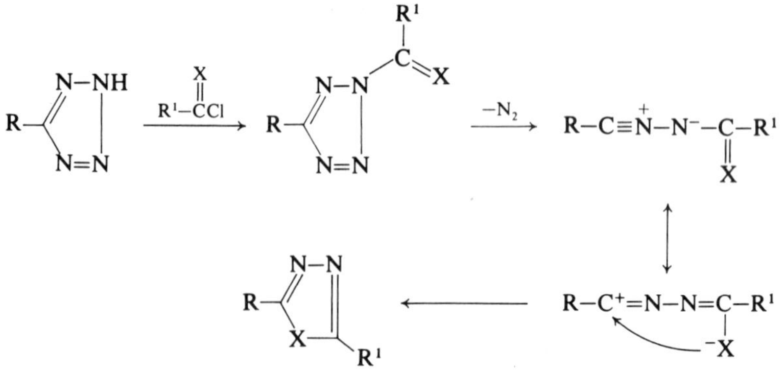

X = O, S, N–

SCHEME 9

tion is employed with a bistetrazole and a monoacylhalide or a monotetrazole and a diacylhalide, bisoxadiazoles are obtained.[305] When the reaction is employed using both a diacylhalide and a bistetrazole, heterocyclic polymers such as **114** are obtained.[304,306]

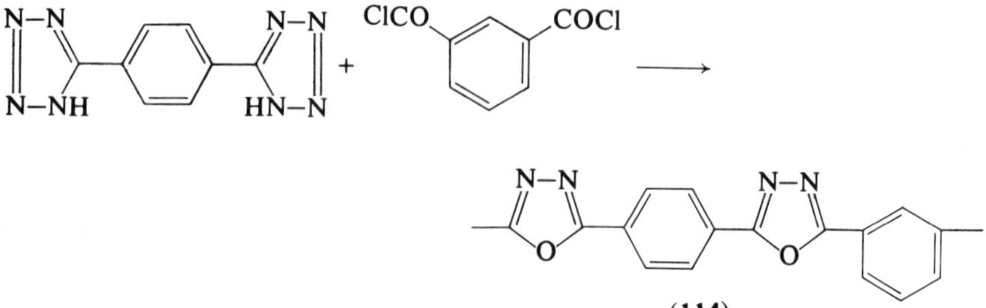

(**114**)

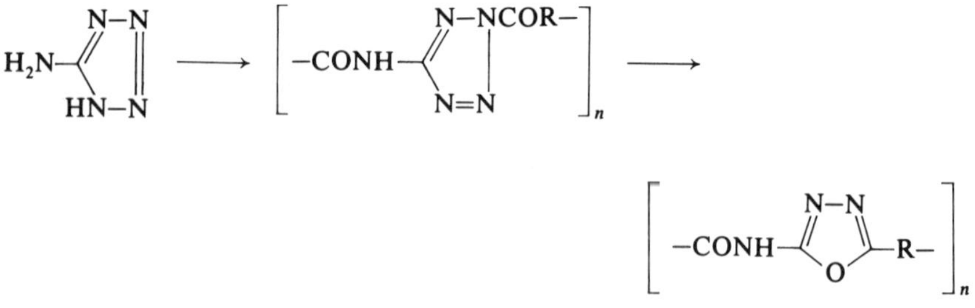

(**115**)

[306] C. J. Abshire and C. S. Marvel, *Makromol. Chem.* **44/46**, 338 (1961) [*CA* **55**, 20487 (1961)].

Treatment of 5-aminotetrazole with diacylchlorides also leads to polymeric products of type **115**.[307] Reimlinger has recently[308,309] classified these reactions as electrocyclic cyclizations involving a 1,5-dipolar nitrilimine intermediate (originally proposed by Huisgen) (Scheme 9). Spontaneous evolution of nitrogen also occurs when 5-(*p*-toluenesulfonamido)tetrazole is treated with *p*-toluenesulfonyl chloride at room temperature in pyridine [Eq. (8)], and compound **116**, which compares with **113**, is formed.[310]

$$\text{Ar-SO}_2\text{NH} \underset{\text{HN-N}}{\overset{\text{N-N}}{\diagdown}} + \text{ArSO}_2\text{Cl} \xrightarrow{\text{pyridine}} \text{ArSO}_2\text{-N-C-N-NHSO}_2\text{Ar} \quad (8)$$

(**116**) Ar = *p*-tolyl-

A new cleavage of 5-alkyltetrazoles leading to substituted ethylenes and acetylenes has been reported by Behringer and Matner.[311] When the tetrazoles **117** were treated with bromine at 150° in a sealed tube, the

$$\text{RR}^1\text{C} \underset{\text{HN-N}}{\overset{\text{H N-N}}{\diagdown}} \xrightarrow{\text{Br}_2} \text{R-CBr=CBrR}^1$$

(**117**)   (**118**)

compounds **118** were formed. The reaction involved initial halogenation to yield compounds such as **119** (X = Br) followed by elimination of HBr generating a tetra-azafulvene intermediate (**120**) which loses N_2 ac-

$$\text{RR}^1\text{C} \underset{\text{HN-N}}{\overset{\text{X N-N}}{\diagdown}} \xrightarrow{-\text{HX}} \left[\text{RR}^1\text{C} \underset{\text{N=N}}{\overset{\text{N=N}}{=}} \right] \xrightarrow{-N_2} \text{R-C≡C-R}^1$$

(**119**)   (**120**)

X = halogen, OH, NH$_2$

---

[307] L. Stoicescu-Crivat and H. Bruma, *Rev. Roum. Chim.* **12**, 1245 (1967) [*CA* **69**, 27943 (1968)].
[308] H. Reimlinger, *Chem. Ber.* **103**, 1900 (1970).
[309] H. Reimlinger, J. J. M. Vanderwalle, G. S. D. King, W. R. F. Lingier, and R. Merenyi, *Chem. Ber.* **103**, 1918 (1970).
[310] K. A. Jensen and C. Christophersen, *Acta Chem. Scand., Ser. B.* **28**, 1 (1974) [*CA* **80**, 133348 (1974)].
[311] H. Behringer and M. Matner, *Tetrahedron Lett.*, 1663 (1966).

companied by migration of one of the substituted groups. Pyrolysis of the compounds **119** separately in the absence of bromine yielded the expected acetylenes.[311] A comparable degradation of substituted 5-aminotetrazoles which yields isocyanide dibromides also occurs when

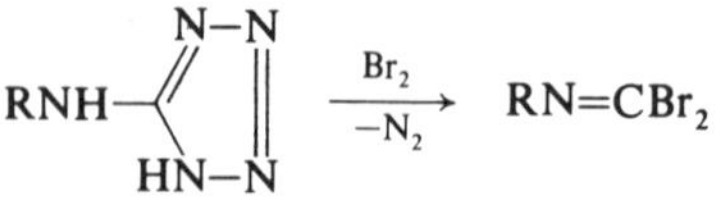

$$\text{RNH}-\underset{\text{HN}-\text{N}}{\overset{\text{N}-\text{N}}{\diagup\!\!\!\diagdown}}\xrightarrow[-\text{N}_2]{\text{Br}_2}\text{RN}=\text{CBr}_2 \qquad (9)$$

these compounds are treated with bromine [Eq. (9)].[312] With tetrazol-5-ylhydrazones the reaction leads to trihalogenodiazabutadienes (**121**) (Scheme 10). Evidence has been cited[313,314] for a mechanism involving initial N-halogenation to yield an intermediate comparable with **119**,

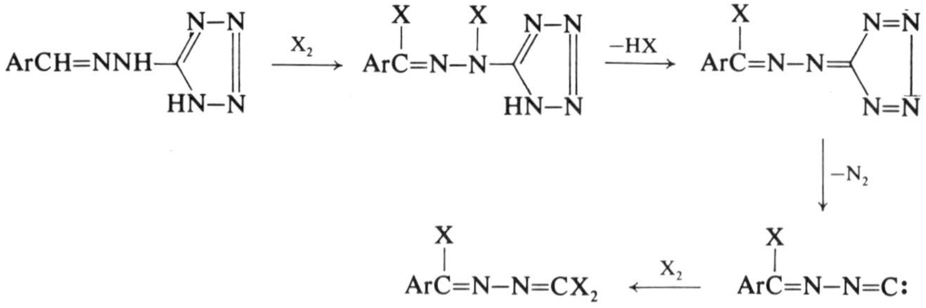

$$\text{ArCH}=\text{NNH}-\underset{\text{HN}-\text{N}}{\overset{\text{N}-\text{N}}{\diagup\!\!\!\diagdown}}\xrightarrow{X_2}\overset{X}{\underset{}{\text{ArC}}}=\text{N}-\overset{X}{\underset{}{\text{N}}}-\underset{\text{HN}-\text{N}}{\overset{\text{N}-\text{N}}{\diagup\!\!\!\diagdown}}\xrightarrow{-HX}\overset{X}{\underset{}{\text{ArC}}}=\text{N}-\text{N}=\underset{\text{N}=\text{N}}{\diagup\!\!\!\diagdown}$$

$$\downarrow -N_2$$

$$\overset{X}{\underset{}{\text{ArC}}}=\text{N}-\text{N}=\text{CX}_2 \xleftarrow{X_2} \overset{X}{\underset{}{\text{ArC}}}=\text{N}-\text{N}=\text{C}:$$
$$(121) \qquad\qquad (122)$$
$$X = \text{Cl, Br}$$

SCHEME 10

which, in turn, yields a tetra-azafulvene intermediate and an isonitrilic species (**122**) (Scheme 10). A new degradation of 5-substituted tetrazoles with Vilsmeier reagents, which gives formamide hydrazone derivatives, has also been reported.[227]

c. *General (5-Aminotetrazole)*. Treatment of 5-aminotetrazole with ethoxycarbonyl- or benzoyl isothiocyanate in pyridine yields the 5-amidotetrazoles **123** and **124** [Eq. (10)].[315] The reaction, which is general for π-deficient azoles, involves an addition at the 5-amino group followed by elimination of HSCN in an intramolecular trans-

[312] E. Kuhle, B. Anders, and G. Zumach, *Angew. Chem., Int. Ed. Engl.* **6**, 654 (1967).
[313] J. A. Cronin, M.Sc. Thesis, National University of Ireland (1966).
[314] F. L. Scott, J. O'Donovan, and J. K. O'Halloran, *Tetrahedron Lett.*, 4079 (1970).
[315] L. Capuano and H. J. Schrepfer, *Chem. Ber.* **104**, 3039 (1971).

Sec. III.A] RECENT ADVANCES IN TETRAZOLE CHEMISTRY 369

$$\underset{\underset{H}{N-N}}{\overset{N-N}{\|}}\!\!\!>\!\!-NH_2 \xrightarrow{RNCS} \left[\underset{\underset{H}{N-N}}{\overset{N-N}{\|}}\!\!\!>\!\!-NH\overset{S}{\overset{\|}{C}}-NHR\right] \xrightarrow{-HSCN} \underset{\underset{H}{N-N}}{\overset{N-N}{\|}}\!\!\!>\!\!-NHR \quad (10)$$

(123) R = CO$_2$Et
(124) R = COPh

acylation.[316] Thermolysis of 5-aminotetrazole to give hydrazoic acid, cyanamide, and polymeric materials has also been reported.[317] Diazotization of 5-aminotetrazole gives a diazonium salt,[318] which has been used to generate atomic carbon by thermal decomposition.[319-321] The tetrazol-5-yltetrazine (126) was obtained when the formazan (125)

(125)   (126)

was treated with formaldehyde in dilute base [Eq. (11)].[322] Treatment of 5-aminotetrazole with methyl propiolate and methyl acrylate resulted in both ring N–H and amino N–H addition across the unsaturated carbon–carbon moiety (Scheme 11), giving the products 127–129.[323] The decomposition of the anion of 5,5'-azoditetrazole with dilute mineral acids has also been reported. Products such as 5-hydrazinotetrazole and 5-azidotetrazole were obtained, depending on the acid used. With dilute hydrochloric and sulfuric acids, 5-hydrazinotetrazole was formed along with some uncharacterized high-nitrogen products that were explosive. Dilute nitric acid gave 5-azidotetrazole and some gaseous products.[324]

[316] H. J. Schrepfer, L. Capuano, and H-L. Schmidt, *Chem. Ber.* **106**, 2925 (1973).
[317] H. Reimlinger, *Chem. Ind. (London)* 294 (1972).
[318] For a review of diazotisation of amino-azoles, cf. R. N. Butler, *Chem. Rev.* **75**, 241 (1975).
[319] P. B. Shevlin, *J. Am. Chem. Soc.* **94**, 1379 (1972).
[320] S. Kammula and P. B. Shevlin, *J. Am. Chem. Soc.* **95**, 4441 (1973).
[321] S. Kammula and P. B. Shevlin, *J. Am. Chem. Soc.* **96**, 7830 (1974).
[322] V. P. Shchipanov and A. A. Skachilova, *Khim. Geterotsikl. Soedin.*, 857 (1974) [*CA* **81**, 105463 (1974)].
[323] H. Reimlinger, M. A. Peiren, and R. Merenyi, *Chem. Ber.* **105**, 103 (1972).
[324] A. J. Barratt, L. R. Bates, J. M. Jenkins, and J. R. White, *U.S. Nat. Tech. Inform. Serv., AD Rep.*, 1971 No. 752370 [*CA* **78**, 124508 (1973)].

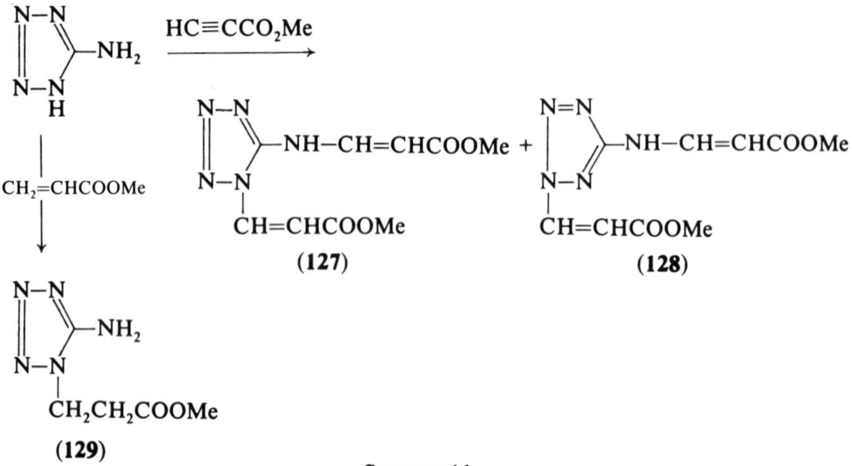

SCHEME 11

An interesting arylation of 5-phenyltetrazolide anion with di-*p*-tolyliodonium bromide gave mixtures of both 1- and 2-*meta* and *para*-tolyl-5-phenyltetrazoles [Eq. (12)]. A mechanism involving a benzyne intermediate was proposed.[325]

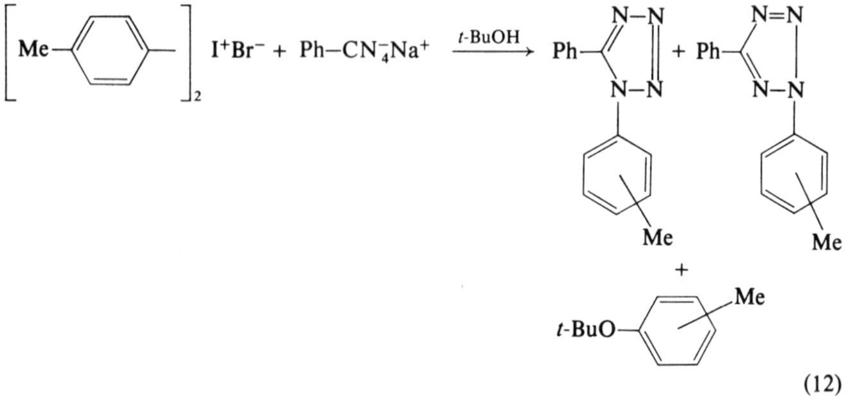

(12)

B. 1-Monosubstituted and 1,5-Disubstituted Tetrazoles

1. Synthesis and Reactions of 1-Substituted Tetrazoles

The reaction of isocyanides with hydrazoic acid has provided one of the main routes to 1-substituted tetrazoles (**130**) for many years.[1] In

---

[325] T. Akiyama, Y. Imasaki, and M. Kawanisi, *Chem. Lett.*, 229 (1974) [*CA* **81**, 3846 (1974)].

general, the reaction has been sluggish and long periods of time were required. Recently, Zimmerman and Olofson[326] have found this reaction to be highly sensitive to acid catalysis and have cut reaction times from days to hours by simply adding one drop of acid to the mixture. These workers postulate that the reaction involves a protonated isocyanide, which is attacked by azide ion (Scheme 12). The catalysis proved

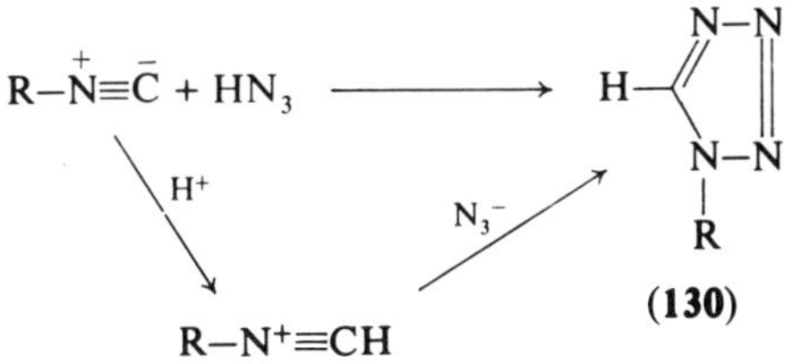

$$R-\overset{+}{N}\equiv\bar{C} + HN_3 \longrightarrow H-\underset{\underset{R}{|}}{\overset{N-N}{\underset{N-N}{||}}}$$

$$\searrow H^+ \qquad \nearrow N_3^-$$

$$R-N^+\equiv CH \qquad (130)$$

SCHEME 12

successful for the preparation of the compound **130** where R = aryl, alkyl, and vinyl. 1-Dialkylaminotetrazoles (**130**; R = NR$_2'$) are also obtained from the appropriate isocyanide and hydrazoic acid.[327] Two routes to 1-aminotetrazole (**130**; R = NH$_2$) have recently been reported. Raap[43] obtained this material by treating tetrazole with hydroxylamine-O-sulfonic acid in weakly alkaline aqueous solutions when both 1- and 2-aminotetrazole were formed. Hagedorn and Winkelmann[328] obtained 1-substituted tetrazoles by treating imidoyl ethers (**131**) with azide ion in acidic solution [Eq. (13)]. With compounds **131** (R = ArCH=N–)

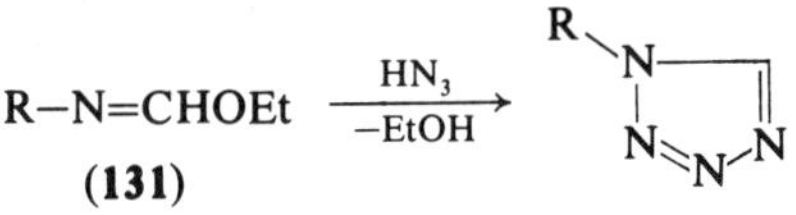

$$R-N=CHOEt \xrightarrow[-EtOH]{HN_3} \underset{(131)}{} \quad \underset{N\underset{N}{\searrow}N}{\overset{R}{\underset{|}{\overset{\diagdown N}{\underset{||}{}}}}} \qquad (13)$$

arylidene derivatives of 1-aminotetrazole were obtained. These could readily be hydrolyzed to the parent amine.[328] Oxidative desulfurization of 1-aryl-5-mercaptotetrazoles has also been used as a means of preparing 1-aryltetrazoles.[329] The 1-pyrazolyltetrazole (**132**) was obtained by treating pyrazole-3-diazonium chloride with an equimolar quantity of diazomethane.[330,331] When an excess of diazomethane was used, methylation of the pyrazole moiety of **132** occurred and mixtures of the 1- and

---

[326] D. H. Zimmerman and R. A. Olofson, *Tetrahedron Lett.*, 5081 (1969).
[327] H. Bredereck, B. Fohlisch, and K. Walz, *Justus Liebigs Ann. Chem.* **688**, 93 (1965).
[328] I. Hagedorn and H. D. Winkelmann, *Chem. Ber.* **99**, 850 (1966).
[329] J. C. Kauer and W. A. Sheppard, *J. Org. Chem.* **32**, 3580 (1967).
[330] H. Reimlinger and R. Merényi, *Chem. Ber.* **103**, 3284 (1970).
[331] H. Reimlinger, G. D. S. King, and M. A. Peiren, *Chem. Ber.* **103**, 2821 (1970).

2-methylated derivatives were formed.[330] The reaction of aromatic diazonium salts with diazomethane has been used previously[1] to a limited extent as a route to 1-aryltetrazoles. The 1-alkanonyltetrazole (133) and the corresponding 2-substituted isomer were obtained, presumably via addition of the anion to mesityl oxide, in a reaction between tetrazole dimethylammonium salt and acetone.[332]

(132)   (133)

1-Aminotetrazole behaves like a normal primary amine in reactions with chloroform in base,[328] phenyl isocyanate,[328] p-toluenesulfonyl chloride,[328] and aromatic aldehydes,[43] and the expected derivatives of the amino moiety are obtained without interference at the tetrazole ring. With acid chlorides and anhydrides acylation of the amino group is accompanied by ring cleavage and yields 2-acylamino-1,3,4-oxadiazoles.[43,328] Flash vacuum pyrolysis of 1-phenyltetrazole at 500° has recently been reported to give phenylcyanamide as the only detectable product.[333]

2. Synthesis of 1,5-Disubstituted Tetrazoles

Many routes to 1,5-disubstituted tetrazoles, the most common of tetrazole derivatives, have been developed in the past.[1,2] The chief of these may be described under the following separate headings, although most of them are intrinsically related since the final step generally involves the cyclization of an imino azide ($-C(N_3)=N-$): (i) the reaction between nitriles and organic azides[334]; (ii) the reaction of nitrilium salts with azide ion; (iii) the reaction of isocyanides with hydrazoic acid including the Ugi reaction; (iv) the reaction of carbodiimides with hydrazoic acid or azide ion; (v) the reaction of azide ion or hydrazoic acid with imidoyl compounds; (vi) azide displacement reactions with α-halocarbonyl compounds, substituted guanidines, and cyclic lactone groups attached to C=N moieties; (vii) the reactions of substituted hydrazidines with nitrous acid[334a]; (viii) cyclization of acyl tetrazenes;

[332] J. E. Oliver and J. B. Stokes, J. Heterocycl. Chem. 7, 961 (1970).
[333] T. L. Gilchrist, C. W. Rees, and C. Thomas, J. Chem. Soc., Perkin Trans. 1, 12 (1975).
[334] G. L. 'Abbe, Chem. Rev. 69, 359 (1969).
[334a] D. G. Neilson, R. Roger, J. W. M. Heatlie, and L. R. Newlands, Chem. Rev. 70, 151 (1970).

Sec. III.B] RECENT ADVANCES IN TETRAZOLE CHEMISTRY 373

(ix) the Schmidt reaction of ketones with 2 mol of hydrazoic acid; (x) the reaction of nitriles with hydrazoic acid; (xi) thermal reactions of *gem*-diazides, and (xii) rearrangement of arylaminothiatriazoles. In recent years these general methods continue to be used for tetrazole synthesis, but main interest appears to have centered on a few of them.

a. From Nitriles and Azides. The phosphorylated tetrazolines (**134**) were obtained by treating phenylacetonitriles with diphenylphosphinic azide under basic conditions [Eq. (14)].[335] The compounds **134** were readily hydrolyzed to 5-benzyltetrazoles.[335]

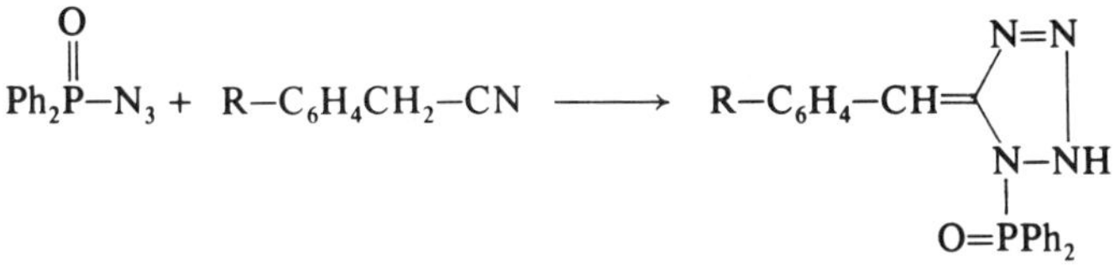

$$Ph_2\overset{O}{\underset{\|}{P}}-N_3 + R-C_6H_4CH_2-CN \longrightarrow R-C_6H_4-CH\underset{\underset{O=PPh_2}{\overset{|}{N-NH}}}{\overset{N=N}{\diagup}} \quad (14)$$

(**134**)

The tetrazole polymers (**135**) were obtained by heating mixtures of organic diazides and perfluoroalkyl dinitriles.[336] In general, organic

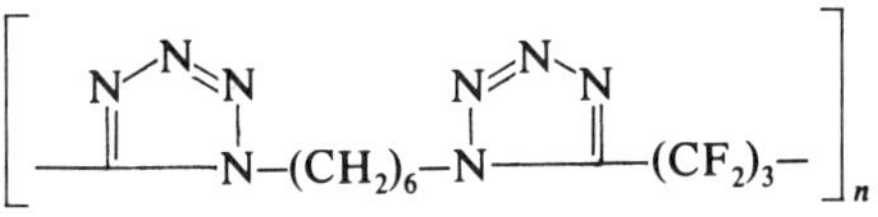

$$\left[\underset{}{\overset{N\diagdown N}{\underset{N}{\diagup}}}-N-(CH_2)_6-N\underset{}{\overset{N\diagdown N}{\underset{N}{\diagup}}}-(CF_2)_3-\right]_n$$

(**135**)

azides react with perfluoronitriles to give 1-substituted 5-perfluoroalkyltetrazoles (**136**) [Eq. (15)].[337,338]

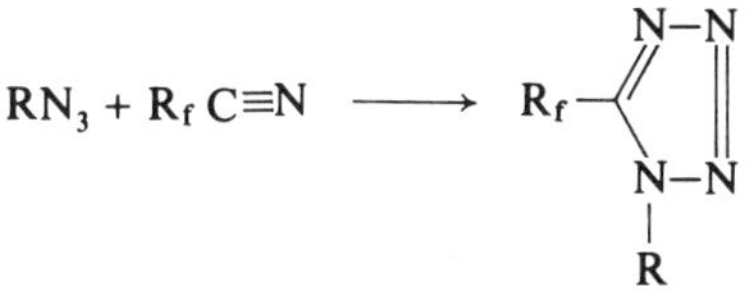

$$RN_3 + R_f C\equiv N \longrightarrow R_f\underset{\underset{R}{\overset{|}{N-N}}}{\overset{N-N}{\diagup}} \quad (15)$$

(**136**)

The preparation of cyanogen azide from cyanogen bromide and sodium azide in anhydrous medium has recently been described.[339] In aqueous solution, further involvement of a second azide ion leads to the sodium salt of 5-azidotetrazole.[339] The influence of solvent on these reactions

---

[335] K. D. Berlin, R. Ranganathan, and H. Haberlein, *J. Heterocycl. Chem.* **5**, 813 (1968).
[336] W. R. Carpenter, U.S. Patent 3,386,968 (1968) [*CA* **69**, 28106 (1968)].
[337] W. R. Carpenter, *J. Org. Chem.* **27**, 2085 (1962).
[338] W. P. Norris and W. G. Finnegan, *J. Org. Chem.* **31**, 3292 (1966).
[339] F. D. Marsh, *J. Org. Chem.* **37**, 2966 (1972).

and those described below, which involve azide cycloadditions, has been reviewed.[5]

b. *From Nitrilium Ions.* Treatment of the nitrilium salts (**137**) with azide ion allows the synthesis of a wide range of 1,5-disubstituted tetrazoles [Eq. (16)].[340,341]

$$R^1-\overset{+}{C}\equiv N-R \quad BF_4^- \xrightarrow{NaN_3} R^1 \underset{\underset{R}{N-N}}{\overset{N-N}{\diagup\diagdown}} \quad (16)$$

(**137**)

Addition of halogens to olefins in a nitrile solvent containing silver perchlorate leads to nitrilium ions, which may be trapped with azide ion to yield 1,5-disubstituted tetrazoles, e.g., **138** [Eq. (17)].[342-345] The

$$Ph-CH=CH_2 \xrightarrow[MeCN]{Br_2/AgClO_4} Ph-CH-CH_2Br \xrightarrow{N_3^-} Ph-CH-CH_2Br \quad (17)$$

with intermediate nitrilium (N⁺≡C–Me) and product Me-substituted tetrazole (**138**)

stereospecific trans addition of this reaction is consistent with attack by the nitrile on an intermediate halonium species.[342,343] An interesting variation of this reaction with α-pinene (Scheme 13) has recently been reported.[346] The same tetrazole product (Scheme 13) has also been obtained by treating α-pinene with $Pb(OAc)_{4-n}[N_3]_n$ in acetonitrile as solvent.[346a] The kinetics of solvolysis of 1-adamantyl arenesulfonates in acetonitrile containing azide ion have been measured.[347] The products were 1-(1-adamantyl)-5-methyltetrazole in high yield accompanied by a low yield of 1-adamantyl azide. The kinetics were consistent with a

---

[340] L. A. Lee, E. V. Grabtree, J. W. Lowe, M. J. Cziela, and R. Evans, *Tetrahedron Lett.*, 2885 (1965).

[341] L. A. Lee, R. Evans, and J. W. Wheeler, *J. Org. Chem.* **37**, 343 (1972).

[342] A. Hassner, *Acc. Chem. Res.* **4**, 10 (1971); A. Hassner, L. A. Levy, and R. Gault, *Tetrahedron Lett.*, 3119 (1966).

[343] A. Hassner and F. Boerwinkle, *Tetrahedron Lett.*, 3309 (1969).

[344] A. Terada and A. Hassner, *Bull. Chem. Soc. Jpn.* **42**, 2666 (1969).

[345] S. N. Moorthy, D. Devaprabhakara, and K. G. Das, *Tetrahedron Lett.*, 257 (1975).

[346] S. Ranganathan, D. Ranganathan, and A. K. Mehrotra, *Tetrahedron Lett.*, 2265 (1973).

[346a] A. Ztütz and E. Zbiral, *Justus Liebigs Ann. Chem.* **765**, 34 (1972); cf. also a review by E. Zbiral, *Synthesis*, **285** (1972).

[347] D. N. Kevill and C.-B. Kim, *J. Org. Chem.* **39**, 3085 (1974).

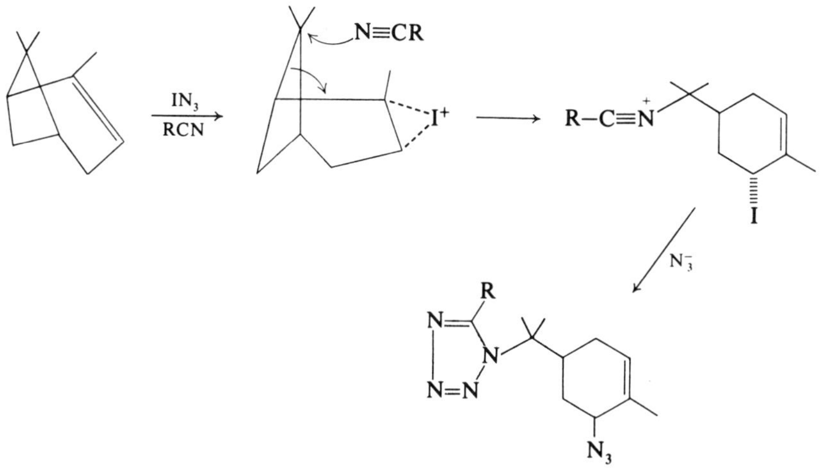

SCHEME 13

mechanism involving competition between azide ion and acetonitrile for an intermediate carbonium ion, the tetrazole product arising from subsequent 1,3-addition of azide ion (Scheme 14).[347]

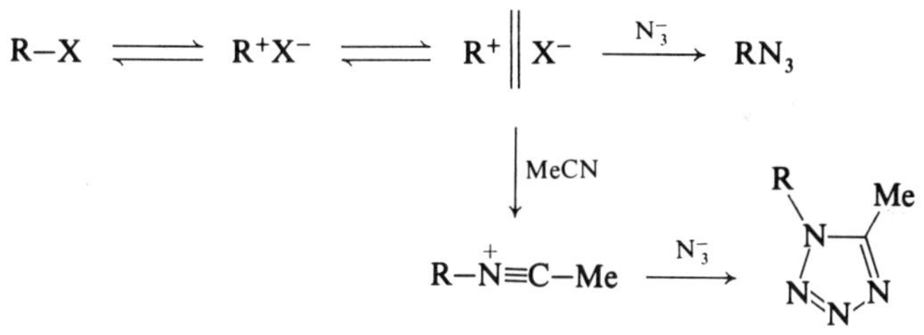

SCHEME 14

c. From Isocyanide Derivatives and Isocyanates. A new synthesis of 1-aryl-5-chloro and -5-azidotetrazoles (**139**) by treating isocyanide dihalides with sodium azide has been reported (Scheme 15).[329,348] Replacement of the 5-chloro substituent by azide ion was facilitated with acetone as solvent.[329] In general, nucleophilic substitution of the 5-chloro atom has led to a range of tetrazoles of type **139** (X = NHR, NH$_2$NH, N$_3$, OR, OAr).[348] Treatment of isothiocyanates with azide ion leads to 1-substituted 5-mercaptotetrazoles (**140**). These have been further converted into the derivatives shown (Scheme 15).[329,349]

[348] C. A. Maggiulli and R. E. Paine, Belg. Patent 671,402 (1966) [*CA* **65**, 8926 (1966)].
[349] R. Neidlein and J. Tauber, *Chem. Ber.* **100**, 736 (1967).

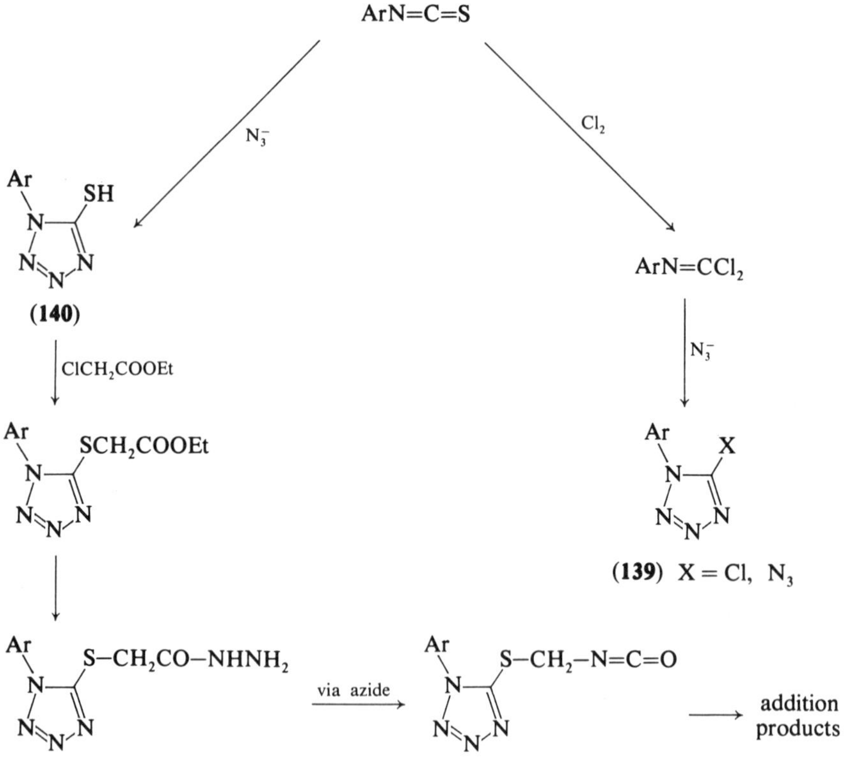

SCHEME 15

Organometallic azides, for example, tri-*n*-butyltin azide and triphenyltin azide, also react with phenylisothiocyanate to give N=C adducts [Eq. (18)].[350,351] The organometallic group could readily be removed to give the 1,5-disubstituted tetrazole by treatment with cold dilute HCl.[350] In

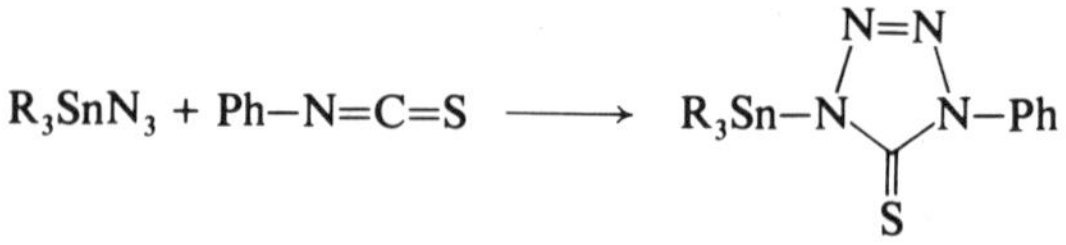

$$R_3SnN_3 + Ph-N=C=S \longrightarrow R_3Sn-N\underset{S}{\overset{N=N}{\diagdown\diagup}}N-Ph \qquad (18)$$

contrast, hydrazoic acid and alkyl azides react with isothiocyanates at the C=S bond and give 5-(substituted amino) 1,2,3,4-thiatriazoles.[352] A range of 1-aryl-5-mercapto tetrazoles has been obtained by treating a

[350] P. Dunn and D. Oldfield, *Aust. J. Chem.* **24**, 645 (1971).
[351] P. Kreutzer, C. Weis, H. Boehme, T. Kemmerich, W. Beck, C. Spencer, and R. Mason, *Z. Naturforsch., Teil B* **27**, 745 (1972) [*CA* **77**, 119436 (1972)].
[352] G. L'abbe, E. Van Loock, R. Albert, S. Toppet, G. Verhelst, and G. Smets, *J. Am. Chem. Soc.* **96**, 3973 (1974).

previously prepared mixture of the aromatic amine, carbon disulfide, and sodium hydroxide with sodium azide and heating under reflux.[353,354] Further syntheses of 1-substituted 5-mercaptotetrazoles[355] and 1-substituted 5-thiocyanatotetrazoles[356] have also been reported. Organic azides add to substituted isocyanates at the —N=C— bond and give 1,4-disubstituted tetrazolin-5-ones (141) [Eq. (19)].[357,358] This reaction has been successful for aryl, acyl, carboalkoxy, and sulfonyl isocyanates

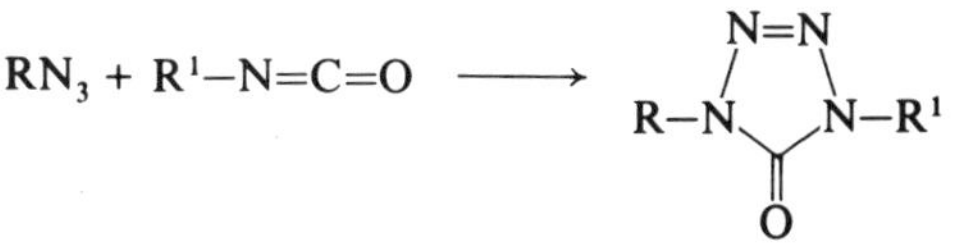

$$RN_3 + R^1-N=C=O \longrightarrow \underset{(141)}{R-N\underset{\underset{O}{\|}}{\underset{|}{\diagdown}}\overset{N=N}{\underset{N-R^1}{\diagup}}} \qquad (19)$$

with alkyl azides.[357] The phosphinylmethyltetrazole (142) was obtained[359] when diphenylphosphinylmethylisocyanide was treated with sodium azide, formaldehyde, and piperidine in acetone, an example of the Ugi reaction which has been widely used[1] in tetrazole synthesis. Hydroxylamines have also been successfully employed in the Ugi

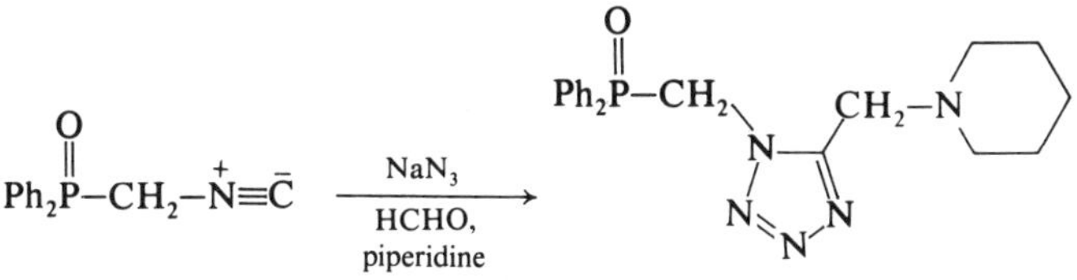

$$Ph_2\overset{O}{\overset{\|}{P}}-CH_2-\overset{+}{N}\equiv\overset{-}{C} \xrightarrow[\text{piperidine}]{NaN_3, HCHO,} (142)$$

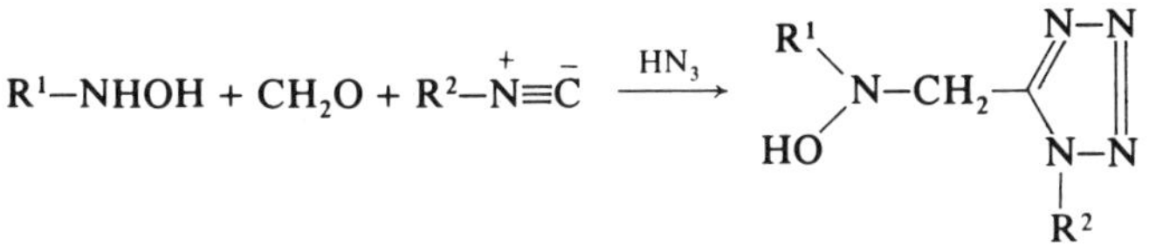

$$R^1-NHOH + CH_2O + R^2-\overset{+}{N}\equiv\overset{-}{C} \xrightarrow{HN_3} (143) \qquad (20)$$

---

[353] H. Loester and E. Lippmann, Ger. (East) Patent 106,645 (1974) [CA **82**, 72998 (1975)].

[354] H. Loester and E. Lippmann, Ger. (East) Patent 105,224 (1974) [CA **81**, 169549 (1974)].

[355] R. G. Dubenko and V. D. Panchenko, Khim. Geterotsikl. Soedin. 199 (1967) [CA **70**, 87687 (1969)].

[365] R. Pohloudek-Fabini, K. Kottke, and F. Friedrich, Pharmazie **24**, 433 (1969) [CA **71**, 10811 (1969)].

[357] J.-M. Vandensavel, G. Smets, and G. L'abbe, J. Org. Chem. **38**, 675 (1973).

[358] G. J. Smets and J. M. Vandensavel, Ger. Offen. 2,226,525 (1972) [CA **78**, 58984 (1973)].

[359] N. Kreutzkamp and K. Lammerhirt, Angew. Chem., Int. Ed. Engl. **7**, 372 (1968).

Reaction [Eq. (20)], and a range of the tetrazoles **143** has been obtained.[360] A similar reaction with hydrazines, $R-NH-NHR^1$, ketones R^2R^3CO, hydrazoic acid, and an isocyanide, $R^4-\overset{+}{N}\equiv\bar{C}$, gave the bistetrazoles (**144**).[361]

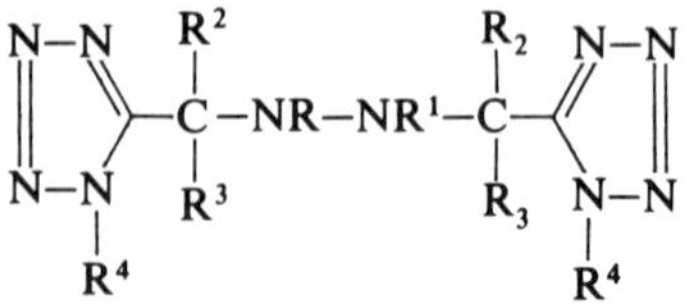

$$\underset{R^4}{\underset{|}{\underset{N-N}{\overset{N-N}{\|}}}}\overset{R^2}{\underset{R^3}{\overset{|}{C}}}-NR-NR^1-\overset{R_2}{\underset{R_3}{\overset{|}{C}}}\underset{R^4}{\underset{|}{\underset{N-N}{\overset{N-N}{\|}}}}$$

(**144**)

d. From Imidoyl Compounds and Imides. Treatment of a range of the imidoyl halides (**145**) with azide ion gives high yields of the corresponding tetrazoles [Eq. (21)].[362-364]

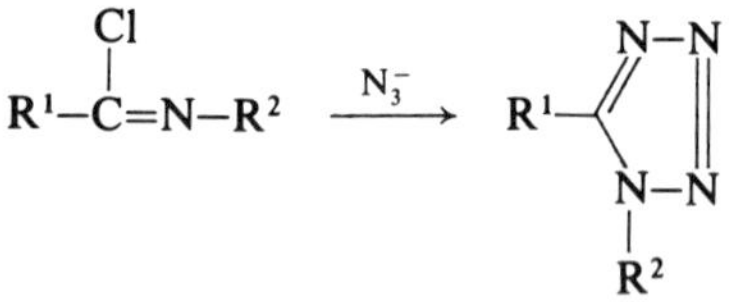

$$R^1-\overset{\overset{Cl}{|}}{C}=N-R^2 \xrightarrow{N_3^-} R^1\underset{\underset{R^2}{|}}{\overset{N-N}{\underset{N-N}{\|}}} \quad (21)$$

(**145**) $R^1 = -COOEt$; $-COCl$; $-CO-Ar$; $-CO-N_3$; $NH-R^3$;
$R^2 = $ alkyl, aryl

The imidoyl ether **146** similarly gave the tetrazoles **147** on treatment with hydrazoic acid.[365] The kinetics of the reaction of the imidoyl ether **148** in water and D_2O containing azide ion in the pH range 2–13 have

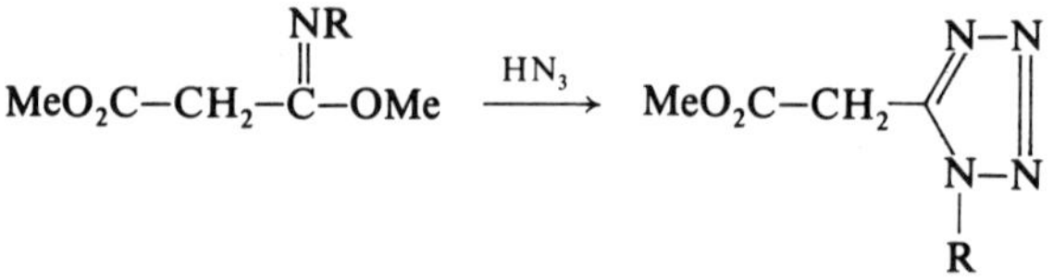

$$MeO_2C-CH_2-\overset{\overset{NR}{\|}}{C}-OMe \xrightarrow{HN_3} MeO_2C-CH_2\underset{\underset{R}{|}}{\overset{N-N}{\underset{N-N}{\|}}}$$

(**146**)   (**147**) $R = Ph, Me_2N$

---

[360] G. Zinner, D. Moderhack, O. Hantelmann, and W. Bock, *Chem. Ber.* **107**, 2947 (1974).

[361] G. Zinner and W. Bock, *Arch. Pharm. (Weinheim)* **304**, 933 (1971) [*CA* **76**, 85764 (1972)].

[362] M. O. Lozinskii, A. F. Shivanyuk, and P. S. Pel'kis, *Ukr. Khim. Zh.* **39**, 1245 (1973) [*CA* **80**, 70753 (1974)]; *Dopov. Akad. Nauk. Ukr. RSR, Ser. B* **31**, 1096 (1969) [*CA* **73**, 14767 (1970)].

[363] M. O. Lozinskii, A. F. Shivanyuk, and P. S. Pel'kis, *Khim. Geterotsikl. Soedin.* **7**, 471 (1971) [*CA* **76**, 25184 (1972)].

[364] N. S. Zefirov, N. K. Chapovskaya, and S. S. Trach, *Zh. Org. Khim.* **8**, 629 (1972) [*CA* **77**, 5408 (1972)].

[365] R. Raap, *Can. J. Chem.* **46**, 2255 (1968).

been reported.[366] The protonated form of the imidoyl ether was the primary reactive species, and the reaction with azide ion gave 1,5-dimethyltetrazole and phenol.[366] Replacement of the halide from

$$\text{Me-N=C(OPh)-Me} \longrightarrow \text{Me-}\overset{+}{\text{NH}}=\text{C(OPh)-Me} \underset{}{\overset{N_3^-}{\rightleftarrows}} \text{Me-}\underset{\underset{\text{Me}}{|}}{\begin{array}{c}N-N\\ \diagup\diagdown \\ N-N\end{array}} + \text{PhOH}$$

(148)

hydrazonyl halides[367,368] and hydroxyamoyl halides[369-371] with azide ion yields the azides **149** and **150**, respectively, rather than the expected tetrazoles. These azides do not cyclize readily to tetrazoles, possibly

$$\begin{array}{cc} \text{R-C=N:} & \text{R-C=N:} \\ |\quad\quad\backslash & |\quad\quad\backslash \\ \text{N}_3\quad\text{NHAr} & \text{N}_3\quad\text{OH} \\ (149) & (150) \end{array}$$

because of an unfavorable orientation of the imino lone pair as in **149** and **150**. Recently, however, the azide **151** has been cyclized by treatment with propionyl chloride [Eq. (22)].[372]

$$\text{Ph}_2\text{CH-C(N}_3\text{)=N-OH} \xrightarrow{\text{EtCOCl}} \text{Ph}_2\text{CH-}\underset{\underset{\text{OCOEt}}{|}}{\begin{array}{c}N-N\\ \diagup\diagdown \\ N-N\end{array}} \xrightarrow{\text{EtOH}} \text{Ph}_2\text{CH-}\underset{\underset{\text{OH}}{|}}{\begin{array}{c}N-N\\ \diagup\diagdown \\ N-N\end{array}} \quad (22)$$

(151)

Treatment of the sulfonylcarbodiimides **152** with hydrazoic acid yielded the 1-substituted 5-sulfonylaminotetrazoles (**153**) via an imino azide [Eq. (23)].[373] In general, such azides cyclize onto the nitrogen bearing the more electron-donating substituent (see Section IV). Compounds

---

[366] Y. Pocker, M. W. Brug, and K. L. Stephens, *J. Am. Chem. Soc.* **96**, 174 (1974).
[367] A. F. Hegarty, J. B. Aylward, and F. L. Scott, *J. Chem. Soc., C*, 2587 (1967).
[368] A. F. Hegarty, J. B. Aylward, and F. L. Scott, *Tetrahedron Lett.*, 1257 (1967).
[369] F. Eloy, *J. Org. Chem.* **26**, 953 (1961).
[370] M. S. Chang and A. J. Matuszko, *J. Org. Chem.* **28**, 2260 (1963).
[371] C. Grundmann and H-D. Frommeld, *J. Org. Chem.* **31**, 157 (1966).
[372] J. Plenkiewicz, *Tetrahedron Lett.*, 341 (1975).
[373] R. Neidlein and E. Heukelbach, *Angew. Chem., Int. Ed. Engl.*, **5**, 520 (1966); L. F. Pronskji, E. A. Abrashanova, and V. N. Sevast'yanov, *Vopr. Khim. Khim. Tekhnol,* **33**, 10 (1974) [*CA* **82**, 156195 (1975)]; R. Neidlein and H. Haussmann, *Tetrahedron Lett.*, 5401 (1966).

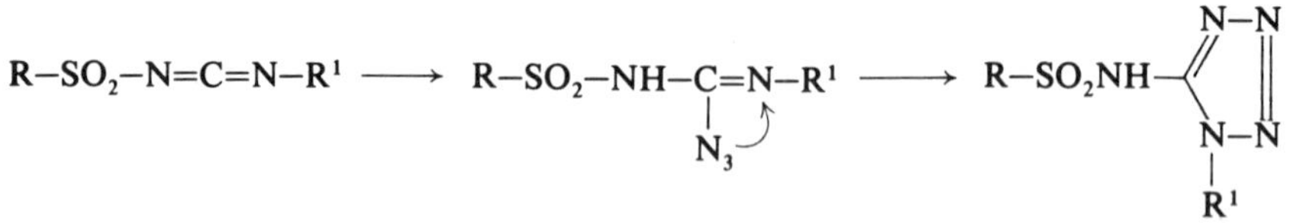

$$R-SO_2-N=C=N-R^1 \longrightarrow R-SO_2-NH-\underset{\underset{N_3}{|}}{C}=N-R^1 \longrightarrow R-SO_2NH-\underset{\underset{R^1}{|}}{\overset{N-N}{\underset{N-N}{\left\langle\right.}}}$$

(152) (153)

(23)

of type **153** have also been obtained by heating 1-substituted 5-aminotetrazoles with *p*-toluenesulfonyl chloride in pyridine.[374] Competing cyclizations were observed when the carbodiimide-azide adducts (**154**) were generated by treating the corresponding thiosemicarbazides with PbO under N_2 in the presence of NaN_3 (Scheme 16).[290] Cyclization by path a giving compounds **155** was the dominant process, and only ca. 1–3% of cyclization by path b was encountered.[290]

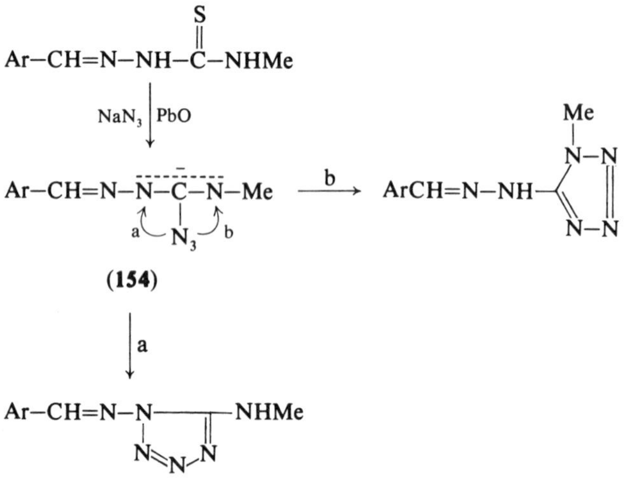

$$Ar-CH=N-NH-\overset{\overset{S}{\|}}{C}-NHMe$$

NaN_3 | PbO

$Ar-CH=N-\overset{a}{N}-\underset{N_3}{\overset{|}{C}}-\overset{b}{N}-Me \xrightarrow{b} ArCH=N-NH-\underset{\underset{N-N}{}}{\overset{\overset{Me}{|}}{\underset{}{\left\langle\right.}}}\overset{N-N}{}$

(154)

↓ a

$Ar-CH=N-\underset{\underset{N\!\!\!\diagdown_{N}\diagup N}{}}{\overset{|}{N}}-NHMe$

(155)

SCHEME 16

A new route to 1,5-disubstituted tetrazoles from the reaction of triphenyliminophosphoranes (**156**) with acid chlorides and sodium azide (Scheme 17) has recently been developed.[375,376]

[374] V. P. Shapinov and I. Ya. Postovskii, *Zh. Org. Chim.* **2**, 1108 (1966) [*CA* **65**, 15368 (1966)].
[375] E. Zbiral and J. Ströh, *Justus Liebigs Ann. Chem.* **725**, 29 (1969).
[376] E. Zbiral, *Synthesis*, 775 (1974).

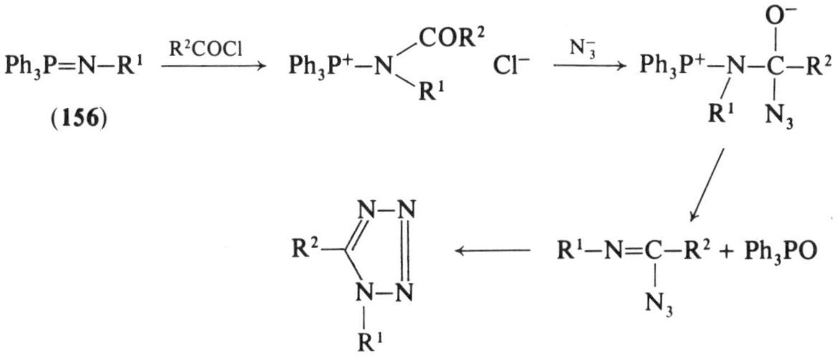

Sec. III.B] RECENT ADVANCES IN TETRAZOLE CHEMISTRY 381

$$Ph_3P=N-R^1 \xrightarrow{R^2COCl} Ph_3P^+-N\begin{smallmatrix}COR^2\\R^1\end{smallmatrix} Cl^- \xrightarrow{N_3^-} Ph_3P^+-N-\underset{N_3}{\overset{O^-}{\underset{|}{C}}}-R^2$$
(156) R^1

$$R^2-\underset{\underset{R^1}{|}}{\overset{N-N}{\underset{N-N}{\|}}} \longleftarrow R^1-N=\underset{N_3}{\overset{}{\underset{|}{C}}}-R^2 + Ph_3PO$$

SCHEME 17

The reaction is applicable to a range of acyl halides and iminophosphoranes and should be particularly useful for introducing selected substituents at the 5-position.

e. From the Schmidt Reaction. Recent studies of this reaction for the synthesis of tetrazoles, e.g., Eq. (24), have been concerned particularly with the mechanism of the cyclization stage.[377,378] Cyclization of the azide is inhibited by increasing acidity of the medium[377a] and only the anti-form of the azide is reported to cyclize.[378] The reaction of

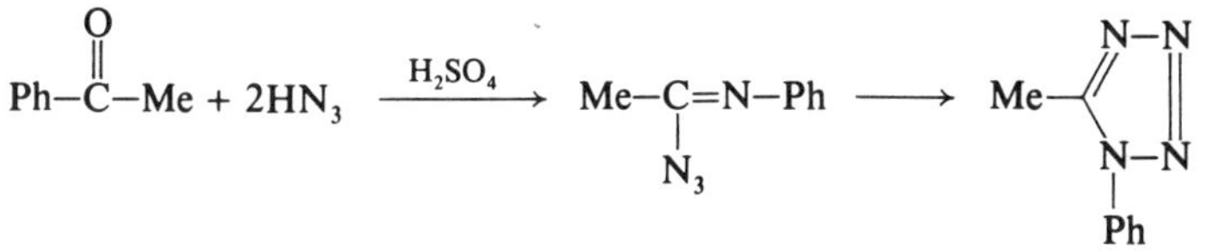

$$Ph-\overset{O}{\underset{\|}{C}}-Me + 2HN_3 \xrightarrow{H_2SO_4} Me-\underset{N_3}{\overset{}{\underset{|}{C}}}=N-Ph \longrightarrow Me-\underset{\underset{Ph}{|}}{\overset{N-N}{\underset{N-N}{\|}}} \quad (24)$$

acetophenone with hydrazoic acid in aqueous sulfuric acid [Eq. (24)] also gives a low yield of MeCONHPh, which can be decreased further by increasing the proportion of hydrazoic acid used. Up to 5% of PhCONHMe and the accompanying 1-methyl-5-phenyltetrazole are also formed.[377b] The main products arise from competitive attack by OH$^-$ and HN$_3$ on the intermediate Me—C≡$\overset{+}{N}$—Ph.[377b] The isolation of a urea from the Schmidt reaction on camphor, from which tetrazoles were previously isolated, has also been reported.[379] The Schmidt reaction with steroids as ketones has been used widely to obtain compounds containing a tetrazole ring fused to a steroid system (cf. Section IV,C).

[377] S. A. Enin, G. I. Koldobskii, and L. I. Bagal, *Zh. Org. Khim.* **7**, (a) 2560, (b) 1672 (1971) [*CA* **76**, 71677 (1972) and **75**, 139961 (1971)].
[378] S. A. Enin, G. I. Koldobskii, V. A. Ostrovskii, and L. I. Bagal, *Zh. Org. Khim.* **8**, 1895 (1972) [*CA* **78**, 42663 (1973)].
[379] J. W. ApSimon and N. R. Hunter, *Tetrahedron Lett.*, 187 (1972).

f. From Azide Displacement of Cyclic Lactone Groups and General. A series of 1-(substituted vinyl) 5-substituted tetrazoles (**157**) was obtained by treating substituted oxazol-5-ones with azide ion followed by decarboxylation with Cu in quinoline [Eq. (25)].[380] The decarboxylation

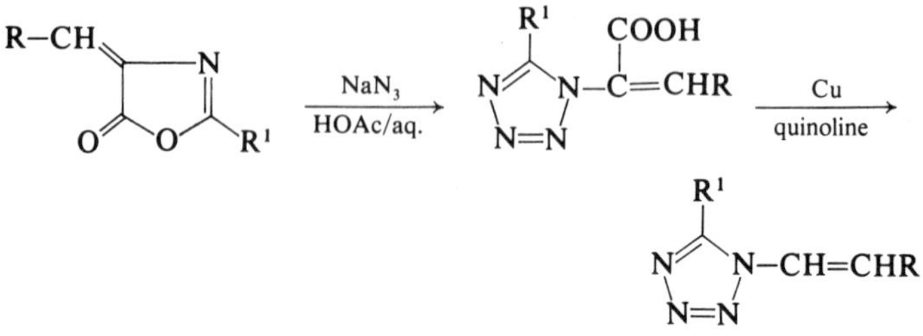

$$\text{(25)}$$

procedure gave rise to mixtures of cis and trans forms of **157**.[380] The kinetics of the reactions of *N*-phenylphthalisoimide (**158**) with a range of nucleophiles including azide ion have been investigated at various pH values.[381]

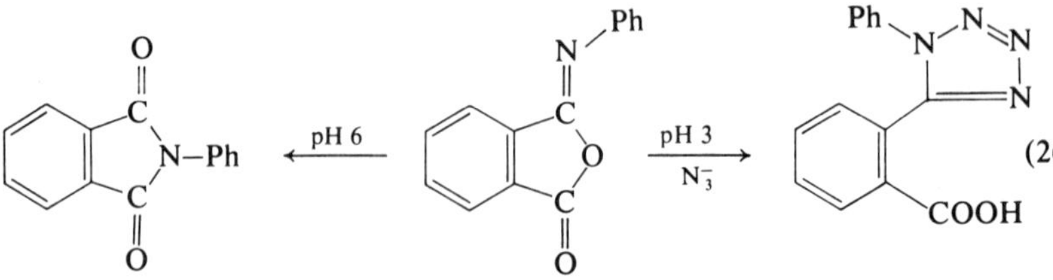

$$\text{(26)}$$

The isoimide gave the tetrazole **159** at pH 3, and at pH 6 extensive rearrangement to *N*-phenylphthalimide occurred [Eq. (26)].[381] At pH 4.5, mixtures of compound **159** and *N*-phenylphthalimide were obtained. An imino azide arising from azide ion attack on the protonated isomide was considered to be a likely intermediate under the acidic conditions.[381] Thermolysis[150] and photolysis[148] of *gem*-diazides have also led to 1,5-diaryl-, 1,5-dimethoxycarbonyl- and 1,5-dicarboxamido-tetrazoles. Treatment of the nitrones (**160**) with hydrazoic acid interestingly led to azides (**161**),[382] not to the expected[383] tetrazoles (**162**). Only in one case was a small yield of a tetrazole (**162**, R = NO$_2$), obtained.[382] The tetrazole **164** was obtained by treating 2-methyl-5-phenylisoxazolium

[380] J. Lykkeberg and N. A. Klitgaard, *Acta Chem. Scand.* **26**, 266 (1972).
[381] M. L. Ernst and G. L. Schmir, *J. Am. Chem. Soc.* **88**, 5001 (1966).
[382] D. Moderhack, *Synthesis*, 299 (1973).
[383] F. Kröhnke, *Angew. Chem.* **75**, 317 (1963).

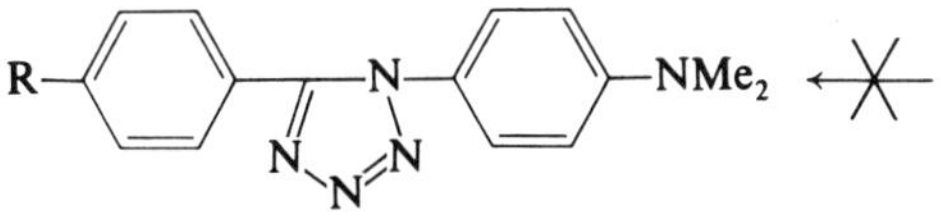

(162) [structure: R-C6H4-tetrazole-N-C6H4-NMe2, with ✗ indicating not formed]

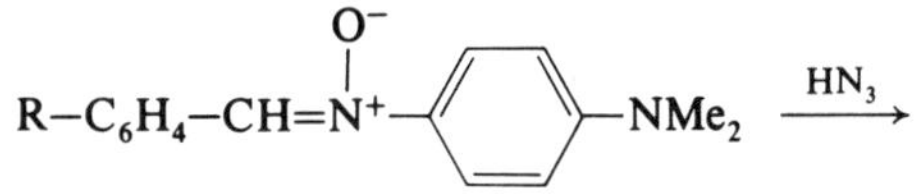

(160) [structure: R-C6H4-CH=N+(O⁻)-C6H4-NMe2] → HN3

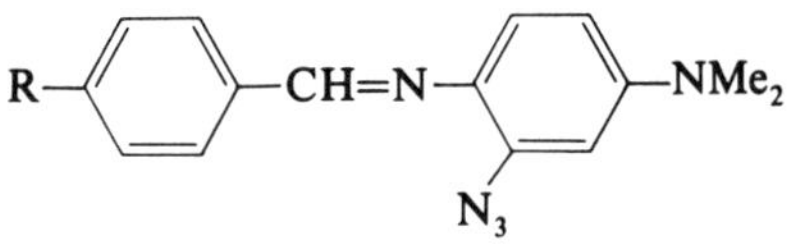

(161) [structure: R-C6H4-CH=N-C6H4(N3)-NMe2]

methylsulfate (**163**) with sodium azide [Eq. (27)], followed by cyclization of the imino azide by heating under reflux in ethanol.[384] It is of interest for the synthesis of substituted tetrazoles that the recently prepared[385-387] iminoazimines (**165**) exist preferentially in acyclic dipolar

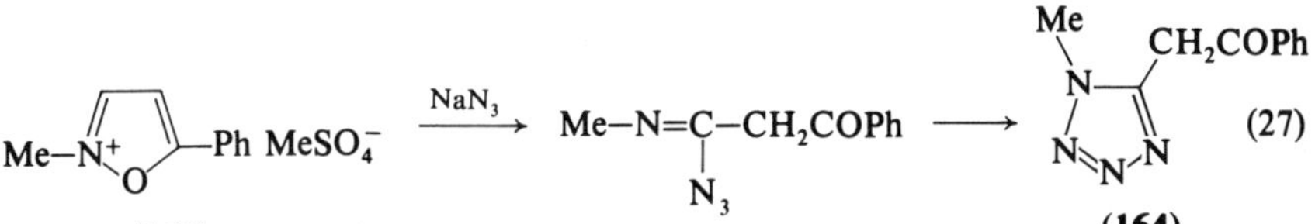

$$\text{Me-N}^+\text{(O)-Ph MeSO}_4^- \xrightarrow{\text{NaN}_3} \text{Me-N=C(N}_3\text{)-CH}_2\text{COPh} \longrightarrow (164) \quad (27)$$

(163) → (164) [1,4-dimethyl-5-(phenacyl)-tetrazole-type structure]

structures rather than in the alternative cyclic tetrazoline form (**166**). Some 1,4-dialkyltetrazolin-5-ones have also been obtained recently from the reaction of nucleophiles with 1,4-dialkoxycarbonyl-1,4-dialkyltetrazenes.[154] An interesting synthesis of 1,5-dimethyltetrazole has been

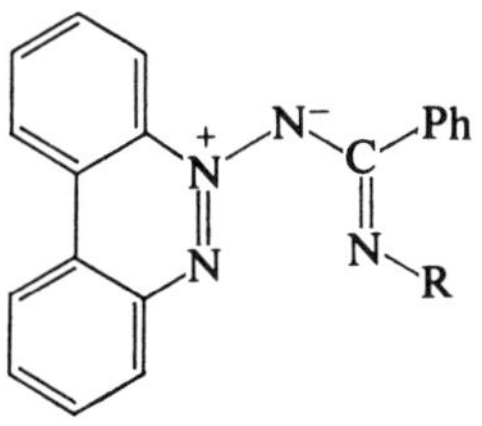

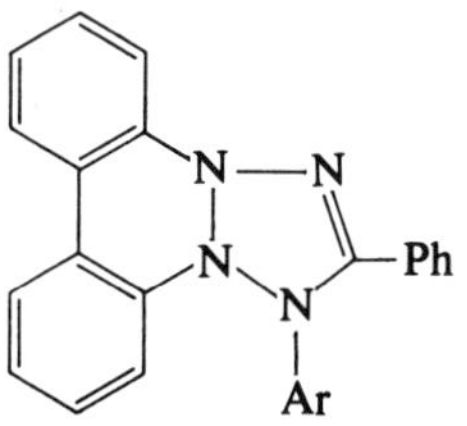

(**165**) vs. (**166**)

[384] R. B. Woodward and R. A. Olofson, *Tetrahedron Suppl.* **7**, 415 (1966).
[385] F. A. Neugebauer and H. Fischer, *Chem. Ber.* **106**, 1589 (1973).
[386] J. J. Barr, R. C. Storr, and J. Rimmer, *Chem. Commun.*, 657 (1974).
[387] M. J. Rance, C. W. Rees, P. Spagnolo, and R. C. Storr, *Chem. Commun.*, 659 (1974).

reported by treating the oxime ester Me$_2$C=N—O—SO$_2$Ph, with NaN$_3$ in aqueous methanol, which presumably involves a Beckmann type rearrangement with subsequent azide attack on nitrilium ion intermediate.[387a]

3. Reactions of 1,5-Disubstituted Tetrazoles

1,5-Disubstituted tetrazoles display a wide range of interesting reactions. A number of these have already been discussed in Sections II,B,2, II,C, II,D,3 and 4, III,A,4, and III,B,2.

a. Exchange and Replacement at C-5. Replacement of halogen atoms at the 5-C position by nucleophiles has been widely used[1,329,348] for the synthesis of substituted tetrazoles. Kinetic studies of the reaction between 1-methyl-5-bromotetrazole and the 2-methyl isomer with piperidine have shown the 1-methyl compound to be considerably more reactive, possibly owing to a greater number of mesomeric canonical forms in the transition state for the bimolecular process.[388] Comparison of these kinetics with similar reactions for a series of azoles has indicated that two to three electron-withdrawing sp^2 nitrogen atoms (—N=) are required to overcome the electron release from one pyrrole-type nitrogen atom in these systems.[388]

Kinetic studies of base-induced deuterium–protium exchange at the 5-C atom of 1-substituted tetrazoles and tetrazolium cations have indicated the presence of carbanionic intermediates in the exchange process.[389] The reaction for 1-substituted tetrazoles had a ρ value of +1.3 in piperidine-MeOD—DMF.[389] The pH profile for protium–deuterium exchange for 1-methyltetrazole-5-d in aqueous solution has also been reported.[390] In the pH range 3–5 the reaction was considered to involve either the free carbanion (**167**) resulting from D$^+$ abstraction, the ylid (**168**) resulting from initial protonation at N-1 or

Me—N⎯N   R—N⎯N$^+$—R^1
  \ /   \ /
  N=N   N=N
 (**167**)   (**168**) R = Me, R^1 = H
              (**169**) R, R^1 = alkyl, aryl, etc.

[387a] F. E. Condon, R. Waldman, N. Kundu, and J. P. Trivedi, *Org. Prep. Proced. Int.* **6**, 135 (1974) [*CA* **81**, 37520 (1974)].
[388] G. B. Barlin, *J. Chem. Soc., B*, 641 (1967).
[389] A. C. Rochat and R. A. Olofson, *Tetrahedron Lett.*, 3377 (1969).
[390] H. Kohn, S. J. Benkovic, and R. A. Olofson, *J. Am. Chem. Soc.* **94**, 5759 (1972).

N-2 followed by D^+ abstraction, or a pathway comparable to an electrophilic substitution with H_2O as the acid and OH^- as the base.[390] A major rate-enhancing effect was observed with Cu^{2+} and Zn^{2+} ions due to σ complexation with the heterocycle.[390] The rate of base-induced protium–deuterium exchange of 1-methyltetrazole was 10^5 times faster than that of 2-methyltetrazole,[391] and, in general, ring protons in positions α to sp^3 pyrrole type nitrogen atoms were found to exchange much faster than those located β to such nitrogens.[392-395] Calculated energies of deprotonation for azolium cations were consistent with the experimental trends and the relative importance of coulombic, inductive, hybridization and resonance effects in governing these energies has been discussed.[396] Species of type **169** are considered to be intermediates in these reactions.[389] Attempts to prepare a carbanionic tetrazolyl Grignard reagent by treating 1-phenyl-5-chlorotetrazole with Mg or n-butyllithium were unsuccessful: decomposition of the tetrazole ring occurred, yielding phenylcyanamide [Eq. (28)].[329] It is not clear whether a

$$\text{Cl-tetrazole(Ph)} \xrightarrow{BuLi} [\text{-tetrazole(Ph)}] \longrightarrow \text{PhNH-CN} \quad (28)$$

carbanion was involved in the reaction or not, since the rate of deuterium–protium exchange was 10^6 times faster than the decomposition.[329,394,395] However, by treatment of 1-methyltetrazole with n-butyllithium in THF at −60°, Raap[397] has obtained 1-methyltetrazol-5-yllithium (**170**). This decomposed to N_2 and lithium methylcyanamide on heating above −50°. 1-Phenyltetrazol-5-yl lithium was also prepared and found to be less stable, decomposing at temperatures as low as −70°.[397] Compound **170** gave useful synthetic reactions (Scheme 18) at temperatures below −60°, which are somewhat comparable with conventional Grignard reactions.[397]

---

[391] R. A. Olofson, H. Kohn, R. V. Kendall, and W. P. Piekielek, *Abstr. 160th Nat. Meeting Am. Chem. Soc., Abstr. ORGN 76, 1970.*
[392] W. P. Norris and R. A. Henry, *Tetrahedron Lett.,* 1213 (1965).
[393] R. A. Olofson, W. R. Thompson, and J. S. Michelman, *J. Am. Chem. Soc.* **86**, 1865 (1964).
[394] R. A. Olofson and J. M. Landesberg, *J. Am. Chem. Soc.* **88**, 4263 (1966).
[395] R. A. Olofson, J. M. Landesberg, K. N. Houk, and J. S. Michelman, *J. Am. Chem. Soc.* **88**, 4265 (1966).
[396] M. A. Schroeder and R. C. Makino, *Tetrahedron* **29**, 3469 (1973).
[397] R. Raap, *Can. J. Chem.* **49**, 2139 (1970).

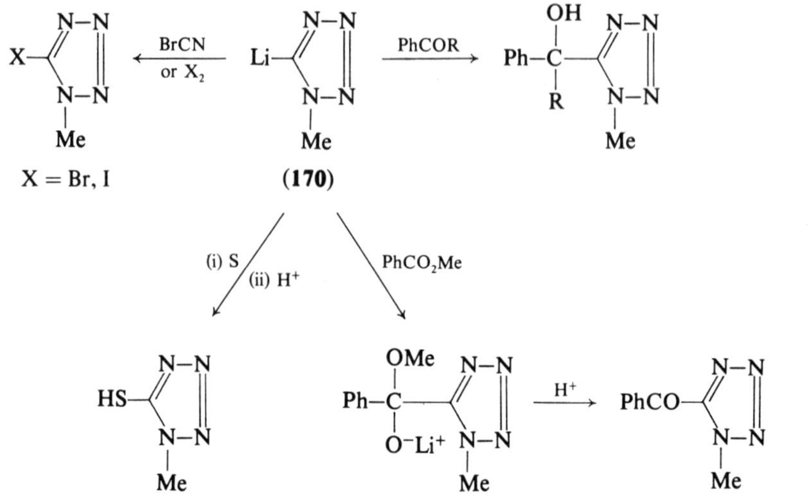

SCHEME 18

b. Cleavage and Rearrangement of Tetrazol-5-yl Ethers. 1-Phenyl-5-chlorotetrazole has proved to be a particularly useful reagent for selective removal of phenolic hydroxy groups. The procedure involves displacement of the halogen atom to yield a tetrazolyl ether [Eq. (29)],

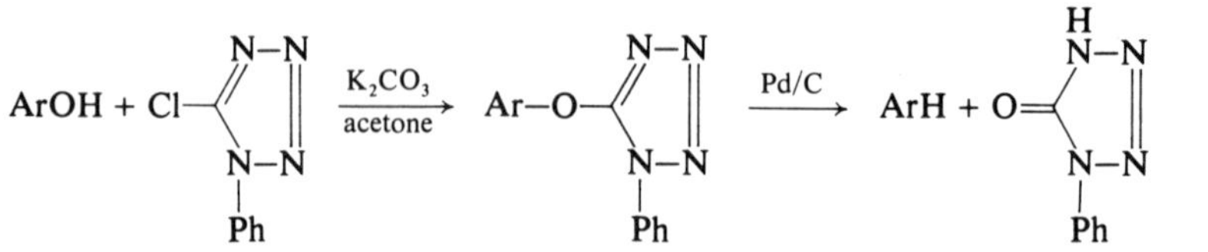

which readily gives the required elusive aryl–oxygen cleavage when treated with 5% Pd on charcoal.[398,399] This reaction was also successful with some other heterocycles, but the tetrazolyl ethers were the most advantageous since they were readily formed, had high melting points, and both they and the 1-phenyltetrazol-5-one by-product had the least poisoning effect on the catalyst.[398,399] The dehydroxylation reaction has proved to be widely applicable and has been used with such diverse systems as substituted pyrimidines,[400] gardneria alkaloids,[401] substituted

[398] W. J. Musliner and J. W. Gates, *J. Am. Chem. Soc.* **88**, 4271 (1966).
[399] W. J. Musliner and J. W. Gates, *Org. Syn.* **51**, 82 (1971) [*CA* **76**, 59083 (1972)].
[400] B. A. Otter, J. Taube, and J. J. Fox, *J. Org. Chem.* **36**, 1251 (1971).
[401] S. Sakai, A. Kubo, T. Hamamoto, M. Wokabayashi, K. Takahashi, Y. Ohtani, and J. Haginiwa, *Tetrahedron Lett.* 1489 (1969).

berbines,[402] dihydroxytoluenes,[403] morphine derivatives,[403a] and a range of substituted phenols.[398,399,404]

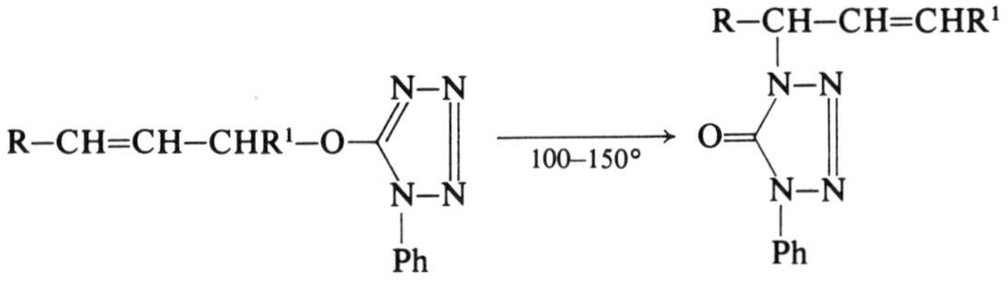

$$R-CH=CH-CHR^1-O-\underset{\underset{Ph}{|}}{\overset{N-N}{\underset{N-N}{\bigg\langle}}} \xrightarrow{100-150°} O=\underset{\underset{Ph}{|}}{\overset{N-N}{\underset{N-N}{\bigg\langle}}}\overset{R-CH-CH=CHR^1}{|}$$

(171)   (172)

$$R-CH=CH-CH_2-O^+=\underset{\underset{Ph}{|}}{\overset{^-N-N}{\underset{N-N}{\bigg\langle}}}$$

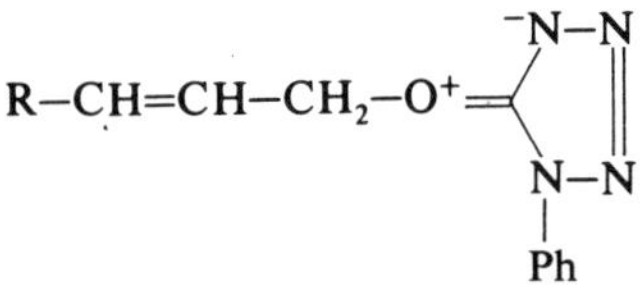

(173)

The 5-alloxy-1-phenyltetrazoles (171) were readily obtained by displacement of chloride from 1-phenyl-5-chlorotetrazole.[405] The compounds 171 underwent Claisen rearrangement with inversion when heated, giving the products 172.[405] The rates of the rearrangement were first order in some cases, but in others the order varied with

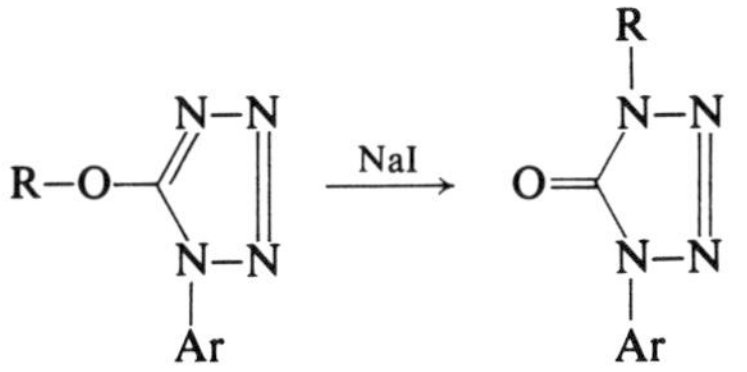

$$R-O-\underset{\underset{Ar}{|}}{\overset{N-N}{\underset{N-N}{\bigg\langle}}} \xrightarrow{NaI} O=\underset{\underset{Ar}{|}}{\overset{N-N}{\underset{N-N}{\bigg\langle}}}\overset{R}{|} \qquad (30)$$

(174)

temperature. It has been suggested that ground-state canonical contributions such as 173 facilitate the rearrangement.[405] Rearrangement of the 5-alkoxy-1-aryl tetrazoles (174) was also observed when these compounds were heated with NaI [Eq. (30)].[406] Small quantities of mesoionic isomers were also formed in this reaction.[406]

---

[402] A. Brossi, H. Bruderer, A. I. Rachlin, and S. Teitel, *Tetrahedron* **24**, 4277 (1968).
[403] C. F. Barfknecht, R. V. Smith, and V. D. Reif, *Can. J. Chem.* **48**, 2138 (1970).
[403a] R. Bognar, Gy. Goal, P. Kerekes, G. Horvath, and M. T. Kovacs, *Org. Prep. Proced. Int.* **6**, 305 (1974) [*CA* **82**, 140346 (1975)].
[404] J. W. Gates and W. J. Musliner, U.S. Patent 3,658,835 (1972) [*CA* **77**, 48474 (1972)].
[405] J. K. Elwood and J. W. Gates, *J. Org. Chem.* **32**, 2956 (1967).
[406] A. Vollmar and A. Hassner, *J. Heterocycl. Chem.* **11**, 491 (1974).

c. *Acylation and Alkylation 5-Amino- and 5-Mercaptotetrazoles.* Acylation of 1-substituted 5-aminotetrazoles gave 1-substituted 5-acylamido derivatives retaining the tetrazole ring.[300] In contrast, acylation of the phosphorylated tetrazolines (134) resulted in ring cleavage and gave phosphorylated oxadiazoles.[335] Acylation of 1-substituted 5-mercaptotetrazoles occurred on both the exocyclic sulfur and ring nitrogen atoms [Eq. (31)]. Generally, for R = aryl[407] or benzyl,[408] products of type 175 and 176 were obtained. However, with R = n-C_3H_7 and n-C_4H_9 and R^1 = Me, compounds of type 177 have also been reported

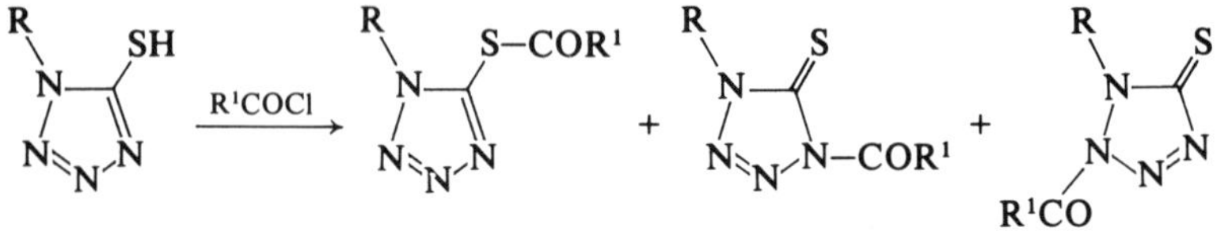

in low yields.[409] Such 1,2-disubstituted tetrazolines are quite rare, and few examples have been noted previously. Alkylation of 1-benzyl-5-mercaptotetrazole with alkyl halides[408] occurred exclusively on the sulfur atom, as did the reaction with trialkoxymethanes [Eq. (32)].[410] Acylation of 1-phenyl-5-hydroxytetrazole generally occurs exclusively at the 4-N-position but *O*-acylation to give the ester 178 [Eq. (33)] has been observed in up to 50% yield when the acylating agent was 2-methylpropanoyl chloride.[411]

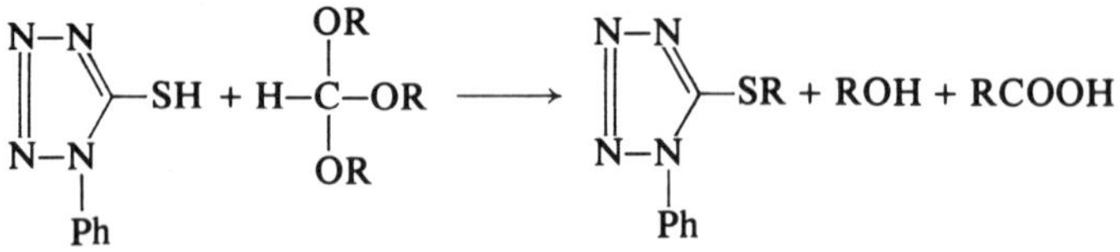

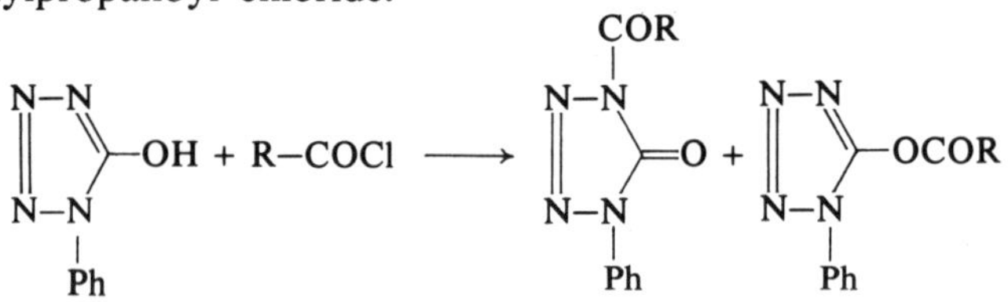

(178) R = $CHMe_2$

---

[407] E. Lippmann, D. Reifegerste, and E. Kleinpeter, *Z. Chem.* **13**, 134 (1973) [*CA* **79**, 66256 (1973)].

[408] G. L'abbe, S. Toppet, G. Verhelst, and C. Martens, *J. Org. Chem.* **39**, 3770 (1974).

[409] E. Lippmann, D. Reifegerste, and E. Kleinpeter, *Z. Chem.*, **14**, 16 (1974) [*CA* **81**, 25615 (1974)].

[410] E. Lippmann and D. Reifegerste, *Z. Chem.* **15**, 54 (1975) [*CA* **82**, 156186 (1975)].

[411] E. Lippmann, R. Widera, and E. Kleinpeter, *Z. Chem.* **13**, 429 (1973) [*CA* **80**, 82829 (1974)].

Treatment of 1-phenyl-5-mercaptotetrazole with acrylonitrile in triethylamine resulted in cyanoethylation of the molecule.[412] This has been compared with the similar reaction of 1-phenyl-5-mercapto-1,2,4-triazole.[412] The sodium salt of 1-phenyl-5-mercaptotetrazole has been used as a nucleophile in studies of the nucleophilic cleavage of epoxides and α-epoxyketones.[413-417] Kinetic studies indicated a ring opening in which the 2-substituent on the epoxide interferes with the approach of the nucleophile and which is not facilitated by α-carbonyl groups.[416] With cyclohexene oxide, 1-phenyl-5-mercaptotetrazole was converted to 1-phenyl-5-tetrazolinone.[416] In the reaction of 1-phenyl-5-mercaptotetrazole with 2,3-epoxides of Diels–Alder adducts of 1,4-benzoquinone, tetrazol-5-ylthioether hydroquinones and tetrazol-5-ylthioether enediones were the products.[413,414,416] 1-Substituted 5-

$$\underset{\underset{R}{\overset{N-N}{\|}}}{\overset{N-N}{\|}}\text{—SH} \xrightarrow[-\text{HCl}]{\text{ClSAr}} \underset{\underset{R}{\overset{N-N}{\|}}}{\overset{N-N}{\|}}\text{—S—S—Ar} \xrightarrow[\text{HOAc}]{\text{H}_2\text{O}_2} \underset{\underset{R}{\overset{N-N}{\|}}}{\overset{N-N}{\|}}\text{—S—SO}_2\text{—Ar}$$

(179)   (180)

mercaptotetrazoles have also been converted to the disulfides (179), and these in turn oxidized to the compounds 180.[418] Thermal rearrangement of 1-aryl-5-mercaptotetrazoles to 5-arylamino-1,2,3,4-thiatriazoles has also been reported.[329]

d. General. Treatment of aryl- and benzyl-substituted 5-aminotetrazoles with sodium nitrite in dilute hydrochloric acid leads to primary nitrosamines (181), in contrast to the diazonium salts obtained

$$\underset{R}{\overset{N-N}{\underset{N=N}{\|}}}\text{—NH}_2 \xrightarrow{\text{HONO}} \underset{R}{\overset{N-N}{\underset{N=N}{\|}}}\text{—NH—NO} \xrightarrow{\text{C}_6\text{H}_6} \underset{R}{\overset{N-N}{\underset{N=N}{\|}}}\text{—Ph} \quad (34)$$

(181)

---

[412] E. G. Kovalev and I. Ya. Postovskii, *Khim. Geterotsikl. Soedin.* **6**, 1138 (1970) [*CA* **74**, 141644 (1971)].

[413] H. S. Wilgus, E. Frauenglass, P. P. Chiesa, G. H. Nawn, F. H. Evans, and J. W. Gates, *Can. J. Chem.* **44**, 603 (1966).

[414] D. F. O'Brien and J. W. Gates, *J. Org. Chem.* **30**, 2593 (1965).

[415] M. J. Youngquist, D. F. O'Brien, and J. W. Gates, *J. Am. Chem. Soc.* **88**, 4960 (1966).

[416] D. F. O'Brien, *J. Org. Chem.* **32**, 262 (1968).

[417] L. A. Vlasova, and I. Ya. Postovskii, *Khim. Geterotsikl. Soedin.* **7**, 700 (1971) [*CA* **76**, 12687 (1972)].

[418] G. Stájer, E. A. Szabo, J. Pintye, F. Klivenyi, and P. Sohar, *Chem. Ber.* **107**, 299 (1974).

directly from 5-aminotetrazole.[318,419-421] When the compounds **181** were heated under reflux in benzene or nitrobenzene, a Gomberg–Bachmann type homolysis occurred leading to 5-phenyltetrazoles or 5-*p*-nitrophenyltetrazoles [Eq. (34)].[419] This reaction, which was general for a series of heterocyclic primary nitrosamines, probably involves homolysis of the nitrosamine moiety in its diazotate (—N=N—OH) form and may involve a tetrazolyl radical.

A number of reactions involving substituents in the α-position of 5-alkyl-substituted tetrazoles have been reported. Thus new 1-substituted tetrazole-5-carbaldehydes (**183**) and their 2-substituted isomers have recently been prepared (Scheme 19) and treated with a range of reagents.[422] Studies of the relative reactivities of 1-aryl- and 2-

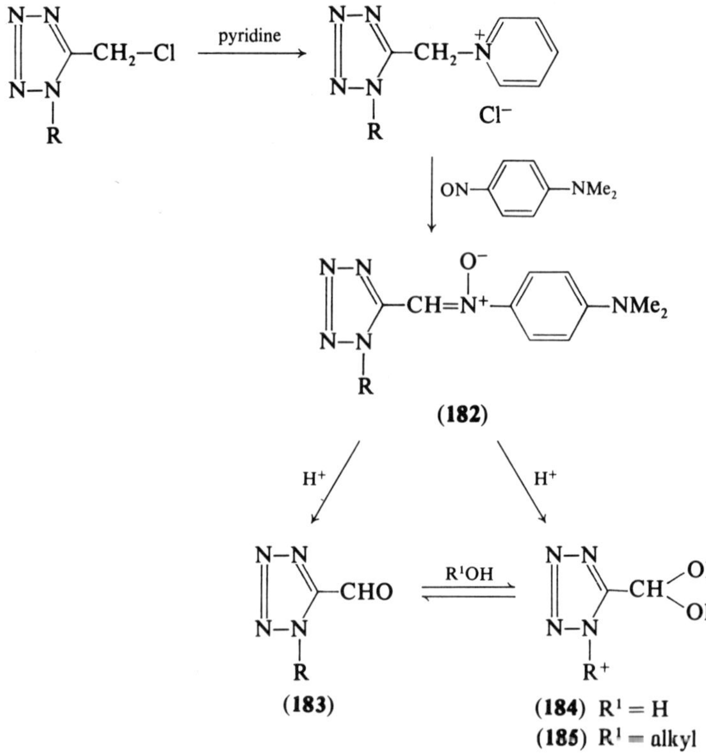

SCHEME 19

[419] R. N. Butler, T. M. Lambe, J. C. Tobin, and F. L. Scott, *Chem. Soc., Perkin Trans.* 1, 1357 (1973).
[420] J. C. Tobin, R. N. Butler, and F. L. Scott, *Chem. Commun.*, 112 (1970).
[421] R. N. Butler, T. M. Lambe, and F. L. Scott, *Chem. Ind. (London)*, 628 (1970).
[422] D. Moderhack, *Justus Liebigs Ann. Chem.* **758**, 29 (1972); D. Moderhack, *Chem. Ber.* **108**, 887 (1975).

aryltetrazole-5-carbaldehydes have shown the 1,5-disubstituted derivatives to be considerably more reactive to nucleophilic attack.[422a] Electrophilic attack on 1- and 2-phenyltetrazoles with such reagents as bromine and mercuric acetate also occurred much more easily with the 1-phenyl isomer[422a] in agreement with the general trends of higher reactivity in 1,5-disubstituted tetrazoles noted already (e.g., Section III,B,3a).

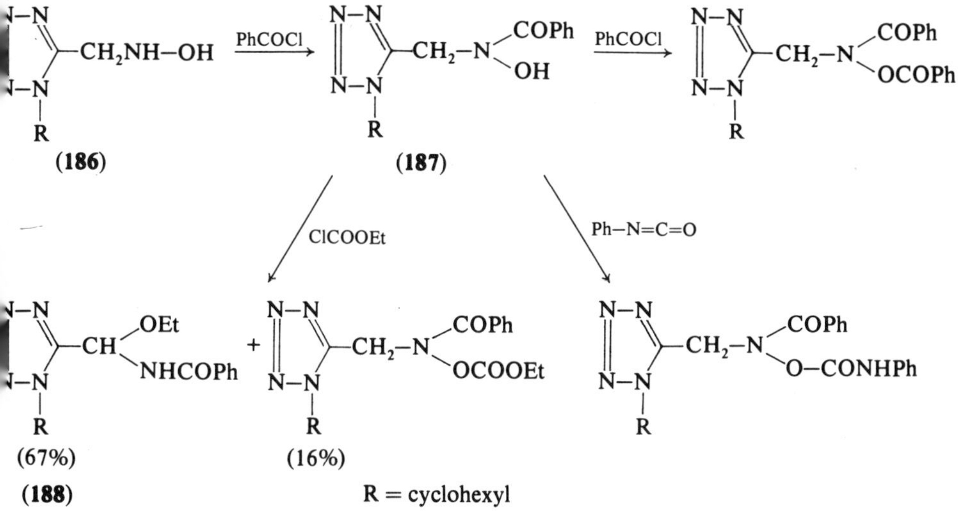

SCHEME 20

The interesting hydrates (**184**) and hemiacetals (**185**) which have been isolated are stabilized apparently by the electron-withdrawing power of the tetrazole ring.[422] Acylation of 1-cyclohexyl-5-hydroxylaminomethyltetrazole (**186**) (Scheme 20) with benzoyl chloride gave the *N*-benzoyl derivative (**187**), which gave the further reactions indicated when treated with benzoyl chloride, phenylisocyanate, and ethyl chloroformate.[423] The reaction with ethyl chloroformate is particularly interesting owing to the formation in high yield of the rearranged product (**188**). This is also a derivative of compound **183**, and from it a 2,4-dinitrophenylhydrazone of 1-cyclohexyltetrazole-5-carbaldehyde was readily obtained.[423] A series of tetrazolyl nitrones of general type **182** has also been reported from a range of the compounds **187** (PhCO— replaced by R^1R^2CH-).[424] An interesting thermolysis of 5-

[422a] E. Lippmann and A. Konnecke, *Abstr. Int. Congr. Heterocycl. Chem.*, 5th, 1975, p. 297.
[423] G. Zinner and O. Hantelmann, *Chem. Zgt.* **97**, 269 (1973) [*CA* **79**, 42419 (1973)].
[424] G. Zinner and O. Hantelmann, *Arch. Pharm.* **307**, 780 (1974) [*CA* **82**, 4182 (1975)].

substituted tetrazole-1-(α)-acrylic acids and amides, which yields oxazolones and imidazolones [Eq. (35)], has been described.[380,425] Thermal

$$\underset{H}{\overset{CHR^1}{\underset{N}{\overset{N}{\bigg|}}}}\overset{R}{\underset{O}{\bigg|}} \xleftarrow{Cu, X=NH_2} \underset{\underset{COX}{\overset{|}{C=CHR^1}}}{\overset{N-N}{\underset{N-N}{\bigg|}}} R \xrightarrow{X=OH} \underset{O}{\overset{CHR^1}{\underset{N}{\overset{N}{\bigg|}}}}\overset{R}{\underset{O}{\bigg|}} \quad (35)$$

decomposition of 1,5-diaryltetrazoles to give carbodiimides and 2-arylbenzimidazoles has been noted in earlier studies.[426,427] By contrast, thermal decomposition of 1-aryl-5-chlorotetrazoles gave completely intractable products in a reaction that was an order of magnitude faster in nonpolar solvents than in polar solvents and in which only small substituent effects on the rate of N_2 evolution were observed.[329]

C. 2-Substituted and 2,5-Disubstituted Tetrazoles

1. Synthesis

The main direct synthetic routes to 2,5-disubstituted tetrazoles which have been developed are (i) diazotization of substituted hydrazidines, (ii) the reaction of aromatic azides with hydrazones, (iii) heterocyclic rearrangements involving preferential ring interconversions, (iv) displacements from formazans or tetrazolium salts. Most of the recent work also falls into these general areas.

a. From the Hydrazone–Organic Azide Reaction. Aldehyde arylhydrazones and aryl azides, when heated in basic solution, yield 2,5-disubstituted tetrazoles [Eq. (36)], in which two nitrogen atoms of the

$$Ar-CH=N-\bar{N}-Ar^1 \xrightarrow{Ar^2N_3} \left[\begin{array}{c} Ar-CH \overset{N^-}{\underset{N=N^+}{\diagdown}} N-Ar^1 \\ \underset{\bar{N}-Ar^2}{\diagdown} \end{array}\right] \longrightarrow Ar{-}\underset{N=N}{\overset{N-N}{\bigg\langle}}\overset{Ar^1}{} + Ar^2NH_2 \quad (36)$$

(189)

[425] J. Lykkeberg and N. A. Klitgaard, *Acta Chem. Scand.* **26**, 2687 (1972).
[426] P. A. S. Smith and J. E. Leon, *J. Am. Chem. Soc.* **80**, 4647 (1958).
[427] J. Vaughan and P. A. S. Smith, *J. Org. Chem.* **23**, 1909 (1958).

tetrazole ring are supplied by the aryl azide.[428] That the amine formed in the reaction arises from the azide rather than the hydrazone has been established by inserting labeling substituents on both the N-aryl ring of the azides and the hydrazones[428] and also from ^{14}C labeling of the N-aryl ring of the hydrazone when 99% of the radioactivity was incorporated into the tetrazole.[429] Kinetic studies have not been reported, but a 3 + 2 cycloaddition between the hydrazone anion and the azide to give an intermediate of type **189**, followed by tautomerism and elimination of Ar^2NH^-, has been suggested.[429]

b. From Heterocyclic Rearrangements. A series of 5-acetonyl-2-aryltetrazoles (**190**) was obtained by heating 3-diazoamino-5-methylisoxazoles with ammonia in aqueous acetone.[430] This reaction

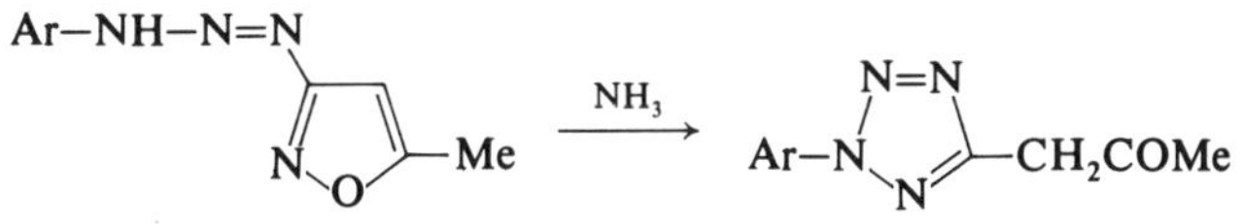

Ar—NH—N=N

$\xrightarrow{NH_3}$

(**190**) Ar—N, N=N, —CH$_2$COMe

represents an example of a general heterocyclic rearrangement [Eq. (37)] recently recognized and categorized for a wide range of compounds.[431]

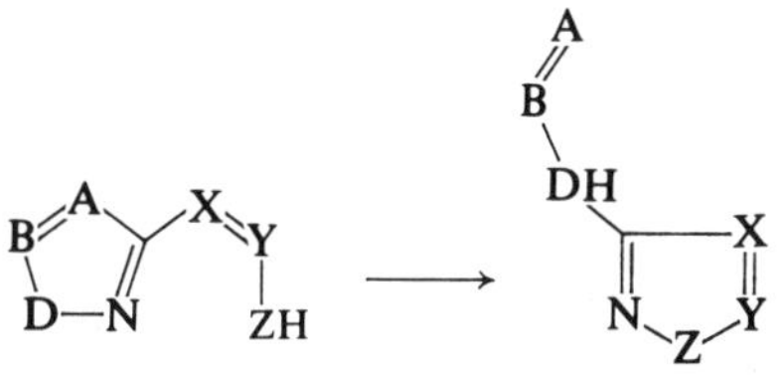

(37)

c. From Formazans and Tetrazolium Salts. Formazans, e.g., **191**, may be readily cyclized to tetrazolium salts (**192**) by treatment with oxidizing agents. However, if one of the N-substituents, e.g., R^1 in **191**, is a good leaving group, cyclization to a 2,5-disubstituted tetrazole may occur with loss of this group. The exact timing of leaving group loss in the reaction is not clear, i.e., whether before, during, or after the cyclization. A series of tetrazoles (**193**, Ar^1 = Ph) has been obtained

[428] S. Y. Hong and J. E. Baldwin, *Tetrahedron* **24**, 3787 (1968).
[429] J. E. Baldwin and S. Y. Hong, *J. Heterocycl. Chem.* **5**, 565 (1968).
[430] H. Kano and E. Yamazaki, *Tetrahedron* **20**, 461 (1964); *Chem. Pharm. Bull.* **10**, 993 (1962).
[431] A. J. Boulton, A. R. Katritzky, and A. Majid Hamid, *J. Chem. Soc., C,* 2005 (1967).

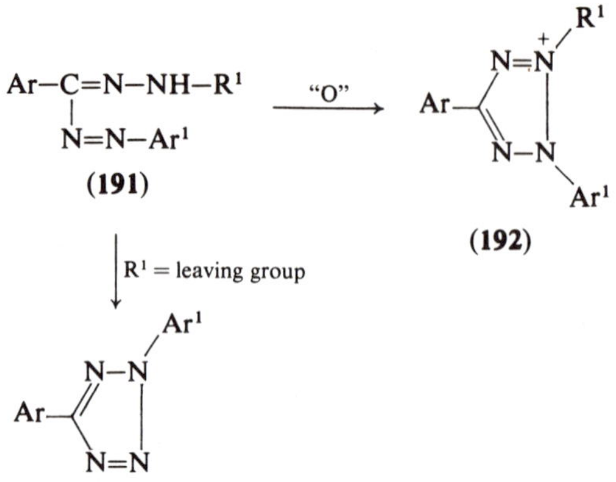

SCHEME 21

(191) Ar—C(=N—NH—R^1)—N=N—Ar^1

(192) Tetrazolium intermediate with Ar, Ar^1, R^1

(193) 2,5-diaryltetrazole with Ar, Ar^1

from formazans of phenylsulfonylhydrazones (191, R$^1_+$ = —SO$_2$Ph) in basic solution, and a 1,5-dipole, Ar^1—N$^-$—N=C(Ar)—N≡N, has been suggested as a possible intermediate.[432] Treatment of formazans of type 191, where the leaving group R^1 was a 2-quinoxalyl substituent, with lead tetraacetate also gave high yields of diaryl tetrazoles (193).[433] 2,5-Diaryltetrazoles have also been obtained by heating formazans (191, R^1 = Acyl, Ar^1 = 4-NO$_2$C$_6$H$_4$) under reflux in acetic acid.[434] In a similar reaction where the NO$_2$ substituent of Ar^1 was replaced by a H atom, the products were 1,3,4-oxadiazoles, not tetrazoles.[434] 2,5-Disubstituted tetrazoles (105, Scheme 8) have been obtained by elimination of alkyl halides from 1,3,5-trisubstituted tetrazolium salts[297,298] and also in cases where the 1-substituent is an alkyl group and the 3- and 5-substituents are both aryl groups.[435] Ring opening of the pyridine ring of 3-aryltetrazolopyridinium salts (194) with nucleophiles such as NaBH$_4$, NaOMe, and morpholine leads to 2-aryltetrazol-5-ylbutadienes (195) [Eq. (38)].[436]

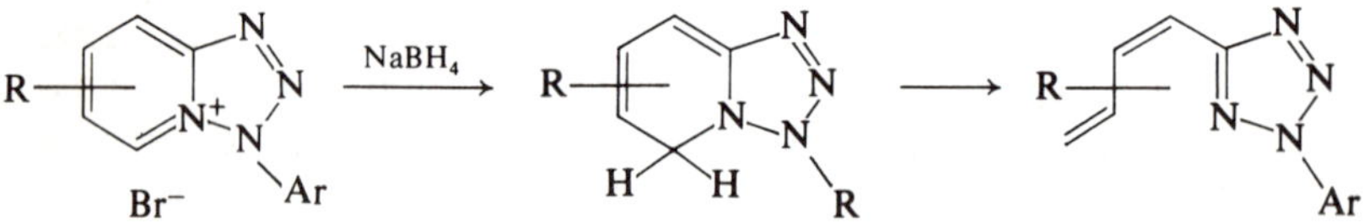

(194) → (via NaBH$_4$) → (195) (38)

---

[432] S. Ito, Y. Tanaka, A. Kakehi, and K. Kondo, *Chem. Lett.*, 1071 (1973) [*CA* **80**, 3437 (1974)].

[433] Yu. A. Sedov, A. I. Zabolotskaya, and N. V. Koba, *Khim. Geterotsikl. Soedin.*, 1705 (1973) [*CA* **80**, 82886 (1974)].

[434] B. I. Buzykin, L. P. Sysoeva, and Yu. P. Kitaev, *Zh. Org. Khim.* **10**, 2200 (1974) [*CA* **82**, 43275 (1975)].

[435] R. Grasbey, M. Baumann, and H. Bauer, *Chem. Ztg.* **96**, 225 (1972) [*CA* **77**, 34427 (1972)].

[436] A. Gelleri and A. Messmer, *Tetrahedron Lett.*, 4295 (1973).

Treatment of compounds **194** with aqueous sodium dithionate results in cleavage of the tetrazole ring and gives pyrid-2-ylaryltriazenes, and cleavage with sodium methoxide gives nitrogen evolution from the tetrazole ring forming *N*-arylaminopyrid-2-ones.[436a] Thus, depending on the nucleophile, selective cleavage of compounds **194** at three different sites can be achieved.[436a]

d. General. An interesting addition of azidosilanes to nitriles has led to a range of 5-substituted 2-silyltetrazoles (**196**).[437,438] The compound **196** (R = Ph) was also obtained by treating 5-phenyltetrazole with hexamethyldisilazane.[439] When the compounds **196** were treated with

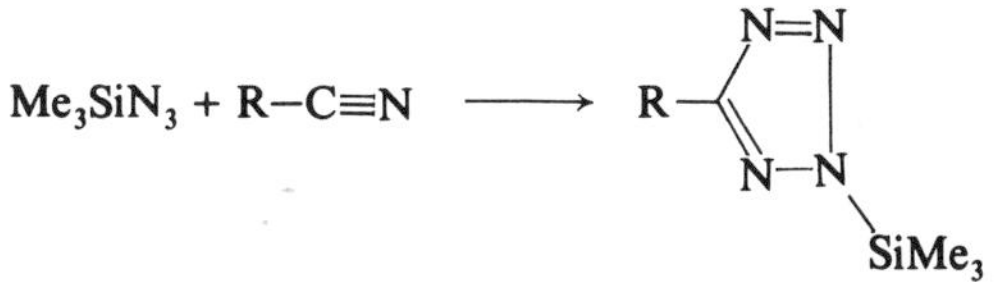

$$Me_3SiN_3 + R-C\equiv N \longrightarrow R-\underset{\underset{SiMe_3}{|}}{\overset{N=N}{\underset{N-N}{\diagdown\diagup}}}$$

(**196**)

aqueous ethanol, cleavage of the silyl moiety readily occurred, yielding a 5-substituted tetrazole.[437] Treatment of 2-diazoacetophenone (**197**) with potassium *t*-butoxide in *t*-butyl alcohol yielded 5-benzoyl-2-phenacyltetrazole (**198**) along with benzoic acid and some 2-phenacyltetrazole resulting from cleavage of **198**.[440] With potassium hydroxide as base and DMSO as solvent, the dihydrotetrazine **199** was obtained (Scheme 22).[440,441] The reaction has been viewed as involving

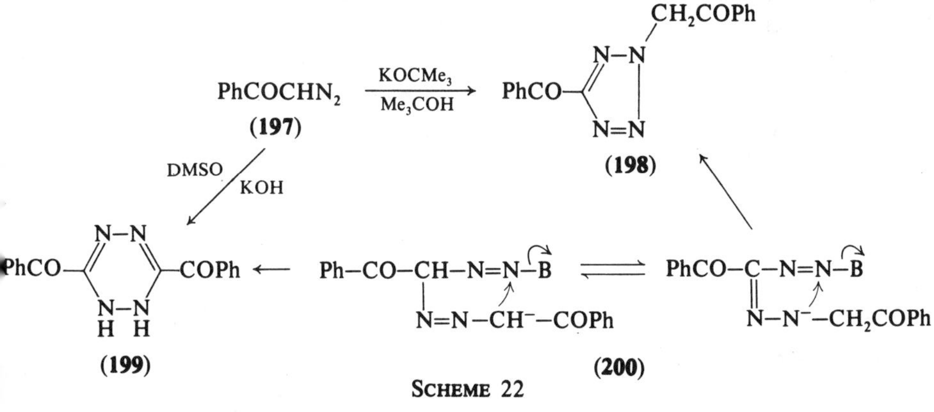

SCHEME 22

[436a] A. Messmer, A. Gelleri, and G. Hajos, *Abstr. Int. Congr. Heterocycl. Chem., 5th.* 1975, p. 167.
[437] E. Ettenhuber and K. Rühlman, *Chem. Ber.* **101**, 743 (1968).
[438] L. Birkofer and P. Wegner, *Chem. Ber.* **99**, 2512 (1966).
[439] L. Birkofer, A. Ritter, and P. Richter, *Chem. Ber.* **96**, 2750 (1963).
[440] P. Yates, R. G. F. Giles, and D. G. Farnum, *Can. J. Chem.* **47**, 3997 (1969).
[441] P. Yates and O. Meresz, *Tetrahedron Lett.*, 77 (1967).

initial terminal attack of base on the diazo moiety of **197** followed by addition of the resulting intermediate to an unreacted molecule of starting material yielding the intermediate **200**, which, in turn, cyclizes to the products (Scheme 22). The influence of base and solvent on the orientation of the products is, however, unclear, and the rate and equilibrium constant for the tautomerism involved in **200** would also seem to merit further investigation. Other routes to 2,5-disubstituted tetrazoles have been discussed in Section III,A,4.

2. Reactions

2,5-Disubstituted tetrazoles generally display reactions at the 5-position which are similar to those of other tetrazoles, but only a limited amount of work has been reported on these compounds.[1] For example, diazotization of 2-benzyl- and 2-aryl-5-aminotetrazoles [Eq. (34)] is similar to that of the 1-substituted compounds.[419-421] Diazotization of 2-methyl-5-aminotetrazole in dilute acid solution gives the triazene **201**.[291]

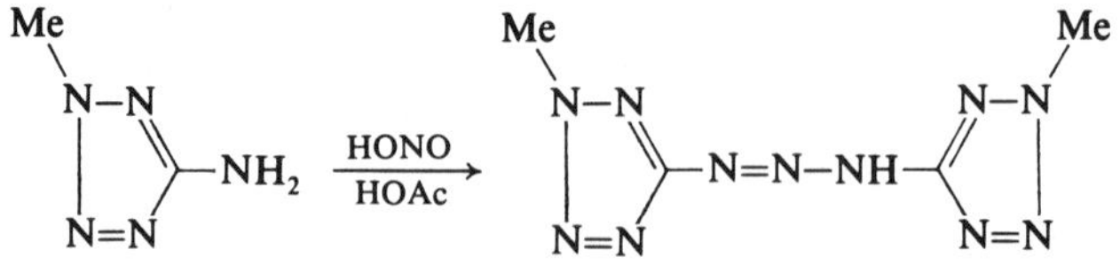

(201)

Recent studies of the chemistry of 2,5-disubstituted tetrazoles have been mainly concerned with reactions that involve loss of nitrogen and generation of nitrilimine intermediates. Examples of this have already been seen in the photolysis of these compounds and in the thermolysis of the unstable 2-acyltetrazoles obtained from acylation of 5-substituted tetrazoles. Direct thermolysis of 2,5-diaryltetrazoles[141,142,442] also leads to nitrilimine intermediates [Eq. (39)]. These undergo a wide variety of

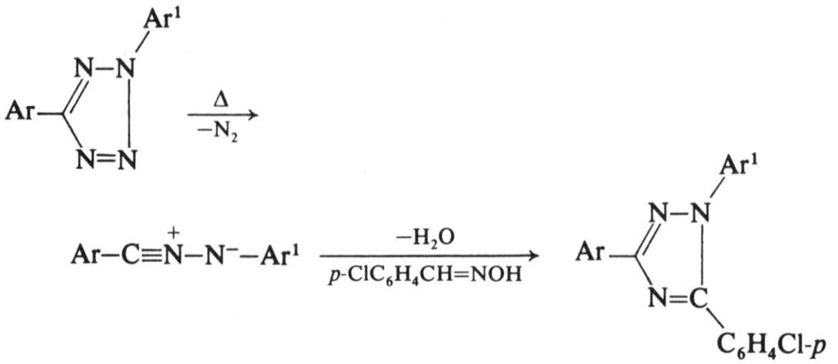

(39)

[442] R. Huisgen, R. Grashey, E. Aufderhaar, and R. Kunz, *Chem. Ber.* **98**, 642 (1965).

1,3-dipolar cycloaddition reactions.[143,443,444] When the liquid 2-silyltetrazoles (**196**) were heated, evolution of nitrogen occurred, yielding the nitrilimine species **202** which self-condensed to the 4-(bistrimethylsilylamino)-1,2,4-triazoles (**203**) (Scheme 23).[437] The inter-

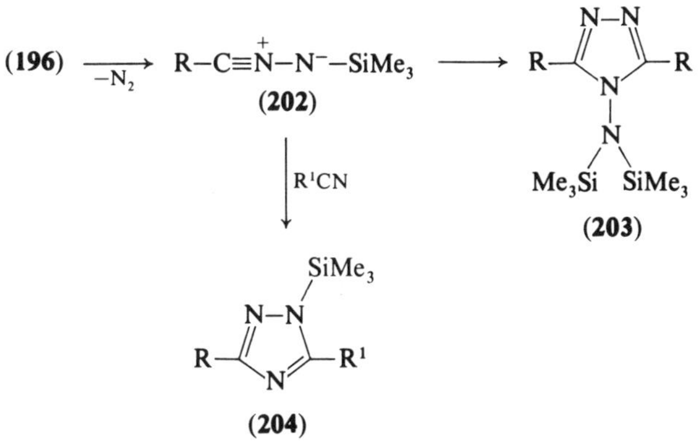

SCHEME 23

mediates **202** were trapped when the reaction was carried out in the presence of nitriles. In this case, the triazoles **204** were obtained in up to 56% yields (Scheme 23).[437] Treatment of the compounds **196** with acetyl chloride resulted in the displacement of the trimethylsilyl group

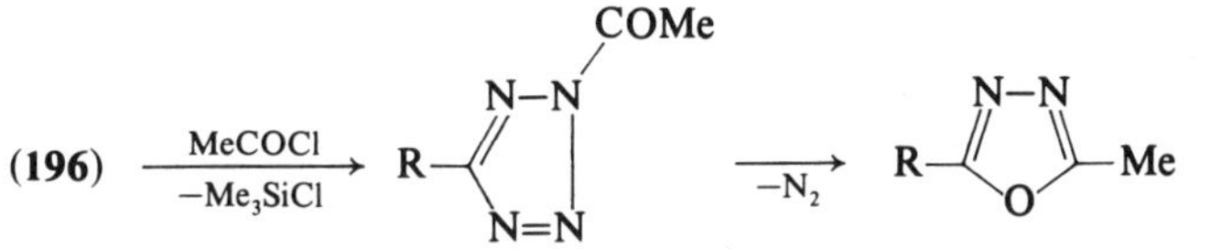

by the acyl moiety [Eq. (40)] followed by oxadiazole formation by the pathway described in Scheme 9 (Section III,A,4,b).

The kinetics of thermolysis of 2,5-diaryltetrazoles (**205**) have been investigated by Hong and Baldwin[428,445] in order to estimate the sensitivity of 1,3-dipolar cycloadditions to the electronic characteristics of the nitrilimine rather than the steric characteristics, which often tend to dominate such cycloadditions. For the species **205** the steric factors are constant, and the reverse of the cleavage reaction was considered to be analogous to the 1,3-cycloadditions with symmetrical olefins. A Hammett ρ value of +1.16 was obtained for the compounds **205** (X = H) and a value of −0.23 for the series **205** (Y = H). The ρ values

---

[443] R. A. Firestone, *J. Org. Chem.* **33**, 2285 (1968).
[444] R. Huisgen, *J. Org. Chem.* **33**, 2291 (1968).
[445] J. E. Baldwin and S. Y. Hong, *Chem. Commun.*, 1136 (1967).

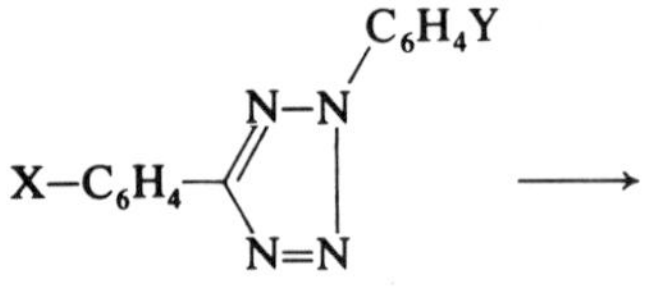

of opposite sign and different magnitude suggested an unsymmetrical type of activated complex (206) for the 1,3-cycloeliminations. A similar pathway may be involved in cycloadditions, but, from this point of view, the results are somewhat indecisive since they do not allow a distinction between a concerted and nonconcerted process, an aspect of these reactions that has proved to be controversial.[443,444]

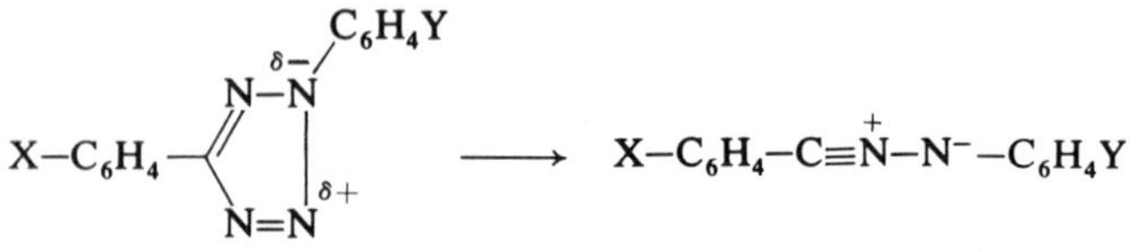

D. MESOIONIC TETRAZOLES

Recent studies of mesoionic tetrazoles have centered mainly on 2,3-diphenyl-2H-tetrazolium-5-thiolate (72), also known as dehydrodithizone, and on the synthesis of some new tetrazoles. The structure of the product formed when dithizone (207) was oxidized with a number of oxidizing agents was correctly formulated as 72 by Ogilvie and Corwin[446] and has since been confirmed by X-ray analysis.[113,134]

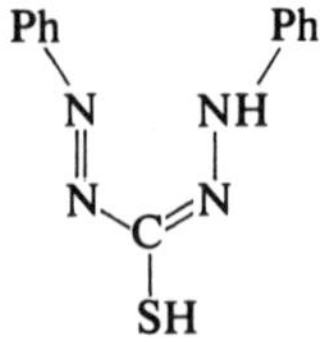

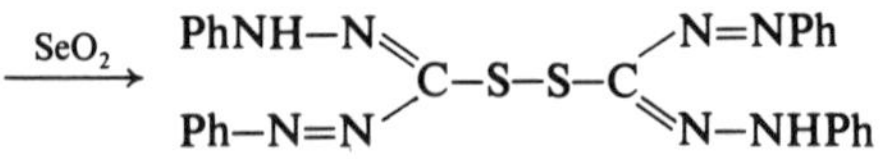

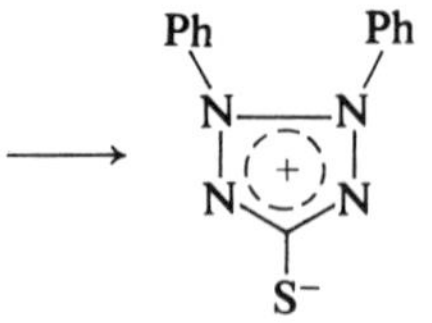

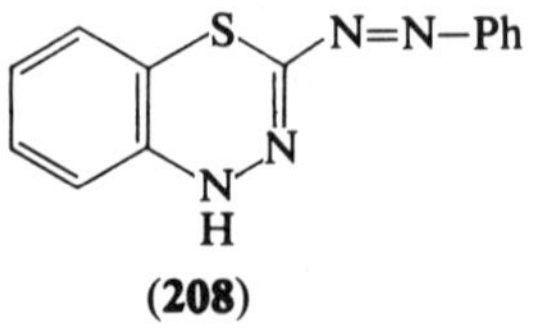

---

[446] J. Ogilvie and A. Corwin, *J. Am. Chem. Soc.* **83**, 5023 (1961).

Oxidation of dithizone with selenium dioxide, gives the disulfide (**209**) which gradually disproportionates to the compound **72**.[447] Under acidic conditions the oxidation may also lead to elimination of an ortho-proton from one of the phenyl rings to give the product **208**.[448] This may also be obtained by heating compound **72** in HOAc.[448] The electronic absorptions of compound **72** in the UV and visible are highly sensitive to solvent variation and exhibit hypsochromic shifts with increasing solvent polarity, suggesting a more polar ground state for the molecule.[449] The compound **72** undergoes a variety of 1,3-dipolar cycloadditions yielding products of type **210**.[450] In these reactions the 1,3-dipolar species is the unusual $-N^+-C-S^-$ moiety.[450] Addition of 1-diethylaminoprop-1-yne to dehydrodithizone yields the pyrazole (**211**) in an interesting reaction that involves extrusion of sulfur.[451]

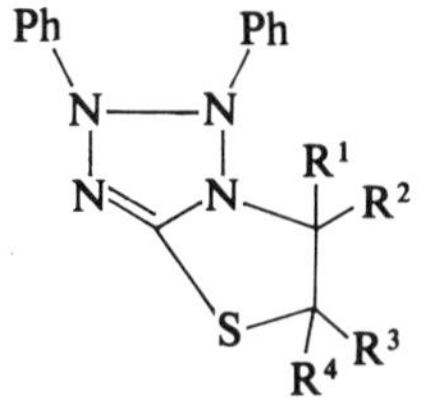

(**210**) 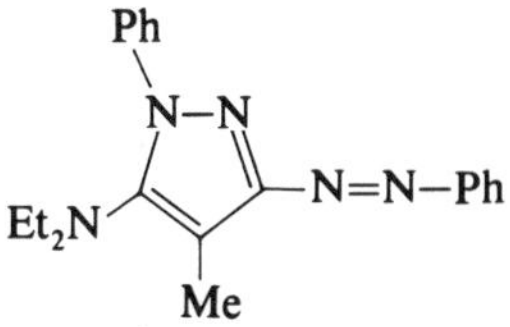   (**211**) 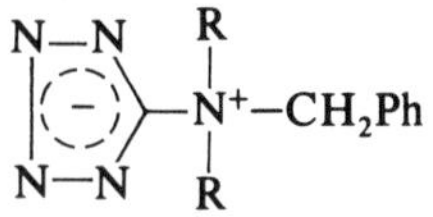   (**212**)

The tetrazolyl ylids (**212**) have been obtained, along with other isomers, from the benzylation of 5-dialkylaminotetrazoles.[452] The structure of these compounds has been established by X-ray analysis.[453] Tetrazolium ylids of type **214** have been obtained from the oxidation of

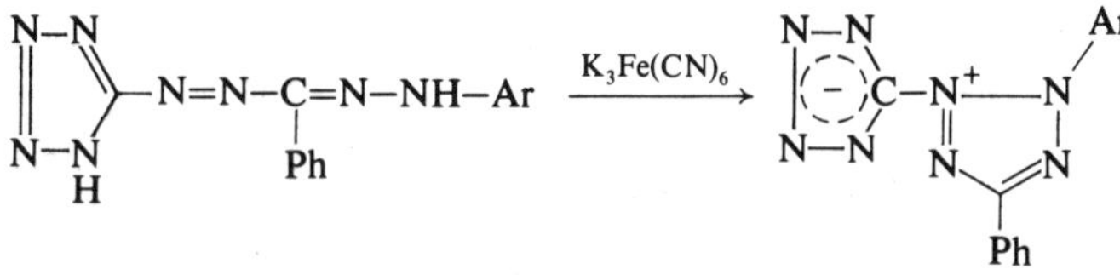

(**213**)   (**214**)   (41)

[447] R. S. Ramakrishna and H. M. N. H. Irving, *Chem. Commun.*, 1356 (1969).
[448] W. S. McDonald, H. M. N. H. Irving, G. Raper, and D. C. Rupainwar, *Chem. Commun.*, 392 (1969).
[449] A. M. Kiwan and H. M. N. H. Irving, *J. Chem. Soc., B*, 898 (1971).
[450] P. Rajagopalan and P. Penov, *Chem. Commun.*, 490 (1971).
[451] G. V. Boyd, T. Norris, and P. F. Lindley, *Chem. Commun.*, 639 (1974).
[452] L. Huff, D. M. Forkey, D. W. Moore, and R. A. Henry, *J. Org. Chem.* **35**, 2074 (1970).
[453] G. B. Ansell, *Chem. Commun.*, 684 (1970).

tetrazolylformazans (**213**) [Eq. (41)].[454-456] Formazans of type **213** are generally obtained by treating tetrazol-5-ylhydrazones with diazonium salts.[454-456] Mesoionic 1,3-dimethyl-5-arylsulfonimidotetrazoles have been obtained by methylation of the parent monomethyl isomers.[457,458] These mesoionic 1,3-dimethyltetrazoles were more stable to alkali than the corresponding 1,4-dimethyltetrazolines. Recently, the new 3-substituted (1,2,3,4-oxatriazolio) amides (**215**) have been obtained by treating 1,4-disubstituted thiosemicarbazides with nitrous acid.[459] These compounds rearrange to the mesoionic 1,3-diaryltetrazolio-5-oxides (**216**) when treated with base.[459] This interesting reaction appears to be the formal oxygen analog of the base-catalyzed rearrangement of 5-arylamino-1,2,3,4-thiatriazoles. Compounds of type **216** have also been obtained directly from aromatic diazonium salts and bis(alkylsulfonyl-)-methanes in basic solution (Scheme 24).[460,461] Hydrazones (**217**) are also obtained from these reactions, and a mechanism involving the tetrazenes (**218**) have been proposed (Scheme 24).[460]

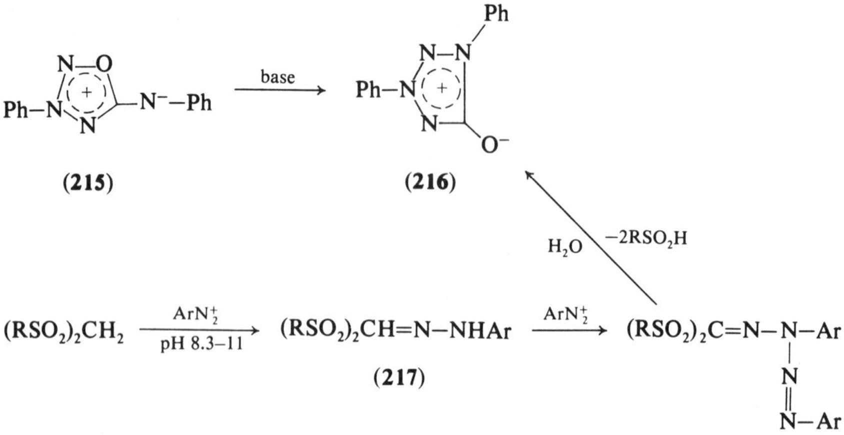

$$(RSO_2)_2CH_2 \xrightarrow[pH\ 8.3-11]{ArN_2^+} (RSO_2)_2CH=N-NHAr \xrightarrow{ArN_2^+} (RSO_2)_2C=N-N-Ar$$

(**217**)

SCHEME 24 (**218**)

[454] V. P. Schipanov, K. I. Krashina, and A. A. Skachilova, *Khim. Geterotsikl. Soedin.* 1570 (1973) [*CA* **80**, 59898 (1974)].

[455] V. P. Schipanov and G. F. Grigor'eva, *Khim. Geterotsikl. Soedin.* 268 (1974) [*CA* **80**, 133352 (1974)].

[456] N. P. Bednyagina, A. P. Novikova, and I. Ya. Postovskii, *Khim. Geterotsikl. Soedin.* 456 (1966) [*CA* **65**, 15369 (1966)].

[457] V. P. Schipanov, *Zh. Org. Chim.* **2**, 1489 (1966) [*CA* **66**, 55441 (1967)].

[458] V. P. Schipanov and I. Ya. Postovskii, *Z. Org. Khim.* **2**, 350–376 (1966) [*CA* **65**, 2248 (1966)].

[459] C. Christophersen and S. Treppendahl, *Acta Chem. Scand.* **25**, 625 (1971).

[460] V. M. Neplynev and P. S. Pel'kis, *Zh. Org. Khim.* **10**, 1725 (1974) [*CA* **81**, 136059 (1974)].

[461] V. Neplynev, R. G. Dubenko, and P. S. Pelkis, *Zh. Org. Khim.* **5**, 1832 (1969) [*CA* **72**, 21461 (1970)].

Oxidation of formazans with a wide range of oxidizing agents leads to 2,3-disubstituted tetrazolium salts. Recently, oxidations with cupric acetate and cobalt chloride have led to mixtures of metal–formazan complexes and tetrazolium salts arising from competing reactions in which the balance depended on the substituents present in the formazans.[462] Reduction of 1,4,5-trisubstituted tetrazolium iodides with $NaBH_4$ has been used to prepare Δ^2-tetrazolines.[463] A similar attempted reduction of 1,3,5-trisubstituted tetrazolium salts gave no reaction.[463]

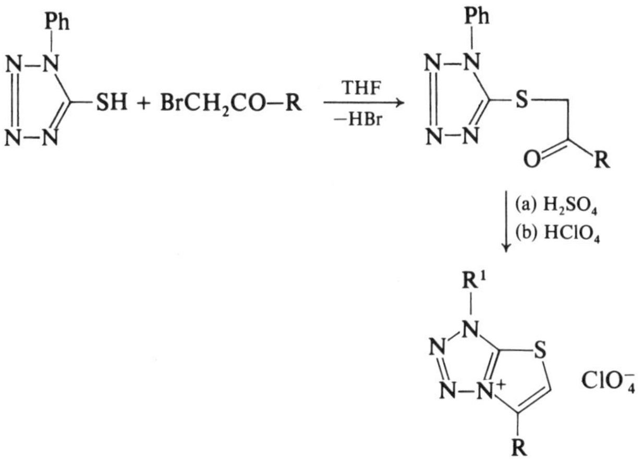

SCHEME 25

Thiazolo[3,2-d] tetrazolium salts (**219**) have been prepared by treatment of 1-phenyl-5-mercaptotetrazole with α-bromoketones followed by cyclization (Scheme 25).[464] Further aspects of the chemistry of tetrazolium salts are discussed in Sections II,D,4 and III,A,4. These compounds represent an area of study distinct from the tetrazoles, and a number of interesting reviews have been published on them separately.[465–467]

A full review of the chemistry of mesoionic compounds including the early work on tetrazoles, has been published recently.[468]

[462] R. Price, *J. Chem. Soc., A*, 3379, 3385 (1971).
[463] T. Isida, T. Akyama, N. Mihara, S. Kozima, and K. Sisido, *Bull. Chem. Soc. Jpn.* **46**, 1250 (1973) [*CA* **78**, 159527 (1973)].
[464] H. Alper and R. W. Stout, *J. Heterocycl. Chem.* **10**, 5, 569 (1973).
[465] A. W. Nineham, *Chem. Rev.* **55**, 355 (1955).
[466] W. Ried, *Angew. Chem.* **64**, 391 (1952).
[467] N. D. Cheronis, *J. Chem. Ed.* **33**, 120 (1956).
[468] W. D. Ollis and C. A. Ramsden, *Adv. Heterocycl. Chem.* **19**, 1 (1976).

IV. Azidoazomethine Tetrazole Isomerism—Fused Tetrazoloheterocycles

A. General Characteristics. Monocyclic Systems

Molecules containing an azide group bonded to a doubly bound carbon may exist in one or both of the acyclic and cyclic forms **220** and **221**. When X is an oxygen atom, the molecule exists in the acyclic form,[469] and when X is a sulfur atom the cyclic form predominates.[329]

$$R-C(=X)-N_3 \rightleftharpoons \text{(tetrazole ring)}$$

(220)   (221)

When X is an N–R¹ moiety, either form may predominate or both forms may exist in equilibrium. The orientation of the equilibrium generally depends on the following factors: (i) The nature of the substituents R and R^1 about the C=N: In general, electron-withdrawing substituents favor the azide form, but electron-donating substituents enhance ring closure and stabilize the tetrazole form.[290,470,471] (ii) The temperature: Cleavage of the tetrazole ring to the azide form is generally an endothermic process (cf. Tisler[8]), and consequently higher temperatures tend to favor the acyclic form. (iii) The solvent system: A basic medium in which the azidoazomethine moiety is likely to exist as an anion strongly favors the tetrazole form by increasing the electron density on the C=N. Acidic solvents, e.g., CF_3COOH, in which the system tends to be protonated strongly favor the azide form due to electron depletion at the C=N. In general, polar solvents tend to favor the tetrazole form, and nonpolar solvents the azide form. Dipolar aprotic solvents have proved particularly useful for synthesizing tetrazoles in reactions involving an azidoazomethine ring closure as the final step.[5]

Another factor that might be expected to influence the azide–tetrazole isomerism is the syn–anti isomerism about the C=N moiety, since presumably for ring closure to occur the nitrogen lone pair must be cis to the azide group. This factor does not appear to have been discussed in any detail in the literature to date. When the azidoazomethine system is present in a cyclic molecule, the geometrical orientation of the azide

---

[469] E. Lieber. R. L. Minnis, and C. N. R. Rao, *Chem. Rev.* **65**, 377 (1965).
[470] W. P. Norris, and R. A. Henry, *J. Org. Chem.* **29**, 650 (1964).
[471] B. Stanovnik and M. Tisler, *Tetrahedron* **25**, 3313 (1969).

group and the lone pair is always favorable and ring closure results in the formation of a fused tetrazoloheterocycle. The isomerism in such systems is again governed by the above factors, but the major substituent effects in this case are the electron-withdrawing or -donating power of the second ring, or the general heterocyclic system if more than one ring is involved. A further factor that has an important influence on the orientation of the isomerism in such heterocyclic systems is the possibility of preferential ring interconversions. In such cases, introduction of a new fused ring to a tetrazoloheterocycle may result in simultaneous formation of the new ring and cleavage of the existing tetrazole ring, thereby giving a heterocycle in which the electron-donating azido substituent[472] exerts a stabilizing influence as against the reverse effect of an electron-withdrawing tetrazole ring. In recent years the azidoazomethine ring closure step has been classified as an electrocyclic 1,5-dipolar cyclization.[143,308]

$$R-C\begin{smallmatrix}+\\ \diagup NH_2 \\ \diagdown N_3 \end{smallmatrix} \quad SbCl_6^-$$

(222)

$$R-SO_2-N=C-SMe \\ | \\ N_3$$

(223)

$$CF_3-C=N-R \\ | \\ N_3$$

(224)

Examples of monocyclic systems which illustrate some of the characteristics of the azidoazomethine–tetrazole isomerism have been given in Section III,B, e.g., Eqs. (22)–(24), Scheme 16, and compounds **149** and **150**. The effect of protonation in favoring the azide form is illustrated by the isolation of imidium azide salts, e.g., **222**.[473] The effect of electron-withdrawing substituents is demonstrated by the compounds **223** and **224**,[375] which exist in the azide form. When the CF_3 group of compound **224** (R = β-naphthyl) is replaced by a CH_3, the molecule exists in the tetrazole form.[375] The carbonylazide substituent in compound **225** causes the azidoazomethine system to exist in the azide form.[362] Interestingly, when this compound is treated with triphenylphosphine, the tetrazole **226** is formed [Eq. (42)].[362]

$$Ph-N=C-C-N_3 \quad \xrightarrow[Et_2O]{Ph_3P} \quad \text{(226)} \tag{42}$$

(225)

[472] P. A. S. Smith, J. H. Hall, and R. O. Kan, *J. Am. Chem. Soc.* **84**, 485 (1962).
[473] A. Schmidt, *Chem. Ber.* **100**, 3319 (1967).

B. Bicyclic Fused Tetrazoloheterocycles

1. Tetrazoles Fused with Five-Membered Rings

The azidoazomethine system has now been inserted into the full range of azoles, and in each case, under normal conditions, the effect of the azole ring is completely to favor the azide form. Thus, 3-azidopyrazoles exist exclusively in the azide form (**227**).[330,474-476] When the molecules are converted into anions, the cyclic pyrazolotetrazole form (**228**) predominates.[475,476] Similarly 2-azidoimidazoles[477-479] exist in the azide form, as do 5-azidoimidazoles.[480] When 2-azidoimidazoles are converted into anions, cyclization of the azidoazomethine moiety again occurs, giving imidazotetrazole anions.[477,478] Treatment of 2-azidoimidazole with acetic anhydride gives 1-acetyl-2-azidoimidazole for which both azide and tetrazole forms exist in equilibrium in the ratio 3:2 azide:tetrazole in DMSO solution.[477]

The 3-azido-1,2,4-triazoles (**230**) exist in the azide form rather than the triazolotetrazole form.[481-484] These compounds were obtained by cyclization of the hydrazonyl bromides (**229**, R = H) and by treating the 3-hydrazinotriazole (**233**) with nitrous acid[482] (Scheme 26). The former reaction involves a preferential ring interconversion of tetrazole to triazole and presumably could be included in the general classification of Eq. (37) An alkyl substituent located in the tetrazole moiety appears to have a stabilizing influence on the triazolotetrazole ring system, since the bicyclic compounds (**232**), not azides (**231**), were obtained when the

---

[474] P. A. Smith and M. Dounchis, *J. Org. Chem.* **38**, 2958 (1973).
[475] E. Alcalde, J. De Mendoza, and J. Elguero, *Chem. Commun.*, 411 (1974).
[476] E. Alcalde, J. De Mendoza, and J. Elguero, *J. Heterocycl. Chem.* **11**, 921 (1974).
[477] E. Alcalde and R. M. Claramunt, *Tetrahedron Lett.*, 1523 (1975).
[478] R. M. Claramunt, R. Granados, and E. Pedroso, *Bull. Soc. Chim. Fr.*, 1854 (1973).
[479] J. A. VanAllan, G. A. Reynolds, and D. P. Maier, *J. Org. Chem.* **34**, 1691 (1969).
[480] V. I. Nifontov, V. S. Mokrushin, Z. V. Pushkareva, and A. A. Mukhina, *Khim. Geterotsikl. Soedin.* 94 (1973) [*CA* **78**, 97552 (1973)].
[481] F. L. Scott, D. A. Cronin, and J. K. O'Halloran, *J. Chem. Soc., C*, 2769 (1971).
[482] H. Gehlen and K. H. Uteg, *Z. Chem.* **9**, 338 (1969); *Chem. Abstr.* **72**, 3440 (1970).
[483] R. N. Butler and F. L. Scott, *J. Chem. Soc., C*, 239 (1967); 1202 (1966).
[484] B. T. Heitke and C. G. McCarty, *J. Org. Chem.* **39**, 1522 (1974).

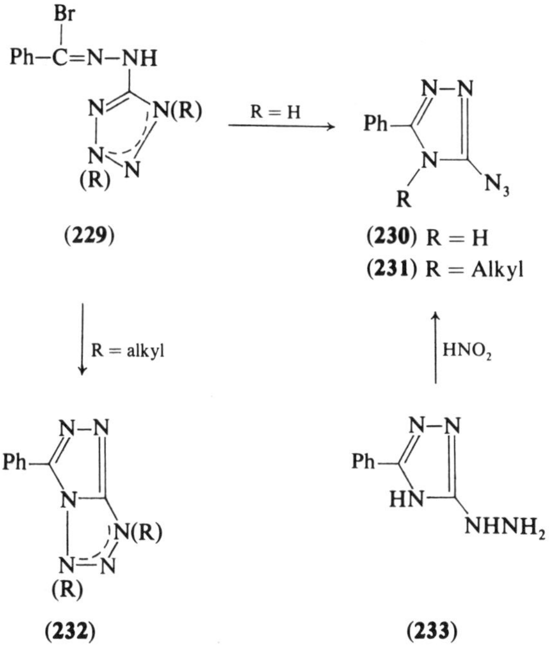

Sec. IV.B] RECENT ADVANCES IN TETRAZOLE CHEMISTRY

SCHEME 26

hydrazonyl bromides (**229**, R = alkyl) were cyclized (Scheme 26).[483] The preparation of a series of 3-azido-1,2,4-triazoles of type **230**, with the 5-phenyl group replaced by electron-donating substituents such as MeO and Me$_2$N and with Me groups at the ring 2-N site, has been reported[484] by diazotization of the corresponding amines followed by treatment with azide ion. A systematic search for a tetrazole tautomer in these systems using polar solvents and temperatures down to −82° showed that even under these favorable conditions the compounds exist exclusively in the azide form.[484] C-Azido-1,2,3-triazoles also exist in the azide form,[474,485,486] as do 5-azidotetrazoles.[329,348] The latter appear to exist in the azide form also when converted to anions,[339] and tetrazolotetrazoles have not been reported. Thermolysis of azidopyrazoles and 4-azido-1,5-diphenyl-1,2,3-triazole involves evolution of nitrogen and gives rise to fragmentation products that may arise from nitrene intermediates.[474]

In contrast to the azoles, an azidoazomethine unit in a thiazole ring, e.g., compounds **234**, may exist in either the cyclized or open-chain form

[485] Y. F. Shealey and C. A. O'Dell, *J. Heterocycl. Chem.* **10**, 839 (1974).
[486] P. A. S. Smith, G. J. W. Breen, M. K. Hajek, and D. C. V. Awang, *J. Org. Chem.* **35**, 2215 (1970).

under normal conditions, and in general both forms are detected.[487-489] In these systems the balance of the isomerism is particularly sensitive to substituent effects, and introduction of a sulfonyl substituent as in compounds (235) (Scheme 27) causes ring cleavage to the azide form.[490]

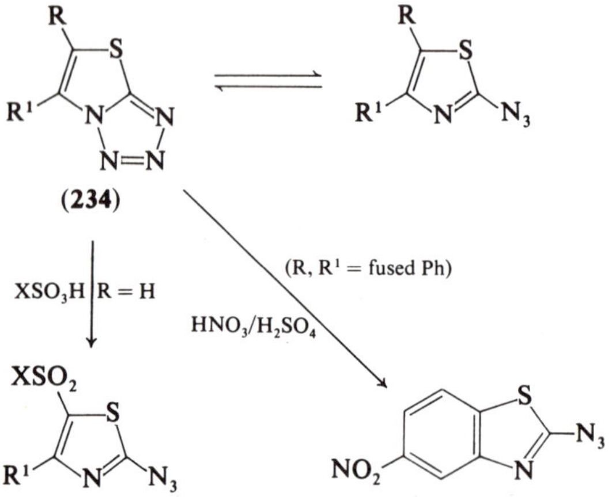

SCHEME 27

Tetrazolo[1,5-b]benzothiazoles generally exist in the tetrazole form in the solid state, but in CHCl$_3$ solution the azido form is also present.[491] Introduction of electron-withdrawing substituents, such as NO$_2$ or Cl, in the phenyl ring results in a preference for the azide form, e.g., 236 (Scheme 27).[488,491] Conversion of the sulfur atom of benzothiazoles into an SO$_2$ group as in compounds 237[492] and 238[493] completely favors the azide form, and the system is comparable to acyclic sulfonyl-substituted

[487] L. F. Avramenko, T. A. Zakharova, V. Ya. Pochinok, and S. Yi. Rozum, *Khim. Geterotsikl. Soedin.* 423 (1968) [*CA* **69**, 96591 (1968)].
[488] L. I. Skripnik and V. Ya. Pochinok, *Khim. Geterotsikl. Soedin.* 474 (1968) [*CA* **69**, 96600 (1968)].
[489] S. Maiorana and G. Pagani, *Chem. Ind. (Milan)* **53**, 470 (1971) [*CA* **75**, 48991 (1971)].
[490] V. N. Skopenko, L. F. Avramenko, V. Ya. Pochinok, and N. E. Smoilovskaya, *Ukr. Khim. Zh. (Russ. Ed.)*, **31**, 60 (1973) [*CA* **78**, 124507 (1973)].
[491] V. N. Skopenko, L. F. Avramenko, V. Ya. Pochinok, and M. I. Svichar, *Ukr. Khim. Zh. (Russ. Ed.)* **39**, 215 (1973) [*CA* **78**, 147878 (1973)]; see also R. Faure, J. P. Galy, G. Giusti, E. J. Vincent, and J. Elguero, *Org. Magn. Reson.* **6**, 485 (1975) [*CA* **82**, 169850 (1975)].
[492] J. H. Boyer and E. J. Miller, *J. Am. Chem. Soc.* **81**, 4671 (1959).
[493] G. A. Reynolds, J. A. Van Allen, and J. F. Tinker, *J. Org. Chem.* **24**, 1205 (1959).

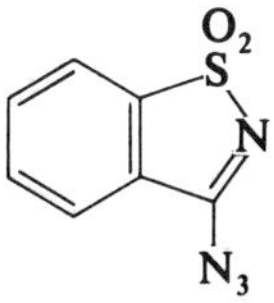

(237)

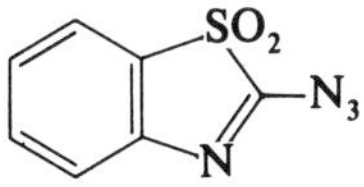

(238)

azidoazomethines, e.g., **223**. Tetrazolothiadiazoles of type **239** exist in the bicyclic form in the solid state and in the azide form in carbon tetrachloride solution.[494] When the substituent R is electron withdrawing as in compound **240**, the system exists in the azide form in the solid

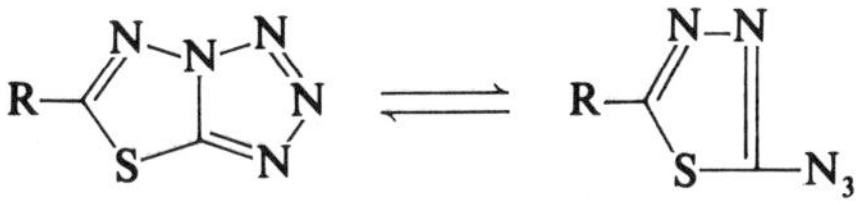

(239)
(240) R = ArCO—

state.[495] 2-Azido-1,3-benzoxazole also exists in the azide form.[479,493] The thiazolo[3,2-d]tetrazolium salts (**219**) represent another example of a fused thiazolotetrazole system.[464,496] In this case, the presence of the substituent R^1 prevents the formation of a stable azido form. Tetrazoles

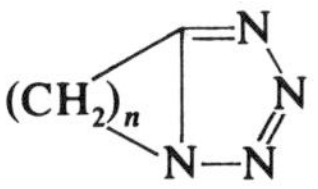

(241)

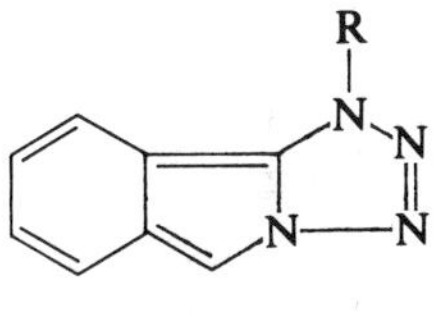

(242) R = H
(243) R = Me

fused to saturate carbocyclic rings (**241**) exist exclusively in the cyclic form. A number of new preparations of these compounds, including halogeno derivatives,[497] have been reported.[498,499] Pyrolysis of the compounds **241** in the gas phase results in elimination of nitrogen-yielding cyanamides which subsequently cyclize to substituted pyrrolidines.[500]

[494] A. Alemagna, T. Bacchetti, and P. Beltrame, *Tetrahedron* **24**, 3209 (1968).
[495] T. Baccheti, A. Alemagna, and B. Danieli, *Ann. Chim. (Rome)* **55**, 615 (1965) [*CA* **63**, 14849 (1965)].
[496] U. Askani, R. Neidlein, and J. Täuber, *Pharmazie* **26**, 463 (1971) [*CA* **75**, 110220 (1971)].
[497] F. M. D'Itri and A. I. Popov. *J. Heterocycl. Chem.* **7**, 221 (1970).
[498] A. Etienne and Y. Correia, *Bull. Soc. Chim. Fr.*, 3704 (1969).
[499] H. D. Torre. German Patent 1,914,553 (1969) [*CA* **72**, 12736 (1970)].
[500] C. Wentrup, *Tetrahedron* **27**, 1281 (1971).

The tetrazoloisoindole **242** has recently been prepared and methylated via a tetrazolium salt to the *N*-methyl derivative **243**.[501]

2. Tetrazoles Fused with Six-Membered Rings

a. Tetrazolopyridines and Pyridazines. By contrast with the π-excessive azole series, and somewhat enigmatically in view of the influence of the electron density at the imino bond on the isomerism, the azidoazomethine unit in the lower π-deficient azine compounds generally exists in the cyclic tetrazole form. Thus tetrazolo[1,5-*a*]-pyridine (**244**) exists in equilibrium with the azide form, but the

equilibrium lies strongly on the side of the tetrazole.[502] The tetrazole form is also dominant for derivatives of **244** bearing substituents, such as trifluoromethyl[503] or thienyl[503] on the pyridine ring, or halogens other than fluoride[504] at the 8-position. Introduction of a number of fluorine atoms into the pyridine ring orients the isomerization to the azide form, and perfluoro-(2-azido-4-isopropylpyridine) and 2-azido-3,5,6-trifluoro-4-methoxypyridine exist preferentially or even exclusively in the azide form.[505] The 5-chloro derivative of compound **244** is reported to exist predominantly in the azide form,[308,506] and the 6-nitro derivative exists as an equilibrium mixture of both forms in chloroform solution.[492] Treatment of tetrazolopyridines with alkali results in degradation of the pyridine ring, giving alkadienyltetrazoles.[507-509] When the 8-cyano derivative **245** was treated with an ethanolic solution of potassium

---

[501] F. S. Babichev and N. N. Romanov, *Ukr. Khim. Zh. (Russ. Ed.)* **39**, 49 (1973) [*CA* **78**, 111229 (1973)].
[502] R. Huisgen and K. F. Fraunberg, *Tetrahedron Lett.*, 2595 (1969).
[503] S. Portnoy, *J. Heterocycl. Chem.* **25**, 272 (1971).
[504] Yu. A. Azev, G. A. Mokrushina, and I. Y. Postovskii, *Khim. Geterotsikl. Soedin.*, 792 (1974) [*CA* **81**, 169406 (1974)].
[505] R. E. Banks and A. Prakash, *J. Chem. Soc., Perkin Trans. 1*, 2479 (1974).
[506] T. Sasaki, K. Kanematsu, and M. Murata, *Tetrahedron* **27**, 5121 (1971).
[507] B. Stanovnik and M. Tisler, *Chimia* **25**, 272 (1971) [*CA* **75**, 110239 (1971)].
[508] B. V. Golomolzin and I. Ya. Postovskii, *Khim. Geterotsikl. Soedin.*, 281 (1970) [*CA* **72**, 111379 (1970)].
[509] A. Pollak, S. Polanc, B. Stanovnik, and M. Tisler, *Monatsh. Chem.* **103**, 1591 (1972) [*CA* **78**, 136179 (1973)].

hydroxide and hydrogen peroxide, the interesting diepoxy compound **246** was formed [Eq. (43)].[510]

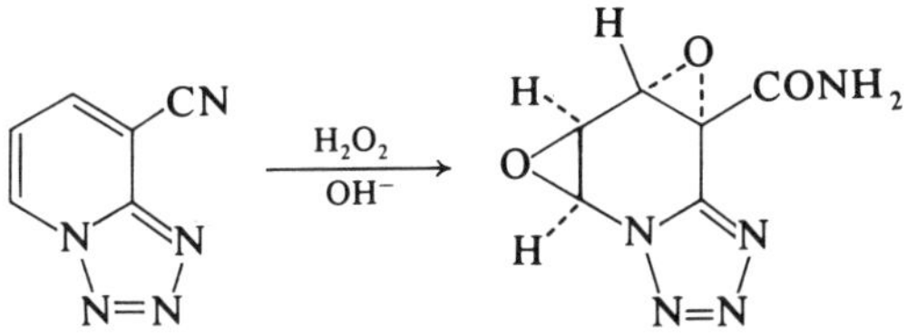

$$\text{(245)} \xrightarrow{\text{H}_2\text{O}_2, \text{OH}^-} \text{(246)} \quad (43)$$

Tetrazolopyridines, when heated to high temperatures with substituted acetylenes and olefin dipolarophiles, yielded a wide range of products arising from cycloaddition of the azide form,[511] and the equilibrium was shifted toward the azide form at the higher temperatures.[512]

The azidoazomethine moiety in a pyridazine ring tends to exist predominantly in the tetrazole form. Compounds of type **247** exist exclusively in the tetrazole form in a range of solvents.[513,514] In strong acids, such as concentrated sulfuric acid and trifluoromethanesulfonic acid, both forms were present and the equilibrium could be studied.[514] The persistence of the tetrazole form in such acidic solutions is unusual, since the azide form is normally dominant in strong acids. When the

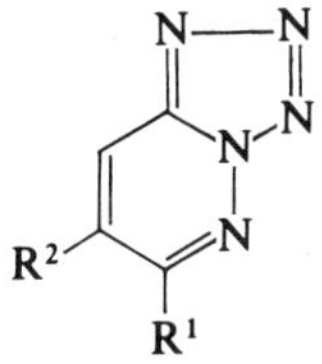

(**247**)

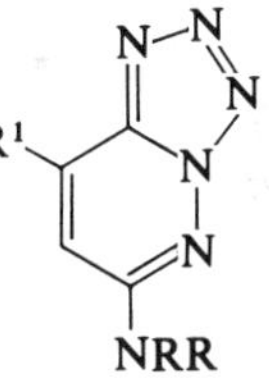

(**248**) $R^1 = H$
(**249**) $R^1 = Me, R = H$

substituent R^1 in **247** is an azide group, this second azide group remains in the open form.[513] Oxidation of compound **247** ($R^1 = R^2 = H$) with hydrogen peroxide resulted in simultaneous N-oxidation and cleavage of the tetrazole ring, giving 3-azidopyridazine 1-oxide.[515] The compounds

[510] J. F. Blount, R. Pitcher, M. R. Uskokovic, B. Stanovnik, and M. Tisler, *J. Org. Chem.* **38**, 2717 (1973).
[511] R. Huisgen, K. V. Fraunberg, and H. J. Sturm, *Tetrahedron Lett.*, 2589 (1969).
[512] W. D. Crow and C. Wentrup, *Chem. Commun.*, 1082 (1968).
[513] T. Sasaki, K. Kanematsu, and M. Murata, *J. Org. Chem.* **36**, 446 (1971).
[514] V. Pirc, B. Stanovnik, and M. Tisler, *Croat. Chem. Acta* **45**, 547 (1973) [*CA* **81**, 104502 (1974)].
[515] B. Stanovnik, M. Tisler, M. Ceglar, and V. Bah, *J. Org. Chem.* **35**, 1138 (1970).

248 and **249** exist in the cyclic form.[471,516] When the substituent NRR is replaced by an azide group, this second azide function does not cyclize,[516,517] but the azido derivative undergoes isomerization in warm DMSO to a new isomer, in which the original azide group has cyclized to a tetrazole ring and the original tetrazole ring has opened to an azide group.[471] Compounds with two tetrazole rings fused to a central pyridazine nucleus have not been observed under normal conditions.[516,517] In mass spectral fragmentations of azidotetrazolopyridazines, a loss of six nitrogen atoms has been accounted for by a path proceeding via a molecular ion with a bistetrazolo structure.[518] Equally, two azido substituents on a pyridazine ring have not been observed, and attempts to generate such a system by treating compounds of type **247** ($R^1 = N_3$) with strong acids at high temperatures did not result in cleavage of the tetrazole ring.[514] Reduction of tetrazolo[1,5-*b*]pyridazines to tetrahydro derivatives with sodium borohydride has been reported[519] as well as the reactions of compounds of type **247** ($R^1 = N_3$) with β-diketones to give the compounds **247** ($R^1 =$ substituted 1,2,3-triazol-1-yl).[520,521] Photolysis of tetrazolopyridazines interestingly gives 3-cyanocyclopropenes and 3-cyanopyrazoles [Eq. (44)].[522]

b. *Tetrazolopyrimidines.* Tetrazolo[1,5-*a*]pyrimidines (**250**) ($R = R^2 = Me$; $R^1 = H$) exist in both the azide and tetrazole forms.[523] When the substituents R and R^2 are electron donating, the tetrazole form predominates.[523-525] The electron-withdrawing character of the

---

[516] I. B. Lundina, N. J. Sheinker, and I. Ya. Postovskii, *Izv. Akad. Nauk. SSSR, Ser. Khim.* 66 (1967) [*CA* **69**, 106647 (1968)].
[517] A. Kovacik, B. Stanovnik, and M. Tisler, *J.Heterocycl. Chem.* **5**, 351 (1968).
[518] V. Pirc, B. Stanovnik, M. Tisler, J. Marcel, and W. W. Paudler, *J. Heterocycl. Chem.* **7**, 639 (1970).
[519] P. K. Kadaba, B. Stanovnik, and M. Tisler, *Tetrahedron Lett.*, 3715 (1974).
[520] A. Karklina and E. Gudriniece, *Latv. PSR Zinat. Akad. Vestis, Khim. Ser.*, 750 (1973) [*CA* **80**, 82860 (1974)].
[521] A. Karklina, E. Gudriniece, and L. K. Bronfman, *Latv. PSR Zinat. Akad. Vestis, Khim. Ser.* 206 (1974) [*CA* **81**, 37518 (1974)].
[522] T. Tsuchiya, H. Arai, and H. Igeta, *Chem. Commun.*, 1059 (1972).
[523] C. Wentrup, *Tetrahedron* **26**, 4969 (1970); **27**, 361 (1971).
[524] C. Temple and J. A. Montgomery, *J. Org. Chem.* **30**, 826 (1965).
[525] C. Temple and J. A. Montgomery, *J. Am. Chem. Soc.*, **86**, 2946 (1964).

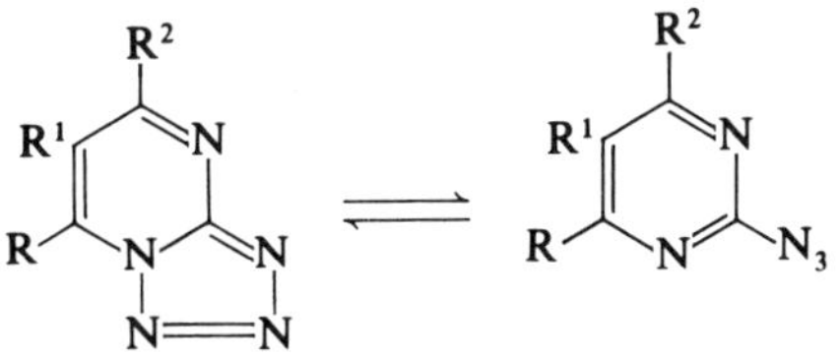

pyrimidine ring in the compound **251** favored the azide form.[308] The compound was obtained by treating 5-aminotetrazole with malonic ester.[308]

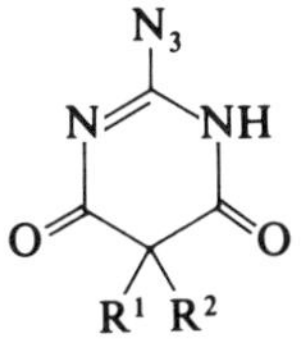

(251) $R^1 = R^2 = H$

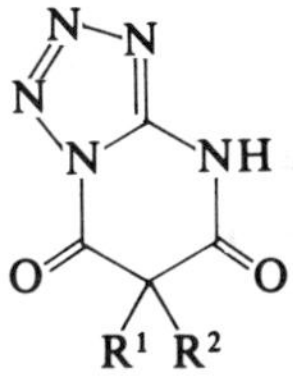

(252) R^1, R^2 = alkyl

When one of the carbonyl groups of compound **251** is replaced by a CH_2 group, the compound exists in a tetrazole form in which the azide function is cyclized to the nitrogen farthest from the remaining carbonyl group.[323] When the substituents R^1 and R^2 in compound **251** are alkyl groups, the tetrazole form is favored. The compounds **252**, which exist as tetrazoles, were obtained by treating the corresponding hydrazines with nitrous acid.[526] The two tetrazole isomers **253** and **255** have been observed in solution.[527] In the solid state, the compound **253**

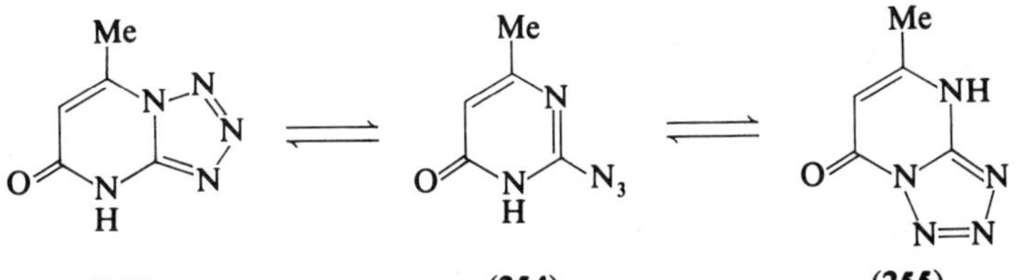

(253)   (254)   (255)

predominates. This is expected to be the preferred form because of the location of both the methyl and carbonyl groups, and the existence of compound **255** even in low concentration is interesting.[527] Treatment of β-cyanoacrylic esters with aluminum azide in THF leads to compounds

[526] A. V. Spasov and Z. Raikov, *Z. Chem.* **11**, 422 (1971) [*CA* **76**, 85785 (1972)].
[527] C. Temple, N. C. Coburn, M. C. Thorpe, and J. A. Montgomery, *J. Org. Chem.* **30**, 2395 (1965).

of type **256**, which have been reported in the tetrazole form, as well as other products [Eq. (45)].[528]

$$\underset{CN}{\overset{Me}{>}}C=CH-COOEt \xrightarrow{Al(N_3)_3} \underset{(256)}{[fused\ tetrazolopyrimidine]} + NC-\underset{H}{C}=C\underset{N=N}{\overset{Me}{<}}\underset{NH}{\overset{O}{>}} \quad (45)$$

Pyrolysis of the compounds **250** yields nitrene intermediates, which undergo a wide range of insertion reactions.[502,529] *N*-Cyanopyrroles, which are obtained from the pyrolysis of tetrazolopyridines, also arise from nitrene intermediates via ring contraction.[512,530] Mass spectral studies have indicated that the first step in the gas-phase pyrolysis of fused tetrazolopyridines and pyrimidines involves tautomerism to the azide with subsequent nitrene formation.[523] Copper compounds have been found to catalyze the generation of nitrene intermediates from the compounds **250** and **244**.[531] The compounds **250** also undergo cycloaddition reactions through the azido form.[511]

(257) ⇌ (258)

The tetrazolo[1,5-*c*]pyrimidines (**257**; R = NR$_2$) exist in both the azide and tetrazole forms.[532,533] The compound **257** (R = H) exists predominantly as the tetrazole form.[533] Some electron-donating substituents in the 5- and 7-positions of this system seem to favor the azide form in contrast to the effects of similar substituents with other systems.[523] With electron-donating substituents in the 8-position, both forms are observed.[532] The compound **259** exists in the tetrazole form and is converted into the (protonated) azide in trifluoroacetic acid.[534,535]

---

[528] E. R. Wagner, *J. Org. Chem.* **38**, 2976 (1973).
[529] C. Wentrup, *Helv. Chim. Acta* **55**, 562 (1972).
[530] W. D. Crow and C. Wentrup, *Chem. Commun.*, 1387 (1969).
[531] K. V. Fraunberg and R. Huisgen, *Tetrahedron Lett.*, 2599 (1969).
[532] I. Ya. Postovskii and N. B. Smirnova, *Dokl. Akad. Nauk. SSSR* **165**, 1136 (1966) [*CA* **64**, 17589 (1966)].
[533] C. Temple, R. L. McKee, and J. A. Montgomery, *J. Org. Chem.* **30**, 829 (1965).
[534] J. A. Hyatt and J. S. Swenton, *J. Heterocycl. Chem.* **9**, 409 (1972).
[535] J. A. Hyatt and J. S. Swenton, *J. Org. Chem.* **37**, 3216 (1972).

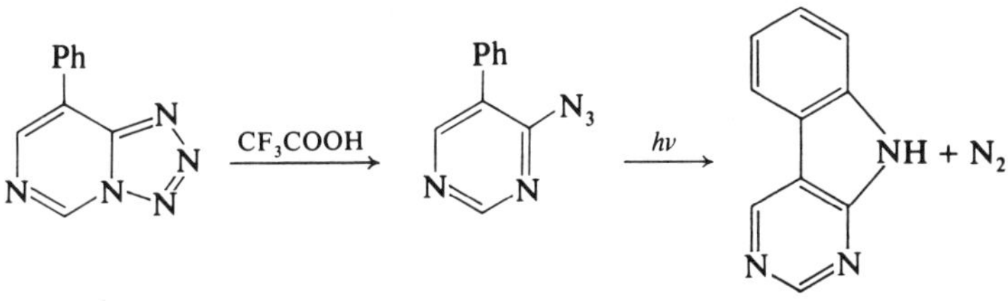

(259) → CF₃COOH → (shown) → hv → (260) NH + N₂ (46)

Photolysis of the compounds **259** in this solvent rapidly gave high yields of the pyrimido[4,5-b]indoles (**260**) resulting from evolution of N_2 from the azide form [Eq. (46)].[534,535] Treatment of 4-chloro-5-nitropyrimidines (**261**) with azide ion yielded the products **263** via the azides **262** [Eq. (47)].[536] Evidence that the azide **262** was in equilibrium with the tetrazole form was obtained, and, when R was NH_2, a tetrazolopyrimidine was isolated.[536]

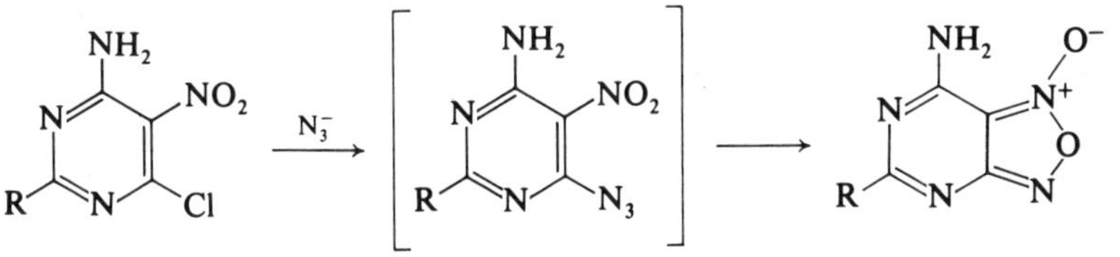

(261) → N_3^- → [(262)] → (263) (47)

c. *Tetrazolopyrazines.* Tetrazolo[1,5-a]pyrazine (**264**) was obtained by treating fluoropyrazine with sodium azide.[537] In the solid state and in carbon tetrachloride it appears to exist in the tetrazole form.[537] The

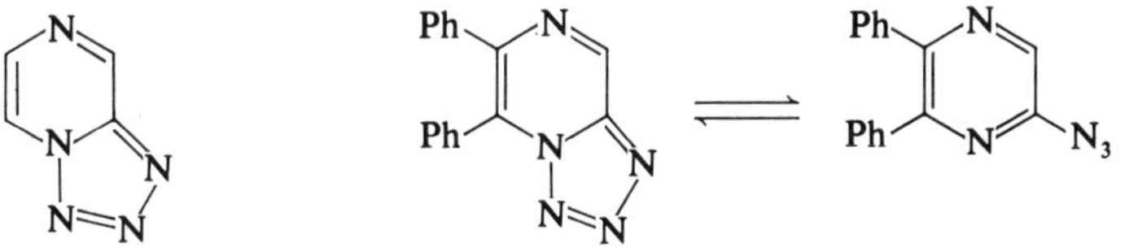

(264)   (265) ⇌ (266)

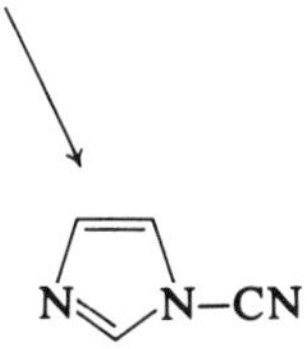

↓

(267)

---

[536] C. Temple, C. L. Kussner, and J. A. Montgomery, *J. Org. Chem.* **33**, 2086 (1968).
[537] H. Rutner and P. E. Spoerri, *J. Heterocycl. Chem.* **3**, 435 (1966).

compound **265** may exist in different forms depending on the conditions. In the solid and in dimethyl sulfoxide solution, only the tetrazole form **265** was detected.[513] In trifluoroacetic acid the compound exists exlusively in the (protonated) azide form **266**, while both forms are present in the ratio 10:3, **265:266**, in chloroform.[513] Thermolysis of compound **264** has recently been reported.[529,538,539] 1-Cyanoimidazole (**267**) was formed via ring contraction in the intermediate pyrazinyl nitrene.[529,538,539] The tetrazolo[1,5-a]quinoxalines **268** and **269** have also been prepared and reported as in the tetrazole form.[540]

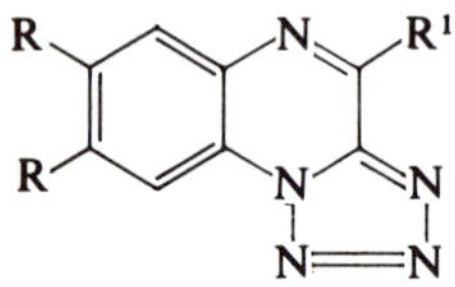

(**268**) 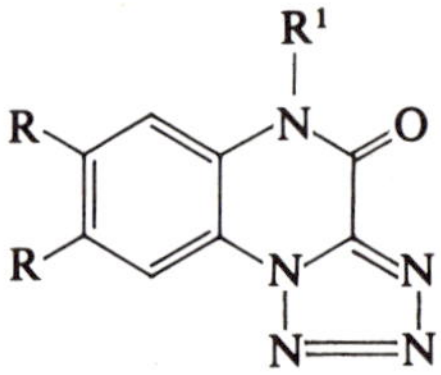   (**269**)

d. Tetrazolotriazines and tetrazines and General. Treatment of bistetrazolylamine with cyanogen bromide yielded 2-amino-4,6-diazido-1,3,5-triazine (**270**) [Eq. (48)].[541] This compound exists in the diazide form, presumably because of the π-deficient nature of the 1,3,5-triazine ring. Other mono- and diazido-1,3,5-triazines which have been prepared also exist in the azide forms.[542] Photolysis of 2-azido-4,6-dimethoxy-1,3,5-triazine in the presence of nitriles gives triazolotriazinones via nitrene attack on the nitrile followed by cyclization.[542a] By contrast with

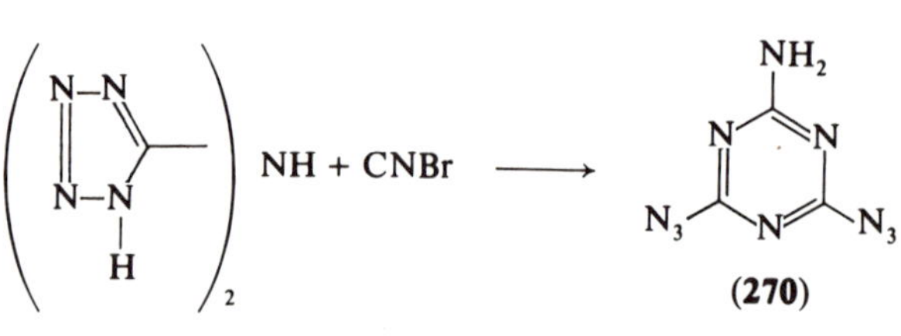

(48)

(**270**)

[538] C. Wentrup, C. Thétaz, and R. Gleiter, *Helv. Chim. Acta* **55**, 2633 (1972).
[539] C. Wentrup and W. D. Crow, *Tetrahedron* **26**, 4915 (1970).
[540] N. G. Koshel, E. G. Kovalev, and I. Ya. Postovskii, *Khim. Geterotsikl. Soedin*, 851 (1970) [*CA* **73**, 120586 (1970)].
[541] R. Henry, *J. Org. Chem.* **31**, 1973 (1966).
[542] J. Kobe, B. Stanovnik, and M. Tisler, *Monatsh. Chem.* **101**, 724 (1970) [*CA* **73**, 45470 (1970)].
[542a] H. Yamada, H. Shizuka, and K. Matsui, *J. Org. Chem.* **40**, 1351 (1975); see also, R. Kayama, S. Hasunuma, S. Sekiguchi, and K. Matsui, *Bull. Chem. Soc. Jpn.* **47**, 2825 (1974) [*CA* **82**, 138795 (1975)].

the 1,3,5-triazine systems, the 3-azido-1,2,4-triazines (271) exist preferentially in the tetrazolo form 272, but the azide form is also present in chloroform solution.[489,513,543] Attempts to obtain the isomer 273 by heating compound 272 in acetic acid and pyridine were unsuccessful.[543] When compound 272 was heated in saturated amine solvents, 3-amino-1,2,4-triazines were formed.[543] The compound 274 has also been reported in the tetrazole form.[544] The compounds 275 and

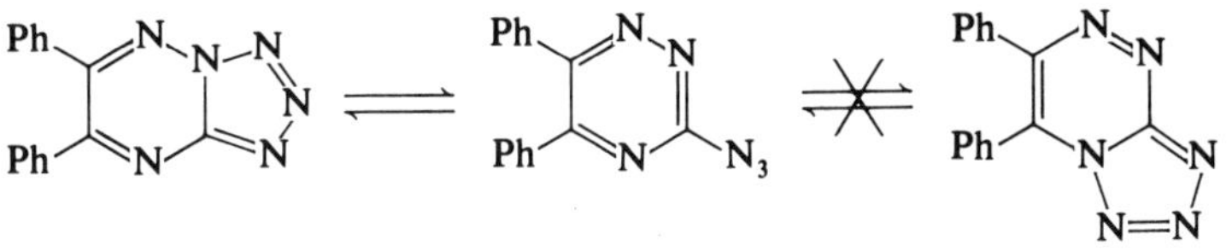

(272)   (271)   (273)

276 exist in an equilibrium in which the azide form predominates in solution and the tetrazole form predominates in the solid.[545] The compounds 277 and 278 have been isolated separately and can be interconverted. The tetrazole form is the more stable in the solid state, and the azide is preferred in solution. In these compounds the azide group cyclizes to the 4-N atom of the 1,2,4-triazine ring. When the 1-N atom of the 1,2,4-triazine ring is N-oxidized, only the azide forms are stable.[545]

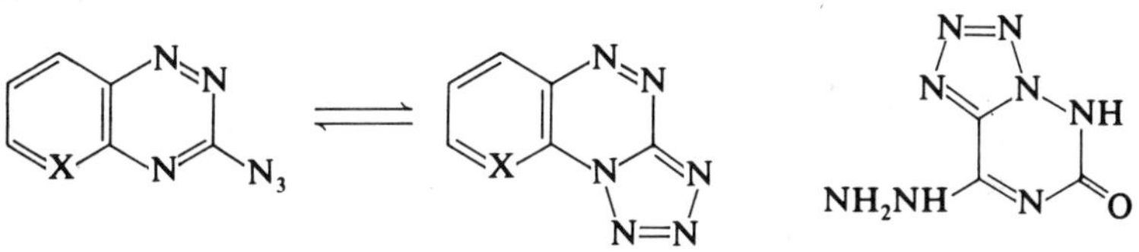

(275) X = CH   (276) X = CH   (274)
(277) X = N   (278) X = N

The azidotetrazines (279) have been obtained by treating 3-bromo-6-aryltetrazines with NaN_3.[546] The azide forms were preferred when Ar

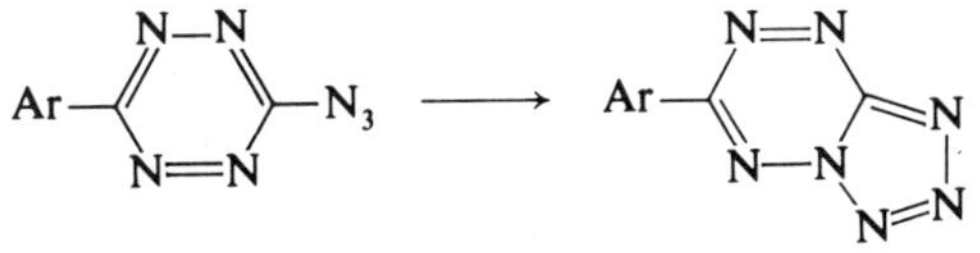

(279)   (280) Ar = Ph

[543] M. F. G. Stevens, *J. Chem. Soc., Perkin Trans. 1*, 1221 (1972).
[544] C. Critescu, *Rev. Roum. Chim.* **16**, 18 (1971) [*CA* **75**, 20358 (1971)].
[545] A. Messmer, Cy. Hajos, P. Benko, and L. Pallos, *J. Heterocycl. Chem.* **10**, 575 (1973); cf. also *Magy. Chem. Foly.* **80**, 527 (1974) [*CA* **82**, 125362 (1975)].
[546] V. A. Ershov and I. Ya. Postovskii, *Khim. Geterotsikl. Soedin.* **7**, 771 (1971) [*CA* **76**, 126947 (1972)].

was a phenyl ring with an electron-withdrawing substituent. With Ar = Ph, the tetrazole form **280** was also detected in polar solvents, although the azide form was dominant in the solid state and in nonpolar solvents.[546]

Tetrazolomorpholines **281**[547] and **282**[548] have also been reported. Like other saturated tetrazoloheterocycles, they exist in the cyclic form.

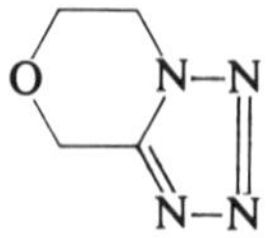

(281) 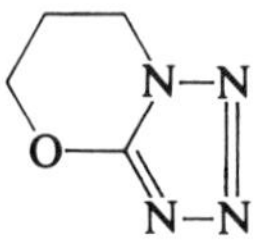   **(282)**

C. Tricyclic and Higher Fused Tetrazoloheterocycles

1. Tetrazolopyridazines Fused with Five-Membered Rings

The imidazo [4,5-*d*]tetrazolo[1,5-*b*]pyridazine (**283**) exists exclusively in the tetrazole form.[549] Both azide and tetrazole forms of the compound **284** are reported to be detected in the solid state, which shows photochromic properties. In dimethylformamide the tetrazole form of

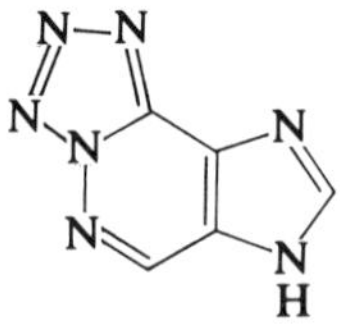

(283) 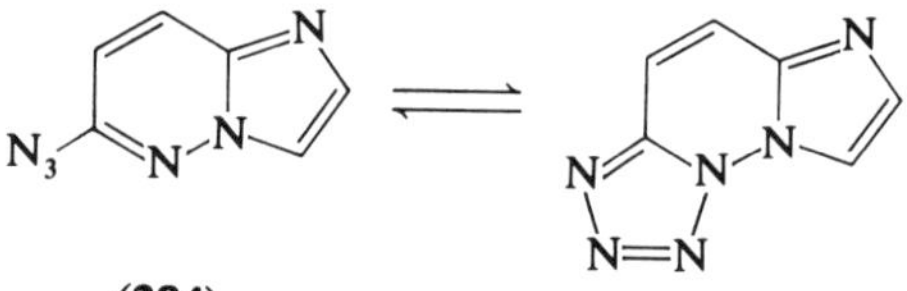   **(284)**

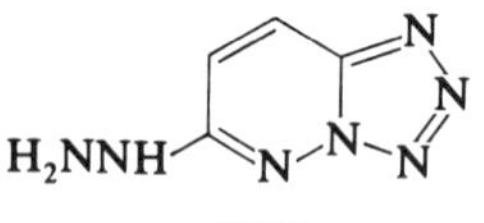

(285) 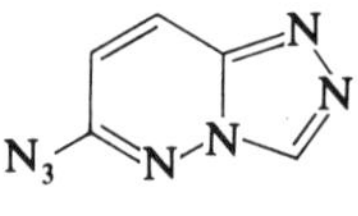   **(286)**

this system predominates.[550,551] Attempts to introduce a fused triazole ring into the tetrazolopyridazine system via compound **285** resulted in

[547] R. G. Glushkov and O. Yu. Magidson, *Khim. Geterotsikl. Soedin.*, 192 (1966) [*CA* **65**, 5460 (1966)].
[548] A. Krantz and B. Hoppe, *Tetrahedron Lett.*, 695 (1975).
[549] S. F. Martin and R. N. Castle, *J. Heterocycl. Chem.* **6** 93 (1969).
[550] B. Stanovnik and M. Tisler, *Tetrahedron* **23**, 387 (1967).
[551] B. Stanovnik and M. Tisler, *Tetrahedron Lett.*, 2403 (1966).

cleavage of the tetrazole ring to give the azido product **286**.[513,517] The influence of a 1,2,4-triazole moiety fused to the pyridazine ring of a tetrazolopyridazine seems consistently to effect cleavage of the tetrazole ring. Thus, cyclization of compound **287** also gave an azide, **288**, which is an isomer of compound **286**.[552,553] A tetrazolopyridazine fused to a

(**287**)   (**288**)

triazole ring, compound **290**, has, however, been reported from the reaction of the hydrazone **289** with lead tetraacetate [Eq. (49)].[554] This contrasts with other attempts to fuse a triazole ring to the tetrazolopyridazine system. When the furan ring of compound **290** was

(**289**)   (**290**)   (49)

replaced by a phenyl group, the azide form only of the compound was detected and the cyclization of the corresponding hydrazone involved a preferential ring interconversion of tetrazole to triazole.[515] Photolytic degradation of compounds of type **286** via nitrene intermediates has recently been reported.[513,555] The addition of azidotetrazolopyridazines to substituted acrylates has also been described.[556]

The tetrazolobenzothienopyridazines **291** and **293** exist in the cyclic forms, while for the system **292** the cyclic form dominates in the solid state, but both forms are present in DMSO solution and the azide form predominates in trifluoroacetic acid.[557-559] The tetrazolofuropyridazines

[552] S. Polanc, B. Vercek, B. Stanovnik, and M. Tisler, *Tetrahedron Lett.*, 1677 (1973).
[553] S. Polanc, B. Vercek, B. Sek, B. Stanovnik, and M. Tisler, *J. Org. Chem.* **39**, 2143 (1974).
[554] H. Berger, K. Stach, and W. Wormal, S. African Patent 6,706,255 (1968) [*CA* **70**, 57869 (1969)].
[555] B. Stanovnik, *Tetrahedron Lett.*, 3211 (1971).
[556] B. Stanovnik, *J. Heterocycl. Chem.* **8**, 1055 (1971).
[557] M. Robba, G. Dore, and M. Bonhomme, *C.R. Hebd. Seances Acad. Sci., Ser. C* **271**, 1990 (1970) [*CA* **74**, 112021 (1971)].
[558] M. Robba, G. Dore, and M. Bonhomme, *C.R. Hebd. Seances Acad. Sci., Ser. C* **271**, 1328 (1970) [*CA* **74**, 53673 (1971)].
[559] M. Robba, M. Bonhomme, and G. Dore, *Tetrahedron* **29**, 2919 (1973).

(295)[560] and the tetrazolo[1,2,5]thiadiazolopyridazines (294)[561] are reported to exist in the cyclic forms.

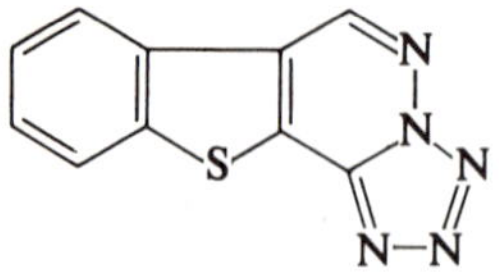

(291) 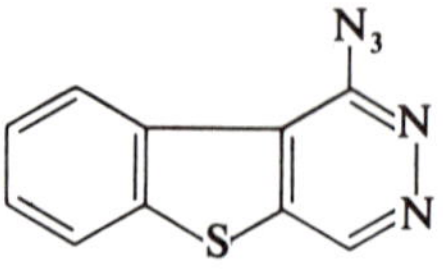   (292) 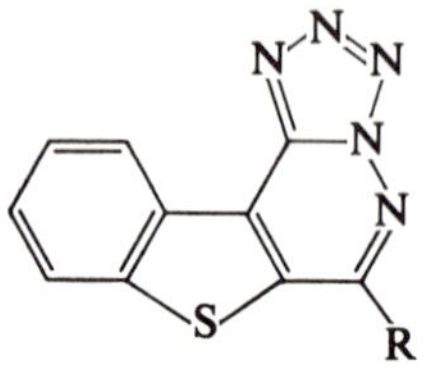   (293)

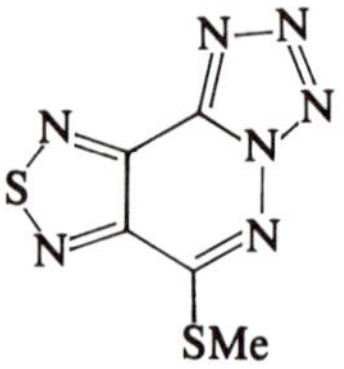

(294) 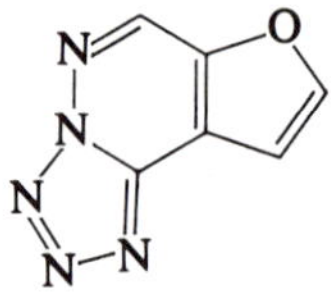   (295)

2. Tetrazolopyridazines Fused with Six-Membered Rings

The compounds **296** and **297** have been prepared from 1,4-dichoro-5,6-dihydrobenzo[f]phthalazines.[562-564] These compounds contain one tetrazole ring, the second azidoazomethine moiety remaining in the

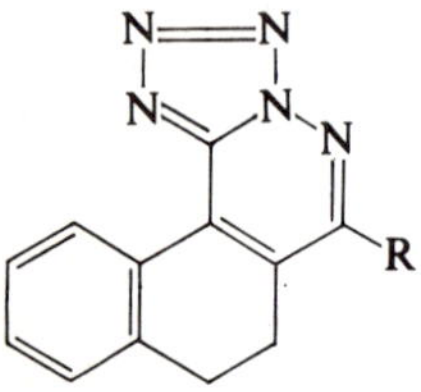

(296) R = N₃
(297) R = NHNH₂

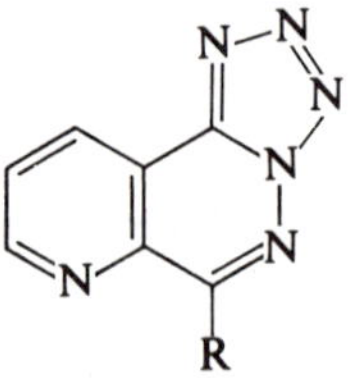

(298) R = H, Cl
(300) R = N₃

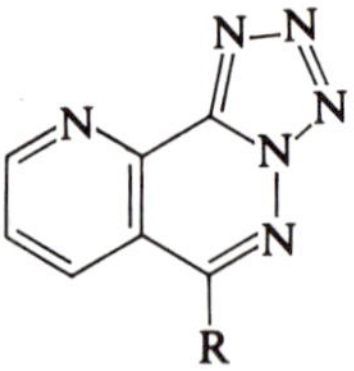

(299) R = H, Cl
(301) R = N₃

---

[560] M. Robba, M. Zaluski, P. Roques, and M. Bonhomme, *Bull. Soc. Chim. Fr.*, 4004 (1969).

[561] J. Marn, B. Stanovnik, and M. Tisler, *Croat. Chem. Acta* **43**, 101 (1971) [*CA* **75**, 98523 (1971)].

[562] B. Stanovnik, M. Tisler, and P. Skufca, *J. Org. Chem.* **33**, 2910 (1968).

[563] B. Stanovnik and M. Tisler, *Chimia* **22**, 141 (1968) [*CA* **69**, 19371 (1968)].

[564] J. Kobe, A. Krbavcic, B. Stanovnik, and M. Tisler, *Croat. Chem. Acta* **41**, 245 (1969) [*CA* **72**, 66884 (1970)].

acyclic form. Insertion of a triazole ring to the system via the hydrazino group in **297** results in cleavage of the tetrazole ring to the azide.[562] The azidoazomethine–tetrazole isomerism in these systems has been used in the synthesis of azasteroids by allowing selective orientation of the rings in the buildup of the polynuclear system.[562] The compounds **298** and **299** exist in the cyclic form, in which the entire polyaza nucleus contains 14 π-electrons and possesses special stability.[565-567] An interesting thermal isomerism of the azido derivative **300** to the higher melting isomer **301** has been observed.[471] Treatment of the compounds **301** with 1,3-dicarbonyl compounds gives triazoles of type **301** (R = substituted-1,2,3-triazol-1-yl).[568] When the azides **301** were heated in the presence of diethylamine, the azido group was replaced by NH_2, NEt_2, and, interestingly, $N=CHNEt_2$.[569] These reactions are considered to involve

(302)

(303)

nitrene intermediates.[569] The analogous pyrazino derivatives of the compounds **300** and **301** have been prepared, and these also contain one tetrazole ring and one azide group.[570] The compound **302** has been reported in the azide form.[515] When the *as*-triazine ring of this compound was cleaved by acid hydrolysis, a fused tetrazole ring was immediately generated. The compound **303** has also been reported in the azide form.[571] The tetrazole ring of the precursory tetrazolopyridazine opened as soon as the 1,2,4-triazine was introduced.[571] Generally, it appears that in tricyclic tetrazolopyridazines the azidoazomethine moiety exists in the tetrazole form unless the third ring is strongly electron withdrawing, for example, triazole, tetrazole, or 1,2,4-triazine. The influence of the 1,2,4-triazine system in this case appears to contrast with its influence in bicyclic systems, where the tetrazole forms predominate.

[565] A. Krbavcic, B. Stanovnik, and M. Tisler, *Croat. Chem. Acta* **40**, 181 (1968).
[566] B. Stanovnik, A. Krbavcic, and M. Tisler, *J. Org. Chem.* **32**, 1139 (1967).
[567] M. Tisler, B. Stanovnik, and B. Stefanov, *J. Org. Chem.* **36**, 3812 (1971).
[568] A. Gorup, M. Kovacic, B. Kranjc-Skraba, B. Mihelcic, S. Simonic, B. Stanovnik, and M. Tisler, *Tetrahedron* **30**, 2251 (1974).
[569] S. Polanc, B. Stanovnik, and M. Tisler, *J. Heterocycl. Chem.* **10**, 565 (1973).
[570] L. Di Stefano and R. N. Castle, *J. Heterocycl. Chem.* **5**, 109 (1968).
[571] B. Stanovnik and M. Tisler, *Synthesis* **2**, 180 (1970).

3. Tetrazolopyrimidines Fused with Five- and Six-Membered Rings

The isomerism in the tetrazolopurine systems **304** and **306** has been directly observed, and all the isomers have been detected.[572,573] Both forms of the systems **304** are present, but the tetrazole forms are generally favored except where the substituent R^1 is electron

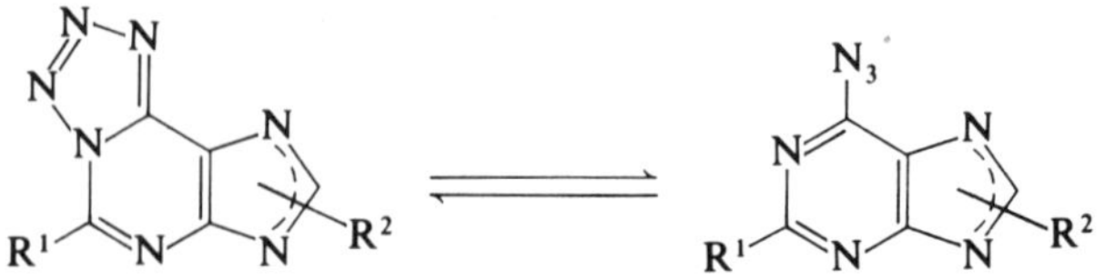

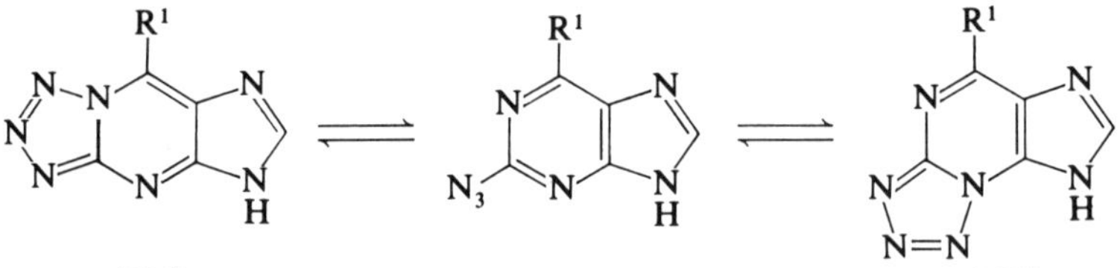

withdrawing, for example, Cl. In this case the azido form (**305**) dominates.[573] Compounds of type **304** (R^1 = H, R^2 = β-D-ribofuranose) have also been prepared and found to exist predominantly in the tetrazole form.[573a] The heat of isomerization of **306** ⇌ **307** was 4.7 kcal mole^{-1}, similar to the value of 5.1 kcal mole^{-1} for the isomerism in tetrazolo[1,5-a]pyrimidine.[572] Full tables of heats of isomerism for these systems in general have been drawn up by Tisler[8] in a recent review. When the substituent R^1 in compound **306** is an azide group, five isomers are possible. Three of these have been detected in DMSO solution.[573] In acidic solutions the diazido form only is present.[573]

Tetrazolothienopyrimidines **309** and **310** have been reported in the

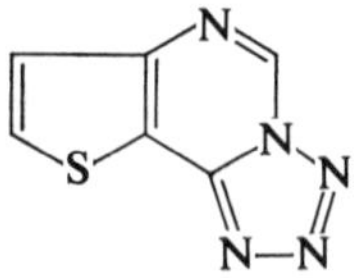

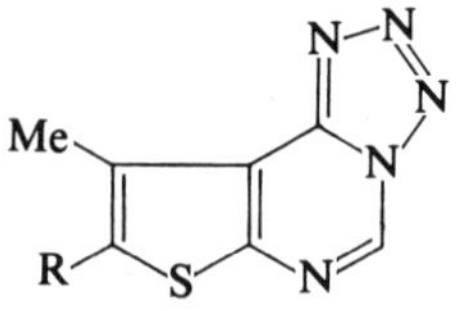

---

[572] C. Temple, M. C. Thorpe, W. C. Coburn, and J. A. Montgomery, *J. Org. Chem.* **31**, 935 (1966).

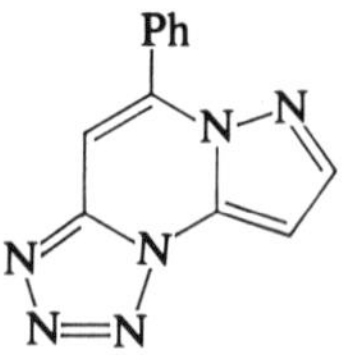

[573] C. Temple, C. L. Kussner, and J. A. Montgomery, *J. Org. Chem.* **31**, 2210 (1966).

[573a] R. Wetzel and F. Eckstein, *J. Org. Chem.* **40**, 658 (1975); J. A. Johnson, H. J. Thomas, and H. J. Schaefer, *J. Am. Chem. Soc.* **80**, 699 (1968).

tetrazole form.[574,575] The tetrazolopyrimidine **311** and the isomeric azide have been prepared by Reimlinger.[576] The tetrazole form **311** was thermostable and could be obtained by cyclization of the azide.

Tetrazoloquinazolines **312** and **313** ($R^1 = Ph$) exist mainly as tetrazoles, but with compounds of type **313** ($R^1 = H$) some of the azido form is also present in solution.[577,578] The tetrazolo[1,5-*a*]-[1,8]naphthyridines (**314**) exist in both tetrazole and azide forms.[579,580]

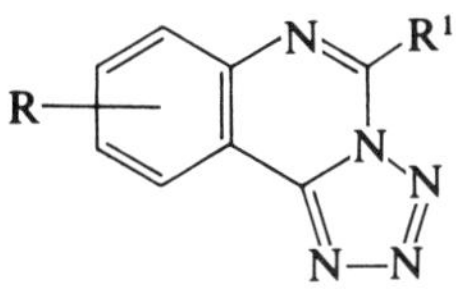

(313)

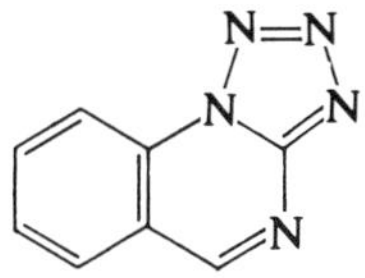

(312)

The tetrazole form is favored in the solid state and in basic solution, and the azido form predominates in acidic solution.[579] The presence of the phenyl group adjacent to the tetrazole ring is reported to stabilize the tetrazole form.[579] When these compounds contain a second azidoazomethine moiety, e.g., **315**, one of the azide groups remains in the open-chain form, and in solution mixtures of isomers are

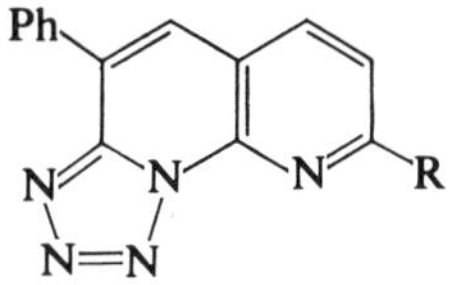

(314) R = NHAc
(315) R = N_3

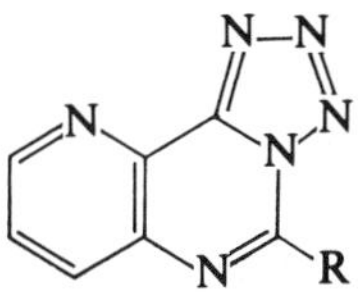

(316) R = alkyl

[574] M. Robba, J. M. Lecomte, and Y. L. Le Guen, *C.R. Hebd. Seances Acad. Sci., Ser. C* **266** 1706 (1968) [*CA* **69**, 96635 (1968)].

[575] M. Robba, J. M. Lecomte, and M. Cugnon de Sevricourt, *C.R. Hebd. Seances Acad. Sci., Ser. C* **267**, 697 (1968) [*CA* **70**, 37767 (1969)].

[576] H. Reimlinger and M. A. Peiren, *Chem. Ber.* **104**, 2237 (1971).

[577] B. Golomolzin and I. Ya. Postovskii, *Khim. Geterotsikl. Soedin*, 855 (1970) [*CA* **73**, 120586 (1970)].

[578] I. Ya. Postovskii and I. N. Goncharova, *Zh. Org. Khim.* **33**, 2334 (1963).

[579] S. Carboni, A. Da Settimo, P. L. Ferrarini, and P. L. Ciantelli, *J. Heterocycl. Chem.* **7**, 1037 (1970).

present.[579-584] Pyridotetrazolopyrimidines (316) have also been reported in the tetrazole form.[585]

Treatment of 2-azido-pyrido[1,2-a]pyrimidone (317), which can exist only in an azide form, with base resulted in simultaneous cleavage of the pyrimidine ring and closure of the azidoazomethine moiety [Eq. (50)].[586] Cycloaddition reactions of the azide group of compound 317 with 1,3-dicarbonyl compounds to form 1,2,3-triazoles has also been described.[586]

4. General Discussion of Fused Tetrazoloheterocycles

The tetrazolo-s-triazolo-as-triazine (318) has been found to exist predominantly in the tetrazole form,[513,587] but the azide form is also present in chloroform solution. In this case, the presence of the singly bound nitrogen atom in the 1,2,4-triazine ring seems to have counteracted the deactivating influence of the fused triazole system. In general, it appears that the tetrazole form is preferred when the azido–azomethine moiety is directly linked to a 1,2,4-triazine ring, e.g., compounds 272, 278, 274, and 318, while the azido form is favored when the 1,2,4-triazine system is present as a fused substituent on the ring containing the azidoazomethine moiety, e.g., compounds 302 and 303. The azidoazomethine moiety in a fused 1,2,3-triazine system, e.g.,

[580] P. Ferrarini, *Ann. Chim. (Rome)* **61**, 318 (1971) [*CA* **75**, 129726 (1971)].
[581] S. Carboni, A. Da Settimo, P. L. Ferrarini, and G. Pirisino, *Gazz. Chim. Ital.* **96**, 1456 (1966) [*CA* **67**, 86 (1967)].
[582] S. Carboni, A. Da Settimo, and P. L. Ferrarini, *Gazz. Chim. Ital.* **97**, 42 (1967) [*CA* **68**, 29647 (1968)].
[583] S. Carboni, A. Da Settimo, and P. L. Ferrarini, *Gazz. Chim. Ital.* **97**, 1061 (1967) [*CA* **68**, 29648 (1968)].
[584] S. Carboni, A. Da Settimo, P. L. Ferrarini, and F. Trusendi, *Gazz. Chim. Ital.* **98**, 1174 (1968) [*CA* **70**, 68265 (1969)].
[585] L. Godefroy, A. Decormeille, G. Queguiner, and P. Pastour, *C.R. Hebd. Seances Acad. Sci., Ser. C* **278**, 1421 (1974) [*CA* **81**, 91461 (1974)].
[586] M. Kovacic, S. Polanc, B. Stanovnik, and M. Tisler, *J. Heterocycl. Chem.* **11**, 949 (1974).
[587] T. Sasaki, K. Minamoto, and M. Murata, *Chem. Ber.* **101**, 3969 (1968).

319[588] and 320,[589] appears to exist exclusively in the azide form. The effects of the various triazine systems on the azidoazomethine moiety

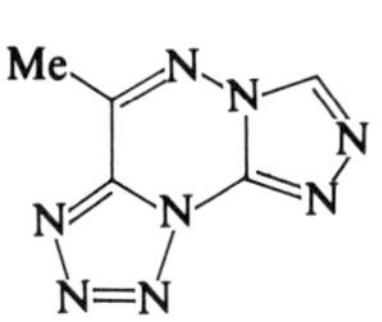

(318) 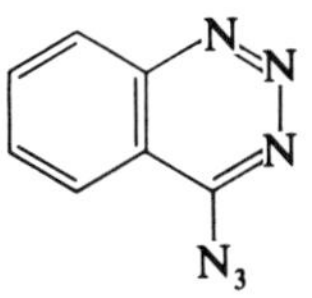   (319) 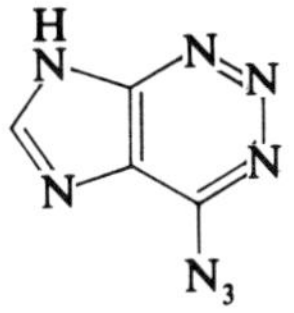   (320)

provides an interesting comparison, e.g., 1,2,3-triazines, compounds 319 and 320; 1,3,5-triazines, compound 270, and 1,2,4-triazines, compounds 272, 274, and 278. Very recently two groups of workers[590] have reported that potential 2,3-diazidopyrazines exist exclusively in the ditetrazolo form (321). This was observed with both parent pyrazine[590a] and quinoxaline systems.[590b]

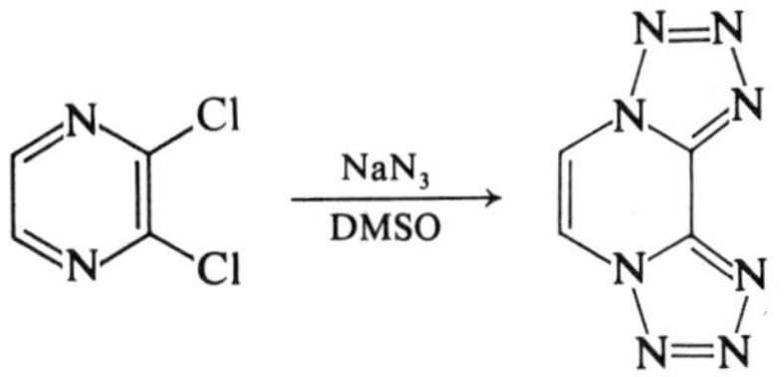

(321)

Other polycyclic systems containing an azidoazomethine moiety which exist in the cyclic form include tetrazoloadamantanes,[591] dihydrobenzoxazepine tetrazole derivatives,[592] and steroid tetrazoles.[593-598] The

[588] B. Stanovnik and M. Tisler, *J. Heterocycl. Chem.* **8**, 785 (1971).
[589] C. Temple, C. L. Kussner, and J. A. Montgomery, *J. Org. Chem.* **32**, 2241 (1967).
[590] (a) M. Maeck and R. Promel, *Abstr. Int. Congr. Heterocycl. Chem., 5th., 1975* p. 169; (b) V. V. Titov and L. F. Kozhokina, *ibid.* p. 177.
[591] T. Sasaki, S. Eguchi, and T. Toru, *J. Org. Chem.* **36**, 2454 (1971).
[592] D. Misiti and V. Rimatori, *Tetrahedron Lett.*, 947 (1970).
[593] A. Cervantes, P. Crabbe, J. Iriarte, and G. Rosenkranz, *J. Org. Chem.* **33**, 4298 (1968).
[594] J. Moural and K. Syhora, *Collect. Czech. Chem. Commun.*, **35**, 2018 (1970) [*CA* **73**, 66803 (1970)].
[595] H. Singh, R. B. Mathur, and P. P. Sharma, *J. Chem. Soc., Perkin Trans. 1*, 990 (1972).
[596] H. Singh, R. K. Malhotra, and N. K. Luhadiya, *J. Chem. Soc., Perkin Trans. 1*, 1480 (1974).
[597] H. Singh and P. Dharam, *Indian J. Chem.* **12**, 1211 (1974) [*CA* **82**, 86475 (1975)].
[598] H. Singh, R. K. Malhotra, and V. R. Parashar, *Tetrahedron Lett.*, 2587 (1973).

latter are generally obtained from the Schmidt reaction with ketones, and recently an interesting azidonitrile product (**322**), arising from

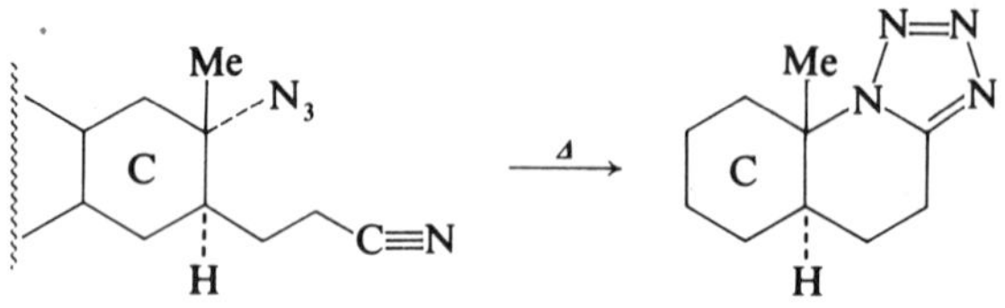

(**322**)   (**323**)

cleavage of the D ring, was obtained by treating 4-androstene-3,17-dione with HN_3.[598] This cyclized to the tetrazole form (**323**) on heating, in an interesting reaction involving intramolecular cycloaddition.[598] Tetrazoloazonines have also been obtained from the Schmidt reaction with 2,4,6-cyclo-octatrien-1-one in trifluoroacetic acid.[599] Compounds of type **324** have also been prepared and found to exist in the tetrazole form despite the influence of the tetrazole-1-N-carbonyl group.[600-602]

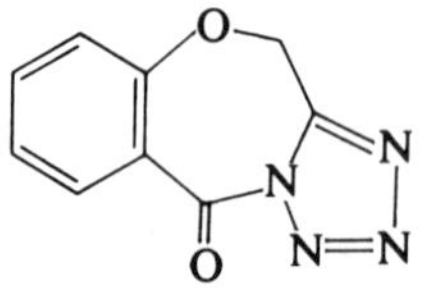

(**324**)

This discussion of the azidoazomethine moiety in cyclic molecules has been concerned mainly with the synthesis and structure of these systems. The reactions of the fused heterocycles at sites removed from the azomethine unit are outside the scope of this work. Some aspects of these general reactions have, however, been discussed in a recent review.[8]

V. Conclusion

It is hoped that this review demonstrates the range and depth of interest shown in the chemistry of the tetrazoles in recent years. The im-

[599] A.-H. Khuthier and J. C. Robertson, *J. Org. Chem.* **35**, 3760 (1970).
[600] J. G. B. Howes and R. A. Selway, German Patent 1,806,546 (1969) [*CA* **71**, 70606 (1969)].
[601] J. G. B. Howes and R. A. Selway, German Patent 1,954,839 (1970) [*CA* **73**, 25546 (1970)].
[602] J. G. B. Howes and R. A, Selway, German Patent 1,806,546 (1969) [*CA* **71**, 112939 (1969)].

portance of tetrazole chemistry is, we believe, further underlined by the emergence of the cyclic tetrazene compounds of type **325** with

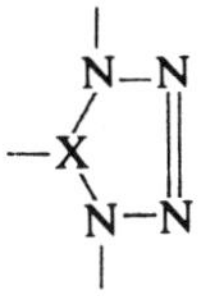

(**325**) 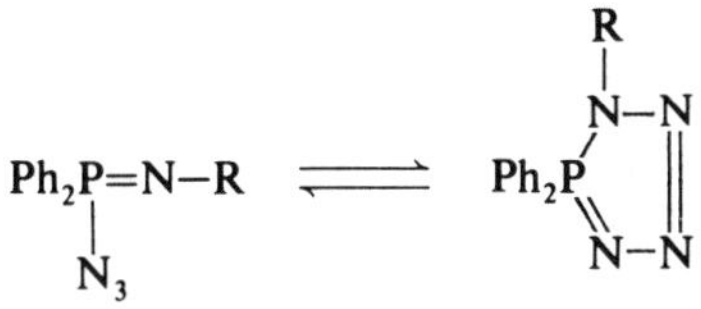   (**326**)

$X = $ boron,[603] and transition metals,[604,605] and by the possibility of developing phosphorus systems of type **326**.[606] All of these represent separate areas of study, which should ultimately relate in various ways to the carbon analogs, the tetrazoles, first discovered over 90 years ago.[607]

Notes Added in Proof

These notes cover the papers published between September 1975 and August 1976. Reviews of the synthesis and reactions of 2-aryltetrazoles[608] and the azidoazomethine tetrazole isomerism[609] have been published during this time.

Section II,A

Comparisons of ^{1}H and ^{13}C NMR spectra of 1-aryl- and 2-aryltetrazoles have indicated a considerably higher extent of interannular conjugation in the 2-aryl derivatives.[610] This has been ascribed to a loss of coplanarity in the 1-aryl derivatives due to steric interaction between the tetrazole 5-H atom and the ortho proton of the aryl substituent.[610] Comparison of J_{CH}^{13} (DMSO) at the 5-position for tetrazole

---

[603] N. N. Greenwood and J. H. Morris, *J. Chem Soc.*, 6205 (1965).
[604] F. W. B. Einstein, A. B. Gilchrist, G. W. Rayner-Canham, and D. Sutton, *J. Am. Chem. Soc.* **93**, 1826 (1971).
[605] R. J. Doedens, *Chem. Commun.*, 1271 (1968).
[606] H. Bock and W. Wiegrabe, *Chem. Ber.* **99**, 1068 (1966).
[607] J. A. Bladin, *Ber.* **18**, 1544 (1885).
[608] E. Lippmann and A. Konnecke, *Z. Chem.* **16**, 90 (1976).
[609] V. Ya. Pochinok, L. F. Avramenko, T. F. Grigorenko, and V. N. Shopenko, *Usp. Khim.* **44**, 1028 (1975) [*CA* **83**, 164021 (1975)].
[610] A. Konnecke, E. Lippmann, and E. Kleinpeter, *Tetrahedron* **32**, 499 (1976).

(216 Hz), 1-phenyltetrazole (216 Hz), and 2-phenyltetrazole (211 Hz) has been quoted as further evidence that tetrazole exists in solution predominantly as the 1-H tautomer.[610] Intermolecular association in concentrated solutions of azoles, including methyltetrazole derivatives, has been detected by Begtrup[611] using ^{13}C NMR. The values for J_{CH}^{13} (acetone) reported by this worker[611] for tetrazole (217.9 Hz), 1-methyltetrazole (218.1 Hz), and 2-methyltetrazole (213.5 Hz) agree well with those of Konnecke et al.[610] ^{13}C NMR has also been applied to the tautomerism of 1-phenyl-5-hydroxy- and 5-thioltetrazoles.[612] The tetrazolinone forms were confirmed as the predominating species in both cases.[612] Proton and ^{13}C NMR spectra of N-acetyltetrazole derivatives have also been reported.[613] Nitrogen-14 chemical shifts for some N-methyltetrazole and triazole derivatives have been compared[614] and ^{13}C NMR spectra of some 1,4-disubstituted tetrazoline-5-thiones have also been described.[615]

SECTION II,B

Mass spectral fragmentation of the 1-phenyltetrazolines (**327**) involved loss of the PhNCX moiety.[616] Thus for X = O and S the fragments PhN=C=O and PhN=C=S were lost while for X = NHR^1

$$\text{Ph}-\text{N} \underset{N=N}{\overset{\overset{X}{\|}}{\underset{}{\diagdown}}} \text{N}-\text{R}$$

(**327**)

phenylcarbodiimide fragments were eliminated.[616] Mass spectral fragmentation of isomeric 1- and 2-N-aryltetrazoles has been found suitable for distinguishing between isomeric pairs.[617] In both cases the fragmentation began by elimination of N$_2$ followed by HCN, but different abundances allowed for a distinction between the isomers.[617] Mass spectral fragmentations of a number of substituted tetrazolo[1,5-c]pyrimidines have also been reported.[618]

[611] M. Begtrup, *J. Chem. Soc., Perkin Trans. 2*, 763 (1976).
[612] A. Konnecke, E. Lippmann, and E. Kleinpeter, *Z. Chem.* **15**, 402 (1975) [*CA* **84**, 73399 (1976)].
[613] A. Konnecke, E. Lippmann, and E. Kleinpeter, *Tetrahedron Lett.*, 533 (1976).
[614] A. Holm, K. Schaumburg, N. Dahlberg, C. Christophersen, and J. P. Snyder, *J. Org. Chem.* **40**, 431 (1975).
[615] G. L'abbe, G. Vermeulen, J. Flemal, and S. Toppet, *J. Org. Chem.* **41**, 1875 (1976).
[616] N. W. Rokke, J. J. Worman, and W. S. Wadsworth, *J. Heterocycl. Chem.* **12**, 1031 (1975).
[617] A. Konnecke and E. Lippmann, *Org. Mass. Spectrosc.* **11**, 167 (1976).
[618] J. C. Tou, *J. Heterocycl. Chem.* **11**, 707 (1974).

Section II,C

The complex (**328**) has been obtained from *trans*-cyanohalogeno-bis-(triphenylphosphine) Pt(II) by halide displacement with azide ion followed by cycloaddition of acetonitrile.[619] Above 150°C fluxional exchange of the coordinated N atom around the tetrazole ring was observed.[619] New dimeric π-allyl complexes of Pd and Pt with tetrazolate groups as bridging ligands bonded at the 2-N and 3-N atoms (**329**) have

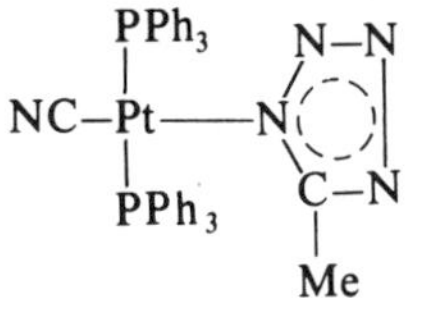

$$\begin{array}{c} \text{PPh}_3 \quad \text{N-N} \\ | \quad \quad / \\ \text{NC-Pt---N} \\ | \quad \quad \backslash \\ \text{PPh}_3 \quad \text{C-N} \\ \quad \quad \quad | \\ \quad \quad \quad \text{Me} \end{array} \qquad [(\pi\text{-C}_3\text{H}_5)\text{PdCN}_4\text{CF}_3]_2$$

(**328**)   (**329**)

been obtained by treating the azido complexes $[(\pi\text{-C}_3\text{H}_5) \text{PdN}_3]_2$ with CF_3CN.[620] Intermolecular association of 2-(tri-*n*-butylstannyl)-tetrazoles in benzene solution has been examined and found to be polymeric in nature involving the tetrazole 1-N and 3-N atoms.[621] The crystal structure data of the ylide 5-[(3-chlorobenzyl)dimethylammonio]-tetrazolide have been published.[622]

Section II,D

Photolysis of 1,4-dimethyltetrazolin-5-ones of type (**327**) (X = S, O, NMe) gave elimination of nitrogen and yielded diaziridinones and the carbodiimide MeN=C=NMe.[623] With X = S, the carbodiimide only was formed along with sulfur.[623]

Section III,A

Tetrazole has been obtained in 94.5–97% yields by hydrolysis and decarboxylation of ethyl tetrazol-5-ylcarboxylate.[624] Series of 5-aryltetrazoles have been prepared by treating arylnitriles with sodium

---

[619] W. Beck and K. Schorpp, *Chem. Ber.* **108**, 3317 (1975).
[620] L. Busetto and A. Palazzi, *Inorg. Chim. Acta* **13**, 233 (1975).
[621] S. Kozima, T. Hitomi, T. Akiyama, and T. Isida, *J. Organmetal. Chem.* **32**, 303 (1975).
[622] G. B. Ansell, *J. Chem. Soc., Perkin Trans. 2*, 1200 (1975).
[623] H. Quast and L. Bieber, *Angew. Chem.* **87**, 422 (1975).
[624] A. Konnecke and E. Lippmann, *Z. Chem.* **16**, 53 (1976) [*CA* **85**, 21225 (1976)].

azide.[625,626] A series of aliphatic and aromatic aldehyde tetrazol-5-ylhydrazones has been reported from diazotization of 5-aminotetrazole to bistetrazol-5-yltriazine followed by reductive cleavage to tetrazol-5-yl hydrazine which was condensed with the aldehydes.[627]

An interesting oxidative degradation of 5-substituted aminotetrazoles, $RNHCN_4H$, to isocyanides, R—N=C: [cf. Eq. (9) and Scheme 10] has also been reported using a variety of oxidizing agents.[628] These included sodium hypobromite in water; lead tetraacetate in dichloromethane containing triethyl amine; bromine-triethylamine; or anodic oxidation.[628] Studies of ring contractions in the carbene–nitrene rearrangement observed in the pyrolysis of 2-(tetrazol-5-yl)-pyridine have been carried out using a ^{13}C label at the tetrazole 5-position.[629] *t*-Butylation of the 1-N and 2-N atoms of 5-substituted tetrazoles has also been achieved.[629a]

Section III,A,2

Tetrazole analogs of D-alanyl-D-alanine have been synthesized as potential inhibitors of bacterial cell wall biogenesis.[630] Tetrazole analogs of phenylalanine have also been reported.[631] The tetrazole ribonucleoside 1-β-D-ribofuranosyltetrazole and the corresponding tetrazol-5-ylcarboxamido and -acetamido derivatives have been synthesized.[632] The β-configuration was confirmed by X-ray crystallography.[632] The compounds have been tested as potential agents against the influenza A2/Asian/J-305 virus.[632] A series of copper chelates has been presented as a unique class of potentially more therapeutically useful antiarthritic agents having both antiinflammatory and anti-ulcer activities.[633] A number of these involve some 1-substituted 5-aminotetrazole compounds as bridging ligands.[633] Further new active tetrazole derivatives that have

---

[625] A. Antonowa and S. Hauptmann, *Z. Chem.* **16**, 17 (1976) [*CA* **85**, 21226 (1976)].
[626] B. Decroix, P. Dubus, J. Morel, and P. Pastour, *Bull. Soc. Chim. Fr.,* 621 (1976).
[627] V. P. Schipanov, A. I. Zabalotskaya, and R. A. Badryzlova, *Khim. Geterotsikl. Soedin.* 850 (1975) [*CA* **83**, 178943 (1975)].
[628] G. Höfle and B. Lange, *Angew. Chem., Int. Ed. Engl.* **15**, 113 (1976)
[629] C. Thetaz and C. Wentrup, *J. Am. Chem. Soc.* **98**, 1258 (1976).
[629a] R. A. Henry, *J. Heterocycl. Chem.* **13**, 391 (1976).
[630] E. E. Smissman, A. Terada, and S. El-Antably, *J. Med. Chem.* **19**, 165 (1976).
[631] K. Brewster and R. M. Pinder, *Eur. J. Med. Chem.-Chim. Ther.* **10**, 117 (1975) [*CA* **83**, 164087 (1975)].
[632] M. S. Poonian, E. F. Nowoswiat, J. F. Blount, T. H. Williams, R. G. Pitcher, and M. J. Kramer, *J. Med. Chem.* **19**, 286 (1976).
[633] J. R. J. Sorensen, *J. Med. Chem.* **19**, 135 (1976).

been reported include diuretics,[634] antiallergic agents,[635] plant growth inhibitors,[636] and antiinflammatory agents.[637]

Section III,B

The reaction of substituted isocyanates with organic azides [Eq. (19)] has been extended to obtain vinyltetrazolines by using β-chloroethyl azide, $N_3CH_2CH_2Cl$, in the cycloaddition followed by elimination of HCl from the tetrazoline product.[638] Treatment of 5-aminotetrazole with vinyl acetate at 80°–100° in acetic acid containing $Hg(OAc)_2$ gave 5-amino-1-vinyltetrazole.[639] Perfluoroalkyltetrazoles have been obtained by treating perfluoroimidoyl chlorides, e.g. $CF_3-CCl=NCF(CF_3)_2$, with

$$Me_2N^+=C-Cl.Cl^- \quad \xrightarrow{RNH_2} \quad Me_2N-C \underset{N-N}{\overset{N-N}{\diagdown\!\!\!\diagup}}$$
$$\underset{N_3}{|} \qquad\qquad\qquad\qquad\qquad \underset{R}{|}$$

(330)   (331)

azide ion.[640] The tetrazoles (**331**) have been obtained by treating the immonium salts (**330**) with primary amines.[641] Some 1-(o-carboxyphenyl)-5-styryltetrazoles have been reported from the reaction of 2-substituted styryl-3,1-benzoxazin-4-ones with sodium azide.[642]

1-Benzoyltetrazole has proved to be a useful benzoylating agent for hydroxy groups in nucleosides.[643] Some selectivity between primary and

---

[634] J. H. Biel, D. S. Bariana, and A. K. L. Fung, Ger. Offen. 2,508,852 (1975) [*CA* **84**, 59478 (1976)].

[635] A. Nohara, T. Umetani, and Y. Sanno. Japan Kokai 75 29566 (1975) [*CA* **83**, 79252 (1975)].

[636] E. F. George and W. D. Riddell, U.S. Patent 3,865,570 (1975) [*CA* **83**, 54601 (1975)]; British Patent 1,381,840 (1975) [*CA* **82**, 170970 (1975)].

[637] J. Hollowood and J. F. Cavalla, British Patent 1,381,860 (1975) [*CA* **82**, 170971 (1975)].

[638] G. Denecker, G. Smets, and G. L'abbe, *Tetrahedron* **31**, 765 (1975); cf. also G. L'abbe, *Bull. Soc. Chim. Fr.*, 1127 (1975).

[639] V. A. Chuiguk, U.S.S.R. Patent 504,772 (1976) [*CA* **85**, 33019 (1976)].

[640] K. E. Peterman and J. M. Shreeve, *J. Fluorine Chem.* **6**, 83 (1975).

[641] H. G. Viehe and P. George, *Chimia* **29**, 209 (1975) [*CA* **83**, 178940 (1975)].

[642] N. N. Messiha, A. M. M. Abdel-Kadar, and M. M. Nosseir, *Indian J. Chem.* **13**, 326 (1975).

[643] J. Stawinski, T. Hozumi, and S. A. Narang, *Chem. Commun.* 243 (1976).

secondary hydroxy groups and with the primary amino groups of deoxycytidine, deoxyadenosine, and deoxyguanosine was also observed.[643] Some 1-arylsulfonyltetrazoles have been prepared and investigated as agents for building the phosphotriester bond between protected oligonucleotides.[644] Thermolysis or photolysis of the 1-(o,o-disubstituted phenyl)-5-phenyltetrazoles (332) interestingly gave the products (333)–(335).[645] N-Arylbenzimidoyl nitrenes, which cyclize to the ortho position giving 3aH-benzimidazole intermediates, have been

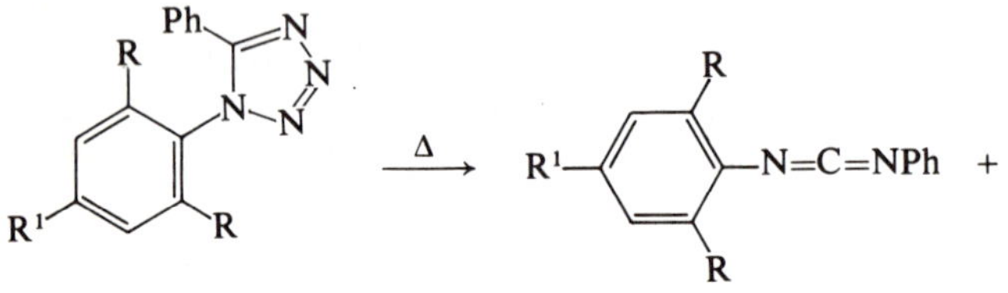

(332)   (333)

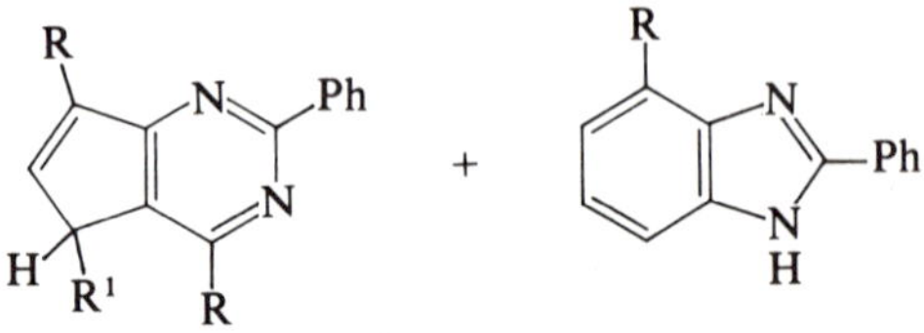

(334)   (335) R = MeO

invoked to explain the formation of products (334) and (335).[645] Thermolysis of β-substituted α-(1-tetrazolyl) acrylamides yielded 2,4-disubstituted imidazole-5-carboxamides.[646] Alkylation of 1-phenyl-5-mercaptotetrazole at the sulfur atom with α-bromocarbonyl compounds has also been reported.[647] Derivatives of 1-phenyltetrazol-5-ylmethylene bromide have been obtained from the parent compound, which was itself prepared by treating the potassium salt of 1-phenyltetrazol-5-ylcarboxylic acid with $BrCH_2COCl$.[648] Addition reactions of 1-phenyl-5-mercaptotetrazole with α,β-unsaturated carbonyl compounds have also been carried out.[649]

[644] J. Stawinski, T. Hozumi, and S. A. Narang, *Can. J. Chem.* **54**, 670 (1976).
[645] T. L. Gilchrist, C. J. Moody, and C. W. Rees, *Chem. Commun.* 414 (1976).
[646] J. Lykkeberg and B. Jerselv, *Acta Chem. Scand., Ser. B* **29**, 793 (1975).
[647] E. Lippmann and D. Reifegerste, *Z. Chem.* **15**, 351 (1975) [*CA* **85**, 21219 (1976)].
[648] M. O. Lozinskii, A. F. Shivanyuk, V. N. Bodnar, and P. S. Pel'kis, *Zh. Org. Khim.* **12**, 915 (1976) [*CA* **85**, 32928 (1976)].
[649] E. Lippmann and D. Reifegerste, *Z. Chem.* **15**, 146 (1975) [*CA* **83**, 97142 (1975)].

Section III,C

Synthesis of 2-substituted tetrazole-5-carbaldehydes has been reported by two groups [650,651] using reactions similar to those described in Scheme 19 and employing the corresponding 2-substituted tetrazole derivatives. A wide series of 2,5-diaryltetrazoles has been prepared in yields of 16–70% by treating phenylsulfonylhydrazonyl chlorides, p-RC_6H_4–CCl=N–$NHSO_2Ph$, with arylhydrazines.[652] The reaction may involve an intermediate formazan which cyclized with elimination of phenylsulfinic acid (cf. Scheme 21).[652]

Section III,D

An interesting thermal isomerization of the *meso*-ionic 2,3-diphenyl compounds (**336**) into the 2,4-diphenyl isomers (**337**) has been reported.[653] The structure of compound (**337b**) was confirmed by X-ray crystallography.[653] Addition reactions of dehydrodithizone (**336b**) with

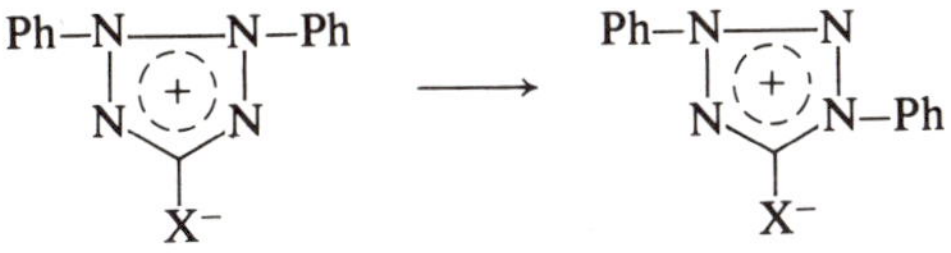

Ph–N——N–Ph   Ph–N——N
    (+)   ⟶   (+)
N⟍⟋N   N⟍⟋N–Ph
  X⁻   X⁻

(**336**)   (**337**)
a: X = O   a: X = O
b: X = S   b: X = S

pentacarbonyliron giving a 1,3,4-thiadiazol-2-one[654] and with the enamines (**338**) giving the compounds (**339**)[655] have been reported. These reactions may involve the acyclic valence isomer of dehydrodithizone. Synthesis of pyridinium 1- and 2-phenyltetrazol-5-ylmethylides and their alkylation and acylation reactions have

---

[650] E. Lippmann, A. Konnecke, and G. Beyer, *Monatsh. Chem.* **106**, 437 (1975) [*CA* **83**, 97139 (1975)]; **106**, 443 (1975) [*CA* **83**, 97140 (1975)]; *Z. Chem.* **15**, 102 (1975) [*CA* **83**, 28159 (1975)].

[651] D. Moderhack, *Chem. Ber.* **108**, 887 (1975).

[652] S. Ito, Y. Tanaka, and A. Kakehi, *Bull. Chem. Soc. Jpn.* **49**, 762 (1976).

[653] P. N. Preston, K. K. Tiwari, K. Turnbull, and T. J. King, *Chem. Commun.* 343 (1976).

[654] P. N. Preston, N. J. Robinson, K. Turnbull, and T. J. King, *Chem. Commun.* 998 (1974).

[655] G. V. Boyd, T. Norris, and P. F. Lindley, *Chem. Commun.* 100 (1975); cf. also *J. Chem. Soc., Perkin Trans. 1*, 1673 (1976).

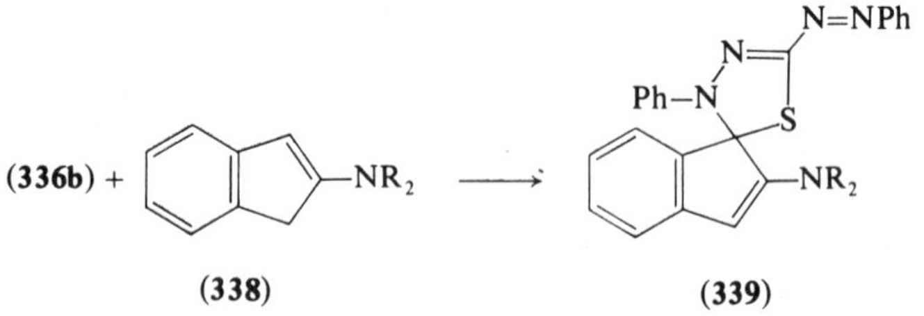

(336b) + [indene-NR₂ (338)] ⟶ [spiro thiadiazole structure (339)]

described.[656] Cycloaddition of alkyl azides with alkyl nitrilium salts gave 1,4-dialkyltetrazolium salts.[657]

Section IV,A

An interesting theoretical study of the azidoazomethine tetrazole isomerism has been published.[658] It has been suggested that in the reaction path the heavy atoms remain in the same plane, and the transition state strongly resembles the starting azide rather than the tetrazole. The formation of the activated complex is considered to involve a movement of the imino lone pair (which forms the new σ bond) toward the terminal N-atom of the azido group, while simultaneously a lone pair is forming on the central nitrogen atom from the N—N π bond. It was predicted that polar solvents would stabilize the transition state which was more polarized than the starting azide.[658] The involvement of azide intermediates in the synthesis of tetrazoles via the reactions of imidoyl chlorides with inorganic azides has been discussed.[659] Solvation effects on this reaction have been examined and dipolar aprotic solvents were found to give clean high yield reactions in contrast to the sluggish low yield reactions observed with aqueous and protic solvents.[659] The greater reactivity of the azide anion, which was poorly solvated relative to the transition state, was considered to facilitate the reaction in dipolar aprotic solvents.[659]

Section IV,B

Fused tetrazolodiazoles, e.g. (**340**) and (**341**), have been obtained from methylation of the anions of azidodiazoles which exist in the

---

[656] A. R. Katritzky and D. Moderhack, *J. Chem. Soc., Dalton Trans.* 909 (1976).
[657] H. Quast and L. Bieber, *Tetrahedron Lett.* 1485 (1976).
[658] L. A. Burke, J. Elguero, G. Leroy, and M. Sana, *J. Am. Chem. Soc.* **98**, 1685 (1976).
[659] P. K. Kadaba, *J. Org. Chem.* **41**, 1073 (1976).

cyclized tetrazole form.[660] The influence of solvents on the azidotetrazole equilibrium of 2-azidothiazoles has been investigated.[661] A new route to heterocyclic azido-azomethines via aza-transfer reactions with heterocyclic hydrazines and benzene diazonium fluoroborate has been developed.[662] ^{15}N-labeling of the tetrazole nitrogen atoms in a range of tetrazoloazines using end-labeled potassium azide, K^{15}NN$_2$ has been achieved[663] and the labeled compounds have been used to study the ring contractions observed with nitrene intermediates derived from such systems.[664]

New fused tetrazoloazines that have been reported include tetrazolo[5,1-a]isoquinolines,[665] tetrazolo[5,1-b]benzothiazoles,[666] and tetrazolopyridines.[667] Reactions of fused tetrazoloazines, or the azido isomers, that have been reported include hydrolytic cleavage of the pyrimidine ring of tetrazolopyrimidines,[668] pyrolysis of 3-azidopyridazine 2-oxides,[669] and photolysis of azido-1,3,5-triazine.[670] Dipole moments of 3-azido-1,2,4-triazole derivatives have also been reported.[671]

---

[660] E. Alcalde and R. M. Claramunt, *J. Heterocycl. Chem.* **13**, 379 (1976).

[661] J. Elguero, R. Faure. J. P. Galy, and E. J. Vincent, *Bull. Soc. Chim. Belg.* **84**, 1189 (1975).

[662] B. Stanovnik, M. Tisler, S. Polanc, V. Kovacic-Bratina, and B. Spicer-Smolnikar, *Tetrahedron Lett.* 3193 (1976).

[663] C. Wentrup and C. Thetaz, *Helv. Chim. Acta* **59**, 256 (1976); C. Thetaz, F. W. Wehrli, and C. Wentrup, *Helv. Chim. Acta.* **59**, 259 (1976).

[664] R. Harder and C. Wentrup, *J. Am. Chem. Soc.* **98**, 1259 (1976).

[665] H. Reimlinger, W. R. F. Lingier, and J. J. M. Vandewalle, *Chem. Ber.* **108**, 3780 (1975).

[666] T. F. Grigorenko, L. F. Avramenko, V. Ya. Pochinok, and T. P. Naidenova, *Ukr. Khim. Zh. (Russ Ed.)* **41**, 1105 and 1222 (1975) [*CA* **84**, 121728, 73205 (1976)]; R. Faure, J. P. Galy, G. Giusti, E. J. Vincent, and J. Elguero, *Org. Magn. Reson.* **6**, 485 (1975) [*CA* **82**, 169850 (1975)].

[667] M. Lacan and K. Tabakovic, *Croat. Chem. Acta* **47**, 127 (1975) [*CA* **84**, 59320 (1976)].

[668] R. Nutiu and A. J. Boulton, *J. Chem. Soc., Perkin Trans. 1*, 1327 (1976).

[669] R. A. Abramovitch and I. Shinkai, *Chem. Commun.* 703 (1975).

[670] H. Yamada. H. Shizuka, and K. Matsui, *J. Org. Chem.* **40**, 1351 (1975).

[671] M. A. Pervozvanskaya, V. V. Mel'nikov, M. S. Pevzner, and B. V. Gidaspov, *Zh. Org. Khim.* **11**, 1974 (1975) [*CA* **84**, 4258 (1976)].

Section IV,C

The ditetrazolopyrazine (**342**) was formed when azidoacetonitrile, N_3CH_2CN, was heated at 125°C.[672] The compounds (**344**) were obtained from an intramolecular 1,3-dipolar cycloaddition in the compounds (**343**).[673] The tetrazolo[1,5-e]pyrazolo[1,5-a]-1,3,5-triazine

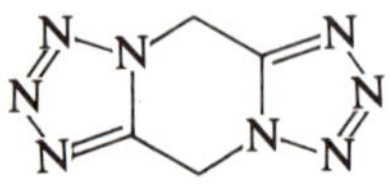

(**342**) 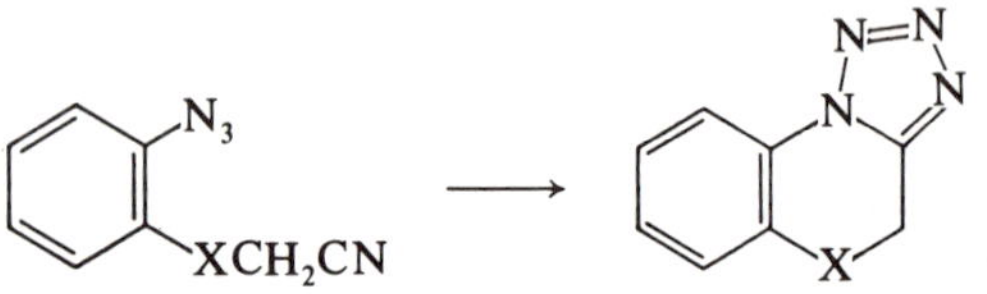   (**343**)   (**344**)

$X = O, CH_2$

system has been reported.[674] The azido form (**345**) was the predominant form but the tetrazolo form was also present in DMSO solution.[674] The preparation of some 1,2,5-thiadiazolo-[3,4-d]-tetrazolo[1,5-b]pyri-

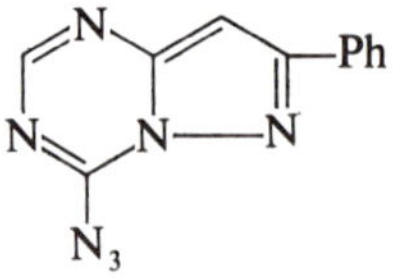

(**345**)

dazines has been described[675] and also the synthesis of 9-methyl-s-triazolo[4,3-c]tetrazolo[1,5-a]pyrimidine via the azide from the precursory hydrazine.[676] The naphthotetrazolo-1,2,4-triazine (**347**) has been obtained by treating β-naphthol with 5-diazotetrazole giving

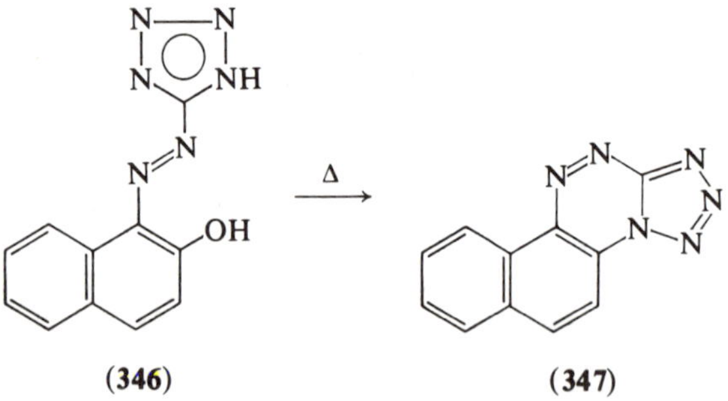

(**346**)   (**347**)

[672] J. H. Boyer, J. Dunn, and J. Kooi, *J. Chem. Soc., Perkins Trans. 1*, 1743 (1975).
[673] R. Fusco, L. Garanti, and G. Zecchi, *J. Org. Chem.* **40**, 1906 (1975).
[674] J. Kobe, D. E. O'Brien, R. K. Robins, and T. Novinson, *J. Heterocycl. Chem.* **11**, 991 (1974).
[675] A. Majcen, B. Stanovnik, and M. Tisler, *Vestn. Slov. Kem. Drus.* **21**, 23 (1974) [*CA* **83**, 79175 (1975)].
[676] T. La Noce and A. M. Giulani, *J. Heterocycl. Chem.* **12**, 551 (1975).

compound (346), which was then dehydrated by heating in a polar medium.[677] The tetrazolo form (347) (ring closure at 4-N)[677a] was the dominant form, but the azido form was also detected in chloroform and DMSO solutions.[677] The preferred cyclic structures of the compounds (345) and (347) are consistent with the trends noted (Section IV,C,4) for the azidoazomethine moiety in a triazine ring. Addition reactions of azidoazolopyridazines to unsymmetrical acetylenes[678] and the reactions of 1,8-naphthyridine azides with ethylacrylate[679] have also been described.

Several new steroidal tetrazoles from the Schmidt reaction of HN_3 with cholestanone derivatives have been reported.[680] The tetrazoles (348) have also been obtained from the Schmidt reaction of the corresponding lactams.[681]

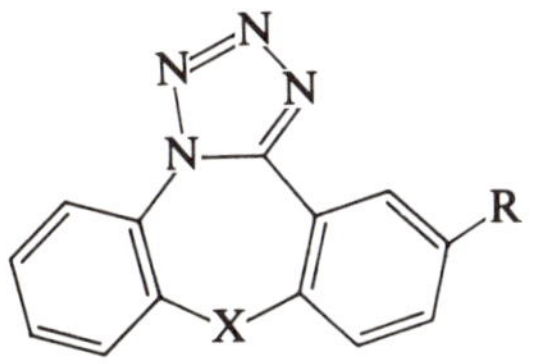

(348)

X = O, S, NMe

---

[677] J. Vilarrasa and R. Granados, *J. Heterocycl. Chem.* **11**, 867 (1974).

[677a] For discussion and controversy concerning the site of ring closure in tetrazolo-1,2,4-triazines, cf. M. M. Goodman, J. L. Atwood, R. Carlin, W. Hunter, and W. W. Paudler, *J. Org. Chem.* **41**, 2860 (1976).

[678] B. Mihelcic, S. Simonic, B. Stanovnik, and M. Tisler, *Croat. Chem. Acta* **46**, 275 (1974).

[679] O. Livi, P. L. Ferrarini, B. Bertini, and I. Tonnetti, *Farmaco, Ed. Sci.* **30**, 1017 (1975).

[680] H. Singh and R. K. Malhotra, *J. Chem. Soc., Perkin Trans. 1*, 1404 (1975); M. S. Ahmad, H. Z. Chaudry, and P. N. Khan, *Aust. J. Chem.* **29**, 447 (1976).

[681] L. S. Crawley and S. R. Safir, *J. Heterocycl. Chem.* **12**, 1075 (1975).

The Chemistry of 1,2-Dioxetanes

WALDEMAR ADAM

*Department of Chemistry, University of Puerto Rico,
Rio Piedras, Puerto Rico*

I. Introduction 438
II. Preparation 439
   A. Eliminative Cyclizations 439
      1. 1,2-Dioxetanes 439
      2. α-Peroxylactones 441
      3. 1,2-Dioxetanediones 443
   B. Cycloaddition 443
III. Characterization 448
   A. Nonspectroscopic Physical Methods 448
   B. Spectroscopic Methods 449
      1. Ultraviolet-Visible Spectra 449
      2. Chemiluminescence 449
      3. Photoelectron Spectra 450
      4. Mass Spectrometry 450
      5. Nuclear Magnetic Resonance (NMR) 450
      6. Infrared Spectra 450
   C. Chemical Methods 451
IV. Dioxetanes as Reactive Intermediates in Oxygenations . . . 451
   A. Chemical Oxygenation 451
      1. Alkenes 451
      2. Enamines 453
      3. Enol Ethers 456
      4. Thioenol Ethers 457
      5. Ketenes 457
   B. Biochemical Oxygenations 458
      1. Bioelectronic Processes 458
      2. Processes Not Involving Electronically Excited States . . . 463
V. Chemical Properties 463
   A. Thermolysis 463
      1. Kinetics 464
      2. Thermochemistry 466
      3. Efficiency and Selectivity of Excited-State Production . . . 467
      4. Mechanism 468
      5. Theoretical Considerations 472
      6. Chemienergized Transformations 473
   B. Photolysis 477
   C. Nucleophilic Substitution 478

I. Introduction

Among the four-membered ring cyclic peroxides, three of the possible types are known in the literature: 1,2-dioxetanes (**1**), 1,2-dioxetanones (**2**) or α-peroxylactones, and 1,2-dioxetanedione (**3**), all derivatives of the 1,2-dioxacyclobutane ring system. Only in recent years has it been possible to prepare, isolate, and characterize these thermally labile compounds. Thus, an authentic 1,2-dioxetane (**1**) was first prepared by Kopecky and Mumford,[1] while Adam and Liu[2] prepared the first α-peroxylactone (**2**). Although 1,2-dioxetanedione (**3**) has been claimed to be prepared[3] and isolated,[4] to date a full characterization is still lacking.

$$
\begin{array}{ccc}
\text{(1)} & \text{(2)} & \text{(3)}
\end{array}
$$

In the 1930s, these elusive compounds were invoked as reaction intermediates in oxygenation reactions; but little if any concrete evidence existed to substantiate these claims other than that they provided a convenient mechanistic interpretation. In fact, a number of the early reports were found to be erroneous.[5] For example, in the autoxidation of olefins the formation of 1,2-dioxetanes (**1**) were claimed by Stephens[6] and by Hock and Schrader.[7] However, subsequently Criegee and co-workers[8] showed that allylic hydroperoxides were produced instead of the originally proposed 1,2-dioxetanes (**1**). On the other hand, several decades earlier, as one of the earliest claims of 1,2-dioxetanes, Kohler[9] interpreted the oxidative cleavage of enol (**4**) to proceed via 1,2-dioxetanes [Eq. (1)]. Recent work[10] confirms these interpretations.

[1] R. R. Kopecky and C. Mumford, *Can. J. Chem.* **47**, 709 (1969).
[2] W. Adam and J. C. Liu, *J. Am. Chem. Soc.* **94**, 2894 (1972).
[3] M. M. Rauhut, *Acc. Chem. Res.* **2**, 80 (1969).
[4] J. Stauff, W. Jaesche, and G. Schlögl, *Z. Naturforsch. B* **27**, 1434 (1972).
[5] D. Swern, "Autoxidation and Antioxidants" (W. O. Lundberg, ed.), Vol. 1, p. 5. Interscience, New York, 1961.
[6] H. N. Stephens, *J. Am. Chem. Soc.* **50**, 568 (1928).
[7] H. Hock and O. Schrader, *Angew. Chem.* **39**, 365 (1936).
[8] (a) R. Criegee, *Justus Liebigs Ann. Chem.* **552**, 75 (1936); (b) R. Criegee, H. Pilz, and H. Flygare, *Chem. Ber.* **72**, 1799 (1939).
[9] E. P. Kohler, *Am. Chem. J.* **36**, 177 (1906).
[10] (a) W. H. Richardson, V. F. Hodge, D. L. Stiggall, M. B. Yelvington, and F. C. Montgomery, *J. Am. Chem. Soc.* **96**, 6652 (1974); (b) W. H. Richardson, G. Ranney, and F. C. Montgomery, *ibid.* **96**, 4688 (1974); (c) A. G. Pinkus, M. Z. Haq, and J. G. Lindberg, *J. Org. Chem.* **35**, 2555 (1970); (d) Y. Sawaki and Y. Ogata, *J. Am. Chem. Soc.* **97**, 6983 (1975).

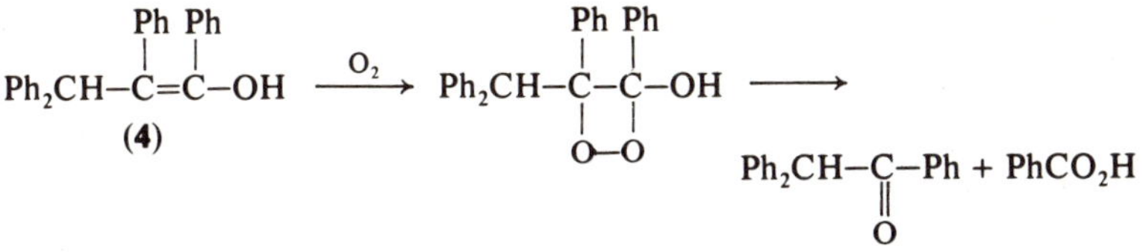

$$\underset{(4)}{Ph_2CH-\overset{\overset{Ph}{|}}{C}=\overset{\overset{Ph}{|}}{C}-OH} \xrightarrow{O_2} Ph_2CH-\overset{\overset{Ph}{|}}{\underset{\underset{O\!-\!O}{|}}{C}}-\overset{\overset{Ph}{|}}{\underset{\underset{}{|}}{C}}-OH \longrightarrow$$

$$Ph_2CH-\underset{\underset{O}{\|}}{C}-Ph + PhCO_2H \quad (1)$$

The most distinctive behavior of the 1,2-dioxetanes is the chemiluminescence that accompanies their thermal decomposition. Electronically excited carbonyl compounds are produced in the thermal decomposition of these "high energy" molecules and de-energization generates light.[11] It is the purpose of this chapter to review the existing literature on 1,2-dioxetanes with particular emphasis on their intriguing chemielectronic decomposition. This review is not intended to be encyclopedic in scope, but rather illustrative of the concepts involved. We shall begin with the preparative aspects of the three 1,2-dioxetane types, which have presented enormous synthetic challenges.

II. Preparation

Two principal methods have been successful in the preparation and isolation of 1,2-dioxetanes: eliminative cyclization [Eq. (2a)] and cycloaddition [Eq. (2b)].

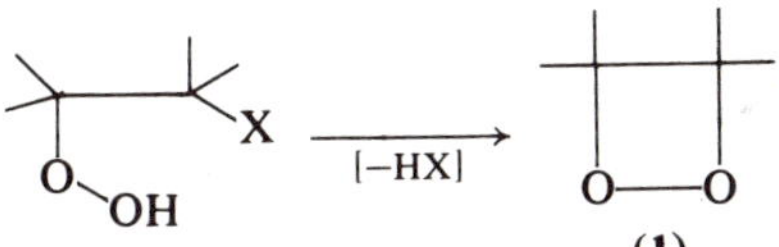

(2a)

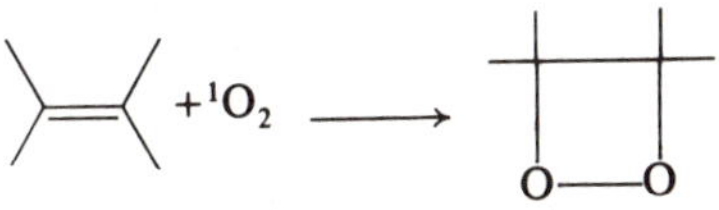

(2b)

A. Eliminative Cyclizations

1. *1,2-Dioxetanes*

This classical method, which is of historical importance since it led to the isolation of the first stable four-membered cyclic peroxide, was

---

[11] (a) W. Adam, *J. Chem. Educ.* **52**, 138 (1975); (b) N. J. Turro, P. Lechtken, N. E. Schore, G. Schuster, H.-C. Steinmetzer, and A. Yekta, *Acc. Chem. Res.* **7**, 97 (1974); (c) N. J. Turro, and P. Lechtken, *Pure Appl. Chem.* **33**, 363 (1973).

developed by Kopecky and co-workers.[1,12] In this synthesis the β-halohydroperoxide (5) is dehydrohalogenated to the 1,2-dioxetane (1) by base or silver ion catalysis, as illustrated for the trimethyl-1,2-dioxetane (1a), the first to be prepared and characterized [Eq. (3)]. The β-

$$\text{Me–C(Me)(O–OH)–CH(Me)–Br} \xrightarrow[\text{or AgOAc}]{\text{NaOH}} \text{Me–C(Me)(O–O)–C(Me)–H} \quad (3)$$

(5a)   (1a)

bromohydroperoxide (5a) is readily available by bromination of 2-methyl-2-butene with 1,3-dibromo-5,5-dimethylhydantoin in the presence of concentrated hydrogen peroxide.[12] Since then a number of 1,2-dioxetanes (1b–m) have been prepared by the Kopecky procedure[12-16] and are collected in Table I.

Significant is the preparation of 1,2-dioxetane (1c), since this system was suggested[17] to be formed in the ozonolysis of ethylidenecyclohexane [Eq. (4)] to explain the chemiluminescence observed on allowing

$$\text{ethylidenecyclohexane} \xrightarrow[\text{pinacolone}]{O_3} \text{(1c)} \longrightarrow \text{MeCHO} + \text{cyclohexanone} + h\nu \quad (4)$$

the ozonated solution to warm up. However, Kopecky's work[12] demonstrated that the dioxetane (1c) is stable at room temperature and should have accumulated in the ozonolysis experiment. Since we and others[18] could not repeat the work of Story and co-workers[17] and of

---

[12] (a) K. R. Kopecky, J. E. Filby, C. Mumford, P. A. Lockwood, and J. Y. Ding, *Can. J. Chem.* **53**, 1103 (1975); (b) K. R. Kopecky, P. A. Lockwood, J. E. Filby, and R. W. Reid, *ibid.* **57**, 468 (1973).

[13] B. S. Campbell, D. B. Denney, D. Z. Denney, and L. S. Shih, *J. Am. Chem. Soc.* **97**, 3850 (1975).

[14] (a) W. H., Richardson and V. F. Hodge, *J. Am. Chem. Soc.* **93**, 3996 (1971); (b) W. H. Richardson, M. B. Yelvington, and H. E. O'Neal, *ibid.* **94**, 1619 (1972); (c) W. H. Richardson, F. C. Montgomery, M. B. Yelvington, and H. E. O'Neal, *ibid.* **96**, 7525 (1974).

[15] T. R. Darling and C. S. Foote, *J. Am. Chem. Soc.* **96**, 1625 (1974).

[16] (a) E. H. White, P. D. Wildes, J. Wiecko, H. Doshan, and C. C. Wei, *J. Am. Chem. Soc.* **95**, 7050 (1973); (b) P. D. Wildes and E. H. White, *ibid.* **93**, 6286 (1971).

[17] P. R. Story, E. A. Whited, and J. A. Alford, *J. Am. Chem. Soc.* **94**, 2143 (1972).

[18] (a) P. S. Bailey, T. P. Carter, C. M. Fischer, and J. A. Thompson, *Can J. Chem.* **51**, 1298 (1973); (b) N. C. Yang and J. Libman, *J. Org. Chem.* **39**, 1782 (1974).

TABLE I
1,2-DIOXETANES (1) PREPARED BY ELIMINATIVE CYCLIZATION

$$\begin{array}{c} \text{O-O} \\ R^1 \underset{R^2}{\overset{}{|}} \underset{R^3}{\overset{}{|}} R^4 \end{array}$$

	Melting point (°C)	References
(1a) $R^1 = H; R^2 = R^3 = R^4 = Me$	5–7	12a, 16
(1b) $R^1 = R^2 = R^3 = R^4 = Me$	77–78	12a
(1c) $R^1 = H; R^2 = Me; R^3 + R^4 = \{CH_2\}_4$	−11	12a,b
(1d) $R^1 = R^4 = Me; R^2 + R^3 = \{CH_2\}_4$	48–50	12a
(1e) $R^1 + R^4 = R^2 + R^3 = \{CH_2\}_4$	34–36	12a
(1f) $R^1 = R^2 = H; R^3 = R^4 = Me$	—	13, 14a
(1g) $R^1 = R^2 = H; R^3 = Me; R^4 = Ph$	—[a]	14b, 16
(1h) $R^1 = R^2 = H; R^3 = R^4 = Ph$	—[a]	14c
(1i) $R^1 = R^2 = H; R^3 = R^4 = CH_2Ph$	60–61	14c
(1j) $R^1 = R^2 = Me; R^3 = R^4 = n\text{-Bu}$	—[a]	15
(1k) $R^1 = R^4 = H; R^2 = R^3 = Me$	—[a]	16
(1l) $R^1 = R^3 = H; R^2 = R^4 = Me$	—[a]	16
(1m) $R^1 = H; R^2 = R^3 = Me; R^4 = Et$	—[a]	16

[a] Not isolated in pure form.

Yang and Carr,[19] it is questionable whether ozonization of olefins constitutes a feasible method of preparing 1,2-dioxetanes.

2. α-Peroxylactones

These exceedingly unstable substances, which are invoked as the active intermediates in bioluminescence,[20] were first prepared and isolated by Adam and Liu.[2] The *t*-butyl system (**2a**) was the originally synthesized derivative by dehydrative cyclization of the α-hydroperoxy acid (**6**) by dicyclohexylcarbodiimide (DCC) [Eq. (5)]. It was not possible to isolate the pure material in view of its great thermal instability; but its characteristic carbonyl band at 1875 cm^{-1} in the infrared (IR) served as unequivocal structure identification of these novel cyclic peroxides.

$$\underset{(6)}{\begin{array}{c} R^1 \\ R^2 \end{array} \underset{O-OH}{\overset{O}{\diagup}} OH} \xrightarrow[{[-H_2O]}]{DCC} \underset{(2)}{\begin{array}{c} R^1 \\ R^2 \end{array} \underset{O-O}{\overset{O}{\square}}} \qquad (5)$$

a: $R^1 = H; R^2 = t\text{-Bu}$ **c:** $R^1 = R^2 = t\text{-Bu}$ **e:** $R^1 + R^2 =$ (adamantyl)
b: $R^1 = R^2 = Me$ **d:** $R_1 = H; R^2 = 1\text{-Ad}$

[19] N. C. Yang and R. V. Carr, *Tetrahedron Lett.* 5143 (1972).
[20] W. Adam, *Chem. Unserer Zeit* **7**, 182 (1973).

Since our original publication,[2] we have improved the method by conducting the cyclization at −78° in methylene chloride followed by flash distillation. Solutions of pure α-peroxylactone without decomposition products (checked by IR) can be obtained under optimized conditions.[21] This method has served to prepare the dimethyl (**2b**) and di-*tert*-butyl (**2c**) systems. However, for relatively involatile derivatives as the adamantyl (**2d**) and adamantylidene (**2e**) derivatives, which have been prepared but not isolated,[22] flash distillation is not feasible. For such cases we are developing[23] polymer-supported carbodiimides[24] as cyclants.

A considerable difficulty in the preparation of α-peroxylactones (**2**) has been the availability of the precursors, i.e., the α-hydroperoxy acids (**6**). In view of their acid and base sensitivity, we devised the novel process shown in Eq. (6).[2] On treatment of the ketene acetal (**7**) with

$$R_2C(H)-CO_2H \xrightarrow[\text{(ii) Me}_3\text{SiCl}]{\text{(i) }i\text{-Pr}_2\text{NLi}} R_2C=C\begin{matrix}OSiMe_3\\OSiMe_3\end{matrix} \xrightarrow{^1O_2}$$

(**7**)

$$R_2C\begin{matrix}O\\|\\O-SiMe_3\end{matrix}\begin{matrix}C\\\end{matrix}OSiMe_3 \xrightarrow{\text{MeOH}} R_2C\begin{matrix}O\\|\\O\\OH\end{matrix}\begin{matrix}C\\\end{matrix}OH \quad (6)$$

(**6**)

singlet oxygen, the oxygenophilic trimethylsilyl group undergoes a silatropic shift,[25] introducing the peroxy substituent adjacent to the carbonyl group, at the same time protecting the carboxy and peroxy functions in the form of silyl derivatives. Desilylation with methanol affords the pure α-hydroperoxy acid in high yield. However, in the case of ketene acetals with allylic hydrogen, the normal ene-reaction[26] competes, reducing the generality of this method. In such cases we recommend[27] direct oxygenation of the α-lithiocarboxylate (**8**) at very low temperature.

[21] W. Adam, A. Alzérreca, and F. Yany, unpublished work.
[22] W. Adam and H.-C. Steinmetzer, *Angew, Chem.* **84**, 590 (1972).
[23] W. Adam and F. Yany, unpublished work.
[24] N. M. Weinschenker and C. M. Shen, *Tetrahedron Lett.* 3285 (1972).
[25] G. M. Rubottom and M. I. Lopez Nieves, *Tetrahedron Lett.* 2423 (1972).
[26] W. Adam, *Chem. Ztg.* **99**, 142 (1975).
[27] W. Adam, O. Cueto, and V. Ehrig, *J. Org. Chem.* **43**, 370 (1976).

3. *1,2-Dioxetanediones*

Treatment of oxalyl chloride with concentrated hydrogen peroxide under base catalysis in the presence of fluorescers leads to bright chemiluminescence.[28] This rather unusual observation was also demonstrated for other oxalyl derivatives (9)[29] and interpreted[3] to involve 1,2-dioxetandione (3) as the intermediate [Eq. (7)]. Although isolation of the carbon dioxide dimer (3) was claimed,[4] no characteristic spectral data could be observed.[16] Furthermore, the claim[30] of detecting

$$X-\underset{\underset{X}{\|}}{C}-\underset{\underset{}{\|}}{C}-X \xrightarrow[\text{base}]{H_2O_2} \begin{array}{c} O \diagdown \diagup O \\ C—C \\ | | \\ O—O \end{array} \quad (7)$$

(9)   (3)

$X = -Cl, -OAr, -NR_2$

the carbon dioxide dimer by mass spectrometry of the effluent gas formed in the perhydrolysis of oxalyl esters was shown to be erroneous.[31] Consequently, it is still tentative whether the chemiluminescence of this system is due to the postulated 1,2-dioxetanedione intermediate.

B. CYCLOADDITION

The direct cycloaddition of singlet oxygen, either photochemically or chemically, to double bonds has so far been successfully demonstrated only for the simple 1,2-dioxetanes (1). The claim[32] that ketenes and singlet oxygen give α-peroxylactones directly is not substantiated with concrete experimental data. Attempts to singlet oxygenate di-*tert*-butyl- and bis(trifluoromethyl)ketenes did not afford isolable α-peroxylactones; but weak chemiluminescence could be observed when the nitrogen purged solutions were allowed to warm up directly in the photometer.[33]

---

[28] E. A. Chandross, *Tetrahedron Lett.* 761 (1963).

[29] (a) A. G. Mohan and N. J. Turro, *J. Chem. Educ.* **51**, 528 (1974); (b) L. J. Bollyky, R. H. Whitman, and B. G. Roberts, *J. Org. Chem.* **33**, 4266 (1968); (c) D. R. Maulding, R. A. Clarke, B. G. Roberts, and M. M. Rauhut, *ibid.* **33**, 250 (1968); (d) L. J. Bollyky, R. H. Whitman, B. G. Roberts, and M. M. Rauhut, *J. Am. Chem. Soc.* **89**, 6523 (1967); (e) M. M. Rauhut, L. J. Bollyky, G. G. Roberts, M. Loy, R. H. Whitman, A. V. Iannotta, A. M. Semsel, and R. A. Clarke, *ibid.* **89**, 6515 (1967).

[30] H. F. Cordes, H. P. Richter, and C. A. Heller, *J. Am. Chem. Soc.* **91**, 2709 (1969).

[31] J. J. De Corp, A. Baronovski, M. V. McDowell, and F. E. Saalfeld, *J. Am. Chem. Soc.* **94**, 2879 (1972).

[32] L. J. Bollyky, *J. Am. Chem. Soc.* **92**, 3230 (1970).

[33] W. Adam, J.-C. Liu, and F. Yany, unpublished work.

The first 1,2-dioxetanes prepared in this way were the *cis*-1,2-diethoxy (**1n**) and *trans*-1,2-diethoxy (**1o**) derivatives, obtained by photosensitized singlet oxygenation of the respective 1,2-diethoxyethylenes.[34] The reaction is a stereospecific *cis*-addition.[34a,c,d] However, when the same substrates are treated with triphenylphosphine ozonide, the *cis*-1,2-dioxetane (**1n**) is formed preferentially from either cis- or trans-olefin.[34] A stepwise ionic mechanism was postulated to explain this stereoselective process [Eq. (8)].

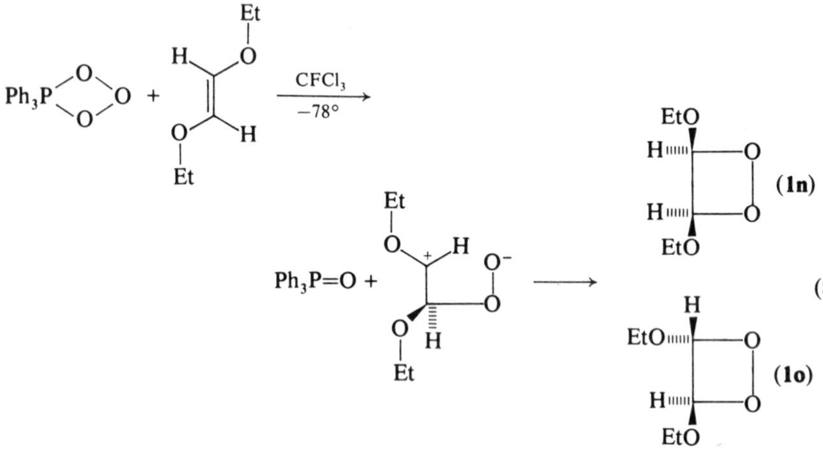

$$\text{(8)}$$

Since the original report of photosensitized singlet oxygenation of olefins, a range of 1,2-dioxetanes (**1n–1ah**) have been prepared, which are summarized in Table II. When allylic hydrogens are present, the classical ene-reaction with singlet oxygen[26] is a serious side reaction. However, when the allylic hydrogens are on bridgeheads, as in diadamantylidene (**10**), the corresponding 1,2-dioxetane (**1z**) is formed

---

[34] (a) P. D. Bartlett and A. P. Schaap, *J. Am. Chem. Soc.* **92**, 3223 (1970); (b) A. P. Schaap and P D. Bartlett, *ibid.* **92**, 6056 (1970); (c) A. P. Schaap and N. Tontapanish, *Pet. Reprints* **16**, A-78 (1971); (d) A. P. Schaap, *Tetrahedron Lett.* 1757 (1971).

[35] G. Rio and J. Berthelot, *Bull. Soc. Chim. Fr.*, 3555 (1971).

[36] (a) S. Mazur and C. S. Foote, *J. Am. Chem. Soc.* **92**, 3225 (1970); (b) C. S. Foote, A. A. Dzakpasu, J. W.-P. Lin, *Tetrahedron Lett.* 1247 (1975); (c) P. A. Burns and C. S. Foote, *J. Am. Chem. Soc.* **96**, 4339 (1974).

[37] N. M. Hasty and D. R. Kearns, *J. Am. Chem. Soc.* **95**, 3380 (1973).

[38] (a) J. H. Wieringa, J. Strating, H. Wynberg, and W. Adam, *Tetrahedron Lett.* 169 (1972); (b) G. B. Schuster, N. J. Turro, H.-C. Steinmetzer, A. P. Schaap, G. Faler, W. Adam, and J. -C. Liu, *J. Am. Chem. Soc.* **97**, 7110 (1975).

[39] P. D. Bartlett and M. S. Ho, *J. Am. Chem. Soc.* **96**, 627 (1975).

[40] H. E. Zimmerman and G. E. Keck, *J. Am. Chem. Soc.* **97**, 3527 (1975).

TABLE II
1,2-DIOXETANES (1) PREPARED BY CYCLOADDITION

$$\underset{R^2 \; R^3}{R^1 \overset{O-O}{\overline{\underline{\quad\quad}}} R^4}$$

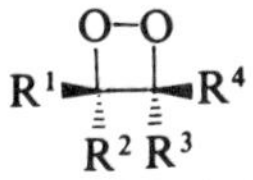		Melting point (°C)	References
(1n)	$R^1 = R^4 = H; R^2 = R^3 = OEt$	—[a]	34a,b
(1o)	$R^1 = R^3 = H; R^2 = R^4 = OEt$	—[a]	34a,b
(1p)	$R^1 = R^4 = H; R^2 = OEt; R^3 = OPh$	—[b]	34c
(1q)	$R^1 = R^3 = H; R^2 = OEt; R^4 = OPh$	—[b]	34c
(1r)	$R^1 = R^4 = H; R^2 + R^3 = -O-CH_2CH_2-O-$	—[b]	34b,d
(1s)	$R^1 = R^4 = H; R^2 + R^3 = -O-CH_2-O-$	—[b]	34d
(1t)	$R^1 = R^4 = Ph; R^2 = R^3 = OMe$	67–68	35
(1u)	$R^1 = R^3 = Ph; R^2 = R^4 = OMe$	59–61 (dec.)	35
(1v)	$R^1 = R^2 = R^3 = R^4 = OMe$	−9 to −8	36a
(1w)	$R^1 = H; R^2 = NMe_2; R^3 = R^4 = Me$	Liquid	36b
(1x)	$R^1 = H; R^2 = R^3 = Me; R^4 = $ 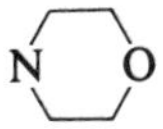 morpholino	−6 (epl.)	36b
(1y)	$R^1 = H; R^2 = $ 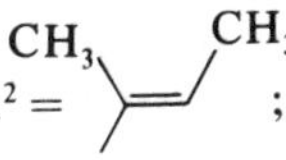 (CH_3)_2C=CH– ; $R^3 = R^4 = Me$	—[a]	37
(1z)	$R^1 + R^2 = R^3 + R^4 = $ adamantylidene	174–176	38
(1aa)	$R^1 + R^2 = R^3 + R^4 = $ 7-norbornylidene	129–130	39
(1ab)	$R^1 = R^4 = Ph; R^2 + R^3 = $ 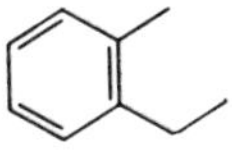 o-xylylene	47–51 (dec)	36c
(1ac)	$R^1 = Me; R^2 + R^3 = $ 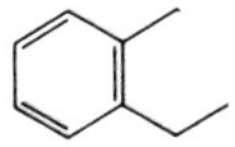 o-xylylene ; $R^4 = Ph$	69–70 (dec)	36c
(1ad)	$R^1 = i$-$Pr; R^2 + R^3 = $ 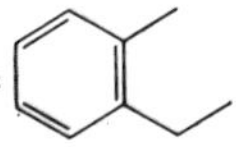 o-xylylene ; $R^4 = H$	49.5–51	36c
(1ae)	$R^1 = Me; R^2 + R^3 = $ 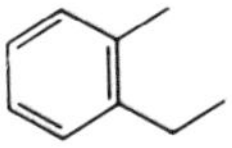 o-xylylene ; $R^4 = H$	56–58	36c
(1af)	$R^1 = Me; R^2 = Ph; R^3 + R^4 = $ 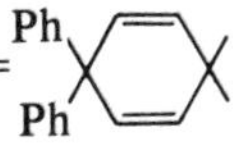 Ph,Ph-cyclohexylidene	88–92 (dec)	40
(1ag)	$R^1 = Me; R^2 = p$-$An; R^3 + R^4 = $ 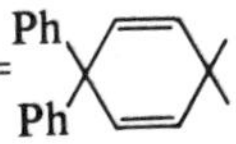 Ph,Ph-cyclohexylidene	Liquid	40
(1ah)	$R^1 = Me; R^2 = \beta$-$Naph; R^3 + R^4 = $ 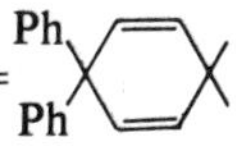 Ph,Ph-cyclohexylidene	89–94 (dec)	40

[a] Crystalline solid, but no melting point reported.
[b] Not isolated in pure form.

in high yield [Eq. (9)].[38] The ene-reaction with an allylic hydrogen at a bridgehead would lead to a highly strained bridgehead olefin, and consequently 1,2-cycloaddition is preferred. A similar case constitutes the singlet oxygenation of 7,7'-dinorbornylidene, affording the 2,2-dioxetane (1aa) in high yield.

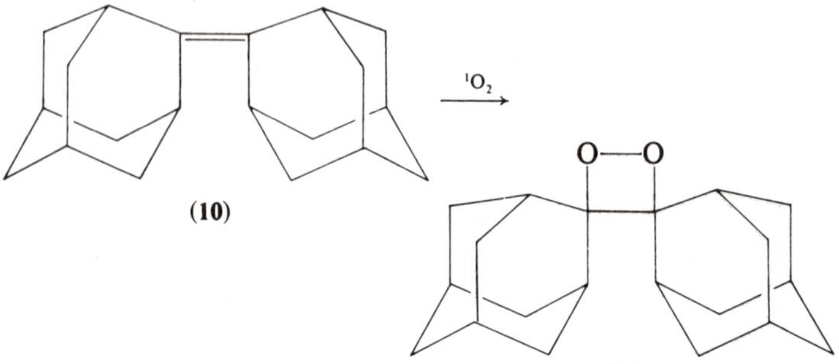

(9)

The claim by Rigaudy and co-workers[41] that diene (11) afforded on singlet oxygenation the unusually stable (m.p. 181°) bis-1,1-dioxetane (12), the only example of its kind, has recently been refuted. It was shown that the tetraoxa-octalin (13) was formed [Eq. (10)].[42]

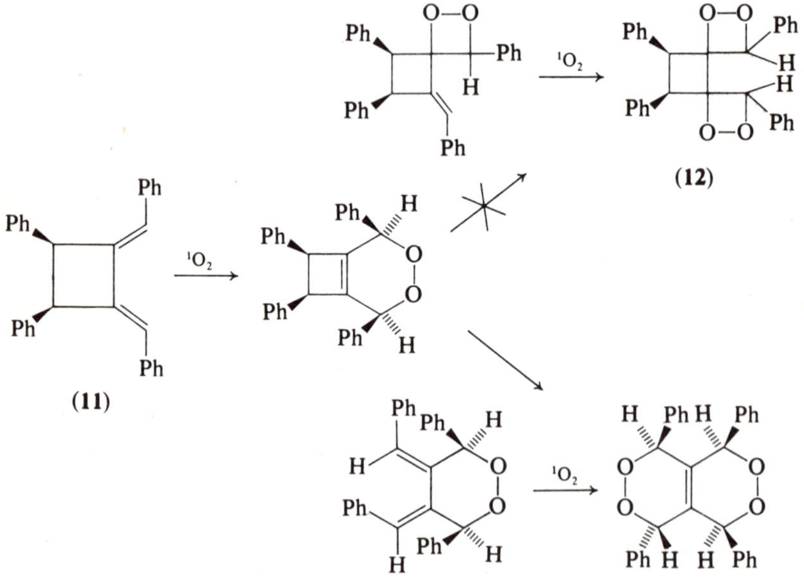

(10)

[41] J. Rigaudy, P. Capdevielle, and M. Maumy, *Tetrahedron Lett.*, 4997 (1972).
[42] (a) E. Friedrich, W. Lutz, and D. Seebach, private communication; (b) E. Friedrich, Diplom-Arbeit. Univ. Giessen, Giessen, Germany, 1975.

Sec. II.B]   THE CHEMISTRY OF 1,2-DIOXETANES

The mechanism of 1,2-dioxetane formation via singlet oxygenation is still rather perplexing despite intensive efforts.[26] Besides the simple concerted 1,2-cycloaddition via the 2s:2a transition state (14) proposed by Bartlett and Schaap,[34] in order to account for the high degree of stereospecificity [Eq. (11a)], alternative serious contenders are the intervention of the perepoxide intermediate (15),[37] shown in Eq. (11b), or the electron transfer process leading to the zwitterion (17) via the caged radical pair (16), as suggested by Foote and co-workers[36b] [Eq. (11c)]. The last mechanism probably applies to electron-rich olefins, such as enamines and vinyl ethers, and is hardly expected to be general.

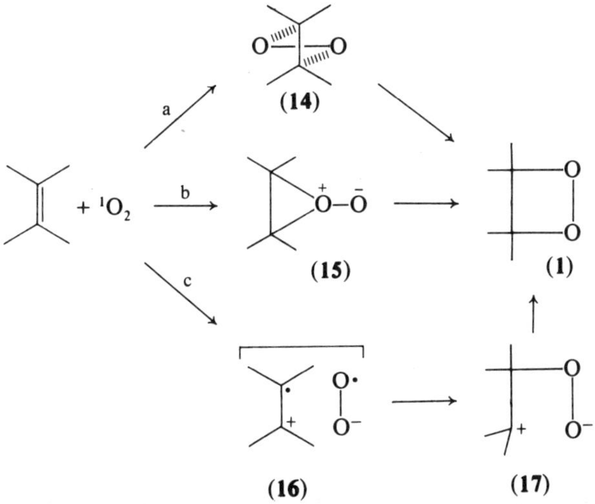

$$ \text{(11)} $$

The trapping experiment of Schaap and Faler[43] in the photooxygenation of diadamantylidene (10) lends support to the perepoxide mechanism [Eq. (11b)]. In the presence of pinacolone, the epoxide (18) is formed in addition to the 1,2-dioxetane (1z), as shown in Eq. (12).

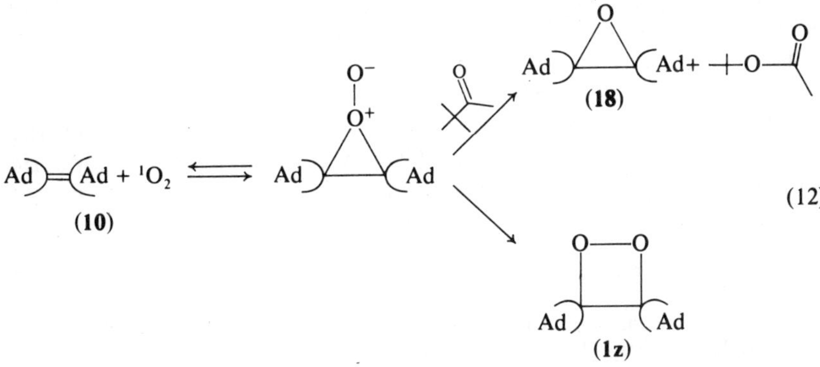

$$ \text{(12)} $$

[43] A. P. Schaap and G. R. Faler, *J. Am. Chem. Soc.* **95**, 3381 (1973).

However, conflicting evidence has been reported by Bartlett and Ho,[39] who observed 1,2-dioxetane as well as epoxide formation in the singlet oxygenation of 7,7'-dinorbornylidene even in the absence of pinacolone. In fact, in benzene as solvent the epoxide:dioxetane ratio was 19. Using MINDO/3 calculations, Dewar and co-workers[44] suggested that the epoxide is formed by direct oxygen atom transfer of the perepoxide to the starting olefin. Recent solvent and secondary isotope effects in enol ether photooxygenation suggest[45] essentially a 2s:2a transition state (**14a**) with dipolar character, which readily converts to a perepoxide transition state (**15a**) by a rocking motion [Eq. (13)].

$$\text{(15a)} \rightleftharpoons \text{(14a)} \equiv \text{(14a)} \rightleftharpoons \text{(15a)} \quad (13)$$

Further challenging work will be necessary to resolve the mechanism of 1,2-dioxetane formation.

III. Characterization

A. Nonspectroscopic Physical Methods

In the case of relatively stable and crystalline 1,2-dioxetanes, which can be isolated by crystallization and elution chromatography on silica gel,[36,40] the melting points (Tables I and II) have served as good criteria of purity. However, in some cases, e.g., derivatives (**1u**), (**1x**), (**1ab**), (**1ac**), (**1af**) and (**1ah**), decomposition points are involved, which limits this criterion for purity and structure confirmation. Where liquid dioxetanes are obtained, purification other than column chromatography is not feasible because of the thermal lability of these materials.

Of crucial importance have been molecular weight determinations in order to differentiate macrocyclic and polymeric compounds from the four-membered ring cyclic product. Usually cryoscopic and osmometric techniques are employed for this purpose,[12,16,34-40] using such diverse solvents as C_6H_6, $CHCl_3$, CH_2Cl_2, Freon-11, CH_2Br_2, and CH_3NO_2.

---

[44] (a) M. J. S. Dewar, A. C. Griffin, W. Thiel, and I. J. Turchi, *J. Am. Chem. Soc.* **97**, 4439 (1975); (b) M. J. S. Dewar and W. Thiel, *ibid.* **97**, 3978 (1975).

[45] (a) A. A. Frimer and P. D. Bartlett, *Abstr. IUPAC Congr. 25th, 1975*; (b) A. A. Frimer, Ph.D. Thesis, Harvard Univ. 1974.

These classical characterization methods have been used even for 1,2-dioxetanes, which cannot be obtained as pure, crystalline solids, for example (**1w**).[36b]

B. SPECTROSCOPIC METHODS

1. Ultraviolet-Visible Spectra

With few exceptions, 1,2-dioxetanes are yellow. They exhibit a λ_{max} at about 280 nm, with low extinction coefficients ($\epsilon \sim 20$).[11b, 12] The origin of this band is still not certain, but it most likely corresponds to the $\pi_y^* \to \pi_z^*$ excitation of the oxygen–oxygen bond, as suggested by theoretical work[108e] and photon–electron spectra.[46] The yellow color is due to tail-end absorption even out to 450 nm. One of the exceptions is the diadamantylidene-1,2-dioxetane (**1z**), which is white and has its λ_{max} at 265 nm ($\epsilon = 21.5$).[38]

2. Chemiluminescence

As pointed out already, one of the most characteristic properties of the four-membered ring cyclic peroxides is their ability to chemiluminesce on thermal decay into carbonyl products [Eq. (14)].

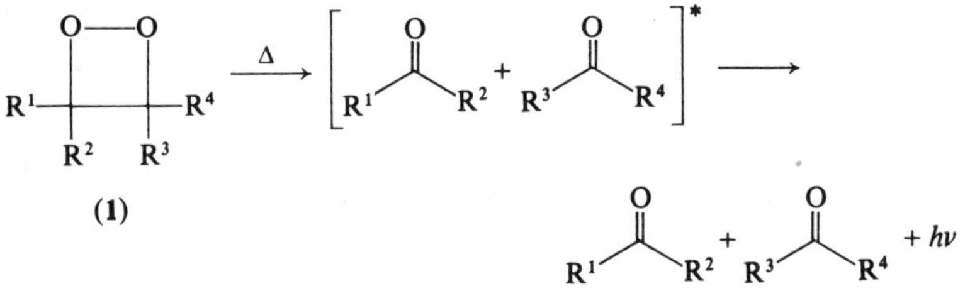

$$\underset{(1)}{\underset{R^2\quad R^3}{R^1 {\vdash\!\!\!\!\!\!\overset{O\!-\!O}{\square}\!\!\!\!\!\!\dashv} R^4}} \xrightarrow{\Delta} \left[\underset{R^1}{\overset{O}{\|}}\!\!\!\diagup\!\!\!\underset{R^2}{} + \underset{R^3}{\overset{O}{\|}}\!\!\!\diagup\!\!\!\underset{R^4}{} \right]^* \longrightarrow \underset{R^1}{\overset{O}{\|}}\!\!\!\diagup\!\!\!\underset{R^2}{} + \underset{R^3}{\overset{O}{\|}}\!\!\!\diagup\!\!\!\underset{R^4}{} + h\nu \quad (14)$$

The light emission, which ranges at about 420 ± 10 nm, corresponds to the fluorescence of the carbonyl product.[1b, 11b, 12a, 14b, 37, 38] Intriguing is the tetramethyl system (**1b**), which exhibits emission at 400 nm in nondegassed and at 430 nm in degassed solutions of acetonitrile. The former has been attributed to fluorescence, the latter to phosphorescence.[11b, 47]

---

[46] R. S. Brown, *Can. J. Chem.* **53**, 3439 (1975).
[47] N. J. Turro, H.-C. Steinmetzer, and A. Yekta, *J. Am. Chem. Soc.* **95**, 6468 (1973).

3. Photoelectron Spectra

Although an attempt was made to obtain a PE spectrum of the diadamantylidene dioxetane (**1z**),[48] it decomposed. However, recently Brown[46] reported the PE spectrum of the 1,2-dioxetane (**1b**). It exhibits vertical ionization potentials at 8.98, 10.94, 11.41, and 12.09 eV. The first IP was assigned to ionization from the π^* level of the peroxide linkage of **1b**, thus corroborating the UV-visible and fluorescence spectra.

4. Mass Spectrometry

Usually 1,2-dioxetanes are thermally too labile to permit observation of the parent ion by mass spectrometry. Under optimized conditions the parent ions have been detected for the dioxetanes **1b**,[12a] **1z**,[38] and **1aa**.[39]

5. Nuclear Magnetic Resonance (NMR)

One of the most powerful techniques to characterize 1,2-dioxetanes has been proton NMR. Thus, resonance of the dioxetane ring protons which are alkyl substituted are observed at 4.9–5.2 ppm.[12-14, 16] Dimethylamino substitution as in **1w** does not alter the chemical shift, since the ring proton resonance still occurs at 4.90 ppm.[36b] However, alkoxy substitution deshields the ring proton significantly, and the resonance occurs at 5.6–6.8 ppm.[34] Aryl and olefinic substitution also shift the resonance to lower field, occurring at 5.6–5.9 ppm.[14,36,37]

Carbon-13 resonance is also useful in structural assignments of 1,2-dioxetanes. The quaternary carbon occurs at 88.5–89.5 ppm downfield from TMS.[12,38] However, the adamantylidine system falls out of line and exhibits the resonance of the quaternary carbon at 95 ppm.[38]

6. Infrared Spectra

For the 1,2-dioxetanes the only characteristic band is the O—O deformation[12,34,36,38] at 845–895 cm^{-1}, but this band is weak and thus of limited value. Of great help is the characteristic carbonyl band at 1870 cm^{-1} for the α-peroxylactones.[2] It was not possible to observe an infrared spectrum of the carbon dioxide dimer **3**.[16a]

---

[48] C. Battich and W. Adam, *Tetrahedron Lett.* 1467 (1974).

C. Chemical Methods

Iodometric titration is one of the most direct methods for identifying 1,2-dioxetanes, since we are specifically analyzing for peroxide content. When the necessary precautions are taken to avoid thermal decomposition, many 1,2-dioxetanes titrate quantitatively.[12] However, the adamantylidene (**17**) does not react with iodide ion since it is too sterically hindered.[38] The product of I^- reduction of **1a** is the corresponding diol.[12]

Combustion analysis has also been applied successfully for characterizing 1,2-dioxetanes.[12,16,34,36,39,40] In view of the explosion hazard, all precautions should be taken.

Reduction with lithium aluminum hydride is a useful method of identification since the respective diols are formed.[12,34] Catalytic hydrogenation usually leads to decomposition of the dioxetanes into the corresponding carbonyl products.[12,38] This is not surprising since transition metals promote fragmentation of these labile materials.[49] In the case of dioxetane **1z**, the use of zinc in acetic acid was the only successful method for reducing it to its diol.[38]

IV. Dioxetanes as Reactive Intermediates in Oxygenations

In the preceding sections we discussed only those dioxetanes that were isolated and characterized. Yet, the literature is generously documented with oxygenations of olefinic substrates proceeding probably through 1,2-dioxetanes (**1**) in order to explain the carbonyl fragmention products and the accompanying chemiluminescence. This does not necessarily prove the intervention of the 1,2-dioxetane, but this interpretation constitutes a convenient mechanistic rationalization. Again, it is not our purpose to be exhaustive in our coverage, but rather selective and illustrative.

First, we will consider oxygenation of chemical substrates, including alkenes, enamines, enol ethers, and ketenes. Subsequently we shall take up the oxygenation of biological substrates.

A. Chemical Oxygenation

1. *Alkenes*

Evidence has finally appeared concerning the existence of the unsubstituted 1,2-dioxetane, i.e., that derived from ethylene. Bogan and co-

---

[49] P. D. Barlett, A. L. Baumstark, and M. E. Landis, *J. Am. Chem. Soc.* **96**, 5557 (1974).

workers,[50] have shown that singlet oxygenation of ethylene in the gas phase leads to formaldehyde fluorescence. It is postulated that the 1,2-dioxetane 1 is formed as an intermediate with excess vibrational energy.

In the photooxygenation of thujopsene (19), the formation of ketoaldehyde 20 is interpreted[51] as proceeding via the 1,2-dioxetane 1ai, as illustrated in Eq. (15). The cleavage of 9,9'-bifluorenylidene into fluorenone by singlet oxygen is thought to pass through the respective 1,2-dioxetane.[52]

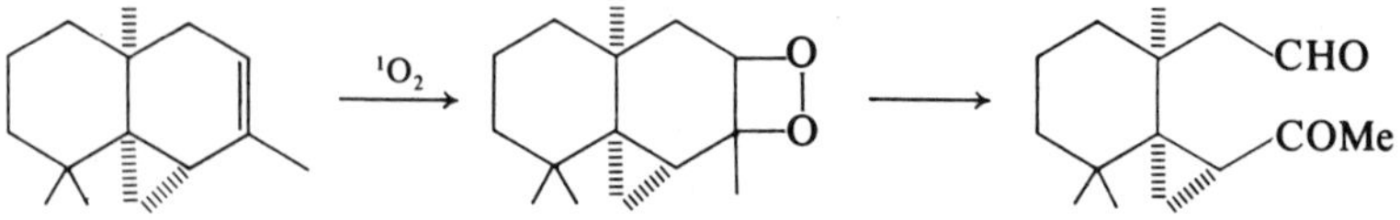

$$\text{(19)} \xrightarrow{^1O_2} \text{(1ai)} \longrightarrow \text{(20)} \quad (15)$$

While the formation of 1,3-dione (22) by singlet oxygenation of the cyclopropene 21a via dioxetane 1aj appears to be quite straightforward, the transformation of cyclopropene 21b into dione 23 has been postulated to involve rearrangement of the initial 1,2-dioxetane 1ak to 1ak', as shown in Eq. (16).[53] Even the bridged annulene 24 affords the

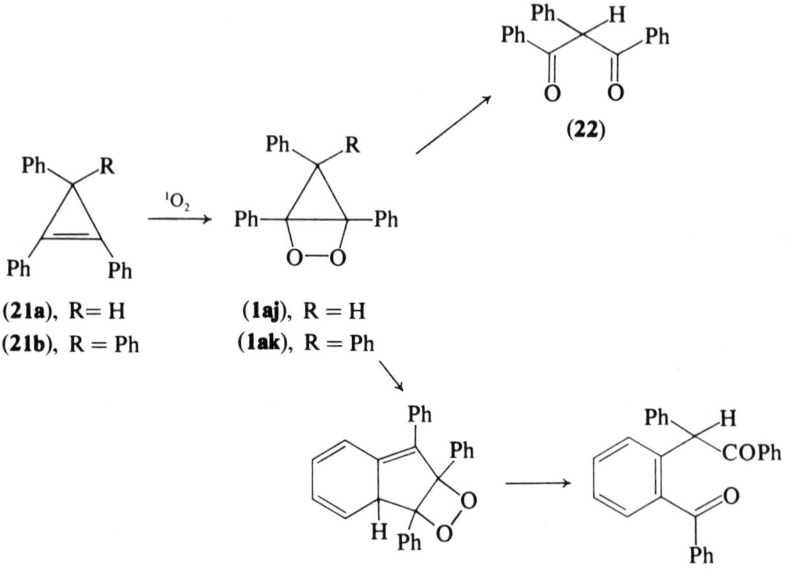

(21a), R = H
(21b), R = Ph

(1aj), R = H
(1ak), R = Ph

(22)

(1ak')

(23)

(16)

[50] D. J. Bogan, R. S. Sheinson, and F. W. William, *J. Am. Chem. Soc.* **98**, 1034 (1976).
[51] S. Ito, H. Takeshita, and M. Hirama, *Tetrahedron Lett.* 1181 (1971).
[52] W. H. Richardson and V. Hodge, *J. Org. Chem.* **35**, 1216 (1970).
[53] I. R. Politzer and G. W. Griffin, *Tetrahedron Lett.* 4775 (1974).

Sec. IV.A] THE CHEMISTRY OF 1,2-DIOXETANES 453

dialdehyde **25** as a side product on singlet oxygenation, presumably via the 1,2-dioxetane **1al**, as illustrated in Eq. (17).[54]

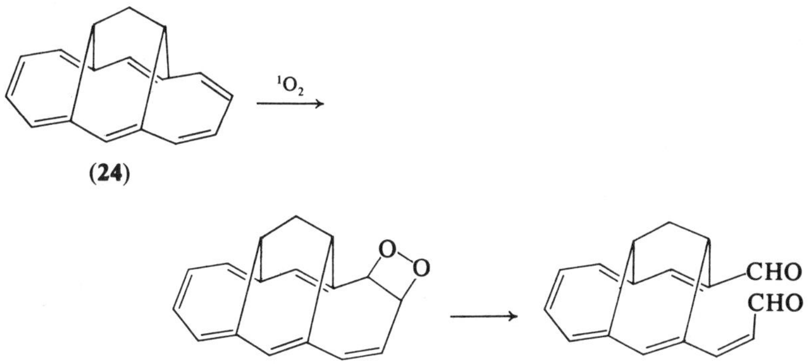

Although norbornene (**26**) was reported to be inert toward singlet oxygenation,[55] recently it has been shown that the cleavage product (**27**) is indeed formed.[55b] Presumably the dioxetane **1am** is the precursor [Eq. (18)]. Puzzling, however, is the formation of substantial quantities of *exo*-norbornene epoxide. Control experiments suggest that a perepoxide (**15**) is not involved.[55b]

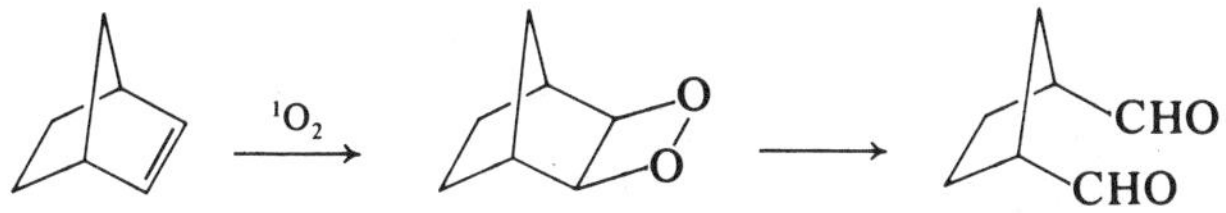

As we have pointed out already, the claim[17,19] that dioxetanes are formed in the ozonization of olefins has been refuted.[18] Recently, however, the chemiluminescence observed in the gas phase ozonization of 2-butene, which leads to complex cleavage products, has been explained in terms of intermediary 1,2-dioxetanes.[56]

2. *Enamines*

One of the early examples was the photosensitized oxygenation of piperidino derivatives (**28**), affording ketones and *N*-formylpiperidine

[54] E. Vogel, A. Alscher, and K. Wilms, *Angew. Chem.* **86**, 407 (1974).
[55] (a) F. A. Litt and A. Nickson, *Adv. Chem. Ser.* **77**, 118 (1968); (b) C. W. Jefford and A. F. Boschung, *Helv. Chim. Acta* **57**, 2257 (1974).
[56] B. J. Finlayson, J. N. Pitts, and R. Atkinson, *J. Am. Chem. Soc.* **96**, 5356 (1974).

(29) as fragmentation products [Eq. (19)]. Also more complex products are formed.[57] Recently Foote and co-workers[36b] were able to isolate dioxetanes such as **1an**. Similarly, the chemiluminescent oxygenation of

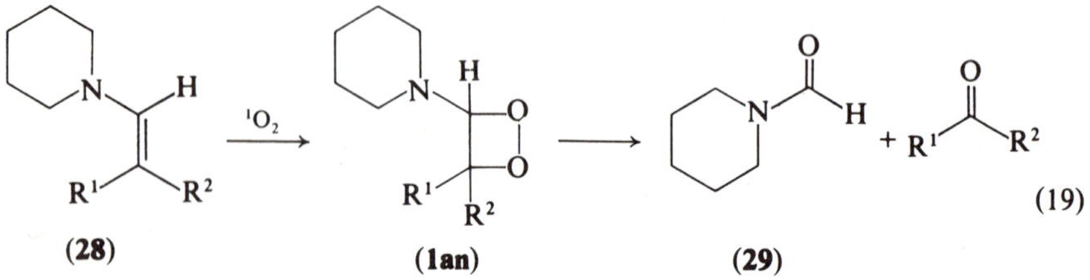

indole derivatives into amidoacetophenones involves 1,2-dioxetanes as intermediates.[58] Photodegradation of the dipyrrylmethene (**30**) and biliverdin (**31**) in the presence of oxygen is also suggested to proceed

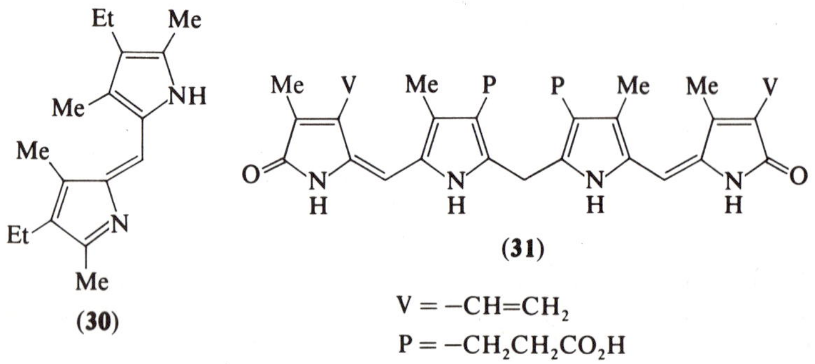

$V = -CH=CH_2$
$P = -CH_2CH_2CO_2H$

via 1,2-dioxetane intermediates.[59] An efficient chemiluminescent system involving dioxetane **1ao** is the singlet oxygenation of the biacridylidene (**32**) leading to N-methylacridone (**33**), as illustrated in Eq. (20).[60] Even 2,4-dimethoxy-N,N-dimethylaniline is singlet oxygenated to the inter-

---

[57] (a) C. S. Foote and J. W.-P. Lin, *Tetrahedron Lett.* 3267 (1968); (b) H. H. Wasserman and S. Terao, *ibid.*, 1735 (1975); (c) W. Ando, T. Saiki, and T. Migita, *J. Am. Chem. Soc.* **97**, 5028 (1975).

[58] (a) N. Sugiyama, M. Akutagawa, and H. Yamamoto, *Bull. Chem. Soc. Jpn.* **41**, 936 (1968); (b) N. A. Evans, *Aust. J. Chem.* **24**, 1971 (1971); (c) I. Saito, M. Imuta, S. Matugo, and T. Matsuura, *J. Am. Chem. Soc.* **97**, 719 (1975).

[59] (a) D. A. Lightner and D. C. Crandall, *Tetrahedron Lett.* 1799 (1973); (b) *Idem.*, 953 (1973); (c) J.-H. Fuhrhop, S. Besecke, J. Subramanian, C. Mengerson, and D. Riesner, *J. Am. Chem. Soc.* **97**, 7141 (1975).

[60] F. McCapra and R. A. Hann, *J. Chem. Soc. Chem. Commun.* 442 (1969).

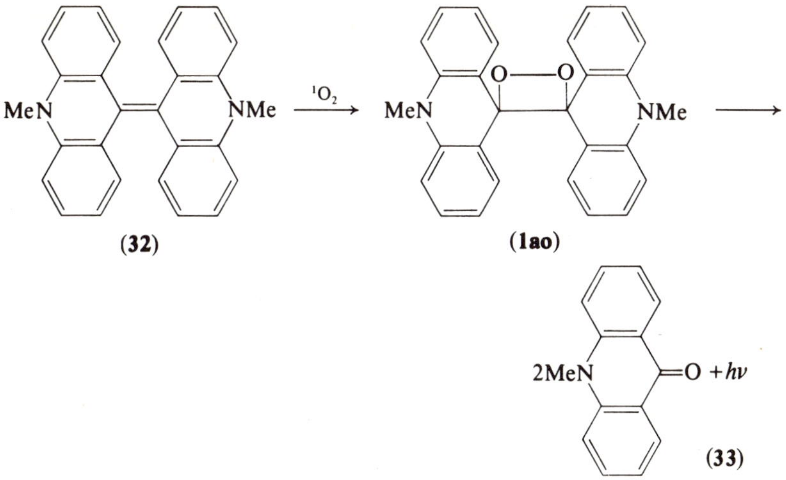

Sec. IV.A] THE CHEMISTRY OF 1,2-DIOXETANES 455

$$(32) \xrightarrow{^1O_2} (1ao) \longrightarrow$$

$$2 \, \text{MeN} \!\!=\!\!\text{O} + h\nu \quad (20)$$

$$(33)$$

mediary dioxetane **1ap**, which cleaves to give **34** [Eq. (21)].[61] This is the first example of 1,2-cycloaddition of singlet oxygen to a benzene derivative.

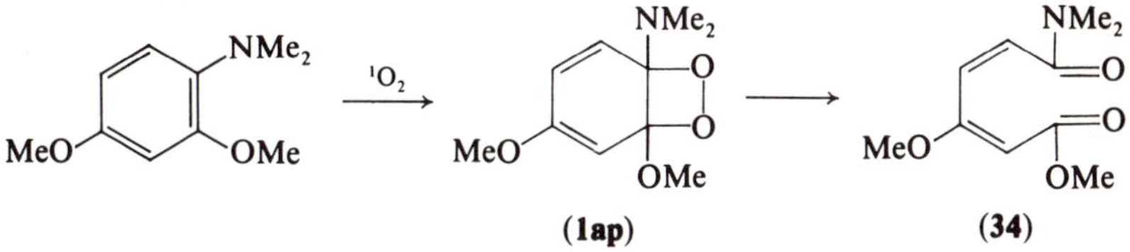

$$\xrightarrow{^1O_2} (1ap) \longrightarrow (34) \quad (21)$$

The chemiluminescent autoxidation of tetrakis(dimethylamino)-ethylene (**35**) is spectacular.[62] Ground-state oxygen is sufficiently reactive to produce the dioxetane **1aq**, presumably via electron transfer, radical cage coupling, and cyclization [Eq. (22)]. Besides tetramethyl-

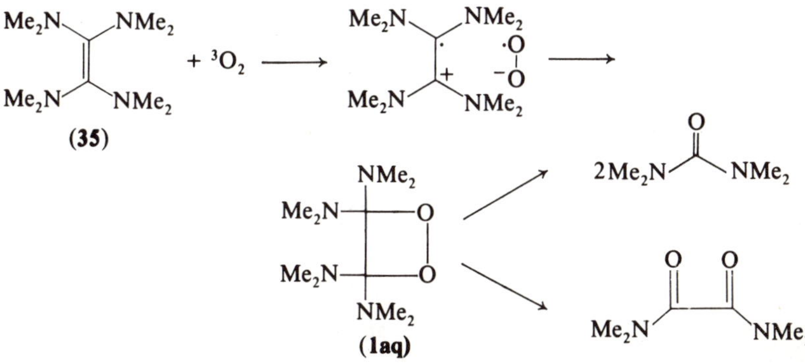

$$(35) + {}^3O_2 \longrightarrow \cdots \longrightarrow (1aq) \longrightarrow \quad (22)$$

---

[61] I. Saito, S. Abe, Y. Takahashi, and T. Matsuura, *Tetrahedron Lett.* 4001 (1974).
[62] (a) W. Carpenter and E. M. Bens, *Tetrahedron* **26**, 59 (1970); (b) W. H. Urry and J. J. Sheeto, *Photochem. Photobiol.* **4**, 1067 (1965); (c) J. P. Paris, *ibid.* **4**, 1059 (1965).

urea, tetramethyloxamide and tetramethylhydrazine are also formed. These are unusual dioxetane cleavage products since instead of ring fragmentation, dimethylamino substituents are cleaved off.

3. Enol Ethers

The gas phase singlet oxygenation of ethyl vinyl ether leads to formaldehyde chemiluminescence via the corresponding dioxetane.[63] The latter is formed in a highly vibrational excited state with a lifetime estimated to be less than 10^{-7} sec.

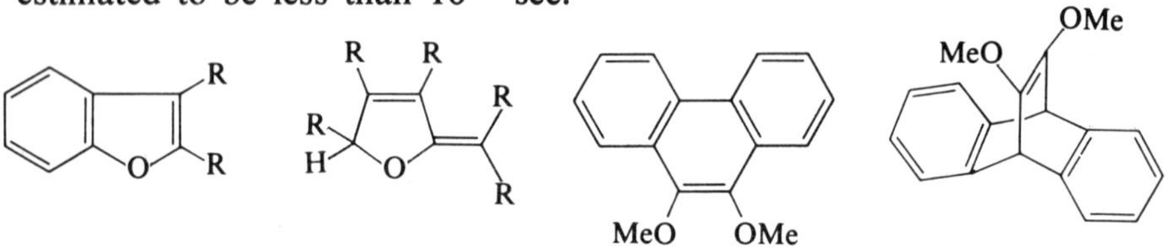

(36)   (37)   (38)   (39)

Numerous examples are known in solution, of which we shall mention only a few. For example, the benzofurans (36),[64a] the alkylidenedihydrofurans (37),[65] the phenanthrene (38),[64b] and the ethenoanthracene (39)[66] all afford dioxetane cleavage products with singlet oxygen. Also worth mentioning is the autoxidation of succinylfluorescein (40), leading to electronically excited xanthone (41) via dioxetane (1ar),[67] as illustrated in Eq. (23).

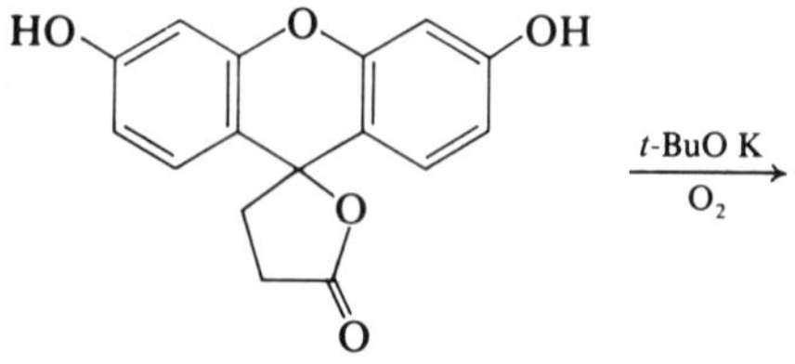

(40)

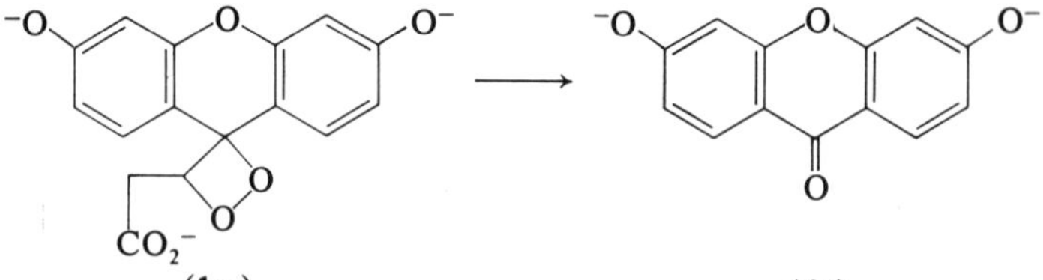

(1ar)   (41)   (23)

[63] P. J. Bogan, R. S. Sheinson, R. G. Gann, and F. W. Williams, *J. Am. Chem. Soc.* **97**, 2560 (1975).
[64] (a) G. Rio and J. Berthelot, *Bull. Soc. Chim. Fr.*, 1705 (1971); (b) 822 (1972).
[65] J. J. Basselier and J. P. Le Roux, *Bull. Soc. Chim. Fr.*, 4443 (1971).
[66] J. Font, F. Serratosa, and L. Vilarrasa, *Tetrahedron Lett.* 4105 (1970).
[67] I. Kamiya and K. Aoki, *Bull. Soc. Chem. Jpn* **47**, 1744 (1974).

4. Thioenol Ethers

Also thioalkyl-substituted olefins undergo 1,2-cycloaddition with singlet oxygen producing the corresponding dioxetanes. Thus, thioenol

$$\underset{(42)}{\overset{RS}{\underset{RS}{>}}\!\!=\!\!\overset{SR}{\underset{SR}{<}}} \xrightarrow{{}^1O_2} \underset{(1as)}{\overset{RS\ \ RS}{\underset{RS\ \ RS}{\square_O^O}}} \longrightarrow \underset{RS}{\overset{RS}{>}}\!\!=\!\!\underset{RS}{\overset{RS}{<}} + \underset{RS}{\overset{RS}{>}}\!\!\overset{O}{\underset{O}{\diagdown\!\!\!\diagup}}\!\!\underset{RS}{\overset{}{<}} \quad (24)$$

ethers[68] and 1,1-dithioethylenes[69] afford the normal dioxetane cleavage products, while the dioxetanes (1as) of the tetrathioethylenes (42) undergo the unusual cleavage observed for tetrakis(dimethylamino)ethylene [Eq. (22)], leading to dithiooxalates and disulfides [Eq. (24)].[69] Furthermore, singlet oxygenation of carbon–sulfur bonds has also been

$$\underset{(43)}{\overset{S}{\underset{R\ X\ R}{\bigcirc}}} \xrightarrow{{}^1O_2} \underset{(1at)}{\overset{S\diagup\!\!{}^O}{\underset{R\ X\ R}{\bigcirc}}} \xrightarrow{[-SO]} \underset{(44)}{\overset{O}{\underset{R\ X\ R}{\bigcirc}}} \quad X = O \text{ or } S \quad (25)$$

achieved. Thus, the thiones (43) are converted into the ketones (44) via the dioxathiete (1at),[70] as shown in Eq. (25).

5. Ketenes

Diphenylketene on singlet oxygenation affords benzophenone and carbon dioxide with light emission, suggesting the intermediacy of α-peroxylactone (2f).[32] We have demonstrated the presence of 2f by its characteristic infrared band and its direct chemiluminescence.[33] An analogous case is the photochemical conversion of friedelin (45) into 3,4-seco-3-nor-friedelan-2-al (47) by in situ singlet oxygenation of the intermediary ketene (46) into α-peroxylactone (2g) and subsequent

---

[68] W. Ando, K. Watanabe, J. Suzuki, and T. Migita, *J. Am. Chem. Soc.* **96**, 6766 (1974).
[69] (a) W. Adam and J.-C. Liu, *J. Chem. Soc., Chem. Commun.*, 73 (1972); (b) W. Adam and J.-C. Liu, *J. Am. Chem. Soc.* **94**, 1206 (1972).
[70] N. Ishibe, M. Odami, and M. Sunami, *J. Chem. Soc., Chem. Commun.* 118 (1971).

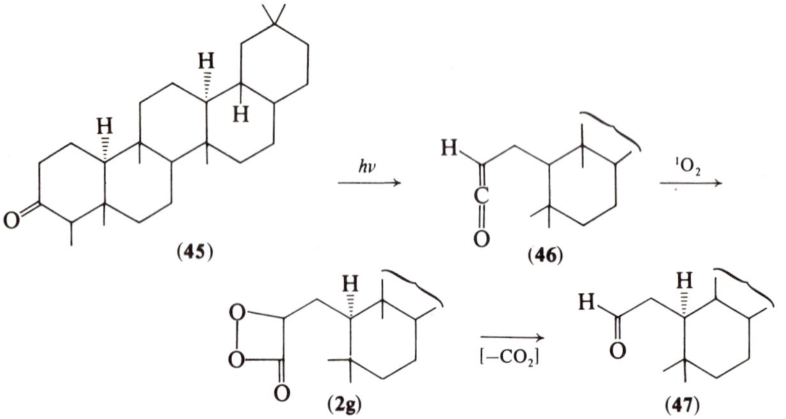

decarboxylation (Eq. 26).[71] Related is the photooxygenation of sulfines (**48**) to the corresponding ketones by SO_2 elimination from the dioxetanone (**2h**).[72] as shown in Eq. (27).

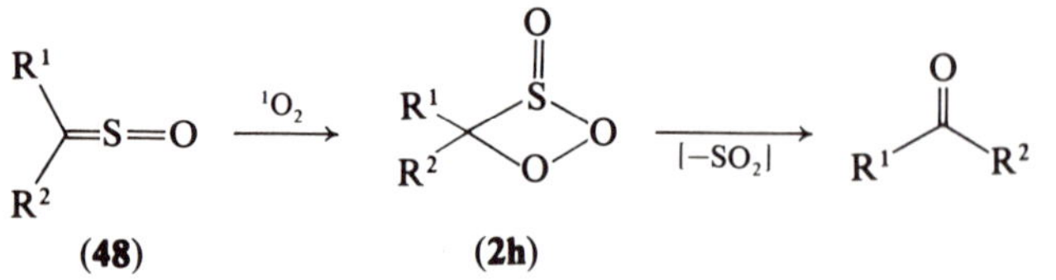

B. Biochemical Oxygenations

Under this heading we feature the oxygenation of organic matter by enzymes that function as dioxygenases, affording dioxetane intermediates. If the latter fragment into excited carbonyl products, we label such processes as bioelectronic. This topic is further subdivided into bioluminescent and dark processes, depending on whether the bioenergized electronically excited carbonyl product emits light or not.

1. Bioelectronic Processes

a. *Bioluminescence.* Numerous reviews have been published on this subject.[11a,20,73] It will suffice to be brief.

[71] R. Aoyagi, T. Tsuyuki, T. Takahashi, and R. Stevenson, *Tetrahedron Lett.* 3397 (1972).
[72] B. Zwanenburg, A. Wagenaar, and J. Strating, *Tetrahedron Lett.* 4683 (1970).
[73] (a) E. N. Harvey, "Living Light." Hafner, New York, 1965; (b) F. McCapra, *Endeavour* **32**, 139 (1973); (c) K.-D. Gundermann, *Topics Curr. Chem.* **46**, 63 (1979); (d) T. Goto and Y. Kishi, *Angew. Chem., Int. Ed. Engl.* **7**, 421 (1968); (e) M. J. Cormier, D. M. Hercules, and J. Lee, "Chemiluminescence and Bioluminescence." Plenum Press, New York, 1973; (f) M. J. Cormier, K. Hori, and J. M. Anderson, *Biochim. Biophys. Acta* **346**, 137 (1974).

Sec. IV.B] THE CHEMISTRY OF 1,2-DIOXETANES 459

In luciferin bioluminescence, the most extensively studied system, the substrate luciferin (**49**) becomes oxygenated by molecular oxygen with the help of the luciferase to afford the intermediary α-peroxylactone.[74] On decarboxylation an electronically excited carbonyl product is obtained which is photodeactivated with light emission, as illustrated in Eq. (28) for the firefly luciferin (**49a**), affording the oxyluciferin (**50**) via α-

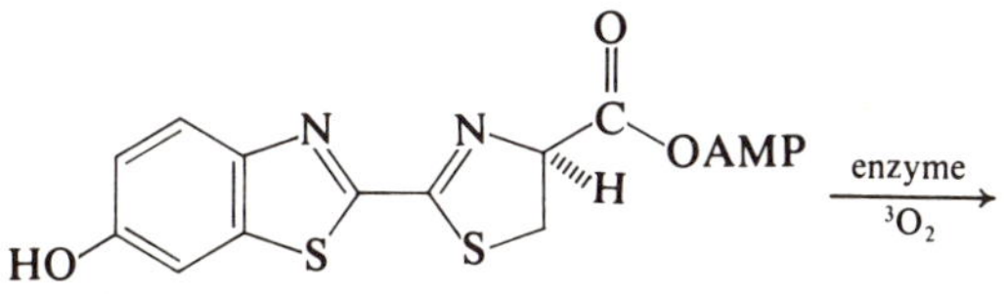

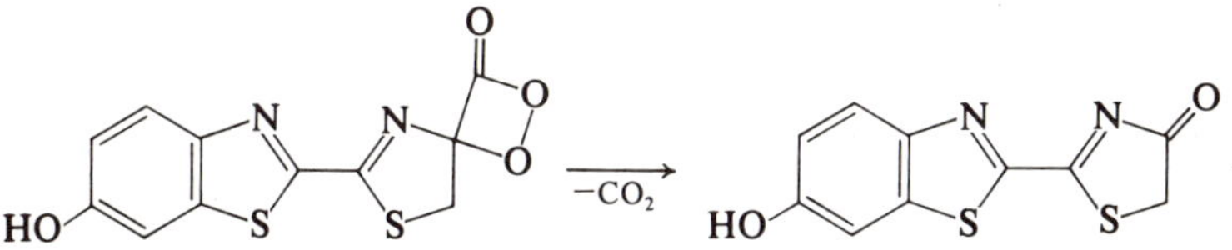

peroxylactone (**2i**). Other luciferins whose structures have been characterized and were postulated to proceed via α-peroxylactones include that isolated from *Cypridina hilgendorfii* (**49b**)[75] and *Renilla reniformis* (**49c**).[76] A number of other bioluminescent organisms have been studied, but structures of the luciferins have not yet been characterized.[73]

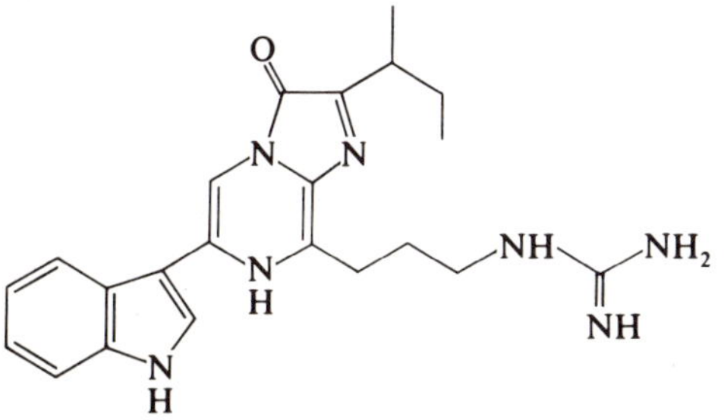

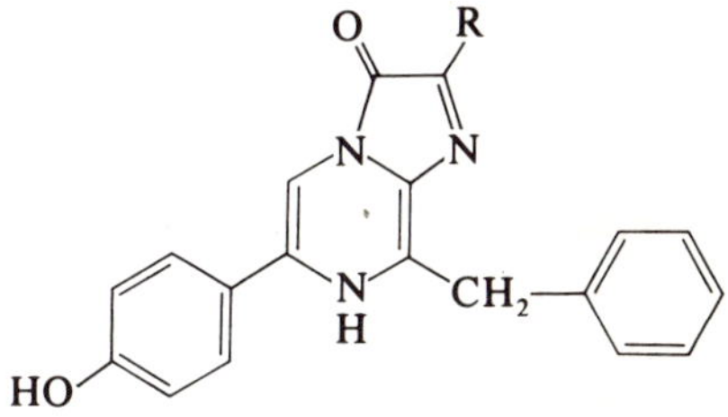

[74] (a) T. A. Hopkins, H. H. Seliger, E. H. White, and M. W. Cass, *J. Am. Chem. Soc.* **89**, 714 (1967); (b) F. McCapra, *J. Chem. Soc., Chem. Commun.*, 155 (1968).
[75] (a) Y. Kishi, S. Inoue, S. Sugiura, and H. Kishimoto, *Tetrahedron Lett.* 3445 (1966); (b) Y. Kishi, T. Goto, Y. Hirata, O. Shimamura, and F. H. Johnson, *ibid.* 3427 (1966).
[76] (a) K. Hori, J. E. Wampler, J. C. Mathews, and M. J. Cormier, *Biochemistry* **12**, 4463 (1973); (b) K Hori, J. E. Wampler, and M. J. Cormier, *J. Chem. Soc., Chem. Commun.* 492 (1973).

Oxygen-18 labeling experiments provide mechanistic confirmation for the intervention of α-peroxylactones (2) in luciferin bioluminescence. The first such experiment was performed by DeLuca and Dempsey.[77] It was concluded that no α-peroxylactone intermediate was involved in firefly bioluminescence. Instead of the cyclic peroxide path [Eq. (29a)], the tetrahedral intermediate path [Eq. (29b)] was postulated. However, White and co-workers[78] observed that in the chemical oxygenation the

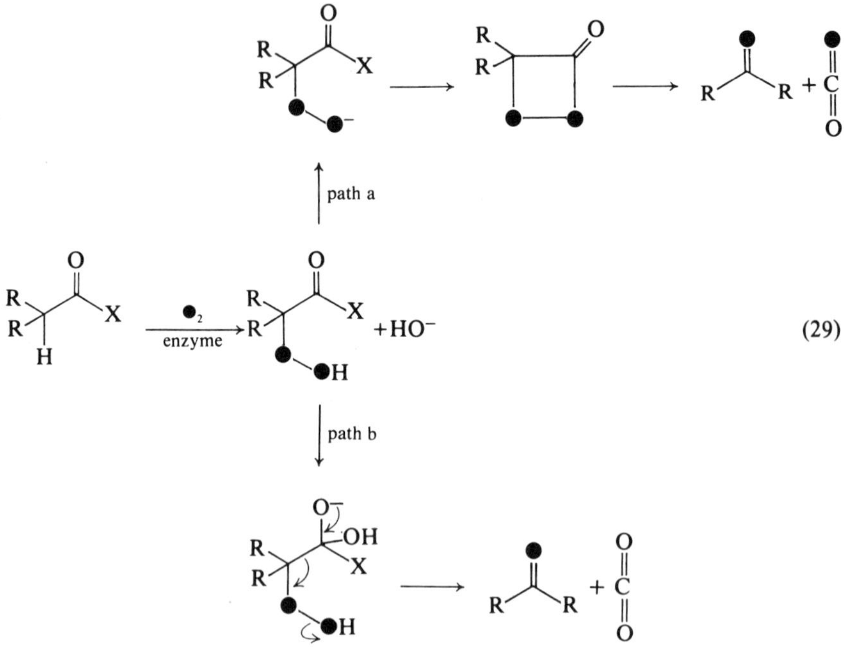

$$(29)$$

α-peroxylactone was formed, by using oxygen-18 labeling as criterion. In the *Cypridina* bioluminescence Shimamura and Johnson[79a] have confirmed by oxygen-18 labeling that the α-peroxylactone intervenes. They observed that at very low substrate concentrations oxygen exchange with the medium might invalidate the labeling results.[79b] Oxygen-18 labeling experiments in the *Renilla reniformis* bioluminescence suggest that the α-peroxylactone mechanism does not apply either.[80] However, in view of the *Cypridina* results,[79a] it is surprising that such structurally similar substrates would engage different chemielectronic paths. It is probable that isotopic exchange with the medium[79b] is responsible for

[77] M. DeLuca and M. E. Dempsey, *Biochem. Biophys. Res. Commun.* **40**, 117 (1970).
[78] E. H. White, J. D. Miano, and M. Umbreit, *J. Am. Chem. Soc.* **97**, 198 (1975).
[79] (a) O. Shimamura and F. H. Johnson, *Biochem. Biophys. Res. Commun.* **44**, 340 (1971); (b) **51**, 558 (1973).
[80] M. DeLuca, M. E. Dempsey, K. Hori, J. E. Wampler, and M. J. Cormier, *Proc. Natl. Acad. Sci. U.S.A.* **68**, 1658 (1971).

b. *Dark Processes.* The two excellent reviews by Cilento[81] and by White and co-workers[82] recognize the importance of this emerging interdisciplinary frontier. The principal difficulty is to recognize such bioelectronic processes, since, as the adjective "dark" implies, the electronically excited state does not emit light, but instead utilizes its excitation energy in performing a chemical transformation. Obvious criteria to look for in the enzymic oxygenation are (i) light leakage by ultrasensitive photodetection; (ii) stoichiometry and products akin to the bioluminescent process, i.e., a carbonyl substrate consumes 1 mole of oxygen and leads to the next lower carbonyl product homolog [Eq. (30)]; (iii) products typical of photochemical origin.

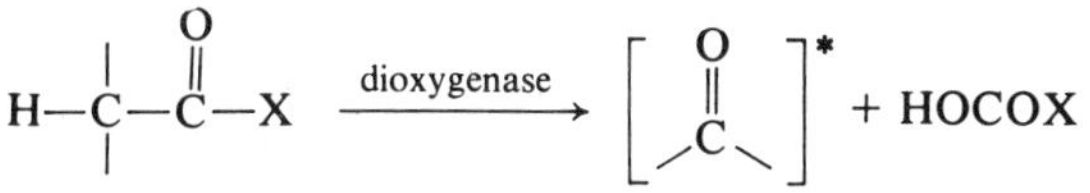

$$H-\underset{|}{\overset{|}{C}}-\overset{O}{\overset{\|}{C}}-X \xrightarrow{\text{dioxygenase}} \left[\overset{O}{\underset{C}{\overset{\|}{}}}\right]^* + HOCOX \tag{30}$$

Several examples have recently been documented by Cilento and co-workers.[83] Thus, in the chemical autoxidation of 4-hydroxy-3,5-diiodophenylpyruvic acid (**51**) electronically excited 4-hydroxy-3,5-diiodobenzaldehyde (**52**) is formed, since its chemiluminescence could be detected.[83a] Presumably the dioxetane (**1au**) intervenes Eq. (31)].

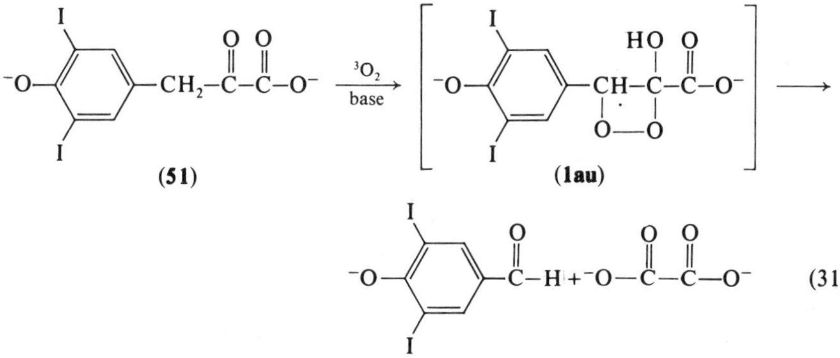

(**51**) → (**1au**) →

(**52**) (31)

---

[81] G. Cilento, *Quart. Rev. Biophys.* **6**, 485 (1973).

[82] E. H. White, J. D. Miano, C. J. Watkins, and E. J. Breaux, *Angew. Chem.* **86**, 292 (1974).

[83] (a) G. Cilento, M. Nakano, H. Fukuyama, K. Sawa, and I. Kamiya, *Biochem. Biophys. Res. Commun.* **58**, 296 (1974); (b) C. C. Vidigal, K. Zinner, N. Duran, E. J. H. Bechara, and G. Cilento, *ibid.* **65**, 138 (1975); (c) C. C. Vidigal and G. Cilento, *ibid.* **62**, 184 (1975); (d) K. Zinner, R. Casadei de Baptista and G. Cilento, *ibid.* **61**, 889 (1974); (e) O. M. M. Faria Oliveira, D. L. Sanioto, and G. Cilento, *ibid.* **56**, 391 (1974); (f) K. Zinner, N. Duran, C. C. Vidigal, Y. Shimizu, and G. Cilento, *Arch. Biophys. Biochem.*, in press.

Since this process is involved in thyroxine biosynthesis,[84] it is likely that the enzymic oxygenation also involves dioxetanes. The horseradish peroxidase (HRP)-catalyzed oxygenation of indole-3-acetic acid (53) affords electronically excited indole-3-aldehyde (54).[83b] The α-peroxylactone (2j) was postulated to be responsible for the feeble light

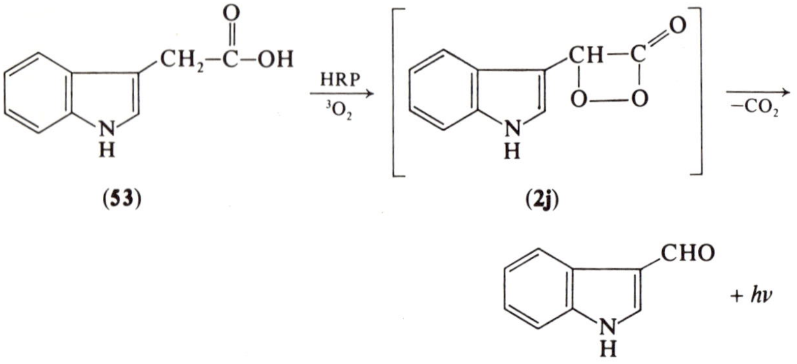

$$\text{(53)} \xrightarrow[{}^3O_2]{HRP} \text{[(2j)]} \xrightarrow{-CO_2}$$

$$\text{(54)} + h\nu \quad (32)$$

emission and enzyme self-damage [Eq. (32)]. In the myoglobin-catalyzed oxygenation of acetoacetate, electronically excited glyoxal is presumably formed via the α-peroxylactone (2k), as illustrated in Eq. (33), in order to reconcile the self-destruction of this hemoprotein.[83c]

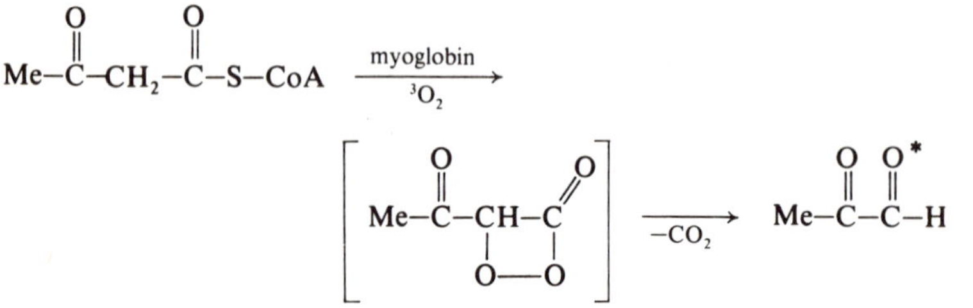

$$\text{Me-CO-CH}_2\text{-CO-S-CoA} \xrightarrow[{}^3O_2]{myoglobin} \text{[(2k)]} \xrightarrow{-CO_2} \text{Me-CO-CO-H}^* \quad (33)$$

Numerous interesting examples worthy of investigation are outlined in the two reviews.[81,82] Among the more intriguing ones, we mention the decarboxylation of retinoic acid by horseradish peroxidase,[85] the peroxidative decarboxylation of long-chain fatty acids by peroxidase,[86] biogenetic lumisantonin formation,[87] enzyme-catalyzed repair of

[84] A. Nishinaga, H. J. Cahnmann, H. Kon, and T. Matsuura, *Biochemistry* **7**, 388 (1968).
[85] E. C. Nelson, M. Mayberry, R. Reid, and K. U. John, *Biochem. J.* **121**, 731 (1971).
[86] R. O. Martin and P. K. Stumpf, *J. Biol Chem.* **234**, 2548 (1959).
[87] I. Satoda, N. Yoshida, and E. Yoshii, *Yakugaku Zasshi* **79**, 269 (1959).

photodamaged DNA,[88] biogenesis of vitamin D in fish,[89] and enzymic oxygenation of quercetin.[90]

2. Processes Not Involving Electronically Excited States

In these enzymic oxygenations no evidence exists that excited-state products are formed. In fact, 1,2-dioxetane intervention has not been soundly established, except that it constitutes a mechanistic convenience. The dioxygenase action entails cleavage of aromatic substrates [Eq. (34a)] or their cis-dihydroxylation [Eq. (34b)].[91] A recent chemical model study on pyrocatechase discounts 1,2-dioxetane formation.[92]

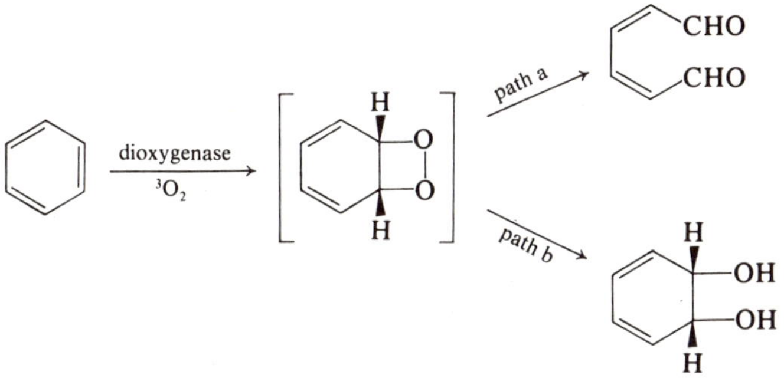

V. Chemical Properties

A. Thermolysis

As pointed out in Section III,B,2, the most characteristic and novel feature of 1,2-dioxetanes is the chemiluminescence that accompanies their thermal decomposition. Electronically excited carbonyl products [Eq. (14)] are photodeactivated either by light emission or energy transfer. We shall now consider the mechanistic details of this unique chemielectronic process, in which a ground-state molecule is converted into an excited-state product.

---

[88] (a) J. S. Cook, in "Photophysiology" (A. C. Giese, ed.), Vol. 5, p. 181. Academic Press, New York, 1970, (b) J. S. Cook, *Photochem. Photobiol.* **6**, 97 (1967).

[89] C. E. Bills, *J. Biol. Chem.* **72**, 751 (1927).

[90] A. Nishinaga and T. Matsuura, *J. Chem. Soc., Chem. Commun.* 11 (1973).

[91] (a) D. T. Gibson, *Science* **161**, 1093 (1968); (b) J. W. Daly, D. M. Jerina, and B. B. Witkop, *Experientia* **28**, 1129 (1972); (c) O. Hayaishi, Y. Ishimura, H. Fujisawa, and M. Nozaki, in "Oxidases and Related Redox Systems" (T. E. King, H. S. Mason, and M. Morrison, eds.), Univ. Park Press, London, 1973; (d) O. Hayaishi, "Molecular Mechanisms of Oxygen Activation." Academic Press, New York, 1974.

[92] J. Tsuji and H. Takayanagi, *J. Am. Chem. Soc.* **96**, 7349 (1974).

1. Kinetics

The thermal decompositions are first order and usually unimolecular. A variety of experimental methods can be used to follow the rates, which include direct chemiluminescence of the excited carbonyl product (A),[14c,50,93,96] activated chemiluminescence by energy transfer of the excited carbonyl to an efficient fluorescer (B),[14c,12a,94-96] dioxetane consumption or carbonyl product formation by NMR spectroscopy (C),[12a,14b,c] iodometric titration of the cyclic peroxide (D),[12a,14b,c] and infrared spectroscopy of α-peroxylactone consumption or carbonyl product formation (E).[2,22,38] The method of choice depends on the particular system, but usually several techniques can be employed.

The activation parameters are collected in Table III for the 1,2-dioxetanes (1) and α-peroxylactones (2). Clearly, variation of substituent structure has a minor effect on the activation parameters. A significant exception is the diadamantylidene system (1z), which is unusually thermally stable.[38] Assuming a diradical mechanism for the thermal decomposition of 1,2-dioxetanes, O'Neal and Richardson[97a] were able to reproduce the experimental activation parameters with good precision, employing thermokinetic calculations developed by Benson.[97d]

Although preliminary quantum-chemical calculations had predicted that the α-peroxylactones should be more stable than the simple 1,2-dioxetanes,[32] the experimental data in Table III indicate the contrary.[2,22] Moreover, thermokinetic calculations are in excellent accord with the experimental data.[97b] Thus, Richardson and co-workers[97] predict activation energies of the order of 24–25 kcal for the 1,2-dioxetanes (1), 21–22 kcal for α-peroxylactones (2), and 17 kcal for the carbon dioxide dimer (3).

Great care must be exercised in interpreting solvent effects. The original claim that the low activation enthalpy and highly negative activation entropy of dioxetane 1b thermolysis in methanol supported a concerted fragmentation of the ring system, was shown to be erroneous.[98] A competing dark reaction catalyzed by traces of transition

---

[93] W. Adam, N. Durán, and G. A. Simpson, *J. Am. Chem. Soc.* **97**, 5464 (1975).

[94] (a) N. J. Turro and P. Lechtken, *J. Am. Chem. Soc.* **95**, 264 (1973); (b) N. J. Turro and P. Lechtken, *Pure Appl. Chem.* **33**, 363 (1973); (c) H.-C. Steinmetzer, A. Yekta, and N. J. Turro, *J. Am. Chem. Soc.* **96**, 282 (1974).

[95] T. Wilson and A. P. Schaap, *J. Am. Chem. Soc.* **93**, 4126 (1971).

[96] W. Adam, G. A. Simpson, and F. Yany, *J. Phys. Chem.* **78**, 2559 (1974).

[97] (a) H. E. O'Neal and W. H. Richardson, *J. Am. Chem. Soc.* **92**, 6553 (1970); (b) W. H. Richardson and H. E. O'Neal, *ibid.* **94**, 8665 (1972); (c) W. H. Richardson, F. C. Montgomery, M. B. Yelvington, and H. E. O'Neal, *ibid.* **96**, 7525 (1974); (d) S. W. Benson, "Thermochemical Kinetics." Wiley, New York, 1968.

[98] (a) T. Wilson, M. E. Landis, A. L. Baumstark, and P. D. Bartlett, *J. Am. Chem. Soc.* **95**, 4765 (1973); (b) W. H. Richardson, F. C. Montgomery, P. Slusser, and M. B. Yelvington, *ibid.* **97**, 2819 (1975).

TABLE III

KINETIC PARAMETERS FOR THE DECOMPOSITION OF 1,2-DIOXETANES AND α-PEROXYLACTONES

$$\begin{array}{c} O-O \\ R^1 \overset{|}{\underset{R^2}{-}} \overset{|}{\underset{R^3}{-}} R^4 \end{array}$$

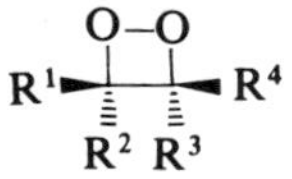	Solvent	Method[a]	ΔH‡ (kcal/mol)	ΔS‡ (e.u.)	Reference
(1) $R^1 = R^2 = R^3 = R^4 = H$	gas phase	A	21.7 ± 1.3	—	50
(1f) $R^1 = R^2 = H; R^3 = R^4 = Me$	CCl_4	C	22.4 ± 0.1	−5 ± 0.2	14b
(1i) $R^1 = R^2 = H; R^3 = R^4 = PhCH_2$	C_6H_6	A, B, C, D	23.6 ± 0.1	−2 ± 0.5	14c
(1g) $R^1 = R^2 = H; R^3 = Me; R^4 = Ph$	CCl_4	C	22.3 ± 0.2	−5.3 ± 0.9	14b
(1h) $R^1 = R^2 = H; R^3 = R^4 = Ph$	C_6H_6	A, B, C, D	22.0 ± 0.1	−4.1 ± 0.3	14c
(1a) $R^1 = H; R^2 = R^3 = R^4 = Me$	CCl_4	B, C, D	24.2 ± 0.5	−5 ± 2	12a
(1b) $R^1 = R^2 = R^3 = R^4 = Me$	C_6H_6, CCl_4	A, B, C, D	26 ± 2	−2 to +2	12a, 93, 94
(1j) $R^1 = R^2 = Me; R^3 = R^4 = n$-Bu	decalin	—	24.8 ± 0.3	−3.6 ± 1.0	15
(1z) $R^1 + R^2 = R^3 + R^4 = $ adamantylidene	o-xylene	E	33.8	2.9 ± 2	38
(1d) $R^1 = R^4 = Me; R^2 + R^3 = -(CH_2)_4-$	C_6H_6	B, C, D	26.2 ± 0.7	2 ± 2	12a
(1e) $R^1 + R^4 = R^2 + R^3 = -(CH_2)_4-$	C_6H_6	B, C, D	23.4 ± 0.9	−3 ± 3	12a
(1n) $R^1 = R^4 = H; R^2 = R^3 = OEt$	C_6H_6	B	23–24	—	95
(2a) $R^1 = H; R^2 = t$-Bu$; R^3 + R^4 = O$	CCl_4	A, B, E	18.8	−8.9	2, 95
(2b) $R^1 = R^2 = CH_3; R^3 + R^2 = O$	CCl_4	E	13.7	−24.1	22

[a] A, direct fluorescence of carbonyl product; B, activated fluorescence by energy transfer; C, NMR spectroscopy; D, iodometry; E, infrared spectroscopy.

metal ions in the methanol causes these suspect activation data. When sequestors of transition metal ions, such as EDTA or CHELEX 100 resin, were utilized, the rates in methanol were normal.

2. Thermochemistry

As a crucial requisite for formation of electronically excited carbonyl product in a chemielectronic process, the reactant must store sufficient energy to chemienergize the product during its thermal decomposition. Since only about 20–25 kcal/mol of thermal activation are available for the 1,2-dioxetanes, the heats of reaction must be highly exothermic. In other words, the condition shown in Eq. (35) must continue, in which E_a

$$E_a - \Delta H° > E^* \tag{35}$$

is the activation energy, $\Delta H°$ the heat of reaction, and E^* the energy of the electronically excited product. Recently this oversimplified criterion of chemienergization has been justly criticized, and it has been pointed out that free energies should be employed instead of enthalpies in order to deal properly with the entropic factor.[99] However, in the absence of reliable entropy data, one is usually obliged to assess the feasibility of the process at the simplified level of Eq. (35).

Experimental measurements of heats of reaction for chemielectronic processes are scarce. However, for the 1,2-dioxetane **1b** the necessary data were determined by Lechtken and Höhne, using DSC calorimetry.[100] Indeed, it was confirmed that the thermal decomposition was highly exothermic, i.e., $\Delta H° = -61$ kcal/mol in solution. Since the activation energy for **1b** is about 25 kcal/mol, a total of 86 kcal/mol are available to energize the acetone product electronically. Thus, sufficient energy is released in the thermolysis of dioxetane **1b** to energize singlet and triplet acetone product, since the respective excitation energies are 84 and 78 kcal/mol.[11b] Perrin concluded that on a free-energy scale this reaction is also sufficiently exergonic to afford electronically excited acetone.[99a]

By means of thermochemical calculations, O'Neal and Richardson[97] showed that sufficient energy is stored in the 1,2-dioxetanes (**1**), α-peroxylactones (**2**), and carbon dioxide dimer (**3**) to chemienergize carbonyl products in their electronically excited states. For **3** the available energy is only sufficient to produce triplet excited carbon dioxide. For the α-peroxylactones **2b**, we concluded that the stored energy is sufficient only to energize singlet or triplet acetone, but not carbon diox-

---

[99] (a) C. L. Perrin, *J. Am. Chem. Soc.* **97**, 4119 (1975); (b) E. Lissi, *ibid.* **98**, 3386 (1976); (c) E. B. Wilson, *ibid.* **98**, 3387 (1976).
[100] P. Lechtken and G. Höhne, *Angew. Chem.* **85**, 7525 (1973).

3. Efficiency and Selectivity of Excited-State Production

From the kinetic and thermochemical data it is clear that sufficient energy is stored in the 1,2-dioxetanes and relatives to generate electronically excited products. Yet, such data tell nothing about the efficiency of chemienergization and how this excitation energy is partitioned among the singlet and triplet excited-state carbonyl products in such "high-energy" molecules. It is, therefore, necessary to determine experimentally the quantum yields of singlet- and triplet-state production. This is a difficult task, but significant progress has been made in developing efficient and selective counting techniques.[11b]

We distinguish between intramolecular and intermolecular counting, since the two methods entail different methodologies. In intramolecular counting, the excited product directly chemiluminesces (physical counting) either by fluorescence or phosphorescence, depending on whether singlet or triplet excited states are involved, or undergoes a chemical transformation (chemical counting). The essential relationships for determining the efficiency of excited-state production are given in Eq. (36a) for intramolecular physical counting and Eq. (36b) for intra-

$$\phi_{DC} = \phi_{ES} \cdot \phi_{PL} \quad (36a)$$

where $\phi_{DC} \equiv$ direct chemiluminescence efficiency of product; $\phi_{ES} \equiv$ efficiency of excited state production, i.e., ϕ_S for singlets and ϕ_T for triplets; $\phi_{PL} \equiv$ photoluminescence efficiency of product, i.e., ϕ_{fl} for fluorescence of singlets and ϕ_{ph} for phosphorescence of triplets.

$$\phi_{CHEM} = \phi_{ES} \cdot \phi_{PHOTO} \quad (36b)$$

where $\phi_{CHEM} \equiv$ efficiency of chemienergization and $\phi_{PHOTO} \equiv$ efficiency of photoenergization.

molecular chemical counting. In intermolecular counting, the excited product transfers its energy to an acceptor, which chemiluminesces efficiently (physical counting) or undergoes a chemical transformation (chemical counting). The fundamental relationships are basically the same [Eqs. (37a) and (37b)], except that the efficiency of energy

$$\phi_{EC} = \phi_{ES} \cdot \phi_A \cdot \phi_{ET} \quad (37a)$$

where $\phi_{EC} \equiv$ enhanced chemiluminescence of efficiency of acceptor, $\phi_A \equiv$ photoluminescence efficiency of acceptor, and $\phi_{ET} \equiv$ energy transfer efficiency.

$$\phi_{CHEM} = \phi_{ES} \cdot \phi_{PHOTO} \cdot \phi_{ET} \tag{37b}$$

transfer must be known. The counting process becomes a good deal more complicated if both singlet and triplets are produced simultaneously and if intersystem crossing is involved. The above counting equations need then to be modified appropriately by deriving the complete steady-state kinetics.

A convenient method for determining relative yields of singlet and triplet excited states has been developed by Turro and co-workers,[101c] by measuring the enhanced chemiluminescence intensities of singlet- and triplet-state specific acceptors, energized by the 1,2-dioxetane via energy transfer. For example, 9,10-diphenylanthracene (DPA) selectively counts singlets and 9,10-dibromoanthracene (DBA) efficiently counts triplets. The method is an extension of the technique originally pioneered by Vassilev[103]; however, our own experience has been discouraging, and caution is advocated.[104]

In Table IV the efficiencies and selectivities of excited state production from 1,2-dioxetanes and α-peroxylactones are summarized. In view of the large errors in the counting methods, these data are at best within 50% accuracy. Even with this large tolerance, glaring discrepancies are evident in Table IV. No doubt, more reliable counting techniques will have to be designed to discern important features such as structural effects, solvent effects, types of excited states. Despite these shortcomings one qualitative feature emerges from the data in Table IV, namely that all systems, except **1j**, preferentially chemienergize triplet excited product. The chemielectronic process associated with the thermolysis of 1,2-dioxetanes is indeed of a unique and rare type in that a singlet state reactant is thermally converted into a triplet excited product.

4. *Mechanism*

It is now our task to amalgamate the experimental data on the thermolysis of 1,2-dioxetanes into a consistent mechanism. This has not been at all easy, as witnessed by the mechanistic turmoil precipitated during the last couple of years. Two mechanisms have been argued.

---

[101] (a) N. J. Turro, P. Lechtken, G. Schuster, J. Orell, H. C. Steinmetzer, and W. Adam, *J. Am. Chem. Soc.* **96**, 1629 (1974); (b) N. J. Turro and P. Lechtken, *ibid.* **94**, 2886 (1972); (c) H.-C. Steinmetzer, P. Lechtken, and N. J. Turro, *Justus Liebigs Ann. Chem.* 1984 (1973); (d) N. J. Turro and H.-C. Steinmetzer, *J. Am. Chem. Soc.* **96**, 4677 (1974); (e) **96**, 4679 (1974).

[102] K. R. Kopecky, private communication.

[103] R. F. Vassilev, *Progr. React. Kinet.* **4**, 305 (1967).

[104] W. Adam, E. Cancio, and O. Rodriguez, unpublished work.

TABLE IV
Quantum Yields of Excited State Production and Spin Multiplicity Dependence

		Quantum yields (%)					
		ϕ_{Total}	ϕ_S	ϕ_T	ϕ_T/ϕ_S	Method[a]	References
(1b)	$R^1 = R^2 = R^3 = R^4 = Me$	50	0.5	50	100	A	47
		—	0.16	—	—	A	101b
		—	0.1	—	—	A	93
		50	0.25	50	200	B	101a,c
		22	0.02	22	1100	B	102
		48	0.28	48	170	C	101b
(1a)	$R^1 = H; R^2 = R^3 = R^4 = Me$	50	0.2	50	>300	B	101a
		12	0.007	12	1700	B	102
		14	—	—	—	B,C	16
(1h)	$R^1 = R^2 = H; R^3 = R^4 = Ph$	3.5	—	—	—	C	14c
(1j)	$R^1 = R^3 = Me; R^2 = R^4 = n\text{-}Bu$	8–11	5–7	3.0–3.5	1.6–2	C	15
(1z)	$R^1 + R^2 = R^3 + R^4 = $ adamantylidene	17	2	15	7.5	C	38b
(1d)	$R^1 = R^2 = Me; R^3 + R^4 = (CH_2)_4$	17	0.02	17	850	B	102
(1e)	$R^1 + R^2 = R^3 + R^4 = (CH_2)_4$	0.8	0.0003	0.8	2600	B	102
(1n)	$R^1 = R^4 = H; R^2 = R^3 = OEt$	—	—	—	100	B	93
(1af)	$R^1 = Me; R^2 = Ph; R^3 + R^4 = $ 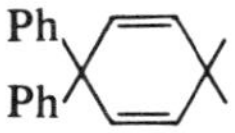 <chemical structure: cyclohexenyl with Ph, Ph substituents>	20	—	—	—	C	40
(1ag)	$R^1 = Me; R^2 = p\text{-}An; R^3 + R^4 = $  <chemical structure: cyclohexenyl with Ph, Ph substituents>	16	—	—	—	C	40
(1ah)	$R^1 = Me; R^2 = \beta\text{-}Naph; R^3 + R^4 = $ 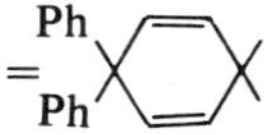 <chemical structure: cyclohexenyl with Ph, Ph substituents>	14	—	—	—	C	40
(2a)	$R^1 = H; R^2 = t\text{-}Bu; R^3 + R^4 = O$	15	0.047	15	320	A,B	96
(2b)	$R^1 = R^2 = Me; R^3 + R^4 = O$	5.3	0.046	5.3	115	A,B	96
		21	1	20	20	B	101a

[a] A, direct chemiluminescence; B, enhanced chemiluminescence; C, chemical counting.

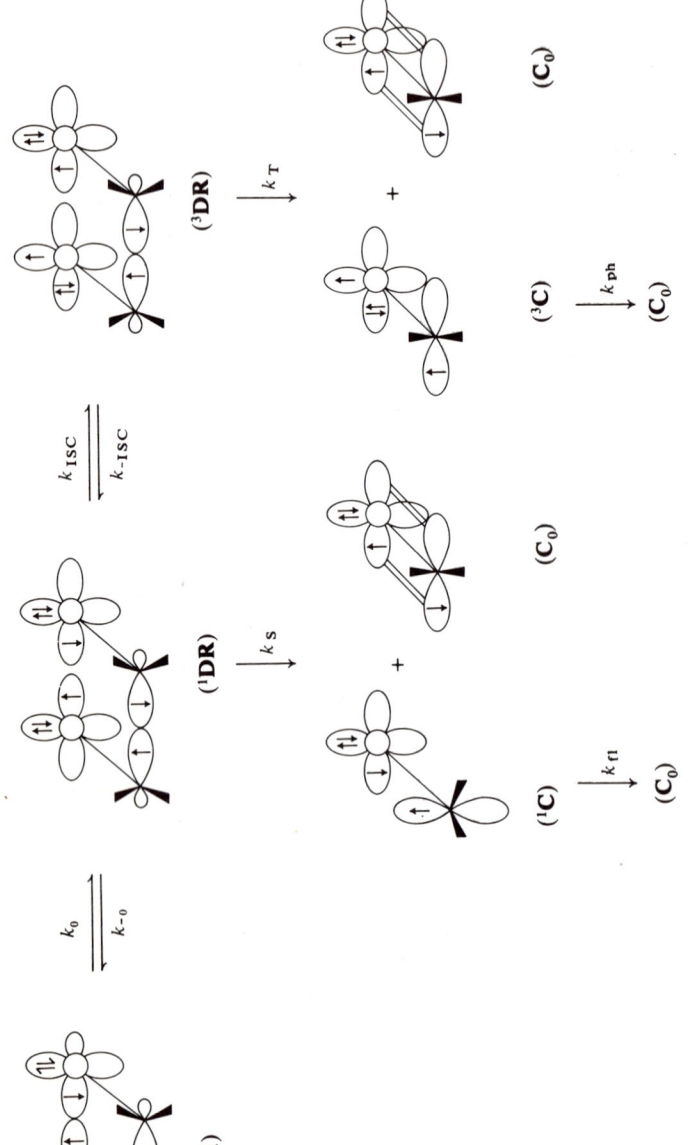

(38)

These are the stepwise diradical path [Eq. (38)] proposed by Richardson[14,97b] and the concerted retrocyclic path [Eq. (39)] suggested by Turro.[11,94]

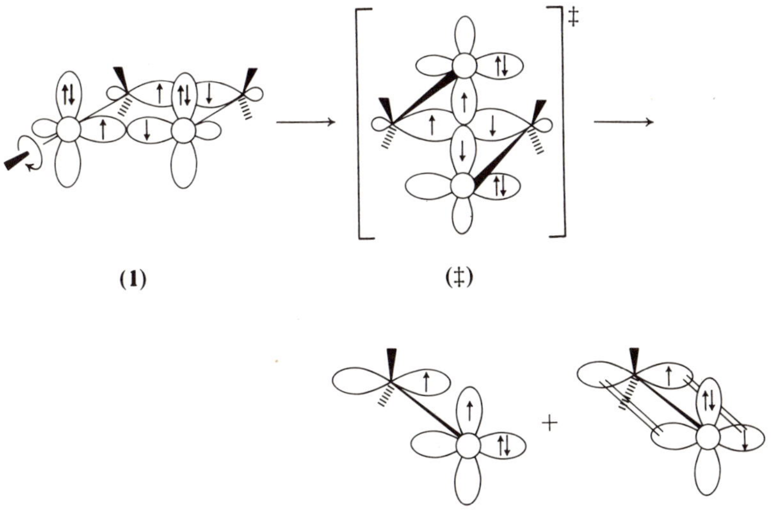

(1)   (‡)

$$({}^3C) \quad + \quad (C_0) \tag{39}$$

In the stepwise diradical paths first the oxygen–oxygen bond is disengaged by stretching (k_0), leading to the singlet diradical (^{1}DR). Recyclization (k_{-0}) generates back the 1,2-dioxetane (1). Rupture of the carbon–carbon bond (k_s) affords the singlet excited (n, π^*) carbonyl product (^{1}C) and singlet ground-state carbonyl product (C_0). Deenergization by fluorescence (k_{fl}) leads to (C_0). Competing with the k_s step is intersystem crossing (k_{ISC}) to the triplet diradical (^{3}DR). The latter can reverse (k_{-ISC}) to the singlet diradical (^{1}DR) or unzip the carbon–carbon bond (k_T) to afford the triplet excited (n, π^*) carbonyl product (^{3}C) and singlet ground-state product (C_0). Deenergization by phosphorescence (k_{ph}) or any other process affords (C_0). If the conditions $k_{ISC} \gg k_S$ and $k_T \gg k_{-ISC}$ continue, triplet-energized product (^{3}C) prevails, as observed experimentally.

In the concerted retrocyclic path, vibrational deformations leading to puckering of the 1,2-dioxetane ring system displaces electron density from its normal distribution by charge acceleration. This acceleration of electronic charge promotes spin-orbit coupling at the oxygen atom of the incipient n, π^* state of the excited carbonyl fragment. The activated complex (‡) disengages into a triplet excited (n, π^*) carbonyl product (^{3}C) and singlet ground state carbonyl product (C_0) by simultaneous rupture of the oxygen–oxygen and carbon–carbon bonds.

Both mechanisms have in common a spin multiplicity change; however, the fundamental difference between them is that in the diradical mechanism [Eq. (38)] a reversible intersystem crossing step is involved, whereas in the concerted retrocyclic mechanism [Eq. (39)] the intersystem crossing step is irreversible. Thus, the classical mechanistic dilemma of distinguishing between normal spin-conserved diradical and concerted paths,[105] which to date has not been unequivocally resolved, is still further aggravated by the fact that in the dioxetane thermolysis distinct spin multiciplicity changes are involved. No information is available on how the usual mechanistic criteria, such as solvent effects, kinetic parameters, substituent effects, isotope effects, etc., influence reversible versus irreversible intersystem crossing phenomena. Consequently, it is our contention that the present experimental data do not differentiate conclusively between the diradical and concerted paths suggested for the dioxetane thermolysis. Certainly thermochemical calculations, which by necessity are independent of spin multiplicity considerations, cannot genuinely distinguish between the two argued mechanisms.

On the more optimistic side of the mechanistic problem, most probably a composite mechanism applies in this novel chemielectronic transformation. In other words, on thermal activation the oxygen–oxygen bond is being stretched and the singlet ground-state dioxetane molecule begins its decomposition on the energy surface for the singlet diradical. However, before the activated complex of the singlet diradical is reached, the singlet diradical energy surface intersects the energy surface of a triplet excited product state via a puckered concerted activated complex by twisting around the carbon–carbon bond with simultaneous stretching. Recent quantum-chemical calculations support this dualistic mechanism, which is discussed next.

5. Theoretical Considerations

On the basis of orbital symmetry arguments, McCapra[106] pointed out that a concerted 2s:2s retrocyclization must lead to electronically excited product. Such a process would afford a doubly excited π,π^* state, and clearly not enough energy is available to generate such an excited carbonyl product. However, this interpretation nicely accounts for the high efficiency (100%) of firefly bioluminescence.[11a]

The theoretical problem of concerted decomposition using orbital symmetry arguments was examined in much greater detail by Kearns.[107]

---

[105] (a) L. Salem and R. S. Rowland, *Angew Chem., Int. Ed. Engl.* **11**, 92 (1972); (b) W. G. Dauben, L. Salem, and N. J. Turro, *Acc. Chem. Res.* **8**, 41 (1975).
[106] F. McCapra, *J. Chem. Soc., Chem. Commun.*, 155 (1968).
[107] D. R. Kearns, *J. Am. Chem. Soc.* **91**, 6554 (1969).

Long before it was experimentally recognized that in most simple 1,2-dioxetanes triplet excited carbonyl products are formed preferentially, he postulated that the energy surface of a triplet excited product intersected the energy surface of the singlet excited product much before the concerted activated complex is reached. It was then clear why triplet (n, π^*) product was formed; however, Kearns's arguments lacked quantitative rigor.

Several CNDO calculations with configuration interaction further clarified the theoretical mechanism.[108] It was concluded that a diradical ring opening was preferred initially; however, before the transition state is reached, a triplet excited product surface intersects the singlet diradical surface. Thus, a merger of the two extreme views, i.e., diradical versus concerted mechanism, was established by these semiempirical calculations. Also MINDO/3 with configuration interaction confirmed this dual mechanism.[109] Finally, a recent detailed analysis of energy surface crossings by Turro and Devaquet[110] substantiated on qualitative grounds crossover of the diradical path to the triplet excited product path, but never completely reaching a stable diradical state, as suggested by the thermochemical calculations.[97]

6. Chemienergized Transformations

A number of photochemical processes have been chemienergized by 1,2-dioxetanes. In fact, most of them have served as chemical counting techniques by titration of the excited states (cf. Section VA3). In this section we recount these reactions in order to sense the usefulness of 1,2-dioxetanes as chemisensitizers. For convenience we shall subdivide this section into intramolecular, intermolecular, and chemisensitized transformations.

a. Intramolecular Transformations. The α-cleavage of excited-state

$$\underset{(1i)}{\underset{\underset{PhCH_2}{|}}{PhCH_2-\overset{\overset{O-O}{||}}{C}-CH_2}} \xrightarrow[C_6H_6]{60°} PhCH_2 \overset{O^*}{\underset{}{\overset{\|}{C}}} CH_2Ph \longrightarrow$$

$$PhCH_2\cdot + \cdot\overset{O}{\overset{\|}{C}}-CH_2Ph \xrightarrow[|-CO|]{} PhCH_2-CH_2Ph \qquad (40)$$

---

[108] (a) E. M. Evleth and G. Feller, *Chem. Phys. Lett.* **22**, 499 (1973); (b) D. R. Roberts, *J. Chem. Soc., Chem. Commun.* 683 (1974); (c) G. Barnett, *Can. J. Chem.* **52**, 3837 (1974).
[109] (a) M. J. S. Dewar and S. Kirschner, *J. Am. Chem. Soc.* **96**, 7578 (1974); (b) M. J. S. Dewar, S. Kirschner, and K. W. Kollmar, *ibid.* **96**, 7579 (1974).
[110] N. J. Turro and A. Devaquet, *J. Am. Chem. Soc.* **97**, 3859 (1975).

ketones was employed by Richardson and co-workers.[111] It was shown that excited dibenzyl ketone, chemigenerated from 1,2-dioxetane **1i**, gave a 2.2% yield of bibenzyl [Eq. (40)]. The β-cleavage and cyclobutane formation of excited ketones [Eq. (41)] with γ-hydrogen

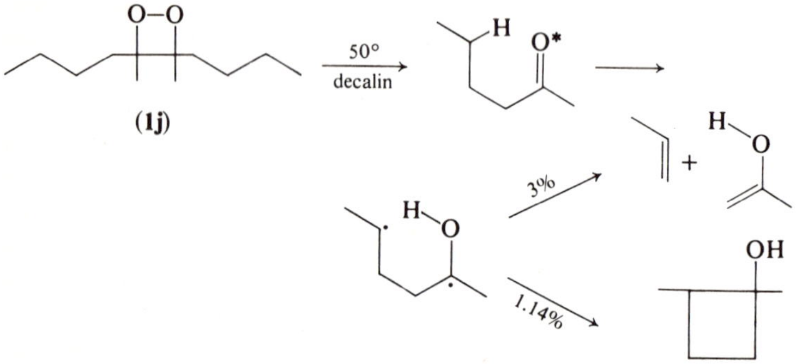

was illustrated by Darling and Foote,[15] utilizing the 1,2-dioxetane **1j**. Finally, the dienone rearrangement [Eq. (42)] was achieved by Zimmerman and Keck[40] via the 1,2-dioxetanes **1af–h**.

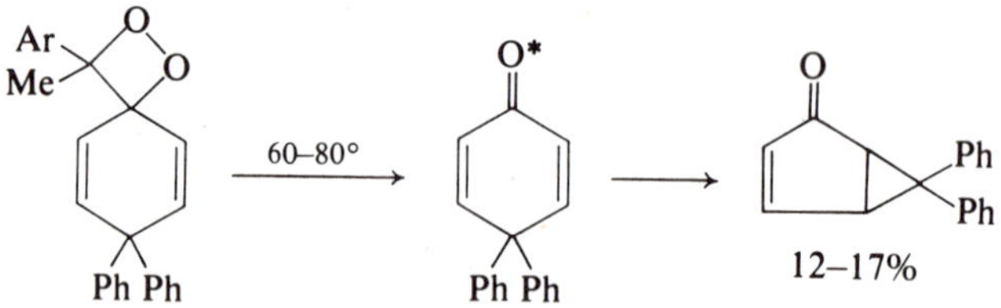

(**1af**) Ar = Ph
(**1ag**) Ar = p-An
(**1ah**) Ar = β-Naph

b. Intermolecular Transformations. Turro and Lechtken[101b,112] were the first to employ the Paterno–Büchi cycloaddition as a chemical titration technique for singlet excited carbonyl product. Thus, *trans*-1,2-dicyanoethylenes stereospecifically and spin-selectively cycloadd to singlet acetone [Eq. (43)] that is chemigenerated by thermolysis of diox-

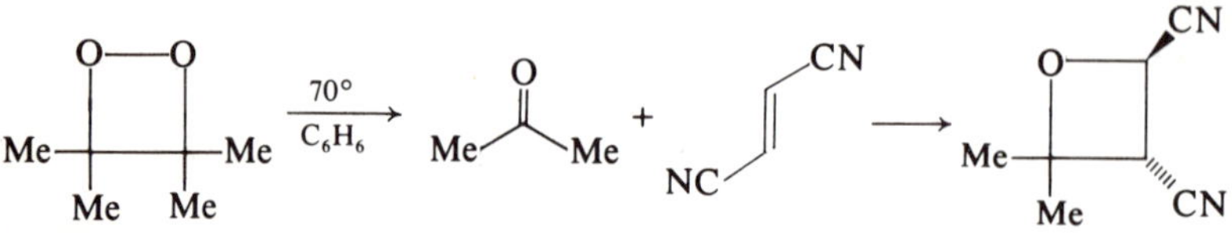

$\phi_s = 0.28\%$ $\phi = 0.022\%$

---

[111] W. H. Richardson, F. C. Montgomery, and M. B. Yelvingon, *J. Am. Chem. Soc.* **94**, 9277 (1972).
[112] P. Lechtken and N. J. Turro, *Angew. Chem.* **85**, 300 (1973).

etane **1b**. Similarly, singlet excited adamantanone derived from dioxetane **12** cycloadded to *trans*-1,2-dicyanoethylene to give the corresponding oxetane.[38b] Excited benzophenone, energized by way of dioxetane **1h**, afforded 2,2-diphenyl-3,3,4-trimethyloxetane with 2-methyl-2-butene ($\phi_{app} = 0.5\%$).[113] Finally, Bartlett and co-workers[98a] observed that the triplet acetone generated from **1b** produced isopropanol and benzene in 7.5% yield by hydrogen abstraction from 1,4-cyclohexadiene.

c. Chemisensitized Transformations. Here the chemienergized excited carbonyl product energy-transfers to an acceptor, which then suffers a photochemical change. Cis–trans isomerization of olefins is the most abundantly documented process. The earliest example was provided by White and co-workers,[16] who chemiisomerized *trans*-stilbene ($\phi_{app} = 4\%$), by heating it with dioxetane **1a** in benzene at 95°–100°. Dioxetane **1h** cis–trans isomerized stilbene ($\phi_{app} = 1.9\%$) when heated in benzene at 45°. With the help of the carbon dioxide dimer, derived from perhydrolysis of diaryl oxalates, Güsten and Ullman[114] isomerized ($\phi_{app} = 4.5\%$) *p*-methoxy-*p'*-nitrostilbene in benzene under ambient conditions. Spin-selective isomerization of *trans*-1,2-dicyanoethylene was

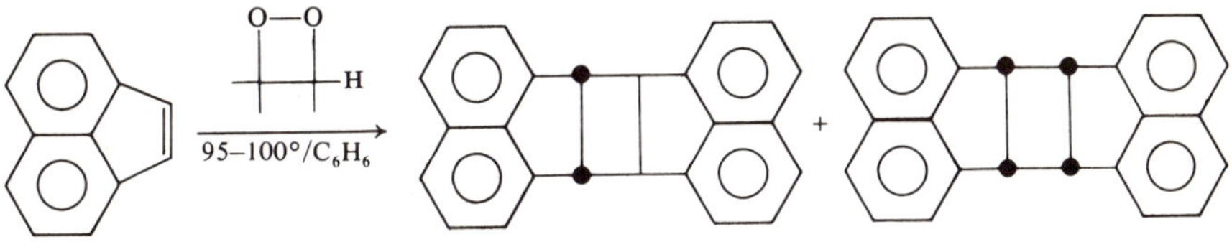

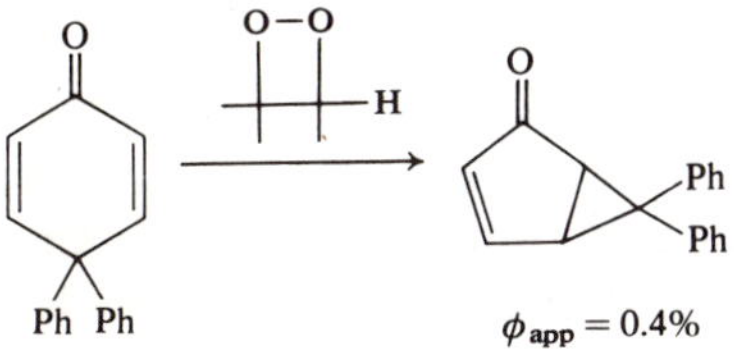

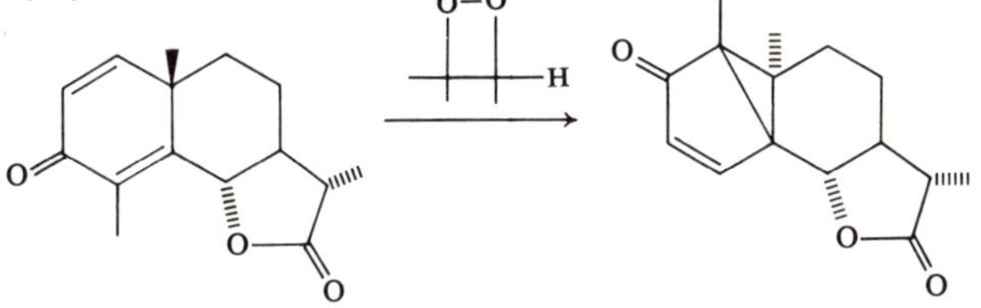

(44)

[113] W. H. Richardson, F. C. Montgomery, M. B. Yelvington, and G. Ranney, *J. Am. Chem. Soc.* **96**, 4045 (1974).
[114] H. Güsten and E. F. Ullman, *J. Chem. Soc., Chem. Commun.* 29 (1970).

effected by triplet acetone derived from **1b**[101b] and triplet adamantanone derived from **1z**.[38b] Chemisensitized self-dimerization of acenaphthene (**55**) and the dienone rearrangement of 4,4-diphenyl-2,6-cyclohexadienone (**56**) and santonin (**57**) were accomplished by White and co-workers,[16] by heating these substrates with dioxetane **1a** at 95–100° in benzene [Eq. (44)]. Photocyclization of *o*-tolylpropane-1,2-dione (**58**) into the indanone (**58a**) and photocyclization of phenanthrene (**59**) with ethyl vinyl ether were effected by the carbon dioxide dimer at room temperature in benzene [Eq. (45)].[114]

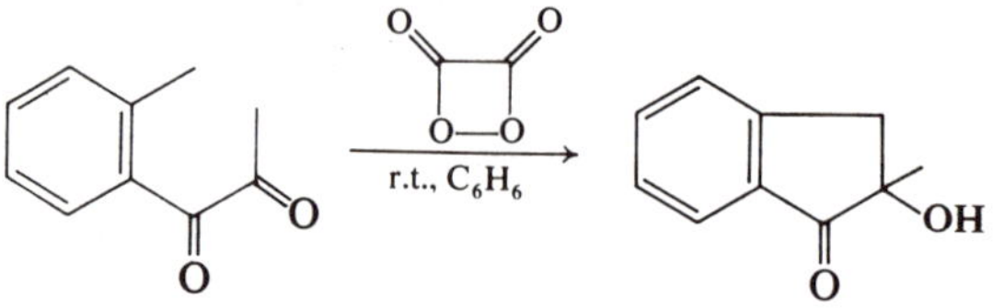

(45)

$\phi_{app}= 0.6\%$

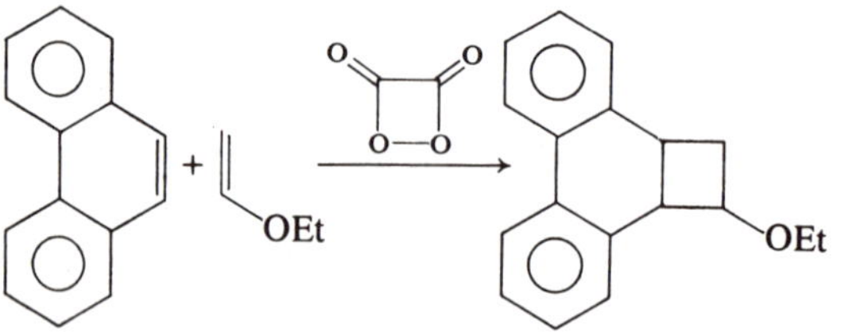

$\phi_{app}= 4\%$

Recently Bartlett and Shimizu[115] demonstrated that triplet acetone derived from dioxetane **1b** chemisensitized the decomposition of benzoyl peroxide in carbon tetrachloride, affording chlorobenzene, which exhibited a photo-CINDP effect. Finally, we observed that the β-peroxylactone (**60**) afforded exclusively tetramethylethylene oxide when chemisensitized by **1b**, as shown in Eq. (46).[116]

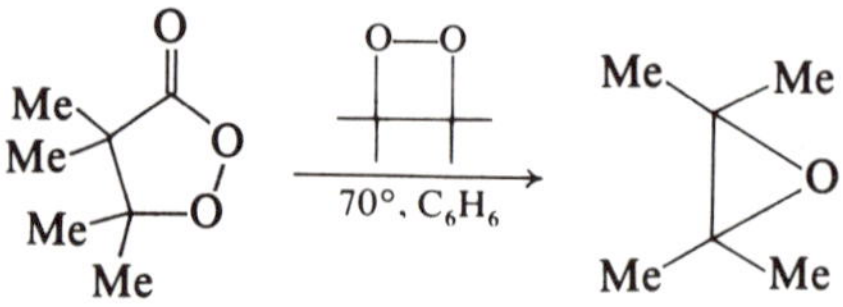

(46)

[115] P. D. Bartlett and N. Shimizu, *J. Am. Chem. Soc.* **97**, 6253 (1975).
[116] W. Adam and L. Guedes, *Abstr. IUPAC Congr., 25th, 1975*.

B. Photolysis

The photolytic decomposition of 1,2-dioxetanes has been studied much less extensively in comparison to their thermal decompositions. Bartlett and Schaap[34] observed that prolonged irradiation times in the preparation of 1,2-dioxetanes **1n** by photooxygenation decreased the yields substantially. Presumably the dioxetane was photosensitive. This was confirmed by Wilson and Schaap[95] in their classical paper on the mechanism of chemiluminescence of dioxetane **1n**, showing that the photoenergized 9,10-diphenyl- and 9,10-dibromoanthracene singlets efficiently induced the decomposition of **1n.**

Turro's group has contributed most significantly on this topic. For example, it was shown[117] that the direct photolysis of dioxetane **1b** at 366 nm gave 10% singlet and 43% triplet excited acetone. This corresponds to a triplet:singlet ratio of efficiencies of 4.3, or ca. 25-fold lower than the thermal process.[47] Furthermore, the more energetic the photons, the more singlet excited acetone is produced. Similarly, direct photolysis of the adamantylidene dioxetane (**1z**) at 280 nm in CH_3CN gave 8% singlet and 40% triplet excited adamantanone.[38b] Again this corresponds to a greater yield of singlet excited product compared to thermolysis results; but the difference is not as dramatic as for **1b**. The direct photolysis of dioxetane **1j** with > 360 nm light in decalin afforded 2.3% of β-cleavage and 0.86% cyclobutanol formation [Eq. (41)]. In this case, the thermal and photolytic singlet/triplet efficiencies are comparable.

Also a number of photosensitized decompositions of dioxetanes have been reported. Thus, Barron and Turro[118] showed that dioxetane **1b** is dissociated by pyrene singlets into 30% acetone triplets, determined by cis–trans isomerization of crotononitrile. Also biacetyl triplets sensitize the decomposition of **1b**, which was used in an ingenious way by Turro and Lechtken[119] to promote the Norrish type II cleavage of butyrophenone. The sequence of events in this ternary system is illustrated in Eq. (47).

The protecting effect of oxygen, which was first observed by Wilson and Schaap,[95] stimulated an interesting series of papers on quantum chain reactions.[120] In brief, when triplet oxygen is present, the triplet-excited carbonyl product is deenergized by energy transfer to molecular

---

[117] N. J. Turro, P. Lechtken, A. Lyons, R. R. Hautala, E. Carnaham, and T. J. Katz, *J. Am. Chem. Soc.* **95**, 2035 (1973).
[118] W. J. Barron and N. J. Turro, *Tetrahedron Lett.* 3515 (1974).
[119] N. J. Turro and P. Lechtken, *Tetrahedron Lett.* 565 (1973).
[120] (a) P. Lechtken, A. Yekta, and N. J. Turro, *J. Am. Chem. Soc.* **95**, 3027 (1973); (b) N. J. Turro, N. E. Schore, H.-C. Steinmetzer, and A Yekta, *ibid.* **96**, 1936 (1974); (c) N. J. Turro and W. H. Waddell, *Tetrahedron Lett.* 2069 (1975).

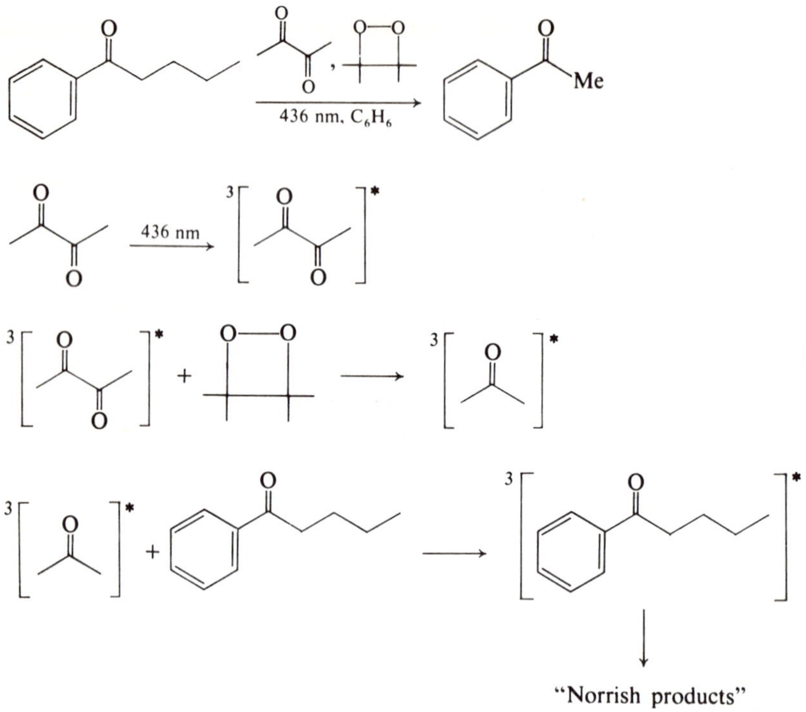

$$\text{(47)}$$

"Norrish products"

oxygen, thereby preventing autocatalytic or photosensitized destruction of the dioxetane. From the kinetics of decomposition of the dioxetane (**1**) in the absence of triplet oxygen, the efficiency of induced decomposition was found to be ca. 100%,[93,120] with chain lengths of the order of 1000.[11b,120c]

C. Nucleophilic Substitution

Besides thermal and photolytic decompositions of 1,2-dioxetanes, there has been relatively little other chemistry explored with these "high energy" molecules. Several nucleophilic substitution reactions have been documented.

In an effort to resolve the mechanistic question of dioxetane or perepoxide trapping by azide ion in singlet oxygenations,[121] Richardson and Hodge[122] showed that sodium azide coverts 1,2-dioxetane **1f** into acetone and nitrogen, instead of the claimed β-azidohydroperoxide (**61**).

---

[121] (a) W. Fenical, D. R. Kearns, and P. Radlick, *J. Am. Chem. Soc.* **91**, 1771 (1969); (b) A. G. Schultz and R. H. Schlessinger, *Tetrahedron Lett.* 2731 (1970).

[122] W. H. Richardson and V. F. Hodge, *Tetrahedron Lett*, 749 (1971).

The mechanism [Eq. (48)] discounts the possibility that 1,2-dioxetane is the precursor of **61**. Later Gollnick and co-workers[123] showed that azide

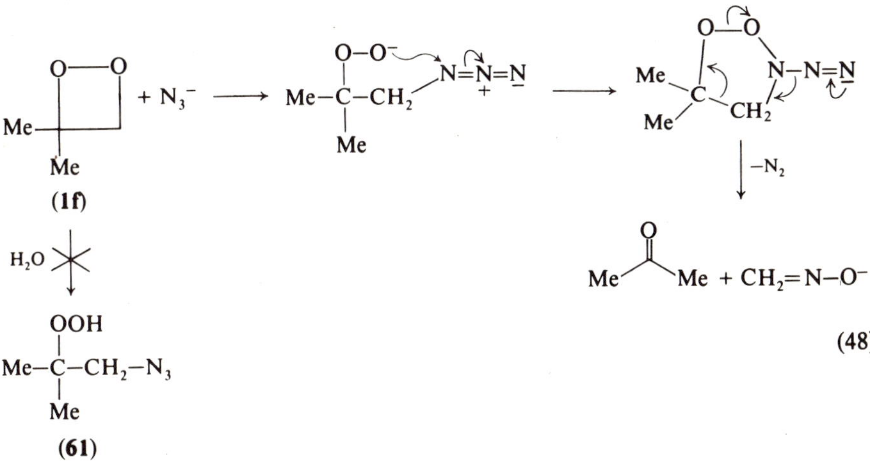

$$\text{(1f)} + N_3^- \longrightarrow \text{Me-C(O-O}^-\text{)-CH}_2\text{-N=N=N} \longrightarrow \cdots \longrightarrow Me_2C=O + CH_2=N-O^-$$

$$\downarrow H_2O$$

Me-C(OOH)(Me)-CH$_2$-N$_3$

(61)

(48)

ion trapping was an artifact, resulting from a free-radical chain reaction of the olefin with the azide radical.

Interesting chemistry has been discovered by Bartlett and co-workers[124] in the reaction of 1,2-dioxetane **1b** with trivalent phosphorus nucleophiles. At low temperature the cyclic phosphorane **62** could be isolated [Eq. (49)], which at 55° smoothly decomposed into

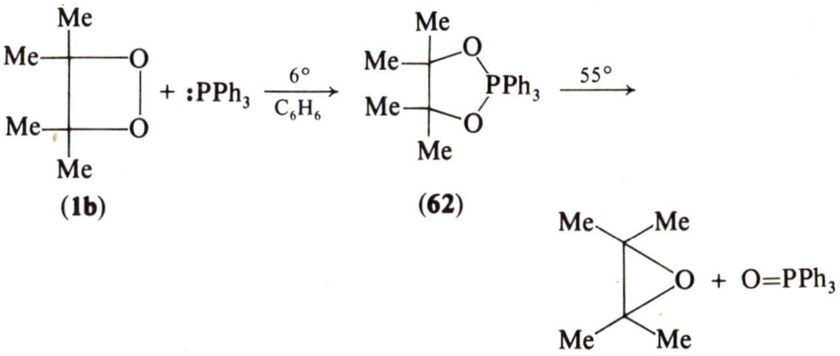

$$\text{(1b)} + :PPh_3 \xrightarrow[C_6H_6]{6°} \text{(62)} \xrightarrow{55°} Me_2C\text{-}CMe_2\text{-}O + O=PPh_3 \quad (49)$$

tetramethylethylene oxide and triphenylphosphine oxide. When the ligands on phosphorus are alkoxy groups, the cyclic phosphoranes **62** show, as expected, a much greater stability.[124b]

---

[123] K. Gollnick, D. Haisch, and G. Schade, *J. Am Chem. Soc.* **94**, 1747 (1972).
[124] (a) P. D. Bartlett, A. L. Baumstark, and M. E. Landis, *J. Am. Chem. Soc.* **95**, 6486 (1973); (b) P. D. Bartlett, A. L. Baumstark, M. E. Landis, and C. L. Lerman, *ibid.* **96**, 5267 (1974).

Analogous chemistry was observed by Denney and co-workers[13] when divalent sulfur nucleophiles were allowed to react with dioxetanes. Thus, diphenyl sulfide gave epoxide and diphenyl sulfoxide, presumably via the sulfurane (**63**), as shown in Eq. (50). When the ligands on the sulfur nucleophile were alkoxy groups, NMR evidence for the intermediary sulfuranes was provided. Similar results were observed by Wasserman and Saito[125] in the photooxygenation of 1,2-dialkoxyethylenes in the presence of diphenyl sulfide. For example, the *p*-dioxene **64** gave the ketal **65** and epoxide **66**, besides diphenyl sulfoxide, under these conditions. The mechanism is rationalized in Eq. (51).

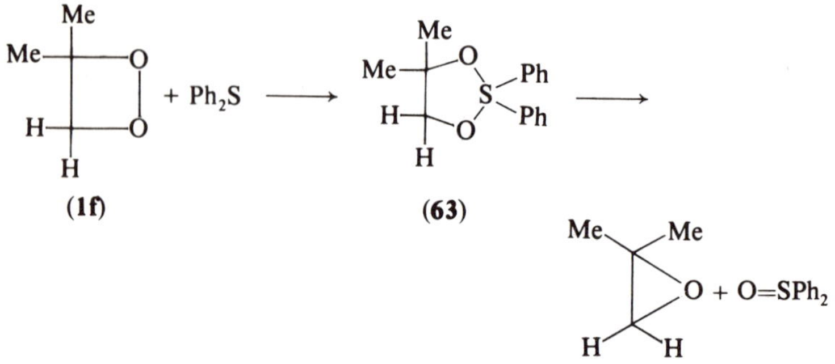

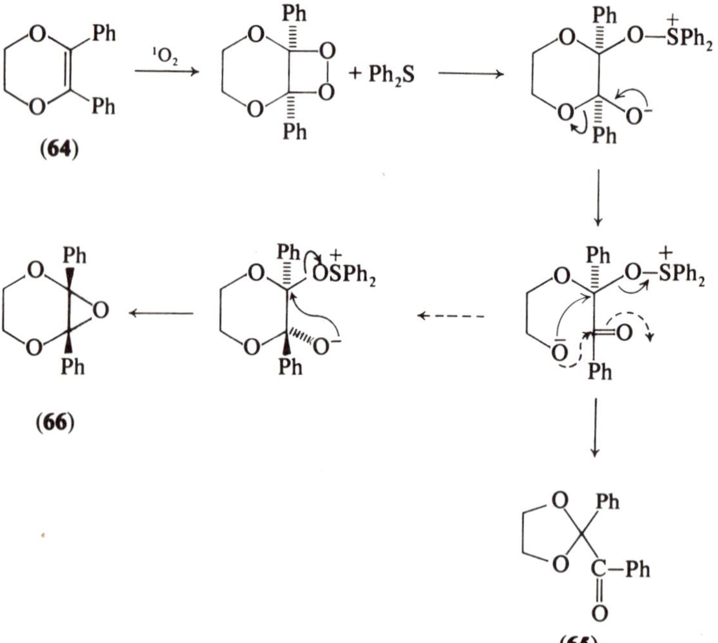

[125] H. Wasserman and I. Saito, *J. Am. Chem. Soc.* **97**, 905 (1975).

A novel reduction was discovered by Baumstark,[126] involving the reaction of dioxetane **1b** with stannous chloride, leading to pinacol and stannic ion on hydrolysis [Eq. (52)]. Formally it could be considered as

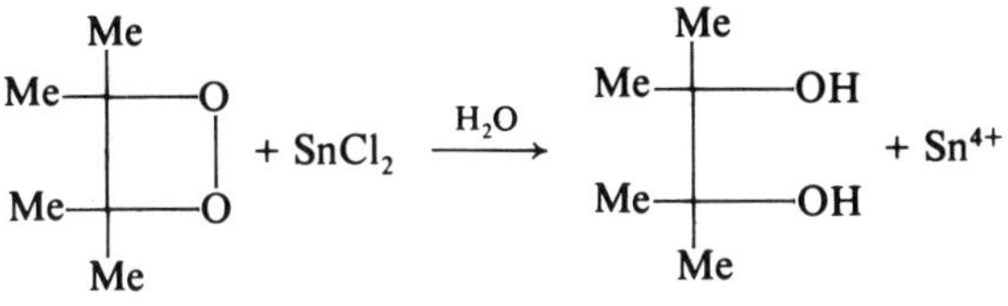

$$\underset{(\mathbf{1b})}{\begin{array}{c}\text{Me} \\ \text{Me}\!-\!\!\!\!-\!\!\!-\!\!\text{O} \\ |\quad\quad| \\ \text{Me}\!-\!\!\!\!-\!\!\!-\!\!\text{O} \\ \text{Me}\end{array}} + \text{SnCl}_2 \xrightarrow{\text{H}_2\text{O}} \underset{\text{Me}}{\begin{array}{c}\text{Me} \\ \text{Me}\!-\!\!\!\!-\!\!\!-\!\!\text{OH} \\ |\quad\quad\quad\;\;| \\ \text{Me}\!-\!\!\!\!-\!\!\!-\!\!\text{OH} \\ \text{Me}\end{array}} + \text{Sn}^{4+} \quad (52)$$

a nucleophilic substitution of $SnCl_2$ on the dioxetane. Varying amounts of acetone are obtained as well.

ACKNOWLEDGMENTS[127]

Acknowledgments are made to the donors of the Petroleum Research Fund (Grant 8341-AC-1,4), administered by the American Chemical Society, the National Science Foundation (Grant CHE 72-04956-A03) and the National Institutes of Health (Grants GM-22119-02, GM-00142-02, and RR-8102-03) for support of our work in this field. Fellowships from the A. P. Sloan Foundation (1968–1972) and J. S. Guggenheim Memorial Foundation (1972–1973), and a Research Career Award from the National Institutes of Health (1975–1980) have been particularly helpful. The diligent and capable collaboration of the author's students is especially appreciated.

---

[126] A. L. Baumstark, Ph.D. Thesis. Harvard Univ., 1974.
[127] This is Paper XLIII in the Cyclic Peroxide Series.

Cumulative Index of Titles

A

Acetylenecarboxylic acids and esters, reactions with N-heterocyclic compounds, **1**, 125
Acetylenic esters, synthesis of heterocycles through nucleophilic additions to, **19**, 297
Acid-catalyzed polymerization of pyrroles and indoles, **2**, 287
t-Amino effect, **14**, 211
Aminochromes, **5**, 205
Anthracen-1,4-imines, **16**, 87
Anthranils, **8**, 277
Applications of NMR spectroscopy to indole and its derivatives, **15**, 277
Applications of the Hammett equation to heterocyclic compounds, **3**, 209; **20**, 1
Aromatic quinolizines, **5**, 291
Aromaticity of heterocycles, **17**, 255
Aza analogs of pyrimidine and purine bases, **1**, 189
7-Azabicyclo[2.2.1]hepta-2,5-dienes, **16**, 87
Azines, reactivity with nucleophiles, **4**, 145
Azines, theoretical studies of, physicochemical properties of reactivity of, **5**, 69
Azinoazines, reactivity with nucleophiles, **4**, 145
1-Azirines, synthesis and reactions of, **13**, 45

B

Base-catalyzed hydrogen exchange, **16**, 1
1-, 2-, and 3-Benzazepines, **17**, 45
Benzisothiazoles, **14**, 43
Benzisoxazoles, **8**, 277
Benzoazines, reactivity with nucleophiles, **4**, 145
1,5-Benzodiazepines, **17**, 27
Benzo[*b*]furan and derivatives, recent advances in chemistry of, Part I, occurrence and synthesis, **18**, 337
Benzofuroxans, **10**, 1
2*H*-Benzopyrans (chrom-3-enes), **18**, 159
Benzo[*b*]thiophene chemistry, recent advances in, **11**, 177
Benzo[*c*]thiophenes, **14**, 331
1,2,3-(Benzo)triazines, **19**, 215

Biological pyrimidines, tautomerism and electronic structure of, **18**, 199

C

Carbenes, reactions with heterocyclic compounds, **3**, 57
Carbolines, **3**, 79
Cationic polar cycloaddition, **16**, 289 (**19**, xi)
Chemistry
  of benzo[*b*]furan, Part I, occurrence and synthesis, **18**, 337
  of benzo[*b*]thiophenes, **11**, 178
  of chrom-3-enes, **18**, 159
  of diazepines, **8**, 21
  of dibenzothiophenes, **16**, 181
  of dioxetanes, **21**, 437
  of furans, **7**, 377
  of isatin, **18**, 1
  of isoxazolidines, **21**, 207
  of lactim ethers, **12**, 185
  of mononuclear isothiazoles, **14**, 1
  of 4-oxy- and 4-keto-1,2,3,4-tetrahydroisoquinolines, **15**, 99
  of phenanthridines, **13**, 315
  of phenothiazines, **9**, 321
  of 1-pyrindines, **15**, 197
  of tetrazoles, **21**, 323
  of 1,3,4-thiadiazoles, **9**, 165
  of thienothiophenes, **19**, 123
  of thiophenes, **1**, 1
Chrom-3-ene chemistry, advances in, **18**, 159
Claisen rearrangements, in nitrogen heterocyclic systems, **8**, 143
Complex metal hydrides, reduction of nitrogen heterocycles with, **6**, 45
Covalent hydration
  in heteroaromatic compounds, **4**, 1, 43
  in nitrogen heterocycles, **20**, 117
Cyclic enamines and imines, **6**, 147
Cyclic hydroxamic acids, **10**, 199
Cyclic peroxides, **8**, 165
Cycloaddition, cationic polar, **16**, 289 (**19**, xi)
(2 + 2)-Cycloaddition and (2 + 2)-cycloreversion reactions of heterocyclic compounds, **21**, 253

D

Development of the chemistry of furans (1952–1963), **7**, 377
2,4-Dialkoxypyrimidines, Hilbert–Johnson reaction of, **8**, 115
Diazepines, chemistry of, **8**, 21
1,4-Diazepines, 2,3-dihydro-, **17**, 1
Diazo compounds, heterocyclic, **8**, 1
Diazomethane, reactions with heterocyclic compounds, **2**, 245
Dibenzothiophenes, chemistry of, **16**, 181
2,3-Dihydro-1,4-diazepines, **17**, 1
1,2-Dihydroisoquinolines, **14**, 279
1,2-Dioxetanes, chemistry of, **21**, 437
Diquinolylmethane and its analogs, **7**, 153
1,2- and 1,3-Dithiolium ions, **7**, 39

E

Electrolysis of N-heterocyclic compounds, **12**, 213
Electronic aspects of purine tautomerism, **13**, 77
Electronic structure of biological pyrimidines, tautomerism and, **18**, 199
Electronic structure of heterocyclic sulfur compounds, **5**, 1
Electrophilic substitutions of five-membered rings, **13**, 235

F

Ferrocenes, heterocyclic, **13**, 1
Five-membered rings, electrophilic substitutions of, **13**, 235
Free radical substitutions of heteroaromatic compounds, **2**, 131
Furans, development of the chemistry of (1952–1963), **7**, 377

G

Grignard reagents, indole, **10**, 43

H

Halogenation of heterocyclic compounds, **7**, 1
Hammett equation, applications to heterocyclic compounds, **3**, 209; **20**, 1
Hetarynes, **4**, 121
Heteroaromatic compounds
    free-radical substitutions of, **2**, 131
    homolytic substitution of, **16**, 123
    nitrogen, covalent hydration in, **4**, 1, 43
    prototropic tautomerism of, **1**, 311, 339; **2**, 1, 27; Suppl. 1
Heteroaromatic N-imines, **17**, 213
Heteroaromatic substitution, nucleophilic, **3**, 285
Heterocycles
    aromaticity of, **17**, 255
    nomenclature of, **20**, 175
    photochemistry of, **11**, 1
    by ring closure of ortho-substituted t-anilines, **14**, 211
    synthesis of, through nucleophilic additions to acetylenic esters, **19**, 279
    thioureas in synthesis of, **18**, 99
Heterocyclic chemistry, literature of, **7**, 225
Heterocyclic compounds
    application of Hammett equation to, **3**, 209; **20**, 1
    (2 + 2)-cycloaddition and (2 + 2)-cycloreversion reactions of, **21**, 253
    halogenation of, **7**, 1
    isotopic hydrogen labeling of, **15**, 137
    mass spectrometry of, **7**, 301
    quaternization of, **3**, 1
    reaction of acetylenecarboxylic acids with, **1**, 125
    reactions of, with carbenes, **3**, 57
    reactions of diazomethane with, **2**, 245
N-Heterocyclic compounds, electrolysis of, **12**, 213
Heterocyclic diazo compounds, **8**, 1
Heterocyclic ferrocenes, **13**, 1
Heterocyclic oligomers, **15**, 1
Heterocyclic pseudo bases, **1**, 167
Heterocyclic sulphur compounds, electronic structure of, **5**, 1
Heterocyclic syntheses, from nitrilium salts under acidic conditions, **6**, 95
Hilbert–Johnson reaction of 2,4-dialkoxy-pyrimidines, **8**, 115
Homolytic substitution of heteroaromatic compounds, **16**, 123
Hydrogen exchange
    base-catalyzed, **16**, 1
    one-step (labeling) methods, **15**, 137
Hydroxamic acids, cyclic, **10**, 199

CUMULATIVE INDEX OF TITLES

I

Imidazole chemistry, advances in, **12**, 103
N-Imines, heteroaromatic, **17**, 213
Indole Grignard reagents, **10**, 43
Indole(s)
    acid-catalyzed polymerization, **2**, 287
    and derivatives, application of NMR spectroscopy to, **15**, 277
Indoxazenes, **8**, 277
Isatin, chemistry of, **18**, 1
Isoindoles, **10**, 113
Isoquinolines
    1,2-dihydro-, **14**, 279
    4-oxy- and 4-keto-1,2,3,4-tetrahydro-, **15**, 99
Isothiazoles, **4**, 107
    recent advances in the chemistry of monocyclic, **14**, 1
Isotopic hydrogen labeling of heterocyclic compounds, one-step methods, **15**, 137
Isoxazole chemistry, recent developments in, **2**, 365
Isoxazolidines, chemistry of, **21**, 207

L

Lactim ethers, chemistry of, **12**, 185
Literature of heterocyclic chemistry, **7**, 225

M

Mass spectrometry of heterocyclic compounds, **7**, 301
Meso-ionic compounds, **19**, 1
Metal catalysts, action on pyridines, **2**, 179
Monoazaindoles, **9**, 27
Monocyclic pyrroles, oxidation of, **15**, 67
Monocyclic sulfur-containing pyrones, **8**, 219
Mononuclear isothiazoles, recent advances in chemistry of, **14**, 1

N

Naphthalen-1,4-imines, **16**, 87
Naphthyridines, **11**, 124
Nitriles and nitrilium salts, heterocyclic syntheses involving, **6**, 95
Nitrogen-bridged six-membered ring systems, **16**, 87
Nitrogen heterocycles
    covalent hydration in, **20**, 117
    reduction of, with complex metal hydrides, **6**, 45
Nitrogen heterocyclic systems, Claisen rearrangements in, **8**, 143
Nomenclature of heterocycles, **20**, 175
Nuclear magnetic resonance spectroscopy, application to indoles, **15**, 277
Nucleophiles, reactivity of azine derivatives with, **4**, 145
Nucleophilic additions to acetylenic esters, synthesis of heterocycles through, **19**, 299
Nucleophilic heteroaromatic substitution, **3**, 285

O

Oligomers, heterocyclic, **15**, 1
1,2,4-Oxadiazoles, **20**, 65
1,3,4-Oxadiazole chemistry, recent advances in, **7**, 183
1,3-Oxazine derivatives, **2**, 311
Oxazole chemistry, advances in, **17**, 99
Oxazolone chemistry
    new developments in, **21**, 175
    recent advances in, **4**, 75
Oxidation of monocyclic pyrroles, **15**, 67
3-Oxo-2,3-dihydrobenz[*d*]isothiazole-1,1-dioxide (Saccharin) and derivatives, **15**, 233
4-Oxy- and 4-keto-1,2,3,4-tetrahydroisoquinolines, chemistry of, **15**, 99

P

Pentazoles, **3**, 373
Peroxides, cyclic, **8**, 165 (*see also* 1,2-Dioxetanes)
Phenanthridine chemistry, recent developments in, **13**, 315
Phenothiazines, chemistry of, **9**, 321
Phenoxazines, **8**, 83
Photochemistry of heterocycles, **11**, 1
Physicochemical aspects of purines, **6**, 1
Physicochemical properties
    of azines, **5**, 69
    of pyrroles, **11**, 383
3-Piperideines, **12**, 43
Polymerization of pyrroles and indoles, acid-catalyzed, **2**, 287
Prototropic tautomerism of heteroaromatic compounds, **1**, 311, 339; **2**, 1, 27; Suppl. 1
Pseudo bases, heterocyclic, **1**, 167
Purine bases, aza analogs, **1**, 189

Purines
    physicochemical aspects of, **6**, 1
    tautomerism, electronic aspects of, **13**, 77
Pyrazine chemistry, recent advances in, **14**, 99
Pyrazole chemistry, progress in, **6**, 347
Pyridazines, **9**, 211
Pyridine(s)
    action of metal catalysts on, **2**, 179
    effect of substituents on substitution in, **6**, 229
    1,2,3,6-tetrahydro-, **12**, 43
Pyridoindoles (the carbolines), **3**, 79
Pyridopyrimidines, **10**, 149
Pyrimidine bases, aza analogs of, **1**, 189
Pyrimidines
    2,4-dialkoxy-, Hilbert–Johnson reaction of, **8**, 115
    tautomerism and electronic structure of biological, **18**, 199
1-Pyrindines, chemistry of, **15**, 197
Pyrones, monocyclic sulfur-containing, **8**, 219
Pyrroles
    acid-catalyzed polymerization of, **2**, 287
    oxidation of monocyclic, **15**, 67
    physicochemical properties of, **11**, 383
Pyrrolizidine chemistry, **5**, 315
Pyrrolodiazines, with a bridgehead nitrogen, **21**, 1
Pyrrolopyridines, **9**, 27
Pyrylium salts, syntheses, **10**, 241

Q

Quaternization of heterocyclic compounds, **3**, 1
Quinazolines, **1**, 253
Quinolizines, aromatic, **5**, 291
Quinoxaline chemistry, recent advances in, **2**, 203
Quinuclidine chemistry, **11**, 473

R

Reduction of nitrogen heterocycles with complex metal hydrides, **6**, 45
Reissert compounds, **9**, 1
Ring closure of ortho-substituted *t*-anilines, for heterocycles, **14**, 211

S

Saccharin and derivatives, **15**, 233
Selenazole chemistry, present state of, **2**, 343
Selenophene chemistry, advances in, **12**, 1
Six membered ring systems, nitrogen bridged, **16**, 87
Substitution(s),
    electrophilic, of five-membered rings, **13**, 235
    homolytic, of heteroaromatic compounds, **16**, 123
    nucleophilic heteroaromatic, **3**, 285
    in pyridines, effect of substituents, **6**, 229
Sulfur compounds, electronic structure of heterocyclic, **5**, 1
Synthesis and reactions of 1-azirines, **13**, 45
Synthesis of heterocycles through nucleophilic additions to acetylenic esters, **19**, 279

T

Tautomerism
    electronic aspects of purine, **13**, 77
    and electronic structure of biological pyrimidines, **18**, 199
    prototropic, of heteroaromatic compounds, **1**, 311, 339; **2**, 1, 27; Suppl. 1
Tellurophene and related compounds, **21**, 119
1,2,3,4-Tetrahydroisoquinolines, 4-oxy- and 4-keto-, **15**, 99
1,2,3,6-Tetrahydropyridines, **12**, 43
Theoretical studies of physicochemical properties and reactivity of azines, **5**, 69
Tetrazole chemistry, recent advances in, **21**, 323
1,2,4-Thiadiazoles, **5**, 119
1,2,5-Thiadiazoles, chemistry of, **9**, 165
Thiathiophthenes (1,6,6aS^{IV}-Trithiapentalenes), **13**, 161
1,2,3,4-Thiatriazoles, **3**, 263; **20**, 145
Thienopyridines, **21**, 65
Thienothiophenes and related systems, chemistry of, **19**, 123
Thiochromanones and related compounds, **18**, 59
Thiophenes, chemistry of, recent advances in, **1**, 1
Thiopyrones (monocyclic sulfur-containing pyrones), **8**, 219
Thioureas in synthesis of heterocycles, **18**, 99
Three-membered rings with two heteroatoms, **2**, 83
1,3,5-, 1,3,6-, 1,3,7-, and 1,3,8-Triazanaphthalenes, **10**, 149
1,2,3-Triazines, **19**, 215
1,2,3-Triazoles, **16**, 33
1,6,6aS^{IV}-Trithiapentalenes, **13**, 161